# Springer Texts in Statistics

T0137633

*Advisors:*
George Casella   Stephen Fienberg   Ingram Olkin

Allan Gut

# Probability:
# A Graduate Course

 Springer

Allan Gut
Department of Mathematics
University of Uppsala
SE-751 06 Uppsala
Sweden

ISBN 978-1-4419-1985-4          e-ISBN 978-0-387-27332-7

Library of Congress Cataloging-in-Publication Data
Gut, Allan, 1944–
    Probability: a graduate course / Allan Gut.
        p. cm. — (Springer texts in statistics)
    Includes bibliographical references and index.

    1. Probabilities.  2. Distribution (Probability theory)  3. Probabilistic number theory.  I.
Title.  II. Series.
QA273.G869   2005
519.2—dc22                                                    2004056469

Printed on acid-free paper.

9 8 7 6 5 4 3 2 1

springeronline.com

# Preface

Toss a symmetric coin twice. What is the probability that both tosses will yield a head?

This is a well-known problem that anyone can solve. Namely, the probability of a head in each toss is $1/2$, so the probability of two consecutive heads is $1/2 \cdot 1/2 = 1/4$.

*BUT!* What did we do? What is involved in the solution? What are the arguments behind our computations? Why did we multiply the two halves connected with each toss?

This is reminiscent of the centipede[1] who was asked by another animal how he walks; he who has so many legs, in which order does he move them as he is walking? The centipede contemplated the question for a while, but found no answer. However, from that moment on he could no longer walk.

This book is written with the hope that we are not centipedes.

There exist two kinds of probabilists. One of them is the mathematician who views probability theory as a purely mathematical discipline, like algebra, topology, differential equations, and so on. The other kind views probability theory as the *mathematical modeling of random phenomena*, that is with a view toward applications, and as a companion to statistics, which aims at finding methods, principles and criteria in order to analyze data emanating from experiments involving random phenomena and other observations from the real world, with the ultimate goal of making wise decisions. I would like to think of myself as both.

What kind of a random process describes the arrival of claims at an insurance company? Is it one process or should one rather think of different processes, such as one for claims concerning stolen bikes and one for houses that have burnt down? How well should the DNA sequences of an accused offender and a piece of evidence match each other in order for a conviction? A

---

[1]Cent is 100, so it means an animal with 100 legs. In Swedish the name of the animal is *tusenfoting*, where "tusen" means 1000 and "fot" is foot; thus an animal with 1000 legs or feet.

milder version is how to order different species in a phylogenetic tree. What are the arrival rates of customers to a grocery store? How long are the service times? How do the clapidemia cells split? Will they create a new epidemic or can we expect them to die out? A classical application has been the arrivals of telephone calls to a switchboard and the duration of calls. Recent research and model testing concerning the Internet traffic has shown that the classical models break down completely and new thinking has become necessary. And, last but (not?) least, there are many games and lotteries.

The aim of this book is to provide the reader with a fairly thorough treatment of the main body of basic and classical probability theory, preceded by an introduction to the mathematics which is necessary for a solid treatment of the material. This means that we begin with basics from measure theory, such as $\sigma$-algebras, set theory, measurability (random variables) and Lebesgue integration (expectation), after which we turn to the Borel-Cantelli lemmas, inequalities, transforms and the three classical limit theorems: the law of large numbers, the central limit theorem and the law of the iterated logarithm. A final chapter on martingales – one of the most efficient, important, and useful tools in probability theory – is preceded by a chapter on topics that could have been included with the hope that the reader will be tempted to look further into the literature. The reason that these topics did not get a chapter of their own is that beyond a certain number of pages a book becomes deterring rather than tempting (or, as somebody said with respect to an earlier book of mine: "It is a nice format for bedside reading").

One thing that is *not* included in this book is a philosophical discussion of whether or not chance exist, whether or not randomness exists. On the other hand, probabilistic modeling is a wonderful, realistic, and efficient way to model phenomena containing uncertainties and ambiguities, regardless of whether or not the answer to the philosophical question is yes or no.

I remember having read somewhere a sentence like "There exist already so many textbooks [of the current kind], so, why do I write another one?" This sentence could equally well serve as an opening for the present book.

Luckily, I can provide an answer to that question. The answer is the short version of the story of the mathematician who was asked how one realizes that the fact he presented in his lecture (because this was really a he) was trivial. After 2 minutes of complete silence he mumbled

**I know it's trivial, but I have forgotten why.**

I strongly dislike the arrogance and snobbism that encompasses mathematics and many mathematicians. Books and papers are filled with expressions such as "it is easily seen", "it is trivial", "routine computations yield", and so on. The last example is sometimes modified into "routine, but tedious, computations yield". And we all know that behind things that are easily seen there may be years of thinking and/or huge piles of scrap notes that lead nowhere, and one sheet where everything finally worked out nicely.

Clearly, things become routine after many years. Clearly, facts become, at least *intuitively*, obvious after some decades. But in writing papers and books we try to help those who do not know yet, those who want to learn. We wish to attract people to this fascinating part of the world. Unfortunately though, phrases like the above ones are repellent, rather than being attractive. If a reader understands immediately that's fine. However, it is more likely that he or she starts off with something that either results in a pile of scrap notes or in frustration. Or both. And nobody is made happier, certainly not the reader. I have therefore avoided, or, at least, tried to avoid, expressions like the above unless they are adequate.

The main aim of a book is to be helpful to the reader, to help her or him to understand, to inform, to educate, and to attract (and not for the author to prove himself to the world). It is therefore essential to keep the flow, not only in the writing, but also in the reading. In the writing it is therefore of great importance to be rather extensive and not to leave too much to the (interested) reader.

A related aspect concerns the style of writing. Most textbooks introduce the reader to a number of topics in such a way that further insights are gained through exercises and problems, some of which are not at all easy to solve, let alone trivial. We take a somewhat different approach in that several such "would have been" exercises are given, together with their solutions as part of the ordinary text – which, as a side effect, reduces the number of exercises and problems at the end of each chapter. We also provide, at times, results for which the proofs consist of variations of earlier ones, and therefore are left as an exercise, with the motivation that doing almost the same thing as somebody else has done provides a much better understanding than reading, nodding and agreeing. I also hope that this approach creates the atmosphere of a dialogue rather than of the more traditional monologue (or sermon).

The ultimate dream is, of course, that this book contains no errors, no slips, no misprints. Henrik Wanntorp has gone over a substantial part of the manuscript with a magnifying glass, thereby contributing immensely to making that dream come true. My heartfelt thanks, Henrik. I also wish to thank Raimundas Gaigalas for several perspicacious remarks and suggestions concerning his favorite sections, and a number of reviewers for their helpful comments and valuable advice. As always, I owe a lot to Svante Janson for being available for any question at all times, and, more particularly, for always providing me with an answer. John Kimmel of Springer-Verlag has seen me through the process with a unique combination of professionalism, efficiency, enthusiasm and care, for which I am most grateful.

Finally, my hope is that the reader who has digested this book is ready and capable to attack any other text, for which a solid probabilistic foundation is necessary or, at least, desirable.

Uppsala                                                                                        *Allan Gut*
November 2004

# Contents

# Outline of Contents

In this extended list of contents, we provide a short expansion of the headings into a quick overview of the contents of the book.

## Chapter 1. Introductory Measure Theory

The mathematical foundation of probability theory is measure theory and the theory of Lebesgue integration. The bulk of the introductory chapter is devoted to measure theory: sets, measurability, $\sigma$-algebras, and so on. We do not aim at a full course in measure theory, rather to provide enough background for a solid treatment of what follows.

## Chapter 2. Random Variables

Having set the scene, the first thing to do is to forget probability spaces (!). More precisely, for modeling random experiments one is interested in certain specific quantities, called *random variables*, rather than in the underlying probability space itself. In Chapter 2 we introduce random variables and present the basic concepts, as well as concrete applications and examples of probability models. In particular, Lebesgue integration is developed in terms of expectation of random variables.

## Chapter 3. Inequalities

Some of the most useful tools in probability theory and mathematics for proving finiteness or convergence of sums and integrals are *inequalities*. There exist many useful ones spread out in books and papers. In Chapter 3 we make an attempt to present a sizable amount of the most important inequalities.

## Chapter 4. Characteristic Functions

Just as there are i.a. Fourier transforms that transform convolution of functions into multiplication of their corresponding transforms, there exist probabilistic transforms that "map" addition of independent random variables into multiplication of their transforms, the most prominent one being the characteristic function.

## Chapter 5. Convergence

Once we know how to add random variables the natural problem is to investigate asymptotics. We begin by introducing some convergence concepts, prove uniqueness, after which we investigate how and when they imply each other. Other important problems are when, and to what extent, limits and expectations (limits and integrals) can be interchanged, and when, and to what extent, functions of convergent sequences converge to the function of the limit.

## Chapter 6. The Law of Large Numbers

The law of large numbers states that (the distribution of) the arithmetic mean of a sequence of independent trials stabilizes around the center of gravity of the underlying distribution (under suitable conditions). There exist *weak* and *strong* laws and several variations and extensions of them. We shall meet some of them as well as some applications.

## Chapter 7. The Central Limit Theorem

The central limit theorem, which (in its simplest form) states that if the variance is finite, then the arithmetic mean, properly rescaled, of a sequence of independent trials approaches a normal distribution as the number of observations increases. There exist many variations and generalizations, of the theorem, the central one being the Lindeberg-Lévy-Feller theorem. We also prove results on moment convergence, and rate results, the foremost one being the celebrated Berry-Esseen theorem.

## Chapter 8. The Law of the Iterated Logarithm

This is a special, rather delicate and technical, and very beautiful, result, which provides precise bounds on the oscillations of sums of the above kind. The name obviously stems from the iterated logarithm that appears in the expression of the parabolic bound.

## Chapter 9. Limit Theorems; Extensions and Generalizations

There are a number of additional topics that would fit well into a text like the present one, but for which there is no room. In this chapter we shall meet a number of them – stable distributions, domain of attraction, infinite divisibility, sums of dependent random variables, extreme value theory, the Stein-Chen method – in a somewhat more sketchy or introductory style. The reader who gets hooked on such a topic will be advised to some relevant literature (more can be found via the Internet).

## Chapter 10. Martingales

This final chapter is devoted to one of the most central topics, not only in probability theory, but also in more traditional mathematics. Following some introductory material on conditional expectations and the definition of a martingale, we present several examples, convergence results, results for stopped martingales, regular martingales, uniformly integrable martingales, stopped random walks, and reversed martingales.

## In Addition

A list of notation and symbols precedes the main body of text, and an appendix with some mathematical tools and facts, a bibliography, and an index conclude the book. References are provided for more recent results, for more nontraditional material, and to some of the historic sources, but in general not to the more traditional material. In addition to cited material, the list of references contains references to papers and books that are relevant without having been specifically cited.

## Suggestions for a Course Curriculum

One aim with the book is that it should serve as a graduate probability course – as the title suggests. In the same way as the sections in Chapter 9 contain materials that no doubt would have deserved chapters of their own, Chapters 6,7, and 8 contain sections entitled "Some Additional Results and Remarks", in which a number of additional results and remarks are presented, results that are not as central and basic as earlier ones in those chapters.

An adequate course would, in my opinion, consist of Chapters 1-8, and 10, except for the sections "Some Additional Results and Remarks", plus a skimming through Chapter 9 at the level of the instructor's preferences.

# Notation and Symbols

| | |
|---|---|
| $\Omega$ | the sample space |
| $\omega$ | an elementary event |
| $\mathcal{F}$ | the $\sigma$-algebra of events |
| | |
| $x^+$ | $\max\{x, 0\}$ |
| $x^-$ | $-\min\{x, 0\}$ |
| $[x]$ | the integer part of $x$ |
| $\log^+ x$ | $\max\{1, \log x\}$ |
| $\sim$ | the ratio of the quantities on either side tends to 1 |
| | |
| $\mathbb{N}$ | the (positive) natural numbers |
| $\mathbb{Z}$ | the integers |
| $\mathbb{R}$ | the real numbers |
| $\mathcal{R}$ | the Borel $\sigma$-algebra on $\mathbb{R}$ |
| $\lambda(\cdot)$ | Lebesgue measure |
| $\mathbb{Q}$ | the rational numbers |
| $\mathbb{C}$ | the complex numbers |
| $C$ | the continuous functions |
| $C_0$ | the functions in $C$ tending to 0 at $\pm\infty$ |
| $C_B$ | the bounded continuous functions |
| $C[a, b]$ | the functions in $C$ with support on the interval $[a, b]$ |
| $D$ | the right-continuous, functions with left-hand limits |
| $D[a, b]$ | the functions in $D$ with support on the interval $[a, b]$ |
| $D^+$ | the non-decreasing functions in $D$ |
| $\mathbb{J}_G$ | the discontinuities of $G \in D$ |
| | |
| $I\{A\}$ | indicator function of (the set) $A$ |
| $\#\{A\}$ | number of elements in (cardinality of) $A$ |
| $|A|$ | number of elements in (cardinality of) $A$ |
| $A^c$ | complement of $A$ |
| $\partial A$ | the boundary of $A$ |
| $P(A)$ | probability of $A$ |

| | |
|---|---|
| $X, Y, Z, \ldots$ | random variables |
| $F(x), F_X(x)$ | distribution function (of $X$) |
| $X \in F$ | $X$ has distribution (function) $F$ |
| $C(F_X)$ | the continuity set of $F_X$ |
| $p(x), p_X(x)$ | probability function (of $X$) |
| $f(x), f_X(x)$ | density (function) (of $X$) |
| $\mathbf{X}, \mathbf{Y}, \mathbf{Z}, \ldots$ | random (column) vectors |
| $\mathbf{X}', \mathbf{Y}', \mathbf{Z}', \ldots$ | the transpose of the vectors |
| $F_{X,Y}(x,y)$ | joint distribution function (of $X$ and $Y$) |
| $p_{X,Y}(x,y)$ | joint probability function (of $X$ and $Y$) |
| $f_{X,Y}(x,y)$ | joint density (function) (of $X$ and $Y$) |
| $E, E X$ | expectation (mean), expected value of $X$ |
| Var, Var $X$ | variance, variance of $X$ |
| Cov $(X,Y)$ | covariance of $X$ and $Y$ |
| $\rho, \rho_{X,Y}$ | correlation coefficient (between $X$ and $Y$) |
| med $(X)$ | median of $X$ |
| | |
| $g(t), g_X(t)$ | (probability) generating function (of $X$) |
| $\psi(t), \psi_X(t)$ | moment generating function (of $X$) |
| $\varphi(t), \varphi_X(t)$ | characteristic function (of $X$) |
| | |
| $X \sim Y$ | $X$ and $Y$ are equivalent random variables |
| $X \stackrel{a.s.}{=} Y$ | $X$ and $Y$ are equal (point-wise) almost surely |
| $X \stackrel{d}{=} Y$ | $X$ and $Y$ are equidistributed |
| $X_n \stackrel{a.s.}{\to} X$ | $X_n$ converges almost surely (a.s.) to $X$ |
| $X_n \stackrel{p}{\to} X$ | $X_n$ converges in probability to $X$ |
| $X_n \stackrel{r}{\to} X$ | $X_n$ converges in $r$-mean ($L^r$) to $X$ |
| $X_n \stackrel{d}{\to} X$ | $X_n$ converges in distribution to $X$ |
| | |
| $X_n \stackrel{a.s.}{\nrightarrow}$ | $X_n$ does not converge almost surely |
| $X_n \stackrel{p}{\nrightarrow}$ | $X_n$ does not converge in probability |
| $X_n \stackrel{r}{\nrightarrow}$ | $X_n$ does not converge in $r$-mean ($L^r$) |
| $X_n \stackrel{d}{\nrightarrow}$ | $X_n$ does not converge in distribution |
| | |
| $\Phi(x)$ | standard normal distribution function |
| $\phi(x)$ | standard normal density (function) |
| $F \in \mathcal{D}(G)$ | $F$ belongs to the domain of attraction of $G$ |
| $g \in \mathcal{RV}(\rho)$ | $g$ varies regularly at infinity with exponent $\rho$ |
| $g \in \mathcal{SV}$ | $g$ varies slowly at infinity |

| | |
|---|---|
| $\mathrm{Be}(p)$ | Bernoulli distribution |
| $\beta(r, s)$ | beta distribution |
| $\mathrm{Bin}(n, p)$ | binomial distribution |
| $C(m, a)$ | Cauchy distribution |
| $\chi^2(n)$ | chi-square distribution |
| $\delta(a)$ | one-point distribution |
| $\mathrm{Exp}(a)$ | exponential distribution |
| $F(m, n)$ | (Fisher's) $F$-distribution |
| $\mathrm{Fs}(p)$ | first success distribution |
| $\Gamma(p, a)$ | gamma distribution |
| $\mathrm{Ge}(p)$ | geometric distribution |
| $H(N, n, p)$ | hypergeometric distribution |
| $L(a)$ | Laplace distribution |
| $\mathrm{LN}(\mu, \sigma^2)$ | log-normal distribution |
| $N(\mu, \sigma^2)$ | normal distribution |
| $N(0, 1)$ | standard normal distribution |
| $\mathrm{NBin}(n, p)$ | negative binomial distribution |
| $\mathrm{Pa}(k, \alpha)$ | Pareto distribution |
| $\mathrm{Po}(m)$ | Poisson distribution |
| $\mathrm{Ra}(\alpha)$ | Rayleigh distribution |
| $t(n)$ | (Student's) $t$-distribution |
| $\mathrm{Tri}(a, b)$ | triangular distribution on $(a, b)$ |
| $U(a, b)$ | uniform or rectangular distribution on $(a, b)$ |
| $W(a, b)$ | Weibull distribution |
| $X \in P(\theta)$ | $X$ has a $P$-distribution with parameter $\theta$ |
| $X \in P(\alpha, \beta)$ | $X$ has a $P$-distribution with parameters $\alpha$ and $\beta$ |
| | |
| a.e. | almost everywhere |
| a.s. | almost surely |
| cf. | *confer*, compare, take counsel |
| i.a. | *inter alia*, among other things, such as |
| i.e. | *id est*, that is |
| i.o. | infinitely often |
| iff | if and only if |
| i.i.d. | independent, identically distributed |
| viz. | *videlicet*, in which |
| w.l.o.g. | without loss of generality |
| ♠ | hint for solving a problem |
| ♣ | bonus remark in connection with a problem |
| □ | end of proof, definitions, exercises, remarks, etc. |

# 1

---

# Introductory Measure Theory

## 1 Probability Theory: An Introduction

The object of *probability theory* is to describe and investigate mathematical models of random phenomena, primarily from a theoretical point of view. Closely related to probability theory is *statistics*, which is concerned with creating principles, methods, and criteria in order to treat data pertaining to such (random) phenomena or data from experiments and other observations of the real world, by using, for example, the theories and knowledge available from the theory of probability.

Probability models thus aim at describing *random experiments*, that is, experiments that can be repeated (indefinitely) and where future outcomes cannot be exactly predicted – due to randomness – even if the experimental situation can be fully controlled.

The basis of probability theory is the *probability space*. The key idea behind probability spaces is the *stabilization of the relative frequencies*. Suppose that we perform "independent" repetitions of a random experiment and that we record each time if some "event" $A$ occurs or not (although we have not yet mathematically defined what we mean by independence or by an event). Let $f_n(A)$ denote the number of occurrences of $A$ in the first $n$ trials, and $r_n(A)$ the relative frequency, $r_n(A) = f_n(A)/n$. Since the dawn of history one has observed the stabilization of the relative frequencies, that is, one has observed that (it seems that)

$$r_n(A) \quad \text{converges to some real number as} \quad n \to \infty.$$

The intuitive interpretation of the probability concept is that if the probability of some event $A$ is 0.6, one should expect that by performing the random experiment "many times" the relative frequency of occurrences of $A$ should be approximately 0.6.

The next step is to axiomatize the theory, to make it mathematically rigorous. Although games of chance have been performed for thousands of years, a mathematically rigorous treatment of the theory of probability only came about in the 1930's by the Soviet/Russian mathematician A.N. Kolmogorov (1903–1987) in his fundamental monograph *Grundbegriffe der Wahrschein-lichkeitsrechnung* [163], which appeared in 1933.

The first observation is that a number of rules that hold for relative frequencies should also hold for probabilities. This immediately calls for the question "which is the minimal set of rules?"

In order to answer this question one introduces the *probability space* or *probability triple* $(\Omega, \mathcal{F}, P)$, where

- $\Omega$ is the *sample space*;
- $\mathcal{F}$ is the collection of *events*;
- $P$ is a probability measure.

The fact that $P$ is a probability measure means that it satisfies the three *Kolmogorov axioms* (to be specified ahead).

In a first course in probability theory one learns that

"the collection of events $=$ the subsets of $\Omega$",

maybe with an additional remark that this is not quite true, but true enough for the purpose of that course.

To clarify the situation we need some definitions and facts from measure theory in order to answer questions such as

"What does it mean for a set to be measurable?"

After this we shall return to a proper definition of the probability space.

## 2 Basics from Measure Theory

In addition, to straighten out the problems raised by such questions, we need rules for how to operate on what we shall define as events. More precisely, a problem may consist of finding the probability of one or the other of two things happening, or for something *not to happen*, and so on. We thus need rules and conventions for how to handle events, how we can combine them or not combine them. This means i.a. that we need to define collections of sets with a certain structure. For example, a collection such that the intersection of two events is an event, or the collection of sets such that the complement of an event is an event, and also rules for how various collections connect. This means that we have to confront ourselves with some notions from measure theory. Since this is rather a tool than a central theme of this book we confine ourselves to an overview of the most important parts of the topic, leaving some of the "routine but tedious calculations" as exercises.

## 2.1 Sets

### Definitions; Notation

A set is a collection of "objects", concrete or abstract, called *elements*. A set is *finite* if the number of elements is finite, and it is *countable* if the number of elements is countable, that is, if one can label them by the positive integers in such a manner that no element remains unlabeled. Sets are usually denoted by capitals from the early part of the alphabet, $A$, $B$, $C$, and so on. If several sets are related, "of the same kind", it is convenient to use the same letter for them, but to add indices; $A_1, A_2, A_3, \ldots$.

The set $A = \{1, 2, \ldots, n\}$ is finite; $|A| = n$. The natural numbers, $\mathbb{N}$ constitute a countable set, and so do the set of rationals, $\mathbb{Q}$, whereas the set of irrationals and the set of reals, $\mathbb{R}$, are uncountable, as is commonly verified by Cantor's diagonal method. Although the natural numbers and the reals belong to different collections of sets they are both infinite in the sense that the number of elements is infinite in both sets, but the infinities are different. The same distinction holds true for the rationals and the irrationals. Infinities are distinguished with the aid of the *cardinal numbers*, which came about after Cantor's proof of the fact that the infinity of the reals is larger than that of the natural numbers, that there are "more" reals than natural numbers. Cardinal numbers are denoted by the Hebrew letter *alef*, where the successively larger cardinal numbers have increasing indices. The first cardinal number is $\aleph_0 = |\mathbb{N}|$, the *cardinality* of $\mathbb{N}$. Moreover, $|\mathbb{R}| = 2^{\aleph_0}$.

Let us mention in passing that a long-standing question, in fact, one of Hilbert's famous problems, has been whether or not there exist infinities between $|\mathbb{N}|$ and $|\mathbb{R}|$. The famous *continuum hypothesis* states that this is not the case, a claim that can be formulated as $\aleph_1 = 2^{\aleph_0}$. The interesting fact is that it has been proved that this claim can neither be proved nor disproved within the usual axiomatic framework of mathematics. Moreover, one may assume it to be true or false, and neither assumption will lead to any contradictory results. The continuum hypothesis is said to be *undecidable*. For more, see [51].

### Set Operations

Just as real (or complex) numbers can be added or multiplied, there exist *operations* on sets. Let $A, A_1, A_2, \ldots$ and $B, B_1, B_2, \ldots$ be sets.

- Union: $A \cup B = \{x : x \in A \text{ or } x \in B\}$;
- Intersection: $A \cap B = \{x : x \in A \text{ and } x \in B\}$;
- Complement: $A^c = \{x : x \notin A\}$;
- Difference: $A \smallsetminus B = A \cap B^c$;
- Symmetric difference: $A \triangle B = (A \smallsetminus B) \cup (B \smallsetminus A)$.

We also use standard notations such as $\cup_{k=1}^{n} A_k$ and $\cap_{j=1}^{\infty} B_j$ for unions and intersections of finitely or countably many sets.

**Exercise 2.1.** Check to what extent the associative and distributive rules for these operations are valid.                                                                                          □

Some additional terminology:

- the empty set: $\emptyset$;
- subset: $A$ is a *subset* of $B$, $A \subset B$, if $x \in A \Longrightarrow x \in B$;
- disjoint: $A$ and $B$ are *disjoint* if $A \cap B = \emptyset$;
- power set: $\mathfrak{P}(\Omega) = \{A : A \subset \Omega\}$;
- $\{A_n, n \geq 1\}$ is *non-decreasing*, $A_n \nearrow$, if $A_1 \subset A_2 \subset \cdots$;
- $\{A_n, n \geq 1\}$ is *non-increasing*, $A_n \searrow$, if $A_1 \supset A_2 \supset \cdots$.

The *de Morgan formulas*,

$$\left( \bigcup_{k=1}^{n} A_k \right)^c = \bigcap_{k=1}^{n} A_k^c \quad \text{and} \quad \left( \bigcap_{k=1}^{n} A_k \right)^c = \bigcup_{k=1}^{n} A_k^c, \qquad (2.1)$$

can be verified by picking $\omega \in \Omega$ belonging to the set made up by the left-hand side and then show that it also belongs to the right-hand side, after which one does the same the other way around (please do that!). Alternatively one realizes that both members express the same fact. In the first case, this is the fact that an element that does not belong to any $A_k$ whatsoever belongs to all complements, and therefore to their intersection. In the second case this is the fact that an element that does not belong to every $A_k$ belongs to at least one of the complements.

### Limits of Sets

It is also possible to define limits of sets. However, not every sequence of sets has a limit.

**Definition 2.1.** *Let $\{A_n, n \geq 1\}$ be a sequence of subsets of $\Omega$. We define*

$$A_* = \liminf_{n \to \infty} A_n = \bigcup_{n=1}^{\infty} \bigcap_{m=n}^{\infty} A_m,$$

$$A^* = \limsup_{n \to \infty} A_n = \bigcap_{n=1}^{\infty} \bigcup_{m=n}^{\infty} A_m.$$

*If the sets $A_*$ and $A^*$ agree, then*

$$A = A_* = A^* = \lim_{n \to \infty} A_n. \qquad □$$

One instance when a limit exists is when the sequence of sets is monotone.

**Proposition 2.1.** *Let $\{A_n, n \geq 1\}$ be a sequence of subsets of $\Omega$.*

(i) *If $A_1 \subset A_2 \subset A_3 \cdots$, then*

$$\lim_{n\to\infty} A_n = \bigcup_{n=1}^{\infty} A_n.$$

(ii) *If $A_1 \supset A_2 \supset A_3 \cdots$, then*

$$\lim_{n\to\infty} A_n = \bigcap_{n=1}^{\infty} A_n.$$

**Exercise 2.2.** Prove the proposition.                                   □

## 2.2 Collections of Sets

Collections of sets, are defined according to a setup of rules. Different rules yield different collections. Certain collections are more easy to deal with than others depending on the property or theorem to prove. We now present a number of rules and collections, as well as results on how they connect. Since much of this is more or less well known to a mathematics student we leave essentially all proofs, which consist of longer or shorter, sometimes routine but tedious, manipulations, as exercises.

Let $\mathcal{A}$ be a non-empty collection of subsets of $\Omega$, and consider the following set relations:

(a)   $A \in \mathcal{A} \Longrightarrow A^c \in \mathcal{A}$;

(b)   $A, B \in \mathcal{A} \Longrightarrow A \cup B \in \mathcal{A}$;

(c)   $A, B \in \mathcal{A} \Longrightarrow A \cap B \in \mathcal{A}$;

(d)   $A, B \in \mathcal{A}, \ B \subset A \Longrightarrow A \setminus B \in \mathcal{A}$;

(e)   $A_n \in \mathcal{A}, \ n \geq 1, \Longrightarrow \bigcup_{n=1}^{\infty} A_n \in \mathcal{A}$;

(f)   $A_n \in \mathcal{A}, \ n \geq 1, \ A_i \cap A_j = \emptyset, \ i \neq j \Longrightarrow \bigcup_{n=1}^{\infty} A_n \in \mathcal{A}$;

(g)   $A_n \in \mathcal{A}, \ n \geq 1, \Longrightarrow \bigcap_{n=1}^{\infty} A_n \in \mathcal{A}$;

(h)   $A_n \in \mathcal{A}, \ n \geq 1, \ A_n \nearrow \Longrightarrow \bigcup_{n=1}^{\infty} A_n \in \mathcal{A}$;

(j)   $A_n \in \mathcal{A}, \ n \geq 1, \ A_n \searrow \Longrightarrow \bigcap_{n=1}^{\infty} A_n \in \mathcal{A}$.

A number of relations among these rules and extensions of them can be established. For example (a) and one of (b) and (c), together with the de Morgan formulas, yield the other; (a) and one of (e) and (g), together with the de Morgan formulas, yield the other; (b) and induction shows that (b) can be extended to any finite union of sets; (c) and induction shows that (c) can be extended to any finite intersection of sets, and so on.

**Exercise 2.3.** Check these statements, and verify some more.          □

Here are now definitions of some collections of sets.

**Definition 2.2.** *Let $\mathcal{A}$ be a collection of subsets of $\Omega$.*

- $\mathcal{A}$ *is an* algebra *or a* field *if $\Omega \in \mathcal{A}$ and properties* (a) *and* (b) *hold;*
- $\mathcal{A}$ *is a* $\sigma$-algebra *or a* $\sigma$-field *if $\Omega \in \mathcal{A}$ and properties* (a) *and* (e) *hold;*
- $\mathcal{A}$ *is a* monotone class *if properties* (h) *and* (j) *hold;*
- $\mathcal{A}$ *is a* $\pi$-system *if property* (c) *holds;*
- $\mathcal{A}$ *is a* Dynkin system *if $\Omega \in \mathcal{A}$, and properties* (d) *and* (h) *hold.* □

*Remark 2.1.* Dynkin systems are also called $\lambda$-*systems.*

*Remark 2.2.* The definition of a Dynkin system varies. One alternative, in addition to the assumption that $\Omega \in \mathcal{A}$, is that (a) and (f) hold. □

**Exercise 2.4.** The obvious exercise is to show that the two definitions of a Dynkin system are equivalent. □

The definitions of the different collections of sets are obviously based on minimal requirements. By manipulating the different properties (a)–(j), for example together with the de Morgan formulas, other properties can be derived. The following relations between different collections of sets are obtained by such manipulations.

**Theorem 2.1.** *The following connections hold:*

1. *Every algebra is a $\pi$-system.*
2. *Every $\sigma$-algebra is an algebra.*
3. *An algebra is a $\sigma$-algebra if and only if it is a monotone class.*
4. *Every $\sigma$-algebra is a Dynkin system.*
5. *A Dynkin system is a $\sigma$-algebra if and only if it is $\pi$-system.*
6. *Every Dynkin system is a monotone class.*
7. *Every $\sigma$-algebra is a monotone class.*
8. *The power set of any subset of $\Omega$ is a $\sigma$-algebra on that subset.*
9. *The intersection of any number of $\sigma$-algebras, countable or uncountable, is, again, a $\sigma$-algebra.*
10. *The countable union of a non-decreasing sequence of $\sigma$-algebras is an algebra, but not necessarily a $\sigma$-algebra.*
11. *If $\mathcal{A}$ is a $\sigma$-algebra, and $B \subset \Omega$, then $B \cap \mathcal{A} = \{B \cap A : A \in \mathcal{A}\}$ is a $\sigma$-algebra on $B$.*
12. *If $\Omega$ and $\Omega'$ are sets, $\mathcal{A}'$ a $\sigma$-algebra in $\Omega'$ and $T : \Omega \to \Omega'$ a mapping, then $T^{-1}(\mathcal{A}') = \{T^{-1}(A') : A' \in \mathcal{A}'\}$ is a $\sigma$-algebra on $\Omega$.*

**Exercise 2.5.** (a) Prove the above statements.

(b) Find two $\sigma$-algebras, the union of which is not an algebra (only very few elements in each suffice).

(c) Prove that if, for the infinite set $\Omega$, $\mathcal{A}$ consists of all $A \subset \Omega$, such that either $A$ or $A^c$ is finite, then $\mathcal{A}$ is an algebra, but not a $\sigma$-algebra. □

## 2.3 Generators

Let $\mathcal{A}$ be a collection of subsets of $\Omega$. Since the power set, $\mathfrak{P}(\Omega)$, is a $\sigma$-algebra, it follows that there exists at least one $\sigma$-algebra containing $\mathcal{A}$. Since, moreover, the intersection of any number of $\sigma$-algebras is, again, a $\sigma$-algebra, there exists a *smallest* $\sigma$-algebra containing $\mathcal{A}$. In fact, let

$$\mathcal{F}^* = \{\sigma\text{-algebras} \supset \mathcal{A}\}.$$

The smallest $\sigma$-algebra containing $\mathcal{A}$ equals

$$\bigcap_{\mathcal{G} \in \mathcal{F}^*} \mathcal{G},$$

and is unique since we have intersected *all* $\sigma$-algebras containing $\mathcal{A}$.

**Definition 2.3.** *Let $\mathcal{A}$ be a collection of subsets of $\Omega$. The smallest $\sigma$-algebra containing $\mathcal{A}$, $\sigma\{\mathcal{A}\}$, is called the $\sigma$-algebra generated by $\mathcal{A}$. Similarly, the smallest Dynkin system containing $\mathcal{A}$, $\mathfrak{D}\{\mathcal{A}\}$, is called the Dynkin system generated by $\mathcal{A}$, and the smallest monotone class containing $\mathcal{A}$, $\mathfrak{M}\{\mathcal{A}\}$, is called the monotone class generated by $\mathcal{A}$. In each case $\mathcal{A}$ is called the generator of the actual collection.* □

*Remark 2.3.* The $\sigma$-algebra generated by $\mathcal{A}$ is also called "the minimal $\sigma$-algebra containing $\mathcal{A}$". Similarly for the other collections.

*Remark 2.4.* Let $\{\mathcal{A}_n, n \geq 1\}$ be $\sigma$-algebras. Even though the union need not be a $\sigma$-algebra, $\sigma\{\bigcup_{n=1}^{\infty} \mathcal{A}_n\}$, that is, the $\sigma$-algebra generated by $\{\mathcal{A}_n, n \geq 1\}$, always exists. □

**Exercise 2.6.** Prove that

(i) If $\mathcal{A} = A$, a single set, then $\sigma\{\mathcal{A}\} = \sigma\{A\} = \{\emptyset, A, A^c, \Omega\}$.
(ii) If $\mathcal{A}$ is a $\sigma$-algebra, then $\sigma\{\mathcal{A}\} = \mathcal{A}$. □

The importance and usefulness of generators is demonstrated by the following two results.

**Theorem 2.2.** *Let $\mathcal{A}$ be an algebra. Then*

$$\mathfrak{M}\{\mathcal{A}\} = \sigma\{\mathcal{A}\}.$$

*Proof.* Since every $\sigma$-algebra is a monotone class (Theorem 2.1) and $\mathfrak{M}\{\mathcal{A}\}$ is the minimal monotone class containing $\mathcal{A}$, we know from the outset that $\mathfrak{M}\{\mathcal{A}\} \subset \sigma\{\mathcal{A}\}$. To prove the opposite inclusion we must, due to the minimality of $\sigma\{\mathcal{A}\}$, prove that $\mathfrak{M}\{\mathcal{A}\}$ is a $\sigma$-algebra, for which it is sufficient to prove that $\mathfrak{M}\{\mathcal{A}\}$ is an algebra (Theorem 2.1 once more). This means that we have to verify that properties (a) and (b) hold;

$$\begin{cases} B \in \mathfrak{M}\{\mathcal{A}\} & \Longrightarrow \quad B^c \in \mathfrak{M}\{\mathcal{A}\}, \quad \text{and} \\ B, C \in \mathfrak{M}\{\mathcal{A}\} & \Longrightarrow \quad B \cup C \in \mathfrak{M}\{\mathcal{A}\}. \end{cases} \tag{2.2}$$

Toward this end, let

$$\mathcal{E}_1 = \{B \in \mathfrak{M}\{\mathcal{A}\} : B \cup C \in \mathfrak{M}\{\mathcal{A}\} \text{ for all } C \in \mathcal{A}\},$$
$$\mathcal{E}_2 = \{B \in \mathfrak{M}\{\mathcal{A}\} : B^c \in \mathfrak{M}\{\mathcal{A}\}\}.$$

We first note that $\mathcal{E}_1$ is a monotone class via the identities

$$\left(\bigcap_{k=1}^{\infty} B_k\right) \cup C = \bigcap_{k=1}^{\infty} (B_k \cup C) \quad \text{and} \quad \left(\bigcup_{k=1}^{\infty} B_k\right) \cup C = \bigcup_{k=1}^{\infty} (B_k \cup C), \tag{2.3}$$

and that $\mathcal{E}_2$ is a monotone class via the de Morgan formulas, (2.1).

Secondly, by definition, $\mathcal{A} \subset \mathfrak{M}\{\mathcal{A}\}$, and by construction,

$$\mathcal{A} \subset \mathcal{E}_k \subset \mathfrak{M}\{\mathcal{A}\}, \quad k = 1, 2,$$

so that, in view of minimality of $\mathfrak{M}\{\mathcal{A}\}$,

$$\mathfrak{M}\{\mathcal{A}\} = \mathcal{E}_1 = \mathcal{E}_2.$$

To finish off, let

$$\mathcal{E}_3 = \{B \in \mathfrak{M}\{\mathcal{A}\} : B \cup C \in \mathfrak{M}\{\mathcal{A}\} \text{ for all } C \in \mathfrak{M}\{\mathcal{A}\}\}.$$

Looking at $\mathcal{E}_1 = \mathfrak{M}\{\mathcal{A}\}$ from another angle, we have shown that for every $B \in \mathfrak{M}\{\mathcal{A}\}$ we know that if $C \in \mathcal{A}$, then $B \cup C \in \mathfrak{M}\{\mathcal{A}\}$, which means that

$$\mathcal{A} \subset \mathcal{E}_3.$$

Moreover, $\mathcal{E}_3$ is a monotone class via (2.3), so that, by minimality again, we must have $\mathfrak{M}\{\mathcal{A}\} = \mathcal{E}_3$.

We have thus shown that $\mathfrak{M}\{\mathcal{A}\}$ obeys properties (a) and (b).    □

By suppressing the minimality of the monotone class, the following corollary emerges (because an arbitrary monotone class contains the minimal one).

**Corollary 2.1.** *If $\mathcal{A}$ is an algebra and $\mathcal{G}$ a monotone class containing $\mathcal{A}$, then*

$$\mathcal{G} \supset \sigma\{\mathcal{A}\}.$$

A related theorem, *the monotone class theorem*, concerns the equality between the Dynkin system and the $\sigma$-algebra generated by the same $\pi$-system.

**Theorem 2.3.** (The monotone class theorem)
*If $\mathcal{A}$ is a $\pi$-system on $\Omega$, then*

$$\mathfrak{D}\{\mathcal{A}\} = \sigma\{\mathcal{A}\}.$$

*Proof.* The proof runs along the same lines as the previous one. Namely, one first observes that $\mathfrak{D}\{\mathcal{A}\} \subset \sigma\{\mathcal{A}\}$, since every $\sigma$-algebra is a Dynkin system (Theorem 2.1) and $\mathfrak{D}\{\mathcal{A}\}$ is the minimal Dynkin system containing $\mathcal{A}$.

For the converse we must show that $\mathfrak{D}\{\mathcal{A}\}$ is a $\pi$-system (Theorem 2.1). In order to achieve this, let

$$\mathcal{D}_C = \{B \subset \Omega : B \cap C \in \mathfrak{D}\{\mathcal{A}\}\} \quad \text{for} \quad C \in \mathfrak{D}\{\mathcal{A}\}.$$

We claim that $\mathcal{D}_C$ is a Dynkin system.

To prove this we check the requirements for a collection of sets to constitute a Dynkin system. In this case we use the following alternative (recall Remark 2.2), namely, we show that $\Omega \in \mathcal{D}_C$, and that (a) and (f) hold.

Let $C \in \mathfrak{D}\{\mathcal{A}\}$.

- Since $\Omega \cap C = C$, it follows that $\Omega \in \mathcal{D}_C$.
- If $B \in \mathcal{D}_C$, then

$$B^c \cap C = (\Omega \setminus B) \cap C = (\Omega \cap C) \setminus (B \cap C),$$

which shows that $B^c \in \mathcal{D}_C$.

- Finally, if $\{B_n, n \geq 1\}$ are disjoint sets in $\mathcal{D}_C$, then

$$\left( \bigcup_{n=1}^{\infty} B_n \right) \cap C = \bigcup_{n=1}^{\infty} (B_n \cap C),$$

which proves that $\bigcup_{n=1}^{\infty} B_n \in \mathcal{D}_C$.

The requirements for $\mathcal{D}_C$ to be a Dynkin system are thus fulfilled. And, since $C$ was arbitrarily chosen, this is true for any $C \in \mathfrak{D}\{\mathcal{A}\}$.

Now, since, by definition, $\mathcal{A} \subset \mathcal{D}_A$ for every $A \in \mathcal{A}$, it follows that

$$\mathfrak{D}\{\mathcal{A}\} \subset \mathcal{D}_A \quad \text{for every} \quad A \in \mathcal{A}.$$

For $C \in \mathfrak{D}\{\mathcal{A}\}$ we now have $C \cap A \in \mathfrak{D}\{\mathcal{A}\}$ for every $A \in \mathcal{A}$, which implies that $\mathcal{A} \subset \mathcal{D}_C$, and, hence, that $\mathfrak{D}\{\mathcal{A}\} \subset \mathcal{D}_C$ for every $C \in \mathfrak{D}\{\mathcal{A}\}$. Consequently,

$$B, C \in \mathfrak{D}\{\mathcal{A}\} \quad \Longrightarrow \quad B \cap C \in \mathfrak{D}\{\mathcal{A}\},$$

that is, $\mathfrak{D}\{\mathcal{A}\}$ is a $\pi$-system. □

By combining Theorems 2.2 and 2.3 (and the exercise preceding the former) the following result emerges.

**Corollary 2.2.** *If $\mathcal{A}$ is a $\sigma$-algebra, then*

$$\mathfrak{M}\{\mathcal{A}\} = \mathfrak{D}\{\mathcal{A}\} = \sigma\{\mathcal{A}\} = \mathcal{A}.$$

### 2.4 A Metatheorem and Some Consequences

A frequent proof technique is to establish some kind of reduction from an infinite setting to a finite one; simple functions, rectangles, and so on. Such proofs can often be identified in that they open by statements such as

"it suffices to check rectangles",

"it suffices to check step functions".

The basic idea behind such statements is that there either exists some approximation theorem that "takes care of the rest", or that some convenient part of Theorem 2.1 can be exploited for the remaining part of the proof. Our next result puts this device into a more stringent form, although in a somewhat metaphoric sense.

**Theorem 2.4.** (A Metatheorem)
(i) *Suppose that some property holds for some monotone class $\mathcal{E}$ of subsets. If $\mathcal{A}$ is an algebra that generates the $\sigma$-algebra $\mathcal{G}$ and $\mathcal{A} \subset \mathcal{E}$, then $\mathcal{E} \supset \mathcal{G}$.*
(ii) *Suppose that some property holds for some Dynkin system $\mathcal{E}$ of subsets. If $\mathcal{A}$ is a $\pi$-system that generates the $\sigma$-algebra $\mathcal{G}$ and $\mathcal{A} \subset \mathcal{E}$, then $\mathcal{E} \supset \mathcal{G}$.*

*Proof.* Let
$$\mathcal{E} = \{E : \text{the property is satisfied}\}.$$

(i): It follows from the assumptions and Theorem 2.2, respectively, that
$$\mathcal{E} \supset \mathfrak{M}\{\mathcal{A}\} = \sigma\{\mathcal{A}\} = \mathcal{G}.$$

(ii): Apply Theorem 2.3 to obtain
$$\mathcal{E} \supset \mathfrak{D}\{\mathcal{A}\} = \sigma\{\mathcal{A}\} = \mathcal{G}. \qquad \square$$

*Remark 2.5.* As the reader may have discovered, the proofs of Theorems 2.2 and 2.3 are of this kind.

*Remark 2.6.* The second half of the theorem is called *Dynkin's $\pi$-$\lambda$ theorem.*$\square$

## 3 The Probability Space

We now have sufficient mathematics at our disposal for a formal definition of the *probability space* or *probability triple*, $(\Omega, \mathcal{F}, P)$.

**Definition 3.1.** *The triple $(\Omega, \mathcal{F}, P)$ is a* probability (measure) space *if*

- $\Omega$ *is the* sample space, *that is, some (possibly abstract) set;*
- $\mathcal{F}$ *is a $\sigma$-algebra of sets (events) – the* measurable *subsets of $\Omega$. The "atoms", $\{\omega\}$, of $\Omega$, are called* elementary events*;*
- $P$ *is a* probability measure,

*that is, $P$ satisfies the following* Kolmogorov axioms*:*

*1. For any $A \in \mathcal{F}$, there exists a number $P(A) \geq 0$; the probability of $A$.*
*2. $P(\Omega) = 1$.*
*3. Let $\{A_n, n \geq 1\}$ be disjoint. Then*

$$P(\bigcup_{n=1}^{\infty} A_n) = \sum_{n=1}^{\infty} P(A_n). \qquad \square$$

*Remark 3.1.* Axiom 3 is called *countable additivity* (in contrast to *finite additivity*).    □

Departing from the axioms (only!) one can now derive various relations between probabilities of unions, subsets, complements and so on. Following is a list of some of them:

Let $A, A_1, A_2, \ldots$ be measurable sets. Then

- $P(A^c) = 1 - P(A)$;
- $P(\emptyset) = 0$;
- $P(A_1 \cup A_2) \leq P(A_1) + P(A_2)$;
- $A_1 \subset A_2 \quad \Longrightarrow \quad P(A_1) \leq P(A_2)$;
- $P(\bigcup_{k=1}^{n} A_k) + P(\bigcap_{k=1}^{n} A_k^c) = 1$.

**Exercise 3.1.** Prove these relations.    □

*Remark 3.2.* There exist non-empty sets which have probability 0.    □

*From now on we assume, unless otherwise stated, that all sets are measurable.*

## 3.1 Limits and Completeness

One of the basic questions in mathematics is to what extent limits of objects carry over to limits of functions of objects. In the present context the question amounts to whether or not probabilities of converging sets converge.

**Theorem 3.1.** *Suppose that $A$ and $\{A_n, \ n \geq 1\}$ are subsets of $\Omega$, such that $A_n \nearrow A$ $(A_n \searrow A)$ as $n \to \infty$. Then*

$$P(A_n) \nearrow P(A) \quad (P(A_n) \searrow P(A)) \qquad as \quad n \to \infty.$$

*Proof.* Suppose that $A_n \nearrow A$, let $B_1 = A_1$ and set $B_n = A_n \cap A_{n-1}^c$, $n \geq 2$. Then $\{B_n, \ n \geq 1\}$ are disjoint sets, and

$$A_n = \bigcup_{k=1}^{n} B_k \quad \text{for all} \quad n \geq 1, \quad \text{and} \quad \bigcup_{n=1}^{\infty} A_n = \bigcup_{n=1}^{\infty} B_n,$$

so that by Proposition 2.1 (and $\sigma$-additivity)

$$P(A_n) = \sum_{k=1}^{n} P(B_k) \nearrow \sum_{k=1}^{\infty} P(B_k) = P\left( \bigcup_{k=1}^{\infty} B_k \right) = P\left( \bigcup_{k=1}^{\infty} A_k \right) = P(A).$$

The case $A_n \searrow A$ follows similarly, or, alternatively, by considering complements (since $A_n^c \nearrow A^c$ as $n \to \infty$).    □

A slight extension of this result yields the following one.

**Theorem 3.2.** *Let $A$ and $\{A_n, n \geq 1\}$ be subsets of $\Omega$, and set, as before, $A_* = \liminf_{n \to \infty} A_n$, and $A^* = \limsup_{n \to \infty} A_n$. Then*

(i)     $P(A_*) \leq \liminf_{n \to \infty} P(A_n) \leq \limsup_{n \to \infty} P(A_n) \leq P(A^*)$;

(ii)     $A_n \to A$ *as* $n \to \infty \implies P(A_n) \to P(A)$ *as* $n \to \infty$.

*Proof.* (i): By definition, for any $n$, we obtain, recalling Proposition 2.1, that

$$A_* \nwarrow \bigcap_{m=n}^{\infty} A_m \subset A_n \subset \bigcup_{m=n}^{\infty} A_m \searrow A^*,$$

where the limits are taken as $n \to \infty$. Joining this with Theorem 3.1, yields

$$P(A_*) \leq \liminf_{n \to \infty} P(A_n) \leq \limsup_{n \to \infty} P(A_n) \leq P(A^*),$$

which proves (i), from which (ii) is immediate, since the extreme members coincide under the additional assumption of set convergence.     □

As a corollary we obtain the following intuitively reasonable result.

**Corollary 3.1.** *Suppose that $A_n \to \emptyset$ as $n \to \infty$. Then*

$$P(A_n) \to 0 \quad as \quad n \to \infty.$$

*Proof.* Immediate from the previous theorem with $A = \emptyset$.     □

To prepare for the next two results we introduce the notion of a *null set*.

**Definition 3.2.** *A set $A$ is a* null set *if there exists $B \in \mathcal{F}$, such that $B \supset A$ with $P(B) = 0$.*     □

In normal prose this means that a set is a null set if it is contained in a *measurable* set which has probability 0. In particular, *null sets need not be measurable*. The concept of *completeness* takes care of that problem.

**Definition 3.3.** *A probability space $(\Omega, \mathcal{F}, P)$ is* complete *if every null set is measurable, that is, if*

$$A \subset B \in \mathcal{F}, \quad P(B) = 0 \implies A \in \mathcal{F}, \quad (and, \ hence, \ P(A) = 0).$$     □

One can show that it is always possible to enlarge a given $\sigma$-algebra, and extend the given probability measure to make the, thus, extended probability space complete; one completes the probability space. It is therefore no restriction really to assume from the outset that a given probability space is complete. *In order to avoid being distracted from the main path, we assume from now on, without further explicit mentioning, that all probability spaces are complete.* Completeness is important in the theory of stochastic processes and for stochastic integration.

We close this subsection by showing that the union of a countable number of null sets remains a null set, and that the intersection of countably many sets with probability 1 also has probability 1. The meaning of the latter is that peeling off sets of probability 0 countably many times still does not reduce the intersection with more than a null set.

**Theorem 3.3.** *Suppose that* $\{A_n, n \geq 1\}$ *are subsets of* $\Omega$ *with* $P(A_n) = 0$ *for all* $n$. *Then*

$$P\Big(\bigcup_{n=1}^{\infty} A_n\Big) = 0.$$

*Proof.* By $\sigma$-sub-additivity,

$$P\Big(\bigcup_{n=1}^{\infty} A_n\Big) \leq \sum_{n=1}^{\infty} P(A_n) = 0.$$  $\square$

**Theorem 3.4.** *Suppose that* $\{B_n, n \geq 1\}$ *are subsets of* $\Omega$ *with* $P(B_n) = 1$ *for all* $n$. *Then*

$$P\Big(\bigcap_{n=1}^{\infty} B_n\Big) = 1.$$

*Proof.* Using the de Morgan formulas (2.1) and Theorem 3.3,

$$P\Big(\bigcap_{n=1}^{\infty} B_n\Big) = 1 - P\Big(\bigcup_{n=1}^{\infty} B_n^c\Big) = 1.$$  $\square$

Having defined the probability space, we prove, as a first result that for two probability measures to coincide it suffices that they agree on a suitable generator. The proof is a nice illustration of the Metatheorem, Theorem 2.4.

**Theorem 3.5.** *Suppose that* $P$ *and* $Q$ *are probability measures defined on the same probability space* $(\Omega, \mathcal{F})$, *and that* $\mathcal{F}$ *is generated by a* $\pi$-*system* $\mathcal{A}$. *If* $P(A) = Q(A)$ *for all* $A \in \mathcal{A}$, *then* $P = Q$, *i.e.,* $P(A) = Q(A)$ *for all* $A \in \mathcal{F}$.

*Proof.* Define

$$\mathcal{E} = \{A \in \mathcal{F} : P(A) = Q(A)\}.$$

Since

- $\Omega \in \mathcal{E}$,
- $A, B \in \mathcal{E},\ A \subset B \implies B \setminus A \in \mathcal{E}$,
- $A_n \in \mathcal{E},\ n \geq 1,\ A_n \nearrow \implies \bigcup_n A_n \in \mathcal{E}$,

where we used Theorem 3.1 in the final step, it follows that $\mathcal{E}$ is a Dynkin system. An application of Theorem 2.4 finishes the proof.  $\square$

## 3.2 An Approximation Lemma

The following result states that any set in a $\sigma$-algebra can be arbitrary well approximated by another set that belongs to an algebra that generates the $\sigma$-algebra. The need for this result is the fact that the infinite union of $\sigma$-algebras is not necessarily a $\sigma$-algebra (recall Theorem 2.1.10).

The general description of the result reveals that it reduces an infinite setting to a finite one, which suggests that the proof builds on the metatheorem technique.

**Lemma 3.1.** *Suppose that $\mathcal{F}_0$ is an algebra that generates the $\sigma$-algebra $\mathcal{F}$, that is, $\mathcal{F} = \sigma\{\mathcal{F}_0\}$. For any set $A \in \mathcal{F}$ and any $\varepsilon > 0$ there exists a set $A_\varepsilon \in \mathcal{F}_0$, such that*

$$P(A \bigtriangleup A_\varepsilon) < \varepsilon.$$

*Proof.* Let $\varepsilon > 0$, and define

$$\mathcal{G} = \{A \in \mathcal{F} : P(A \bigtriangleup A_\varepsilon) < \varepsilon \quad \text{for some} \quad A_\varepsilon \in \mathcal{F}_0\}.$$

(i): If $A \in \mathcal{G}$, then $A^c \in \mathcal{G}$, since $A^c \bigtriangleup (A_\varepsilon)^c = A \bigtriangleup A_\varepsilon$.
(ii): If $A_n \in \mathcal{G}$, $n \geq 1$, then so does the union. Namely, set $A = \bigcup_{n=1}^\infty A_n$, let $\varepsilon$ be given and choose $n_*$, such that

$$P\left(A \setminus \bigcup_{k=1}^{n_*} A_k\right) < \varepsilon. \qquad (3.1)$$

Next, let $\{A_{k,\varepsilon} \subset \mathcal{F}_0, 1 \leq k \leq n_*\}$ be such that

$$P(A_k \bigtriangleup A_{k,\varepsilon}) < \varepsilon \quad \text{for} \quad 1 \leq k \leq n_*. \qquad (3.2)$$

Since

$$\left(\bigcup_{k=1}^{n_*} A_k\right) \bigtriangleup \left(\bigcup_{k=1}^{n_*} A_{k,\varepsilon}\right) \subset \bigcup_{k=1}^{n_*}(A_k \bigtriangleup A_{k,\varepsilon}),$$

it follows that

$$P\left(\left(\bigcup_{k=1}^{n_*} A_k\right) \bigtriangleup \left(\bigcup_{k=1}^{n_*} A_{k,\varepsilon}\right)\right) \leq \sum_{k=1}^{n_*} P(A_k \bigtriangleup A_{k,\varepsilon}) < n_*\varepsilon,$$

so that, finally,

$$P\left(A \bigtriangleup \left(\bigcup_{k=1}^{n_*} A_{k,\varepsilon}\right)\right) < (n_* + 1)\varepsilon.$$

This proves the second claim – the claim would have followed with an approximation error $\varepsilon$ instead of $(n_* + 1)\varepsilon$ if we had chosen $\varepsilon$ to be $\varepsilon/2$ in (3.1) and as $\varepsilon/(2n_*)$ in (3.2). But that's cheating.

To summarize: $\mathcal{G}$ is non-empty, since $\mathcal{G} \supset \mathcal{F}_0$ by construction (choose $A_\varepsilon = A$ whenever $A \in \mathcal{F}_0$), and $\mathcal{G}$ obeys properties (a) and (e), so that $\mathcal{G}$ is a $\sigma$-algebra. Moreover, $\mathcal{G} \supset \mathcal{F}$, since $\mathcal{F}$ is the minimal $\sigma$-algebra containing $\mathcal{F}_0$, and, since, trivially, $\mathcal{G} \subset \mathcal{F}$, it finally follows that $\mathcal{G} = \mathcal{F}$. $\qquad \square$

## 3.3 The Borel Sets on $\mathbb{R}$

**Definition 3.4.** *A set $\Omega$ together with an associated $\sigma$-algebra, $\mathcal{A}$, i.e., the pair $(\Omega, \mathcal{A})$, is called a* measurable space. $\qquad \square$

In this subsection we shall find out what the terminology we have introduced above means for $\Omega = \mathbb{R}$, and characterize $\mathcal{R}$, the $\sigma$-algebra of Borel sets on the real line.

A set $F$ is *open* if, for every point $x \in F$, there exists an $\varepsilon$-ball, $B(x, \varepsilon) \subset F$. This means that the boundary points do not belong to the set; $\partial F \not\subset F$. A set $G$ is *closed* if its complement $G^c$ is open. If $G$ is closed, then $\partial G \subset G$.

The sets in $\mathcal{R}$ are called *Borel sets*, and the space $(\mathbb{R}, \mathcal{R})$ is called the *Borel space*. The $\sigma$-algebra of Borel sets, or the Borel-$\sigma$-algebra, is defined as the $\sigma$-algebra generated by the open subsets of $\mathbb{R}$;

$$\mathcal{R} = \sigma\{F : F \text{ is open}\}. \tag{3.3}$$

An important fact is that the Borel sets can, equivalently, be generated by intervals as follows.

**Theorem 3.6.** *We have*

$$\begin{aligned}
\mathcal{R} &= \sigma\{(a, b], \ -\infty \leq a \leq b < \infty\} \\
&= \sigma\{[a, b), \ -\infty < a \leq b \leq \infty\} \\
&= \sigma\{(a, b), \ -\infty \leq a \leq b \leq \infty\} \\
&= \sigma\{[a, b], \ -\infty < a \leq b < \infty\} \\
&= \sigma\{(-\infty, b], \ -\infty < b < \infty\}.
\end{aligned}$$

*Proof.* We confine ourselves by providing a sketch. The equivalences build on relations such as

$$(a, b) = \bigcup_{n=1}^{\infty} \left(a, b - \frac{1}{n}\right] \quad \text{and} \quad (a, b] = \bigcap_{n=1}^{\infty} \left(a, b + \frac{1}{n}\right),$$

and so on, or, more generally, by choosing a sequence $\{x_n, n \geq 1\}$, such that, if $x_n \downarrow 0$ as $n \to \infty$, then

$$(a, b) = \bigcup_{n=1}^{\infty} (a, b - x_n] \quad \text{and} \quad (a, b] = \bigcap_{n=1}^{\infty} (a, b + x_n).$$

Once these relations have been established one shows that a given $\sigma$-algebra is contained in another one and vice versa (which proves that they coincide), after which one of them is proven to be equivalent to (3.3). We leave the (boring?) details to the reader. $\qquad \square$

For probability measures on the real line Theorem 3.5 becomes

**Theorem 3.7.** *Suppose that $P$ and $Q$ are probability measures on $(\mathbb{R}, \mathcal{R})$ that agree on all intervals $(a, b], \ -\infty \leq a \leq b < \infty$, say. Then $P = Q$.*

*Proof.* The collection of intervals constitutes a $\pi$-system, that generates $\mathcal{R}$. $\square$

*Remark 3.3.* The theorem obviously remains true for all kinds of intervals mentioned in Theorem 3.6.                                                    □

The statement of Theorem 3.7 amounts to the fact that, if we know a probability measure on all intervals (of one kind) we know it on any Borel set. Knowledge on the intervals thus *determines* the measure. The intervals are said to form a *determining class*.

For comparison with higher dimensions we interpret intervals as one-dimensional rectangles.

## 3.4 The Borel Sets on $\mathbb{R}^n$

For Borel sets in higher (finite) dimensions one extends Theorem 3.6 to higher-dimensional rectangles. The extension of Theorem 3.7 tells us that two probability measures on $(\mathbb{R}^n, \mathcal{R}^n)$ agree if and only if they agree on the rectangles.

In infinite dimensions things become much harder. Existence follows from the famous Kolmogorov extension theorem. Moreover, by using the metatheorem technique one can show that if the finite-dimensional distributions of two probability measures agree, then the measures agree. The finite-dimensional distributions constitute a determining class. We omit all details.

# 4 Independence; Conditional Probabilities

One of the most central concepts of probability theory is *independence*, which means that successive experiments do not influence each other, that the future does not depend on the past, that knowledge of the outcomes so far does not provide any information about future experiments.

**Definition 4.1.** *The events $\{A_k, 1 \leq k \leq n\}$ are* independent *iff*

$$P\left(\bigcap A_{i_k}\right) = \prod P(A_{i_k}),$$

*where intersections and products, respectively, are to be taken over all subsets of $\{1, 2, \ldots, n\}$.*

*The events $\{A_n, n \geq 1\}$ are independent if $\{A_k, 1 \leq k \leq n\}$ are independent for all $n$.*                                           □

**Exercise 4.1.** How many equations does one have to check in order to establish that $\{A_k, 1 \leq k \leq n\}$ are independent?                           □

The classical examples are coin-tossing and throwing dice where successive outcomes are independent, and sampling without replacement from a finite population, where this is not the case, the typical example being drawing cards from a card deck.

**Exercise 4.2.** Prove that, if $A$ and $B$ are independent, then so are $A$ and $B^c$, $A^c$ and $B$, and $A^c$ and $B^c$.    □

A suggestive way to illustrate independence is to introduce *conditional probabilities*.

**Definition 4.2.** *Let $A$ and $B$ be two events, and suppose that $P(A) > 0$. The conditional probability of $B$ given $A$ is defined as*

$$P(B \mid A) = \frac{P(A \cap B)}{P(A)}.$$    □

The conditional probability thus measures the probability of $B$ given that we know that $A$ has occurred. The numerator is the probability that both of them occur, and the denominator rescales this number in order for conditional probabilities to satisfy the Kolmogorov axioms.

**Exercise 4.3.** Prove that, given $A$ with $P(A) > 0$, $P(\cdot \mid A)$ satisfies the Kolmogorov axioms, and, hence, is a bona fide probability measure.    □

If, in particular, $A$ and $B$ are independent, then

$$P(B \mid A) = \frac{P(A) \cdot P(B)}{P(A)} = P(B),$$

which means that knowing that $A$ has occurred does not change the probability of $B$ occurring. As expected.

*Remark 4.1.* It is also possible to begin by defining conditional probabilities, after which one "discovers" that for events satisfying $P(A \cap B) = P(A) \cdot P(B)$ one has $P(B \mid A) = P(B)$, which implies that for such events, the fact that $A$ has occurred does not change the probability of $B$ occurring, after which one introduces the notion of independence. In order to take care of sets with measure 0 one observes that

$$0 \leq P(A \cap B) \leq P(A) = 0,$$

i.e., null sets are independent of everything.

Note, in particular, that a null set is independent of itself.    □

## 4.1 The Law of Total Probability; Bayes' Formula

Having introduced conditional probabilities, the following facts are just around the corner.

**Definition 4.3.** *A* partition *of $\Omega$ is a collection of disjoint sets, the union of which equals $\Omega$. Technically, $\{H_k, 1 \leq k \leq n\}$ is a partition of $\Omega$ if*

$$\Omega = \bigcup_{k=1}^{n} H_k, \quad where \quad H_i \cap H_j = \emptyset \quad for \quad 1 \leq i, j \leq n, \ i \neq j.$$    □

**Proposition 4.1.** (The law of total probability)
*Let $\{H_k, 1 \leq k \leq n\}$ be a partition of $\Omega$. Then, for any event $A \subset \Omega$,*

$$P(A) = \sum_{k=1}^{n} P(A \mid H_k) \cdot P(H_k).$$

**Proposition 4.2.** (Bayes' formula)
*Let $\{H_k, 1 \leq k \leq n\}$ be a partition of $\Omega$. Then, for any event $A \subset \Omega$, such that $P(A) > 0$,*

$$P(H_k \mid A) = \frac{P(A \mid H_k) \cdot P(H_k)}{P(A)} = \frac{P(A \mid H_k) \cdot P(H_k)}{\sum_{i=1}^{n} P(A \mid H_i) \cdot P(H_i)}.$$

**Exercise 4.4.** Prove these two results.                                   □

## 4.2 Independence of Collections of Events

Next we extend the definition of independence to independence between collections, in particular $\sigma$-algebras, of events.

**Definition 4.4.** *Let $\{\mathcal{A}_k\}$ be a finite or infinite collection. The collections are independent iff, for any $k \in \mathbb{N}$, and non-empty subset of indices $i_1, i_2, \ldots, i_k$, the events $\{A_{i_j} \in \mathcal{A}_{i_j}, j = 1, 2, \ldots, k\}$ are independent.*                                   □

*Remark 4.2.* It follows from the definition that:

- If every collection of events contains exactly one event, the definition reduces to Definition 4.1.
- An infinite collection of events is independent if and only if every finite sub-collection is independent.
- Every sub-collection of independent events is independent.
- Disjoint sub-collections of independent events are independent.                                   □

**Exercise 4.5.** Check the statements of the remark.                                   □

From the metatheorem and some of its consequences we already know that it frequently suffices to "check rectangles", which, more stringently speaking, amounts to the fact that it suffices to check some generator. This is also true for independence.

**Theorem 4.1.** (i) *If $\{\mathcal{A}_n, n \geq 1\}$ are independent non-empty collections of events, then so are $\{\mathfrak{D}\{\mathcal{A}_n\}, n \geq 1\}$, the Dynkin systems generated by $\{\mathcal{A}_n, n \geq 1\}$.*
(ii) *If, in addition, $\{\mathcal{A}_n, n \geq 1\}$ are $\pi$-systems, then $\{\sigma\{\mathcal{A}_n\}, n \geq 1\}$, the $\sigma$-algebras generated by $\{\mathcal{A}_n, n \geq 1\}$, are independent.*                                   □

*Proof.* Since, as was noticed before, an infinite collection is independent if and only if every finite subcollection is independent, it is no restriction to depart from a finite collection $\{\mathcal{A}_k, 1 \leq k \leq n\}$. Moreover, once the result is established for $n = 2$, the general case follows by induction.

(i): Let $n = 2$, and define

$$\mathcal{E}_C = \{B \in \mathfrak{D}\{\mathcal{A}_1\} : P(B \cap C) = P(B) \cdot P(C)\} \quad \text{for} \quad C \in \mathcal{A}_2.$$

We claim that $\mathcal{E}_C$ is a Dynkin system for every $C \in \mathcal{A}_2$.
  To see this we argue as in the proof of Theorem 2.3:
  Let $C \in \mathcal{A}_2$.

- Since $P(\Omega \cap C) = P(C) = P(\Omega) \cdot P(C)$, we have $\Omega \in \mathcal{E}_C$.
- For $B \in \mathcal{E}_C$, we have $B^c \in \mathcal{E}_C$, since

$$\begin{aligned}
P(B^c \cap C) &= P((\Omega \smallsetminus B) \cap C) = P((\Omega \cap C) \smallsetminus (B \cap C)) \\
&= P(C \smallsetminus (B \cap C)) = P(C) - P(B \cap C) \\
&= P(C) - P(B)P(C) = (1 - P(B))P(C) \\
&= P(B^c) \cdot P(C).
\end{aligned}$$

- If $B_n \in \mathcal{E}_C$, $n \geq 1$, are disjoint sets, then $\bigcup_{n=1}^{\infty} B_n \in \mathcal{E}_C$, because

$$P\Big(\Big(\bigcup_{n=1}^{\infty} B_n\Big) \cap C\Big) = P\Big(\bigcup_{n=1}^{\infty}(B_n \cap C)\Big) = \sum_{n=1}^{\infty} P(B_n \cap C)$$

$$= \sum_{n=1}^{\infty} P(B_n) \cdot P(C) = P\Big(\bigcup_{n=1}^{\infty} B_n\Big) \cdot P(C).$$

This concludes the proof of the fact that $\mathcal{E}_C$ is a Dynkin system for every $C \in \mathcal{A}_2$.
  Next, since $\mathcal{A}_1 \subset \mathcal{E}_C$ for every $C \in \mathcal{A}_2$, it follows that, $\mathfrak{D}\{\mathcal{A}_1\} \subset \mathcal{E}_C$ for every $C \in \mathcal{A}_2$, which, by definition, means that $\mathfrak{D}\{\mathcal{A}_1\}$ and $\mathcal{A}_2$ are independent.
  Repeating the same arguments with

$$\mathcal{F}_C = \{B \in \mathfrak{D}\{\mathcal{A}_2\} : P(B \cap C) = P(B) \cdot P(C)\} \quad \text{for} \quad C \in \mathfrak{D}\{\mathcal{A}_1\},$$

that is, with $\mathfrak{D}\{\mathcal{A}_2\}$ and $\mathfrak{D}\{\mathcal{A}_1\}$ playing the roles of $\mathfrak{D}\{\mathcal{A}_1\}$ and $\mathcal{A}_2$, respectively, shows that $\mathfrak{D}\{\mathcal{A}_1\}$ and $\mathfrak{D}\{\mathcal{A}_2\}$ are independent as desired.
  This completes the proof of the first part of the theorem.
(ii): The second half is immediate from the first part and Theorem 2.3, according to which the $\sigma$-algebras and Dynkin systems coincide because of the assumption that $\{\mathcal{A}_n, n \geq 1\}$ are $\pi$-systems.    $\square$

### 4.3 Pair-wise Independence

This is an independence concept which is slightly weaker than independence as defined in Definition 4.1.

**Definition 4.5.** *The events $\{A_k, 1 \leq k \leq n\}$ are* pair-wise independent *iff*

$$P(A_i \cap A_j) = P(A_i) \cdot P(A_j) \quad \text{for all} \quad 1 \leq i \neq j \leq n.$$    $\square$

**Exercise 4.6.** How many equations does one have to check?    □

Having defined pair-wise independence the obvious request at this point is an example which shows that this is something different from independence as defined earlier.

*Example 4.1.* Pick one of the four points $(1,0,0)$, $(0,1,0)$, $(0,0,1)$, and $(1,1,1)$ at random, that is, with probability $1/4$ each, and let

$$A_k = \{\text{the } k\text{th coordinate equals } 1\}, \quad k = 1, 2, 3.$$

An easy calculation shows that

$$P(A_k) = \frac{1}{2} \quad \text{for} \quad k = 1, 2, 3,$$

$$P(A_i \cap A_j) = \frac{1}{4} \quad \text{for} \quad i, j = 1, 2, 3, \ i \neq j,$$

$$P(A_i)P(A_j) = \frac{1}{4} \quad \text{for} \quad i, j = 1, 2, 3, \ i \neq j,$$

$$P(A_1 \cap A_2 \cap A_3) = \frac{1}{4},$$

$$P(A_1)P(A_2)P(A_3) = \frac{1}{8}.$$

The sets $A_1$, $A_2$, $A_3$ are thus pair-wise independent, but not independent.    □

## 5 The Kolmogorov Zero-one Law

One magic result in probability theory concerns situations in which the probability of an event can only be 0 or 1. Theorems that provide criteria for such situations are called *zero-one laws*. We shall encounter one such result here, another one in Section 2.18, and then reprove the first one in Chapter 10.

In order to state the theorem we need to define the $\sigma$-algebra that contains information about "what happens at infinity".

Let $\{A_n, n \geq 1\}$ be arbitrary events and, set

$$\mathcal{A}_n = \sigma\{A_1, A_2, \ldots, A_n\} \quad \text{for} \quad n \geq 1,$$
$$\mathcal{A}'_n = \sigma\{A_{n+1}, A_{n+2}, \ldots\} \quad \text{for} \quad n \geq 0.$$

**Definition 5.1.** *The tail-$\sigma$-field is defined as*

$$\mathcal{T} = \bigcap_{n=0}^{\infty} \mathcal{A}'_n.$$    □

*Remark 5.1.* Since $\mathcal{T}$ is defined as an intersection of $\sigma$-algebras it is, indeed, itself a $\sigma$-algebra.    □

If we think of $n$ as time, then $\mathcal{A}'_n$ contains the information beyond time $n$ and $\mathcal{T}$ contains the information "beyond time $n$ for all $n$", that is, loosely speaking, the information at infinity. Another name for the tail-$\sigma$-field is *the $\sigma$-algebra of remote events*.

The famous Kolmogorov zero-one law states that if $\{A_n, \ n \geq 1\}$ are independent events, then the tail-$\sigma$-field is trivial, which means that it only contains sets of probability measure 0 or 1.

**Theorem 5.1.** (The Kolmogorov zero-one law)
*Suppose that $\{A_n, \ n \geq 1\}$ are independent events. If $A \in \mathcal{T}$, then*

$$P(A) = 0 \quad or \quad 1.$$

There exist several proofs of this theorem. We shall provide one "proof" and one proof.

*"Proof".* This is only a sketch which has to be rectified. The essential idea is that if $A \in \mathcal{T}$, then $A \in \mathcal{A}'_n$ for all $n$ and, hence, is independent of $\mathcal{A}_n$ for all $n$. This implies, on the one hand, that

$$A \in \sigma\{A_1, A_2, A_3, \ldots\}, \tag{5.1}$$

and, on the other hand, that

$$A \text{ is independent of } \sigma\{A_1, A_2, A_3, \ldots\}. \tag{5.2}$$

By combining the two it follows that $A$ *is independent of itself*, which leads to the very special equation,

$$P(A) = P(A \cap A) = P(A) \cdot P(A) = (P(A))^2, \tag{5.3}$$

the solutions of which are $P(A) = 0$ and $P(A) = 1$.                          "$\square$"

This is not completely stringent; the passages to infinity have to be performed with greater care, remember, for example, that the infinite union of $\sigma$-algebras need not be a $\sigma$-algebra.

*Proof.* Suppose that $A \in \mathcal{T}$. This means that $A \in \mathcal{A}'_n$ for every $n$, and, hence, that $A$ and $\mathcal{A}_n$ are independent for every $n$.

To turn the "proof" into a proof we need to rectify the transition to (5.2). This is achieved by the following slight elaboration of Lemma 3.1, namely, that there exists a sequence $E_n \in \mathcal{A}_n$, such that

$$P(E_n \triangle A) \to 0 \quad \text{as} \quad n \to \infty.$$

This tells us that

$$P(E_n \cap A) \to P(A) \quad \text{as} \quad n \to \infty, \tag{5.4}$$

$$P(E_n) \to P(A) \quad \text{as} \quad n \to \infty. \tag{5.5}$$

Moreover, since $A \in \mathcal{A}'_n$ for *all* $n$, it follows that $A$ is independent of *every* $E_n$, so that (5.4) can be rewritten as

$$P(A) \cdot P(E_n) \to P(A) \quad \text{as} \quad n \to \infty. \tag{5.6}$$

However, by (5.5) we also have

$$P(A) \cdot P(E_n) \to (P(A))^2 \quad \text{as} \quad n \to \infty, \tag{5.7}$$

which, in view of (5.6) forces the equation $P(A) = (P(A))^2$, and we rediscover (5.3) once again. □

**Exercise 5.1.** Write down the details of the elaboration of Lemma 3.1 that was used in the section proof. □

We shall return to the theorem in Chapter 10 and provide another, very elegant, proof (which, admittedly, rests on *a lot* more background theory).

## 6 Problems

Throughout, $A, B, \{A_n, n \geq 1\}$, and $\{B_n, n \geq 1\}$ are subsets of $\Omega$.

1. Show that

$$A \triangle B = (A \cap B^c) \cup (B \cap A^c) = (A \cup B) \setminus (A \cap B),$$
$$A^c \triangle B^c = A \triangle B,$$
$$\{A_1 \cup A_2\} \triangle \{B_1 \cup B_2\} \subset \{A_1 \triangle B_1\} \cup \{A_2 \triangle B_2\}.$$

2. Which of the following statements are (not) true?

$$\limsup_{n \to \infty}\{A_n \cup B_n\} = \limsup_{n \to \infty} A_n \cup \limsup_{n \to \infty} B_n ;$$
$$\limsup_{n \to \infty}\{A_n \cap B_n\} = \limsup_{n \to \infty} A_n \cap \limsup_{n \to \infty} B_n ;$$
$$\liminf_{n \to \infty}\{A_n \cup B_n\} = \liminf_{n \to \infty} A_n \cup \liminf_{n \to \infty} B_n ;$$
$$\liminf_{n \to \infty}\{A_n \cap B_n\} = \liminf_{n \to \infty} A_n \cap \liminf_{n \to \infty} B_n ;$$
$$A_n \to A, \quad B_n \to B \Longrightarrow A_n \cup B_n \to A \cup B \quad \text{as} \quad n \to \infty ;$$
$$A_n \to A, \quad B_n \to B \Longrightarrow A_n \cap B_n \to A \cap B \quad \text{as} \quad n \to \infty .$$

3. Let $\Omega = \mathbb{R}$, set $I_k = [k-1, k)$ for $k \geq 1$, and let $\mathcal{A}_n = \sigma\{I_k, k = 1, 2, \dots, n\}$. Show that $\bigcup_{n=1}^{\infty} \mathcal{A}_n$ is *not* a $\sigma$-algebra.

4. Suppose that $B_k = \bigcup_{j=1}^{k} A_j$ for $k = 1, 2, \dots, n$. Prove that

$$\sigma\{B_1, B_2, \dots, B_n\} = \sigma\{A_1, A_2, \dots, A_n\}.$$

5. Let $\{A_n, n \geq 1\}$ be a sequence of sets.
   (a) Suppose that $P(A_n) \to 1$ as $n \to \infty$. Prove that there exists a subsequence $\{n_k, k \geq 1\}$, such that

$$P\left(\bigcap_{k=1}^{\infty} A_{n_k}\right) > 0.$$

(b) Show that this is *not* true if we only assume that

$$P(A_n) \geq \alpha \quad \text{for all} \quad n. \tag{6.1}$$

(c) Show that, if $A_n$, $n \geq 1$, are non-decreasing and such that (6.1) holds for some $\alpha \in (0,1)$, then

$$P\left(\bigcap_{n=1}^{\infty} A_n\right) \geq \alpha.$$

(d) Show that, if $A_n$, $n \geq 1$, are non-increasing and

$$P(A_n) \leq \alpha \quad \text{for all} \quad n$$

for some $\alpha \in (0,1)$, then

$$P\left(\bigcap_{n=1}^{\infty} A_n\right) \leq \alpha.$$

6. Show that

$$I\{A \cup B\} = \max\{I\{A\}, I\{B\}\} = I\{A\} + I\{B\} - I\{A\} \cdot I\{B\};$$
$$I\{A \cap B\} = \min\{I\{A\}, I\{B\}\} = I\{A\} \cdot I\{B\};$$
$$I\{A \triangle B\} = I\{A\} + I\{B\} - 2I\{A\} \cdot I\{B\} = (I\{A\} - I\{B\})^2.$$

7. Show that

$$A_n \to A \text{ as } n \to \infty \quad \Longleftrightarrow \quad I\{A_n\} \to I\{A\} \text{ as } n \to \infty.$$

8. Let, for $n \geq 1$, $0 \leq a_n \nearrow 1$, and $1 < b_n \searrow 1$ as $n \to \infty$. Show that

$$\sup_n [0, a_n) = [0, 1) \quad \text{and/but} \quad \sup_n [0, a_n] \neq [0, 1];$$
$$\inf_n [0, b_n] = [0, 1] \quad \text{and} \quad \inf_n [0, b_n) = [0, 1].$$

9. *The Bonferroni inequalities.* Show that

$$P\left(\bigcup_{k=1}^{n} A_k\right) \leq \sum_{k=1}^{n} P(A_k);$$

$$P\left(\bigcup_{k=1}^{n} A_k\right) \geq \sum_{k=1}^{n} P(A_k) - \sum_{1 \leq i < j \leq n} P(A_i \cap A_j);$$

$$P\left(\bigcup_{k=1}^{n} A_k\right) \leq \sum_{k=1}^{n} P(A_k) - \sum_{1 \leq i < j \leq n} P(A_i \cap A_j)$$
$$+ \sum_{1 \leq i < j < k \leq n} P(A_i \cap A_j \cap A_k).$$

10. *The inclusion-exclusion formula.* Show that

$$P\left(\bigcup_{k=1}^{n} A_k\right) = \sum_{k=1}^{n} P(A_k) - \sum_{1\leq i<j\leq n} P(A_i \cap A_j)$$

$$+ \sum_{1\leq i<j<k\leq n} P(A_i \cap A_j \cap A_k)$$

$$+\cdots - (-1)^n P(A_1 \cap A_2 \cap \cdots \cap A_n).$$

11. Suppose that the events $\{A_n, n \geq 1\}$ are independent. Show that the inclusion-exclusion formula reduces to

$$P\left(\bigcup_{k=1}^{n} A_k\right) = 1 - \prod_{k=1}^{n} (1 - P(A_k)).$$

12. Show, in the setup of the previous problem, that

$$P\left(\bigcup_{k=1}^{n} A_k\right) \geq 1 - \exp\left\{-\sum_{k=1}^{n} P(A_k)\right\},$$

$$\sum_{n=1}^{\infty} P(A_n) = \infty \implies P\left(\bigcup_{n=1}^{\infty} A_n\right) = 1.$$

13. Let $A_n$, $n \geq 1$, be Borel sets on the Lebesgue space $([0,1], \mathcal{F}(0,1), \lambda)$. Show that, if there exists $\eta > 0$, such that $\lambda(A_n) \geq \eta$ for all $n$, then there exists at least one point that belongs to infinitely many sets $A_n$.

14. At the end of a busy day $n$ fathers arrive at kindergarten to pick up their kids. Each father picks a child to take home uniformly at random. Use the inclusion-exclusion formula to show that the probability that at least one father picks his own child equals

$$1 - \frac{1}{2!} + \frac{1}{3!} - \frac{1}{4!} + \cdots - (-1)^n \frac{1}{n!},$$

and that this probability tends to $1 - 1/e \approx 0.6321$ as $n \to \infty$.

♣ The mathematical formulation of this problem, which is called the *rencontre problem* or the *mathcing problem*, is that we seek the probability that a random permutation of the numbers $1, 2, \ldots, n$ leaves at least one position unchanged. The traditional (outdated) formulation is that $n$ men pick a hat at random from the hat rack when they leave a party, and one seeks the probability that at least one of them picks his own hat.

# 2
## Random Variables

The standard situation in the modeling of a random phenomenon is that the quantities of interest, rather than being defined on the underlying probability space, are *functions* from the probability space to some other (measurable) space. These functions are called *random variables*. Strictly speaking, one uses the term random variable when they are functions from the probability space to $\mathbb{R}$. If the image is in $\mathbb{R}^n$ for some $n \geq 2$ one talks about $n$-dimensional random variables or simply random vectors. If the image space is a general abstract one, one talks about random elements.

## 1 Definition and Basic Properties

Let $(\Omega, \mathcal{F}, P)$ be a probability space.

**Definition 1.1.** *A random variable $X$ is a measurable function from the sample space $\Omega$ to $\mathbb{R}$;*

$$X : \Omega \to \mathbb{R},$$

*that is, the inverse image of any Borel set is $\mathcal{F}$-measurable:*

$$X^{-1}(A) = \{\omega : X(\omega) \in A\} \in \mathcal{F} \quad \text{for all} \quad A \in \mathcal{R}.$$

*We call $X$ simple if, for some $n$,*

$$X = \sum_{k=1}^{n} x_k I\{A_k\},$$

*where $\{x_k, 1 \leq k \leq n\}$ are real numbers, and $\{A_k, 1 \leq k \leq n\}$ is a* finite *partition of $\Omega$, that is $A_i \cap A_j = \emptyset$ if $i \neq j$ and $\cup_{k=1}^{n} A_k = \Omega$.*
*We call $X$ elementary if*

$$X = \sum_{n=1}^{\infty} x_n I\{A_n\},$$

*where* $\{x_n, \, n \geq 1\}$ *are real numbers, and* $\{A_n, \, n \geq 1\}$ *is an* infinite *partition of* $\Omega$.

*If* $X : \Omega \to [-\infty, +\infty]$ *we call* $X$ *an* extended *random variable.* □

Random variables are traditionally denoted by large capitals toward the end of the alphabet; $X$, $Y$, $Z$, $U$, $V$, $W$, …. For sequences of "similar kinds" it is convenient to use indices; $X_1$, $X_2$, …, and so on.

We do not distinguish between random variables that differ on a null set.

**Definition 1.2.** *Random variables which only differ on a null set are called* equivalent

*The* equivalence class *of a random variable* $X$ *is the collection of random variables that differ from* $X$ *on a null set.*

*If* $X$ *and* $Y$ *are equivalent random variables we write* $X \sim Y$. □

So far we have described the map from $(\Omega, \mathcal{F})$ to $(\mathbb{R}, \mathcal{R})$. In order to complete the picture we must find a third component in the triple – the appropriate probability measure.

To each random variable $X$ we associate an *induced probability measure*, $\mathbb{P}$, through the relation

$$\mathbb{P}(A) = P(X^{-1}(A)) = P(\{\omega : X(\omega) \in A\}) \quad \text{for all} \quad A \in \mathcal{R}. \quad (1.1)$$

In words this means that we define the probability (on $(\mathbb{R}, \mathcal{R})$) that the random variable $X$ falls into a Borel set as the probability (on $(\Omega, \mathcal{F})$) of the inverse image of this Borel set. This is the motivation for the measurability assumption.

That the definition actually works is justified by the following result.

**Theorem 1.1.** *The induced space* $(\mathbb{R}, \mathcal{R}, \mathbb{P})$ *with* $\mathbb{P}$ *defined by (1.1) is a probability space – the* induced *probability space.*

*Proof.* The proof amounts to checking the Kolmogorov axioms, which, in turn, amounts to going back and forth between the two probability spaces.
1.  $\mathbb{P}(A) = P(\{\omega : X(\omega) \in A\}) \geq 0$ for any $A \in \mathcal{R}$.
2.  $\mathbb{P}(X) = P(\{\omega : X(\omega) \in \Omega\}) = 1$.
3.  Suppose that $\{A_n, \, n \geq 1\}$ are disjoint subsets of $\mathcal{R}$. Then

$$\mathbb{P}\left( \bigcup_{n=1}^{\infty} A_n \right) = P\left( \left\{ \omega : X(\omega) \in \bigcup_{n=1}^{\infty} A_n \right\} \right) = P\left( \bigcup_{n=1}^{\infty} \{\omega : X(\omega) \in A_n\} \right)$$

$$= \sum_{n=1}^{\infty} P(\{\omega : X(\omega) \in A_n\}) = \sum_{n=1}^{\infty} \mathbb{P}(A_n). \qquad \square$$

*Remark 1.1.* Once one gets used to the fact that only the random variables are of interest one need no longer worry about the exact probability space behind the random variables. In the remainder of this book we therefore refrain from distinguishing between the probability measures $P$ and $\mathbb{P}$, and we omit the brackets { and } to emphasize that $\{X \in A\}$ actually is a *set*. So, instead of writing $\mathbb{P}(A)$ we shall write $P(X \in A)$. There is absolutely no danger of confusion!                                                                                   □

**Definition 1.3.** *A degenerate random variable is constant with probability 1. Thus, X is degenerate if, for some $a \in \mathbb{R}$, $P(X = a) = 1$. A random variable that is not degenerate is called* non-degenerate.                                                                                   □

There are different ways to interpret the equality $X = Y$. The random variables $X$ and $Y$ are *equal in distribution* iff they are governed by the same probability measure:

$$X \stackrel{d}{=} Y \quad \Longleftrightarrow \quad P(X \in A) = P(Y \in A) \quad \text{for all} \quad A \in \mathcal{R}$$

and they are *point-wise equal*, iff they agree for almost all elementary events:

$$X \stackrel{a.s.}{=} Y \quad \Longleftrightarrow \quad P(\{\omega : X(\omega) = Y(\omega)\}) = 1,$$

i.e., $X$ and $Y$ are equivalent random variables, $X \sim Y$.

Next we provide an example to illustrate that there is a clear difference between the two equality concepts. The following example shows that two random variables, in fact, may well have the same distribution, and at the same time there is no elementary event where they agree.

*Example 1.1.* Toss a fair coin once and set

$$X = \begin{cases} 1, & \text{if the outcome is heads,} \\ 0, & \text{if the outcome is tails,} \end{cases}$$

and

$$Y = \begin{cases} 1, & \text{if the outcome is tails,} \\ 0, & \text{if the outcome is heads.} \end{cases}$$

Clearly, $P(X = 1) = P(X = 0) = P(Y = 1) = P(Y = 0) = 1/2$, in particular, $X \stackrel{d}{=} Y$. But $X(\omega)$ and $Y(\omega)$ differ for every $\omega$.                                                                                   □

**Exercise 1.1.** Prove that if $X(\omega) = Y(\omega)$ for almost all $\omega$, then $X \stackrel{d}{=} Y$.                                                                                   □

For $X$ to be a random variable one has to check that the set $\{\omega : X(\omega) \in A\} \in \mathcal{F}$ for all $A \in \mathcal{R}$. However, as a consequence of Theorem 1.3.6 it suffices to check measurability for (e.g.) all sets of the form $(-\infty, x]$; why? This important fact deserves a separate statement.

**Theorem 1.2.** *X is a random variable iff*

$$\{\omega : X(\omega) \leq x\} \in \mathcal{F} \quad \text{for all} \quad x \in \mathbb{R}.$$

## 1.1 Functions of Random Variables

A random variable is, as we have seen, a function from one space ($\Omega$) to another space ($\mathbb{R}$). What can be said of a (real valued) function of a random variable? Since, we know from analysis that a function of a function is a function, the following result does not come to us as a surprise.

**Theorem 1.3.** *A Borel measurable function of a random variable is a random variable, viz., if g is a real, Borel measurable function and X a random variable, then $Y = g(X)$ is a random variable.*

*Proof.* The proof follows, in fact, from the verbal statement, since $Y$ is a composite mapping from $\Omega$ "via $\mathbb{R}$" to $\mathbb{R}$. A more detailed proof of this is that, for any $A \in \mathcal{R}$,

$$\{Y \in A\} = \{\omega : Y(\omega) \in A\} = \{\omega : g(X(\omega)) \in A\}$$
$$= \{\omega : X(\omega) \in g^{-1}(A)\} \in \mathcal{F}. \qquad \square$$

By taking advantage of Theorem 1.2 we can prove measurability of the following functions, that is, we can prove that the following objects are, indeed, random variables.

**Proposition 1.1.** *Suppose that $X_1$, $X_2$, ... are random variables. The following quantities are random variables:*

(a)   $\max\{X_1, X_2\}$ *and* $\min\{X_1, X_2\}$;
(b)   $\sup_n X_n$ *and* $\inf_n X_n$;
(c)   $\limsup_{n \to \infty} X_n$ *and* $\liminf_{n \to \infty} X_n$.
(d)   *If $X_n(\omega)$ converges for every $\omega$ as $n \to \infty$, then* $\lim_{n \to \infty} X_n$ *is a random variable.*

*Proof.* (a): For any $x$,

$$\{\omega : \max\{X_1, X_2\}(\omega) \le x\} = \{\omega : \max\{X_1(\omega), X_2(\omega)\} \le x\}$$
$$= \{\omega : X_1(\omega) \le x\} \cap \{\omega : X_2(\omega) \le x\}$$

and

$$\{\omega : \min\{X_1, X_2\}(\omega) \le x\} = \{\omega : \min\{X_1(\omega), X_2(\omega)\} \le x\}$$
$$= \{\omega : X_1(\omega) \le x\} \cup \{\omega : X_2(\omega) \le x\},$$

which proves (a), since an intersection and a union, respectively, of two measurable sets are measurable.
(b): Similarly,

$$\{\omega : \sup_n X_n(\omega) \le x\} = \bigcap_n \{\omega : X_n(\omega) \le x\} \in \mathcal{F},$$

since a countable intersection of measurable sets is measurable, and

$$\{\omega : \inf_n X_n(\omega) < x\} = \bigcup_n \{\omega : X_n(\omega) < x\} \in \mathcal{F},$$

since a countable union of measurable sets is measurable.

(c): In this case,

$$\{\omega : \limsup_{n \to \infty} X_n(\omega) \le x\} = \{\omega : \inf_n \sup_{m \ge n} X_m(\omega) \le x\}$$

$$= \bigcup_{n=1}^{\infty} \bigcap_{m=n}^{\infty} \{\omega : X_m(\omega) \le x\} \in \mathcal{F},$$

and

$$\{\omega : \liminf_{n \to \infty} X_n(\omega) < x\} = \{\omega : \sup_n \inf_{m \ge n} X_m(\omega) < x\}$$

$$= \bigcap_{n=1}^{\infty} \bigcup_{m=n}^{\infty} \{\omega : X_m(\omega) < x\} \in \mathcal{F},$$

since, once again, we have only performed legal operations.

Alternatively, since $\sup_{n \ge m} X_n(\omega)$ is a random variable by (b), it follows, also by (b), that $\inf_m \left( \sup_{n \ge m} X_n(\omega) \right)$ is a random variable. Similarly for $\liminf_{n \to \infty} X_n(\omega)$.

(d): This is true, because in this case lim sup and lim inf coincide (and both are measurable). Moreover, the limit exists and is equal to the common value of lim sup and lim inf. $\square$

**Exercise 1.2.** Prove that a continuous function of a random variable is a random variable. $\square$

The usual construction procedure for properties of functions, or for constructions of new (classes of) functions, is to proceed from non-negative simple functions, sometimes via elementary functions, to non-negative functions, to the general case. This works for random variables too. The following lemma is closely related to Lemma A.9.3 which deals with the approximation of real valued functions.

**Lemma 1.1.** (i) *For every non-negative random variable $X$ there exists a sequence $\{X_n, n \ge 1\}$ of non-negative simple variables, such that*

$$X_n(\omega) \uparrow X(\omega) \quad \text{for all} \quad \omega \in \Omega.$$

(ii) *For every random variable $X$ there exists a sequence $\{X_n, n \ge 1\}$ of simple variables, such that*

$$X_n(\omega) \to X(\omega) \quad \text{for all} \quad \omega \in \Omega.$$

*Proof.* (i): Let $n \ge 1$, and set

$$X_n(\omega) = \begin{cases} \frac{k-1}{2^n}, & \text{for} \quad \frac{k-1}{2^n} \le X(\omega) < \frac{k}{2^n}, \quad k = 1, 2, \ldots, n2^n, \\ n, & \text{for} \quad X(\omega) \ge n. \end{cases} \quad (1.2)$$

The sequence thus constructed has the desired property because of the dyadic construction and since

$$X(\omega) - X_n(\omega) < \frac{1}{2^n} \quad \text{for } n \text{ sufficiently large.}$$

The limit is a random variable by Proposition 1.1. This proves (i).

To prove (ii) we use the mirrored approximation and the fact that $X = X^+ - X^-$. □

# 2 Distributions

In analogy with the arguments that preceded Theorem 1.2, the complete description of the distribution of a random variable $X$ would require knowledge about $P(X \in A)$ for all sets $A \in \mathcal{R}$. And, once again, the fact that the intervals $(-\infty, x]$ generate $\mathcal{R}$ comes to our rescue. This fact is manifested by introducing the concept of *distribution functions*.

## 2.1 Distribution Functions

**Definition 2.1.** *Let $X$ be a real valued random variable. The* distribution function *of $X$ is*

$$F(x) = P(X \le x), \quad x \in \mathbb{R}.$$

*The continuity set of $F$ is*

$$C(F) = \{x : F(x) \text{ is continuous at } x\}.$$   □

Whenever convenient we index a distribution function by the random variable it refers to; $F_X$, $F_Y$, and so on.

**Definition 2.2.** *The distribution function of a degenerate random variable is a* degenerate *distribution function. If $F$ is degenerate, then, for some $a \in \mathbb{R}$,*

$$F(x) = \begin{cases} 0, & \text{for} \quad x < a, \\ 1, & \text{for} \quad x \ge a. \end{cases}$$

*A distribution function that is not degenerate is called* non-degenerate.   □

In order to describe the properties of distribution functions it is convenient to introduce the following class of functions.

**Definition 2.3.** *The class $D$ is defined as the set of right-continuous, real valued functions with left-hand limits;*

$$F(x-) = \lim_{\substack{x_n \nearrow x \\ x_n < x}} F(x_n).$$

*The class $D^+$ is defined as the set of non-decreasing functions in $D^+$.*   □

**Proposition 2.1.** *Let $F$ be a distribution function. Then*

(a)    $F \in D^+$;
(b)    $\lim_{x \to -\infty} F(x) = 0$ *and* $\lim_{x \to +\infty} F(x) = 1$;
(c)    $F$ *has at most a countable number of discontinuities.*

*Proof.* (a): Let $X$ be a random variable associated with $F$. Boundedness follows from the fact that $0 \le F(x) = P(X \le x) \le 1$ for all $x$. To see that $F$ is non-decreasing, let $x \le y$. Then $\{X \le x\} \subset \{X \le y\}$, so that

$$F(x) = P(X \le x) \le P(X \le y) = F(y).$$

Next, let $x_n$, $n \ge 1$, be reals, $x_n \searrow x$ as $n \to \infty$. Then $\{X \le x_n\} \searrow \{X \le x\}$, so that by monotonicity (Theorem 1.3.1)

$$F(x_n) = P(X \le x_n) \searrow P(X \le x) = F(x),$$

which establishes right-continuity.

In order to verify left-continuity, we let $y_n$, $n \ge 1$, be reals, such that $y_n \nearrow x$ as $n \to \infty$. Then $\{X \le y_n\} \nearrow \{X < x\}$, so that by monotonicity (Theorem 1.3.1),

$$F(y_n) = P(X \le y_n) \nearrow P(X < x) = F(x-).$$

This concludes the proof of the fact that $F \in D^+$.
(b): This follows from the set convergences $\{X \le x\} \searrow \emptyset$ as $x \to -\infty$, and $\{X \le x\} \nearrow \Omega$ as $x \to +\infty$, respectively, together with Theorem 1.3.1.
(c): Immediate from Lemma A.9.1(i).                                        □

*Remark 2.1.* We shall, at times, encounter non-negative functions in $D^+$ with total mass *at most* equal to 1. We shall call such functions *sub-probability distribution functions.* They can be described as distribution functions, except for the fact that the total mass need not be equal to 1.          □

To complement the proposition, we find that with $x_n$ and $y_n$ as given there, we have

$$P(y_n < X \le x_n) \to \begin{cases} F_X(x) - F_X(x-) \\ P(X = x) \end{cases} \quad \text{as} \quad n \to \infty.$$

**Proposition 2.2.** *Let $X$ be a random variable. Then*

$$P(X = x) = F_X(x) - F_X(x-).$$

In order to prove uniqueness it sometimes suffices to check a generator or a dense set. Following are some results of this kind.

**Proposition 2.3.** *Suppose that $F$ and $G$ are distribution functions, and that $F = G$ on a dense subset of the reals. Then $F = G$ for all reals.*

*Proof.* Combine Proposition 2.1 and Lemma A.9.1(ii).                    □

In Chapter 1 we discussed probability measures. In this chapter we have introduced distribution functions of random variables. Now, to any given probability measure $\mathbb{P}$ on $(\mathbb{R}, \mathcal{R})$ we can associate a distribution function $F$ via the relation

$$F(b) - F(a) = \mathbb{P}((a, b])  \quad \text{for all } a, b, \quad -\infty < a \leq b < \infty,$$

and since $\mathbb{P}$ as well as $F$ is uniquely defined by their values on the rectangles we have established the following equivalence.

**Theorem 2.1.** *Every probability measure on $(\mathbb{R}, \mathcal{R})$ corresponds uniquely to the distribution function (of some random variable(s)).*

*Remark 2.2.* Recall from Example 1.1 that different random variables may have coinciding distribution functions.                    □

Having defined distribution functions, a natural challenge is to determine how many different kinds or types there may exist. For example, the number of dots resulting after throwing dice, the number of trials until a first success, the number of customers that visit a given store during one day, all those experiments have non-negative integers as outcomes. Waiting times, durations, weights, and so on are continuous quantities. So, there are at least two kinds of random variables or distributions. Are there any others?

Well, there also exist mixtures. A simple mixture is the waiting time at a traffic light. With some given probability the waiting time is 0, namely if the light is green upon arrival. If the light is red the waiting time is some continuous random quantity. So, there exist at least two kinds of random variables and mixtures of them. Are there any others?

The main decomposition theorem states that there exist exactly three kinds of random variables, and mixtures of them. However, before we turn to that problem in Subsection 2.2.3 we need some additional terminology.

## 2.2 Integration: A Preview

The classical integral is the Riemann integral, which was later generalized to the Riemann-Stieltjes integral. However, it turns out that the Riemann-Stieltjes integral has certain deficiencies that are overcome by another integral, the Lebesgue integral. The problem is that we need to be able to integrate certain wild functions that the Riemann-Stieltjes integral cannot handle. After having defined the Lebesgue integral we shall exhibit a perverse example that is Lebesgue integrable, but not Riemann-Stieltjes integrable.

There exists a probabilistic analog to integration, called *expectation*, denoted by the letter $E$. Now, instead of describing and proving a number of properties for functions and integrals and then translating them into statements about random variables and expectations (which basically amounts to replacing $f$ by $X$ and $\int$ by $E$), we shall develop the theory in a probabilistic framework, beginning in Section 2.4. However, since we need some (not much) terminology earlier, we present a few definitions and facts without proof in the language of mathematics already here. The reader who is eager for proofs is referred to standard books on measure theory and/or function theory. The amount of details and the choice of which statements to prove and which to "leave as exercises" varies between books.

**The Riemann-Stieltjes Integral**

From analysis we remember that the *Riemann integral* of a function $g$ on a bounded interval $(a, b]$ is defined via a partition $\boldsymbol{\Delta}$ of the interval into disjoint subintervals;

$$a = x_0 < x_1 < x_2 < \cdots < x_n = b,$$

and the Riemann sums

$$R(n) = \sum_{j=1}^{n} g(t_j)\Delta_j,$$

where $\Delta_j = x_j - x_{j-1}$, and $t_j \in (x_{j-1}, x_j]$. The *mesh* of the partition is $\|\boldsymbol{\Delta}\| = \max_{1 \le k \le n}\{\Delta_k\}$.

The integral exists iff there exists a number $A$, such that

$$\lim_{\|\boldsymbol{\Delta}\| \to 0} |R(n) - A| \to 0,$$

for any partition and arbitrary intermediate points. The limit is denoted with the aid of the integral sign:

$$A = \int_a^b g(x)\,dx.$$

If the integral exists we may, in particular, select the $t_j$'s so that $g$ always assumes its maximum in the subintervals, and also such that the minimum is attained. As a consequence the actual value $A$ is sandwiched between those two special sums, the upper sum and the lower sum.

We also note that, for simple functions, the Riemann integral coincides with the Riemann sum (let the partition coincide with the steps).

In the definition of the *Riemann-Stieltjes integral* of a function $f$ on a bounded interval one replaces the $\Delta$-differences along the $x$-axis by differences of a function. Thus, let, in addition, $\gamma$ be a real valued function on the interval $(a, b]$, let the partition be defined as before, and (or but) set $\Delta_j = \gamma(x_j) - \gamma(x_{j-1})$. The *Riemann-Stieltjes sum* is

$$RS(n) = \sum_{j=1}^{n} g(t_j)\Delta_j = \sum_{j=1}^{n} g(t_j)(\gamma(x_j) - \gamma(x_{j-1})),$$

and the Riemann-Stieltjes integral exists iff there exists a number $A$, such that

$$\lim_{\|\Delta\| \to 0} |RS(n) - A| \to 0$$

for any partition and arbitrary intermediate points. The notation is

$$A = \int_a^b g(x)\,\mathrm{d}\gamma(x).$$

Once again, we may select the points of the partition in such a way that the actual value $A$ can be sandwiched between an upper sum and a lower sum.

As for existence criteria we mention without proof that the Riemann-Stieltjes integral exists if (for example) $g$ is continuous and $\gamma$ is bounded and non-decreasing. The integral is then suitably extended to all of $\mathbb{R}$. The interesting example is that distribution functions fit this requirement for $\gamma$.

An inspection of the definition and the limiting procedure shows that

- if $\gamma$ is discrete with point masses $\{x_j\}$, then

$$\int g(x)\,\mathrm{d}\gamma(x) = \sum_k g(x_k)\gamma(\{x_k\});$$

- if $\gamma$ is absolutely continuous with density $f(x)$, then

$$\int g(x)\,\mathrm{d}\gamma(x) = \int g(x)f(x)\,\mathrm{d}x.$$

For the latter conclusion we also lean on the mean value theorem.

In addition, by departing from the approximating Riemann-Stieltjes sum and partial summation, one obtains a formula for partial integration:

$$\int_a^b g(x)\,\mathrm{d}\gamma(x) = g(b)\gamma(b) - g(a)\gamma(a) - \int_a^b \gamma(x)\,\mathrm{d}g(x).$$

And, needless to say, if $\gamma(x) = x$ the Riemann-Stieltjes integral reduces to the ordinary Riemann integral.

## The Lebesgue Integral

Paralleling the notion of a simple random variable (Definition 1.1) we say that $f$ is *simple* real valued function if, for some $n$,

$$f = \sum_{k=1}^n x_k I\{A_k\},$$

where $\{x_k,\ 1 \leq k \leq n\}$ are real numbers, and $\{A_k,\ 1 \leq k \leq n\}$ is a *finite* partition of $\mathbb{R}$.

We call $f$ *elementary* if

$$f = \sum_{n=1}^{\infty} x_n I\{A_n\},$$

where $\{x_n, n \geq 1\}$ are real numbers, and $\{A_n, n \geq 1\}$ is an *infinite* partition of $\mathbb{R}$.

The following lemma is a translation of Lemma 1.1; cf. also Lemma A.9.3.

**Lemma 2.1.** (i) *For every non-negative function $f$ there exists a sequence $\{f_n, n \geq 1\}$ of non-negative simple functions, such that*

$$f_n \uparrow f \quad point\text{-}wise.$$

(ii) *For every function $f$ there exists a sequence $\{f_n, n \geq 1\}$ of simple functions, such that*

$$f_n \to f \quad point\text{-}wise.$$

*Proof.* For (i) we set

$$f_n(x) = \begin{cases} \frac{k-1}{2^n}, & \text{for} \quad \frac{k-1}{2^n} \leq f(x) < \frac{k}{2^n}, \quad k = 1, 2, \ldots, n2^n, \\ n, & \text{for} \quad f(x) \geq n, \end{cases}$$

and for (ii) we add the mirrored version, and apply $f = f^+ - f^-$. □

The Lebesgue integral is an integral with respect to Lebesgue measure.

**Definition 2.4.** *The* Lebesgue measure, $\lambda$, *is a measure on $(\mathbb{R}, \mathcal{R})$, satisfying*

$$\lambda((a, b]) = b - a \quad for\ all \quad a < b, \ a, b \in \mathbb{R}.$$

**Definition 2.5.** *For the simple function $f = \sum_{k=1}^{n} x_k I\{A_k\}$ we define the Lebesgue integral with respect to a probability measure $\lambda$ as*

$$\int f \, d\lambda = \sum_{k=1}^{n} x_k \lambda(A_k).$$ □

After proving several properties such as additivity and monotonicity one defines the Lebesgue integral of arbitrary non-negative functions as the limit of the integrals of the simple functions defined in the proof of Lemma 1.1:

$$\int f \, d\lambda = \lim_{n\to\infty} \sum_{k=1}^{n2^n} \frac{k-1}{2^n} \lambda\left(\frac{k-1}{2^n} \leq f(x) < \frac{k}{2^n}\right).$$

Since, as mentioned in the introduction of this section, we shall traverse the theory in the probabilistic language in a moment, we close the discussion in mathematical terms with some comments on how the Lebesgue integral and the Riemann-Stieltjes integral relate to each other.

**Theorem 2.2.** *If the Riemann-Stieltjes integral of a function exists, then so does the Lebesgue integral, and both agree.*

We close this preview (i) by recalling that what has been stated so far will soon be justified, and (ii) with an example of a function that is Lebesgue integrable but *not* Riemann-Stieltjes integrable.

*Example 2.1.* Let $f(x)$ be defined as follows on the unit interval:

$$f(x) = \begin{cases} 1, & \text{for } x \in [0,1] \setminus \mathbb{Q}, \\ 0, & \text{for } x \in [0,1] \cap \mathbb{Q}, \end{cases}$$

that is, $f$ equals 1 on the irrationals and 0 on the rationals.

This function is Lebesgue integrable – the integral equals 1 – but not Riemann integrable, the reason for the latter being that the upper and lower sums equal 1 and 0, respectively, for any partition of the unit interval.

The explanation for the difference in integrability is that the "slices" in the definition of the Lebesgue integral are horizontal, whereas those of the Riemann integral are vertical. □

As for a converse we mention without proof that any Lebesgue integrable function can be arbitrarily well approximated by Riemann integrable functions, and refer the reader, once again, to specialized literature.

**Theorem 2.3.** *If $f$ is Lebesgue integrable, then, for any $\varepsilon > 0$, there exists,*
(a) *a simple function $g$, such that $\int_{-\infty}^{\infty} |f(x) - g(x)| \, dx < \varepsilon$.*
(b) *a continuous, integrable function $h$, such that $\int_{-\infty}^{\infty} |f(x) - h(x)| \, dx < \varepsilon$.*

## 2.3 Decomposition of Distributions

In this subsection we show that that every distribution function can be decomposed into a convex combination of three "pure" kinds.

**Definition 2.6.** *A distribution function $F$ is*

• discrete *iff for some countable set of numbers $\{x_j\}$ and point masses $\{p_j\}$,*

$$F(x) = \sum_{x_j \leq x} p_j, \quad \text{for all } x \in \mathbb{R}.$$

*The function $p$ is called* probability function.
• continuous *iff it is continuous for all $x$.*
• absolutely continuous *iff there exists a non-negative, Lebesgue integrable function $f$, such that*

$$F(b) - F(a) = \int_a^b f(x) \, dx \quad \text{for all } a < b.$$

*The function $f$ is called the* density *of $F$.*
• singular *iff $F \not\equiv 0$, $F'$ exists and equals 0 a.e.* □

The ultimate goal of this subsection is to prove the following decomposition theorem.

**Theorem 2.4.** *Every distribution function can be decomposed into a convex combination of three pure types, a* discrete *one, an* absolutely continuous *one, and a* continuous singular *one. Thus, if* $F$ *is a distribution function, then*

$$F = \alpha F_{ac} + \beta F_d + \gamma F_{cs},$$

*where* $\alpha, \beta, \gamma \geq 0$ *and* $\alpha + \beta + \gamma = 1$. *This means that*

- $F_{ac}(x) = \int_{-\infty}^{x} f(y)\, dy$, *where* $f(x) = F'_{ac}(x)$ *a.e.;*
- $F_d$ *is a pure jump function with at most a countable number of jumps;*
- $F_{cs}$ *is continuous and* $F'_{cs}(x) = 0$ *a.e.*

For the proof we have to accept the following (rather natural) facts. For the proof we refer the reader to his or her favourite book on measure theory or function theory.

**Lemma 2.2.** *Let* $F$ *be a distribution function. Then*

(a)  $F'(x)$ *exists a.e., and is non-negative and finite.*
(b)  $\int_{a}^{b} F'(x)\, dx \leq F(b) - F(a)$ *for all* $a, b \in \mathbb{R}$.
(c)  *Set* $F_{ac}(x) = \int_{-\infty}^{x} F'(y)\, dy$, *and* $F_s(x) = F(x) - F_{ac}(x)$ *for all* $x \in \mathbb{R}$. *Then* $F'_{ac}(x) = F'(x)$ *a.e. and* $F'_s = 0$ *a.e. In particular,* $F_s \equiv 0$ *or* $F_s$ *is singular.*

*Remark 2.3.* The components $F_{ac}(x)$ and $F_s(x)$ in Lemma 2.2 are, in contrast to those of Theorem 2.4, sub-distribution functions in that the total mass is only *at most* equal to 1.  □

Discrete distributions are obviously singular. But, as we shall see, there also exist *continuous* singular distributions. We shall exhibit one, the Cantor distribution, in Subsection 2.2.6 below, and later, in more detail, in Section 2.11.

The first step is the *Lebesgue decomposition theorem*, in which the distribution function is split into an absolutely continuous component and a singular one.

**Theorem 2.5.** *Every distribution function can be decomposed into a convex combination of an absolutely continuous distribution function and a singular one. Thus, if* $F$ *is a distribution function, then*

$$F = \alpha F_{ac} + (1 - \alpha) F_s,$$

*where* $0 \leq \alpha \leq 1$.

*Proof.* Let $f(x) = F'(x)$, which, according to Lemma 2.2 exists a.e., and, moreover, equals $F^*_{ac}(x)$ a.e., where

$$F_{ac}^*(x) = \int_{-\infty}^x f(y)\,dy.$$

In order to see that $F_{ac}^*$ is a distribution function, except, possibly, for the fact that $F_{ac}^*(+\infty) \le 1$, we observe that $F_{ac}^*$ is non-decreasing since $f \ge 0$ a.e., and that $F_{ac}^*(-\infty) = 0$ since $F_{ac}^*(x) \le F(x)$. Continuity is obvious, since the integral is continuous.

Next, set $F_s^*(x) = F(x) - F_{ac}^*$. Then, $F_s^*$ is non-decreasing by Lemma 2.2, and $F_s^*(-\infty) = 0$, since $F_s^*(x) \le F(x)$, which shows that $F_s^*$ is also a distribution function, except, possibly, for having total mass less than one.

If $\alpha = 0$ or 1 we are done. Otherwise, set

$$F_{ac}(x) = \frac{F_{ac}^*(x)}{F_{ac}^*(+\infty)} \quad \text{and} \quad F_s(x) = \frac{F_s^*(x)}{F_s^*(+\infty)}. \qquad \square$$

The following theorem provides a decomposition of a distribution function into a discrete component and a continuous component.

**Theorem 2.6.** *Every distribution function $F$ can be written as a convex combination of a discrete distribution function and a continuous one:*

$$F = \beta F_d + (1 - \beta) F_c,$$

*where $0 \le \beta \le 1$.*

*Proof.* By Proposition 2.1 we know that $F$ may have at most a countable number of jumps. Let $\{x_j\}$ be those jumps (if they exist), let $p(j) = F(x_j+) - F(x_j-) = F(x_j) - F(x_j-)$ for all $j$ (recall that $F$ is right-continuous), and define

$$F_d^*(x) = \sum_{x_j \le x} p_j, \quad x \in \mathbb{R}.$$

By construction, $F_d^*$, being equal to the sum of all jumps to the left of $x$, is discrete, and has all properties of a distribution function, except that we only know that $\lim_{x \to \infty} F_d^*(x) \le 1$.

Next, let $F_c^*(x) = F(x) - F_d^*(x)$ for all $x$. Since $F_d^*(x)$ increases at $x_j$ by $p_j$ and stays constant between jumps, and since $F$ is non-decreasing, it follows that $F_c^*(x)$ must be non-negative and non-decreasing. Moreover,

$$\lim_{x \to -\infty} F_c^*(x) = \lim_{x \to -\infty} (F(x) - F_d^*(x)) = 0 - 0 = 0,$$

and

$$0 \le \lim_{x \to \infty} F_c^*(x) = \lim_{x \to \infty} (F(x) - F_d^*(x)) = 1 - \lim_{x \to \infty} F_d^*(x) \le 1,$$

so that $F_c^*(x)$ also has all properties of a distribution function, except that the total mass may be less than 1. In particular, $F_c^* \in D^+$.

The next thing to prove is that $F_c^*(x)$ is continuous, which seems rather obvious, since we have reduced $F$ by its jumps. Nevertheless,

$$F_c^*(x) - F_c^*(x-) = F(x) - F_d^*(x) - (F(x-) - F_d^*(x-))$$
$$= F(x) - F(x-) - (F_d^*(x) - F_d^*(x-))$$
$$= \begin{cases} p_j - p_j = 0, & \text{when} \quad x = x_j \text{ for some } j, \\ 0, & \text{otherwise,} \end{cases}$$

which shows that $F_c^*$ is left-continuous. This tells us that $F_c^*$ is continuous, since, as we have already seen, $F_c^* \in D^+$. A final rescaling, if necessary, finishes the proof. □

*Proof of Theorem 2.4.* Let $F$ be a distribution function. Then, by the Lebesgue decomposition theorem, we know that

$$F = \alpha F_{ac} + (1 - \alpha)F_s,$$

and by Theorem 2.6 applied to the singular part we know that

$$F_s = \beta F_d + (1 - \beta)F_{cs}.$$

□

**Theorem 2.7.** *The decompositions are unique.*

**Exercise 2.1.** Prove the uniqueness (by contradiction). □

### 2.4 Some Standard Discrete Distributions

Following is a list of some of the most common *discrete* distributions. The domains of the parameters below are $a \in \mathbb{R}$, $0 \le p = 1 - q \le 1$, $n \in \mathbb{N}$, and $m > 0$.

| Distribution | Notation | Probability function | Domain |
|---|---|---|---|
| One point | $\delta(a)$ | $p(a) = 1$ | |
| Symmetric Bernoulli | | $p(-1) = p(1) = \frac{1}{2}$ | |
| Bernoulli | $\mathrm{Be}(p)$ | $p(0) = q, \ p(1) = p$ | |
| Binomial | $\mathrm{Bin}(n,p)$ | $p(k) = \binom{n}{k}p^k q^{n-k}$ | $k = 0, 1, \ldots, n$ |
| Geometric | $\mathrm{Ge}(p)$ | $p(k) = pq^k$ | $k \in \mathbb{N} \cup 0$ |
| First success | $\mathrm{Fs}(p)$ | $p(k) = pq^{k-1}$ | $k \in \mathbb{N}$ |
| Poisson | $\mathrm{Po}(m)$ | $p(k) = e^{-m}\frac{m^k}{k!}$ | $k \in \mathbb{N} \cup 0$ |

**Table 2.1.** Some discrete distributions

The $\mathrm{Be}(p)$-distribution describes the outcome of one "coin-tossing" experiment, and the $\mathrm{Bin}(n,p)$-distribution the number of successes in $n$ trials. The $\mathrm{Ge}(p)$-distribution describes the number of failures prior to the first success, and the $\mathrm{Fs}(p)$-distribution the number of trails required to succeed once. Finally, the typical experiment for a Poisson distribution is a coin-tossing experiment where the probability of success is "small". Vaguely speaking, $\mathrm{Bin}(n,p) \approx \mathrm{Po}(np)$ if $p$ is small (typically "small" means $< 0.1$). This can, of course, be rigorously demonstrated.

## 2.5 Some Standard Absolutely Continuous Distributions

In this subsection we list some of the most common *absolutely continuous* distributions. The parameters $p, \theta, \sigma, r, s, \alpha, \beta$ below are all non-negative, and $a, b, \mu \in \mathbb{R}$.

| Distribution | Notation | Density function | Domain |
|---|---|---|---|
| Uniform | $U(a,b)$ | $f(x) = \frac{1}{b-a}$ | $a < x < b$ |
| | $U(0,1)$ | $f(x) = 1$ | $0 < x < 1$ |
| | $U(-1,1)$ | $f(x) = \frac{1}{2}$ | $|x| < 1$ |
| Triangular | $\mathrm{Tri}(-1,1)$ | $f(x) = 1 - |x|$ | $|x| < 1$ |
| Exponential | $\mathrm{Exp}(\theta)$ | $f(x) = \frac{1}{\theta} e^{-x/\theta}$ | $x > 0$ |
| Gamma | $\Gamma(p,\theta)$ | $f(x) = \frac{1}{\Gamma(p)} x^{p-1} \frac{1}{\theta^p} e^{-x/\theta}$ | $x > 0$ |
| Beta | $\beta(r,s)$ | $f(x) = \frac{\Gamma(r+s)}{\Gamma(r)\Gamma(s)} x^{r-1}(1-x)^{s-1}$ | $0 < x < 1$ |
| Normal | $N(\mu,\sigma^2)$ | $f(x) = \frac{1}{\sigma\sqrt{2\pi}} e^{-\frac{1}{2}(x-\mu)^2/\sigma^2}$ | $x \in \mathbb{R}$ |
| | $N(0,1)$ | $f(x) = \frac{1}{\sqrt{2\pi}} e^{-x^2/2}$ | $x \in \mathbb{R}$ |
| Log-normal | $LN(\mu,\sigma^2)$ | $f(x) = \frac{1}{\sigma x \sqrt{2\pi}} e^{-\frac{1}{2}(\log x - \mu)^2/\sigma^2}$ | $x > 0$ |
| Cauchy | $C(0,1)$ | $f(x) = \frac{1}{\pi} \cdot \frac{1}{1+x^2}$ | $x \in \mathbb{R}$ |
| Pareto | $\mathrm{Pa}(\beta,\alpha)$ | $f(x) = \frac{\alpha k^\alpha}{x^{\alpha+1}}$ | $x > \beta$ |

**Table 2.2.** Some absolutely continuous distributions

We have listed two special uniform distributions and the standard normal distribution because of their frequent occurrences, and confined ourselves to the special triangular distribution which has support on $[-1, 1]$ and the standard Cauchy distribution for convenience.

Uniform distributions typically describe phenomena such as picking a point "at random" in the sense that the probability that the resulting point belongs to an interval only depends on the length of the interval and not on its position.

Exponential and gamma distributed random variables typically are used to model waiting times, life lengths, and so on, in particular in connection with the so-called Poisson process.

The normal distribution, also called the Gaussian distribution, models cumulative or average results of "many" repetitions of an experiment; the formal result is the central limit theorem, which we shall meet in Chapter 7. The *multivariate* normal distribution, that we shall encounter in Subsection 4.5.1, plays, i.a., an important role in many statistical applications.

## 2.6 The Cantor Distribution

The third kind of distributions, the continuous singular ones, are the most special or delicate ones. In this subsection we shall define the Cantor distribution and prove that it belongs to that class.

The standard Cantor *set* is constructed on the interval $[0, 1]$ as follows. One successively removes the open middle third of each subinterval of the previous set. The Cantor set itself is the infinite intersection of all remaining sets. More precisely, let $C_0 = [0, 1]$, and, successively,

$$C_1 = \left[0, \frac{1}{3}\right] \cup \left[\frac{2}{3}, 1\right], \quad C_2 = \left[0, \frac{1}{9}\right] \cup \left[\frac{2}{9}, \frac{1}{3}\right] \cup \left[\frac{2}{3}, \frac{7}{9}\right] \cup \left[\frac{8}{9}, 1\right],$$

and so on.

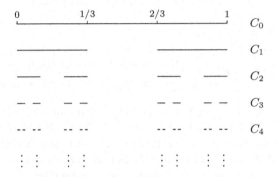

**Figure 2.1.** The Cantor set on $[0, 1]$

The Cantor set is

$$C = \bigcap_{n=0}^{\infty} C_n,$$

and the Cantor distribution is the distribution that is uniform on the Cantor set.

Having thus defined the distribution we now show that it is continuous singular.

(i):  The Lebesgue measure of the Cantor set equals 0, since $C \subset C_n$ for all $n$, so that

$$\lambda(C) \leq \lambda(C_n) = \left(\frac{2}{3}\right)^n \quad \text{for every } n \quad \implies \quad \lambda(C) = 0.$$

Alternatively, in each step we remove the middle thirds. The Lebesgue measure of the pieces we *remove* thus equals

$$\frac{1}{3} + 2\left(\frac{1}{3}\right)^2 + 4\left(\frac{1}{3}\right)^3 + \cdots = \sum_{n=1}^{\infty} 2^{n-1} \left(\frac{1}{3}\right)^n = 1.$$

The Cantor set is the complement, hence $\lambda(C) = 0$.

(ii):  The Cantor distribution is singular, since its support is a Lebesgue null set.

(iii):   The distribution function is continuous. Namely, let $F_n$ be the distribution function corresponding to the distribution that is uniform on $C_n$. This means that $F_n(0) = 0$, $F_n$ is piecewise constant with $2^n$ jumps of size $2^{-n}$ and $F_n(1) = 1$. Moreover, $F'_n(x) = 0$ for all $x$ except for the end-points of the $2^n$ intervals.

The distribution function of the Cantor distribution is

$$F(x) = \lim_{n \to \infty} F_n(x).$$

Now, let $x, y \in C_n$. Every subinterval of $C_n$ has length $3^{-n}$. Therefore,

$$0 < x - y < \frac{1}{3^n} \implies$$

$$F(y) - F(x) \begin{cases} = 0, & \text{when } x, y \text{ are in the same subinterval,} \\ \leq \frac{1}{2^n} & \text{when } x, y \text{ are in adjacent subintervals.} \end{cases}$$

This proves that $F$, in fact, is uniformly continuous on $C$.

(iv)   $F'(x) = 0$ for almost all $x$, because $F'(x) = 0$ for all $x \in C_n$ for all $n$.

This finishes the proof of the fact that the Cantor distribution is continuous singular. We shall return to this distribution in Section 2.11, where an elegant representation in terms of an infinite sum will be given.

## 2.7 Two Perverse Examples

In Example 2.1 we met the function on the unit interval, which was equal to 1 on the irrationals and 0 on the rationals:

$$f(x) = \begin{cases} 1, & \text{for } x \in [0, 1] \setminus \mathbb{Q}, \\ 0, & \text{for } x \in [0, 1] \cap \mathbb{Q}. \end{cases}$$

Probabilistically this function can be interpreted as the density of a random variable, $X$, which is uniformly distributed on the irrationals in $[0, 1]$.

Note that, if $U \in U(0, 1)$, then the probability that the two random variables differ equals

$$P(X \neq U) = P(X \in \mathbb{Q}) = 0,$$

so that $X \sim U$.

An extreme variation of this example is the following:

*Example 2.2.* Let $\{r_k, k \geq 1\}$ be an enumeration of the rationals in the unit interval, and define

$$p(r_k) = \begin{cases} \frac{6}{\pi^2 k^2}, & \text{for } r_k \in (0, 1) \cap \mathbb{Q}, \\ 0, & \text{otherwise.} \end{cases}$$

Since $\sum_{k=1}^{\infty} 1/k^2 = \pi^2/6$, this is a bona fide discrete distribution.

This may seem as a somewhat pathological distribution, since it is defined along an enumeration of the rationals, which by no means is neither unique nor "chronological". □

# 3 Random Vectors; Random Elements

Random vectors are the same as multivariate random variables. Random elements are "random variables" in (more) abstract spaces.

## 3.1 Random Vectors

Random vectors are elements in the Euclidean spaces $\mathbb{R}^n$ for some $n \in \mathbb{N}$.

**Definition 3.1.** *An n-dimensional random vector* $\mathbf{X}$ *is a measurable function from the sample space* $\Omega$ *to* $\mathbb{R}^n$*;*

$$\mathbf{X} : \Omega \to \mathbb{R}^n,$$

*that is, the inverse image of any Borel set is* $\mathcal{F}$*-measurable:*

$$X^{-1}(A) = \{\omega : X(\omega) \in A\} \in \mathcal{F} \quad \text{for all} \quad A \in \mathcal{R}^n.$$

*Random vectors are considered column vectors;*

$$\mathbf{X} = (X_1, X_2, \ldots, X_n)',$$

*where* ' *denotes transpose (i.e.,* $\mathbf{X}'$ *is a row vector).*
    *The* joint distribution function *of* $\mathbf{X}$ *is*

$$F_{X_1, X_2, \ldots, X_n}(x_1, x_2, \ldots, x_n) = P(X_1 \le x_1, X_2 \le x_2, \ldots, X_n \le x_n),$$

*for* $x_k \in \mathbb{R}$, $k = 1, 2, \ldots, n$. □

*Remark 3.1.* A more compact way to express the distribution function is

$$F_{\mathbf{X}}(\mathbf{x}) = P(\mathbf{X} \le \mathbf{x}), \quad \mathbf{x} \in \mathbb{R}^n,$$

where the event $\{\mathbf{X} \le \mathbf{x}\}$ is to be interpreted component-wise, that is,

$$\{\mathbf{X} \le \mathbf{x}\} = \{X_1 \le x_1, X_2 \le x_2, \ldots, X_n \le x_n\} = \bigcap_{k=1}^{n} \{X_k \le x_k\}. \quad □$$

For discrete distributions the *joint probability function* is defined by

$$p_{\mathbf{X}}(\mathbf{x}) = P(\mathbf{X} = \mathbf{x}), \quad \mathbf{x} \in \mathbb{R}^n.$$

In the absolutely continuous case we have a *joint density*;

$$f_{\mathbf{X}}(\mathbf{x}) = \frac{\partial^n F_{\mathbf{X}}(\mathbf{x})}{\partial x_1 \partial x_2 \cdots \partial x_n} \quad \mathbf{x} \in \mathbb{R}^n.$$

The following example illuminates a situation where a problem intrinsically is defined in a "high" dimension, but the object of interest is "low-dimensional" (in the example, high = 2 and low = 1).

*Example 3.1.* Let $(X, Y)$ be a point that is uniformly distributed on the unit disc, that is,

$$f_{X,Y}(x,y) = \begin{cases} \frac{1}{\pi}, & \text{for} \quad x^2 + y^2 \le 1, \\ 0, & \text{otherwise.} \end{cases}$$

Determine the distribution of the $x$-coordinate.                                    □

In order to solve this problem we consider, as a preparation, the discrete analog, which is easier to handle. Let $(X, Y)$ be a given two-dimensional random variable whose joint probability function is $p_{X,Y}(x, y)$ and that we are interested in finding $p_X(x)$. By the law of total probability, Proposition 1.4.1,

$$p_X(x) = P(X = x) = P\left(\{X = x\} \cap \left\{\bigcup_y \{Y = y\}\right\}\right)$$
$$= P\left(\bigcup_y \{\{X = x\} \cap \{Y = y\}\}\right)$$
$$= \sum_y P(X = x, Y = y) = \sum_y p_{X,Y}(x, y).$$

The distribution of one of the random variables thus is obtained by adding the joint probabilities along "the other" variable.

Distributions thus obtained are called *marginal distributions*, and the corresponding probability functions are called *marginal probability functions*.

The *marginal distribution function* of $X$ at the point $x$ is obtained by adding the values of the marginal probabilities to the left of $x$:

$$F_X(x) = \sum_{u \le x} p_X(u) = \sum_{u \le x} \sum_v p_{X,Y}(u, v).$$

Alternatively,

$$F_X(x) = P(X \le x, Y < \infty) = \sum_{u \le x} \sum_v p_{X,Y}(u, v).$$

In the absolutely continuous case we depart from the distribution function

$$F_X(x) = P(X \le x, Y < \infty) = \int_{-\infty}^x \int_{-\infty}^\infty f_{X,Y}(u, v)\, du dv,$$

and differentiate to obtain the *marginal density function*,

$$f_X(x) = \int_{-\infty}^\infty f_{X,Y}(x, y)\, dy,$$

that is, this time we *integrate* along "the other" variable.

Marginal distribution functions are integrals of the marginal densities.

Analogous formulas hold in higher dimensions, and for more general distributions. Generally speaking, marginal distributions are obtained by integrating in the generalized sense ("getting rid of") those components that are not relevant for the problem at hand.

Let us now solve the problem posed in Example 3.1. The joint density was given by

$$f_{X,Y}(x,y) = \begin{cases} \frac{1}{\pi}, & \text{for} \quad x^2 + y^2 \leq 1, \\ 0, & \text{otherwise}, \end{cases}$$

from which we obtain

$$f_X(x) = \int_{-\infty}^{\infty} f_{X,Y}(x,y)\,dy = \int_{-\sqrt{1-x^2}}^{\sqrt{1-x^2}} \frac{1}{\pi}\,dy = \frac{2}{\pi}\sqrt{1-x^2}, \qquad (3.1)$$

for $-1 < x < 1$ (and $f_X(x) = 0$ otherwise).

**Exercise 3.1.** Let $(X, Y, Z)$ be a point chosen uniformly within the three-dimensional unit sphere. Find the marginal distributions of $(X, Y)$ and $X$.  □

We have seen how a problem might naturally be formulated in a higher dimension than that of interest. The converse concerns to what extent the marginal distributions determine the joint distribution. Interesting applications are computer tomography and satellite pictures; in both cases one departs from two-dimensional pictures from which one wishes to make conclusions about three-dimensional objects (the brain and the Earth).

A multivariate distribution of special importance is the normal one, for which some facts and results will be presented in Chapter 4.

## 3.2 Random Elements

Random elements are random variables on abstract spaces.

**Definition 3.2.** *A random element is a measurable mapping from a measurable space $(\Omega, \mathcal{F})$ to a measurable, metric space $(S, \mathcal{S})$:*

$$X : (\Omega, \mathcal{F}) \to (S, \mathcal{S}).$$  □

In this setting measurability thus means that

$$X^{-1}(A) = \{\omega : X(\omega) \in A\} \in \mathcal{F} \quad \text{for all} \quad A \in \mathcal{S}.$$

The meaning thus is the same as for ordinary random variables. With a slight exaggeration one may say that the difference is "notational".

The distribution of a random element is "the usual one", namely the induced one, $\mathbb{P} = P \circ X^{-1}$;

$$\mathbb{P}(A) = P(\{\omega : X(\omega) \in A\}) \quad \text{for} \quad A \in \mathcal{S}.$$

A typical example is the space $C[0, 1]$ of continuous functions on the unit interval, endowed with the uniform topology or metric

$$d(x, y) = \sup_{0 \leq t \leq 1} |x(t) - y(t)| \quad \text{for} \quad x, y \in C[0, 1].$$

For more on this and on the analog for the space $D[0, 1]$ of right-continuous functions with left-hand limits on the unit interval, endowed with the Skorohod $J_1$- or $M_1$-topologies [224], see also [20, 188].

# 4 Expectation; Definitions and Basics

Just as random variables are "compressed versions" of events from a probability space, one might be interested in compressed versions of random variables. The typical one is the *expected value*, which is the probabilistic version of the center of gravity of a physical body. Another name for expectation is *mean*.

Mathematically, expectations are integrals with respect to distribution functions or probability measures. We must therefore develop the theory of integration, more precisely, the theory of *Lebesgue integration*. However, since this is a book on probability theory we prefer to develop the theory of Lebesgue integration in terms of expectations. We also alert the reader to the small integration preview in Subsection 2.2.2, and recommend a translation of what is to come into the traditional mathematics language – remember that rewriting is much more profitable than rereading.

Much of what follows next may seem like we are proving facts that are "completely obvious" or well known (or both). For example, the fact that the tails of convergent integrals tend to 0 just as the terms in a convergent series do. We must, however, remember that we are introducing a new integral concept, namely the Lebesgue integral, and for that concept we "do not yet know" that the results are "trivial". So, proofs and some care are required. Along the way we also obtain the promised justifications of facts from Subsection 2.2.2.

## 4.1 Definitions

We begin with the simple case.

### Simple Random Variables

We remember from Definition 1.1 that a random variable $X$ is *simple* if, for some $n$,

$$X = \sum_{k=1}^{n} x_k I\{A_k\},$$

where $\{x_k, 1 \leq k \leq n\}$ are real numbers, and $\{A_k, 1 \leq k \leq n\}$ is a *finite* partition of $\Omega$.

4 Expectation; Definitions and Basics     47

**Definition 4.1.** *For the simple random variable $X = \sum_{k=1}^{n} x_k I\{A_k\}$, we define the* expected value *as*

$$E X = \sum_{k=1}^{n} x_k P(A_k).$$ □

## Non-negative Random Variables

In the first section of this chapter we found that if $X$ is a non-negative random variable, then the sequence of *simple* non-negative random variables $X_n$, $n \geq 1$, defined by

$$X_n(\omega) = \begin{cases} \frac{k-1}{2^n}, & \text{for} \quad \frac{k-1}{2^n} \leq X(\omega) < \frac{k}{2^n}, \quad k = 1, 2, \ldots, n2^n, \\ n, & \text{for} \quad X(\omega) \geq n, \end{cases}$$

converges monotonically from below to $X$ as $n \to \infty$. With this in mind we make the following definition of the expected value of arbitrary, non-negative random variables.

**Definition 4.2.** *Suppose that $X$ is a non-negative* random variable. *The expected value of $X$ is defined as*

$$E X = \lim_{n \to \infty} \sum_{k=1}^{n2^n} \frac{k-1}{2^n} P\left( \frac{k-1}{2^n} \leq X < \frac{k}{2^n} \right).$$

*Note that the limit may be infinite.* □

The definition is particularly appealing for bounded random variables. Namely, suppose that $X$ is a non-negative random variable, such that

$$X \leq M < \infty, \quad \text{for some} \quad M > 0,$$

and set,

$$Y_n(\omega) = \begin{cases} \frac{k}{2^n}, & \text{for} \quad \frac{k-1}{2^n} \leq X(\omega) < \frac{k}{2^n}, \quad k = 1, 2, \ldots, n2^n, \\ n, & \text{for} \quad X(\omega) \geq n, \end{cases}$$

for $n \geq M$, where we pretend, for simplicity only, that $M$ is an integer. Then $Y_n \searrow X$ as $n \to \infty$, and moreover,

$$X_n \leq X \leq Y_n, \quad \text{and} \quad Y_n - X_n = \frac{1}{2^n}.$$

Thus, by the consistency property that we shall prove in Theorem 4.2 below,

$$E X_n \leq E X \leq E Y_n, \quad \text{and} \quad E(Y_n - X_n) = \frac{1}{2^n} \to 0 \quad \text{as} \quad n \to \infty.$$

**The General Case**

**Definition 4.3.** *For an arbitrary random variable $X$ we define*

$$E X = E X^+ - E X^-,$$

*provided at least one of $E X^+$ and $E X^-$ is finite (thus prohibiting $\infty - \infty$).*
*We write*

$$E X = \int_\Omega X(\omega) \, dP(\omega) \quad or, \ simply, \quad \int X \, dP.$$

*If both values are finite, that is, if $E|X| < \infty$, we say that $X$ is* integrable.☐

Throughout our treatment, $P$ is a probability measure, and assumptions about integrability are with respect to $P$. Recall that a.s. means almost surely, that is, if a property holds a.s. then the set where it does *not* hold is a null set. If $X$ and $Y$ are random variables, such that $X = Y$ a.s., this means that $P(X = Y) = 1$, or, equivalently, that $P(X \neq Y) = 0$.

During the process of constructing the concept of expected values, we shall need the concept of almost sure convergence, which means that we shall meet situations where we consider sequences $X_1, X_2, \ldots$ of random variables such that $X_n(\omega) \to X(\omega)$ as $n \to \infty$, not for *every* $\omega$, but for *almost all* $\omega$. This, as it turns out, is sufficient, (due to equivalence; Definition 1.2), since integrals over sets of measure 0 are equal to 0.

As a, somewhat unfortunate, consequence, the introduction of the concept of almost sure convergence cannot wait until Chapter 5.

**Definition 4.4.** *Let $X, X_1, X_2, \ldots$ be random variables. We say that $X_n$ converges almost surely (a.s.) to the random variable $X$ as $n \to \infty$, $X_n \overset{a.s.}{\to} X$ as $n \to \infty$, iff*

$$P\big(\{\omega : X_n(\omega) \to X(\omega) \ as \ n \to \infty\}\big) = 1,$$

*or, equivalently, iff*

$$P\big(\{\omega : X_n(\omega) \not\to X(\omega) \ as \ n \to \infty\}\big) = 0. \qquad\qquad ☐$$

## 4.2 Basic Properties

The first thing to prove is that the definition of expectation is consistent, after which we turn our attention to a number of properties, such as additivity, linearity, domination, and so on.

### Simple Random Variables

We thus begin with a lemma proving that the expected value of a random variable is independent of the partition.

**Lemma 4.1.** *If $\{A_k, 1 \le k \le n\}$ and $\{B_j, 1 \le j \le m\}$ are partitions of $\Omega$, such that*

$$X = \sum_{k=1}^{n} x_k I\{A_k\} \quad and \quad X = \sum_{j=1}^{m} y_j I\{B_j\},$$

*Then*

$$\sum_{k=1}^{n} x_k P(A_k) = \sum_{j=1}^{m} y_j P(B_j).$$

*Proof.* The fact that $\{A_k, 1 \le k \le n\}$ and $\{B_k, 1 \le k \le m\}$ are partitions implies that

$$P(A_k) = \sum_{j=1}^{m} P(A_k \cap B_j) \quad and \quad P(B_j) = \sum_{k=1}^{n} P(A_k \cap B_j),$$

and, hence, that

$$\sum_{k=1}^{n} x_k P(A_k) = \sum_{k=1}^{n}\sum_{j=1}^{m} x_k P(A_k \cap B_j),$$

and

$$\sum_{j=1}^{m} y_j P(B_j) = \sum_{j=1}^{m}\sum_{k=1}^{n} y_j P(A_k \cap B_j).$$

Since the sets $\{A_k \cap B_j, 1 \le k \le n, 1 \le j \le m\}$ also form a partition of $\Omega$ it follows that $x_k = y_j$ whenever $A_k \cap B_j \ne \emptyset$, which proves the conclusion.  $\square$

Next we show that intuitively obvious operations are permitted.

**Theorem 4.1.** *Let $X, Y$ be non-negative simple random variables. Then:*

(a)  *If $X = 0$ a.s., then $E X = 0$;*
(b)  *$E X \ge 0$;*
(c)  *If $E X = 0$, then $X = 0$ a.s.;*
(d)  *If $E X > 0$, then $P(X > 0) > 0$;*
(e)  *Linearity: $E(aX + bY) = aE X + bE Y$ for any $a, b \in \mathbb{R}^+$;*
(f)  *$E XI\{X > 0\} = E X$;*
(g)  *Equivalence: If $X = Y$ a.s., then $E Y = E X$;*
(h)  *Domination: If $Y \le X$ a.s., then $E Y \le E X$.*

*Proof.* (a): If $X(\omega) = 0$ for all $\omega \in \Omega$, then, with $A_1 = \{X = 0\} (= \Omega)$, we have $X = 0 \cdot I\{A_1\}$, so that $E X = 0 \cdot P(A_1) = 0 \cdot 1 = 0$.

If $X = 0$ a.s., then $X = \sum_{k=1}^{n} x_k I\{A_k\}$, where $x_1 = 0$, and $x_2, x_3, \ldots, x_n$ are finite numbers, $A_1 = \{X = 0\}$, and $A_2, A_3, \ldots, A_n$ are null sets. It follows that

$$E X = 0 \cdot P(A_1) + \sum_{k=2}^{n} x_k \cdot 0 = 0.$$

(b): Immediate, since the sum of non-negative terms is non-negative.

(c): By assumption,

$$\sum_{k=1}^{n} x_k P(A_k) = 0.$$

The fact that the sum of non-negative terms can be equal to 0 if and only if all terms are equal to 0, forces one of $x_k$ and $P(A_k)$ to be equal to 0 for every $k \geq 2$ ($A_1 = \{X = 0\}$ again). In particular, we must have $P(A_k) = 0$ for any nonzero $x_k$, which shows that $P(X = 0) = 1$.

(d): The assumption implies that at least one of the terms $x_k P(A_k)$, and therefore both factors of this term must be positive.

(e): With $X = \sum_{k=1}^{n} x_k I\{A_k\}$ and $Y = \sum_{j=1}^{m} y_j I\{B_j\}$, we have

$$X + Y = \sum_{k=1}^{n} \sum_{j=1}^{m} (x_k + y_j) I\{A_k \cap B_j\},$$

so that

$$E(aX + bY) = \sum_{k=1}^{n} \sum_{j=1}^{m} (ax_k + by_j) P(A_k \cap B_j)$$

$$= a \sum_{k=1}^{n} \sum_{j=1}^{m} x_k P(A_k \cap B_j) + b \sum_{k=1}^{n} \sum_{j=1}^{m} y_j P(A_k \cap B_j)$$

$$= a \sum_{k=1}^{n} x_k P\left( A_k \cap \left( \bigcup_{j=1}^{m} B_j \right) \right) + b \sum_{j=1}^{m} y_j P\left( \left( \bigcup_{k=1}^{n} A_k \right) \cap B_j \right)$$

$$= a \sum_{k=1}^{n} x_k P(A_k) + b \sum_{j=1}^{m} y_j P(B_j) = aE\,X + bE\,Y.$$

(f): Joining (a) and (e) yields

$$E\,X = E\,XI\{X > 0\} + E\,XI\{X = 0\} = E\,XI\{X > 0\} + 0 = E\,XI\{X > 0\}.$$

(g): If $X = Y$ a.s., then $X - Y = 0$ a.s., so that, by (a), $E(X - Y) = 0$, and by (e),

$$E\,X = E((X - Y) + Y) = E(X - Y) + E\,Y = 0 + E\,Y.$$

(h): The proof is similar to that of (g). By assumption, $X - Y \geq 0$ a.s., so that, by (b), $E(X - Y) \geq 0$, and by linearity,

$$E\,Y = E\,X - E(X - Y) \leq E\,X. \qquad \square$$

## Non-negative Random Variables

Once again, the first thing to prove is consistency.

**Theorem 4.2.** (Consistency)
*Let $X$ be a non-negative random variable, and suppose that $\{Y_n, n \geq 1\}$ and $\{Z_n, n \geq 1\}$ are sequences of simple random variables, such that*

$$Y_n \nearrow X \quad \text{and} \quad Z_n \nearrow X \quad \text{as} \quad n \to \infty.$$

*Then*

$$\lim_{n \to \infty} E\, Y_n = \lim_{n \to \infty} E\, Z_n \quad (= E\, X).$$

*Proof.* The first remark is that *if* the limits are equal, then they must be equal to $E\, X$ because of the definition of the expected value for non-negative random variables (Definition 4.2).

To prove equality between the limits it suffices to show that if $0 \leq Y_n \nearrow X$ as $n \to \infty$, and $X \geq Z_m$, then

$$\lim_{n \to \infty} E\, Y_n \geq E\, Z_m, \tag{4.1}$$

because by switching roles between the two sequences, we similarly obtain

$$\lim_{m \to \infty} E\, Z_m \geq E\, Y_n,$$

and the desired equality follows.

To prove (4.1) we first suppose that

$$Z_m > c > 0.$$

Next we note that there exists $M < \infty$, such that $Z_m \leq M$ (because $Z_m$ is simple, and therefore has only a finite number of supporting points). Let $\varepsilon < M$, set $A_n = \{Y_n \geq Z_m - \varepsilon\}$, and observe that, by assumption, $A_n \nearrow \Omega$ a.s. as $n \to \infty$. Moreover,

$$Y_n \geq Y_n I\{A_n\} \geq (Z_m - \varepsilon) I\{A_n\}.$$

By domination we therefore obtain (all random variables are simple)

$$E\, Y_n \geq E\, Y_n I\{A_n\} \geq E(Z_m - \varepsilon) I\{A_n\} = E\, Z_m I\{A_n\} - \varepsilon P(A_n)$$
$$= E\, Z_m - E\, Z_m I\{A_n^c\} - \varepsilon \geq E\, Z_m - M P(A_n^c) - \varepsilon,$$

so that,

$$\liminf_{n \to \infty} E\, Y_n \geq E\, Z_m - \varepsilon,$$

since $P(A_n^c) \to 0$ as $n \to \infty$. The arbitrariness of $\varepsilon$ concludes the proof for that case. Since $c$ was arbitrary, (4.1) has been verified for $Z_m$ strictly positive.

If $c = 0$, then, by domination, and what has already been shown,

$$\liminf_{n \to \infty} E\, Y_n \geq \liminf_{n \to \infty} E\, Y_n I\{Z_m > 0\} \geq E\, Z_m I\{Z_m > 0\} = E\, Z_m,$$

where, to be precise, we used Theorem 4.1(f) in the last step.  □

We have thus shown consistency and thereby that the definition of the expected value is in order.

A slight variation to prove consistency runs as follows.

**Theorem 4.3.** *Suppose that $X$ is a non-negative random variable, and that $\{Y_n,\ n \geq 1\}$ are non-negative simple random variables, such that $0 \leq Y_n \nearrow X$ as $n \to \infty$. Suppose further, that $Y$ is a simple random variable, such that $0 \leq Y \leq X$. Then*

$$\lim_{n \to \infty} EY_n \geq EY.$$

**Exercise 4.1.** Prove the theorem by showing that it suffices to consider indicator functions, $Y = I\{A\}$ for $A \in \mathcal{F}$.
Hint: Think metatheorem. □

The next point in the program is to show that the basic properties we have provided for simple random variables carry over to general non-negative random variables.

**Theorem 4.4.** *Let $X, Y$ be non-negative random variables. Then*

(a)   *If $X = 0$ a.s., then $E X = 0$;*
(b)   $E X \geq 0$;
(c)   *If $E X = 0$, then $X = 0$ a.s.;*
(d)   *If $E X > 0$, then $P(X > 0) > 0$;*
(e)   *Linearity: $E(aX + bY) = aE X + bE Y$ for any $a, b \in \mathbb{R}^+$;*
(f)   $E XI\{X > 0\} = E X$;
(g)   *Equivalence: If $X = Y$ a.s., then $EY = E X$;*
(h)   *Domination: If $Y \leq X$ a.s., then $EY \leq E X$;*
(j)   *If $E X < \infty$, then $X < \infty$ a.s., that is, $P(X < \infty) = 1$.*

*Remark 4.1.* Note that infinite expected values are allowed. □

*Proof.* The properties are listed in the same order as for simple random variables, but verified in a different order (property (j) is new).
   The basic idea is that there exist sequences $\{X_n,\ n \geq 1\}$ and $\{Y_n,\ n \geq 1\}$ of non-negative simple random variables converging monotonically to $X$ and $Y$, respectively, as $n \to \infty$, and which obey the basic rules for each $n$. The conclusions then follow by letting $n \to \infty$.
   For (a) there is nothing new to prove.
   To prove linearity, we know from Theorem 4.1(e) that

$$E(aX_n + bY_n) = aE X_n + bE Y_n \quad \text{for any} \quad a, b \in \mathbb{R}^+,$$

which, by letting $n \to \infty$, shows that

$$E(aX + bY) = aE X + bE Y \quad \text{for any} \quad a, b \in \mathbb{R}^+.$$

The proof of (h), domination, follows exactly the same pattern. Next, (b) follows from (h) and (a): Since $X \geq 0$, we obtain $E X \geq E 0 = 0$. In order to prove (c), let $A_n = \{\omega : X(\omega) \geq \frac{1}{n}\}$. Then

$$\frac{1}{n}I\{A_n\} \leq X_n I\{A_n\} \leq X,$$

so that

$$\frac{1}{n}P(A_n) \leq E X_n I\{A_n\} \leq E X = 0,$$

which forces $P(A_n) = 0$ for all $n$, that is $P(X < \frac{1}{n}) = 1$ for all $n$.

Moreover, (d) follows from (a), and (f) follows from (e) and (a), since

$$E X = E X I\{X = 0\} + E X I\{X > 0\} = 0 + E X I\{X > 0\}.$$

Equivalence follows as in Theorem 4.1, and (j), finally, by linearity,

$$\infty > E X = E X I\{X < \infty\} + E X I\{X = \infty\} \geq E X I\{X = \infty\},$$

from which there is no escape except $P(X = \infty) = 0$.  □

## The General Case

Recall that the expected value of a random variable $X$ is defined as the difference between the expected values of the positive and negative parts, $E X = E X^+ - E X^-$, provided at least one of them is finite, and that the expected value is finite if and only if $E|X| < \infty$, in which case we call the random variable integrable.

By reviewing the basic properties we find that (a) remains (nothing is added), that (b) disappears, and that (c) is no longer true, since, e.g., symmetric random variables whose mean exists have mean 0 – one such example is $P(X = 1) = P(X = -1) = 1/2$. The remaining properties remain with minor modifications.

**Theorem 4.5.** *Let $X, Y$ be integrable random variables. Then*

(a)   *If $X = 0$ a.s., then $E X = 0$;*
(b)   *$|X| < \infty$ a.s., that is, $P(|X| < \infty) = 1$;*
(c)   *If $E X > 0$, then $P(X > 0) > 0$;*
(d)   *Linearity: $E(aX + bY) = aE X + bE Y$ for any $a, b \in \mathbb{R}$;*
(e)   *$E X I\{X \neq 0\} = E X$;*
(f)   *Equivalence: If $X = Y$ a.s., then $E Y = E X$;*
(g)   *Domination: If $Y \leq X$ a.s., then $E Y \leq E X$;*
(h)   *Domination: If $|Y| \leq X$ a.s., then $E|Y| \leq E X$.*

*Proof.* For the proofs one considers the two tails separately. Let us illustrate this by proving linearity.

Since, by the triangle inequality, $|aX + bY| \leq |a||X| + |b||Y|$ it follows, by domination and linearity for non-negative random variables, Theorem 4.4(h) and (e), that

$$E|aX + bY| \leq E|a||X| + E|b||Y| = |a|E|X| + |b|E|Y| < \infty,$$

so that the sum is integrable. Next we split the sum in two different ways:

$$aX + bY = \begin{cases} (aX + bY)^+ - (aX + bY)^-, \\ (aX)^+ - (aX)^- + (bY)^+ - (bY)^-. \end{cases}$$

Because of linearity it suffices to prove additivity.

Since all random variables to the right are non-negative we use linearity to conclude that

$$E(X + Y)^+ + E(X^-) + E(Y^-) = E(X + Y)^- + E(X^+) + E(Y^+),$$

which shows that

$$E(X + Y) = E(X + Y)^+ - E(X + Y)^-$$
$$= E(X^+) - E(X^-) + E(Y^+) - E(Y^-) = E X + E Y. \qquad \square$$

**Exercise 4.2.** Complete the proof of the theorem. $\qquad \square$

## 5 Expectation; Convergence

In addition to the basic properties one is frequently faced with an infinite sequence of functions and desires information about the limit. A well-known fact is that it is not permitted in general to reverse the order of taking a limit and computing an integral; in technical probabilistic terms the problem amounts to the question

$$\lim_{n \to \infty} E X_n \ ? \ = ? \ E \lim_{n \to \infty} X_n. \tag{5.1}$$

We shall encounter this problem in greater detail in Chapter 5 which is devoted to various convergence modes. We therefore provide just one illustration here, the full impact of which will be clearer later.

*Example 5.1.* Let $\alpha > 0$, and set

$$P(X_n = 0) = 1 - \frac{1}{n^2} \quad \text{and} \quad P(X_n = n^\alpha) = \frac{1}{n^2}, \quad n \geq 1.$$

Taking only two different values these are certainly simple random variables, but we immediately observe that one of the points slides away toward infinity as $n$ increases.

One can show (this will be done in Chapter 5) that $X_n(\omega) \to 0$ as $n \to \infty$ for almost every $\omega$, which means that $P(\lim_{n \to \infty} X_n = 0) = 1$ – at this point we may at least observe that $P(X_n = 0) \to 1$ as $n \to \infty$.

As for the limit of the expected values,

$$E X_n = 0 \cdot \left(1 - \frac{1}{n^2}\right) + n^\alpha \cdot \frac{1}{n^2} = n^{\alpha - 2} \to \begin{cases} 0, & \text{for} \quad 0 < \alpha < 2, \\ 1, & \text{for} \quad \alpha = 2, \\ \infty, & \text{for} \quad \alpha > 2. \end{cases}$$

The answer to the question addressed in (5.1) thus may vary. $\qquad \square$

Typical conditions that yield positive results are uniformity, monotonicity or domination conditions. All of these are tailored in order to prevent masses to escape, "to pop up elsewhere".

A first positive result concerns random variables that converge monotonically.

**Theorem 5.1.** (Monotone convergence)
*Let $\{X_n,\ n \geq 1\}$ be non-negative random variables. If $X_n \nearrow X$ as $n \to \infty$, then*

$$E X_n \nearrow E X \quad as \quad n \to \infty.$$

*Remark 5.1.* The limit may be infinite.     □

*Proof.* From the consistency proof we know that the theorem holds if $\{X_n,\ n \geq 1\}$ are non-negative *simple* random variables. For the general case we therefore introduce non-negative, simple random variables $\{Y_{k,n},\ n \geq 1\}$ for every $k$, such that

$$Y_{k,n} \nearrow X_k \quad as \quad n \to \infty.$$

Such sequences exist by definition and consistency.

In addition, we introduce the non-negative simple random variables

$$Z_n = \max_{1 \leq k \leq n} Y_{k,n}, \quad n \geq 1.$$

By construction, and domination, respectively,

$$Y_{k,n} \leq Z_n \leq X_n, \quad and \quad E Y_{k,n} \leq E Z_n \leq E X_n. \tag{5.2}$$

Letting $n \to \infty$ and then $k \to \infty$ in the point-wise inequality yields

$$X_k \leq \lim_{n \to \infty} Z_n \leq \lim_{n \to \infty} X_n = X \quad and \ then \quad X \leq \lim_{n \to \infty} Z_n \leq X,$$

respectively, so that,

$$\lim_{n \to \infty} E Z_n = E X = E \lim_{n \to \infty} Z_n, \tag{5.3}$$

where the first equality holds by definition (and consistency), and the second one by equivalence (Theorem 4.4(g)).

The same procedure in the inequality between the expectations in (5.2) yields

$$E X_k \leq \lim_{n \to \infty} E Z_n \leq \lim_{n \to \infty} E X_n,$$

and then

$$\lim_{k \to \infty} E X_k \leq \lim_{n \to \infty} E Z_n \leq \lim_{n \to \infty} E X_n.$$

Combining the latter one with (5.3) finally shows that

$$\lim_{n \to \infty} E X_n = \lim_{n \to \infty} E Z_n = E \lim_{n \to \infty} Z_n = E X. \qquad \square$$

The following variation for non-increasing sequences immediately suggests itself.

**Corollary 5.1.** *Let $\{X_n, n \geq 1\}$ be non-negative random variables and suppose that $X_1$ is integrable. If $X_n \searrow X$ as $n \to \infty$, then*

$$E X_n \searrow E X \quad as \quad n \to \infty.$$

*Proof.* Since $0 \leq 2X - X_n \nearrow X$ as $n \to \infty$, the conclusion is, indeed, a corollary of the monotone convergence theorem.                           □

A particular case of importance is when the random variables $X_n$ are partial sums of other random variables. The monotone convergence theorem then translates as follows:

**Corollary 5.2.** *Suppose that $\{Y_n, n \geq 1\}$ are non-negative random variables. Then*

$$E\left(\sum_{n=1}^{\infty} Y_n\right) = \sum_{n=1}^{\infty} E Y_n.$$

**Exercise 5.1.** Please write out the details of the translation.          □

In Example 5.1 we found that the limit of the expected values coincided with the expected value in some cases and was larger in others. This is a common behavior.

**Theorem 5.2.** (Fatou's lemma)
(i) *If $\{X_n, n \geq 1\}$ are non-negative random variables, then*

$$E \liminf_{n \to \infty} X_n \leq \liminf_{n \to \infty} E X_n.$$

(ii) *If, in addition, $Y$ and $Z$ are integrable random variables, such that $Y \leq X_n \leq Z$ a.s. for all $n$, then*

$$E \liminf_{n \to \infty} X_n \leq \liminf_{n \to \infty} E X_n \leq \limsup_{n \to \infty} E X_n \leq E \limsup_{n \to \infty} X_n.$$

*Proof.* (i): Set $Y_n = \inf_{k \geq n} X_k$, $n \geq 1$. Since

$$Y_n = \inf_{k \geq n} X_k \nearrow \liminf_{n \to \infty} X_n \quad as \quad n \to \infty,$$

the monotone convergence theorem yields

$$E Y_n \nearrow E \liminf_{n \to \infty} X_n.$$

Moreover, since $Y_n \leq X_n$, Theorem 4.4(h) tells us that

$$E Y_n \leq E X_n \quad for \ all \quad n.$$

Combining the two proves (i).

To prove (ii) we begin by noticing that

$$\liminf_{n\to\infty}(X_n - Y) = \liminf_{n\to\infty} X_n - Y \quad \text{and} \quad \liminf_{n\to\infty}(Z - X_n) = Z - \limsup_{n\to\infty} X_n,$$

after which (ii) follows from (i) and additivity, since $\{X_n - Y,\, n \geq 1\}$ and $\{Z - X_n,\, n \geq 1\}$ are non-negative random variables.    $\square$

*Remark 5.2.* The right-hand side of (i) may be infinite.

*Remark 5.3.* If the random variables are are indicators, the result transforms into an inequality for probabilities and we rediscover Theorem 1.3.2. Technically, if $X_n = I\{A_n\}$, $n \geq 1$, then (i) reduces to $P(\liminf_{n\to\infty} A_n) \leq \liminf_{n\to\infty} P(A_n)$, and so on.

A typical use of Fatou's lemma is in cases where one knows that a pointwise limit exists, and it is enough to assert that the expected value of the limit is finite. This situation will be commonplace in Chapter 5. However, if, in addition, the sequence of random variables is dominated by another, integrable, random variable, we obtain another celebrated result.

**Theorem 5.3.** (The Lebesgue dominated convergence theorem)
*Suppose that $|X_n| \leq Y$, for all $n$, where $EY < \infty$, and that $X_n \to X$ a.s. as $n \to \infty$. Then*

$$E|X_n - X| \to 0 \quad \text{as} \quad n \to \infty,$$

*In particular,*

$$E X_n \to E X \quad \text{as} \quad n \to \infty.$$

*Proof.* Since also $|X| \leq Y$ it follows that $|X_n - X| \leq 2Y$, so that by replacing $X_n$ by $|X_n - X|$, we find that the proof reduces to showing that if $0 \leq X_n \leq Y \in L^1$, and $X_n \to 0$ almost surely as $n \to \infty$, then $E X_n \to 0$ as $n \to \infty$. This, however, follows from Theorem 5.2(ii).    $\square$

*Remark 5.4.* If, in particular, $Y$ is constant, that is, if the random variables are uniformly bounded, $|X_n| \leq C$, for all $n$ and some constant $C$, the result is sometimes called the *bounded* convergence theorem.

*Remark 5.5.* In the special case when the random variables are indicators of measurable sets we rediscover the last statement in Theorem 1.3.2:

$$A_n \to A \quad \Longrightarrow \quad P(A_n) \to P(A) \quad \text{as} \quad n \to \infty.    \qquad \square$$

The following corollary, the verification of which we leave as an exercise, parallels Corollary 5.2.

**Corollary 5.3.** *Suppose that $\{Y_n,\, n \geq 1\}$ are random variables, such that $\left|\sum_{n=1}^{\infty} Y_n\right| \leq X$, where $X$ is integrable. If $\sum_{n=1}^{\infty} Y_n$ converges a.s. as $n \to \infty$, then $\sum_{n=1}^{\infty} Y_n$, as well as every $Y_n$, are integrable, and*

$$E\left(\sum_{n=1}^{\infty} Y_n\right) = \sum_{n=1}^{\infty} E Y_n.$$

This concludes our presentation of expected values. Looking back we find that the development is rather sensitive in the sense that after having traversed elementary random variables, the sequence of results, that is, extensions, convergence results, uniqueness, and so on, have to be pursued in the correct order. Although many things, such as linearity, say, are intuitively "obvious" we must remember that when the previous section began we *knew* nothing about expected values – everything had to be verified.

Let us also mention that one can define expected values in different, albeit equivalent ways. Which way one chooses is mainly a matter of taste.

**Exercise 5.2.** Prove that the definition

$$E X = \sup_{0 \leq Y \leq X} \{E Y : Y \text{ is a simple random variable}\}$$

is equivalent to Definition 4.2.

**Exercise 5.3.** Review the last two sections in the language of Subsection 2.2.2, i.e., "translate" the results (and the proofs) into the language of mathematics.     □

## 6 Indefinite Expectations

In mathematical terminology one integrates over sets. In probabilistic terms we suppose that $X$ is an integrable random variable, and consider expressions of the form

$$\mu_X(A) = E X I\{A\} = \int_A X \, dP = \int_\Omega X I\{A\} \, dP, \quad \text{where} \quad A \in \mathcal{F}.$$

In other words, $\mu_X(\cdot)$ is an "ordinary" expectation applied to the random variable $X I\{\cdot\}$. In order to justify the definition and the equalities it therefore suffices to consider indicator variables, for which the equalities reduce to equalities between probabilities – note that $\mu_{I\{A\}}(A) = P(A \cap A)$ for $A \in \mathcal{F}$ –, after which one proceeds via non-negative simple random variables, monotone convergence, and $X = X^+ - X^-$ according to the usual procedure.

The notation $\mu_X(\cdot)$ suggests that we are confronted with a *signed measure* with respect to the random variable $X$, that is, a measure that obeys the properties of a probability measure except that it can take negative values, and that the total mass need not be equal to 1. If $X$ is non-negative and integrable the expression suggests that $\mu$ is a non-negative, finite measure, and if $E X = 1$ a probability measure.

**Theorem 6.1.** *Suppose that $X$ is a non-negative, integrable random variable. Then:*

(a)     $\mu_X(\emptyset) = 0$.
(b)     $\mu_X(\Omega) = E X$.
(c)     $P(A) = 0 \implies \mu_X(A) = 0$.

(d)   If $\mu_X(A) = 0$ for all $A \in \mathcal{F}$, then $X = 0$ a.s.
(e)   If $\{A_n, n \geq 1\}$ are disjoint sets, then $\mu_X(\bigcup_{n=1}^{\infty} A_n) = \sum_{n=1}^{\infty} \mu_X(A_n)$.
(f)   If $\mu_X(A) = 0$ for all $A \in \mathcal{A}$, where $\mathcal{A}$ is a $\pi$-system that generates $\mathcal{F}$, then $X = 0$ a.s.
(g)   If $\mu_X(A) = 0$ for all $A \in \mathcal{A}$, where $\mathcal{A}$ is an algebra that generates $\mathcal{F}$, then $X = 0$ a.s.

*Proof.* The conclusions follow, essentially, from the definition and the different equivalent forms of $\mu_X(\cdot)$. For (a)–(e) we also need to exploit some of the earlier results from this chapter, and for (f) and (g) we additionally need Theorems 1.2.3 and 1.2.2, respectively.

**Exercise 6.1.** Spell out the details.                                        □

*Remark 6.1.* The theorem thus verifies that $\mu_X$ is a finite measure whenever $X$ is a non-negative integrable random variable.                       □

It is now possible to extend the theorem to arbitrary integrable random variables by considering positive and negative parts separately, and to compare measures, corresponding to different random variables, by paralleling the development for ordinary expectations.

**Theorem 6.2.** *Suppose that $X$ and $Y$ are integrable random variables. Then:*

(i)   If $\mu_X(A) = \mu_Y(A)$ for all $A \in \mathcal{F}$, then $X = Y$ a.s.
(ii)  If $\mu_X(A) = \mu_Y(A)$ for all $A \in \mathcal{A}$, where $\mathcal{A}$ is a $\pi$-system that generates $\mathcal{F}$, then $X = Y$ a.s.
(iii) If $\mu_X(A) = \mu_Y(A)$ for all $A \in \mathcal{A}$, where $\mathcal{A}$ is an algebra that generates $\mathcal{F}$, then $X = Y$ a.s.

**Exercise 6.2.** Once again we urge the reader to fill in the proof.          □

The following result is useful for integrals over tails or small, shrinking sets of integrable random variables.

**Theorem 6.3.** *Let $X$ be a random variable with finite mean, and $A$ and $A_n$, $n \geq 1$, be arbitrary measurable sets (events). Then:*

(i)   $\left| \mu_X(\{|X| > n\}) \right| \leq \mu_{|X|}(\{|X| > n\}) \to 0$ as $n \to \infty$.

(ii)  If $P(A_n) \to 0$ as $n \to \infty$, then $\left| \mu_X(A_n) \right| \leq \mu_{|X|}(A_n) \to 0$ as $n \to \infty$.

*Proof.* Since the inequalities are consequences of the basic properties it suffices to prove the conclusion for non-negative random variables.

Thus, suppose that $X \geq 0$. The first claim follows from monotone convergence, Theorem 5.1(i) and linearity. Namely, since $XI\{X \leq n\} \nearrow X$, which is integrable, it follows that

$$E\,XI\{X \leq n\} \nearrow E\,X < \infty \quad \text{as} \quad n \to \infty,$$

so that

$$\mu_X(\{X > n\}) = E\,XI\{X > n\} = E\,X - E\,XI\{X \le n\} \searrow 0 \quad \text{as} \quad n \to \infty.$$

As for (ii), let $M > 0$. Then

$$\mu_X(A_n) = E\,XI\{A_n\} = E\,XI\{A_n \cap \{X \le M\}\} + E\,XI\{A_n \cap \{X > M\}\}$$
$$\le MP(A_n) + E\,XI\{X > M\},$$

so that

$$\limsup_{n \to \infty} E\,XI\{A_n\} \le E\,XI\{X > M\}.$$

The conclusion now follows from (i), since $E\,XI\{X > M\}$ can be made arbitrarily small by choosing $M$ large enough. $\qquad\square$

*Remark 6.2.* Note the idea in (ii) to split the set $A_n$ into a "nice" part which can be handled in more detail, and a "bad" part which is small. This device is used abundantly in probability theory (and in analysis in general) and will be exploited several times as we go on. $\qquad\square$

# 7 A Change of Variables Formula

We have seen that random variables are functions from the sample space to the real line, and we have defined expectations of random variables in terms of integrals over the sample space. Just as the probability space behind the random variables sinks into the background, once they have been properly defined by the induced measure (Theorem 1.1), one would, in the same vein, prefer to compute an integral on the real line rather than over the probability space. Similarly, since measurable functions of random variables are new random variables (Theorem 1.3), one would also like to find the relevant integral corresponding to expectations of functions of random variables. The following theorem, which we might view as the establishing of "induced expectations", settles the problem.

**Theorem 7.1.** (i) *Suppose that $X$ is integrable. Then*

$$E\,X = \int_\Omega X\,dP = \int_{\mathbb{R}} x\,dF_X(x).$$

(ii) *Let $X$ be a random variable, and suppose that $g$ is a measurable function, such that $g(X)$ is an integrable random variable. Then*

$$E\,g(X) = \int_\Omega g(X)\,dP = \int_{\mathbb{R}} g(x)\,dF_X(x).$$

*Proof.* We follow the usual procedure.

(i) If $X$ is an indicator random variable, $X = I\{A\}$ for some $A \in \mathcal{F}$, then the three members all reduce to $P(X \in A)$. If $X$ is a simple random variable, $X = \sum_{k=1}^{n} x_k I\{A_k\}$, where $\{A_k, 1 \leq k \leq n\}$ is a partition of $\Omega$, then the three members reduce to $\sum_{k=1}^{n} P(A_k)$. If $X$ is non-negative, the conclusion follows by monotone convergence, and for the general case we use $X = X^+ - X^-$ and additivity.

(ii) We proceed as in (i) with $g$ playing the role of $X$. If $g(x) = I_A(x)$, then

$$\{\omega : g(X(\omega)) = 1\} = \{\omega : X(\omega) \in A\},$$

so that

$$E\, g(X) = P(X \in A) = \int_A \mathrm{d}F_X(x) = \int_{\mathbb{R}} g(x)\, \mathrm{d}F_X(x).$$

If $g$ is simple, the conclusion follows by linearity, if $g$ is non-negative by monotone convergence, and, finally, in the general case by decomposition into positive and negative parts. □

**Exercise 7.1.** As always, write out the details. □

By analyzing the proof we notice that if $X$ is discrete, then $X$ is, in fact, an elementary random variable (recall Definition 1.1), that is, an infinite sum $\sum_{k=1}^{\infty} x_k I\{A_k\}$. If $X$ is non-negative, then, by monotonicity,

$$E X = \sum_{k=1}^{\infty} x_k P(A_k),$$

and in the general case this holds by the usual decomposition. This, and the analogous argument for $g(X)$, where $g$ is measurable proves the following variation of the previous result in the discrete and absolutely continuous cases, respectively.

**Theorem 7.2.** *If $X$ is a discrete random variable with probability function $p_X(x)$, $g$ is a measurable function, and $E|g(X)| < \infty$, then*

$$E\, g(X) = \int_{\Omega} g(X)\, \mathrm{d}P = \sum_{k=1}^{\infty} g(x_k) p_X(x_k) = \sum_{k=1}^{\infty} g(x_k) P(X = x_k).$$

*Proof.* We use the decomposition $A_k = \{X = x_k\}$, $k = 1, 2, \ldots$, and $A_0 = \left( \bigcup_{n=1}^{\infty} A_k \right)^c$, observing that $P(A_0) = 0$. □

**Theorem 7.3.** *If $X$ is an absolutely continuous random variable, with density function $f_X(x)$, $g$ is a measurable function, and $E|g(X)| < \infty$, then*

$$E\, g(X) = \int_{\Omega} g(X)\, \mathrm{d}P = \int_{-\infty}^{\infty} g(x) f_X(x)\, \mathrm{d}x.$$

*Proof.* If $g(x) = I_A(x)$ is an indicator, of $A \in \mathcal{R}$, say, then

$$E\, g(X) = \int_\Omega I\{A\}\, dP = P(A) = \int_A f_X(x)\, dx$$
$$= \int_{-\infty}^\infty I_A(x) f_X(x)\, dx = \int_{-\infty}^\infty g(x) f_X(x)\, dx,$$

after which one proceeds along the usual scheme. □

In addition to being a computational vehicle, the formula for computing $E\, g(X)$ shows that we do not need to know the distribution of $g(X)$ in order to find its mean.

*Example 7.1.* Let $X \in U(0,1)$, and suppose, for example, that $g(x) = \sin x$. Then

$$E \sin X = \int_0^1 \sin x\, dx = 1 - \cos 1,$$

whereas one has to turn to the arcsin function in order to find the density of $\sin X$. And, ironically, if one then computes $E \sin X$, one obtains the same integral as three lines ago after a change of variable. □

# 8 Moments, Mean, Variance

Expected values measure the center of gravity of a distribution; they are measures of *location*. In order to describe a distribution in brief terms there exist additional measures, such as the variance which measures the *dispersion* or spread, and moments.

**Definition 8.1.** *Let $X$ be a random variable. The*

- moments *are $E\, X^n$, $n = 1, 2, \ldots$;*
- central moments *are $E(X - E\,X)^n$, $n = 1, 2, \ldots$;*
- absolute moments *are $E|X|^n$, $n = 1, 2, \ldots$;*
- absolute central moments *are $E|X - E\,X|^n$, $n = 1, 2, \ldots$.*

*The first moment, $E\,X$, is the* mean. *The second central moment is called* variance*:*

$$\mathrm{Var}\, X = E(X - E\,X)^2 \quad (= E\,X^2 - (E\,X)^2).$$

*All of this, provided the relevant quantities exist.* □

Following are tables which provide mean and variance for the standard discrete and absolutely continuous distributions listed earlier in this chapter. The reader is advised to check that the entries have been correctly inserted in both tables.

Mean and variance for the Cantor distribution will be given in Section 2.11 ahead.

| Distribution | Notation | Mean | Variance |
|---|---|---|---|
| One point | $\delta(a)$ | $a$ | $0$ |
| Symmetric Bernoulli | | $0$ | $1$ |
| Bernoulli | $\mathrm{Be}(p)$ | $p$ | $pq$ |
| Binomial | $\mathrm{Bin}(n,p)$ | $np$ | $npq$ |
| Geometric | $\mathrm{Ge}(p)$ | $\frac{q}{p}$ | $\frac{q}{p^2}$ |
| First success | $\mathrm{Fs}(p)$ | $\frac{1}{p}$ | $\frac{q}{p^2}$ |
| Poisson | $\mathrm{Po}(m)$ | $m$ | $m$ |

**Table 2.3.** Mean and variance for some discrete distributions

| Distribution | Notation | Mean | Variance |
|---|---|---|---|
| Uniform | $U(a,b)$ | $\frac{a+b}{2}$ | $\frac{(b-a)^2}{12}$ |
| | $U(0,1)$ | $\frac{1}{2}$ | $\frac{1}{12}$ |
| | $U(-1,1)$ | $0$ | $\frac{1}{3}$ |
| Triangular | $\mathrm{Tri}(-1,1)$ | $0$ | $\frac{1}{6}$ |
| Exponential | $\mathrm{Exp}(\theta)$ | $\theta$ | $\theta^2$ |
| Gamma | $\Gamma(p,\theta)$ | $p\theta$ | $p\theta^2$ |
| Beta | $\beta(r,s)$ | $\frac{r}{r+s}$ | $\frac{rs}{(r+s)^2(r+s+1)}$ |
| Normal | $N(\mu,\sigma^2)$ | $\mu$ | $\sigma^2$ |
| | $N(0,1)$ | $0$ | $1$ |
| Log-normal | $LN(\mu,\sigma^2)$ | $e^{\mu+\frac{1}{2}\sigma^2}$ | $e^{2\mu}(e^{2\sigma^2}-e^{\sigma^2})$ |
| Cauchy | $C(0,1)$ | $-$ | $-$ |
| Pareto | $\mathrm{Pa}(\beta,\alpha)$ | $\frac{\alpha\beta}{\alpha-1}$ | $\frac{\alpha\beta^2}{(\alpha-2)(\alpha-1)}$ |

**Table 2.4.** Mean and variance for some absolutely continuous distributions

The Cauchy distribution possesses neither mean nor variance. The expected value and variance for the Pareto distribution only exist for $\alpha > 1$ and $\alpha > 2$, respectively (as is suggested by the formulas).

If we think of the physical interpretation of mean and variance it is reasonable to expect that a linear transformation of a random variable changes the center of gravity linearly, and that a translation does not change the dispersion. The following exercise puts these observations into formulas.

**Exercise 8.1.** Prove the following properties for linear transformations: Let $X$ be a random variable with $EX = \mu$ and $\mathrm{Var}\,X = \sigma^2$, and set $Y = aX + b$, where $a, b \in \mathbb{R}$. Prove that

$$EY = a\mu + b \quad \text{and that} \quad \mathrm{Var}\,Y = a^2\sigma^2. \qquad \square$$

## Two Special Examples Revisited

In Subsection 2.2.7 we presented two examples, the first of which was a random variable $X$ which was uniformly distributed on the irrationals in $[0,1]$, that is, with density

$$f(x) = \begin{cases} 1, & \text{for} \quad x \in [0,1] \setminus \mathbb{Q}, \\ 0, & \text{for} \quad x \in [0,1] \cap \mathbb{Q}. \end{cases}$$

The random variable was there seen to be equivalent to a standard $U(0,1)$-distributed random variable, so that a direct computation shows that $EX = 1/2$ and that $\operatorname{Var} X = 1/12$.

The other example was a discrete random variable with probability function

$$p(r_k) = \begin{cases} \frac{6}{\pi^2 k^2}, & \text{for} \quad r_k \in (0,1) \cap \mathbb{Q}, \\ 0, & \text{otherwise}, \end{cases}$$

where $\{r_k, \; k \geq 1\}$ was an enumeration of the rationals in the unit interval. We also pointed out that this is a somewhat pathological situation, since the enumeration of the rationals is not unique. This means that all moments, in particular the expected value and the variance, are ambiguous quantities in that they depend on the actual enumeration of $\mathbb{Q}$.

# 9 Product Spaces; Fubini's Theorem

Expectations of functions of random vectors are defined in the natural way as the relevant multidimensional integral. The results from Section 2.7 carry over, more or less by notation, that is, by replacing appropriate roman letters by boldface ones.

For example, if $(X,Y)'$ is a random vector and $g$ a measurable function, then

$$E\,g(X,Y) = \int_\Omega g(X,Y)\,\mathrm{d}P = \int_{\mathbb{R}^2} g(x,y)\,\mathrm{d}F_{X,Y}(x,y).$$

In the discrete case,

$$E\,g(X,Y) = \sum_{i=1}^\infty \sum_{j=1}^\infty g(x_i,x_j)p_{X,Y}(x_i,x_j),$$

and in the absolutely continuous case

$$E\,g(X,Y) = \int_{\mathbb{R}^2} g(x,y)f_{X,Y}(x,y)\,\mathrm{d}x\mathrm{d}y.$$

In each case the proviso is absolute convergence.

Expectations of functions of random variables take special and useful forms when the probability spaces are product spaces.

## 9.1 Finite-dimensional Product Measures

Let $(\Omega_k, \mathcal{F}_k, P_k)$, $1 \leq k \leq n$, be probability spaces. We introduce the notation

$$\mathcal{F}_1 \times \mathcal{F}_2 \times \cdots \times \mathcal{F}_n = \sigma\{F_1 \times F_2 \times \cdots \times F_n : F_k \in \mathcal{F}_k, \; k = 1, 2, \ldots, n\}.$$

Given this setup one can now construct a product space, $(\times_{k=1}^n \Omega_k, \times_{k=1}^n \mathcal{F}_k)$, with an associated probability measure $\mathbb{P}$, such that

$$\mathbb{P}(A_1 \times A_2 \times \cdots \times A_n) = \prod_{k=1}^n P_k(A_k) \quad \text{for} \quad A_k \in \mathcal{F}_k, \; 1 \le k \le n.$$

Note that the probability measure has a built-in independence.

Moreover, the probability space $(\times_{k=1}^n \Omega_k, \times_{k=1}^n \mathcal{F}_k, \times_{k=1}^n P_k)$ thus obtained is unique. We refer to the literature on measure theory for details.

As for infinite dimensions we confine ourselves to mentioning the existence of a theory. A prominent example is the space of continuous functions on the unit interval and the associated $\sigma$-algebra – $(C[0, 1], \mathcal{C}[0, 1])$. For this and more we recommend [20].

## 9.2 Fubini's Theorem

Fubini's theorem is a result on integration, which amounts to the fact that an expectation, which in its general form is a double integral, can be evaluated as iterated single integrals.

**Theorem 9.1.** *Let* $(\Omega_1, \mathcal{F}_1, P_1)$ *and* $(\Omega_2, \mathcal{F}_2, P_2)$ *be probability spaces, and consider the product space* $(\Omega_1 \times \Omega_2, \mathcal{F}_1 \times \mathcal{F}_2, P)$, *where* $P = P_1 \times P_2$ *is the product measure as defined above, suppose that* $\mathbf{X} = (X_1, X_2)'$ *is a two-dimensional random variable, and that* $g$ *is* $\mathcal{F}_1 \times \mathcal{F}_2$-*measurable, and* (i) *non-negative or* (ii) *integrable. Then*

$$E\, g(\mathbf{X}) = \int_\Omega g(\mathbf{X}) \, dP = \int_{\Omega_1 \times \Omega_2} g(X_1, X_2) \, d(P_1 \times P_2)$$

$$= \int_{\Omega_1} \left( \int_{\Omega_2} g(\mathbf{X}) \, dP_2 \right) dP_1 = \int_{\Omega_2} \left( \int_{\Omega_1} g(\mathbf{X}) \, dP_1 \right) dP_2.$$

*Proof.* For indicators the theorem reduces to the construction of product measure, after which one proceeds via simple functions, monotone convergence and non-negative functions and the usual decomposition. We omit all details.   □

A change of variables (recall Section 2.7) applied to Fubini's theorem yields the following computationally more suitable variant.

**Theorem 9.2.** *Suppose that* $(X, Y)'$ *is a two-dimensional random variable, and* $g$ *is* $\mathcal{R}^2 = \mathcal{R} \times \mathcal{R}$-*measurable, and non-negative or integrable. Then*

$$E\, g(X, Y) = \iint_{\mathbb{R}^2} g(x, y) \, dF_X(x) \, dF_Y(y)$$

$$= \int_{\mathbb{R}} \left( \int_{\mathbb{R}} g(x, y) \, dF_Y(y) \right) dF_X(x)$$

$$= \int_{\mathbb{R}} \left( \int_{\mathbb{R}} g(x, y) \, dF_X(x) \right) dF_Y(y).$$

**Exercise 9.1.** Write down the analogous formulas in the absolutely continuous and discrete cases, respectively.    □

## 9.3 Partial Integration

A first application of Fubini's theorem is to show that the usual formula for partial integration carries over to the present context.

**Theorem 9.3.** *Let* $a < b \in \mathbb{R}$, *and suppose that* $F, G \in D^+$ *have no common points of discontinuity on* $(a, b]$. *Then*

$$\int_a^b G(x)\,\mathrm{d}F(x) = G(b)F(b) - G(a)F(a) - \int_a^b F(x)\,\mathrm{d}G(x).$$

*If, in addition, G is absolutely continuous with density g, then*

$$\int_a^b G(x)\,\mathrm{d}F(x) = G(b)F(b) - G(a)F(a) - \int_a^b F(x)g(x)\,\mathrm{d}x.$$

*Proof.* We first note that if the formula holds for $F$ and $G$, then, by linearity, it also holds for linear transformations; $\alpha F + \beta$ and $\gamma G + \delta$, since then

$$\int_a^b \gamma G(x)\,\mathrm{d}(\alpha F(x)) = \gamma\alpha \int_a^b G(x)\mathrm{d}F(x)$$

$$= \gamma\alpha \left( G(b)F(b) - G(a)F(a) - \int_a^b F(x)\,\mathrm{d}G(x) \right)$$

$$= \left(\gamma G(b)\right)\left(\alpha F(b)\right) - \left(\gamma G(a)\right)\left(\alpha F(a)\right) - \int_a^b \left(\alpha F(x)\right)\mathrm{d}(\gamma G(x)),$$

and

$$\int_a^b (G(x) + \delta)\,\mathrm{d}(F(x) + \beta) = \int_a^b G(x)\,\mathrm{d}F(x) + \delta(F(b) - F(a))$$

$$= G(b)F(b) - G(a)F(a) - \int_a^b F(x)\,\mathrm{d}G(x) + \delta(F(b) - F(a))$$

$$= (G(b) + \delta)F(b) - (G(a) + \delta)F(a) - \int_a^b F(x)\,\mathrm{d}(G(x) + \delta).$$

It is therefore no restriction to assume that $F$ and $G$ are true distribution functions, which we associate with the random variables $X$ and $Y$, respectively, the point being that we can express the integrals as probabilities. Namely, by an appeal to Fubini's theorem, we obtain, on the one hand, that

$$P(a < X \le b, a < Y \le b) = \int_a^b \int_a^b \mathrm{d}(F \times G)(x, y) = \int_a^b \int_a^b \mathrm{d}F(x)\mathrm{d}G(y)$$

$$= \int_a^b \mathrm{d}F(x) \int_a^b \mathrm{d}G(y) = \left(F(b) - F(a)\right)\left(G(b) - G(a)\right),$$

and, by splitting the probability that the point $(X, Y)$ lies inside the square $(a, b] \times (a, b]$ into three pieces, on the other hand (via product measure and Fubini), that

$$P(a < X \le b, a < Y \le b) = P(a < X < Y \le b) + P(a < Y < X \le b)$$
$$+ P(a < Y = X \le b)$$

$$= \int_a^b \int_a^x \mathrm{d}(F \times G)(x, y) + \int_a^b \int_a^x \mathrm{d}(G \times F)(x, y) + 0$$

$$= \int_a^b \left( \int_a^x \mathrm{d}F(y) \right) \mathrm{d}G(x) + \int_a^b \left( \int_a^x \mathrm{d}G(y) \right) \mathrm{d}F(x)$$

$$= \int_a^b (F(x) - F(a)) \mathrm{d}G(x) + \int_a^b (G(x) - G(a)) \mathrm{d}F(x)$$

$$= \int_a^b F(x) \, \mathrm{d}G(x) + \int_a^b G(x) \, \mathrm{d}F(x) - F(a)(G(b) - G(a))$$
$$- G(a)(F(b) - F(a)).$$

The formula for partial integration now follows by equating the two expressions for $P(a < X \le b, a < Y \le b)$.

The conclusion for the special case when $G$ is absolutely continuous follows from the fact that

$$\int_a^b F(x) \, \mathrm{d}G(x) = \int_a^b F(x) g(x) \, \mathrm{d}x. \qquad \square$$

*Remark 9.1.* The interval $(a, b]$ can be replaced by infinite intervals provided enough integrability is available. $\qquad \square$

## 9.4 The Convolution Formula

Consider once again the usual product space $(\Omega_1 \times \Omega_2, \mathcal{F}_1 \times \mathcal{F}_2, P_1 \times P_2)$, and suppose that $(X_1, X_2)'$ is a two-dimensional random variable whose marginal distribution functions are $F_1$ and $F_2$, respectively. The convolution formula provides the distribution of $X_1 + X_2$.

**Theorem 9.4.** *In the above setting*

$$F_{X_1 + X_2}(u) = \int_{-\infty}^{\infty} F_1(u - y) \, \mathrm{d}F_2(y).$$

*If, in addition, $X_2$ is absolutely continuous with density $f_2$, then*

$$F_{X_1 + X_2}(u) = \int_{-\infty}^{\infty} F_1(u - y) f_2(y) \, \mathrm{d}y.$$

*If $X_1$ is absolutely continuous with density $f_1$, the density of the sum equals*

$$f_{X_1+X_2}(u) = \int_{-\infty}^{\infty} f_1(u-y)\, dF_2(y).$$

*If both are absolutely continuous, then*

$$f_{X_1+X_2}(u) = \int_{-\infty}^{\infty} f_1(u-y) f_2(y)\, dy.$$

*Proof.* Once again, an application of Fubini's theorem does the job for us.

$$F_{X_1+X_2}(u) = P(X_1 + X_2 \le u) = \iint_{x+y \le u} d(F_1 \times F_2)(x,y)$$

$$= \int_{-\infty}^{\infty} \int_{-\infty}^{u-y} d(F_1 \times F_2)(x,y) = \int_{-\infty}^{\infty} \left( \int_{-\infty}^{u-y} dF_1(x) \right) dF_2(y)$$

$$= \int_{-\infty}^{\infty} F_1(u-y)\, dF_2(y).$$

The remaining parts are immediate.    □

## 10 Independence

One of the central concepts in probability theory is independence. The outcomes of repeated tosses of coins and throws of dice are "independent" in a sense of normal language, meaning that coins and dice do not have a memory. The successive outcomes of draws *without replacements* of cards from a deck are not independent, since a card that has been drawn cannot be drawn again. The mathematical definition of independence differs from source to source. Luckily the two following ones are equivalent.

**Definition 10.1.** *The random variables $X_1$, $X_2$, ..., $X_n$ are independent iff, for arbitrary Borel measurable sets $A_1, A_2, \ldots, A_n$,*

$$P\left( \bigcap_{k=1}^{n} \{X_k \in A_k\} \right) = \prod_{k=1}^{n} P(X_k \in A_k).$$

**Definition 10.2.** *The random variables $X_1$, $X_2$, ..., $X_n$ or, equivalently, the components of the random vector $\mathbf{X}$ are independent iff*

$$F_{\mathbf{X}}(\mathbf{x}) = \prod_{k=1}^{n} F_{X_k}(x_k) \quad \text{for all} \quad \mathbf{x} \in \mathbb{R}^n.$$    □

Independence according to the first definition thus means that all possible joint events are independent, and according to the second definition that the joint distribution function equals the product of the marginal ones.

**Theorem 10.1.** *The two definitions are equivalent.*

*Proof.* The second definition obviously is implied by the first one, since the half-open infinite sets are a subclass of all measurable sets. For the converse we note that this subclass is a $\pi$-system that generates the $\sigma$-algebra of Borel measurable sets (Theorem 1.3.6). An application of Theorem 1.3.5 finishes the proof. $\qquad\square$

*Remark 10.1.* Independence implies that the joint measure is product measure (due to uniqueness). $\qquad\square$

**Exercise 10.1.** Prove that it is, in fact, enough to check any class of sets that generates the Borel sets to assert independence. $\qquad\square$

For discrete and absolutely continuous distributions independence is equivalent to the factorization of joint probability functions and joint densities, respectively.

**Theorem 10.2.** (i) *If $X$ and $Y$ are discrete, then $X$ and $Y$ are independent iff the joint probability function is equal to the product of the marginal ones, that is iff*

$$p_{X,Y}(x,y) = p_X(x) \cdot p_Y(y) \quad \text{for all} \quad x,y \in \mathbb{R}.$$

(ii) *If $X$ and $Y$ are absolutely continuous, then $X$ and $Y$ are independent iff the joint density is equal to the product of the marginal ones, that is iff*

$$f_{X,Y}(x,y) = f_X(x) \cdot f_Y(y) \quad \text{for all} \quad x,y \in \mathbb{R}.$$

*Proof.* The discrete case follows immediately by taking differences.

As for the absolutely the continuous case, if factorization holds, then, via Fubini's Theorem, Theorem 9.1,

$$F_{X,Y}(x,y) = \int_{-\infty}^{x} \int_{-\infty}^{y} f_{X,Y}(u,v)\, du\, dv = \int_{-\infty}^{x} \int_{-\infty}^{y} f_X(u) f_Y(v)\, du\, dv$$

$$= \int_{-\infty}^{x} f_X(u)\, du \int_{-\infty}^{y} f_Y(v)\, dv = F_X(x) \cdot F_Y(y).$$

To prove the converse, we use the metatheorem approach. Suppose that $X$ and $Y$ are independent and define, for $C = A \times B$, where $A, B \in \mathcal{R}$,

$$\mathcal{E} = \left\{ C : \iint_C f_{X,Y}(u,v)\, du\, dv = \iint_C f_X(u) f_Y(v)\, du\, dv. \right\}$$

Let, for $x,y \in \mathbb{R}$, $A = (-\infty, x]$ and $B = (-\infty, y]$. Then, by definition, the independence assumption, and Fubini's theorem,

$$\iint_C f_{X,Y}(u,v)\, du\, dv = P(A \cap B) = P(A)P(B)$$

$$= \int_A f_X(u)\, du \int_B f_Y(v)\, dv = \iint_{A \times B} f_X(u) f_Y(v)\, du\, dv$$

$$= \iint_C f_X(u) f_Y(v)\, du\, dv.$$

This shows that $\mathcal{E}$ contains all rectangles. Since the class of rectangles constitutes a $\pi$-system and generate the Borel $\sigma$-algebra, Theorem 1.2.3 tells us that $\mathcal{E} = \mathcal{R}$.    □

A more modern (but less common) definition (which we state for $n = 2$) is that $X$ and $Y$ are independent iff

$$E\, g(X)h(Y) = E\, g(X) \cdot E\, h(Y) \quad \text{for all} \quad g, h \in C_B,$$

where $C_B$ is the class of bounded continuous functions. For details and equivalences, see [145], Chapter 10.

**Exercise 10.2.** Prove, via simple functions, non-negative functions, monotone convergence, and differences of non-negative functions, that this definition is equivalent to the other ones.    □

**Exercise 10.3.** Prove that if $X_1, X_2, \ldots, X_n$ are independent, then

$$E \prod_{k=1}^{n} |X_k|^{s_k} = \prod_{k=1}^{n} E|X_k|^{s_k},$$

where $s_1, s_2, \ldots, s_n$ are positive reals, and that

$$E \prod_{k=1}^{n} X_k^{j_k} = \prod_{k=1}^{n} EX_k^{j_k},$$

where $j_1, j_2, \ldots, j_n$ are positive integers.    □

Two of the basic properties of expectations were additivity and linearity. A related question concerns variances; if $X$ and $Y$ are random variables with finite variances, is it true that the variance of the sum equals the sum of the variances? Do variances have the linearity property? These questions are (partially) answered next.

**Theorem 10.3.** *Let* $X$ *and* $Y$ *be* independent *random variables with finite variances, and* $a, b \in \mathbb{R}$. *Then*

$$\text{Var}\, aX = a^2 \text{Var}\, X,$$
$$\text{Var}\, (X + Y) = \text{Var}\, X + \text{Var}\, Y,$$
$$\text{Var}\, (aX + bY) = a^2 \text{Var}\, X + b^2 \text{Var}\, Y.$$

**Exercise 10.4.** Prove the theorem.    □

*Remark 10.2.* Independence is sufficient for the variance of the sum to be equal to the sum of the variances, but not necessary.

*Remark 10.3.* Linearity should not hold, since variance is a quadratic quantity.

*Remark 10.4.* Note, in particular, that $\text{Var}\, (-X) = \text{Var}\, X$. This is as expected, since switching the sign should not alter the spread of the distribution.    □

## 10.1 Independence of Functions of Random Variables

The following theorem puts the natural result that functions of independent random variables are independent into print.

**Theorem 10.4.** *Let $X_1$, $X_2$, ..., $X_n$ be random variables and $h_1, h_2, \ldots, h_n$, be measurable functions. If $X_1$, $X_2$, ..., $X_n$ are independent, then so are $h_1(X_1), h_2(X_2), \ldots, h_n(X_n)$.*

*Proof.* Let $A_1, A_2, \ldots, A_n$ be Borel measurable sets. Then, by turning to inverse images and the Definition 10.1, we find that

$$P\Big( \bigcap_{k=1}^{n} \{h_k(X_k) \in A_k\} \Big) = P\Big( \bigcap_{k=1}^{n} \{X_k \in h_k^{-1}(A_k)\} \Big)$$

$$= \prod_{k=1}^{n} P(X_k \in h_k^{-1}(A_k)) = \prod_{k=1}^{n} P(h_k(X_k) \in A_k). \qquad \square$$

## 10.2 Independence of $\sigma$-Algebras

As an analog to Theorem 1.4.1, independence of random variables implies independence of the $\sigma$-algebras generated by them.

**Theorem 10.5.** *If $X_1$, $X_2$, ..., $X_n$ are independent, then so are*

$$\sigma\{X_1\}, \ \sigma\{X_2\}, \ \ldots, \ \sigma\{X_n\}.$$

**Exercise 10.5.** Write out the details of the proof. $\qquad \square$

## 10.3 Pair-wise Independence

Recall the distinction between independence and *pair-wise* independence of sets from Section 1.4. The same distinction exists for random variables.

**Definition 10.3.** *The random variables $X_1$, $X_2$, ..., $X_n$ are pair-wise independent iff all pairs are independent.* $\qquad \square$

Independence obviously implies pair-wise independence, since there are several additional relations to check in the former case. The following example shows that there exist random variables that are pair-wise independent, but not (completely) independent.

*Example 10.1.* Pick one of the points $(1,0,0)$, $(0,1,0)$, $(0,0,1)$, and $(1,1,1)$ uniformly at random, and set, for $k = 1, 2, 3$,

$$X_k = \begin{cases} 1, & \text{if coordinate } k = 1, \\ 0, & \text{otherwise.} \end{cases}$$

Then, with $A_k = \{X_k = 1\}$, we rediscover Example 1.4.1, which proves the desired assertion. In addition,

$$E\,X_k = \frac{1}{2}, \quad \text{for} \quad k = 1, 2, 3,$$

$$E(X_1 X_2 X_3) = \frac{1}{4} \neq E\,X_1 E\,X_2 E\,X_3 = \frac{1}{8}.$$

However, since $X_i X_j = 1$ if the point $(1, 1, 1)$ is chosen, and $X_i X_j = 0$ otherwise, we obtain

$$P(X_i X_j = 1) = \frac{1}{4}, \quad \text{and} \quad P(X_i X_j = 0) = \frac{3}{4}.$$

for all pairs $(i, j)$, where $(i \neq j)$, which implies that

$$E\,X_i X_j = \frac{1}{4} = E\,X_i E\,X_j.$$

In other words, moment factorization holds for pairs but not for triplets.   □

**Exercise 10.6.** Prove that if $X_1, X_2, \ldots, X_n$ are *pair-wise* independent, then

$$\mathrm{Var}\,(X_1 + X_2 + \cdots + X_n) = \mathrm{Var}\,X_1 + \mathrm{Var}\,X_2 + \cdots + \mathrm{Var}\,X_n.$$   □

## 10.4 The Kolmogorov Zero-one Law Revisited

The proof of the following Kolmogorov zero-one law for random variables amounts to a translation of the proof of the zero-one law for events, Theorem 1.5.1.

Let $\{X_n, n \geq 1\}$ be arbitrary random variables, and set

$$\mathcal{F}_n = \sigma\{X_1, X_2, \ldots, X_n\} \quad \text{for} \quad n \geq 1,$$
$$\mathcal{F}'_n = \sigma\{X_{n+1}, X_{n+2}, \ldots\} \quad \text{for} \quad n \geq 0.$$

Then

$$\mathcal{T} = \bigcap_{n=0}^{\infty} \mathcal{F}'_n$$

is the tail-$\sigma$-field (with respect to $\{X_n, n \geq 1\}$).

**Theorem 10.6.** (The Kolmogorov zero-one law)
*Suppose that $\{X_n, n \geq 1\}$ are independent random variables. If $A \in \mathcal{T}$, then*

$$P(A) = 0 \quad or \quad 1.$$

**Exercise 10.7.** Prove the theorem, that is, rewrite (e.g.) the second proof of Theorem 1.5.1 into the language of random variables.   □

**Corollary 10.1.** *If, in the setting of the Theorem 10.6, $X$ is a $\mathcal{T}$-measurable random variable, then $X$ is a.s. constant.*

*Proof.* The event $\{X \leq x\} \in \mathcal{T}$ for all $x \in \mathbb{R}$. Thus,

$$F_X(x) = P(X \leq x) = 0 \quad \text{or} \quad 1 \quad \text{for all} \quad x \in \mathbb{R},$$

which, in view of the properties of distribution functions, implies that there exists $c \in \mathbb{R}$, such that

$$F_X(x) = P(X \leq x) = \begin{cases} 0, & \text{for} \quad x < c, \\ 1, & \text{for} \quad x \geq c. \end{cases} \qquad \square$$

A consequence of the corollary is that random variables, such as limits, limit superior and limit inferior of sequences of independent random variables must be constant a.s. if they converge at all.

## 11 The Cantor Distribution

A beautiful way to describe a random variable that has the Cantor distribution on the unit interval is the following: Let $X, X_1, X_2, \ldots$ be independent identically distributed random variables such that

$$P(X = 0) = P(X = 2) = \frac{1}{2}.$$

Then

$$Y = \sum_{n=1}^{\infty} \frac{X_n}{3^n} \in \text{Cantor}(0, 1).$$

Namely, the random variables $X_1, X_2, \ldots$ are the successive decimals of a number whose decimals in the base 3 expansion are 0 or 2, and never 1. Moreover, since the decimals each time have a 50-50 chance of being 0 or 2, the infinite sum that constitutes $Y$ is uniformly distributed over the Cantor set.

To compute the mean we use additivity and monotone convergence for series to obtain

$$EY = E\left(\sum_{n=1}^{\infty} \frac{X_n}{3^n}\right) = \sum_{n=1}^{\infty} E\left(\frac{X_n}{3^n}\right) = \sum_{n=1}^{\infty} \frac{EX_n}{3^n} = \sum_{n=1}^{\infty} \frac{1}{3^n} = \frac{1}{2},$$

which coincides with intuition.

To verify the result for the variance we also need the fact that the summands are independent (and an additional argument due to the fact that we are faced with an infinite series) to obtain

$$\text{Var } Y = \text{Var}\left(\sum_{n=1}^{\infty} \frac{X_n}{3^n}\right) = \sum_{n=1}^{\infty} \text{Var}\left(\frac{X_n}{3^n}\right) = \sum_{n=1}^{\infty} \frac{\text{Var } X_n}{(3^n)^2} = \sum_{n=1}^{\infty} \frac{1}{9^n} = \frac{1}{8}.$$

By letting $X$ be equal to 0 with probability $1/2$, and equal to some other positive integer with probability $1/2$, and by modifying $Y$ accordingly, we can construct other Cantor-type distributions. For example,

$$Z = \sum_{n=1}^{\infty} \frac{X_n}{4^n}, \quad \text{where} \quad P(X = 0) = P(X = 3) = \frac{1}{2},$$

is a random variable corresponding to a number that is uniform over the subset of the interval $[0, 1]$ which consists of the numbers whose base 4 decimal expansion contains only 0's and 3's, no 1's or 2's.

We have thus exhibited two different Cantor-type distributions.

**Exercise 11.1.** We have not explicitly *proved* that the base 4 example produces a continuous singular distribution. Please check that this is the case (although this seems pretty clear since the constructions is the same as that of the Cantor distribution). □

**Exercise 11.2.** Compute $E\,Z$ and $\text{Var}\,Z$. □

Although Cantor sets have Lebesgue measure 0 they are, somehow, of different "sizes" in the sense that some are more "nullish" than others. After all, in the classical, first case, we delete one-third of the support in each step, whereas, in the second case we delete halves. The null set in the first case therefore seems larger than in the second case.

There exists, in fact, a means to classify such sets, namely the *Hausdorff dimension*, which can be used to measure the dimension of sets (such as fractals), whose topological dimension is not a natural number. One can show that the Hausdorff dimension of the classical Cantor set on the unit interval is $\log 2/\log 3$, and that the Hausdorff dimension pertaining to our second example is $\log 2/\log 4 = 1/2 < \log 2/\log 3 \approx 0.631$, and, hence smaller than the classical Cantor set.

We close by mentioning that the same argument with 3 (or 4) replaced by 2, and $X_k$ being 0 or 1 with equal probabilities for all $k$, generates a number that is $U(0, 1)$-distributed (and, hence, an absolutely continuous distribution), since it is the binary expansion of such a number. Its Hausdorff dimension is, in fact, equal to $\log 2/\log 2 = 1$, (which coincides with the topological dimension).

# 12 Tail Probabilities and Moments

The existence of an integral or a moment clearly depends on how quickly tails decay. It is therefore not far-fetched to guess that there exist precise results concerning this connection.

**Theorem 12.1.** *Let* $r > 0$, *and suppose that* $X$ *is a non-negative random variable. Then:*

(i)    $E X = \int_0^\infty (1 - F(x)) \, dx = \int_0^\infty P(X > x) \, dx$,
       *where both members converge or diverge simultaneously;*

(ii)   $E X^r = r \int_0^\infty x^{r-1}(1 - F(x)) \, dx = r \int_0^\infty x^{r-1} P(X > x) \, dx$,
       *where both members converge or diverge simultaneously;*

(iii)  $E X < \infty \quad \Longleftrightarrow \quad \sum_{n=1}^\infty P(X \geq n) < \infty$.
       *More precisely,*

$$\sum_{n=1}^\infty P(X \geq n) \leq E X \leq 1 + \sum_{n=1}^\infty P(X \geq n).$$

(iv)   $E X^r < \infty \quad \Longleftrightarrow \quad \sum_{n=1}^\infty n^{r-1} P(X \geq n) < \infty$.
       *More precisely,*

$$\sum_{n=1}^\infty n^{r-1} P(X \geq n) \leq E X^r \leq 1 + \sum_{n=1}^\infty n^{r-1} P(X \geq n).$$

*Proof.* (i) and (ii): Let $A > 0$. By partial integration,

$$\int_0^A x^r \, dF(x) = -A^r(1 - F(A)) + \int_0^A r x^{r-1}(1 - F(x)) \, dx$$

$$= -A^r P(X > A) + r \int_0^A x^{r-1} P(X > x) \, dx.$$

If $E X^r < \infty$, then

$$A^r(1 - F(A)) \leq \int_A^\infty x^r \, dF(x) \to 0 \quad \text{as} \quad A \to \infty,$$

which shows that the integral on the right-hand side converges. If, on the other hand, the latter converges, then so does the integral on the left-hand side since it is smaller.

As for (iii),

$$E X = \sum_{n=1}^\infty \int_{n-1}^n x \, dF(x) \leq \sum_{n=1}^\infty n P(n - 1 < |X| \leq n)$$

$$= \sum_{n=1}^\infty \sum_{k=1}^n P(n - 1 < |X| \leq n) = \sum_{k=1}^\infty \sum_{n=k}^\infty P(n - 1 < |X| \leq n)$$

$$= \sum_{k=1}^\infty P(X > k - 1) \leq 1 + \sum_{k=1}^\infty P(X > k) \leq 1 + \sum_{k=1}^\infty P(X \geq k).$$

The other half follows similarly, since

$$EX \geq \sum_{n=1}^{\infty}(n-1)P(n-1 < |X| \leq n),$$

after which the computations are the same as before, and (iv) follows by "slicing" the corresponding integral similarly. □

*Remark 12.1.* Alternatively, it suffices to prove (i), because

$$EX^r = \int_0^{\infty} P(X^r > x)\,dx = \int_0^{\infty} P(X > x^{1/r})\,dx,$$

after which the change of variable $y = x^{1/r}$ establishes the claim. □

If $X$ is integer valued one can be a little more precise.

**Theorem 12.2.** *If $X$ is a non-negative, integer valued random variable, then*

$$EX = \sum_{n=1}^{\infty} P(X \geq n).$$

*Proof.* The conclusion can be obtained from Theorem 12.1, or, else, directly:

$$EX = \sum_{n=1}^{\infty} nP(X=n) = \sum_{n=1}^{\infty}\left(\sum_{k=1}^{n}1\right)P(X=n)$$
$$= \sum_{k=1}^{\infty}\sum_{n=k}^{\infty} P(X=n) = \sum_{k=1}^{\infty} P(X \geq k).$$

Interchanging the order of summation is no problem since all terms are non-negative. □

**Exercise 12.1.** Let $X$ and $Y$ be random variables and suppose that $E|Y| < \infty$. Show that, if there exists $x_0 > 0$, such that

$$P(|X| > x) \leq P(|Y| > x) \quad \text{for all} \quad x > x_0,$$

then $E|X| < \infty$. □

By modifying the proof of Theorem 12.1 one can obtain the following more general results.

**Theorem 12.3.** *Let $X$ be a non-negative random variable, and $g$ a non-negative, strictly increasing, differentiable function. Then,*

(i) $E\,g(X) = g(0) + \int_0^{\infty} g'(x)P(X > x)\,dx$, *where both members converge or diverge simultaneously;*

(ii) $E\,g(X) < \infty \iff \sum_{n=1}^{\infty} g'(n)P(X > n) < \infty$.

**Exercise 12.2.** Prove the theorem. □

**Exercise 12.3.** Let $X$ be a non-negative random variable. Prove that

$$E \log^+ X < \infty \quad \Longleftrightarrow \quad \sum_{n=1}^{\infty} \frac{1}{n} P(X > n) < \infty;$$

$$E \log^+ \log^+ X < \infty \quad \Longleftrightarrow \quad \sum_{n=1}^{\infty} \frac{1}{n \log n} P(X > n) < \infty;$$

$$E X^r (\log^+ X)^p < \infty \quad \Longleftrightarrow \quad \sum_{n=1}^{\infty} n^{r-1} (\log n)^p P(X > n) < \infty, \quad r > 1, \; p > 0;$$

$$E(\log^+ X)^p < \infty \quad \Longleftrightarrow \quad \sum_{n=1}^{\infty} \frac{(\log n)^{p-1}}{n} P(X > n) < \infty, \quad p > 1. \qquad \square$$

A common proof technique is to begin by proving a desired result for some subsequence. In such cases one sometimes runs into sums of the above kind for subsequences. The following results may then be useful.

**Theorem 12.4.** *Let $X$ be a non-negative random variable, and $\lambda > 1$. Then,*

$$E X < \infty \iff \int_0^{\infty} \lambda^x P(X > \lambda^x) \, dx < \infty \iff \sum_{n=1}^{\infty} \lambda^n P(X > \lambda^n) < \infty.$$

*Proof.* By a change of variable, $y = \lambda^x$,

$$\int_0^{\infty} \lambda^x P(X > \lambda^x) \, dx = \log \lambda \int_0^{\infty} P(X > y) \, dy,$$

which, together with Theorem 12.1 proves the conclusion. $\qquad \square$

More general subsequences can be handled as follows.

**Theorem 12.5.** *Suppose that $\{n_k, \; k \geq 1\}$ is a strictly increasing subsequence of the positive integers, and set*

$$m(x) = \#\{k : n_k \leq x\} \quad and \quad M(x) = \sum_{k=1}^{[x]} n_k, \quad x > 0.$$

*Finally, let $X$ be a non-negative random variable. Then*

$$\sum_{k=1}^{\infty} n_k P(X \geq n_k) = E M(m(X)),$$

*where both sides converge and diverge together.*

*Proof.* The conclusion follows from the fact that

$$\{X \geq n_k\} = \{m(X) \geq k\},$$

partial summation and Theorem 12.1. $\qquad \square$

**Exercise 12.4.** Verify the following special cases:

$$EX^{3/2} < \infty \quad \Longleftrightarrow \quad \sum_{k=1}^{\infty} k^2 P(X \geq k^2) < \infty;$$

$$EX^{1+(1/d)} < \infty \quad \Longleftrightarrow \quad \sum_{k=1}^{\infty} k^d P(X \geq k^d) < \infty \quad \text{for} \quad d \in \mathbb{N}.$$

**Exercise 12.5.** Show that Theorem 12.5 reduces to Theorem 12.4 for $n_k = \lambda^k$ where $\lambda > 1$.                                                                    □

The subsequences we have dealt with so far were at most geometrically increasing. For more rapidly increasing subsequences we have the following special case.

**Theorem 12.6.** *Suppose that $\{n_k,\ k \geq 1\}$ is a strictly increasing subsequence of the positive integers, such that*

$$\limsup_{k \to \infty} \frac{n_k}{n_{k+1}} < 1,$$

*and let $X$ be a non-negative random variable. Then*

$$EX < \infty \quad \Longrightarrow \quad \sum_{k=1}^{\infty} n_k P(X \geq n_k) < \infty.$$

*Proof.* Set $\Sigma = \sum_{k=1}^{\infty} n_k P(X \geq n_k)$. A consequence of the growth condition is that there exists $\lambda > 1$, such that $n_{k+1} \geq \lambda n_k$ for all $k$, so that

$$\Sigma = \sum_{k=1}^{\infty} \left( n_{k-1} + (n_k - n_{k-1}) \right) P(X \geq n_k)$$

$$\leq \sum_{k=1}^{\infty} \lambda^{-1} n_k + \sum_{k=1}^{\infty} \sum_{j=n_{k-1}+1}^{n_k} P(X \geq j)$$

$$\leq \lambda^{-1} \Sigma + \sum_{j=1}^{\infty} P(X \geq j) = \lambda^{-1} \Sigma + EX,$$

so that

$$\Sigma \leq \frac{\lambda}{\lambda - 1} EX < \infty.$$                                                 □

*Remark 12.2.* Combining this with Theorem 12.5 shows that

$$EM(m(X)) \leq \frac{\lambda}{\lambda - 1} EX.$$

The last result is, in general, weaker than the previous one, although frequently sufficient. If, in particular, $M(m(x)) \geq Cx$ as $x \to \infty$, the results coincide. One such example is $n_k = 2^{2^k}$, $k \geq 1$.                                          □

Another variation involves double sums.

**Theorem 12.7.** *Let $X$ be a non-negative random variable. Then*

$$E\,X\log^+ X < \infty \quad\Longleftrightarrow\quad \sum_{m=1}^{\infty}\sum_{n=1}^{\infty} P(X > nm) < \infty.$$

*Proof.* By modifying the proof of Theorem 12.1(i) we find that the double sum converges iff

$$\int_1^{\infty}\int_1^{\infty} P(X > xy)\,dx\,dy < \infty.$$

Changing variables $u = x$ and $v = xy$ transforms the double integral into

$$\int_1^{\infty}\int_1^{v}\frac{1}{u}P(X > v)\,du\,dv = \int_1^{\infty}\log v\,P(X > v)\,dv,$$

and the conclusion follows from Theorem 12.3.                                  □

## 13 Conditional Distributions

Conditional distributions in their complete generality involve some rather delicate mathematical complications. In this section we introduce this concept for pairs of purely discrete and purely absolutely continuous random variables. Being an essential ingredient in the theory of martingales, *conditional expectations* will be more thoroughly discussed in Chapter 10.

**Definition 13.1.** *Let $X$ and $Y$ be discrete, jointly distributed random variables. For $P(X = x) > 0$, the* conditional probability function *of $Y$ given that $X = x$ equals*

$$p_{Y|X=x}(y) = P(Y = y \mid X = x) = \frac{p_{X,Y}(x,y)}{p_X(x)},$$

*and the* conditional distribution function *of $Y$ given that $X = x$ is*

$$F_{Y|X=x}(y) = \sum_{z \le y} p_{Y|X=x}(z).$$                                  □

**Exercise 13.1.** Show that $p_{Y|X=x}(y)$ is a probability function of a true probability distribution.                                  □

This definition presents no problems. It is validated by the definition of conditional probability; just put $A = \{X = x\}$ and $B = \{Y = y\}$. If, however, $X$ and $Y$ are jointly absolutely continuous, expressions like $P(Y = y \mid X = x)$ have no meaning, since they are of the form $\frac{0}{0}$. However, a glance at the previous definition suggests the following one.

**Definition 13.2.** *Let $X$ and $Y$ have a joint absolutely continuous distribution. For $f_X(x) > 0$, the* conditional density function *of $Y$ given that $X = x$ equals*

$$f_{Y|X=x}(y) = \frac{f_{X,Y}(x,y)}{f_X(x)},$$

*and the* conditional distribution function *of $Y$ given that $X = x$ is*

$$F_{Y|X=x}(y) = \int_{-\infty}^{y} f_{Y|X=x}(z)\, dz. \qquad \square$$

**Exercise 13.2.** Show that $f_{Y|X=x}(y)$ is the density function of a true probability distribution

**Exercise 13.3.** Prove that if $X$ and $Y$ are independent then the conditional distributions and the unconditional distributions are the same. Explain why this is reasonable. $\qquad \square$

*Remark 13.1.* The definitions can (of course) be extended to situations with more than two random variables. $\qquad \square$

By combining the expression for the marginal density with the definition of conditional density we obtain the following density version of the law of total probability, Proposition 1.4.1:

$$f_Y(y) = \int_{-\infty}^{\infty} f_{Y|X=x}(y) f_X(x)\, dx. \qquad (13.1)$$

We also formulate, leaving the details to the reader, the following mixed version, in which $Y$ is discrete and $X$ absolutely continuous:

$$P(Y = y) = \int_{-\infty}^{\infty} p_{Y|X=x}(y) f_X(x)\, dx. \qquad (13.2)$$

*Example 13.1.* In Example 3.1 a point was chosen uniformly on the unit disc. The joint density was $f_{X,Y}(x,y) = \frac{1}{\pi}$, for $x^2 + y^2 \leq 1$, and 0 otherwise, and we found that the marginal densities were $f_X(x) = f_Y(x) = \frac{2}{\pi}\sqrt{1-x^2}$, for $|x| < 1$ and 0 otherwise.

Using this we find that the conditional density of the $y$-coordinate given the $x$-coordinate equals

$$f_{Y|X=x}(y) = \frac{f_{X,Y}(x,y)}{f_X(x)} = \frac{1/\pi}{\frac{2}{\pi}\sqrt{1-x^2}} = \frac{1}{2\sqrt{1-x^2}} \qquad \text{for} \quad |y| \leq \sqrt{1-x^2},$$

and 0 otherwise. This shows that the conditional distribution is uniform on the interval $(-\sqrt{1-x^2}, \sqrt{1-x^2})$.

This should not be surprising, since we can view the joint distribution in the three-dimensional space as a homogeneous, circular cake with a thickness equal to $1/\pi$. The conditional distributions can then be viewed as the profile of a face after a vertical cut across the cake. And this face, which is a picture of the marginal distribution is a rectangle.

Note also that the conditional density is a function of the $x$-coordinate, which means that the coordinates are not independent (as they would have been if the cake were a square and we make a cut parallel to one of the coordinate axes).

The conditional density of the $x$-coordinate given the $y$-coordinate is the same, by symmetry. □

A simple example involving discrete distributions is that we pick a digit randomly among $0, 1, 2, \ldots, 9$, and then a second one among those that are smaller than the first one. The corresponding continuous analog is to break a stick of length 1 randomly at some point, and then break one of the remaining pieces randomly.

# 14 Distributions with Random Parameters

Random variables with random parameters are very natural objects. For example, suppose that $X$ follows a Poisson distribution, but in such a way that the parameter itself is random. An example could be a particle counter that emits particles of different kinds. For each kind the number of particles emitted during one day, say, follows a Poisson distribution, However, the parameters for the different kinds are different. Or the intensity depends on temperature or air pressure, which, in themselves, are random. Another example could be an insurance company that is subject to claims according to some distribution, the parameter of which depends on the kind of claim: is it a house on fire? a stolen bicycle? a car that has been broken into? Certainly, the intensities with which these claims occur can be expected to be different.

It could also be that the parameter is unknown. The so-called Bayesian approach is to consider the parameter as a random variable with a so-called prior distribution.

Let us for computational convenience consider the following situation:

$$X \in \text{Po}(M) \quad \text{where} \quad M \in \text{Exp}(1).$$

This is an abusive way of writing that

$$X \mid M = m \in \text{Po}(m) \quad \text{with} \quad M \in \text{Exp}(1).$$

What is the "real" (that is, the unconditional) distribution of $X$? Is it a Poisson distribution? Is it definitely not a Poisson distribution?

By use of the mixed version (13.2) of the law of total probability, the following computation shows tells us that $X$ is geometric; $X \in \text{Ge}(\frac{1}{2})$. Namely, for $k = 0, 1, 2, \ldots$ we obtain

$$P(X = k) = \int_0^\infty P(X = k \mid M = x) \cdot f_M(x)\, dx = \int_0^\infty e^{-x}\frac{x^k}{k!} \cdot e^{-x}\, dx$$

$$= \int_0^\infty \frac{x^k}{k!} e^{-2x}\, dx = \frac{1}{2^{k+1}} \cdot \int_0^\infty \frac{1}{\Gamma(k+1)} 2^{k+1} x^{k+1-1} e^{-2x}\, dx$$

$$= \frac{1}{2^{k+1}} \cdot 1 = \frac{1}{2} \cdot \left(\frac{1}{2}\right)^k,$$

which establishes the geometric distribution as claimed.

**Exercise 14.1.** Determine the distribution of $X$ if

- $M \in \mathrm{Exp}(a)$;
- $M \in \Gamma(p, a)$. □

Suppose that a radioactive substance emits $\alpha$-particles in such a way that the number of particles emitted during one hour, $N \in \mathrm{Po}(\lambda)$. Unfortunately, though, the particle counter is unreliable in the sense that an emitted particle is registered with probability $p \in (0,1)$, whereas it remains unregistered with probability $q = 1 - p$. All particles are registered independently of each other. Let $X$ be the number of particles that are registered during one hour.

This means that our model is

$$X \mid N = n \in \mathrm{Bin}(n, p) \quad \text{with} \quad N \in \mathrm{Po}(\lambda).$$

So, what is the unconditional distribution of $X$? The following computation shows that $X \in \mathrm{Po}(\lambda p)$. Namely, for $k = 0, 1, 2, \ldots$,

$$P(X = k) = \sum_{n=0}^\infty P(X = k \mid N = n) P(N = n) = \sum_{n=k}^\infty \binom{n}{k} p^k q^{n-k} e^{-\lambda} \frac{\lambda^n}{n!}$$

$$= e^{-\lambda}\frac{(\lambda p)^k}{k!} \sum_{n=k}^\infty \frac{(\lambda q)^{n-k}}{(n-k)!} = e^{-\lambda}\frac{(\lambda p)^k}{k!} e^{\lambda q} = e^{-\lambda p}\frac{(\lambda p)^k}{k!}.$$

Note that the sum starts at $n = k$; there must be at least as many particles emitted as there are registered ones.

The following two exercises may or may not have anything to do with everyday life.

**Exercise 14.2.** Susan has a coin with $P(\text{head}) = p_1$ and John has a coin with $P(\text{head}) = p_2$. Susan tosses her coin $m$ times. Each time she obtains heads, John tosses his coin (otherwise not). Find the distribution of the total number of heads obtained by John.

**Exercise 14.3.** Toss a coin repeatedly, and let $X_n$ be the number of heads after $n$ coin tosses, $n \geq 1$. Suppose now that the coin is completely unknown to us in the sense that we have no idea of whether or not it is fair. Suppose, in fact, the following, somewhat unusual situation, namely, that

$$X_n \mid P = p \in \mathrm{Bin}(n, p) \quad \text{with} \quad P \in U(0, 1),$$

that is, we suppose that the probability of heads is $U(0,1)$-distributed.

- Find the distribution of $X_n$.
- Explain why the answer is reasonable.
- Compute $P(X_{n+1} = n + 1 \mid X_n = n)$.
- Are the outcomes of the tosses independent?                    $\square$

A special family of distributions is the family of *mixed normal*, or *mixed Gaussian*, distributions. These are normal distributions with a random variance, namely,

$$X \mid \Sigma^2 = y \in N(\mu, y) \quad \text{with} \quad \Sigma^2 \in F,$$

where $F$ is some distribution (on $(0, \infty)$).

As an example, consider a production process where some measurement of the product is normally distributed, and that the production process is not perfect in that it is subject to rare disturbances. More specifically, the observations might be $N(0, 1)$-distributed with probability 0.99 and $N(0, 100)$-distributed with probability 0.01. We may write this as

$$X \in N(0, \Sigma^2), \quad \text{where} \quad P(\Sigma^2 = 1) = 0.99 \text{ and } P(\Sigma^2 = 100) = 0.01.$$

What is the "real" distribution of $X$? A close relative is the next section.

# 15 Sums of a Random Number of Random Variables

In many applications involving processes that evolve with time, one is interested in the state of affairs at some given, fixed, *time* rather than after a given, fixed, number of steps, which therefore amounts to checking the random process or sequence after a *random number of events*. With respect to what we have discussed so far this means that we are interested in the state of affairs of the sum of *a random number* of independent random variables. In this section we shall always assume that *the number of terms is independent of the summands*. More general random indices or "times" will be considered in Chapter 10.

Apart from being a theory in its own right, there are several interesting and important applications; let us, as an appetizer, mention branching processes and insurance risk theory which we shall briefly discuss in a subsection following the theory.

Thus, let $X, X_1, X_2, \ldots$ be independent, identically distributed random variables with partial sums $S_n = \sum_{k=1}^{n} X_k$, $n \geq 1$, and let $N$ be a non-negative, integer valued random variable which is independent of $X_1, X_2, \ldots$. Throughout, $S_0 = 0$.

The object of interest is $S_N$, that is, the sum of $N$ $X$'s. We may thus interpret $N$ as a random index.

For any Borel set $A \subset (-\infty, \infty)$,

$$P(S_N \in A \mid N = n) = P(S_n \in A \mid N = n) = P(S_n \in A), \quad (15.1)$$

where the last equality, being a consequence of the additional independence, is the crucial one.

Here is an example in which the index is *not* independent of the summands.

*Example 15.1.* Let $N = \min\{n : S_n > 0\}$. Clearly, $P(S_N > 0) = 1$. This implies that if the summands are allowed to assume negative values (with positive probability) then so does $S_n$, whereas $S_N$ is always positive. Hence, $N$ is not independent of the summands, on the contrary, $N$ is, in fact, defined in terms of the summands. $\square$

By (15.1) and the law of total probability, Proposition 1.4.1, it follows that

$$P(S_N \in A) = \sum_{n=1}^{\infty} P(S_N \in A \mid N = n)P(N = n)$$

$$= \sum_{n=1}^{\infty} P(S_n \in A)P(N = n), \tag{15.2}$$

in particular,

$$P(S_N \leq x) = \sum_{n=1}^{\infty} P(S_n \leq x)P(N = n), \quad -\infty < x < \infty, \tag{15.3}$$

so that, by changing the order of integration and summation,

$$E\, h(S_N) = \sum_{n=1}^{\infty} E\big(h(S_n)\big)P(N = n), \tag{15.4}$$

provided the integrals are absolutely convergent.

By letting $h(x) = x$ and $h(x) = x^2$ we obtain expressions for the mean and variance of $S_N$.

**Theorem 15.1.** *Suppose that $X, X_1, X_2, \ldots$ are independent, identically distributed random variables with partial sums $S_n = \sum_{k=1}^{n} X_k$, $n \geq 1$, and that $N$ is a non-negative, integer valued random variable which is independent of $X_1, X_2, \ldots$.*
(i) *If*

$$E\,N < \infty \quad \text{and} \quad E\,|X| < \infty,$$

*then*

$$E\,S_N = E\,N \cdot E\,X.$$

(ii) *If, in addition,*

$$\text{Var}\,N < \infty \quad \text{and} \quad \text{Var}\,X < \infty,$$

*then*

$$\text{Var}\,S_N = E\,N \cdot \text{Var}\,X + (E\,X)^2 \cdot \text{Var}\,N.$$

*Proof.* (i): From (15.4) we know that

$$E\,S_N = \sum_{n=1}^{\infty} E\,S_n P(N=n) = \sum_{n=1}^{\infty} nE\,XP(N=n)$$

$$= E\,X \sum_{n=1}^{\infty} nP(N=n) = E\,X E\,N.$$

(ii): Similarly

$$E\big(S_N^2\big) = \sum_{n=1}^{\infty} E\big(S_n^2\big)P(N=n) = \sum_{n=1}^{\infty} \big(\mathrm{Var}\,S_n + (E\,S_n)^2\big)P(N=n)$$

$$= \sum_{n=1}^{\infty} \big(n\,\mathrm{Var}\,X + n^2 (E\,X)^2\big)P(N=n)$$

$$= \mathrm{Var}\,X \sum_{n=1}^{\infty} nP(N=n) + (E\,X)^2 \sum_{n=1}^{\infty} n^2 P(N=n)$$

$$= \mathrm{Var}\,X E\,N + (E\,X)^2 E\,N^2.$$

By inserting the conclusion from (i) we find that

$$\mathrm{Var}\,S_N = E\big(S_N^2\big) - (E\,S_N)^2 = E\,N\,\mathrm{Var}\,X + (E\,X)^2 E\,N^2 - \big(E\,N E\,X\big)^2$$
$$= E\,N\,\mathrm{Var}\,X + (E\,X)^2\,\mathrm{Var}\,N. \qquad \square$$

## 15.1 Applications

Applications of this model are ubiquitous. In this subsection we first illustrate the theory with what might be called a toy example, after which we mention a few more serious applications. It should also be mentioned that in some of the latter examples the random index is not necessarily independent of the summands (but this is of no significance in the present context).

## A "Toy" Example

*Example 15.2.* Suppose that the number of customers that arrive at a store during one day is $Po(\lambda)$-distributed and that the probability that a customer buys something is $p$ and just browses around without buying is $q = 1 - p$. Then the number of customers that buy something can be described as $S_N$, where $N \in Po(\lambda)$, and $X_k = 1$ if customer $k$ shops and 0 otherwise.

Theorem 15.1 then tells us that

$$E\,S_N = E\,N E\,X = \lambda \cdot p,$$

and that

$$\text{Var } S_N = E N \text{ Var } X + (E X)^2 \text{ Var } N = \lambda \cdot pq + p^2 \lambda = \lambda p.$$

We have thus found that $E S_N = \text{Var } S_N = \lambda p$, which makes it tempting to guess that, in fact $S_N \in \text{Po}(\lambda p)$. This may seem bold, but knowing that the Poisson process has many "nice" features, this may seem reasonable. After all, the new process can be viewed as the old process after having run through a "filter", which makes it seem like a thinner version of the old one. And, in fact, there is a concept, the *thinned Poisson process*, which is precisely this, and which is Poisson distributed with a parameter that is the product of the old one and the thinning probability.

And, in fact, by (15.2), we have, for $k = 0, 1, 2, \ldots$,

$$P(S_N = k) = \sum_{n=1}^{\infty} P(S_n = k) P(N = n) = \sum_{n=k}^{\infty} \binom{n}{k} p^k q^{n-k} e^{-\lambda} \frac{\lambda^n}{n!}$$

$$= e^{-\lambda} \frac{(\lambda p)^k}{k!} \sum_{n=k}^{\infty} \frac{(\lambda q)^{n-k}}{(n-k)!} = e^{-\lambda} \frac{(\lambda p)^k}{k!} e^{\lambda q} = e^{-\lambda p} \frac{(\lambda p)^k}{k!},$$

so that, indeed $S_n \in \text{Po}(\lambda p)$.

*Remark 15.1.* The computations for determining the distribution of $S_N$ are the same as in the previous section. The reason for this is that, instead of introducing an indicator random variable to each customer, we may consider the total number of customers as some random variable $X$, say, and note that $X \mid N = n \in \text{Bin}(n, p)$, after which we proceed as before. This is no surprise, since if we identify every customer with a particle, then a shopping customer is identified with a registered particle. So, they are conceptually the same problem, just modeled or interpreted somewhat differently.     □

More generally, let $Y_k$ be the amount spent by the $k$th customer. The sum $S_N = \sum_{k=1}^{N} Y_k$ then describes the total amount spent by the customers during one day.

If, for example, $Y_1, Y_2, \ldots \in \text{Exp}(\theta)$, and $N \in \text{Fs}(p)$, then

$$E S_N = \frac{1}{p} \cdot \theta \quad \text{and} \quad \text{Var } S_N = \frac{1}{p} \theta^2 + \theta^2 \frac{q}{p^2} = \frac{\theta^2}{p^2}.$$     □

**Exercise 15.1.** Find the distribution of $S_N$ and check that mean and variance agree with the above ones.     □

## Branching Processes

The most basic kind of *branching processes*, the *Galton-Watson process*, can be described as follows:

At time $t = 0$ there exists one (or many) founding members $X(0)$. During its life span, every individual gives birth to a random number of children, who during their life spans give birth to a random number of children, who during their life spans . . ..

The reproduction rules in this model are the same for all individuals:

- all individuals give birth according to the same probability law, independently of each other;
- the number of children produced by an individual is independent of the number of individuals in his or her generation.

Let, for $n \geq 0$, $X(n) = \#$ individuals in generation $n$, and $\{Y_k, \ k \geq 1\}$ and $Y$ be generic random variables denoting the number of children obtained by individuals. We also suppose that $X(0) = 1$, and exclude the degenerate case $P(Y = 1) = 1$.

It follows from the assumptions that

$$X(2) = Y_1 + \cdots + Y_{X(1)},$$

and, recursively, that

$$X(n+1) = Y_1 + \cdots + Y_{X(n)}.$$

Thus, by identifying $Y_1, Y_2, \ldots$ with $X_1, X_2, \ldots$ and $X(n)$ with $N$ it follows that $X(n+1)$ is an "$S_N$-sum".

One simple example is cells that split or die, in other words, with probability $p$ they get two children and with probability $1-p$ they die. What happens after many generations? Will the cells spread all over the universe or is the cell culture going to die out? If the cells are antrax cells, say, this question may be of some interest.

### Insurance Risk Theory

Consider an insurance company whose business runs as follows:

- Claims arrive at random time points according to some random process;
- Claim sizes are (can be considered as being) independent, identically distributed random variables;
- The gross premium rate, that is, the premium paid by the policy holders, arrive at a constant rate $\beta$/month (which is probably not realistic since people pay their bills at the end of the month, just after payday).

Let us denote the number of claims during one year by $N$, and the successive claims by $X_1, X_2, \ldots$. If the initial capital, called the risk reserve, is $v$, then the capital at the end of the first year equals

$$v + 12\beta - \sum_{k=1}^{N} X_k.$$

Relevant questions are probabilities of ruin, of ruin in 5 years, and so on. Another important issue is the deciding of premiums, which means that one wishes to estimate parameters from given data, and, for example, investigate if parameters have changed or not.

**A Simple Queueing Model**

Consider a store to which customers arrive, one at a time, according to some random process (and that the service times, which are irrelevant here, are, say, i.i.d. exponentially distributed random variables). If $X_1$, $X_2$, ... denotes the amount of money spent by the customers and there are $M$ customers during one day, then

$$\sum_{k=1}^{M} X_k$$

depicts the amount of money in the cash register at the end of the day. The toy example above falls into this category.

# 16 Random Walks; Renewal Theory

An important assumption in Theorem 15.1 was the *independence of the random index $N$ and the random summands $X_1$, $X_2$, ....* There obviously exist many situations where such an assumption is unrealistic. It suffices to imagine examples where a process is observed until something "special" occurs. The number of summands at that moment is random and, by construction, defined via the summands. In this section we present some applications where more general random indices are involved.

## 16.1 Random Walks

A *random walk* $\{S_n, n \geq 0\}$ is a sequence of random variables, starting at $S_0 = 0$, with independent, identically distributed increments $X_1$, $X_2$, ....

The classical example is the *simple random walk*, for which the increments, or steps, assume the values $+1$ or $-1$. The standard notation is

$$P(X = 1) = p, \quad P(X = -1) = q, \quad \text{where} \quad 0 \leq p, q \leq 1, \quad p + q = 1,$$

and where $X$ is a generic random variable.

The following figure illustrates the situation.

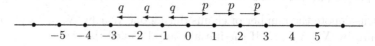

**Figure 2.2.** The simple random walk

If the values are assumed with equal probabilities, $p = q = 1/2$, we call it a *symmetric simple random walk*. Another example is the *Bernoulli random walk*, where the steps are $+1$ or $0$ with probabilities $p$ and $q$, respectively.

Random walk theory is a classical topic. For an introduction and background we refer to the second edition of Spitzer's legendary 1964 book, [234]. Applications are abundant: Sequential analysis, insurance risk theory, queueing theory, reliability theory, just to name a few.

## 16.2 Renewal Theory

Renewal processes are random walks with non-negative increments. The canonical application is a light bulb that fails after a random time and is instantly replaced by a new, identical one, which, upon failure is replaced by another one, which, in turn, .... The central object of interest is the number of replacements during a given time.

In order to model a renewal process we let $X_1$, $X_2$, ... be the individual life times and set $S_n = \sum_{k=1}^{n} X_k$, $n \geq 1$. The number of replacements in the time interval $(0, t]$ then becomes

$$N(t) = \max\{n : S_n \leq t\}.$$

The following figure depicts a typical realization of a renewal process.

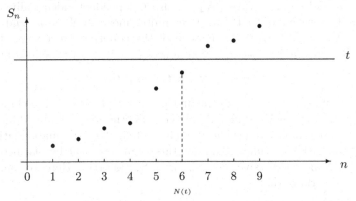

**Figure 2.3.** The realization of a renewal process

The main process of interest is the *renewal counting process*,

$$\{N(t), \, t \geq 0\}.$$

Some classical references are [53, 65, 201, 229, 230]. A summary of results can be found in [110], Chapter II. A discrete version called *recurrent events* dates back to [85] see also [87]. If, in particular, the life times are exponential, then $\{N(t), \, t \geq 0\}$, is a *Poisson process*.

A more general model which allows for repair times is the *alternating renewal process*, a generalization of which is a two-dimensional random walk, stopped when the second component reaches a given level after which the first component is evaluated at that time point. For more on this, see [118] and/or [110], Chapter IV (and Problem 7.8.17).

Classical proofs for the renewal counting process are based on the inversion

$$\{N(t) \geq n\} = \{S_n \leq t\}. \tag{16.1}$$

The idea is that a limit theorem for one of the processes may be derived from the corresponding limit theorem for the other one via inversion by letting $t$ and $n$ tend to infinity jointly in a suitable manner.

### 16.3 Renewal Theory for Random Walks

Instead of considering a random walk after a fixed number of steps, that is, at a random time point, one would rather inspect or observe the process at fixed time points, which means after *a random number of steps*. For example, the closing time of a store is fixed, but the number of customers during a day is random. The number of items produced by a machine during an 8-hour day is random, and so on. A typical random index is "the first $n$, such that ...". With reference to renewal theory in the previous subsection, we also note that it seems more natural to consider a random process at the *first* occurrence of some kind rather than the *last* one, defined by the counting process, let alone, how does one know that a given occurrence really is the last one before having information about the future of the process?

For this model we let $X, X_1, X_2, \ldots$ be independent, identically distributed random variables, with positive, finite, mean $EX = \mu$, and set $S_n = \sum_{k=1}^{n} X_k$, $n \geq 1$. However, instead of the counting process we shall devote ourselves to *the first passage time process*, $\{\tau(t), t \geq 0\}$, defined by

$$\tau(t) = \min\{n : S_n > t\}, \quad t \geq 0.$$

Although the counting process and the first passage time process are close on average, they have somewhat different behaviors in other respects. In addition, first passage times have, somewhat vaguely stated, "better" mathematical properties than last exit times. Some of this vagueness will be clarified in Section 10.14. A more extensive source is [110], Section III.3. Here we confine ourselves by remarking that

- whereas $N(t) + 1 = \tau(t)$ for renewal processes, this is not necessarily the case for random walks;
- the inversion relation (16.1) does not hold for random walks, since the random walk may well fall below the level $t$ after having crossed it.

Both facts may be observed in the next figure.

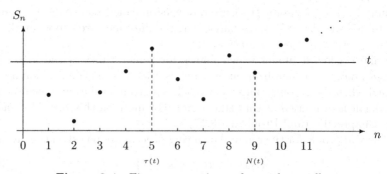

**Figure 2.4.** First passage times of a random walk

Proofs of subsequent limit theorems for first passage time processes will be based on limit theorems for randomly indexed random walks, $\{S_{N(t)}, t \geq 0\}$.

A special feature is that those proofs cover renewal processes as well as random walks. In addition, no distinction is necessary between the continuous cases and the discrete ones. A specialized reference on this topic is [110].

## 16.4 The Likelihood Ratio Test

Let $X_1, X_2, \ldots, X_n$ be a sample from an absolutely continuous distribution with a characterizing parameter $\theta$ of interest, and suppose that we wish to test the null hypothesis $H_0 : \theta = \theta_0$ against the alternative $H_1 : \theta = \theta_1$. The Neyman-Pearson lemma in statistics tells us that such a test should be based on the likelihood ratio statistic

$$L_n = \prod_{k=1}^{n} \frac{f(X_k; \theta_1)}{f(X_k; \theta_0)},$$

where $f_{\theta_0}$ and $f_{\theta_1}$ are the densities under the null and alternative hypotheses, respectively.

The factors $\frac{f(X_k;\theta_1)}{f(X_k;\theta_0)}$ are independent, identically distributed random variables, and, under the null hypothesis, the mean equals 1;

$$E_0\left(\frac{f(X_k; \theta_1)}{f(X_k; \theta_0)}\right) = \int_{-\infty}^{\infty} \frac{f(x; \theta_1)}{f(x; \theta_0)} f(x; \theta_0)\, dx = \int_{-\infty}^{\infty} f(x; \theta_1)\, dx = 1,$$

so that $L_n$ equals a product of independent, identically distributed random variables with mean 1.

For technical reasons it is sometimes more convenient to investigate the log-likelihood, $\log L_n$, which is a *sum* of independent, identically distributed random variables, however, *not* with mean $\log 1 = 0$.

## 16.5 Sequential Analysis

This is one of the most important statistical applications within the renewal theoretic framework. The idea is that, instead of basing a log-likelihood test on a sample of a *fixed* predetermined size, one performs the test sequentially.

The typical sequential procedure then would be to continue sampling until, depending on the circumstances, the likelihood ratio $L_n$ or the log-likelihood ratio, $\log L_n$, falls outside a given strip, at which time point one takes a decision. Technically, this means that one defines

$$\tau_{a,b} = \min\{n : L_n \notin (a, b)\}, \quad \text{where} \quad 0 < a < b < \infty,$$

or, equivalently,

$$\tau_{A,B} = \min\{n : \log L_n \notin (A, B)\}, \quad \text{where} \quad -\infty < A < B < \infty.$$

and continues sampling until the likelihood ratio (or, equivalently, the log-likelihood ratio) escapes from the interval and rejects the null hypothesis if $L_{\tau_{a,b}} > b$ ($\log L_{\tau_{A,B}} > B$), and accepts the null hypothesis if $L_{\tau_{a,b}} < a$ ($\log L_{\tau_{A,B}} < A$).

Although one can show that the procedure stops after a finite number of steps, that is, that the sample size will be finite almost surely, one may introduce a "time horizon", $m$, and stop sampling at $\min\{\tau, m\}$ (and accept $H_0$ if the (log)likelihood-ratio has not escaped from the strip at time $m$).

The classic here is the famous book by Wald [250]. A more recent one is [223].

## 16.6 Replacement Based on Age

Let $X_1$, $X_2$, ... be the independent, identically distributed lifetimes of some component in a larger machine. The simplest replacement policy is to change a component as soon as it fails. In this case it may be necessary to call a repairman at night, which might be costly. Another policy, called *replacement based on age*, is to replace at failure or at some given age, $a$, say, whichever comes first. The inter-replacement times are

$$W_n = \min\{X_n, a\}, \quad n \geq 1,$$

in this case. A quantity of interest would be the number of replacements due to failure during some given time unit.

In order to describe this quantity we define

$$\tau(t) = \min\left\{n : \sum_{k=1}^{n} W_k > t\right\}, \quad t > 0.$$

The quantity $\tau(t)$ equals the number of components that have been in action at time $t$.

Next, let

$$Z_n = I\{X_n \leq a\}, \quad n \geq 1,$$

that is, $Z_n = 1$ if the $n$th component is replaced because of failure, and $Z_n = 0$ if replacement is due to age. The number of components that have been replaced because of failure during the time span $(0, t]$ is then described by

$$\sum_{k=1}^{\tau(t)} Z_k.$$

If we attach a cost $c_1$ to replacements due to failure and a cost $c_2$ to replacements due to age, then

$$\sum_{k=1}^{\tau(t)} \left(c_1 I\{X_k \leq a\} + c_2 I\{X_k > a\}\right)$$

provides information about the replacement cost during the time span $(0, t]$.

For detailed results on this model, see [110, 118].

*Remark 16.1.* Replacement based on age applied to humans is called retirement, where $a$ is the retirement age. □

# 17 Extremes; Records

The central results in probability theory are limit theorems for sums. However, in many applications, such as strength of materials, fatigue, flooding, oceanography, and "shocks" of various kinds, *extremes* rather than sums are of importance. A flooding is the result of one single extreme wave, rather than the cumulative effect of many small ones.

In this section we provide a brief introduction to the concept of extremes – "the largest observation so far" and a more extensive one to the theory of records – "the extreme observations at their first appearance".

## 17.1 Extremes

Let $X_1$, $X_2$, ... be independent, identically distributed random variables. The quantities in focus are the *partial maxima*

$$Y_n = \max_{1 \leq k \leq n} X_k \quad \text{or, at times,} \quad \max_{1 \leq k \leq n} |X_k|.$$

Typical results are analogs to the law of large numbers and the central limit theorem for sums. For the latter this means that we wish to find normalizing sequences $\{a_n > 0,\, n \geq 1\}$, $\{b_n \in \mathbb{R},\, n \geq 1\}$, such that

$$\frac{Y_n - b_n}{a_n} \quad \text{possesses a limit distribution,}$$

a problem that will be dealt with in Chapter 9.

## 17.2 Records

Let $X$, $X_1$, $X_2$, ... be independent, identically distributed, continuous random variables. The *record times* are $L(1) = 1$ and, recursively,

$$L(n) = \min\{k : X_k > X_{L(n-1)}\}, \quad n \geq 2,$$

and the *record values* are
$$X_{L(n)}, \quad n \geq 1.$$

The associated *counting process* $\{\mu(n),\, n \geq 1\}$ is defined by

$$\mu(n) = \# \text{records among } X_1, X_2, \ldots, X_n = \max\{k : L(k) \leq n\}.$$

The reason for assuming continuity is that we wish to avoid ties. And, indeed, in this case we obtain, by monotonicity (Lemma 1.3.1),

$$P\left( \bigcup_{\substack{i,j=1 \\ i \neq j}}^{\infty} \{X_i = X_j\} \right) = \lim_{n \to \infty} P\left( \bigcup_{\substack{i,j=1 \\ i \neq j}}^{n} \{X_i = X_j\} \right)$$

$$\leq \lim_{n \to \infty} \sum_{\substack{i,j=1 \\ i \neq j}}^{n} P(X_i = X_j) = 0.$$

The pioneering paper in the area is [205]. For a more recent introduction and survey of results, see [187, 207, 208].

Whereas the sequence of partial maxima, $Y_n$, $n \geq 1$, describe "the largest value so far", the record values pick these values the first time they appear. The sequence of record values thus constitutes a subsequence of the partial maxima. Otherwise put, the sequence of record values behaves like a compressed sequence of partial maxima, as is depicted in the following figure.

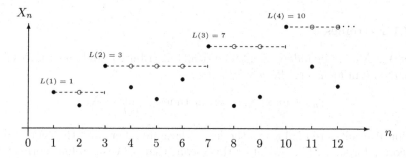

**Figure 2.5.** Partial maxima ∘

A preliminary observation is that the record times and the number of records are distribution independent. This is a consequence of the fact that given $X$ with distribution function $F$, then $F(X)$ is $U(0,1)$-distributed, so that there is a $1-1$ map from every (absolutely continuous) random variable to every other one. And, by monotonicity, record times are preserved under this transformation – however, not the record *values*.

Next, set

$$I_k = \begin{cases} 1, & \text{if } X_k \text{ is a record,} \\ 0, & \text{otherwise,} \end{cases}$$

so that $\mu(n) = \sum_{k=1}^{n} I_k$, $n \geq 1$.

By symmetry, all permutations between $X_1$, $X_2$, ..., $X_n$ are equally likely. Taking advantage of this fact, we introduce *ranks*, so that $X_n$ has rank $j$ if $X_n$ is the $j$th largest among $X_1$, $X_2$, ..., $X_n$. Notationally, $R_n = j$. This means, in particular, that if $X_n$ is the largest among them, then $R_n = 1$, and if $X_n$ is the smallest, then $R_n = n$. Moreover,

$$P(R_1 = r_1, R_2 = r_2, \ldots, R_n = r_n) = \frac{1}{n!},$$

in particular,

$$P(I_k = 1) = 1 - P(I_k = 0) = \frac{1}{k}, \quad k = 1, 2, \ldots, n.$$

The marginal probabilities are

$$P(R_n = r_n) = \sum_{\{r_1, r_2, \ldots, r_{n-1}\}} P(R_1 = r_1, R_2 = r_2, \ldots, R_n = r_n),$$

where the summation thus extends over all possible values of $r_1, r_2, \ldots, r_{n-1}$. By symmetry, the summation involves $(n-1)!$ terms, all of which are the same, namely $1/n!$, so that

$$P(R_n = r_n) = \frac{(n-1)!}{n!} = \frac{1}{n}.$$

Since the same argument is valid for all $n$, we have, in fact, shown that

$$P(R_1 = r_1, R_2 = r_2, \ldots, R_n = r_n) = \frac{1}{n!} = \prod_{k=1}^{n} \frac{1}{k} = \prod_{k=1}^{n} P(R_k = r_k),$$

which proves the independence of the ranks. Moreover, since $\{I_n = 1\} = \{R_n = 1\}$ it follows, in particular, that $\{I_k, \ k \geq 1\}$ are independent random variables.

Joining the above conclusions yields the following result.

**Theorem 17.1.** *Let $X_1, X_2, \ldots, X_n$ be independent, identically distributed, absolutely continuous, random variables, $n \geq 1$. Then*

(i) *The ranks $R_1, R_2, \ldots, R_n$ are independent, and $P(R_k = j) = 1/k$ for $j = 1, 2, \ldots, k$, where $k = 1, 2, \ldots, n$;*
(ii) *The indicators $I_1, I_2, \ldots, I_n$ are independent, and $P(I_k = 1) = 1/k$ for $k = 1, 2, \ldots, n$.*

As a corollary it is now a simple task to compute the mean and the variance of $\mu(n)$, and their asymptotics.

**Theorem 17.2.** *Let $\gamma = 0.5772\ldots$ denote Euler's constant. We have*

$$m_n = E\,\mu(n) = \sum_{k=1}^{n} \frac{1}{k} = \log n + \gamma + o(1) \quad as \quad n \to \infty;$$

$$\operatorname{Var}\mu(n) = \sum_{k=1}^{n} \frac{1}{k}\left(1 - \frac{1}{k}\right) = \log n + \gamma - \frac{\pi^2}{6} + o(1) \quad as \quad n \to \infty.$$

*Proof.* That $E\,\mu(n) = \sum_{k=1}^{n} \frac{1}{k}$, and that $\operatorname{Var}\mu(n) = \sum_{k=1}^{n} \frac{1}{k}(1 - \frac{1}{k})$, is clear. The remaining claims follow from Remark A.3.1, and the (well-known) fact that $\sum_{n=1}^{\infty} 1/n^2 = \pi^2/6$. □

# 18 Borel-Cantelli Lemmas

This section is devoted to an important tool frequently used in connection with questions concerning almost sure convergence – a concept that we shall meet in detail in Chapter 5 – the *Borel-Cantelli lemmas* [26].

We begin by recalling the definitions of limsup and liminf of sets from Chapter 1 and by interpreting them in somewhat greater detail.

Let $\{A_n, n \geq 1\}$ be a sequence of events, that is, measurable subsets of $\Omega$. Then, recalling Definition 1.2.1,

$$A_* = \liminf_{n \to \infty} A_n = \bigcup_{n=1}^{\infty} \bigcap_{m=n}^{\infty} A_m, \quad \text{and} \quad A^* = \limsup_{n \to \infty} A_n = \bigcap_{n=1}^{\infty} \bigcup_{m=n}^{\infty} A_m.$$

Thus, if $\omega \in \Omega$ belongs to the set $\liminf_{n \to \infty} A_n$, then $\omega$ belongs to $\bigcap_{m=n}^{\infty} A_m$ for some $n$, that is, there exists an $n$ such that $\omega \in A_m$ for *all* $m \geq n$. In particular, if $A_n$ is the event that something special occurs at "time" $n$, then $\liminf_{n \to \infty} A_n^c$ means that from some $n$ on this property never occurs.

Similarly, if $\omega \in \Omega$ belongs to the set $\limsup_{n \to \infty} A_n$, then $\omega$ belongs to $\bigcup_{m=n}^{\infty} A_m$ for every $n$, that is, no matter how large we choose $n$ there is always some $m \geq n$ such that $\omega \in A_m$, or, equivalently, $\omega \in A_m$ for infinitely many values of $m$ or, equivalently, for arbitrarily large values of $m$. A convenient way to express this is

$$\omega \in A^* \quad \Longleftrightarrow \quad \omega \in \{A_n \text{ i.o.}\} = \{A_n \text{ infinitely often}\}.$$

If the upper and lower limits coincide the limit exists, and

$$A = A^* = A_* = \lim_{n \to \infty} A_n.$$

## 18.1 The Borel-Cantelli Lemmas 1 and 2

We now present the standard Borel-Cantelli lemmas, after which we prove a zero-one law and provide an example to illustrate the applicability of the results.

**Theorem 18.1.** (The first Borel-Cantelli lemma)
*Let $\{A_n, n \geq 1\}$ be arbitrary events. Then*

$$\sum_{n=1}^{\infty} P(A_n) < \infty \quad \Longrightarrow \quad P(A_n \text{ i.o.}) = 0.$$

*Proof.* We have

$$P(A_n \text{ i.o.}) = P(\limsup_{n \to \infty} A_n) = P\left(\bigcap_{n=1}^{\infty} \bigcup_{m=n}^{\infty} A_m\right)$$

$$\leq P\left(\bigcup_{m=n}^{\infty} A_m\right) \leq \sum_{m=n}^{\infty} P(A_m) \to 0 \quad \text{as} \quad n \to \infty. \qquad \square$$

The converse does not hold in general. The easiest accessible one is obtained under the additional assumption of independence.

**Theorem 18.2.** (The second Borel-Cantelli lemma)
Let $\{A_n, \, n \geq 1\}$ be independent events. Then

$$\sum_{n=1}^{\infty} P(A_n) = \infty \quad \Longrightarrow \quad P(A_n \text{ i.o.}) = 1.$$

*Proof.* By independence,

$$P(A_n \text{ i.o.}) = P\left( \bigcap_{n=1}^{\infty} \bigcup_{m=n}^{\infty} A_m \right) = 1 - P\left( \bigcup_{n=1}^{\infty} \bigcap_{m=n}^{\infty} A_m^c \right)$$

$$= 1 - \lim_{n \to \infty} P\left( \bigcap_{m=n}^{\infty} A_m^c \right) = 1 - \lim_{n \to \infty} \prod_{m=n}^{\infty} P(A_m^c)$$

$$= 1 - \lim_{n \to \infty} \prod_{m=n}^{\infty} \left( 1 - P(A_m) \right) = 1 - 0 = 1,$$

since, by Lemma A.4.1, the divergence of $\sum_{n=1}^{\infty} P(A_n)$ is equivalent to the divergence of $\prod_{m=1}^{\infty} (1 - P(A_m))$. □

By combining the two results we note, in particular, that if the events $\{A_n, \, n \geq 1\}$ are independent, then $P(A_n \text{ i.o.})$ can only assume the values 0 or 1, and that the convergence or divergence of $\sum_{n=1}^{\infty} P(A_n)$ is the decisive factor.

**Theorem 18.3.** (A zero-one law)
If the events $\{A_n, \, n \geq 1\}$ are independent, then

$$P(A_n \text{ i.o.}) = \begin{cases} 0, & \text{when } \sum_{n=1}^{\infty} P(A_n) < \infty, \\ 1, & \text{when } \sum_{n=1}^{\infty} P(A_n) = \infty. \end{cases} \qquad \square$$

A consequence of this zero-one law is that it suffices to prove that $P(A_n \text{ i.o.}) > 0$ in order to conclude that the probability equals 1 (and that $P(A_n \text{ i.o.}) < 1$ in order to conclude that it equals 0).

Here is an example to illuminate the results.

*Example 18.1.* Let $X_1, X_2, \ldots$ be a sequence of arbitrary random variables and let $A_n = \{|X_n| > \varepsilon\}$, $n \geq 1$, $\varepsilon > 0$. Then $\omega \in \liminf_{n \to \infty} A_n^c$ means that $\omega$ is such that $|X_n(\omega)| \leq \varepsilon$, for all sufficiently large $n$, and $\omega \in \limsup_{n \to \infty} A_n$ means that $\omega$ is such that there exist arbitrarily large values of $n$ such that $|X_n(\omega)| > \varepsilon$. In particular, every $\omega$ for which $X_n(\omega) \to 0$ as $n \to \infty$ must be such that, for every $\varepsilon > 0$, only finitely many of the real numbers $X_n(\omega)$ exceed $\varepsilon$ in absolute value. Hence,

$$P(\{\omega : \lim_{n\to\infty} X_n(\omega) = 0\}) = 1 \quad \Longleftrightarrow \quad P(|X_n| > \varepsilon \text{ i.o.}) = 0 \text{ for all } \varepsilon > 0.$$

If convergence holds as in the left-hand side we recall from Definition 4.4 that the conclusion may be rephrased as

$$X_n \overset{a.s.}{\to} 0 \text{ as } n \to \infty \quad \Longleftrightarrow \quad P(|X_n| > \varepsilon \text{ i.o.}) = 0 \text{ for all } \varepsilon > 0. \qquad \square$$

Summarizing our findings so far, we have seen that the first Borel-Cantelli lemma tells us that if $\sum_{n=1}^{\infty} P(|X_n| > \varepsilon) < \infty$, then $X_n \overset{a.s.}{\to} 0$ as $n \to \infty$, and the second Borel-Cantelli lemma tells us that the converse holds if, in addition, $X_1$, $X_2$, ... are independent random variables. In the latter case we obtain the following zero-one law, which we state for easy reference.

**Corollary 18.1.** *Suppose that $X_1$, $X_2$, ... are independent random variables. Then*

$$X_n \overset{a.s.}{\to} 0 \text{ as } n \to \infty \quad \Longleftrightarrow \quad \sum_{n=1}^{\infty} P(|X_n| > \varepsilon) < \infty \text{ for all } \varepsilon > 0.$$

*Remark 18.1.* Convergence is a tail event, since convergence or not is independent of $X_1$, $X_2$, ..., $X_n$ for any $n$. The zero-one law therefore is also a consequence of the Kolmogorov zero-one law, Theorem 1.5.1. However, the present, alternative, derivation is more elementary and direct. $\qquad \square$

A common method in probability theory is to begin by considering subsequences. A typical case in the present context is when one wishes to prove that $P(A_n \text{ i.o.}) = 1$ and the events are not independent, but a suitable subsequence consists of independent events. In such cases the following rather immediate result may be helpful.

**Theorem 18.4.** *Let $\{A_n, n \geq 1\}$ be arbitrary events. If $\{A_{n_k}, k \geq 1\}$ are independent events for some subsequence $\{n_k, k \geq 1\}$, and*

$$\sum_{k=1}^{\infty} P(A_{n_k}) = \infty,$$

*then $P(A_n \text{ i.o.}) = 1$.*

*Proof.* This is immediate from the fact that $\{A_n \text{ i.o.}\} \supset \{A_{n_k} \text{ i.o.}\}$, and the second Borel-Cantelli lemma:

$$P(A_n \text{ i.o.}) \geq P(A_{n_k} \text{ i.o.}) = 1 \qquad \square$$

## 18.2 Some (Very) Elementary Examples

We first present a simple coin-tossing example, which is then expanded via a monkey and a typewriter to the more serious problem of the so-called Bible code, where *serious* is not to be interpreted mathematically, but as an example of the dangerous impact of what is believed to be paranormal phenomena on society. In a following subsection we provide examples related to records and random walks.

## Coin Tossing

Toss a *fair* coin repeatedly (independent tosses) and let

$$A_n = \{\text{the } n\text{th toss yields a head}\}, \quad n \geq 1.$$

Then
$$P(A_n \text{ i.o.}) = 1.$$

To prove this we note that $\sum_{n=1}^{\infty} P(A_n) = \sum_{n=1}^{\infty} \frac{1}{2} = \infty$, and the conclusion follows from Theorem 18.2.

For an arbitrary coin, one could imagine that if the probability of obtaining heads is "very small," then it might happen that, with some "very small" probability, only finitely many heads appear. However, set $P(\text{heads}) = p$, where $0 < p < 1$. Then $\sum_{n=1}^{\infty} P(A_n) = \sum_{n=1}^{\infty} p = \infty$, and we conclude, once again, that $P(A_n \text{ i.o.}) = 1$.

Finally, suppose that the tosses are performed with different coins, let $A_n$ be defined as before, and set $p_n = P(A_n)$. Then

$$P(A_n \text{ i.o.}) = 1 \quad \Longleftrightarrow \quad \sum_{n=1}^{\infty} p_n = +\infty.$$

The following exercises can be solved similarly, but a little more care is required, since the corresponding events are no longer independent.

**Exercise 18.1.** Toss a coin repeatedly as before and let

$$A_n = \{\text{the } (n-1)\text{th and the } n\text{th toss both yield a head}\}, \quad n \geq 2.$$

Show that
$$P(A_n \text{ i.o.}) = 1.$$

In other words, the event "two heads in a row" will occur infinitely often with probability 1. (Remember Theorem 18.4.)

**Exercise 18.2.** Toss another coin. Show that any finite pattern occurs infinitely often with probability 1.

**Exercise 18.3.** Toss a fair die with one face for every letter from A to Z repeatedly. Show that any finite word will appear infinitely often with probability 1.    □

## The Monkey and the Typewriter

A classical, more humorous, example states that if one puts a monkey at a typewriter he (or she) will "some day" all of a sudden have produced the complete works of Shakespeare, and, in fact, repeat this endeavor infinitely many times. In between successes the monkey will also complete the Uppsala telephone directory and lots of other texts.

Let us prove that this is indeed the case. Suppose that the letters the monkey produces constitute an independent sequence of identically distributed random variables. Then, by what we have just shown for coins, and extended in the exercises, every finite sequence of letters will occur (infinitely often!) with probability 1. And since the complete works of Shakespeare (as well as the Uppsala telephone directory) are exactly that, a finite sequence of letters, the proof is complete – under these model assumptions, which, of course, can be debated. After all, it is not quite obvious that the letters the monkey will produce are independent of each other . . ..

Finally, by the same argument it follows that the same texts also will appear if we spell out only every second letter or every 25th letter or every 37,658th letter.

## The Bible Code

Paranormal or supernatural phenomena and superstition have always been an important ingredient in the lives of many persons. Unfortunately a lot of people are fooled and conned by this kind of mumbo-jumbo or by others who exploit their fellow human beings.

In 1997 there appeared a book, *The Bible Code* [67], which to a large extent is based on the paper [254]. In the book it is claimed that the Hebrew Bible contains a code that reveals events that will occur thousands of years later. The idea is that one writes the 304,805 letters in an array, after which one reads along lines backward or forward, up or down, and looks for a given word. It is also permitted to follow every $n$th letter for any $n$. By doing so one finds all sorts of future events. One example is that by checking every 4772nd letter one finds the name of Yitzhak Rabin, which shows that one could already in the Bible find a hint concerning his murder in November 1995. An additional comment is that it is claimed that only the Hebrew version contains the code, no translation of it.

Although the "problem" is not exactly the same as the problem with the monkey and the typewriter, the probabilistic parallel is that one faces a (random) very long list of letters, among which one looks for a given word. Here we do not have an infinite sequence, but, on the other hand, we do not require a given word to appear infinitely often either.

If we look for a word of, say, $k$ letters in an alphabet of, say, $N$ letters, the probability of this word appearing at any given spot is $p = 1/N^k$, under the assumption that letters occur independently of each other and with the same distribution at every site. Barring all model discussions, starting at letters $m(k+1)$, for $m = 1, 2, \ldots$ (in order to make occurrences independent of each other), the number of repetitions before a hit is geometric with mean $1/p = N^k$, which is a finite number.

With the Borel-Cantelli lemmas in our mind it is thus not surprising that one can find almost anything one wishes with this program. More about the book can be found in the article [244], where, among other things, results from the same search method applied to translations of the bible as well as to other books are reported.

Apart from all of this one might wonder: If G-d really has put a code into the Bible, wouldn't one expect a more sophisticated one? And if the code really is a code, why did nobody discover the WTC attack on September 11, 2001, ahead of time? And the subway bombing in Madrid 2-1/2 years later?

Admittedly these examples may seem a bit elementary. On the other hand, they illustrate to what extent such examples are abundant in our daily lives; one may wonder how many fewer copies of books of this kind would be sold if everybody knew the Borel-Cantelli lemmas . . ..

## 18.3 Records

Recall the setting from Subsection 2.17.2: $X_1$, $X_2$, ... are independent, identically distributed, continuous random variables; the record times are

$$L(n) = \min\{k : X_k > X_{L(n-1)}\}, \quad n \geq 2, \quad L(1) = 1;$$

and the associated counting variables are

$$\mu(n) = \#\text{records among } X_1, X_2, \ldots, X_n = \sum_{k=1}^{n} I_k, \quad n \geq 1,$$

where $P(I_k = 1) = P(X_k \text{ is a record}) = 1 - P(I_k = 0) = 1/k$, and the indicators are independent.

Our concern for now is the "intuitively obvious(?)" fact that, one should obtain infinitely many records if we continue sampling indefinitely, the reason being that there is always room for a larger value than the largest one so far. But, intuition is not enough; we require a proof.

Mathematically we thus wish to prove that

$$P(I_n = 1 \text{ i.o.}) = 1.$$

Now,

$$\sum_{n=1}^{\infty} P(I_n = 1) = \sum_{n=1}^{\infty} \frac{1}{n} = \infty,$$

so that, the second Borel-Cantelli lemma tells us that our intuition was, indeed, a good one. Note that independence was important.

Let us also consider the number of *double records*, that is, two records in a row. What about our intuition? Is it equally obvious that there will be infinitely many double records? If there are infinitely many records, why not infinitely many times two of them following immediately after each other?

Let $D_n = 1$ if $X_n$ produces a double record, that is, if $X_{n-1}$ and $X_n$ both are records. Let $D_n = 0$ otherwise. Then, for $n \geq 2$,

$$P(D_n = 1) = P(I_n = 1, I_{n-1} = 1) = P(I_n = 1) \cdot P(I_{n-1} = 1) = \frac{1}{n} \cdot \frac{1}{n-1},$$

because of the independence of the indicators. Alternatively, by symmetry and combinatorics, $D_n = 1$ precisely when $X_n$ is the largest and $X_{n-1}$ is the second largest among the first $n$ observations. Thus,

$$\sum_{n=2}^{\infty} P(D_n = 1) = \sum_{n=2}^{\infty} \frac{1}{n(n-1)} = \lim_{m\to\infty} \sum_{n=2}^{m} \left(\frac{1}{n-1} - \frac{1}{n}\right) = \lim_{m\to\infty} \left(1 - \frac{1}{m}\right) = 1,$$

so that by the first Borel-Cantelli lemma

$$P(D_n = 1 \text{ i.o.}) = 0,$$

that is, the probability of infinitely many double records is 0. Note that $\{D_n, n \geq 2\}$ are *not* independent, which, however, is no problem since the sum was convergent.

The expected number of double records equals

$$E \sum_{n=2}^{\infty} D_n = \sum_{n=2}^{\infty} E D_n = \sum_{n=2}^{\infty} P(D_n = 1) = 1,$$

in other words, we can expect *one* double record.

Moreover, since double records seem to be rare events, one might guess that the total number of double records, $\sum_{n=2}^{\infty} D_n$, has a Poisson distribution, and if so, with parameter 1. That this is a correct guess has been proved independently in [125] and [42], Theorem 1.

### 18.4 Recurrence and Transience of Simple Random Walks

Consider a simple random walk, $\{S_n, n \geq 1\}$, starting at 0, and the probabilities that

- the random walk eventually returns to 0;
- doing so infinitely often.

A return can only occur after an even number of steps, equally many to the left and the right. It follows that

$$P(S_{2n} = 0) = \binom{2n}{n} p^n q^n \sim \begin{cases} \frac{1}{\sqrt{\pi n}}(4pq)^n, & \text{for } p \neq q, \\ \frac{1}{\sqrt{\pi n}}, & \text{for } p = q, \end{cases}$$

so that

$$\sum_{n=1}^{\infty} P(S_n = 0) \begin{cases} < +\infty, & \text{for } p \neq q, \\ = +\infty, & \text{for } p = q. \end{cases}$$

The first Borel-Cantelli lemma therefore tells us that

$$P(S_n = 0 \text{ i.o.}) = 0 \quad \text{for} \quad p \neq q.$$

One can, in fact, show that the probability of returning eventually equals $\min\{p,q\}/\max\{p,q\}$, when $p \neq q$. In this case the random walk is called *transient*.

The case $p = q = 1/2$ is called the *recurrent* case, since the probability of eventually returning to 0 equals 1. However, this is *not* a consequence of the second Borel-Cantelli lemma, since the events $\{S_n = 0\}$ are not independent. So, in order to prove this we must use different arguments.

Thus, suppose that $p = q = 1/2$, let $x$ be the probability we seek, namely, that a random walk starting at 0 eventually returns to 0, and let $y$ be the probability that a random walk starting at 0 eventually reaches the point $+1$. By symmetry, $y$ also equals the probability that a random walk starting at 0 eventually reaches the point $-1$, and by translation invariance, $y$ also equals the probability of eventually being one step to the left (or right) of the current state. Conditioning on the first step we obtain, with the aid of these properties,

$$x = \frac{1}{2}y + \frac{1}{2}y,$$
$$y = \frac{1}{2} + \frac{1}{2}y^2,$$

which has the solution $x = y = 1$.

We have thus shown that the probability of eventually returning to 0 equals 1. Now, having returned once, the probability of returning again equals 1, and so on, so that the probability of returning infinitely often equals 1, as claimed.

*Remark 18.2.* Note that the hard part is to show that the random walk returns *once*; that it returns infinitely often follows as an immediate consequence! □

**Exercise 18.4.** If $p \neq q$ an analogous argument also requires $z =$ the probability that a random walk starting at 0 eventually reaches the point $-1$. In the symmetric case $y = z$, but not here. Find the analogous system of (three) equations. □

*Remark 18.3.* A natural extension would be to consider the two-dimensional variant, in which the random walk is performed in the plane in such a way that transitions occur with probability $1/4$ in each of the four directions. The answer is that the probability of eventually returning to 0 equals 1 also in this case. So, what about three dimensions? Well, in this case even the symmetric random walk is transient. This is true for any dimension $d \geq 3$. The mathematical reason is that

$$\sum_{n=1}^{\infty} \left(\frac{1}{\sqrt{n}}\right)^d \begin{cases} = +\infty, & \text{for} \quad d = 1, 2, \\ < +\infty, & \text{for} \quad d \geq 3. \end{cases}$$

Note that for $d \geq 3$ transience is a consequence of this and the first Borel-Cantelli lemma. □

## 18.5 $\sum_{n=1}^{\infty} P(A_n) = \infty$ and $P(A_n \text{ i.o.}) = 0$

In the previous example with $p = q$ we found that $P(A_n \text{ i.o.}) = 1$, but not because the Borel-Cantelli sum was divergent; the events were not independent, so we had to use a different argument. In the following example the Borel-Cantelli sum diverges too, but in this case the conclusion is that $P(A_n \text{ i.o.}) = 0$. In other words, anything can happen for dependent events.

We ask the reader to trust the following claim, and be patient until Chapter 6 where everything will be verified.

*Example 18.2.* Let $X, X_1, X_2, \ldots$ be a sequence of independent, identically distributed random variables and set $S_n = X_1 + X_2 + \cdots + X_n$, $n \geq 1$.

The two facts we shall prove in Chapter 6 are that

$$P\left(\left|\frac{S_n}{n} - \mu\right| > \varepsilon \text{ i.o.}\right) = 0 \text{ for all } \varepsilon > 0 \quad \Longleftrightarrow \quad E|X| < \infty \text{ and } E X = \mu,$$

and that

$$\sum_{n=1}^{\infty} P\left(\left|\frac{S_n}{n} - \mu\right| > \varepsilon\right) < \infty \text{ for all } \varepsilon > 0 \quad \Longleftrightarrow \quad E X = \mu \text{ and } \operatorname{Var} X < \infty.$$

This means that if the mean is finite, but the variance is infinite, then the Borel-Cantelli sum diverges, and, yet, $P(|\frac{S_n}{n} - \mu| > \varepsilon \text{ i.o.}) = 0$. $\qquad \square$

The remainder of this section deals with how to handle cases without (total) independence.

### 18.6 Pair-wise Independence

We know from Subsection 2.10.3 that independence is a more restrictive assumption than *pair-wise* independence. However, if sums of random variables are involved it frequently suffices to assume pair-wise independence; for example, because the variance of a sum is equal to the sum of the variances.

Our first generalization is, basically, a consequence of that fact.

**Theorem 18.5.** *Let* $\{A_n, n \geq 1\}$ *be pair-wise independent events. Then*

$$\sum_{n=1}^{\infty} P(A_n) = \infty \quad \Longrightarrow \quad P(A_n \text{ i.o.}) = 1.$$

*Proof.* It is convenient to introduce indicator random variables. Let

$$I_n = I\{A_n\}, \quad n \geq 1.$$

Then $E I_n = P(A_n)$, $\operatorname{Var} I_n = P(A_n)(1 - P(A_n))$, the pair-wise independence translates into

$$E(I_i I_j) = E I_i \cdot E I_j \quad \text{for} \quad i \neq j,$$

and the statement of the theorem into

$$\sum_{n=1}^{\infty} E\,I_n = \infty \quad \Longrightarrow \quad P\Big(\sum_{n=1}^{\infty} I_n = \infty\Big) = 1.$$

Now, by Chebyshev's inequality,

$$P\Big(\Big|\sum_{k=1}^{n}(I_k - E\,I_k)\Big| > \frac{1}{2}\sum_{k=1}^{n}P(A_k)\Big) \leq \frac{\mathrm{Var}\Big(\sum_{k=1}^{n} I_k\Big)}{(\frac{1}{2}\sum_{k=1}^{n}P(A_k))^2}$$

$$= \frac{4\sum_{k=1}^{n}P(A_k)(1 - P(A_k))}{\Big(\sum_{k=1}^{n}P(A_k)\Big)^2} \leq \frac{4}{\sum_{k=1}^{n}P(A_k)} \to 0 \quad \text{as} \quad n \to \infty.$$

Recalling that $E\,I_k = P(A_k)$, it follows, in particular, that

$$P\Big(\sum_{k=1}^{n} I_k > \frac{1}{2}\sum_{k=1}^{n} E\,I_k\Big) \to 1 \quad \text{as} \quad n \to \infty.$$

Since both sums increase with $n$ we may let $n$ tend to infinity in $\sum_{k=1}^{n} I_k$ and then in $\sum_{k=1}^{n} E\,I_k$, to conclude that

$$P\Big(\sum_{n=1}^{\infty} I_n = \infty\Big) = 1. \qquad \Box$$

An immediate consequence is that the zero-one law, Theorem 18.3, remains true for pair-wise independent random variables. For convenience we state this fact as a theorem of its own.

**Theorem 18.6.** (A second zero-one law)
*If $\{A_n,\ n \geq 1\}$ are pair-wise independent events, then*

$$P(A_n \text{ i.o.}) = \begin{cases} 0, & when \quad \sum_{n=1}^{\infty} P(A_n) < \infty, \\ 1, & when \quad \sum_{n=1}^{\infty} P(A_n) = \infty. \end{cases}$$

## 18.7 Generalizations Without Independence

The following result is due to Barndorff-Nielsen, [10].

**Theorem 18.7.** *Let $\{A_n,\ n \geq 1\}$ be arbitrary events satisfying*

$$P(A_n) \to 0 \quad as \quad n \to \infty, \tag{18.1}$$

*and*

$$\sum_{n=1}^{\infty} P(A_n \cap A_{n+1}^c) < \infty. \tag{18.2}$$

*Then*

$$P(A_n \text{ i.o.}) = 0.$$

*Remark 18.4.* Note that $\sum_{n=1}^{\infty} P(A_n)$ may be convergent as well as divergent under the present assumptions. In particular, the convergence of the sum is not *necessary* in order for $P(A_n$ i.o.) to equal 0.    □

*Proof.* A glance at Theorem 18.1 shows that the second assumption alone implies that $P(A_n \cap A_{n+1}^c$ i.o.) = 0, that is, that there are almost surely only a finite number of switches between the sequences $\{A_n\}$ and $\{A_n^c\}$, so that one of them occurs only a finite number of times, after which the other one takes over for ever. To prove the theorem it therefore suffices to prove that

$$P(A_n^c \text{ i.o.}) = 1.$$

Now,

$$P(A_n^c \text{ i.o.}) = \lim_{m \to \infty} P\left( \bigcup_{n \geq m} A_n^c \right) \geq \lim_{m \to \infty} P(A_m^c) \to 1 \quad \text{as} \quad m \to \infty,$$

where the convergence to 1 follows from the first assumption.    □

Continuing the discussion at the beginning of the proof we note that if $\{A_n, n \geq 1\}$ are independent events, we may, in addition, conclude that one of $\{A_n$ i.o.$\}$ and $\{A_n^c$ i.o.$\}$ has probability 1 and the other one has probability 0, since by the zero-one law in Theorem 18.3, the probabilities of these events can only assume the values 0 or 1. For ease of future reference we collect these facts separately. Note also that the conclusions are true whether (18.1) holds or not.

**Theorem 18.8.** *Let* $\{A_n, n \geq 1\}$ *be arbitrary events, and suppose (18.2) holds.*
(i) *Then*

$$P(A_n \cap A_{n+1}^c \text{ i.o.}) = 0.$$

(ii) *If, in addition,* $\{A_n, n \geq 1\}$ *are independent, then*

$$P(A_n \text{ i.o.}) = 0 \quad \text{and} \quad P(A_n^c \text{ i.o.}) = 1 \quad \text{or vice versa.}$$

To exploit the crossing concept further we formulate the following result.

**Theorem 18.9.** *Let* $\{A_n, n \geq 1\}$ *and* $\{B_n, n \geq 1\}$ *be arbitrary events, and suppose that the pairs* $A_n$ *and* $B_{n+1}$ *are independent for all* $n$. *If*

$$\sum_{n=1}^{\infty} P(A_n \cap B_{n+1}) < \infty,$$

*then*

$$P(A_n \text{ i.o.}) = 0 \quad \text{and} \quad P(B_n \text{ i.o.}) = 1 \quad \text{or vice versa.}$$

*Proof.* The arguments for Theorem 18.8 were given prior to its statement, and those for Theorem 18.9 are the same.    □

In his paper Barndorff-Nielsen applied this result in order to prove a theorem on the rate of growth of partial maxima of independent, identically distributed random variables. In order to illustrate the efficiency of his result we apply the idea to the partial maxima of standard exponentials. The computations are based on a more general result in [124].

## 18.8 Extremes

Suppose that $X_1$, $X_2$, ... are independent, standard exponential random variables, and set $Y_n = \max\{X_1, X_2, \ldots, X_n\}$, $n \geq 1$.

We begin by considering the original sequence, after which we turn our attention to the sequence of partial maxima.

Since

$$P(X_n > \varepsilon \log n) = \frac{1}{n^\varepsilon},$$

it follows that

$$\sum_{n=1}^{\infty} P(X_n > \varepsilon \log n) \begin{cases} < +\infty & \text{for} \quad \varepsilon > 1, \\ = +\infty & \text{for} \quad \varepsilon \leq 1. \end{cases}$$

An appeal to the Borel-Cantelli lemmas asserts that

$$P(\{X_n > \varepsilon \log n\} \text{ i.o.}) = \begin{cases} 0 & \text{for} \quad \varepsilon > 1, \\ 1 & \text{for} \quad \varepsilon \leq 1, \end{cases} \tag{18.3}$$

and, consequently, that

$$\limsup_{n \to \infty} \frac{X_n}{\log n} = 1 \quad \text{a.s.}$$

Moreover, since

$$\sum_{n=1}^{\infty} P(X_n < \varepsilon \log n) = \sum_{n=1}^{\infty} \left(1 - \frac{1}{n^\varepsilon}\right) = +\infty \quad \text{for all} \quad \varepsilon > 0,$$

the second Borel-Cantelli lemma yields

$$\liminf_{n \to \infty} \frac{X_n}{\log n} = 0 \quad \text{a.s.}$$

This means, roughly speaking, that the sequence $\{X_n/\log n, n \geq 1\}$ oscillates between 0 and 1.

Since $Y_n = \max\{X_1, X_2, \ldots, X_n\}$ is non-decreasing in $n$ (and, hence cannot oscillate) it is tempting to guess that

$$\lim_{n \to \infty} \frac{Y_n}{\log n} = 1 \quad \text{a.s.}$$

This is not only a guess as we shall show next.

The crucial observation is that

$$\{Y_n > \varepsilon \log n \text{ i.o.}\} \iff \{X_n > \varepsilon \log n \text{ i.o.}\},$$

since $\log n$ is increasing in $n$; although $Y_n$ exceeds $\varepsilon \log n$ more often that $X_n$, the whole sequences do so *infinitely often* simultaneously. It follows that

$$P(Y_n > \varepsilon \log n \text{ i.o.}) = 1 \quad \text{for} \quad \varepsilon < 1,$$

and that

$$P\left( \limsup_{n \to \infty} \frac{Y_n}{\log n} = 1 \right) = 1. \tag{18.4}$$

In order to show that the limit actually equals 1 we have a problem, since $Y_n$, $n \geq 1$, are not independent, and this is where Theorem 18.9 comes to our rescue.

Let $0 < \varepsilon < 1$, and set

$$A_n = \{Y_n \leq \varepsilon \log n\} \quad \text{and} \quad B_n = \{X_n > \varepsilon \log n\}, \quad n \geq 1.$$

Then

$$\sum_{n=1}^{\infty} P(A_n \cap A_{n+1}^c) = \sum_{n=1}^{\infty} P(A_n \cap B_{n+1}) = \sum_{n=1}^{\infty} P(A_n) \cdot P(B_{n+1})$$

$$= \sum_{n=1}^{\infty} \left( 1 - \frac{1}{n^\varepsilon} \right)^n \cdot \frac{1}{(n+1)^\varepsilon} \leq \sum_{n=1}^{\infty} \exp\{-n^{1-\varepsilon}\} \cdot \frac{1}{n^\varepsilon}$$

$$= \sum_{n=1}^{\infty} \int_{n-1}^{n} \exp\{-n^{1-\varepsilon}\} \cdot \frac{1}{n^\varepsilon} \, dx \leq \sum_{n=1}^{\infty} \int_{n-1}^{n} \exp\{-x^{1-\varepsilon}\} \cdot \frac{1}{x^\varepsilon} \, dx$$

$$= \int_0^\infty \exp\{-x^{1-\varepsilon}\} \cdot \frac{1}{x^\varepsilon} \, dx = \left[ \frac{-\exp\{-x^{1-\varepsilon}\}}{1-\varepsilon} \right]_0^\infty = \frac{1}{1-\varepsilon} < \infty.$$

Since $P(B_n \text{ i.o.}) = 1$ by (18.3), Theorem 18.9 tells us that we must have

$$P(A_n \text{ i.o.}) = 0 \quad \text{for} \quad \varepsilon < 1,$$

which implies that

$$P\left( \liminf_{n \to \infty} \frac{Y_n}{\log n} \geq 1 \right) = 1. \tag{18.5}$$

Joining this with (18.4) establishes that

$$P\left( \lim_{n \to \infty} \frac{Y_n}{\log n} = 1 \right) = 1,$$

or, equivalently, that $\frac{Y_n}{\log n} \overset{a.s.}{\to} 1$ as $n \to \infty$, as desired.

If, instead, the random variables have a standard normal distribution, Mill's ratio, Lemma A.2.1, yields

$$P(X > x) \sim \frac{1}{x\sqrt{2\pi}} \exp\{-x^2/2\} \quad \text{as} \quad x \to \infty,$$

so that, for $N$ large,

$$\sum_{n \geq N} P(X_n > \varepsilon\sqrt{2\log n}) \sim \sum_{n \geq N} \frac{1}{\varepsilon\sqrt{2\pi \log n}} \cdot \frac{1}{n^{\varepsilon^2}} \begin{cases} < +\infty & \text{for } \varepsilon > 1, \\ = +\infty & \text{for } \varepsilon \leq 1, \end{cases}$$

from which it similarly follows that

$$P(\{X_n > \varepsilon\sqrt{2\log n}\} \text{ i.o.}) = \begin{cases} 0 & \text{for } \varepsilon > 1, \\ 1 & \text{for } \varepsilon \leq 1, \end{cases}$$

and that

$$\limsup_{n \to \infty} \frac{X_n}{\sqrt{2\log n}} = 1 \quad \text{a.s.}$$

Since the standard normal distribution is symmetric around 0, it follows, by considering the sequence $\{-X_n, n \geq 1\}$, that

$$\liminf_{n \to \infty} \frac{X_n}{\sqrt{2\log n}} = -1 \quad \text{a.s.}$$

**Exercise 18.5.** Prove the analog for partial maxima of independent standard normal random variables. $\square$

## 18.9 Further Generalizations

For notational convenience we set, throughout the remainder of this section,

$$p_k = P(A_k) \quad \text{and} \quad p_{ij} = P(A_i \cap A_j), \quad \text{for all} \quad k, i, j,$$

in particular, $p_{kk} = p_k$.

Inspecting the proof of Theorem 18.5, we find that the variance of the sum of the indicator becomes

$$\text{Var}\left(\sum_{k=1}^n I_k\right) = \sum_{k=1}^n p_k(1 - p_k) + \sum_{\substack{i=1 \\ i \neq j}}^n \sum_{j=1}^n (p_{ij} - p_i p_j)$$

$$= \sum_{i=1}^n \sum_{j=1}^n p_{ij} - \sum_{i=1}^n \sum_{j=1}^n p_i p_j = \sum_{i=1}^n \sum_{j=1}^n p_{ij} - \left(\sum_{k=1}^n p_k\right)^2,$$

so that, in this case, the computation turns into

$$P\left(\left|\sum_{k=1}^{n}(I_k - E\,I_k)\right| > \frac{1}{2}\sum_{k=1}^{n} p_k\right) \le \frac{\mathrm{Var}\left(\sum_{k=1}^{n} I_k\right)}{\left(\frac{1}{2}\sum_{k=1}^{n} p_k\right)^2}$$

$$= 4\left(\frac{\sum_{i=1}^{n}\sum_{j=1}^{n} p_{ij}}{\left(\sum_{k=1}^{n} p_k\right)^2} - 1\right),$$

which suggests the following strengthening.

**Theorem 18.10.** *Let* $\{A_n,\ n \ge 1\}$ *be arbitrary events, such that*

$$\liminf_{n\to\infty} \frac{\sum_{i=1}^{n}\sum_{j=1}^{n} P(A_i \cap A_j)}{\left(\sum_{k=1}^{n} P(A_k)\right)^2} = 1.$$

*Then*

$$\sum_{n=1}^{\infty} P(A_n) = \infty \quad \Longrightarrow \quad P(A_n \text{ i.o.}) = 1.$$

*Proof.* By arguing as in the proof of Theorem 18.5, it follows from the computations preceding the statement of Theorem 18.10 that

$$\liminf_{n\to\infty} P\left(\sum_{k=1}^{n} I_k \le \frac{1}{2}\sum_{k=1}^{n} E\,I_k\right) = 0.$$

We may therefore select a subsequence $\{n_j,\ j \ge 1\}$ of the integers in such a way that

$$\sum_{j=1}^{\infty} P\left(\sum_{k=1}^{n_j} I_k \le \frac{1}{2}\sum_{k=1}^{n_j} E\,I_k\right) < \infty,$$

which, by the first Borel-Cantelli lemma, shows that

$$P\left(\sum_{k=1}^{n_j} I_k \le \frac{1}{2}\sum_{k=1}^{n_j} E\,I_k \text{ i.o.}\right) = 0,$$

so that,

$$P\left(\sum_{k=1}^{n_j} I_k > \frac{1}{2}\sum_{k=1}^{n_j} E\,I_k \text{ i.o.}\right) = 1.$$

Finally, since this is true for any $j$ and the sum of the expectations diverges, we may, as in the proof of Theorem 18.5, let $j$ tend to infinity in the sum of the indicators, and then in the sum of the expectations to conclude that

$$P\left(\sum_{k=1}^{\infty} I_k = \infty\right) = 1. \qquad \square$$

With some additional work one can prove the following, stronger, result.

**Theorem 18.11.** *Let* $\{A_n, n \geq 1\}$ *be arbitrary events, such that*

$$\limsup_{n \to \infty} \frac{\left( \sum_{k=m+1}^{n} P(A_k) \right)^2}{\sum_{i=m+1}^{n} \sum_{j=m+1}^{n} P(A_i \cap A_j)} \geq \alpha,$$

*for some* $\alpha > 0$ *and* $m$ *large. Then*

$$\sum_{n=1}^{\infty} P(A_n) = \infty \implies P(A_n \text{ i.o.}) \geq \alpha.$$

An early related paper is [49], from which we borrow the following lemma, which, in turn, is instrumental for the proof of the theorem. We also refer to [234], P3, p. 317, where the result is used in connection with a three-dimensional random walk, and to [195], Section 6.1 where also necessary and sufficient conditions for ensuring that $P(A_n \text{ i.o.}) = \alpha$ are given.

**Lemma 18.1.** *Let* $\{A_n, n \geq 1\}$ *be arbitrary events. For* $m \geq 1$,

$$P\left( \bigcup_{k=m+1}^{n} A_k \right) \geq \frac{\left( \sum_{k=m+1}^{n} P(A_k) \right)^2}{\sum_{i=m+1}^{n} \sum_{j=m+1}^{n} P(A_i \cap A_j)}.$$

*Proof.* Set $I_n = I\{A_n\}$, $n \geq 1$. Then

$$E\left( \sum_{k=m+1}^{n} I_k \right)^2 = \sum_{i,j=m+1}^{n} E\, I_i I_j = \sum_{k=m+1}^{n} E\, I_k + \sum_{\substack{i,j=m+1 \\ i \neq j}}^{n} E\, I_i I_j$$

$$= \sum_{k=m+1}^{n} p_k + \sum_{\substack{i,j=m+1 \\ i \neq j}}^{n} p_{ij} = \sum_{i,j=m+1}^{n} p_{ij}.$$

Secondly, via Cauchy's inequality,

$$\left( \sum_{k=m+1}^{n} p_k \right)^2 = \left( E \sum_{k=m+1}^{n} I_k \right)^2 = \left( E \sum_{k=m+1}^{n} I_k \cdot I\{ \sum_{k=m+1}^{n} I_k > 0 \} \right)^2$$

$$\leq E\left( \sum_{k=m+1}^{n} I_k \right)^2 \cdot E\left( I\{ \sum_{k=m+1}^{n} I_k > 0 \} \right)^2$$

$$= E\left( \sum_{k=m+1}^{n} I_k \right)^2 E\left( I\{ \sum_{k=m+1}^{n} I_k > 0 \} \right) = E\left( \sum_{k=m+1}^{n} I_k \right)^2 P\left( \bigcup_{k=m+1}^{n} A_k \right).$$

The conclusion follows by joining the extreme members from the two calculations. $\qquad\square$

*Proof of Theorem 18.11.* Choosing $m$ sufficiently large, and applying the lemma, we obtain

$$P(A_n \text{ i.o.}) = \lim_{m \to \infty} P\left( \bigcup_{k=m+1}^{\infty} A_k \right) \geq \lim_{m \to \infty} \limsup_{n \to \infty} P\left( \bigcup_{k=m+1}^{n} A_k \right)$$

$$\geq \lim_{m \to \infty} \limsup_{n \to \infty} \frac{\left( \sum_{k=m+1}^{n} p_k \right)^2}{\sum_{i=m+1}^{n} \sum_{j=m+1}^{n} p_{ij}} \geq \alpha. \qquad \square$$

Our final extension is a recent result in which the assumption about the ratio of the sums is replaced by the same condition applied to the individual terms; see [196]. For a further generalization we refer to [197].

**Theorem 18.12.** *Let* $\{A_n, n \geq 1\}$ *be arbitrary events, such that, for some* $\alpha \geq 1$,

$$P(A_i \cap A_j) \leq \alpha P(A_i) P(A_j) \quad \text{for all} \quad i, j > m, \ i \neq j. \tag{18.6}$$

*Then*

$$\sum_{n=1}^{\infty} P(A_n) = \infty \quad \Longrightarrow \quad P(A_n \text{ i.o.}) \geq 1/\alpha.$$

*Proof.* We first consider the denominator in right-hand side of Lemma 18.1. Using the factorizing assumption and the fact that $\alpha \geq 1$, we obtain

$$\sum_{i,j=m+1}^{n} p_{ij} = \sum_{k=m+1}^{n} p_k + \sum_{\substack{i,j=m+1 \\ i \neq j}}^{n} p_{ij} \leq \sum_{k=m+1}^{n} p_k + \alpha \sum_{\substack{i,j=m+1 \\ i \neq j}}^{n} p_{ij}$$

$$\leq \sum_{k=m+1}^{n} p_k - \alpha \sum_{k=m+1}^{n} p_k^2 + \alpha \left( \sum_{k=m+1}^{n} p_k \right)^2$$

$$\leq \alpha \sum_{k=m+1}^{n} p_k \left( 1 + \sum_{k=m+1}^{n} p_k \right),$$

so that, by the lemma,

$$P\left( \bigcup_{k=m+1}^{n} A_k \right) \geq \frac{\left( \sum_{k=m+1}^{n} p_k \right)^2}{\sum_{i,j=m+1}^{n} p_{ij}} \geq \frac{\sum_{k=m+1}^{n} p_k}{\alpha \left( 1 + \sum_{k=m+1}^{n} p_k \right)}.$$

The divergence of the sum, finally, yields

$$P(A_n \text{ i.o.}) = \lim_{m \to \infty} P\left( \bigcup_{k=m+1}^{\infty} A_k \right) \geq \lim_{m \to \infty} \limsup_{n \to \infty} P\left( \bigcup_{k=m+1}^{n} A_k \right)$$

$$\geq \lim_{m \to \infty} \limsup_{n \to \infty} \frac{\sum_{k=m+1}^{n} p_k}{\alpha \left( 1 + \sum_{k=m+1}^{n} p_k \right)} = 1/\alpha. \qquad \square$$

We close by connecting Theorem 18.12 to some of the earlier results.

- Lamperti [167] proves that $P(A_n \text{ i.o.}) > 0$ under the assumption (18.6).
- If there exists some kind of zero-one law, the conclusion of the theorem (and of that of Lamperti) becomes $P(A_n \text{ i.o.}) = 1$; cf. Theorem 18.3.
- If (18.6) holds with $\alpha = 1$, then $P(A_n \text{ i.o.}) = 1$. One such case is when $\{A_n, n \geq 1\}$ are (pair-wise) independent, in which case we rediscover Theorems 18.2 and 18.5, respectively.

## 19 A Convolution Table

Let $X$ and $Y$ be independent random variables and set $Z = X + Y$. What type of distribution does $Z$ have if $X$ and $Y$ are absolutely continuous, discrete, or continuous singular, respectively? If both are discrete one would guess that so is $Z$. But what if $X$ has a density and $Y$ is continuous singular? What if $X$ is continuous singular and $Y$ is discrete?

The convolution formula for densities, cf. Subsection 2.9.4, immediately tells us that if both distributions are absolutely continuous, then so is the sum. However, by inspecting the more general result, Theorem 9.4, we realize that it suffices for one of them to be absolutely continuous.

If both distributions are discrete, then so is the sum; the support of the sum is the direct sum of the respective supports:

$$\operatorname{supp}(F_{X+Y}) = \{x + y : x \in \operatorname{supp}(F_X) \text{ and } y \in \operatorname{supp}(F_Y)\}.$$

If $X$ is discrete and $Y$ is continuous singular, the support of the sum is a Lebesgue null set, so that the distribution is singular. The derivative of the distribution function remains 0 almost everywhere, the new exceptional points are contained in the support of $X$, which, by Proposition 2.1(iii), is at most countable.

Our findings so far may be collected in the following diagram:

| X \ Y | AC | D | CS |
|-------|----|----|----|
| AC | AC | AC | AC |
| D | AC | D | CS |
| CS | AC | CS | ?? |

**Figure 2.6.** The distribution of $X + Y$

It remains to investigate the case when $X$ and $Y$ both are continuous singular. However, this is a more sophisticated one. We shall return to that slot in Subsection 4.2.2 with the aid of (rescaled versions of) the Cantor type random variables $Y$ and $Z$ from Subsection 2.2.6.

# 20 Problems

1. Let $X$ and $Y$ be random variables and suppose that $A \in \mathcal{F}$. Prove that
$$Z = XI\{A\} + YI\{A^c\} \quad \text{is a random variable.}$$

2. Show that if $X$ is a random variable, then, for every $\varepsilon > 0$, there exists a bounded random variable $X_\varepsilon$, such that
$$P(X \neq X_\varepsilon) < \varepsilon.$$

   ♣ Observe the difference between a *finite* random variable and a *bounded* random variable.

3. Show that
   (a) if $X$ is a random variable, then so is $|X|$;
   (b) the converse does not necessarily hold.

   ♠ Don't forget that there exist non-measurable sets.

4. Let $X$ be a random variable with distribution function $F$.
   (a) Show that
   $$\lim_{h \searrow 0} P(x - h < X \leq x + h) = \lim_{h \searrow 0} (F(x+h) - F(x-h))$$
   $$= \begin{cases} P(X = x), & \text{if } x \in \mathbb{J}_F, \\ 0, & \text{otherwise.} \end{cases}$$

   (b) A point $x \in \text{supp}\,(F)$ if and only if
   $$F(x+h) - F(x-h) > 0 \quad \text{for every} \quad h > 0.$$
   Prove that
   $$x \in \mathbb{J}_F \implies x \in \text{supp}\,(F).$$

   (c) Prove that the converse holds for isolated points.
   (d) Prove that the support of any distribution function is closed.

5. Suppose that $X$ is an integer valued random variable, and let $m \in \mathbb{N}$. Show that
$$\sum_{n=1}^{\infty} P(n < X \leq n + m) = m.$$

6. Show that, for any random variable, $X$, and $a \in \mathbb{R}$,
$$\int_{-\infty}^{\infty} P(x < X \leq x + a)\,dx = a.$$

   ♣ An extension to two random variables will be given in Problem 20.15.

7. Let $(\Omega, \mathcal{F}, P)$ be the Lebesgue measure space $([0,1], \mathcal{B}([0,1]), \lambda)$, and let $\{X_t, 0 \le t \le 1\}$ be a *family* of random variables defined as

$$X_t(\omega) = \begin{cases} 1, & \text{for} \quad \omega = t, \\ 0, & \text{otherwise.} \end{cases}$$

Show that

$$P(X_t = 0) = 1 \text{ for all } t \qquad \text{and/but} \qquad P(\sup_{0 \le t \le 1} X_t = 1) = 1.$$

♣ Note that there is no contradiction, since the supremum is taken over an uncountable set of $t$ values.

8. Show that, if $\{X_n, n \ge 1\}$ are independent random variables, then

$$\sup_n X_n < \infty \quad \text{a.s} \quad \Longleftrightarrow \quad \sum_{n=1}^{\infty} P(X_n > A) < \infty \quad \text{for some } A.$$

9. The name of the log-normal distribution comes from the fact that its logarithm is a normal random variable. Prove that the name is adequate, that is, let $X \in N(\mu, \sigma^2)$, and set $Y = e^X$. Compute the distribution function of $Y$, differentiate, and compare with the entry in Table 2.2.

10. Let $X \in U(0,1)$, and $\theta > 0$. Verify, by direct computation, that

$$Y = -\theta \log X \in \text{Exp}(\theta).$$

♦ This is useful for generating exponential random numbers, which are needed in simulations related to the Poisson process.

11. Compute the expected number of trials needed in order for all faces of a symmetric die to have appeared at least once.

12. *The coupon collector's problem.* Each time one buys a bag of cheese doodles one obtains as a bonus a picture (hidden inside the package) of a soccer player. Suppose there are $n$ different pictures which are equally likely to be inside every package. Find the expected number of packages one has to buy in order to get a complete collection of players.

♣ For $n = 100$ the numerical answer is 519, i.e., "a lot" more than 100.

13. Let $X_1, X_2, \ldots, X_n$ be independent, identically distributed random variables with $E X^4 < \infty$, and set $\mu = E X$, $\sigma^2 = \text{Var } X$, and $\mu_4 = E(X - \mu)^4$. Furthermore, set

$$\bar{X}_n = \frac{1}{n} \sum_{k=1}^{n} X_k \quad \text{and} \quad m_n^2 = \frac{1}{n} \sum_{k=1}^{n} (X_k - \bar{X}_n)^2.$$

Prove that

$$E(m_n^2) = \sigma^2 \frac{n-1}{n},$$

$$\text{Var}(m_n^2) = \frac{\mu_4 - \sigma^4}{n} - \frac{2\mu_4 - 4\sigma^4}{n^2} + \frac{\mu_4 - 3\sigma^4}{n^3}.$$

♣ Observe that the *sample variance*, $s_n^2 = \frac{1}{n-1}\sum_{k=1}^{n}(X_k - \bar{X}_n)^2 = \frac{n-1}{n}m_n^2$, is *unbiased*, which means that $E\,s_n^2 = \sigma^2$. On the other hand, the expression for $\mathrm{Var}\,s_n^2$ is more involved.

14. Let, for $k \geq 1$, $\mu_k = E(X - E\,X)^k$ be the $k$th central moment of the random variable $X$. Prove that the matrix

$$\begin{pmatrix} 1 & 0 & \mu_2 \\ 0 & \mu_2 & \mu_3 \\ \mu_2 & \mu_3 & \mu_4 \end{pmatrix}$$

(a) has a non-negative determinant;
(b) is non-negative definite.
♠ Assume w.l.o.g. that $E\,X = 0$ and investigate $E(a_0 + a_1 X + a_2 X^2)^2$, where $a_0, a_1, a_2 \in \mathbb{R}$.
(c) Generalize to higher dimensions.

15. This problem extends Problem 20.6. Let $X, Y$ be random variables with finite mean. Show that

$$\int_{-\infty}^{\infty} \big(P(X < x \leq Y) - P(Y < x \leq X)\big)\mathrm{d}x = E\,Y - E\,X.$$

16. Show that, if $X$ and $Y$ are independent random variables, such that $E|X| < \infty$, and $B$ is an arbitrary Borel set, then

$$E\,XI\{Y \in B\} = E\,X \cdot P(Y \in B).$$

17. Suppose that $X_1, X_2, \ldots, X_n$ are random variables, such that $E|X_k| < \infty$ for all $k$, and set $Y_n = \max_{1 \leq k \leq n} X_k$.
(a) Prove that $E\,Y_n < \infty$.
(b) Prove that $E\,X_k \leq E\,Y_n$ for all $k$.
(c) Prove that $E|Y_n| < \infty$.
(d) Show that the analog of (b) for absolute values (i.e. $E|X_k| \leq E|Y_n|$ for all $k$) need *not* be true.
♣ Note the distinction between the random variables $|\max_{1 \leq k \leq n} X_k|$ and $\max_{1 \leq k \leq n}|X_k|$.

18. Let $X_1, X_2, \ldots$ be random variables, and set $Y = \sup_n |X_n|$. Show that

$$E|Y|^r < \infty \quad \Longleftrightarrow \quad |Y| \leq Z \quad \text{for some} \quad Z \in L^r, \quad r > 0.$$

19. Let $X$ be a non-negative random variable. Show that

$$\lim_{n \to \infty} nE\Big(\frac{1}{X}I\{X > n\}\Big) = 0,$$

$$\lim_{n \to \infty} \frac{1}{n}E\Big(\frac{1}{X}I\{X > \frac{1}{n}\}\Big) = 0.$$

♠ A little more care is necessary for the second statement.

20. Let $\{A_n,\ n \geq 1\}$ be independent events, and suppose that $P(A_n) < 1$ for all $n$. Prove that

$$P(A_n \text{ i.o.}) = 1 \quad \Longleftrightarrow \quad P\left(\bigcup_{n=1}^{\infty} A_n\right) = 1.$$

Why is $P(A_n) = 1$ forbidden?

21. Consider the dyadic expansion of $X \in U(0,1)$, and let $l_n$ be the *run length of zeroes* from the $n$th decimal and onward. This means that $l_n = k$ if decimals $n, n+1, \ldots, n+k-1$ are all zeroes. In particular, $l_n = 0$ if the $n$th decimal equals 1.

(a) Prove that $P(l_n = k) = \frac{1}{2^{k+1}}$ for all $k \geq 0$.

(b) Prove that $P(l_n = k \text{ i.o.}) = 1$ for all $k$.

♠ Note that the events $\{l_n = k,\ n \geq 1\}$ are *not* independent (unless $k = 0$).

♣ The result in (b) means that, with probability 1, there will be infinitely many arbitrarily long stretches of zeroes in the decimal expansion of $X$.

(c) Prove that $P(l_n = n \text{ i.o.}) = 0$.

♣ This means that if we require the run of zeroes that starts at $n$ to have length $n$, then, almost surely, this will happen only finitely many times. (There exist stronger statements.)

22. Let $X, X_1, X_2, \ldots$ be independent, identically distributed random variables, such that $P(X = 0) = P(X = 1) = 1/2$.

(a) Let $N_1$ be the number of 0's and 1's until the first appearance of the pattern 10. Find $E\,N_1$.

(b) Let $N_2$ be the number of 0's and 1's until the first appearance of the pattern 11. Find $E\,N_2$.

(c) Let $N_3$ be the number of 0's and 1's until the first appearance of the pattern 100. Find $E\,N_3$.

(d) Let $N_4$ be the number of 0's and 1's until the first appearance of the pattern 101. Find $E\,N_4$.

(e) Let $N_5$ be the number of 0's and 1's until the first appearance of the pattern 111. Find $E\,N_5$.

(f) Solve the same problem if $X \in Be(p)$, for $0 < p < 1$.

♣ No two answers are the same (as one might think concerning (a) and (b)).

# 3

# Inequalities

Inequalities play an important role in probability theory, because much work concerns the estimation of certain probabilities by others, the estimation of moments of sums by sums of moments, and so on.

In this chapter we have collected a number of inequalities of the following kind:

- tail probabilities are estimated by moments;
- moments of sums are estimated by sums of moments and vice versa;
- the expected value of the product of two random variables is estimated by a suitable product of higher-order moments;
- moments of low order are estimated by moments of a higher order;
- a moment inequality for convex functions of random variables is provided;
- relations between random variables and *symmetrized* versions;
- the probability that a maximal partial sum of random variables exceeds some given level is related to the probability that the *last* partial sum does so.

## 1 Tail Probabilities Estimated via Moments

We begin with a useful and elementary inequality.

**Lemma 1.1.** *Suppose that $g$ is a non-negative, non-decreasing function such that $E\,g(|X|) < \infty$, and let $x > 0$. Then,*

$$P(|X| > x) \leq \frac{E\,g(|X|)}{g(x)}.$$

*Proof.* We have

$$E\,g(|X|) \geq E\,g(|X|)I\{|X| > x\} \geq g(x)E\,I\{|X| > x\} = g(x)P(|X| > x). \quad \square$$

Specializing $g$ yields the following famous named inequality.

**Theorem 1.1.** (Markov's inequality) *Suppose that $E|X|^r < \infty$ for some $r > 0$, and let $x > 0$. Then,*

$$P(|X| > x) \leq \frac{E|X|^r}{x^r}.$$

Another useful case is the exponential function applied to bounded random variables.

**Theorem 1.2.** (i) *Suppose that $P(|X| \leq b) = 1$ for some $b > 0$, that $E\,X = 0$, and set $\operatorname{Var} X = \sigma^2$. Then, for $0 < t < b^{-1}$, and $x > 0$.*

$$P(X > x) \leq e^{-tx + t^2 \sigma^2},$$
$$P(|X| > x) \leq 2e^{-tx + t^2 \sigma^2}.$$

(ii) *Let $X_1, X_2, \ldots, X_n$ be independent random variables with mean 0, suppose that $P(|X_k| \leq b) = 1$ for all $k$, and set $\sigma_k^2 = \operatorname{Var} X_k$. Then, for $0 < t < b^{-1}$, and $x > 0$.*

$$P(|S_n| > x) \leq 2 \exp\left\{ -tx + t^2 \sum_{k=1}^{n} \sigma_k^2 \right\}.$$

(iii) *If, in addition, $X_1, X_2, \ldots, X_n$ are identically distributed, then*

$$P(|S_n| > x) \leq 2\exp\{-tx + nt^2 \sigma_1^2\}.$$

*Proof.* (i): Applying Lemma 3.1.1 with $g(x) = e^{tx}$, for $0 \leq t \leq b^{-1}$, and formula (A.A.1) yields

$$P(X > x) = \frac{E\,e^{tX}}{e^{tx}} \leq e^{-tx}(1 + E\,tX + E\,(tX)^2)$$
$$= e^{-tx}(1 + t^2 \sigma^2) \leq e^{-tx} e^{t^2 \sigma^2},$$

which proves the first assertion. The other one follows by considering the negative tail and addition. Statements (ii) and (iii) are then immediate.    □

The following inequality for bounded random variables, which we state without proof, is due to Hoeffding, [138], Theorem 2.

**Theorem 1.3.** (Hoeffding's inequality) *Let $X_1, X_2, \ldots, X_n$ be independent random variables, such that $P(a_k \leq X_k \leq b_k) = 1$ for $k = 1, 2, \ldots, n$, and let $S_n$, $n \geq 1$, denote the partial sums. Then*

$$P(S_n - E\,S_n > x) \leq \exp\left\{ -\frac{2x^2}{\sum_{k=1}^{n}(b_k - a_k)^2} \right\},$$
$$P(|S_n - E\,S_n| > x) \leq 2\exp\left\{ -\frac{2x^2}{\sum_{k=1}^{n}(b_k - a_k)^2} \right\}.$$

The next result is a special case of Markov's inequality and has a name of its own.

**Theorem 1.4.** (Chebyshev's inequality)
(i) *Suppose that* $\operatorname{Var} X < \infty$. *Then*

$$P(|X - E\,X| > x) \le \frac{\operatorname{Var} X}{x^2}, \quad x > 0.$$

(ii) *If* $X_1, X_2, \ldots, X_n$ *are independent with mean 0 and finite variances, then*

$$P(|S_n| > x) \le \frac{\sum_{k=1}^{n} \operatorname{Var} X_k}{x^2}, \quad x > 0.$$

(iii) *If, in addition,* $X_1, X_2, \ldots, X_n$ *are identically distributed, then*

$$P(|S_n| > x) \le \frac{n \operatorname{Var} X_1}{x^2}, \quad x > 0.$$

The Chebyshev inequality presupposes finite variance. There exists, however, a variation of the inequality which is suitable when variances do not exist. Namely, for a given sequence $X_1, X_2, \ldots$ we define *truncated* random variables $Y_1, Y_2, \ldots$ as follows. Let

$$Y_n = \begin{cases} X_n, & \text{when} \quad |X_n| \le b_n, \\ c, & \text{otherwise.} \end{cases} \tag{1.1}$$

Here $\{b_n, n \ge 1\}$ is a sequence of positive reals and $c$ some constant.

Typical cases are $b_n = b$, $b_n = n^\alpha$ for some $\alpha > 0$, and $c = 0$, $c = M$ for some suitable constant $M$. We shall be more specific as we encounter truncation methods.

**Theorem 1.5.** (The truncated Chebyshev inequality) *Let* $X_1, X_2, \ldots, X_n$ *be independent random variables, let* $Y_1, Y_2, \ldots, Y_n$ *be the truncated sequence, and set* $S_n' = \sum_{k=1}^{n} Y_k$. *Suppose also, for simplicity, that* $c = 0$.
(i) *Then, for* $x > 0$,

$$P(|S_n - E\,S_n'| > x) \le \frac{\sum_{k=1}^{n} \operatorname{Var} Y_k}{x^2} + \sum_{k=1}^{n} P(|X_k| > b_k).$$

(ii) *In particular, if* $X_1, X_2, \ldots, X_n$ *are identically distributed, and* $b_k = b$ *for all* $k$, *then*

$$P(|S_n - E\,S_n'| > x) \le \frac{n \operatorname{Var} Y_1}{x^2} + nP(|X_1| > b).$$

*Proof.* Since the second half of the theorem is a particular case of the first half, we only have to prove the latter.

Toward that end,

$$P(|S_n - E\,S_n'| > x) = P\Big(\{|S_n - E\,S_n'| > x\} \cap \Big\{ \bigcap_{k=1}^{n} \{|X_k| \le b_k\}\Big\}\Big)$$

$$+ P\Big(\{|S_n - E\,S_n'| > x\} \cap \Big\{ \bigcup_{k=1}^{n} \{|X_k| > b_k\}\Big\}\Big)$$

$$\le P(|S_n' - E\,S_n'| > x) + P\Big( \bigcup_{k=1}^{n} \{|X_k| > b_k\}\Big)$$

$$\le \frac{\sum_{k=1}^{n} \operatorname{Var} Y_k}{x^2} + \sum_{k=1}^{n} P(|X_k| > b_k). \qquad \square$$

*Remark 1.1.* If $E\,X = 0$ then, in general, it is *not* true that $E\,Y = 0$. It is, however, true for symmetric random variables.

*Remark 1.2.* Since independence is only used to assert that the variance of a sum equals the sum of the variances, the last two theorems remain true under the weaker assumption that $X_1$, $X_2$, ... are *pair-wise* independent. $\qquad \square$

With a little more work one can obtain the following extension of the Chebyshev inequality pertaining to maximal sums; note that the bound remains the same as for the sums themselves.

**Theorem 1.6.** (The Kolmogorov inequality) *Let $X_1$, $X_2$, ..., $X_n$ be independent random variables with mean $0$ and suppose that $\operatorname{Var} X_k < \infty$ for all $k$. Then, for $x > 0$,*

$$P(\max_{1 \le k \le n} |S_k| > x) \le \frac{\sum_{k=1}^{n} \operatorname{Var} X_k}{x^2}.$$

*In particular, if $X_1$, $X_2$, ..., $X_n$ are identically distributed, then*

$$P(\max_{1 \le k \le n} |S_k| > x) \le \frac{n \operatorname{Var} X_1}{x^2}.$$

*Proof.* For $k = 1, 2, \ldots, n$, set

$$A_k = \{\max_{1 \le j \le k-1} |S_j| \le x, |S_k| > x\}.$$

The idea behind this is that

$$\{\max_{1 \le k \le n} |S_k| > x\} = \bigcup_{k=1}^{n} A_k,$$

and that the sets $\{A_k\}$ are disjoint.

Now,

$$\sum_{k=1}^{n} \mathrm{Var}\, X_k = E(S_n^2) \geq \sum_{k=1}^{n} E(S_n^2 I\{A_k\})$$

$$= \sum_{k=1}^{n} E\big((S_k^2 + 2S_k(S_n - S_k) + (S_n - S_k)^2)I\{A_k\}\big)$$

$$\geq \sum_{k=1}^{n} E\big((S_k^2 + 2S_k(S_n - S_k))I\{A_k\}\big)$$

$$= \sum_{k=1}^{n} E(S_k^2 I\{A_k\}) + 2\sum_{k=1}^{n} E\big((S_n - S_k)S_k I\{A_k\}\big)$$

$$= \sum_{k=1}^{n} E(S_k^2 I\{A_k\}) \geq x^2 \sum_{k=1}^{n} E\, I\{A_k\} = x^2 \sum_{k=1}^{n} P(A_k)$$

$$= x^2 P\Big(\bigcup_{k=1}^{n} A_k\Big) = x^2 P(\max_{1\leq k\leq n} |S_k| > x).$$

The expectation of the double product equals 0 because $S_n - S_k$ and $S_k I\{A_k\}$ are independent random variables. □

*Remark 1.3.* Note that a direct application of Chebyshev's inequality to the left-hand side of the statement yields

$$P(\max_{1\leq k\leq n} |S_k| > x) \leq \frac{E(\max_{1\leq k\leq n} |S_k|)^2}{x^2},$$

which is something different. However, with the aid of results for martingales in Chapter 10 (more precisely Theorem 10.9.4) one can show that, in fact, $E(\max_{1\leq k\leq n} |S_k|^2) \leq 4E(S_n^2)$, that is, one is a factor 4 off the Kolmogorov inequality. □

**Theorem 1.7.** (The "other" Kolmogorov inequality) *Let $X_1$, $X_2$, ..., $X_n$ be independent random variables with mean 0, and such that, for some constant $A > 0$,*

$$\sup_{n} |X_n| \leq A. \tag{1.2}$$

*Then*

$$P(\max_{1\leq k\leq n} |S_k| > x) \geq 1 - \frac{(x + A)^2}{\sum_{k=1}^{n} \mathrm{Var}\, X_k}.$$

*Proof.* Let $\{A_k, 1 \leq k \leq n\}$ be given as in the previous proof, set

$$B_k = \{\max_{1\leq j\leq k} |S_k| \leq x\}, \quad \text{for} \quad k = 1, 2, \ldots, n,$$

and note that, for all $k$,

$$A_k \bigcap B_k = \emptyset \quad \text{and} \quad \bigcup_{j=1}^{k} A_j = B_k^c.$$

Thus,

$$S_{k-1}I\{B_{k-1}\} + X_k I\{B_{k-1}\} = S_k I\{B_{k-1}\} = S_k I\{B_k\} + S_k I\{A_k\}. \quad (1.3)$$

Squaring and taking expectations in the left-most equality yields

$$\begin{aligned}
E(S_k I\{B_{k-1}\})^2 &= E(S_{k-1}I\{B_{k-1}\} + X_k I\{B_{k-1}\})^2 \\
&= E(S_{k-1}I\{B_{k-1}\})^2 + E(X_k I\{B_{k-1}\})^2 \\
&\quad + 2E(S_{k-1}I\{B_{k-1}\}X_k I\{B_{k-1}\}) \\
&= E(S_{k-1}I\{B_{k-1}\})^2 + \operatorname{Var} X_k P(B_{k-1}) \quad (1.4)
\end{aligned}$$

by independence. The same procedure with the right-most equality yields

$$\begin{aligned}
E(S_k I\{B_{k-1}\})^2 &= E(S_k I\{B_k\} + S_k I\{A_k\})^2 \\
&= E(S_k I\{B_k\})^2 + E(S_k I\{A_k\})^2 \\
&\quad + 2E(S_k I\{B_k\}S_k I\{A_k\}) \\
&= E(S_k I\{B_k\})^2 + E\big(S_{k-1}I\{A_k\} + X_k I\{A_k\}\big)^2 \\
&\leq E(S_k I\{B_k\})^2 + (x + A)^2 P(A_k), \quad (1.5)
\end{aligned}$$

where the last inequality is due to the fact that $|S_{k-1}|I\{A_k\} < x$ and (1.2).

Joining (1.4) and (1.5), upon noticing that $B_k \supset B_n$ for all $k$, now shows that

$$P(B_n)\operatorname{Var} X_k \leq E(S_k I\{B_k\})^2 - E(S_{k-1}I\{B_{k-1}\})^2 + (x + A)^2 P(A_k),$$

so that, after summation and telescoping,

$$P(B_n) \sum_{k=1}^{n} \operatorname{Var} X_k \leq E(S_n I\{B_n\})^2 + (x + A)^2 P\Big( \bigcup_{k=1}^{n} A_k \Big)$$

$$\leq x^2 P(B_n) + (x + A)^2 P(B_n^c) \leq (x + A)^2.$$

The conclusion follows.    □

*Remark 1.4.* If $E X_n \neq 0$, then $\sup_n |X_n - E X_n| \leq 2A$, so that,

$$P(\max_{1 \leq k \leq n} |S_k - E S_k| > x) \geq 1 - \frac{(x + 2A)^2}{\sum_{k=1}^{n} \operatorname{Var} X_k}. \qquad \square$$

The common feature with the results above is that tail probabilities are estimated by moments or sums of variances. In certain convergence results in which one aims at necessary and sufficient conditions, converse inequalities will be of great help. A rewriting of the conclusion of Theorem 1.7 produces one such result. However, by picking from the last step in the proof of the theorem, omitting the final inequality, we have

$$P(\max_{1\leq k\leq n}|S_k|\leq x)\sum_{k=1}^{n}\operatorname{Var}X_k \leq x^2 P(\max_{1\leq k\leq n}|S_k|\leq x)$$
$$+(x+A)^2 P(\max_{1\leq k\leq n}|S_k|>x),$$

which we reshuffle into an estimate of the sum of the variances, and register as a separate result.

**Corollary 1.1.** *Let* $X_1, X_2, \ldots, X_n$ *be independent random variables with* $E\,X_n = 0$, *such that, for some constant* $A > 0$,

$$\sup_{n}|X_n|\leq A. \tag{1.6}$$

*Then*

$$\sum_{k=1}^{n}\operatorname{Var}X_k \leq x^2 + \frac{(x+A)^2 P(\max_{1\leq k\leq n}|S_k|>x)}{P(\max_{1\leq k\leq n}|S_k|\leq x)},$$

*In particular, if* $P(\max_{1\leq k\leq n}|S_k|>x)<\delta$, *for some* $\delta \in (0,1)$, *then*

$$\sum_{k=1}^{n}\operatorname{Var}X_k \leq x^2 + (x+A)^2\frac{\delta}{1-\delta}. \qquad \square$$

Following is a generalization of the Kolmogorov inequality to weighted sums.

**Theorem 1.8.** (The Hájek-Rényi inequality) *Let* $X_1, X_2, \ldots, X_n$ *be independent random variables with* $E\,X_n = 0$, *and let* $\{c_k, 0\leq k\leq n\}$ *be positive, non-increasing real numbers. Then*

$$P(\max_{1\leq k\leq n}c_k|S_k|>x)\leq \frac{\sum_{k=1}^{n}c_k^2\operatorname{Var}X_k}{x^2}, \qquad x>0.$$

*Remark 1.5.* If $c_k = 1$ for all $k$, the inequality reduces to the Kolmogorov inequality.

*Remark 1.6.* Note the difference between

$$\max_{1\leq k\leq n}c_k|S_k| \quad \text{and} \quad \max_{1\leq k\leq n}\sum_{j=1}^{n}c_j|X_j|,$$

and that a direct application of the Kolmogorov inequality to the latter maximum provides the same upper bound. $\qquad \square$

*Proof.* The proof follows the basic pattern of the proof of Theorem 1.6. Thus, for $k = 1, 2, \ldots, n$, we set

$$A_k = \{\max_{1 \leq j \leq k-1} |c_j S_j| \leq x, |c_k S_k| > x\},$$

so that

$$\{\max_{1 \leq k \leq n} |c_k S_k| > x\} = \bigcup_{k=1}^{n} A_k,$$

with $\{A_k\}$ being disjoint sets.

By partial summation and independence,

$$\sum_{k=1}^{n} c_k^2 \operatorname{Var} X_k = \sum_{k=1}^{n} c_k^2 (\operatorname{Var} S_k - \operatorname{Var} S_{k-1})$$

$$= \sum_{k=1}^{n-1} (c_k^2 - c_{k+1}^2) \operatorname{Var} S_k + c_n^2 \operatorname{Var} S_n$$

$$= \sum_{k=1}^{n-1} (c_k^2 - c_{k+1}^2) E\, S_k^2 + c_n^2 E\, S_n^2.$$

Moreover, $E(S_k^2 \{A_j\}) \geq E(S_j^2 \{A_j\})$ for $k \geq j$, precisely as in the proof of Theorem 1.6. Rerunning that proof with minor modifications we thus arrive at

$$\sum_{k=1}^{n} c_k^2 \operatorname{Var} X_k \geq \sum_{j=1}^{n} \sum_{k=1}^{n} c_k^2 \operatorname{Var} X_k I\{A_j\}$$

$$= \sum_{j=1}^{n} \sum_{k=1}^{n-1} (c_k^2 - c_{k+1}^2) E(S_k^2 I\{A_j\}) + \sum_{j=1}^{n} c_n^2 E(S_n^2 I\{A_j\})$$

$$\geq \sum_{j=1}^{n} \sum_{k=j}^{n-1} (c_k^2 - c_{k+1}^2) E(S_k^2 I\{A_j\}) + \sum_{j=1}^{n} c_n^2 E(S_n^2 I\{A_j\})$$

$$\geq \sum_{j=1}^{n} \sum_{k=j}^{n-1} (c_k^2 - c_{k+1}^2) E(S_j^2 I\{A_j\}) + \sum_{j=1}^{n} c_n^2 E(S_j^2 I\{A_j\})$$

$$\geq \sum_{j=1}^{n} \sum_{k=j}^{n-1} (c_k^2 - c_{k+1}^2) \frac{x^2}{c_j^2} E I\{A_j\} + \sum_{j=1}^{n} c_n^2 \frac{x^2}{c_j^2} E I\{A_j\}$$

$$= \sum_{j=1}^{n} \sum_{k=j}^{n-1} (c_k^2 - c_{k+1}^2) \frac{x^2}{c_j^2} P(A_j) + \sum_{j=1}^{n} c_n^2 \frac{x^2}{c_j^2} P(A_j)$$

$$= x^2 \sum_{j=1}^{n} P(A_j) = x^2 P\left(\bigcup_{j=1}^{n} A_j\right) = x^2 P(\max_{1 \leq k \leq n} |c_k S_k| > x). \qquad \square$$

2 Moment Inequalities    127

# 2 Moment Inequalities

Next, some inequalities that relate moments of sums to sums of moments. Note that we do not assume independence between the summands.

**Theorem 2.1.** *Let $r > 0$. Suppose that $E|X|^r < \infty$ and $E|Y|^r < \infty$. Then*

$$E|X + Y|^r \leq 2^r(E|X|^r + E|Y|^r).$$

*Proof.* Set $x = X(\omega)$ and $y = Y(\omega)$. The triangle inequality and Lemma A.5.1 together yield

$$E|X + Y|^r \leq E(|X| + |Y|)^r \leq 2^r(E|X|^r + E|Y|^r). \qquad \square$$

Although the inequality is enough for many purposes, a sharper one can be obtained as follows.

**Theorem 2.2.** (The $c_r$-inequality) *Let $r > 0$. Suppose that $E|X|^r < \infty$ and $E|Y|^r < \infty$. Then*

$$E|X + Y|^r \leq c_r(E|X|^r + E|Y|^r),$$

*where $c_r = 1$ when $r \leq 1$ and $c_r = 2^{r-1}$ when $r \geq 1$.*

*Proof.* Set $x = X(\omega)$ and $y = Y(\omega)$ for $\omega \in \Omega$. By the triangle inequality and the second inequality of Lemma A.5.1,

$$E|X + Y|^r \leq E(|X| + |Y|)^r \leq E|X|^r + E|Y|^r,$$

which establishes the inequality for the case $0 < r \leq 1$.

For $r \geq 1$ the desired inequality follows the same procedure with the second inequality of Lemma A.5.1 replaced by the third one. $\qquad \square$

The last two results tell us that the if the summands are integrable, then so is the sum. The integrability assumption is of course superfluous, but without it the right-hand side would be infinite, and the result would be void.

There is no general converse to that statement. If both variables are non-negative the converse is trivial, since, then, $X \leq X + Y$. However, let $X$ and $Y$ be Cauchy-distributed, say, and such that $Y = -X$. Then $X + Y$ equals 0 and thus has moments of all orders, but $X$ and $Y$ do not.

However, for independent summands independent there exists a converse.

**Theorem 2.3.** *Let $r > 0$. If $E|X + Y|^r < \infty$ and $X$ and $Y$ are independent, then $E|X|^r < \infty$ and $E|Y|^r < \infty$.*

*Proof.* By assumption,

$$\int_{-\infty}^{\infty} \int_{-\infty}^{\infty} |x + y|^r \, dF(x) dF(y) < \infty.$$

The finiteness of the double integral implies that the inner integral must be finite for at least one $y$ (in fact, for almost all $y$). Therefore, pick $y$, such that

$$\int_{-\infty}^{\infty} |x+y|^r \, dF(x) < \infty,$$

which means that

$$E|X+y|^r < \infty.$$

An application of the $c_r$-inequality we proved a minute ago asserts that

$$E|X|^r \leq E|X+y|^r + |y|^r < \infty.$$

The integrability of $Y$ follows similarly, or, alternatively, via another application of the $c_r$-inequality.

An alternative is to argue with the aid of Theorem 2.12.1. If both variables are non-negative or non-positive the converse is trivial as mentioned above. Thus, suppose that $X$, say, takes values with positive probability on both sides of the origin. Then there exists $\alpha \in (0,1)$, such that

$$P(X > 0) \geq \alpha \quad \text{and} \quad P(X < 0) \geq \alpha.$$

Moreover,

$$\begin{aligned}
P(X+Y > n) &\geq P(\{Y > n\} \cap \{X > 0\}) \\
&= P(Y > n) \cdot P(X > 0) \geq P(Y > n) \cdot \alpha,
\end{aligned}$$

which, together with the analog for the negative tail, yields

$$P(|X+Y| > n) \geq P(|Y| > n) \cdot \alpha,$$

so that

$$\sum_{n=1}^{\infty} n^{r-1} P(|Y| > n) \leq \frac{1}{\alpha} \sum_{n=1}^{\infty} n^{r-1} P(|X+Y| > n) < \infty.$$

An application of Theorem 2.12.1 proves that $E|Y|^r < \infty$, after which we can lean on the $c_r$-inequality to conclude that $E|X|^r < \infty$.    □

Before proceeding we introduce another piece of notation. For a random variable $X$ whose moment of order $r > 0$ is finite we set

$$\|X\|_r = \left(E|X|^r\right)^{1/r}. \tag{2.1}$$

The notation indicates that $\|X\|_r$ is a norm in some space. That this is, in fact, the case when $r \geq 1$ will be seen after the proof of the Minkowski inequality in Theorem 2.6. For convenience we still keep the notation also when $0 < r < 1$, but in that case it is *only* notation.

**Theorem 2.4.** (The Hölder inequality) *Let $p^{-1} + q^{-1} = 1$. If $E|X|^p < \infty$ and $E|Y|^q < \infty$, then*

$$|E\,XY| \leq E|XY| \leq \|X\|_p \cdot \|Y\|_q.$$

*Proof.* Only the second inequality requires a proof. Once again the point of departure is an elementary inequality for positive reals.

Let $\omega \in \Omega$, put $x = |X(\omega)|/\|X\|_p$ and $y = |Y(\omega)|/\|Y\|_q$, insert this into Lemma A.5.2, and take expectations. Then

$$E\frac{|X|}{\|X\|_p} \cdot \frac{|Y|}{\|Y\|_q} \leq \frac{1}{p}E\Big(\frac{|X|^p}{\|X\|_p^p}\Big) + \frac{1}{q}E\Big(\frac{|Y|^q}{\|Y\|_q^q}\Big) = \frac{1}{p} + \frac{1}{q} = 1. \qquad \square$$

A particular case, or a corollary, is obtained by putting $Y = 1$ a.s.

**Theorem 2.5.** (The Lyapounov inequality) *For $0 < r \leq p$,*

$$\|X\|_r \leq \|X\|_p.$$

*Proof.* We use the Hölder inequality with $X$ replaced by $|X|^p$, $Y$ by 1, and $p$ by $r/p$ (and $q$ by $1 - r/p$). Then

$$E|X|^p = E|X|^p \cdot 1 \leq \||X|^p\|_{r/p} \cdot 1$$
$$= \big(E(|X|^p)^{r/p}\big)^{p/r} = (E|X|^r)^{p/r} = (\|X\|_r)^p,$$

which yields the desired conclusion. $\qquad \square$

The Hölder inequality concerns moments of products. The following, triangular type of inequality concerns sums.

**Theorem 2.6.** (The Minkowski inequality) *Let $p \geq 1$. Suppose that $X$ and $Y$ are random variables, such that $E|X|^p < \infty$ and $E|Y|^p < \infty$. Then*

$$\|X + Y\|_p \leq \|X\|_p + \|Y\|_p.$$

*Proof.* If $\|X + Y\|_p = 0$ there is nothing to prove. We therefore suppose in the following that $\|X + Y\|_p > 0$. By the triangular and Hölder inequalities, and by noticing that $(p-1)q = p$ and $p/q = p - 1$, we obtain

$$\|X + Y\|_p^p = E|X + Y|^{p-1}|X + Y|$$
$$\leq E|X + Y|^{p-1}|X| + E|X + Y|^{p-1}|Y|$$
$$\leq \||X + Y|^{p-1}\|_q \|X\|_p + \||X + Y|^{p-1}\|_q \|Y\|_p$$
$$= \big(E|X + Y|^{(p-1)q}\big)^{1/q}(\|X\|_p + \|Y\|_p)$$
$$= \|X + Y\|_p^{p/q}(\|X\|_p + \|Y\|_p).$$

Dividing the extreme members by $\|X + Y\|_p^{p/q}$ finishes the proof. $\qquad \square$

## 3 Covariance; Correlation

The Hölder inequality with $p = q = 2$ yields another celebrated inequality.

**Theorem 3.1.** (The Cauchy-Schwarz inequality) *Suppose that $X$ and $Y$ have finite variances. Then*

$$|E\,XY| \le E|XY| \le \|X\|_2 \cdot \|Y\|_2 = \sqrt{E\,X^2 \cdot E\,Y^2}.$$

Recall from Chapter 2 that the variance is a measure of dispersion. The *covariance*, which we introduce next, measures a kind of "joint spread"; it measures the extent of covariation of two random variables.

**Definition 3.1.** *Let $(X,Y)'$ be a random vector. The* covariance *of $X$ and $Y$ is*

$$\mathrm{Cov}\,(X,Y) = E(X - E\,X)(Y - E\,Y) \quad (= E\,XY - E\,X E\,Y). \qquad \square$$

The covariance measures the interdependence of $X$ and $Y$ in the sense that it is large and positive when $X$ and $Y$ are both large and of the same sign; it is large and negative if $X$ and $Y$ are both large and of opposite signs. Since, as is easily checked, $\mathrm{Cov}\,(aX, bY) = ab\,\mathrm{Cov}\,(X,Y)$, for $a, b \in \mathbb{R}$, the covariance is not scale invariant, which implies that the covariance may be "large" only because the variables themselves are large. Changing from millimeters to kilometers changes the covariance drastically. A better measure of interdependence is the *correlation coefficient*.

**Definition 3.2.** *Let $(X,Y)'$ be a random vector. The* correlation coefficient *of $X$ and $Y$ is*

$$\rho_{X,Y} = \frac{\mathrm{Cov}\,(X,Y)}{\sqrt{\mathrm{Var}\,X \cdot \mathrm{Var}\,Y}}.$$

*The random variables $X$ and $Y$ are* uncorrelated *iff*

$$\rho_{X,Y} = 0. \qquad \square$$

An application of the Cauchy-Schwarz inequality shows that

$$|\mathrm{Cov}\,(X,Y)| \le \sqrt{\mathrm{Var}\,X \cdot \mathrm{Var}\,Y} \quad \text{or, equivalently, that} \quad |\rho_{X,Y}| \le 1.$$

In particular, the covariance is well defined whenever the variances are finite.

**Exercise 3.1.** Check that the correlation coefficient is scale invariant. $\qquad \square$

The next result tells us that uncorrelatedness is a weaker concept than independence.

**Theorem 3.2.** *Let $(X,Y)'$ be a random vector. If $X$ and $Y$ are independent, then they are uncorrelated, viz.,*

$$E\,XY = E\,X \cdot E\,Y \quad \Longrightarrow \quad \mathrm{Cov}\,(X,Y) = 0.$$

*Proof.* By independence and Fubini's theorem, Theorem 2.9.1,

$$E\, XY = \int_{-\infty}^{\infty} \int_{-\infty}^{\infty} xy \, dF_{X,Y}(x,y) = \int_{-\infty}^{\infty} \int_{-\infty}^{\infty} xy \, dF_X(x) dF_Y(y)$$

$$= \int_{-\infty}^{\infty} x \, dF_X(x) \cdot \int_{-\infty}^{\infty} y \, dF_Y(y) = E\,X \cdot E\,Y. \qquad \square$$

In order to see that the implication is strict we return for a moment to Example 2.3.1 – picking a point uniformly on the unit disc. By symmetry (or by direct computation) we find that $E\,X = E\,Y = E\,XY = 0$, which shows that the $X$- and $Y$-coordinates are uncorrelated. On the other hand, since a large value of $X$ *forces* $Y$ to be small in order for the point to stay inside the circle, it seems reasonable to guess that the coordinates are *not* independent. That this is, indeed, the case follows from the fact that the joint density is not equal to the product of the marginal ones (recall (2.3.1));

$$\frac{1}{\pi} \neq \frac{2}{\pi}\sqrt{1-x^2} \cdot \frac{2}{\pi}\sqrt{1-y^2}.$$

We have thus shown that the coordinates are uncorrelated, but *not* independent.

# 4 Interlude on $L^p$-spaces

The $L^p$-spaces are defined as the set of measurable functions $f$ such that the integral $\int |f(x)|^p \, dx$ is convergent. One defines a *norm* on the space by $\|f\|_p = (\int |f(x)|^p \, dx)^{1/p}$. However, this only works for $p \geq 1$, since for $0 < p < 1$ the object $\|f\|_p$ does not fulfill the requirements of a norm, namely, the "triangle inequality" fails to hold.

The probabilistic version of these spaces is to consider the set of random variables with a finite moment of order $p$ and to define the norm of a random variable $\| \cdot \|_p$ as

$$\|X\|_p = \left(E|X|^p\right)^{1/p}.$$

We notice that $\|X\|_p \geq 0$ with equality only when $X = 0$ a.s., and that the Minkowski inequality, Theorem 2.6 is the desired "triangle inequality"; however, only when $p \geq 1$. Moreover, we have homogeneity in the sense that if $\lambda \in \mathbb{R}$, then $\|\lambda X\|_p = |\lambda| \|X\|_p$. The linearity and the Minkowski inequality together show that if $X$ and $Y$ belong to $L^p$ for some $p \geq 1$, then so does $\lambda_1 X + \lambda_2 Y$. Thus, $\| \cdot \|_p$ is, indeed, a norm (for $p \geq 1$).

In addition, one can define a *distance* between two random variables $X$ and $Y$ by $d(X,Y) = \|X - Y\|_p$. It is readily checked that $d(X,Y) \geq 0$ with equality only when $X = Y$ a.s., and that $d(X,Z) \leq d(X,Y) + d(Y,Z)$, the latter, once again, being a consequence of the Minkowski inequality, so that the $L^p$-spaces are *metric spaces* (when $p \geq 1$). In Section 5.12 we shall show that the so-called Cauchy convergence always implies convergence in this space, which means that the spaces are *complete*.

The $L^p$-spaces are examples of *Banach spaces*. All of them have a *dual space* that turns out to be another such space: the dual of $L^p$ is $L^q$, where $p$ and $q$ are what is called *conjugate exponents*, which means that they are related via $p^{-1} + q^{-1} = 1$ (as, for example, in the Hölder inequality, Theorem 2.4 above).

An inspection of the relation between the exponents reveals that the case $p = q = 2$ is special; $L^2$ is self-dual. Moreover, one can define an inner product with certain properties, among them that the inner product of an element with itself equals the square of its norm. Banach spaces with a norm that can be derived via an inner product are called *Hilbert spaces*. The covariance defined in the previous section plays the role of the inner product.

All of this and much more belongs to the area of functional analysis, which is an important branch of mathematics. The aim of this subsection was merely to illustrate the connection to probability theory, how the theory can be described in a probabilistic context, with random variables instead of measurable functions, and moments instead of integrals, and so on.

# 5 Convexity

For the definition of convexity and some basic facts we refer to Section A.5.

**Theorem 5.1.** (Jensen's inequality) *Let $X$ be a random variable, $g$ a convex function, and suppose that $X$ and $g(X)$ are integrable. Then*

$$g(E\,X) \leq E\,g(X).$$

*Proof.* If, in addition, $g$ is twice differentiable we know that the second derivative is always positive. Therefore, by Taylor expansion, for any $x$,

$$g(x) \geq g(E\,X) + (x - E\,X)g'(E\,X).$$

Putting $x = X(\omega)$, and taking expectations yields the desired conclusion.

In the general case one notices that the chord between $g(x)$ and $g(E\,X)$ lies above the curve joining these points, after one proceeds similarly.     □

**Exercise 5.1.** Complete the details in the general case.     □

The first example is $g(x) = |x|^p$ for $p \geq 1$, in which case the inequality states that $(E|X|)^p \leq E|X|^p$. A variation is $(E|X|^r)^p \leq E|X|^{rp}$, for $r > 0$, which ran be restated as $\|X\|_r \leq \|X\|_{rp}$, and thus reproves Lyapounov's inequality, Theorem 2.5.

A simple-minded example is obtained for $g(x) = x^2$, in which case Jensen's inequality amounts to the statement that $\operatorname{Var} X \geq 0$.

# 6 Symmetrization

For various reasons that will be more apparent later it is often easier to prove certain theorems for symmetric random variables. A common procedure in such cases is to begin by proving the theorem under the additional assumption of symmetry and then to remove that assumption – to desymmetrize. Important tools in this context are, as a consequence, relations between ordinary random variables and symmetric ones, more precisely between ordinary random variables and a kind of associated symmetric ones. The connection is made clear via the following definitions, which is then followed by a number of properties.

**Definition 6.1.** *Let $X$ and $X' \stackrel{d}{=} X$ be independent random variables. We call $X^s = X - X'$ the* symmetrized *random variable.*

**Definition 6.2.** *The* median, med $(X)$, *of a random variable $X$ is a real number satisfying*

$$P(X \leq \mathrm{med}\,(X)) \geq \frac{1}{2} \quad \text{and} \quad P(X \geq \mathrm{med}\,(X)) \geq \frac{1}{2}. \qquad \square$$

A median is a kind of center of the distribution in the sense that (at least) half of the probability mass lies to the left of it and (at least) half of it to the right. Medians *always exist* in contrast to expected values which need not.

The median is unique for absolutely continuous random variables. However, it need not be unique in general, and typically not for several discrete distributions.

If moments exist one can obtain bounds for the median as follows.

**Proposition 6.1.** (i) *Let $a > 0$. If $P(|X| > a) < 1/2$, then $|\mathrm{med}\,(X)| \leq a$.*
(ii) *If $E|X|^r < \infty$ for some $r \in (0, 1)$, then*

$$|\mathrm{med}\,(X)| \leq 2^{1/r}\|X\|_r.$$

(iii) *If $E|X|^r < \infty$ for some $r \geq 1$, then*

$$|\mathrm{med}\,(X) - E\,X| \leq 2^{1/r}\|X\|_r.$$

*In particular, if* $\mathrm{Var}\,X = \sigma^2 < \infty$, *then*

$$|\mathrm{med}\,(X) - E\,X| \leq \sigma\sqrt{2}.$$

*Proof.* (i): By assumption, $P(X < -a) < 1/2$, so that, by definition, the median must be $\geq -a$. Similarly for the other tail; the median must be $\leq a$.
(ii): Using Markov's inequality we find that

$$P(|X| > 2^{1/r}\|X\|_r) \leq \frac{E|X|^r}{(2^{1/r}\|X\|_r)^r} = \frac{1}{2}.$$

(iii): The same proof with $X$ replaced by $X - E\,X$.                    $\square$

Next we present two propositions that relate tail probabilities of random variables to tail probabilities for their symmetrizations.

**Proposition 6.2.** (Weak symmetrization inequalities) *For every x and a,*

$$\frac{1}{2}P(X - \text{med}\,(X) \geq x) \leq P(X^s \geq x),$$

$$\frac{1}{2}P(|X - \text{med}\,(X)| \geq x) \leq P(|X^s| \geq x) \leq 2P(|X - a| \geq x/2).$$

*In particular,*

$$\frac{1}{2}P(|X - \text{med}\,(X)| \geq x) \leq P(|X^s| \geq x) \leq 2P(|X - \text{med}\,(X)| \geq x/2).$$

*Proof.* Since $X \overset{d}{=} X'$, med $(X)$ is also a median for $X'$. Thus,

$$\begin{aligned}
P(X^s \geq x) &\geq P(\{X - \text{med}\,(X) \geq x\} \cap \{X' - \text{med}\,(X') \leq 0\}) \\
&= P(X - \text{med}\,(X) \geq x)P(X' - \text{med}\,(X') \leq 0) \\
&\geq P(X - \text{med}\,(X) \geq x) \cdot \frac{1}{2},
\end{aligned}$$

which proves the first assertion. The left-hand inequality in the second assertion follows by applying the first one to $-X$ and addition. The right-most inequality follows via

$$\begin{aligned}
P(|X^s| \geq x) &= P(|X - a - (X' - a)| \geq x) \\
&\leq P(|X - a| \geq x/2) + P(|X' - a| \geq x/2) \\
&= 2P(|X - a| \geq x/2). \qquad \square
\end{aligned}$$

The adjective *weak* in the name of the symmetrization inequalities suggests that there exist strong ones too. The proofs of these consist of a modification of the proof of the weak symmetrization inequalities in a manner related to the extension from Chebyshev's inequality to the Kolmogorov inequality.

**Proposition 6.3.** (Strong symmetrization inequalities) *For every x and all sequences $\{a_k, 1 \leq k \leq n\}$,*

$$\frac{1}{2}P(\max_{1 \leq k \leq n}(X_k - \text{med}\,(X_k)) \geq x) \leq P(\max_{1 \leq k \leq n} X_k^s \geq x)$$

$$\frac{1}{2}P(\max_{1 \leq k \leq n}|X_k - \text{med}\,(X_k)| \geq x) \leq P(\max_{1 \leq k \leq n}|X_k^s| \geq x)$$

$$\leq 2P(\max_{1 \leq k \leq n}|X_k - a_k| \geq x/2).$$

*In particular*

$$\frac{1}{2}P(\max_{1 \leq k \leq n}|X_k - \text{med}\,(X_k)| \geq x) \leq P(\max_{1 \leq k \leq n}|X_k^s| \geq x)$$

$$\leq 2P(\max_{1 \leq k \leq n}|X_k - \text{med}\,(X_k)| \geq x/2).$$

*Proof.* Set

$$A_k = \{\max_{1 \le j \le k-1}(X_j - \mathrm{med}\,(X_j)) < x, X_k - \mathrm{med}\,(X_k) \ge x\},$$
$$B_k = \{X_k' - \mathrm{med}\,(X_k') \le 0\}, \quad \text{and}$$
$$C_k = \{\max_{1 \le j \le k-1} X_j^s < x, X_k^s \ge x\}.$$

Then,

- $\{A_k,\, 1 \le k \le n\}$ are disjoint;
- $\bigcup_{k=1}^n A_k = \{\max_{1 \le k \le n}(X_k - \mathrm{med}\,(X_k)) \ge x\}$;
- $\{C_k,\, 1 \le k \le n\}$ are disjoint;
- $\bigcup_{k=1}^n C_k = \{\max_{1 \le k \le n} X_k^s \ge x\}$;
- $A_k \cap B_k \subset C_k, \quad k = 1, 2, \ldots, n$;
- $A_k$ and $B_k$ are independent.

The conclusion follows upon observing that

$$P\Big(\bigcup_{k=1}^n C_k\Big) = \sum_{k=1}^n P(C_k) \ge \sum_{k=1}^n P(A_k \cap B_k) = \sum_{k=1}^n P(A_k)P(B_k)$$
$$\ge \sum_{k=1}^n P(A_k) \cdot \frac{1}{2} = \frac{1}{2} P\Big(\bigcup_{k=1}^n A_k\Big). \qquad \square$$

*Remark 6.1.* Whereas the weak symmetrization inequalities are weak in the sense that they concern distributional properties, the strong ones are called strong, since the whole history of the process so far is involved. $\qquad \square$

As a first application we use the weak symmetrization inequalities to relate moments of random variables to their symmetrized counterparts.

**Proposition 6.4.** *For any $r > 0$ and $a$,*

$$\frac{1}{2} E|X - \mathrm{med}\,(X)|^r \le E|X^s|^r \le 2c_r E|X - a|^r.$$

*In particular,*

$$\frac{1}{2c_r} E|X|^r \le E|X^s|^r + |\mathrm{med}\,(X)|^r \le 2c_r E|X|^r + |\mathrm{med}\,(X)|^r,$$

*so that*

$$E|X|^r < \infty \quad \Longleftrightarrow \quad E|X^s|^r < \infty.$$

*Proof.* To prove the first double inequality we suppose that $E|X|^r < \infty$; otherwise there is nothing to prove.

The left-hand inequality is obtained via Proposition 6.2 and Theorem 2.12.1. The right-hand inequality follows by applying the $c_r$-inequality, viz.

$$E|X^s|^r = E|X - a - (X' - a)|^r$$
$$\leq c_r(E|X - a|^r + E|X' - a|^r) = 2c_r E|X - a|^r.$$

The second double inequality follows from the first one via the $c_r$-inequality, and by putting $a = 0$, from which the equivalence between the integrability of $X$ and $X^s$ is now immediate. Note also that this, alternatively, follows directly from the $c_r$-inequality and Theorem 2.3.                    □

A useful variation of the relation between moments of random variables and their symmetrizations can be obtained from the following result in which the assumption that the mean equals 0 is crucial; note that no assumptions about centerings were made in the previous proposition.

**Proposition 6.5.** *Let $r \geq 1$, let $X$ and $Y$ be independent random variables, suppose that $E|X|^r < \infty$, $E|Y|^r < \infty$, and that $EY = 0$. Then*

$$E|X|^r \leq E|X + Y|^r.$$

*In particular, if $X$ has mean 0, and $X^s$ is the symmetrized random variable, then*

$$E|X|^r \leq E|X^s|^r.$$

*Proof.* Let $x$ be a real number. Then, since $Y$ has mean 0,

$$|x|^r = |x + EY|^r \leq E|x + Y|^r = \int_{-\infty}^{\infty} |x + y|^r \, dF_Y(y),$$

so that

$$E|X|^r \leq \int_{-\infty}^{\infty} \int_{-\infty}^{\infty} |x + y|^r \, dF_Y(y) \, dF_X(x) = E|X + Y|^r.$$

The particular case follows by letting $Y = -X'$, where $X' \overset{d}{=} X$ and $X'$ is independent of $X$.                    □

As mentioned before, symmetric random variables are, in general, (much) easier to handle. As an example, suppose that $E|X + Y|^r < \infty$ for some $r > 0$ and, in addition, that $X$ and $Y$ are independent and that $Y$ (say) *is symmetric*. Then, since $Y \overset{d}{=} -Y$, it follows that $X + Y \overset{d}{=} X + (-Y) = X - Y$. Exploiting this fact, together with the elementary identity

$$X = \frac{1}{2}((X + Y) + (X - Y)),$$

the Minkowski inequality when $r \geq 1$, and the $c_r$-inequality when $0 < r < 1$, shows that

$$\|X\|_r \le \frac{1}{2}\big(\|X+Y\|_r + \|X-Y\|_r\big) = \|X+Y\|_r,$$

and that

$$E|X|^r \le \frac{1}{2}\big(E|X+Y|^r + E|X-Y|^r\big) = E|X+Y|^r,$$

respectively.

This provides, under slightly different assumptions, an alternative proof of Theorem 2.3 and Proposition 6.5.

The last two inequalities suggest a connection to some kind of a generalized parallelogram identity.

**Theorem 6.1.** *Let* $1 \le r \le 2$.
(i) *Let* $X$ *and* $Y$ *be random variables with finite moments of order* $r$. *Then*

$$E|X+Y|^r + E|X-Y|^r \le 2\big(E|X|^r + E|Y|^r\big).$$

*If, in addition,* $X$ *and* $Y$ *are independent and* $Y$, *say, has a symmetric distribution, then*

$$E|X+Y|^r \le E|X|^r + E|Y|^r.$$

(ii) *If* $X_1, X_2, \ldots, X_n$ *are independent, symmetric random variables with finite moments of order* $r$, *then*

$$E\left|\sum_{k=1}^{n} X_k\right|^r \le \sum_{k=1}^{n} E|X_k|^r.$$

(iii) *If* $X_1, X_2, \ldots, X_n$ *are independent random variables with mean* $0$ *and finite moments of order* $r$, *then*

$$E\left|\sum_{k=1}^{n} X_k\right|^r \le 2^r \sum_{k=1}^{n} E|X_k|^r.$$

*Let* $r \ge 2$.
(iv) *Let* $X$ *and* $Y$ *be random variables with finite moments of order* $r$. *Then*

$$E|X|^r + E|Y|^r \le \frac{1}{2}\big(E|X+Y|^r + E|X-Y|^r\big).$$

*If, in addition,* $X$ *and* $Y$ *are independent and* $Y$, *say, has a symmetric distribution, then*

$$E|X|^r + E|Y|^r \le E|X+Y|^r.$$

(v) *If* $X_1, X_2, \ldots, X_n$ *are independent, symmetric random variables with finite moments of order* $r$, *then*

$$\sum_{k=1}^{n} E|X_k|^r \le E\left|\sum_{k=1}^{n} X_k\right|^r.$$

(vi) *If* $X_1$, $X_2$, ..., $X_n$ *are independent random variables with mean* 0 *and finite moments of order* $r$, *then*

$$\sum_{k=1}^{n} E|X_k|^r \le 2^r E \left| \sum_{k=1}^{n} X_k \right|^r.$$

*Proof.* As before, put $x = X(\omega)$ and $y = Y(\omega)$, insert this into Clarkson's inequality, Lemma A.5.3, and take expectations. The first statement then follows. The second one exploits, in addition, the fact that $X + Y \stackrel{d}{=} X - Y$. Conclusion (ii) follows from (i) and induction.

As for (iii), we use Proposition 6.5, the $c_r$-inequality, and (ii) to obtain

$$E \left| \sum_{k=1}^{n} X_k \right|^r \le E \left| \left( \sum_{k=1}^{n} X_k \right)^s \right|^r = E \left| \sum_{k=1}^{n} X_k^s \right|^r \le \sum_{k=1}^{n} E|X_k^s|^r$$

$$\le \sum_{k=1}^{n} 2^{r-1} \left( E|X_k|^r + E|X_k'|^r \right) = 2^r \sum_{k=1}^{n} E|X_k|^r.$$

The proof of the second half of the theorem follows the same procedure with obvious modifications; (iv) and (v) by using the second inequality in Lemma A.5.3 instead of the first one, and (vi) via (v), Proposition 6.5, and the $c_r$-inequality:

$$\sum_{k=1}^{n} E|X_k|^r \le \sum_{k=1}^{n} E|X_k^s|^r \le E \left| \sum_{k=1}^{n} X_k^s \right|^r = E \left| \left( \sum_{k=1}^{n} X_k \right)^s \right|^r$$

$$\le 2^{r-1} \left( E \left| \sum_{k=1}^{n} X_k \right|^r + E \left| \sum_{k=1}^{n} X_k' \right|^r \right) = 2^{r-1} 2 E \left| \sum_{k=1}^{n} X_k \right|^r.$$

Recall that $X_k' \stackrel{d}{=} X_k$ for all $k$ and that they are independent.   □

*Remark 6.2.* Via Fourier analytic methods it has been shown in [7] (in a somewhat more general situation) that the constant $2^r$ can be replaced by 2 in (iii) for $1 \le r \le 2$ ($2^r \le 4$).

*Remark 6.3.* The fact that the inequalities in (i)–(iii) and (iv)–(vi) are the same, except for the reversal of the inequality sign is a consequence of the duality between the spaces $L^p$ and $L^q$, where $p^{-1} + q^{-1} = 1$, which was mentioned in Section 3.4.   □

# 7 Probability Inequalities for Maxima

Another important kind of inequalities relates tail probabilities for the maximal partial sum so far to tail probabilities of the *last* partial sum. This and some related facts are the topic of the present section.

Throughout, $X_1$, $X_2$, ... are independent random variables with partial sums $S_n$, $n \ge 1$.

**Theorem 7.1.** (The Lévy inequalities) *For any $x$,*

$$P(\max_{1\leq k\leq n} (S_k - \operatorname{med}(S_k - S_n)) > x) \leq 2P(S_n > x),$$

$$P(\max_{1\leq k\leq n} |S_k - \operatorname{med}(S_k - S_n)| > x) \leq 2P(|S_n| > x).$$

*In particular, in the symmetric case,*

$$P(\max_{1\leq k\leq n} S_k > x) \leq 2P(S_n > x),$$

$$P(\max_{1\leq k\leq n} |S_k| > x) \leq 2P(|S_n| > x).$$

*Proof.* For $k = 1, 2, \ldots, n$, set

$$A_k = \{\max_{1\leq j\leq k-1} (S_j - \operatorname{med}(S_j - S_n)) \leq x, S_k - \operatorname{med}(S_k - S_n) > x\},$$

$$B_k = \{S_n - S_k - \operatorname{med}(S_n - S_k) \geq 0\}.$$

The sets $\{A_k\}$ are disjoint, $A_k$ and $B_k$ are independent since they contain no common summands, $P(B_k) \geq 1/2$, and

$$\{S_n > x\} \supset \bigcup_{k=1}^{n} \{A_k \bigcap B_k\}.$$

Consequently,

$$P(S_n > x) \geq \sum_{k=1}^{n} P(A_k \bigcap B_k) = \sum_{k=1}^{n} P(A_k)P(B_k)$$

$$\geq \sum_{k=1}^{n} P(A_k)\frac{1}{2} = \frac{1}{2}P\Big(\bigcup_{k=1}^{n} A_k\Big) = \frac{1}{2}P(\max_{1\leq k\leq n} S_k > x),$$

which proves the first assertion, after which the other one follows by considering the other tail and addition.    □

*Remark 7.1.* Note that, for symmetric random variables with finite variances the Lévy inequality, together with Chebyshev's inequality, yields

$$P(\max_{1\leq k\leq n} S_k > x) \leq \frac{2\sum_{k=1}^{n} \operatorname{Var} X_k}{x^2},$$

thus the Kolmogorov inequality with an additional factor 2.    □

An immediate and rather useful consequence runs as follows.

**Corollary 7.1.** *If $X_1, X_2, \ldots, X_n$ are symmetric, then, for all $x$ and $1 \leq k \leq n$,*

$$P(S_k > x) \leq 2P(S_n > x),$$

$$P(|S_k| > x) \leq 2P(|S_n| > x).$$

*Proof.* Immediate from the Lévy inequalities, since

$$\{S_k > x\} \subset \{\max_{1 \le k \le n} S_k > x\} \quad \text{and} \quad \{|S_k| > x\} \subset \{\max_{1 \le k \le n} |S_k| > x\}. \quad \square$$

If one replaces the symmetry assumption by the assumption that the random variables have mean 0 and finite variance, the following variation of the Lévy inequalities emerges via an application of Proposition 6.1.

**Theorem 7.2.** *Suppose that* $E X_k = 0$ *for all* $k$, *that* $\operatorname{Var} X_k = \sigma_k^2 < \infty$, *and set* $s_n^2 = \sum_{k=1}^n \sigma_k^2$. *Then, for any* $x$,

$$P(\max_{1 \le k \le n} S_k > x) \le 2P(S_n > x - s_n\sqrt{2}),$$

$$P(\max_{1 \le k \le n} |S_k| > x) \le 2P(|S_n| > x - s_n\sqrt{2}).$$

*Remark 7.2.* The usefulness of this version is, of course, mainly when $x$ itself is of the order of magnitude of at least $s_n$. $\square$

The median was the point(s) such that (at least) half of the probability mass is on either side. Analogously one can define *quantiles* as follows; cf. [195], Theorem 2.1.

**Definition 7.1.** *The* $\alpha$-*quantile*, $\lambda_\alpha(X)$, *of a random variable* $X$ *is a real number satisfying*

$$P(X \ge \lambda_\alpha(X)) \ge \alpha. \qquad \square$$

*Remark 7.3.* The median thus is a $\frac{1}{2}$-quantile. $\square$

A suitable modification of the proof of the Lévy inequalities yields the following extension.

**Theorem 7.3.** (Extended Lévy inequalities) *Let* $\alpha \in (0,1)$. *For any* $x$,

$$P(\max_{1 \le k \le n} (S_k - \lambda_\alpha(S_k - S_n)) > x) \le \frac{1}{\alpha}P(S_n > x),$$

$$P(\max_{1 \le k \le n} |S_k - \lambda_\alpha(S_k - S_n)| > x) \le \frac{1}{\alpha}P(|S_n| > x).$$

*Remark 7.4.* For $\alpha = 1/2$ we rediscover the Lévy inequalities. $\square$

**Exercise 7.1.** Prove Theorem 7.3. $\square$

Since a bound on the tails of a random variable provides a bound on the quantiles, the following variation of Theorem 7.3 can be obtained; cf. [195], Theorem 2.3.

**Theorem 7.4.** *If* $X_1, X_2, \ldots, X_n$ *are independent random variables such that*

$$\max_{1 \leq k \leq n} P(S_n - S_k \geq -\gamma) \geq \alpha,$$

*for some constants* $\gamma \geq 0$ *and* $\alpha > 0$, *then, for any* $x$,

$$P(\max_{1 \leq k \leq n} S_k \geq x) \leq \frac{1}{\alpha} P(S_n \geq x - \gamma). \qquad \square$$

*Proof.* The additional assumption implies, in particular, that

$$\max_{1 \leq k \leq n} \lambda_\alpha(S_n - S_k) \geq -\gamma, \quad \text{so that} \quad \max_{1 \leq k \leq n} \lambda_\alpha(S_k - S_n) \leq \gamma,$$

from which the conclusion follows via an application of Theorem 7.3. $\qquad \square$

*Remark 7.5.* If the random variables are symmetric we may take $\gamma = 0$ and $\alpha = 1/2$, and the inequality reduces to the Lévy inequality. $\qquad \square$

Another, extremely efficient inequality that, however, has not found its way into textbooks so far is the following one. The first version was given by Kahane in his celebrated book [150]. The inequality was later extended by Hoffmann-Jørgensen, [139]. The iterated version in (iii) below is from [146].

**Theorem 7.5.** (The Kahane-Hoffmann-Jørgensen (KHJ) inequality)
*Suppose that* $X_1, X_2, \ldots, X_n$ *have a symmetric distribution.*
(i) *For any* $x, y > 0$,

$$P(|S_n| > 2x + y) \leq P(\max_{1 \leq k \leq n} |X_k| > y) + 4\big(P(|S_n| > x)\big)^2$$

$$\leq \sum_{k=1}^{n} P(|X_k| > y) + 4\big((P(|S_n| > x)\big)^2.$$

*In particular, if* $X_1, X_2, \ldots, X_n$ *are identically distributed (and* $x = y$*), then*

$$P(|S_n| > 3x) \leq nP(|X_1| > x) + 4\big((P(|S_n| > x)\big)^2.$$

(ii) *For any* $x, y > 0$,

$$P(\max_{1 \leq k \leq n} |S_k| > 2x + y) \leq 2P(\max_{1 \leq k \leq n} |X_k| > y) + 8\big((P(|S_n| > x)\big)^2$$

$$\leq 2\sum_{k=1}^{n} P(|X_k| > y) + 8\big((P(|S_n| > x)\big)^2.$$

*In particular, if* $X_1, X_2, \ldots, X_n$ *are identically distributed (and* $x = y$*), then*

$$P(\max_{1 \leq k \leq n} |S_k| > 3x) \leq 2nP(|X_1| > x) + 8\big((P(|S_n| > x)\big)^2.$$

(iii) *For any integer $j \geq 1$,*

$$P(|S_n| > 3^j x) \leq C_j P(\max_{1 \leq k \leq n} |X_k| > x) + D_j \left( (P(|S_n| > x))^{2^j} \right),$$

*where $C_j$ and $D_j$ are numerical constants depending only on $j$.*
*In particular, if $X_1, X_2, \ldots, X_n$ are identically distributed, then*

$$P(|S_n| > 3^j x) \leq C_j n P(|X_1| > x) + D_j \left( (P(|S_n| > x))^{2^j} \right).$$

*Proof.* Set $Y_n = \max_{1 \leq k \leq n} |X_k|$, and, following the usual pattern,

$$A_k = \{ \max_{1 \leq j \leq k-1} |S_j| \leq x, |S_k| > x \}, \quad k = 1, 2, \ldots, n.$$

The sets $A_k$ are disjoint as always. Moreover,

$$\{|S_n| > 2x + y\} \subset \bigcup_{k=1}^{n} A_k,$$

so that

$$P(|S_n| > 2x + y) = P\left( \{|S_n| > 2x + y\} \cap \left\{ \bigcup_{k=1}^{n} A_k \right\} \right)$$

$$= \sum_{k=1}^{n} P(\{|S_n| > 2x + y\} \cap A_k). \tag{7.1}$$

Next, since by the triangular inequality,

$$|S_k| \leq |S_{k-1}| + |X_k| + |S_n - S_k| \quad \text{for} \quad 1 \leq k \leq n,$$

it follows that, on the set $\{|S_n| > 2x + y\} \cap A_k$, we must have

$$|S_n - S_k| > |S_n| - |S_{k-1}| - |X_k| > 2x + y - x - Y_n = x + y - Y_n,$$

so that, noticing that $S_n - S_k$ and $A_k$ are independent,

$$P(\{|S_n| > 2x + y\} \cap A_k) \leq P(\{|S_n - S_k| > x + y - Y_n\} \cap A_k)$$
$$= P(\{|S_n - S_k| > x + y - Y_n\} \cap A_k \cap \{Y_n > y\})$$
$$+ P(\{|S_n - S_k| > x + y - Y_n\} \cap A_k \cap \{Y_n \leq y\})$$
$$\leq P(A_k \cap \{Y_n > y\}) + P(\{|S_n - S_k| > x\} \cap A_k)$$
$$= P(A_k \cap \{Y_n > y\}) + P(|S_n - S_k| > x) \cdot P(A_k)$$
$$\leq P(A_k \cap \{Y_n > y\}) + 2P(|S_n| > x) \cdot P(A_k),$$

the last inequality being a consequence of Corollary 7.1.
Joining this with (7.1) finally yields

$$P(|S_n| > 2x + y) \leq \sum_{k=1}^{n} P(A_k \cap \{Y_n > y\}) + 2 \sum_{k=1}^{n} P(|S_n| > x) \cdot P(A_k)$$

$$= P\left(\left\{\bigcup_{k=1}^{n} A_k\right\} \cap \{Y_n > y\}\right) + 2P(|S_n| > x) \cdot P\left(\bigcup_{k=1}^{n} A_k\right)$$

$$\leq P(Y_n > y) + 2P(|S_n| > x) \cdot P(\max_{1 \leq k \leq n} |S_k| > x)$$

$$\leq P(Y_n > y) + 4(P(|S_n| > x))^2,$$

where we exploited the Lévy inequality in the final step.

Since $P(Y_n > y) \leq \sum_{k=1}^{n} P(|X_k| > y)$, which, in turn equals $nP(|X_1| > y)$ in the i.i.d. case, the proof of (i) is complete, from which (ii) follows via the Lévy inequality.

The proof of (iii) follows by induction. Consider the case $j = 2$. Iterating the first inequality with $y = x$, and exploiting the fact that $(a+b)^2 \leq 2a^2 + 2b^2$ for positive reals $a, b$, we obtain

$$P(|S_n| > 9x) \leq P(Y_n > 3x) + 4(P(|S_n| > 3x))^2$$

$$\leq P(Y_n > 3x) + 4\left(P(Y_n > x) + 4(P(|S_n| > x))^2\right)^2$$

$$\leq P(Y_n > 3x) + 8(P(Y_n > x))^2 + 128(P(|S_n| > x))^4$$

$$\leq 9P(Y_n > x) + 128(P(|S_n| > x))^4.$$

Here we also used the fact powers of probabilities are smaller than the probabilities themselves.

This establishes the first relation (with $C_2 = 9$ and $D_2 = 128$). The second one follows as in (i). Continuing the same procedure proves the conclusion for arbitrary $j$.    □

*Remark 7.6.* We leave it to the reader(s) to formulate and prove the obvious statement (iv).    □

As was mentioned before, it is often convenient to prove things via symmetric random variables, but it may still be of interest to have more general inequalities available. For example, for the law of the iterated logarithm (Chapter 8) one cannot use symmetrization procedures if one aims at best results.

In the proof of the KHJ inequality the symmetry property was used in order to take care of $S_n - S_k$ via the Lévy inequality. Reviewing the proof (omitting details) we find that without symmetry we would obtain

$$P(|S_n| > 2x + y) \leq P(\max_{1 \leq k \leq n} |X_k| > y) + 2P(|S_n| > x)$$

$$\times \left(P(|S_n| > x/2) + \max_{1 \leq k \leq n} P(|S_k| > x/2)\right).$$

In the following inequality symmetry is not assumed.

**Theorem 7.6.** (Etemadi's inequality) *Let* $X_1, X_2, \ldots, X_n$ *be independent random variables. Then, for all* $x > 0$,

$$P(\max_{1 \le k \le n} |S_k| > 3x) \le 3 \max_{1 \le k \le n} P(|S_k| > x).$$

*Proof.* The proof resembles somewhat the proof of Theorem 7.5. Let

$$A_k = \{\max_{1 \le j \le k-1} |S_j| \le 3x, |S_k| > 3x\}, \quad k = 1, 2, \ldots, n.$$

Once again, the sets $A_k$ are disjoint, but in the present case,

$$\bigcup_{k=1}^{n} A_k = \{\max_{1 \le k \le n} |S_k| > 3x\}.$$

Now,

$$P(\max_{1 \le k \le n} |S_k| > 3x) = P(\{\max_{1 \le k \le n} |S_k| > 3x\} \cap \{|S_n| > x\})$$

$$+ P(\{\max_{1 \le k \le n} |S_k| > 3x\} \cap \{|S_n| \le x\})$$

$$\le P(|S_n| > x) + \sum_{k=1}^{n} P(A_k \cap \{|S_n - S_k| > 2x\})$$

$$= P(|S_n| > x) + \sum_{k=1}^{n} P(A_k) \cdot P(|S_n - S_k| > 2x)$$

$$\le P(|S_n| > x) + \max_{1 \le k \le n} P(|S_n - S_k| > 2x) \cdot P\left(\bigcup_{k=1}^{n} A_k\right)$$

$$\le P(|S_n| > x) + \max_{1 \le k \le n} P(|S_n - S_k| > 2x)$$

$$\le P(|S_n| > x) + \max_{1 \le k \le n} \left(P(|S_n| > x) + P(|S_k| > x)\right)$$

$$= 2P(|S_n| > x) + \max_{1 \le k \le n} P(|S_k| > x)$$

$$\le 3 \max_{1 \le k \le n} P(|S_k| > x). \qquad \square$$

*Remark 7.7.* If, in addition, we assume symmetry, then an application of Corollary 7.1 to Etemadi's inequality yields

$$P(\max_{1 \le k \le n} |S_k| > 3x) \le 6P(|S_n| > x).$$

However, the corollary alone tells us that

$$P(\max_{1 \le k \le n} |S_k| > 3x) \le 2P(|S_n| > 3x).$$

The strength of Etemadi's inequality therefore is the avoidance of the symmetry assumption. $\qquad \square$

Some inequalities have many names; sometimes it is not clear who discovered it first. The next one is such a case.

**Theorem 7.7.** (Skorohod's or Ottaviani's inequality)
*Suppose that $X_1, X_2, \ldots, X_n$ are independent random variables, and let $x$ and $y$ be positive reals. If*

$$\beta = \max_{1 \leq k \leq n} P(|S_n - S_k| > y) < 1,$$

*then*

$$P(\max_{1 \leq k \leq n} |S_k| > x + y) \leq \frac{1}{1 - \beta} P(|S_n| > x).$$

*Proof.* This time we set

$$A_k = \{ \max_{1 \leq j \leq k-1} |S_j| \leq x + y, |S_k| > x + y \}.$$

Then, with one eye sneaking at the above proofs,

$$P(|S_n| > x) = \sum_{k=1}^{n} P(\{|S_n| > x\} \cap A_k) + \sum_{k=1}^{n} P(\{|S_n| > x\} \cap A_k^c)$$

$$\geq \sum_{k=1}^{n} P(\{|S_n| > x\} \cap A_k) \geq \sum_{k=1}^{n} P(\{|S_n - S_k| \leq y\} \cap A_k)$$

$$= \sum_{k=1}^{n} P(|S_n - S_k| \leq y) \cdot P(A_k) \geq (1 - \beta) P\left( \bigcup_{k=1}^{n} A_k \right)$$

$$= (1 - \beta) P(\max_{1 \leq k \leq k} |S_k| > x + y). \qquad \square$$

By inspecting the proofs concerning maximal partial sums we observe that a common feature is the introduction of the sets $\{A_k\}$, with the aid of which events like $\{\max_{1 \leq k \leq n} S_k \geq x\}$ are sliced into disjoint sets, so that the probability of a union can be decomposed into the sum of probabilities; a common and efficient tool in probability.

Another and very efficient way to produce this decomposition is to exploit the notion of a *stopping time*, that is, vaguely speaking, a positive integer valued random variable which is independent of the future. Stopping times play an important role in martingale theory, and we shall return to them in more detail in Chapter 10. Let us, for now, content ourselves by introducing, for example,

$$\tau = \min\{k : |S_k| > x\}.$$

Then, and these are the main features,

$$\{\tau \leq n\} = \{ \max_{1 \leq k \leq n} |S_k| > x \},$$

$$\{\tau = k\} = \{ \max_{1 \leq j \leq k-1} |S_j| \leq x, |S_k| > x \} = A_k,$$

$$\{\tau \leq k\} \in \sigma\{X_1, X_2, \ldots, X_k\}, \quad k = 1, 2, \ldots, n,$$

$$\{\tau \leq k\} \quad \text{is independent of } S_n - S_k, \quad k = 1, 2, \ldots, n.$$

It is an instructive exercise to rewrite (some of) the above proofs in the stopping time language.

## 8 The Marcinkiewics-Zygmund Inequalities

We close this chapter with two deep inequalities between moments of sums and moments of summands.

The point of departure in both cases is *Khintchine's inequality* [158]. In order to state (and prove) this inequality, we need to introduce the so-called Rademacher functions, which, probabilistically interpreted, are a kind of rescaled and iterated coin-tossing random variables. Namely, for $t \in \mathbb{R}$, let $r(t)$ be the periodically continued function defined by

$$r(t) = \begin{cases} 1, & \text{for } 0 \le t < \frac{1}{2}, \\ -1, & \text{for } \frac{1}{2} \le t < 1, \end{cases} \tag{8.1}$$

and set, for $0 \le t \le 1$, $r_n(t) = r(2^{n-1}t)$, $n = 1, 2, \ldots$. The sequence $\{r_n, n \ge 1\}$ thus defined is the sequence of *Rademacher functions*. By construction, they are piece-wise linear functions jumping between the values $+1$ and $-1$, in such a way that they cut every stretch of the predecessor into two halves, assigning the value $+1$ to the left half and the value $-1$ to the right half. Figure 3.1 depicts this effect for $n = 2$ and $n = 3$.

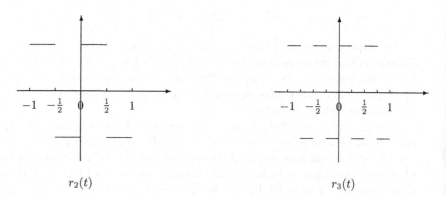

$r_2(t)$ $\qquad\qquad\qquad\qquad\qquad\qquad\qquad$ $r_3(t)$

**Figure 3.1.**

An alternative way to describe the sequence $\{r_n(t), n \ge 1\}$ is that it provides the successive decimals in the binary expansion of $t \in [-1, 1]$.

As for elementary properties, it is easily checked that

$$\int_0^1 r_n(t)\, dt = 0 \quad \text{and that} \quad \int_0^1 r_n(t) r_m(t)\, dt = \begin{cases} 1, & \text{for } m = n, \\ 0, & \text{for } m \ne n, \end{cases}$$

that is, the Rademacher functions form an orthonormal system of functions.

Talking probability theory, the Rademacher functions constitute a sequence of independent random variables, assuming the two values $+1$ and $-1$ with probability $1/2$ each, which implies that the mean is 0 and the variance equals 1.

Here is now the promised lemma.

**Lemma 8.1.** (Khintchine's inequality) *Let $p > 0$, let $c_1, c_2, \ldots, c_n$ be arbitrary reals, and set*

$$f_n(t) = \sum_{k=1}^{n} c_k r_k(t).$$

*There exist constants $A_p$ and $B_p$, depending only on $p$, such that*

$$A_p \Big( \sum_{k=1}^{n} c_k^2 \Big)^{p/2} \leq \int_0^1 |f_n(t)|^p \, dt \leq B_p \Big( \sum_{k=1}^{n} c_k^2 \Big)^{p/2}.$$

*Proof.* Via rescaling it is no loss of generality to assume that $\sum_{k=1}^{n} c_k^2 = 1$. Moreover, it suffices to prove the lemma for integral values of $p$.

We thus wish to prove that, for $p = 1, 2, \ldots$,

$$A_p \leq \int_0^1 |f_n(t)|^p \, dt \leq B_p.$$

*The right-hand inequality:* We first consider $f_n$.

$$\int_0^1 (f_n(t))^p \, dt \leq p! \int_0^1 \exp\{f_n(t)\} \, dt$$

$$= p! \prod_{k=1}^{n} \int_0^1 \exp\{c_k r_k(t)\} \, dt = p! \prod_{k=1}^{n} \frac{1}{2} \big( e^{c_k} + e^{-c_k} \big)$$

$$\leq p! \prod_{k=1}^{n} \exp\{c_k^2\} = p! \exp\Big\{ \sum_{k=1}^{n} c_k^2 \Big\} = p! e.$$

In this argument we have used the fact that one term in the Taylor expansion of the exponential function is smaller than the function itself in the first inequality, independence in the second equality, and, finally, the fact that $\frac{1}{2}(e^x + e^{-x}) \leq \exp\{x^2\}$ (use Taylor expansion; the odd terms cancel and the even ones double).

By symmetry,

$$\int_0^1 |f_n(t)|^p \, dt = 2 \int_0^1 f_n(t) I\{f_n(t) > 0\} \, dt \leq 2p! \int_0^1 e^{f_n(t)} \, dt \leq 2p! e.$$

This establishes the right-hand inequality (with $B_p = 2p! e$ when $p$ is an integer).

*The left-hand inequality:* For $p \geq 2$ we note that, by Lyapounov's inequality, Theorem 2.5 (translated to functions),

$$\left(\int_0^1 |f_n(t)|^p \, dt\right)^{1/p} \geq \left(\int_0^1 |f_n(t)|^2 \, dt\right)^{1/2} = \left(\sum_{k=1}^n c_k^2\right)^{1/2} = 1, \quad (8.2)$$

so that $A_p = 1$ does it in this case.

Now, let $0 < p < 2$, and set $\gamma(p) = p \log(\int_0^1 |f_n(t)|^{1/p} \, dt)$. This is a convex function (please, check this fact!), that is,

$$\gamma(\alpha p_1 + (1 - \alpha)p_2) \leq \alpha \gamma(p_1) + (1 - \alpha)\gamma(p_2) \quad \text{for} \quad 0 < \alpha < 1.$$

Let $\alpha = \frac{p}{4-p}$, so that

$$\alpha \cdot \frac{1}{p} + (1 - \alpha) \cdot \frac{1}{4} = \frac{1}{2}.$$

The convexity of $\gamma$ yields

$$\frac{1}{2} \log \left(\int_0^1 |f_n(t)|^2 \, dt\right) \leq \frac{1}{4-p} \log \left(\int_0^1 |f_n(t)|^p \, dt\right)$$
$$+ \frac{2-p}{8-2p} \log \left(\int_0^1 |f_n(t)|^4 \, dt\right).$$

By exponentiation and (8.2) it follows, recalling that the left-most quantity equals 0, that

$$1 \leq \left(\int_0^1 |f_n(t)|^p \, dt\right)^{1/(4-p)} \cdot \left(\int_0^1 |f_n(t)|^4 \, dt\right)^{(2-p)/(8-2p)},$$

which shows that

$$\int_0^1 |f_n(t)|^p \, dt \geq A_p,$$

with $A_p = \left(\int_0^1 |f_n(t)|^4 \, dt\right)^{(p/2)-1}$.

This completes the proof of the lemma.                                      □

*Remark 8.1.* The main purpose here was to show the *existence* of constants of the desired kind. Much research has been devoted to finding best constants, optimal rates, and so on.

A closer look at Khintchine's inequality suggests the interpretation that it is a relation between quantities of order $p$ in some sense. We therefore introduce the *quadratic variation* of a sequence of reals, functions, or random variables, which we denote by the capital letter $Q$ as follows.

**Definition 8.1.** *Let $\{g_k\}$ be a sequence of real numbers, functions or random variables. The* quadratic variation *of the sequence is*

$$Q(g) = \left(\sum_k g_k^2\right)^{1/2},$$

*where the summation extends over the indices under consideration.*        □

*Remark 8.2.* Note that $Q$ is homogeneous: $Q(cg) = |c|Q(g)$ for $c \in \mathbb{R}$.    □

Letting $Q_n(c)$ be the quadratic variation (the index $n$ referring to the dependence on $n$), Khintchine's inequality becomes

$$A'_p Q_n(c) \leq \left( \int_0^1 |f_n(t)|^p \, dt \right)^{1/p} \leq B'_p Q_n(c),$$

where the primed constants are the unprimed ones raised to the power $1/p$.

By noticing that the middle term is a norm when $p \geq 1$ we can (abusing the notation for $0 < p < 1$) reformulate the inequality succinctly as

$$A'_p Q_n(c) \leq \|f_n(t)\|_p \leq B'_p Q_n(c).$$

In case $\sum_{n=1}^{\infty} c_n^2 < \infty$, Fatou's lemma applied to the first part of the proof, and the fact that the infinite sum dominates any finite sum, together yield the following corollary.

**Corollary 8.1.** *Let $p > 0$, let $c_1, c_2, \ldots$ be real numbers such that*

$$Q(c) = \left( \sum_{n=1}^{\infty} c_n^2 \right)^{1/2} < \infty,$$

*and set*

$$f(t) = \sum_{n=1}^{\infty} c_n r_n(t) \quad and \quad \|f\|_p = \left( \int_0^1 |f(t)|^p \, dt \right)^{1/p}.$$

*There exist constants $A_p$ and $B_p$, depending only on $p$, such that*

$$A_p \left( \sum_{n=1}^{\infty} c_n^2 \right)^{p/2} \leq \int_0^1 |f(t)|^p \, dt \leq B_p \left( \sum_{n=1}^{\infty} c_n^2 \right)^{p/2},$$

*or, equivalently, such that*

$$A_p^{1/p} Q(c) \leq \|f\|_p \leq B_p^{1/p} Q(c).$$

**Exercise 8.1.** Check the details of the proof of the corollary.    □

*Remark 8.3.* A functional analytic interpretation of this is that, for $p \geq 1$, the closed linear subspace of the Rademacher functions in $L^p$ is isomorphic to the space $\ell^2$ of square-summable sequences.    □

Next we turn to the extension of Khintchine's inequality to sums of random variables by Marcinkiewicz and Zygmund, [180, 182].

Let $X_1, X_2, \ldots, X_n$ be independent random variables with mean 0, suppose that $E|X_k|^p < \infty$, for $k = 1, 2, \ldots, n$ and some $p \geq 1$, set $S_n = \sum_{k=1}^{n} X_k$, and let, as always, random variables with superscript $s$ denote symmetrized versions. Further, set

$$T_n(t) = \sum_{k=1}^{n} X_k r_k(t), \quad \text{for} \quad 0 \le t \le 1, \quad n \ge 1.$$

By symmetrization, integration, and Khintchine's inequality and a bit more we are now prepared to prove the following celebrated inequalities.

**Theorem 8.1.** (The Marcinkiewicz-Zygmund inequalities)
*Let $p \ge 1$. Suppose that $X_1$, $X_2$, ..., $X_n$ are independent random variables with mean 0, such that $E|X_k|^p < \infty$, for all $k$, and let $\{S_n, n \ge 1\}$ denote the partial sums. Then there exist constants $A_p^*$ and $B_p^*$ depending only on $p$, such that*

$$A_p^* E\Big( \sum_{k=1}^{n} X_k^2 \Big)^{p/2} \le E|S_n|^p \le B_p^* E\Big( \sum_{k=1}^{n} X_k^2 \Big)^{p/2},$$

*or, equivalently,*

$$(A_p^*)^{1/p} \|Q_n(X)\|_p \le \|S_n\|_p \le (B_p^*)^{1/p} \|Q_n(X)\|_p,$$

*where*

$$Q_n(X) = \Big( \sum_{k=1}^{n} X_k^2 \Big)^{1/2}$$

*is the quadratic variation of the summands.*

*Proof.* Consider $S_n^s = \sum_{k=1}^{n} X_k^s$ and $T_n^s(t) = \sum_{k=1}^{n} X_k r_k(t)$. By the cointossing property of the Rademacher functions and symmetry it follows that

$$S_n^s \overset{d}{=} T_n^s(t).$$

Invoking Proposition 6.5 and the $c_r$-inequality we therefore obtain

$$E|S_n|^p \le E|S_n^s|^p = E|T_n^s(t)|^p \le 2^p E|T_n(t)|^p. \tag{8.3}$$

Integrating the extreme members, changing the order of integration, and applying Khintchine's lemma (for fixed $\omega$) yields

$$E|S_n|^p \le 2^p E \int_0^1 |T_n(t)|^p \, dt \le 2^p B_p \Big( E \sum_{k=1}^{n} X_k^2 \Big)^{p/2},$$

which proves the right-hand inequality with $B_p^* = 2^p B_p$, where $B_p$ is the constant in Khinchine's inequality.

By running (8.3) backward, with $S_n$ and $T_n(t)$ playing reversed roles, we obtain

$$E|T_n(t)|^p \le 2^p E|S_n|^p,$$

which, after the same integration procedure and the other half of Khintchine's inequality, proves the left-hand inequality with $A_p^* = 2^{-p} A_p$, where $A_p$ is the constant in Khinchine's inequality. □

If, in addition, $X_1, X_2, \ldots, X_n$ are identically distributed the right-hand inequality can be further elaborated.

**Corollary 8.2.** *Let $p \geq 1$. Suppose that $X, X_1, X_2, \ldots, X_n$ are independent, identically distributed random variables with mean 0 and $E|X|^p < \infty$. Set $S_n = \sum_{k=1}^n X_k$, $n \geq 1$. Then there exists a constant $B_p$ depending only on $p$, such that*

$$E|S_n|^p \leq \begin{cases} B_p n E|X|^p, & when \quad 1 \leq p \leq 2, \\ B_p n^{p/2} E|X|^{p/2}, & when \quad p \geq 2. \end{cases}$$

*Proof.* Let $1 \leq p \leq 2$. Since $p/2 < 1$ we apply the $c_r$-inequality to the Marcinkiewicz-Zygmund inequalities:

$$E|S_n|^p \leq B_p^* E\Big(\sum_{k=1}^n X_k^2\Big)^{p/2} \leq B_p^* E\Big(\sum_{k=1}^n (X_k^2)^{p/2}\Big) = B_p^* n E|X|^p.$$

For $p \geq 2$ we use the convexity of $|x|^{p/2}$:

$$E|S_n|^p \leq B_p^* E\Big(\sum_{k=1}^n X_k^2\Big)^{p/2} = B_p^* n^{p/2} E\Big(\frac{1}{n}\sum_{k=1}^n X_k^2\Big)^{p/2}$$

$$\leq B_p^* n^{p/2} \frac{1}{n}\Big(\sum_{k=1}^n E(X_k^2)^{p/2}\Big) = B_p^* n^{p/2} E|X|^p. \qquad \square$$

In view of the central limit theorem for sums of independent, identically distributed random variables it is interesting to observe that the bounds in the corollary are of the correct order of magnitude for $p \geq 2$; note that the statement amounts to

$$E\Big|\frac{S_n}{\sqrt{n}}\Big|^p \leq B_p^* E|X|^p.$$

The natural question is, of course, whether or not there is actual *convergence* of moments. A positive answer to this question will be given in Chapter 7.

## 9 Rosenthal's Inequality

This is an inequality with an atmosphere of the Marcinkieiwicz-Zygmund inequalities, and, in part, a consequence of them. The original reference is [214]. We begin with an auxiliary result.

**Lemma 9.1.** *Let $p \geq 1$. Suppose that $X_1, X_2, \ldots, X_n$ are independent random variables such that $E|X_k|^p < \infty$ for all $k$. Then*

$$E|S_n|^p \leq \max\Big\{2^p \sum_{k=1}^n E|X_k|^p, 2^{p^2}\Big(\sum_{k=1}^n E|X_k|\Big)^p\Big\}.$$

*Proof.* Since $S_n \leq \sum_{k=1}^n |X_k|$ it is no restriction to assume that all summands are non-negative.

Set $S_n^{(j)} = \sum_{k=1, k \neq j}^{n} X_k$. Using the $c_r$-inequality (Theorem 2.2), independence, non-negativity and Lyapounov's inequality (Theorem 2.5), we obtain

$$E(S_n)^p = \sum_{j=1}^{n} E(S_n)^{p-1} X_j \leq 2^{p-1} \sum_{j=1}^{n} E\big((X_j^{p-1} + (S_n^{(j)})^{p-1}) X_j\big)$$

$$= 2^{p-1} \sum_{j=1}^{n} \big(E X_j^p + E(S_n^{(j)})^{p-1} \cdot E X_j\big)$$

$$\leq 2^{p-1} \sum_{j=1}^{n} \big(E X_j^p + E(S_n)^{p-1} \cdot E X_j\big)$$

$$= 2^{p-1} \bigg(\sum_{j=1}^{n} E X_j^p + E(S_n)^{p-1} \sum_{j=1}^{n} E X_j\bigg)$$

$$\leq 2^{p-1} \bigg(\sum_{j=1}^{n} E X_j^p + (E(S_n)^p)^{(p-1)/p} \sum_{j=1}^{n} E X_j\bigg)$$

$$\leq 2^p \max\bigg\{\sum_{j=1}^{n} E X_j^p, (E(S_n)^p)^{(p-1)/p} \sum_{j=1}^{n} E X_j\bigg\}.$$

Thus,

$$E(S_n)^p \leq 2^p \sum_{j=1}^{n} E X_j^p \quad \text{and} \quad E(S_n)^p \leq 2^p (E(S_n)^p)^{(p-1)/p} \sum_{j=1}^{n} E X_j.$$

The conclusion follows.     □

With the aid of this lemma, the Marcinkiewicz-Zygmund inequalities of the previous section, and Theorem 6.1, we are now ready for Rosenthal's inequality. Recall the quadratic variation $Q_n(X) = (\sum_{k=1}^{n} X_k^2)^{1/2}$.

**Theorem 9.1.** (Rosenthal's inequality) *For $p > 2$, let $X_1, X_2, \ldots, X_n$ be independent random variables with mean $0$, and suppose that $E|X_k|^p < \infty$ for all $k$. Then*

$$E|S_n|^p \begin{cases} \leq D_p \max\big\{\sum_{k=1}^{n} E|X_k|^p, \big(\sum_{k=1}^{n} E X_k^2\big)^{p/2}\big\}, \\ \geq 2^{-p} \max\big\{\sum_{k=1}^{n} E|X_k|^p, \big(\sum_{k=1}^{n} E X_k^2\big)^{p/2}\big\}, \end{cases}$$

*or, equivalently,*

$$\max\bigg\{\sum_{k=1}^{n} E|X_k|^p, (Q_n(X))^p\bigg\} \leq 2^p E|S_n|^p$$

$$\leq D_p^* \max\bigg\{\sum_{k=1}^{n} E|X_k|^p, (Q_n(X))^p\bigg\},$$

*where $D_p$ is a constant that depends only on $p$, and $D_p^* = 2^p D_p$.*

*Proof.* Applying the Marcinkiewicz-Zygmund inequalities and Lemma 9.1 yields

$$E|S_n|^p \le B_p^* E \left( \sum_{k=1}^n X_k^2 \right)^{p/2}$$

$$\le B_p^* \max \left\{ 2^{p/2} \sum_{k=1}^n E\left( (X_k^2)^{p/2} \right), 2^{(p/2)^2} \left( \sum_{k=1}^n E X_k^2 \right)^{p/2} \right\}$$

$$= D_p \max \left\{ \sum_{k=1}^n E|X_k|^p, \left( \sum_{k=1}^n E X_k^2 \right)^{p/2} \right\},$$

with $D_p = B_p^* 2^{(p/2)^2}$.

As for the lower bound, the Marcinkiewicz-Zygmund inequalities tell us that

$$E|S_n|^p \ge 2^{-p} \left( \sum_{k=1}^n E X_k^2 \right)^{p/2},$$

and Theorem 6.1(vi) tells us that

$$E|S_n|^p \ge 2^{-p} \sum_{k=1}^n E|X_k|^p,$$

which together establish the lower bound.

Note also that the constant $2^{-p}$ can be replaced by $1/2$ if the random variables are symmetric, since, in that case, we lean on Theorem 6.1(v) instead of (vi) in the final inequality.  $\square$

# 10 Problems

1. Show that, for any non-negative random variable $X$,

$$E X + E \frac{1}{X} \ge 2,$$

$$E \max \left\{ X, \frac{1}{X} \right\} \ge 1.$$

2. Let $X$ be a random variable.
   (a) Show that, if $E X^2 = 1$ and $E X^4 < \infty$, then

$$E|X| \ge \frac{1}{\sqrt{E X^4}}.$$

   ♠ Write $X^2 = |X|^r |X|^{2-r}$, choose $r$ conveniently, and exploit the Hölder inequality.
   (b) Suppose that $E X^{2m} < \infty$, where $m \in \mathbb{N}$. State and prove an analogous inequality.

3. Prove that, for any random variable $X$, the minimum of $E(X - a)^2$ is attained for $a = EX$.

4. Prove that, for any random variable $X$, the minimum of $E|X - a|$ is attained for $a = \text{med}(X)$.

5. Let $X$ be a positive random variable with finite variance, and let $\lambda \in (0, 1)$. Prove that

$$P(X \geq \lambda E X) \geq (1 - \lambda)^2 \frac{(EX)^2}{EX^2}.$$

6. Let, for $p \in (0, 1)$, and $x \in \mathbb{R}$, $X$ be a random variable defined as follows:

$$P(X = -x) = P(X = x) = p, \quad P(X = 0) = 1 - 2p.$$

Show that there is *equality* in Chebyshev's inequality for $X$.

   ♦  This means that Chebyshev's inequality, in spite of being rather crude, cannot be improved without additional assumptions.

7. *Cantelli's inequality.* Let $X$ be a random variable with finite variance, $\sigma^2$.
   (a) Prove that, for $x \geq 0$,

$$P(X - EX \geq x) \leq \frac{\sigma^2}{x^2 + \sigma^2},$$

$$P(|X - EX| \geq x) \leq \frac{2\sigma^2}{x^2 + \sigma^2}.$$

   (b) Find $X$ assuming two values where there is equality.
   (c) When is Cantelli's inequality better than Chebyshev's inequality?
   (d) Use Cantelli's inequality to show that $|\text{med}(X) - EX| \leq \sigma\sqrt{3}$; recall, from Proposition 6.1, that an application of Chebyshev's inequality yields the bound $\sigma\sqrt{2}$.
   (e) Generalize Cantelli's inequality to moments of order $r \neq 1$.

8. Recall, from Subsection 2.16.4, the likelihood ratio statistic, $L_n$, which was defined as a product of independent, identically distributed random variables with mean 1 (under the so-called null hypothesis), and the, sometimes more convenient, log-likelihood, $\log L_n$, which was a sum of independent, identically distributed random variables, which, however, do *not* have mean $\log 1 = 0$.
   (a) Verify that the last claim is correct, by proving the more general statement, namely that, if $Y$ is a non-negative random variable with finite mean, then

$$E(\log Y) \leq \log(EY).$$

   (b) Prove that, in fact, there is strict inequality:

$$E(\log Y) < \log(EY),$$

   unless $Y$ is degenerate.
   (c) Review the proof of Jensen's inequality, Theorem 5.1. Generalize with a glimpse on (b).

9. The *concentration function* of a random variable $X$ is defined as

$$Q_X(h) = \sup_x P(x \leq X \leq x + h), \quad h > 0.$$

(a) Show that $Q_{X+b}(h) = Q_X(h)$.

(b) Is it true that $Q_{aX}(h) = Q_X(h/a)$?

(c) Show that, if $X$ and $Y$ are independent random variables, then

$$Q_{X+Y}(h) \leq \min\{Q_X(h), Q_Y(h)\}.$$

♣ To put the concept in perspective, if $X_1, X_2, \ldots, X_n$ are independent, identically distributed random variables, and $S_n = \sum_{k=1}^{n} X_k$, then there exists an absolute constant, $A$, such that

$$Q_{S_n}(h) \leq \frac{A}{\sqrt{n}}.$$

Some references: [79, 80, 156, 213], and [195], Section 1.5.

# 4

# Characteristic Functions

Adding independent random variables is a frequent occupation in probability theory. Mathematically this corresponds to convolving functions. Just as there are Fourier transforms and Laplace transforms which transform convolution into multiplication, there are transforms in probability theory that transform addition of independent random variables into multiplication of transforms. Although we shall mainly use one of them, the characteristic function, we shall, in this chapter, briefly also present three others – the cumulant generating function, which is the logarithm of the characteristic function; the probability generating function; and the moment generating function. In Chapter 5 we shall prove so-called continuity theorems, which permit limits of distributions to be determined with the aid of limits of transforms.

Uniqueness is indispensable in order to make things work properly. This means that if we replace the adding of independent random variables with the multiplication of their transforms we must be sure that the resulting transform corresponds uniquely to the distribution of the sum under investigation. The first thing we therefore have to do in order to see that we are on the right track is to prove that

- summation of independent random variables corresponds to multiplication of their transforms;
- the transformation is 1 to 1; there is a uniqueness theorem to the effect that if two random variables have the same transform then they also have the same distribution.

## 1 Definition and Basics

In this first section we define characteristic functions and prove some basic facts, including uniqueness, inversion, and the "multiplication property", in other words, verify that characteristic functions possess the desired features.

**Definition 1.1.** *The* characteristic function *of the random variable $X$ is*

$$\varphi_X(t) = E\,\mathrm{e}^{\mathrm{i}tx} = \int_{-\infty}^{\infty} \mathrm{e}^{\mathrm{i}bx}\,\mathrm{d}F_X(x).$$    □

*Remark 1.1.* Apart from a minus sign in the exponent (and, possibly, a factor $\sqrt{1/2\pi}$), characteristic functions coincide with Fourier transforms in the absolutely continuous case and with Fourier series in the lattice case.    □

Before we prove uniqueness and the multiplication theorem we present some basic facts. Note that the first property tells us that *characteristic functions exist for all random variables.*

**Theorem 1.1.** *Let $X$ be a random variable. Then*

(a)    $|\varphi_X(t)| \leq \varphi_X(0) = 1;$

(b)    $\overline{\varphi_X(t)} = \varphi_X(-t) = \varphi_{-X}(t);$

(c)    $\varphi_X(t)$ *is uniformly continuous.*

*Proof.* (a): We have

$$|E\,\mathrm{e}^{\mathrm{i}ti}| \leq E\,|\mathrm{e}^{\mathrm{i}tX}| = E\,1 = 1 = E\,\mathrm{e}^{\mathrm{i}\cdot 0\cdot X} = \varphi_X(0).$$

To prove (b) we simply let the minus sign wander through the exponent:

$$\begin{aligned}
\overline{\mathrm{e}^{\mathrm{i}xt}} &= \overline{\cos xt + \mathrm{i}\sin xt} = \cos xt - \mathrm{i}\sin xt \quad (= \mathrm{e}^{(-\mathrm{i})xt}) \\
&= \cos(x(-t)) + \mathrm{i}\sin(x(-t)) \quad (= \mathrm{e}^{\mathrm{i}x(-t)}) \\
&= \cos((-x)t) + \mathrm{i}\sin((-x)t) \quad (= \mathrm{e}^{\mathrm{i}(-x)t}).
\end{aligned}$$

As for (c), let $t$ be arbitrary and $h > 0$ (a similar argument works for $h < 0$). Apart from the trivial estimate $|\mathrm{e}^{\mathrm{i}x} - 1| \leq 2$, we know from Lemma A.1.2 that $|\mathrm{e}^{\mathrm{i}x} - 1| \leq |x|$.

Using the cruder estimate in the tails and the more delicate one in the center (as is common practise), we obtain, for $A > 0$,

$$\begin{aligned}
|\varphi_X(t+h) - \varphi_X(t)| &= |E\,\mathrm{e}^{\mathrm{i}(t+h)X} - E\,\mathrm{e}^{\mathrm{i}tX}| = |E\,\mathrm{e}^{\mathrm{i}tX}(\mathrm{e}^{\mathrm{i}hX} - 1)| \\
&\leq E|\mathrm{e}^{\mathrm{i}tX}(\mathrm{e}^{\mathrm{i}hX} - 1)| = E|\mathrm{e}^{\mathrm{i}hX} - 1| \\
&= E|\mathrm{e}^{\mathrm{i}hX} - 1|I\{|X| \leq A\} + E|\mathrm{e}^{\mathrm{i}hX} - 1|I\{|X| > A\} \\
&\leq E|hX|I\{|X| \leq A\} + 2P(|X| > A) \leq hA + 2P(|X| > A) < \varepsilon
\end{aligned}$$

for any $\varepsilon > 0$ if we first choose $A$ so large that $2P(|X| \geq A) < \varepsilon/2$, and then $h$ so small that $hA < \varepsilon/2$. This proves that $\varphi_X$ is uniformly continuous, since the estimate does not depend on $t$.    □

Following are two tables listing the characteristic functions of some standard distributions. We advise the reader to verify (some of) the entries in the tables.

First, some discrete distributions.

| Distribution | Notation | Characteristic function |
|---|---|---|
| One point | $\delta(a)$ | $e^{ita}$ |
| Symmetric Bernoulli | | $\cos t$ |
| Bernoulli | $\mathrm{Be}(p)$ | $q + pe^{it}$ |
| Binomial | $\mathrm{Bin}(n,p)$ | $(q + pe^{it})^n$ |
| Geometric | $\mathrm{Ge}(p)$ | $\frac{p}{1-qe^{it}}$ |
| First success | $\mathrm{Fs}(p)$ | $\frac{pt}{1-qe^{it}}$ |
| Poisson | $\mathrm{Po}(m)$ | $e^{m(e^{it}-1)}$ |

**Table 4.1.** Some discrete distributions

As for the absolutely continuous distributions, the expressions for the Pareto-, log-normal, and beta distributions are too complicated to put into print.

| Distribution | Notation | Characteristic function |
|---|---|---|
| Uniform | $U(a,b)$ | $\frac{e^{itb}-e^{ita}}{it(b-a)}$ |
| | $U(0,1)$ | $\frac{e^{it}-1}{it}$ |
| | $U(-1,1)$ | $\frac{\sin t}{t}$ |
| Triangular | $\mathrm{Tri}(-1,1)$ | $\left(\frac{\sin t/2}{t/2}\right)^2$ |
| Exponential | $\mathrm{Exp}(\theta)$ | $\frac{1}{1-\theta it}$ |
| Gamma | $\Gamma(p,\theta)$ | $\left(\frac{1}{1-\theta it}\right)^p$ |
| Normal | $N(\mu,\sigma^2)$ | $e^{it\mu-\frac{1}{2}t^2\sigma^2}$ |
| | $N(0,1)$ | $e^{-\frac{1}{2}t^2}$ |
| Cauchy | $C(0,1)$ | $e^{-|t|}$ |

**Table 4.2.** Some absolutely continuous distributions

The most special distribution we have encountered so far (probably) is the Cantor distribution. So, what's the characteristic function of this particular one?

Being prepared for this question, here is the answer: The characteristic function of the Cantor distribution on the interval $[-\frac{1}{2}, -\frac{1}{2}]$ (for simplicity) equals

$$\varphi(t) = \prod_{k=1}^{\infty} \cos\left(\frac{t}{3^k}\right).$$

The verification of this fact will be provided in Subsection 4.2.1.

## 1.1 Uniqueness; Inversion

We begin by stating the uniqueness theorem.

**Theorem 1.2.** *Let $X$ and $Y$ be random variables. If $\varphi_X = \varphi_Y$, then $X \overset{d}{=} Y$ and conversely.* □

Instead of providing a proof here we move directly on to inversion theorems, from which uniqueness is immediate. After all, a uniqueness theorem is an *existence* result (only), whereas an inversion theorem provides a formula for explicitly computing the distribution.

**Theorem 1.3.** *Let $X$ be a random variable with distribution function $F$ and characteristic function $\varphi$. For $a < b$,*

$$F(b) - F(a) + \frac{1}{2}P(X = a) - \frac{1}{2}P(X = b) = \lim_{T\to\infty} \frac{1}{2\pi} \int_{-T}^{T} \frac{e^{-itb} - e^{-ita}}{-it} \cdot \varphi(t)\, dt.$$

*In particular, if $a, b \in C(F)$, then*

$$F(b) - F(a) = \lim_{T\to\infty} \frac{1}{2\pi} \int_{-T}^{T} \frac{e^{-itb} - e^{-ita}}{-it} \cdot \varphi(t)\, dt.$$

*Proof.* By Lemma A.1.2 with $n = 0$,

$$\left| \frac{e^{-itb} - e^{-ita}}{t} \right| = \left| e^{-ita} \right| \cdot \left| \frac{e^{-it(b-a)} - 1}{t} \right| \leq b - a, \tag{1.1}$$

which shows that

$$\left| \int_{-T}^{T} \frac{e^{-itb} - e^{-ita}}{-it} \cdot \varphi(t)\, dt \right| \leq \int_{-T}^{T} \left| \frac{e^{-itb} - e^{-ita}}{t} \right| \cdot 1\, dt \leq 2T(b - a).$$

We may therefore apply Fubini's theorem, the Euler formulas, and symmetry, to obtain

$$\begin{aligned}
I_T &= \frac{1}{2\pi} \int_{-T}^{T} \frac{e^{-itb} - e^{-ita}}{-it} \cdot \varphi(t)\, dt \\
&= \frac{1}{2\pi} \int_{-T}^{T} \frac{e^{-itb} - e^{-ita}}{-it} \left( \int_{-\infty}^{\infty} e^{itx}\, dF(x) \right) dt \\
&= \frac{1}{\pi} \int_{-\infty}^{\infty} \left( \int_{-T}^{T} \frac{e^{it(x-a)} - e^{it(x-b)}}{2it}\, dt \right) dF(x) \\
&= \frac{1}{\pi} \int_{-\infty}^{\infty} \left( \int_{0}^{T} \frac{\sin t(x - a)}{t} - \frac{\sin t(x - b)}{t}\, dt \right) dF(x) \\
&= \frac{1}{\pi} \int_{-\infty}^{\infty} H(a, b, t, x, T)\, dF(x) \\
&= \frac{1}{\pi} E\, H(a, b, t, X, T),
\end{aligned}$$

where, thus, $H(a, b, t, x, T)$ is the inner integral. The expected value is merely a probabilistic interpretation of the preceding line.

From Lemma A.1.3 we know that

$$\int_0^T \frac{\sin x}{x}\,\mathrm{d}x \begin{cases} \leq & \int_0^\pi \frac{\sin x}{x}\,\mathrm{d}x \leq \pi \quad \text{for all} \quad T > 0, \\ \to & \frac{\pi}{2} \quad \text{as} \quad T \to \infty, \end{cases}$$

so that

$$\lim_{T\to\infty} H(a,b,t,x,T) = \begin{cases} 0, & \text{for} \quad x < a, \\ \frac{\pi}{2}, & \text{for} \quad x = a, \\ \pi, & \text{for} \quad a < x < b, \\ \frac{\pi}{2}, & \text{for} \quad x = b, \\ 0, & \text{for} \quad x > b. \end{cases}$$

By dominated convergence, Theorem 2.5.3, we therefore obtain

$$\lim_{T\to\infty} \frac{1}{\pi} E\,H(a,b,t,X,T) = \frac{1}{2}P(X = a) + P(a < X < b) + \frac{1}{2}P(X = b),$$

which proves the first inversion formula, from which the other one is immediate. $\qquad\square$

As for the name of the transform, we have just seen that every random variable possesses a unique characteristic function; the characteristic function *characterizes* the distribution uniquely.

Theorem 1.3 is a general result. If we know more we can say more. This is, for example, the case when the characteristic function is absolutely integrable.

**Theorem 1.4.** *If $\int_{-\infty}^{\infty} |\varphi(t)|\,\mathrm{d}t < \infty$, then $X$ has an absolutely continuous distribution with a bounded, continuous density $f = F'$, given by*

$$f(x) = \frac{1}{2\pi}\int_{-\infty}^{\infty} e^{-itx} \cdot \varphi(t)\,\mathrm{d}t.$$

*Proof.* Let $h > 0$ and set $a = x$ and $b = x + h$ in the inversion formula. First of all, by (1.1),

$$F(x + h) - F(x) + \frac{1}{2}P(X = x) - \frac{1}{2}P(X = x + h) \leq \lim_{T\to\infty} \frac{1}{2\pi}\int_{-T}^{T} h|\varphi(t)|\mathrm{d}t$$

$$\leq \frac{h}{2\pi}\int_{-\infty}^{\infty} |\varphi(t)|\mathrm{d}t \to 0 \quad \text{as} \quad h \to 0,$$

so that

- the inversion formula holds for all of $\mathbb{R}$;
- there cannot be any point masses;
- the limit of the integral exists as $T \to \infty$.

Division by $h$ therefore shows that

$$
\frac{F(x+h) - F(x)}{h} = \frac{1}{2\pi} \int_{-\infty}^{\infty} \frac{e^{-it(x+h)} - e^{-itx}}{-ith} \cdot \varphi(t)\, dt
$$

$$
= \frac{1}{2\pi} \int_{-\infty}^{\infty} e^{-itx} \frac{1 - e^{-ith}}{ith} \cdot \varphi(t)\, dt,
$$

where $h$ may be positive as well as negative. The conclusion follows by letting $h \to 0$ in the right-hand side under the integration sign. To see that this is permitted, we observe that the integrand converges to 1 as $h \to 0$, is bounded in absolute value by 1 (once again be Lemma A.1.2), and that the upper bound is integrable (by assumption), so that the dominated convergence theorem (in its mathematical formulation) applies.

The boundedness of the density follows from the assumption that, for all $x$, $0 \le f(x) \le \frac{1}{2\pi} \int_{-\infty}^{\infty} |\varphi(t)|\, dt < \infty$, and the continuity from the continuity of the defining integral.    □

*Remark 1.2.* Although uniqueness follows from the inversion theorem, we shall present a proof of the uniqueness theorem, as well as an alternative proof of Theorem 1.4 in Subsection 5.11.1 when other tools are available, proofs that are more elegant, primarily because of their more probabilistic nature (although elegance may be considered as a matter of taste). On the other hand, these proofs presuppose a certain amount of additional knowledge. So, as ever so often, the swiftness of a proof or its elegance may be an effect of additional work that is hidden, in the sense that it only appears as a reference; "by Theorem such and such we obtain ...".    □

Theorem 1.4 provides us with a sufficient condition for the distribution to be absolutely continuous and a recipe for how to find the density in that case. The $U(-1, 1)$-distribution, whose characteristic function equals $\sin t / t$, illustrates three facts:

- Absolute integrability is not necessary for absolute continuity.
- The density is *not everywhere* continuous.
- The characteristic function converges to 0 at $\pm\infty$ in spite of the fact that it is not absolutely integrable.

The following result tells us that the latter property is universal for absolutely continuous random variables.

**Theorem 1.5.** (The Riemann-Lebesgue lemma)
*If $X$ is an absolutely continuous random variable with characteristic function $\varphi$, then*

$$
\lim_{t \to \pm\infty} |\varphi(t)| = 0.
$$

*Proof.* For $X \in U(a, b)$ the conclusion follows by recalling that

$$\varphi(t) = \frac{e^{itb} - e^{ita}}{it(b-a)}.$$

The final statement follows the usual way via simple random variables, and approximation (Lemma A.9.3).    □

If the distribution has point masses these can be recovered as follows.

**Theorem 1.6.** *If* $P(X = a) > 0$, *then*

$$P(X = a) = \lim_{T \to \infty} \frac{1}{2T} \int_{-T}^{T} e^{-ita} \cdot \varphi(t) \, dt.$$

*Proof.* By proceeding along the lines of the proof of Theorem 1.3 we obtain

$$\frac{1}{2T} \int_{-T}^{T} e^{-ita} \cdot \varphi(t) \, dt = \frac{1}{2T} \int_{-T}^{T} e^{-ita} \left( \int_{-\infty}^{\infty} e^{itx} \, dF(x) \right) dt$$

$$= \frac{1}{2T} \int_{-\infty}^{\infty} \left( \int_{-T}^{T} e^{it(x-a)} \, dt \right) dF(x)$$

$$= \frac{1}{2T} \int_{-\infty}^{\infty} \left( \int_{-T}^{T} \left( \cos(t(x-a)) + i\sin(t(x-a)) \, dt \right) dF(x) \right.$$

$$= \frac{1}{T} \int_{-\infty}^{\infty} \left( \frac{\sin(T(x-a))}{x-a} + 0 \right) dF(x)$$

$$= \int_{\mathbb{R}\setminus a} \frac{\sin(T(x-a))}{T(x-a)} \, dF(x) + 1 \cdot P(X = a)$$

$$= E\left( \frac{\sin(T(X-a))}{T(X-a)} I\{X \ne a\} \right) + P(X = a)$$

$$\to 0 + P(X = a) = P(X = a) \quad \text{as} \quad T \to \infty,$$

where convergence of the expectation to 0 is justified by the fact that the random variable $\frac{\sin(T(X-a))}{T(X-a)} I\{X \ne a\} \overset{a.s.}{\to} 0$ as $T \to \infty$, and is bounded by 1, so that dominated convergence is applicable.    □

Discrete distributions have their mass concentrated on a countable set of points. A special kind are the *lattice* distributions for which the set of point masses is concentrated on a lattice, that is, on a set of the form

$$\{kd + \lambda : k = 0, \pm 1, \pm 2, \dots, \quad \text{for some } d > 0 \text{ and } \lambda \in \mathbb{R}\}. \tag{1.2}$$

The smallest such $d$ is called the *span*. The characteristic function for lattice distributions reduces to a sum. More precisely, if $X$ has its support on the set (1.2), and $p_k = P(X = kd + \lambda)$, $k \in \mathbb{Z}$, then

$$\varphi_X(t) = \sum_{k \in \mathbb{Z}} p_k e^{i(kd+\lambda)t}.$$

A particular case is the degenerate distribution, that is, the case when $X$ equals a constant, $c$, almost surely, in which case $\varphi_X(t) = e^{itc}$. The following result provides a converse.

**Theorem 1.7.** *Let $X$ be a random variable with characteristic function $\varphi$. The distribution of $X$ is*

(a) *degenerate iff $|\varphi(t)| = |\varphi(s)| = 1$ for two values $s, t$, such that $s/t$ is irrational;*

(b) *a lattice distribution iff $\varphi$ is periodic.*

*Proof.* Only sufficiencies remain to be proved. Thus, suppose that $|\varphi(t)| = 1$, so that $\varphi(t) = e^{ita}$ for some $a \in \mathbb{R}$. Then

$$1 = e^{-ita} \varphi(t) = E \exp\{it(x - a)\} = E \cos(t(X - a)),$$

where the last equality is a consequence of the fact that the expectation is real (equal to 1). Restating this we find that

$$E\left(1 - \cos t(X - a)\right) = 0.$$

Since the integrand is non-negative and the expectation equals 0 we must have $\cos(t(x - a)) = 1$ for all $x$ with $P(X = x) > 0$. Because of the periodicity of the cosine function these points must be situated on a lattice with a span proportional to $2\pi/t$, which proves (b).

If, in addition, $|\varphi(s)| = 1$ where $s/t$ is irrational, the same argument shows that the collection of mass points must be situated on a lattice with a span proportional to $2\pi/s$, which is impossible, unless there is only a single mass point, in which case the distribution is degenerate.                        □

We also mention, without proof, that if $X$ has its support on the set (1.2), and $p_k = P(X = kd + \lambda)$, $k \in \mathbb{Z}$, then the inversion formula reduces to

$$p_k = \frac{d}{2\pi} \int_{-\pi/d}^{\pi/d} e^{-it(kd + \lambda)} \varphi_X(t) \, dt.$$

In particular, if $\lambda = 0$ and $d = 1$, then

$$\varphi_X(t) = \sum_{k \in \mathbb{Z}} p_k e^{ikt} \quad \text{and} \quad p_k = P(X = k) = \frac{1}{2\pi} \int_{-\pi}^{\pi} e^{-itk} \varphi_X(t) \, dt,$$

**Exercise 1.1.** Check the binomial, geometric and Poisson distributions.          □

## 1.2 Multiplication

Next in line is the multiplication theorem.

**Theorem 1.8.** *Let $X_1, X_2, \ldots, X_n$ be independent random variables, and set $S_n = X_1 + X_2 + \cdots + X_n$. Then*

$$\varphi_{S_n}(t) = \prod_{k=1}^{n} \varphi_{X_k}(t).$$

*If, in addition, $X_1, X_2, \ldots, X_n$ are equidistributed, then*

$$\varphi_{S_n}(t) = \left(\varphi_{X_1}(t)\right)^n.$$

*Proof.* Since $X_1$, $X_2$, ..., $X_n$ are independent, we know from Theorem 2.10.4 that same is true for $e^{itX_1}$, $e^{itX_2}$, ..., $e^{itX_n}$, so that

$$\varphi_{S_n}(t) = E\, e^{it(X_1 + X_2 + \ldots + X_n)} = E \prod_{k=1}^{n} e^{itX_k} = \prod_{k=1}^{n} E\, e^{itX_k} = \prod_{k=1}^{n} \varphi_{X_k}(t).$$

The second part is immediate, since all factors are the same.    □

There are several integrals that one can easily compute by identifying a relation between densities and characteristic functions. As an example, recall that the integral $\int_0^\infty \sin x / x \, dx$ was instrumental in the proof of Theorem 1.3. Moreover, we also found (in Lemma A.1.3) that the integral was convergent but not absolutely convergent. Integrating $(\sin x / x)^2$ is a lot easier, and provides a nice application of the inversion theorem for densities and the multiplication theorem.

We first note that the integral is absolutely convergent:

$$\int_0^\infty \left(\frac{\sin x}{x}\right)^2 dx \leq \int_0^1 1 \, dx + \int_1^\infty \frac{1}{x^2} \, dx = 2.$$

Moreover, since $\sin t / t$ is the characteristic function of a $U(-1,1)$-distributed random variable, the square is the characteristic function of a $\text{Tri}(-2,2)$-distributed random variable. The inversion theorem for densities therefore tells us that, for $|x| \leq 2$,

$$\frac{1}{2}\left(1 - \frac{1}{2}|x|\right) = \frac{1}{2\pi} \int_{-\infty}^{\infty} e^{-itx}\left(\frac{\sin t}{t}\right)^2 dt,$$

so that, by putting $x = 0$, switching from $t$ to $x$, and exploiting symmetry, we obtain

$$\frac{1}{2} = \frac{1}{2\pi} \int_{-\infty}^{\infty} \left(\frac{\sin x}{x}\right)^2 dx = \frac{1}{4\pi} \int_0^{\infty} \left(\frac{\sin x}{x}\right)^2 dx,$$

in other words,

$$\int_0^\infty \left(\frac{\sin x}{x}\right)^2 dx = \frac{\pi}{2}.$$    □

## 1.3 Some Further Results

Here is another useful result.

**Theorem 1.9.** *Let $X$ be a random variable. Then*

$$\varphi_X \text{ is real} \quad \Longleftrightarrow \quad X \stackrel{d}{=} -X,$$

*(i.e., iff the distribution of $X$ is symmetric).*

*Proof.* From Theorem 1.1(b) we recall that

$$\varphi_{-X}(t) = \varphi_X(-t) = \overline{\varphi_X(t)}.$$

Now, if $\varphi_X$ is real valued, then $\overline{\varphi_X(t)} = \varphi_X(t)$, and it follows that $\varphi_{-X}(t) = \varphi_X(t)$, which means that $X$ and $-X$ have the same characteristic function, and, hence, by uniqueness, the same distribution.

If, on the other hand, they are equidistributed, then $\varphi_X(t) = \varphi_{-X}(t)$, which, together with Theorem 1.1(b), yields $\varphi_X(t) = \overline{\varphi_X(t)}$, that is, $\varphi_X$ is real valued. □

**Exercise 1.2.** Show that if $X$ and $Y$ are independent, identically distributed random variables then $X - Y$ has a symmetric distribution. □

Another useful tool concerns how to derive the characteristic function of a linearly transformed random variable from the original one.

**Theorem 1.10.** *Let $X$ be a random variable, and let $a, b \in \mathbb{R}$ and b. Then*

$$\varphi_{aX+b}(t) = e^{ibt} \cdot \varphi_X(at).$$

*Proof.* $\varphi_{aX+b}(t) = E\,e^{it(aX+b)} = e^{itb} \cdot E\,e^{i(at)X} = e^{itb} \cdot \varphi_X(at).$ □

**Exercise 1.3.** Let $X \in N(\mu, \sigma^2)$. Use the fact that the characteristic function of the standard normal distribution equals $\exp\{-t^2/2\}$ and the above theorem to show that $\varphi_X(t) = e^{it\mu - \frac{1}{2}\sigma^2 t^2}$.

## 2 Some Special Examples

In addition to the standard examples for which we listed the characteristic functions there are some special ones that are of interest. First in line is the Cantor distribution because of its special features. The Cantor distribution also provides the means to complete the convolution table we started in Section 2.19. After that we determine the characteristic function of the Cauchy distribution, partly because it permits us to introduce a pleasant device, partly because the moment generating function and the probability generating function that we introduce toward the end of this chapter do not exist for the Cauchy distribution.

### 2.1 The Cantor Distribution

In Subsection 2.2.6 we found that a Cantor distributed random variable on the unit interval could be represented as the infinite sum

$$\sum_{n=1}^{\infty} \frac{X_n}{3^n},$$

where $X$, $X_1$, $X_2$, ... are independent, identically distributed random variables such that $P(X = 0) = P(X = 2) = \frac{1}{2}$. Moreover, since $\varphi_X(t) = \frac{1}{2} + \frac{1}{2}e^{2it}$, we have

$$\varphi_{X_n/3^n}(t) = \varphi_X(t/3^n) = \frac{1}{2} + \frac{1}{2}e^{2it/3^n},$$

so that, by independence, the characteristic function of the Cantor distribution on $(0, 1)$ equals the infinite product

$$\prod_{n=1}^{\infty} \left( \frac{1}{2} + \frac{1}{2}e^{2it/3^n} \right).$$

However, for mathematical convenience and beauty we prefer, in this subsection, to consider the Cantor distribution on $[-\frac{1}{2}, \frac{1}{2}]$. We thus consider a number in $[-\frac{1}{2}, \frac{1}{2}]$ whose decimals in base 3 are $-1$ and $+1$ with probability $1/2$ each, and never 0. The analogous representation in this case is

$$Y = \sum_{n=1}^{\infty} \frac{X_n}{3^n},$$

where $X$, $X_1$, $X_2$, ... are independent, identically distributed random variables such that

$$P(X = -1) = P(X = 1) = \frac{1}{2}.$$

The following figure depicts the situation.

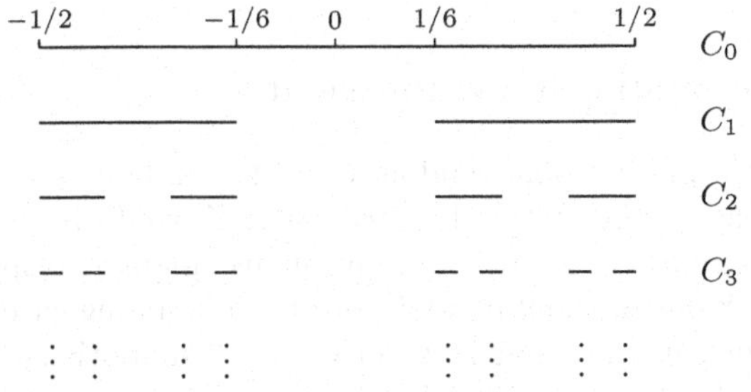

**Figure 4.1.** The Cantor set on $[-\frac{1}{2}, \frac{1}{2}]$

The characteristic function of a decimal is $\varphi_X(t) = \cos t$, so that the characteristic function of the Cantor distribution on $[-\frac{1}{2}, \frac{1}{2}]$ becomes

$$\varphi_Y(t) = \prod_{n=1}^{\infty} \varphi_{X_n/3^n}(t) = \prod_{n=1}^{\infty} \varphi_X\left(\frac{t}{3^n}\right) = \prod_{n=1}^{\infty} \cos\left(\frac{t}{3^n}\right).$$

By the same arguments we may consider the analog of the random variable corresponding to a number that is uniform over the subset of the interval $[0, 1]$, whose base 4 decimal expansion contains only 0's and 3's and no 1's or 2's, and the analogous random variable on the interval $[-\frac{1}{3}, \frac{1}{3}]$.

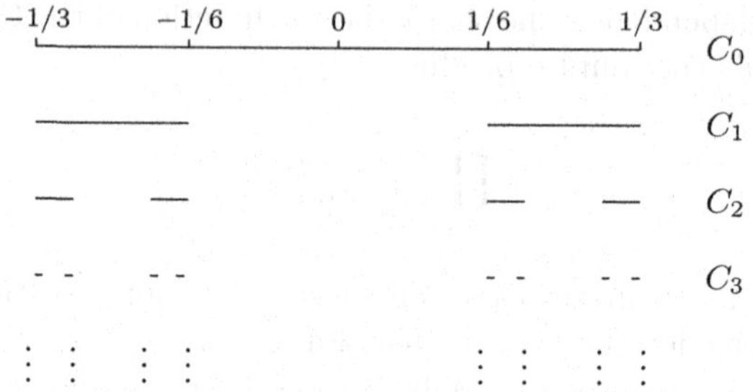

**Figure 4.2.**  A Cantor type set on $[-\frac{1}{3}, \frac{1}{3}]$

The characteristic function of the latter equals

$$\prod_{n=1}^{\infty} \cos\left(\frac{t}{4^n}\right).$$

**Exercise 2.1.** Check this fact.                                    □

We shall make use of these random variables in the following subsection.

### 2.2 The Convolution Table Revisited

Let $X$ and $Y$ be independent random variables, and set $Z = X + Y$. In Chapter 2 we investigated what type of distribution $Z$ would have if $X$ and $Y$ were absolutely continuous, discrete or continuous singular, respectively. We were able to check 8 out of the 9 possibilities, the remaining case being when both are continuous singular, that is, Cantor-type distributions. Having found the characteristic function of such distributions, we are now able to fill the last slot in the diagram.

Remember from Subsection 2.2.6 that the uniform distribution on $[0, 1]$ can be represented as the infinite sum

$$U = \sum_{n=1}^{\infty} \frac{X_n}{2^n},$$

where $X, X_1, X_2, \ldots$ are independent identically distributed random variables such that

$$P(X = 0) = P(X = 1) = \frac{1}{2},$$

If, instead, we consider the $U(-1,1)$-distribution, the representation as an infinite sum is the same, except that now

$$P(X = -1) = P(X = 1) = \frac{1}{2}.$$

By arguing precisely as for the Cantor distribution in the previous subsection, we find that

$$\varphi_U(t) = \prod_{n=1}^{\infty} \cos\left(\frac{t}{2^n}\right).$$

On the other hand, we also know that $\varphi_U(t) = \frac{\sin t}{t}$, which establishes the relation

$$\frac{\sin t}{t} = \prod_{n=1}^{\infty} \cos\left(\frac{t}{2^n}\right), \tag{2.3}$$

by a *purely probabilistic argument*.

Now, splitting the product into two factors, with the even terms in one factor and the odd ones in the other one, yields

$$\prod_{k=1}^{\infty} \cos\left(\frac{t}{2^k}\right) = \prod_{k=1}^{\infty} \cos\left(2 \cdot \frac{t}{4^k}\right) \cdot \prod_{k=1}^{\infty} \cos\left(\frac{t}{4^k}\right).$$

By the uniqueness theorem for characteristic functions,

$$U \stackrel{d}{=} 2V_1 + V_2,$$

where $V_1$ and $V_2$ are independent random variables with the common characteristic function $\prod_{k=1}^{\infty} \cos(\frac{t}{4^k})$ mentioned at the end of the previous subsection.

The punch line is that, although $V_1$ and $V_2$ both are *continuous singular* (and so is $2V_1$), the sum $2V_1 + V_2$ is uniformly distributed on the interval $[-1, 1]$, and, hence, *absolutely continuous*. This shows that it may happen that the sum of two independent, continuous singular random variables is absolutely continuous.

By arguing exactly in the same manner with the Cantor distribution,

$$\prod_{k=1}^{\infty} \cos\left(\frac{t}{3^k}\right) = \prod_{k=1}^{\infty} \cos\left(3 \cdot \frac{t}{9^k}\right) \cdot \prod_{k=1}^{\infty} \cos\left(\frac{t}{9^k}\right),$$

we find that

$$Y \stackrel{d}{=} 3W_1 + W_2,$$

where $W_1$ and $W_2$ are continuous singular, that is, it is also possible to exhibit two independent, continuous singular random variables whose sum is continuous singular.

This means that we have two options for the final slot in our diagram.

| X \ Y | AC | D | CS |
|-------|-----|-----|-----|
| AC | AC | AC | AC |
| D | AC | D | CS |
| CS | AC | CS | AC CS |

**Figure 4.3.** The distribution of $X + Y$

*Remark 2.1.* The traditional mathematical trick for (2.3) is to consider a finite product, $\Pi_N = \prod_{n=1}^{N} \cos\left(\frac{t}{2^n}\right)$, to multiply both members by $\sin(t/2^N)$, and then to apply the double angle formula for the sine function $N - 1$ times to obtain

$$\sin\left(\frac{t}{2^N}\right)\Pi_N = \frac{\sin t}{2^N}.$$

The conclusion then follows by letting $N \to \infty$ (and exploiting the fact that $\frac{\sin \alpha}{\alpha} \to 1$ as $\alpha \to 0$). □

### 2.3 The Cauchy Distribution

In order to compute the characteristic function for the Cauchy distribution we shall use a kind of distribution joining device. We call two distributions *married* if, except for a multiplicative constant, the density of one of them equals the characteristic function of the other. In this terminology the standard normal distribution is married to itself, the density being $\frac{1}{\sqrt{2\pi}}e^{-x^2/2}$ and the characteristic function being $e^{-t^2/2}$.

Let $Y_1$ and $Y_2$ be independent standard exponential random variables. Using the convolution formula one can show that the difference, $Y_1 - Y_2$, has a standard Laplace distribution, which means that

$$f_{Y_1-Y_2}(x) = \frac{1}{2}e^{-|x|} \quad \text{for} \quad -\infty < x < \infty,$$

$$\varphi_{Y_1-Y_2}(t) = \varphi_{Y_1}(t)\varphi_{Y_2}(-t) = \frac{1}{1-it} \cdot \frac{1}{1+it} = \frac{1}{1+t^2}.$$

Since this characteristic function is integrable we may apply the inversion formula for densities, Theorem 1.4, to obtain

$$\frac{1}{2}e^{-|x|} = \frac{1}{2\pi}\int_{-\infty}^{\infty} e^{-itx}\frac{1}{1+t^2}\,dt.$$

A change of variables, $x \to t$ and $t \to x$, and a deletion of 2 from the denominators yields

$$e^{-|t|} = \frac{1}{\pi}\int_{-\infty}^{\infty} e^{-ixt}\frac{1}{1+x^2}\,dx.$$

The imaginary part vanishes by symmetry, so that

$$e^{-|t|} = \int_{-\infty}^{\infty} e^{ixt}\frac{1}{\pi(1+x^2)}\,dx.$$

Inspecting the right-hand side we realize that it defines the characteristic function of the standard Cauchy distribution, which therefore must be equal to the left-hand side. We have thus shown that

$$\varphi_{C(0,1)}(t) = e^{-|t|},$$

and that the Cauchy- and Laplace distributions are a married to each other.

## 2.4 Symmetric Stable Distributions

The Cauchy distribution actually belongs to a special class of distributions, the *symmetric stable distributions*, the characteristic functions of which are

$$e^{-c|t|^{\alpha}}, \quad \text{where} \quad 0 < \alpha \le 2,$$

and where $c$ is some positive constant.

We notice immediately that the standard Cauchy distribution is symmetric stable with index 1, and that the normal distribution with mean 0 is symmetric stable with index 2.

The reason that they are called stable is that the class of distributions is closed under convolution: If $X_1, X_2, \ldots, X_n$ are symmetric with index $\alpha$, then, for any $n$, $\sum_{k=1}^{n} X_k/n^{1/\alpha}$ is also symmetric stable with index $\alpha$.

**Exercise 2.2.** Check this statement with the aid of characteristic functions.    □

That this is true for the centered normal distribution is, of course, no news. For the Cauchy distribution this means that the arithmetic mean, $\frac{S_n}{n}$, has the same distribution as an individual summand.

We shall describe these distributions in a somewhat greater detail in Section 9.1.

## 2.5 Parseval's Relation

Let $X$ and $Y$ be random variables with distribution functions $F$ and $G$, respectively, and characteristic functions $\varphi$ and $\gamma$, respectively. Thus,

$$\varphi(y) = \int_{-\infty}^{\infty} e^{iyx} \, dF(x).$$

Multiplying both members with $e^{-ity}$, integrating with respect to $G$, and applying Fubini's theorem yields

$$\int_{-\infty}^{\infty} e^{-iuy} \varphi(y) \, dG(y) = \int_{-\infty}^{\infty} e^{-iuy} \left( \int_{-\infty}^{\infty} e^{iyx} \, dF(x) \right) dG(y)$$

$$= \int_{-\infty}^{\infty} \left( \int_{-\infty}^{\infty} e^{iy(x-u)} \, dG(y) \right) dF(x)$$

$$= \int_{-\infty}^{\infty} \gamma(x - u) \, dF(x).$$

The equality between the extreme members is (one form of what is) called *Parseval's relation*.

By letting $u = 0$ we obtain the following useful formula:

$$\int_{-\infty}^{\infty} \varphi(y) \, dG(y) = \int_{-\infty}^{\infty} \gamma(x) \, dF(x). \tag{2.4}$$

The idea is to join two distributions, where the left-hand side in (2.4 is a "difficult" integral, but the right-hand side is an "easy" integral.

*Example 2.1.* Show that

$$\int_{-\infty}^{\infty} \frac{\cos y}{1 + y^2} \, dy = \frac{\pi}{e}.$$

In order to solve this problem probabilistically, we let $X$ be a coin-tossing random variable, and $Y$ a standard Cauchy-distributed random variable. By identifying the integral as $\pi \int \varphi_X(y) f_Y(y) \, dy$, we obtain, using (2.4), that

$$\int_{-\infty}^{\infty} \frac{\cos y}{1 + y^2} \, dy = \pi \int_{-\infty}^{\infty} \varphi_X(y) f_Y(y) \, dy = \pi \int_{-\infty}^{\infty} \varphi_Y(y) \, dF_X(y)$$

$$= \pi \int_{-\infty}^{\infty} e^{-|y|} \, dF_X(y) = \pi \left( e^{-|-1|} \cdot \frac{1}{2} + e^{-|1|} \cdot \frac{1}{2} \right) = \frac{\pi}{e}. \quad \square$$

**Exercise 2.3.** (a) Exploit $X$ such that $P(X = 1) = P(X = -1) = 1/2$, and $Y_1 + Y_2$, where $Y_1, Y_2 \in U(-\frac{1}{2}, \frac{1}{2})$ are independent random variables to prove that

$$\int_{-1}^{1} (1 - |y|) \cos y \, dy = 2(1 - \cos 1).$$

(b) Define two convenient random variables and prove that

$$\int_{-\infty}^{\infty} \frac{1 - \cos y}{y^2(1 + y^2)} \, dy = \frac{\pi}{e}. \qquad \square$$

# 3 Two Surprises

Here are two examples that prevent false conclusions.

*Example 3.1.* The uniqueness theorem states that if two characteristic functions coincide then so do their distributions. It is, however, *not true* that it suffices for them to coincide on some finite interval.

To see this, let

$$\varphi(t) = \begin{cases} 1 - |t|, & \text{for } |t| < 1, \\ 0, & \text{otherwise}, \end{cases} \tag{3.5}$$

which, as will be shown at the end of this example, is the characteristic function of the absolutely continuous distribution with density

$$f(x) = \frac{1 - \cos x}{\pi x^2}, \quad -\infty < x < \infty, \tag{3.6}$$

and set

$$\tilde{\varphi}(t) = \frac{1}{3}\varphi(2t) + \frac{2}{3}\varphi(t/2). \tag{3.7}$$

This is the characteristic function of a convex combination of two distributions of the above kind. More precisely, let $F_1$ and $F_2$ be the distribution functions corresponding to the characteristic functions $\varphi_1(t) = \varphi(2t)$, and $\varphi_2(t) = \varphi(t/2)$, respectively. Then $F = \frac{1}{3}F_1 + \frac{2}{3}F_2$ has characteristic function

$$\int_{-\infty}^{\infty} e^{itx}\, dF(x) = \frac{1}{3}\int_{-\infty}^{\infty} e^{itx}\, dF_1(x) + \frac{2}{3}\int_{-\infty}^{\infty} e^{itx}\, dF_2(x)$$

$$= \frac{1}{3}\varphi_1(t) + \frac{2}{3}\varphi_2(t) = \tilde{\varphi}(t).$$

The characteristic functions $\varphi$ and $\tilde{\varphi}$ clearly coincide for $|t| \leq 1/2$, but equally clearly they are not the same function, so that, by the uniqueness theorem, they must correspond to different distributions.

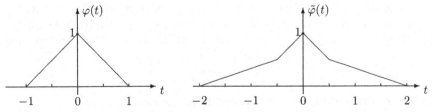

**Figure 4.4.** Two characteristic functions coinciding on an interval

Before closing this example, let us prove that the density $f$ is as claimed in (3.6). This can be done by direct computation, but also by noticing that

$\varphi$ has the same form as the density of the Tri$(-1, 1)$-distribution, that is, we have another married pair.

The triangular density is the result of $X_1 + X_2$, where $X_1$ and $X_2$ are independent $U(-\frac{1}{2}, \frac{1}{2})$-distributed random variables. Since the characteristic function of this uniform distribution equals $\frac{\sin(t/2)}{t/2}$, the characteristic function of the triangular distribution equals the square of the uniform one, which, by definition, means that

$$\int_{-1}^{1} e^{itx}(1 - |x|) \, dx = \left(\frac{\sin(t/2)}{t/2}\right)^2.$$

With $x$ and $t$ switching places, using the symmetry of the integrand and the formula for the double angle, we can rewrite this as

$$\frac{1}{2\pi}\int_{-1}^{1} e^{-ixt}(1 - |t|) \, dt = \frac{1}{2\pi}\left(\frac{\sin(x/2)}{x/2}\right)^2 = \frac{1 - \cos x}{\pi x^2}.$$

At this point we use the inversion formula for densities, Theorem 1.4, to conclude that the expression in the right-hand side is the density corresponding to the characteristic function $\varphi$ of our example.

*Example 3.2.* For real numbers it is true that if $ab = ac$ for $a \neq 0$, then $b = c$. However, the corresponding analog for characteristic functions is *not true*, that is, if $\varphi_i$ are characteristic functions ($i = 1, 2, 3$), then

$$\varphi_1 \cdot \varphi_2 = \varphi_1 \cdot \varphi_3 \quad \not\Longrightarrow \quad \varphi_2 = \varphi_3.$$

In the previous example we found that two characteristic functions may coincide on an interval without being identical. By exploiting this, let $\varphi_2$ and $\varphi_3$ be the characteristic functions $\varphi$ and $\tilde{\varphi}$ in (3.5) and (3.7), respectively. From our previous example we know that $\varphi_2$ and $\varphi_3$ coincide on the interval $(-1/2, 1/2)$. Now, let $\varphi_1(t) = \varphi(2t)$, that is, the characteristic function which is triangular with base $(-1/2, 1/2)$. Then, clearly, $\varphi_1 \cdot \varphi_2 = \varphi_1 \cdot \varphi_3$, since $\varphi$ vanishes outside the interval $(-1/2, 1/2)$. However, from our previous example we know that $\varphi_2 \neq \varphi_3$.

This fact can also be formulated in terms of random variables: Let $X$, $Y$, and $Z$ have characteristic functions $\varphi_1$, $\varphi_2$, and $\varphi_3$, respectively, and suppose that $X$ and $Z$ are independent as well as $Y$ and $Z$. Then we have shown that

$$X + Z \overset{d}{=} Y + Z \quad \not\Longrightarrow \quad X \overset{d}{=} Y,$$

in other words, we cannot "subtract" $Z$ from the equality.

The problem is that $\varphi_1$ vanishes outside the interval where $\varphi_2$ and $\varphi_3$ coincide. If this is not the case the situation is different. For example, a common procedure is "smoothing", which for mathematicians means that one convolves a function with a "nice" one in order to get a better, smoother, behavior. A probabilistic analog is to add an independent normal random variable to some random variable of interest, the point being that the sum will be absolutely continuous regardless of the random variable of interest; remember the convolution table in Section 2.19. The general idea is to end in some limiting procedure to the effect that the contribution of the normal variable vanishes asymptotically.

Some examples of this procedure will be given in Subsection 5.11.1.    □

# 4 Refinements

This section is devoted to connections between existence of moments and differentiability of the characteristic function.

**Theorem 4.1.** *Let $X$ be a random variable with distribution function $F$ and characteristic function $\varphi$.*
(i) *If $E\,|X|^n < \infty$ for some $n = 1, 2, \ldots$, then*

$$\left| \varphi(t) - \sum_{k=0}^{n} \frac{(it)^k}{k!} E\,X^k \right| \leq E \min \left\{ 2 \frac{|t|^n |X|^n}{n!}, \frac{|t|^{n+1} |X|^{n+1}}{(n+1)!} \right\}$$

*In particular,*

$$|\varphi(t) - 1| \leq E \min\{2, |tX|\},$$

*if $E|X| < \infty$, then*

$$|\varphi(t) - 1 - it E\,X| \leq E \min\{2|tX|, t^2 X^2/2\},$$

*and if $E\,X^2 < \infty$, then*

$$|\varphi(t) - 1 - it E\,X + t^2 E\,X^2/2| \leq E \min\{t^2 X^2, |tX|^3/6\}.$$

(ii) *If $E\,|X|^n < \infty$ for all $n$, and $\frac{|t|^n}{n!} E|X|^n \to 0$ as $n \to \infty$ for all $t \in \mathbb{R}$, then*

$$\varphi(t) = 1 + \sum_{k=1}^{\infty} \frac{(it)^k}{k!} E\,X^k.$$

*Proof.* Replace $y$ by $tX$ in Lemma A.1.2 and take expectations. This proves (i), from which (ii) is immediate.    □

The theorem provides us with upper bounds for the difference between the characteristic function and the first terms of the Taylor expansion when a given number of moments exist. A variation of the theme is asymptotics for small values of $t$. Such results prove most useful in proofs of convergence theorems as we shall see in some subsequent chapters.

**Theorem 4.2.** *Let $X$ be a random variable with distribution function $F$ and characteristic function $\varphi$. If $E|X|^n < \infty$ for some $n = 1, 2, \ldots,$ then $\varphi^{(k)}$, $k = 1, 2, \ldots, n$, exist and are uniformly continuous, and*

$$\varphi^{(k)}(t) = \int_{-\infty}^{\infty} (\mathrm{i}x)^k e^{\mathrm{i}tx}\, \mathrm{d}F(x),$$

$$\varphi^{(k)}(0) = \mathrm{i}^k \cdot E X^k,$$

$$\varphi(t) = 1 + \sum_{k=1}^{n} \frac{(\mathrm{i}t)^k}{k!} \cdot E X^k + o(|t|^n) \quad as \quad t \to 0.$$

*In particular, if $EX = 0$ and $\operatorname{Var} X = 1$, then*

$$\varphi(t) = 1 - \frac{1}{2}t^2 + o(t^2) \quad as \quad t \to 0.$$

*Proof.* Let $k = 1$. We thus suppose that $E|X| < \infty$. Then

$$\frac{\varphi(t+h) - \varphi(t)}{h} = \int_{-\infty}^{\infty} \frac{e^{\mathrm{i}(t+h)x} - e^{\mathrm{i}tx}}{h}\, \mathrm{d}F(x) = \int_{-\infty}^{\infty} e^{\mathrm{i}tx} \cdot \frac{e^{\mathrm{i}hx} - 1}{h}\, \mathrm{d}F(x)$$

$$\to \int_{-\infty}^{\infty} \mathrm{i}x e^{\mathrm{i}tx}\, \mathrm{d}F(x) \quad as \quad h \to 0,$$

by dominated convergence, since, by Lemma A.1.2, the integrand converges to $\mathrm{i}x \cdot e^{\mathrm{i}tx}$ as $h \to 0$, is bounded by $|x|$ for $|h| < 1$, and $E|X| < \infty$.

Since the left-hand side converges to $\varphi'(t)$ as $h \to 0$ by definition, the existence of the first derivative has been established, from which the second formula follows by putting $t = 0$.

To prove uniform continuity of the derivative we argue as in the proof of Theorem 1.1(c):

$$|\varphi'(t+h) - \varphi'(t)| = \left| \int_{-\infty}^{\infty} \mathrm{i}x e^{\mathrm{i}tx}(e^{\mathrm{i}hx} - 1)\mathrm{d}F_X(x) \right| \le \int_{-\infty}^{\infty} |x||e^{\mathrm{i}hx} - 1|\mathrm{d}F(x)$$

$$\le \int_{|x| \le A} hx^2\, \mathrm{d}F(x) + 2\int_{|x| > A} |x|\mathrm{d}F(x)$$

$$\le hA^2 + 2\int_{|x| > A} |x|\mathrm{d}F(x),$$

and so on.

This closes the case $k = 1$. The general case follows by induction. Namely, by the same procedure,

$$\frac{\varphi^{(k)}(t+h) - \varphi^{(k)}(t)}{h} = \int_{-\infty}^{\infty} e^{\mathrm{i}tx} \cdot (\mathrm{i}x)^k \frac{e^{\mathrm{i}hx} - 1}{h}\, \mathrm{d}F_X(x)$$

$$\to \int_{-\infty}^{\infty} (\mathrm{i}x)^{k+1} e^{\mathrm{i}tx}\, \mathrm{d}F(x) \quad as \quad h \to 0,$$

by dominated convergence, since, by Lemma A.1.2, the integrand converges to $(ix)^{k+1} \cdot e^{itx}$ as $h \to 0$, is bounded by $|x|^{k+1}$ for $|h| < 1$, and $E|X|^{k+1} < \infty$. Moreover, since the left-hand side converges to the derivative of order $k+1$ as $h \to 0$, the formula for the derivative has been proved, and putting $t = 0$ as before proves the second formula.

To prove uniform continuity a modification of the proof for the first derivative (please check) yields

$$|\varphi^{(k+1)}(t+h) - \varphi^{(k+1)}(t)| = \left| \int_{-\infty}^{\infty} (ix)^{k+1} e^{itx} (e^{ihx} - 1) dF_X(x) \right|$$

$$\leq hA^{k+2} + 2 \int_{|x|>A} |x|^{k+1} dF(x),$$

and so on.

It remains to establish the correct order of magnitude of the remainder in the Taylor expansion. By Theorem 4.1(i) we have

$$\left| \varphi(t) - \sum_{k=0}^{n} \frac{(it)^k}{k!} E X^k \right| \leq |t|^n E \min \left\{ 2 \frac{|X|^n}{n!}, \frac{|t||X|^{n+1}}{(n+1)!} \right\}.$$

Now, $\min \left\{ 2\frac{|X|^n}{n!}, \frac{|t||X|^{n+1}}{(n+1)!} \right\}$ converges to 0 as $t \to 0$, and is bounded by $2\frac{|X|^n}{n!}$, which is integrable by assumption, so that, by dominated convergence,

$$E \min \left\{ 2 \frac{|X|^n}{n!}, \frac{|t||X|^{n+1}}{(n+1)!} \right\} \to 0 \quad \text{as} \quad n \to \infty,$$

that is, the upper bound equals $o(|t|^n)$ as $t \to 0$.    □

**Exercise 4.1.** Find the mean and variance of the binomial, Poisson, uniform, exponential, gamma, and standard normal distributions.    □

The theorem tells us that if moments of a given order exist, then the characteristic function is differentiable as stated above. The converse, however, only holds for moments of *even* order.

**Theorem 4.3.** *Let $X$ be a random variable. If, for some $n = 0, 1, 2, \ldots$, the characteristic function $\varphi$ has a finite derivative of order $2n$ at $t = 0$, then $E|X|^{2n} < \infty$ (and the conclusions of the previous theorem hold).*

*Proof.* The proof is similar to the previous one.

First, let $k = 2$, that is, suppose that $\varphi''(0)$ is finite. Then

$$\frac{\varphi(h) - 2\varphi(0) + \varphi(-h)}{h^2} = \int_{-\infty}^{\infty} \frac{e^{ihx} - 2 + e^{-ihx}}{h^2} dF(x)$$

$$= -\int_{-\infty}^{\infty} \frac{2(1 - \cos hx)}{h^2} dF(x).$$

Since $0 \leq \frac{2(1-\cos hx)}{h^2} \to x^2$ as $h \to 0$, an application of Fatou's lemma tells us that

$$E\, X^2 \leq \liminf_{h \to 0} -\frac{\varphi(h) - 2\varphi(0) + \varphi(-h)}{h^2} = -\varphi''(0),$$

so that the variance is finite.

Higher-order results follow, once again, by induction. Thus, suppose that $\varphi^{(2n+2)}(0)$ is finite. By the induction hypothesis, Theorem 4.2 is at our disposal at the level $2n$. Therefore,

$$\frac{\varphi^{(2n)}(h) - 2\varphi^{(2n)}(0) + \varphi^{(2n)}(-h)}{h^2} = \int_{-\infty}^{\infty} (ix)^{2n} \frac{e^{ihx} - 2 + e^{-ihx}}{h^2}\, dF(x)$$

$$= -(-1)^n \int_{-\infty}^{\infty} x^{2n} \frac{2(1 - \cos hx)}{h^2}\, dF(x),$$

so that, by Fatou's lemma,

$$E\, X^{2n+2} \leq (-1)^n \varphi^{(2n+2)}(0) < \infty. \qquad \Box$$

**Exercise 4.2.** Show that $e^{-t^4}$, $\frac{1}{1+t^4}$ and $|\cos t|$ are *not* characteristic functions.$\Box$

A consequence of Theorem 4.3 is that if, for example, one wishes to find $E\, X$ via differentiation of the characteristic function one must know that the variance exists, which may be a problem if it does not.

For example, consider the symmetric stable distribution with characteristic function

$$\varphi(t) = e^{-|t|^\alpha}, \quad \text{for some} \quad \alpha \in (1, 2).$$

This distribution has a finite mean, but infinite variance. Differentiating the characteristic function twice for $t > 0$ yields

$$\varphi''(t) = -\alpha(\alpha - 1)t^{\alpha-2}e^{-t^\alpha} + \alpha^2 t^{2(\alpha-1)}e^{-t^\alpha} \nearrow +\infty \quad \text{as} \quad t \searrow 0,$$

(and similarly for $t$ negative). The second derivative does not exist at 0, so that Theorem 4.3 cannot be used in order to prove existence of a finite mean, let alone, determine its value (which must be 0 by symmetry!).

There exists, however, a way to find *absolute* moments of order $r \in (0, 2)$ via characteristic functions. The (sufficiency of the) following result is Lemma 2 of [7].

**Theorem 4.4.** *Let $X$ be a random variable with characteristic function $\varphi$. If $E|X|^r < \infty$ for some $r \in (0, 2)$, then*

$$E|X|^r = C(r) \int_{-\infty}^{\infty} \frac{1 - \Re(\varphi(t))}{|t|^{r+1}}\, dt, \qquad (4.8)$$

*where*

$$C(r) = \left( \int_{-\infty}^{\infty} \frac{1 - \cos y}{|y|^{r+1}}\, dy \right)^{-1} = \frac{\Gamma(r+1)}{\pi} \sin \frac{r\pi}{2}.$$

*Conversely, if the integral is finite, then $E|X|^r < \infty$, and (4.8) holds.*

*Proof.* The point of departure is the relation

$$|x|^r = C(r) \int_{-\infty}^{\infty} \frac{1 - \cos xt}{|t|^{r+1}} \, dt,$$

which is obtained from the definition of $C(r)$ by the change of variable $y = xt$.

Putting $x = X(\omega)$ for $\omega \in \Omega$, integrating and changing the order of integration (everything is non-negative), yields

$$E|X|^r = E\, C(r) \int_{-\infty}^{\infty} \frac{1 - \cos Xt}{|t|^{r+1}} \, dt = C(r) \int_{-\infty}^{\infty} \frac{1 - E\cos Xt}{|t|^{r+1}} \, dt$$

$$= C(r) \int_{-\infty}^{\infty} \frac{1 - \Re(\varphi(t))}{|t|^{r+1}} \, dt.$$

The converse follows by running the proof in the reverse.    □

Next we turn our attention to some useful results that connect tail probabilities and truncated moments with characteristic functions.

**Lemma 4.1.** *Let $X$ be a random variable with distribution function $F$ and characteristic function $\varphi$. For $h > 0$,*

$$P(|X| > 2/h) \leq \frac{1}{h} \int_{|t| < h} (1 - \varphi(t)) \, dt.$$

*Proof.* By Fubini's theorem, and the fact that $\frac{\sin u}{u} \leq 1/2$ for $|u| \geq 2$, hence $2(1 - \frac{\sin u}{u}) \geq 1$, we obtain

$$\frac{1}{h} \int_{|t|<h} (1 - \varphi(t)) \, dt = \frac{1}{h} \int_{|t|<h} \left(1 - \int_{-\infty}^{\infty} e^{itx} \, dF(x)\right) dt$$

$$= \frac{1}{h} \int_{|t|<h} \left(\int_{-\infty}^{\infty} (1 - e^{itx}) \, dF(x)\right) dt$$

$$= \int_{-\infty}^{\infty} \left(\frac{1}{h} \int_{|t|<h} (1 - e^{itx}) \, dt\right) dF(x) = 2 \int_{-\infty}^{\infty} \left(1 - \frac{\sin hx}{hx}\right) dF(x)$$

$$\geq 2 \int_{|x| \geq 2/h} \left(1 - \frac{\sin hx}{hx}\right) dF(x) \geq \int_{|x| \geq 2/h} dF(x) = P(|X| > 2/h).    □$$

*Remark 4.1.* Observe that the left-hand side is real, and that the integral of the imaginary part vanishes. Indeed, it *must* vanish ....    □

A corresponding inequality holds for truncated second moments.

**Lemma 4.2.** *Let $X$ be a random variable with distribution function $F$ and characteristic function $\varphi$. For $h > 0$,*

$$E\, X^2 I\{|X| < 2/h\} \leq \frac{3}{h^2}(1 - \Re(\varphi(h))).$$

*In particular, if X is symmetric, then*

$$E\,X^2 I\{|X| < 2/h\} \leq \frac{3}{h^2}(1 - \varphi(h)).$$

*Proof.* This time we exploit the cosine function; for $|u| \leq 2$,

$$\cos u \leq 1 - \frac{u^2}{2} + \frac{u^4}{4!} \leq 1 - \frac{u^2}{2} + \frac{4u^2}{4!} = 1 - \frac{u^2}{3},$$

so that

$$E\,X^2 I\{|X| < 2/h\} \leq \frac{3}{h^2} E\big(1 - \cos(hX)\big)I\{|X| < 2/h\}$$
$$\leq \frac{3}{h^2} E\big(1 - \cos(hX)\big) = \frac{3}{h^2}(1 - \mathfrak{Re}\big(\varphi(h)\big)),$$

which establishes the first inequality, from which the second one is immediate, since characteristic functions of symmetric random variables are real (Theorem 1.9).  □

## 5 Characteristic Functions of Random Vectors

Characteristic functions can naturally be defined also for random vectors, by replacing the product $tx$ in the definition by the scalar product $\mathbf{t}'\mathbf{X}$.

**Definition 5.1.** *Let $\mathbf{X} = (X_1, X_2 \ldots, X_n)'$ be a random vector. The characteristic function of $\mathbf{X}$ is*

$$\varphi_{X_1,X_2,\ldots,X_n}(t_1, t_2, \ldots, t_n) = E\,e^{i(t_1 X_1 + t_2 X_2 + \cdots + t_n X_n)},$$

*or, in the more compact form,*

$$\varphi_{\mathbf{X}}(\mathbf{t}) = E\,e^{i\mathbf{t}'\mathbf{X}}.$$  □

In particular, the following special formulas, which are useful at times, can be obtained:

$$\varphi_{\mathbf{X}}(\mathbf{t}) = \varphi_{\mathbf{t}'\mathbf{X}}(1),$$
$$\varphi_{\mathbf{X}}(t, t, \ldots, t) = \varphi_{X_1 + X_2 + \cdots + X_n}(t),$$
$$\varphi_{\mathbf{X}}(t, 0, \ldots, 0) = \varphi_{X_1}(t).$$

### 5.1 The Multivariate Normal Distribution

This is probably the most important multivariate distribution, which really would be worth a chapter of its own (for such a chapter, see, e.g., [113], Chapter V). In this subsection we provide some basic facts and interesting observations. In order to do so, however, we first need to extend the notions expected value and covariance to random vectors.

**Definition 5.2.** *The* mean vector *of* $\mathbf{X}$ *is* $\boldsymbol{\mu} = E\mathbf{X}$, *the components of which are* $\mu_k = EX_k$, $k = 1, 2, \ldots, n$.

*The* covariance matrix *of* $\mathbf{X}$ *is*

$$\boldsymbol{\Lambda} = E(\mathbf{X} - \boldsymbol{\mu})(\mathbf{X} - \boldsymbol{\mu})',$$

*whose elements are* $\lambda_{ij} = E(X_i - \mu_i)(X_j - \mu_j)$, $i, j = 1, 2, \ldots, n$, *that is,* $\lambda_{kk} = \operatorname{Var} X_k$, $k = 1, 2, \ldots, n$, *and* $\lambda_{ij} = \operatorname{Cov}(X_i, X_j) = \lambda_{ji}$, $i, j = 1, 2, \ldots, n$, $i \neq j$.  □

Since $\operatorname{Cov}(X_i, X_j) = \operatorname{Cov}(X_j, X_i)$ it follows that every covariance matrix is symmetric, and since, by linear algebra,

$$\mathbf{y}'\boldsymbol{\Lambda}\mathbf{y} = \mathbf{y}'E(\mathbf{X} - \boldsymbol{\mu})(\mathbf{X} - \boldsymbol{\mu})'\mathbf{y} = \operatorname{Var}(\mathbf{y}'(\mathbf{X} - \boldsymbol{\mu})) \geq 0 \quad \text{for any} \quad \mathbf{y} \in \mathbb{R}^n,$$

we have, moreover, shown that

**Theorem 5.1.** *Every covariance matrix is non-negative definite.*

*Remark 5.1.* If $\det \boldsymbol{\Lambda} > 0$, the probability distribution of $\mathbf{X}$ is truly $n$-dimensional in the sense that the row vectors of the covariance matrix span all of $\mathbb{R}^n$. In this case we call the distribution non-degenerate. If $\det \boldsymbol{\Lambda} = 0$, the distribution is degenerate in the sense that the row vectors span a space of a lower dimension. This is also called the *singular* case.  □

Following is a vector analog of Exercise 2.8.1.

**Theorem 5.2.** *If* $\mathbf{X}$ *has mean vector* $\boldsymbol{\mu}$ *and covariance matrix* $\boldsymbol{\Lambda}$, *then* $\mathbf{Y} = \mathbf{a} + \mathbf{BX}$ *has mean vector* $\mathbf{a} + \mathbf{B}\boldsymbol{\mu}$ *and covariance matrix* $\mathbf{B}\boldsymbol{\Lambda}\mathbf{B}'$.

**Exercise 5.1.** Prove this.  □

Now we are ready for the multivariate normal distribution.

**Definition 5.3.** *The random* $n$-vector $\mathbf{X}$ *is* normal *iff, for every vector* $\mathbf{a} \in \mathbb{R}^n$, *the (one-dimensional) random variable* $\mathbf{a}'\mathbf{X}$ *is normal. If* $\mathbf{X}$ *is normal with mean vector* $\boldsymbol{\mu}$ *and covariance matrix* $\operatorname{Cov}(\mathbf{X}) = \boldsymbol{\Lambda}$ *we write* $\mathbf{X} \in N(\boldsymbol{\mu}, \boldsymbol{\Lambda})$.  □

*Remark 5.2.* The degenerate normal distribution, $N(0,0)$, is also included as a possible distribution of $\mathbf{a}'\mathbf{X}$.

*Remark 5.3.* Note that no assumption about independence between the components of $\mathbf{X}$ is involved in the definition.  □

**Exercise 5.2.** Suppose that $\mathbf{X} = (X_1, X_2, \ldots, X_n)'$ is normal. Prove that

(a)  every component is normal;
(b)  $X_1 + X_2 + \cdots + X_n$ is normal;
(c)  every marginal distribution is normal.  □

In order to compute the characteristic function $\varphi_{\mathbf{X}}(\mathbf{t}) = E\,e^{i\mathbf{t}'\mathbf{X}}$ of a normal vector we exploit the first of the three special cases, together with the fact that $Z = \mathbf{t}'\mathbf{X}$ has a one-dimensional normal distribution with mean $m = E\,Z = \mathbf{t}'\boldsymbol{\mu}$ and variance $\sigma^2 = \mathbf{t}'\boldsymbol{\Lambda}\mathbf{t}$; please check this. Thus,

$$\varphi_{\mathbf{X}}(\mathbf{t}) = \varphi_Z(1) = \exp\{im - \sigma^2/2\} = \exp\{i\mathbf{t}'\boldsymbol{\mu} - \tfrac{1}{2}\mathbf{t}'\boldsymbol{\Lambda}\mathbf{t}\}.$$

Conversely, one can show that, given a non-negative definite matrix $\boldsymbol{\Lambda}$ and a vector $\boldsymbol{\mu}$, the function $\exp\{i\mathbf{t}'\boldsymbol{\mu} - \tfrac{1}{2}\mathbf{t}'\boldsymbol{\Lambda}\mathbf{t}\}$ is a characteristic function of a random vector, with the property that any linear combination of its components has a one-dimensional normal distribution.

In the non-singular case, that is, when $\det \boldsymbol{\Lambda} > 0$, the inverse exists, and one can show (via a set of independent, identically standard normal random variables and a change of variables) that the density of $\mathbf{X} \in N(\boldsymbol{\mu}, \boldsymbol{\Lambda})$ equals

$$f_{\mathbf{X}}(\mathbf{x}) = \left(\frac{1}{2\pi}\right)^{n/2} \frac{1}{\sqrt{\det \boldsymbol{\Lambda}}} \exp\left\{ -\tfrac{1}{2}(\mathbf{x} - \boldsymbol{\mu})'\boldsymbol{\Lambda}^{-1}(\mathbf{x} - \boldsymbol{\mu})\right\}, \quad \mathbf{x} \in \mathbb{R}^n.$$

Here is now a beautiful and impressive result.

**Theorem 5.3.** *The components of a normal random vector* $\mathbf{X}$ *are independent iff they are uncorrelated.*

*Proof.* The only thing to show is that uncorrelatedness implies independence.

By assumption, $\operatorname{Cov}(X_i, X_j) = 0$, $i \neq j$, that is, the covariance matrix is diagonal, the diagonal elements being $\sigma_1^2, \sigma_2^2, \ldots, \sigma_n^2$. Now, if a variance equals 0, then that component is degenerate and hence independent of the others. We therefore assume in the following that all variances are positive. This implies that the inverse $\boldsymbol{\Lambda}^{-1}$ is also diagonal, with diagonal elements $1/\sigma_1^2, 1/\sigma_2^2, \ldots, 1/\sigma_n^2$. The density function therefore simplifies into

$$f_{\mathbf{X}}(\mathbf{x}) = \left(\frac{1}{2\pi}\right)^{n/2} \frac{1}{\prod_{k=1}^n \sigma_k} \cdot \exp\left\{-\frac{1}{2}\sum_{k=1}^n \frac{(x_k - \mu_k)^2}{\sigma_k^2}\right\}$$

$$= \prod_{k=1}^n \frac{1}{\sqrt{2\pi}\sigma_k} \cdot \exp\left\{-\frac{(x_k - \mu_k)^2}{2\sigma_k^2}\right\},$$

which shows that the joint density factorizes, and, hence, by Theorem 2.10.2, that the components are independent.    □

Let us re-emphasize that independence is a stronger concept than uncorrelatedness in general, but that the concepts, thus, are equivalent for normal vectors. This means that the statement "$X_1, X_2, \ldots, X_n$ are independent, standard normal random variables" is equivalent to the statement "the random $n$-vector $\mathbf{X} \in N(\mathbf{0}, \mathbf{I})$".

Nevertheless, there exist normal random variables that are *not* jointly normal, for which the concepts are *not* equivalent.

**Exercise 5.3.** Let $X \in N(0,1)$, let $Z$ be a coin-tossing random variable, independent of $X$; $P(Z = 1) = P(Z = -1) = \frac{1}{2}$. Set $Y = Z \cdot X$. Prove that $X$ and $Y$ are both (standard) normal, but that the random variables are *not* jointly normal. Hint: Compute $P(X + Y = 0)$. □

Theorem 5.3 can be extended to covariance matrices that have uncorrelated blocks:

**Theorem 5.4.** *Suppose that* $\mathbf{X} \in N(\boldsymbol{\mu}, \boldsymbol{\Lambda})$, *where* $\boldsymbol{\Lambda}$ *can be partitioned as follows:*

$$\boldsymbol{\Lambda} = \begin{pmatrix} \Lambda_1 & 0 & 0 & 0 \\ 0 & \Lambda_2 & 0 & 0 \\ 0 & 0 & \ddots & 0 \\ 0 & 0 & 0 & \Lambda_k \end{pmatrix}$$

*(possibly after reordering the components), where* $\Lambda_1, \Lambda_2, \ldots, \Lambda_k$ *are matrices along the diagonal of* $\boldsymbol{\Lambda}$*. Then* $\mathbf{X}$ *can be partitioned into vectors* $\mathbf{X}^{(1)}, \mathbf{X}^{(2)}, \ldots, \mathbf{X}^{(k)}$ *with* $\mathrm{Cov}\,(\mathbf{X}^{(i)}) = \Lambda_i$, $i = 1, 2, \ldots, k$, *in such a way that these random vectors are independent.*

**Exercise 5.4.** Prove the theorem (by modifying the proof of the previous one). □

*Example 5.1.* Suppose that $\mathbf{X} \in N(\mathbf{0}, \boldsymbol{\Lambda})$, where

$$\boldsymbol{\Lambda} = \begin{pmatrix} 3 & 0 & 0 \\ 0 & 5 & 2 \\ 0 & 2 & 4 \end{pmatrix}.$$

Then $X_1$ and $(X_2, X_3)'$ are independent. □

## 5.2 The Mean and the Sample Variance Are Independent

This is, in fact, a result that characterizes the normal distribution. We shall (only) prove one half.

Let $X_1, X_2, \ldots, X_n$ be independent standard normal random variables. The sample mean and sample variance are

$$\bar{X}_n = \frac{1}{n} \sum_{k=1}^n X_k \quad \text{and} \quad s_n^2 = \frac{1}{n-1} \sum_{k=1}^n (X_k - \bar{X}_n)^2,$$

respectively.

We claim that $\bar{X}_n$ and $s_n^2$ are independent. The beauty of the proof is that it basically relies on Theorem 5.4.

In order to prove this, let $\mathbf{X_n} = (X_1, X_2, \ldots, X_n)' \in N(\mathbf{0}, \mathbf{I})$, where $\mathbf{I}$ is the identity matrix, and set

$$\mathbf{Y}_n = (\bar{X}_n, X_1 - \bar{X}_n, X_2 - \bar{X}_n, \ldots, X_n - \bar{X}_n)'.$$

In matrix notation $\mathbf{Y}_n = \mathbf{B}\mathbf{X}_n$, where

$$
\mathbf{B} = \begin{pmatrix}
\frac{1}{n} & \frac{1}{n} & \frac{1}{n} & \cdots & \frac{1}{n} \\
1-\frac{1}{n} & -\frac{1}{n} & -\frac{1}{n} & \cdots & -\frac{1}{n} \\
-\frac{1}{n} & 1-\frac{1}{n} & -\frac{1}{n} & \cdots & -\frac{1}{n} \\
-\frac{1}{n} & \frac{1}{n} & 1-\frac{1}{n} & \cdots & -\frac{1}{n} \\
\vdots & \vdots & \vdots & \ddots & \vdots \\
-\frac{1}{n} & -\frac{1}{n} & -\frac{1}{n} & \cdots & 1-\frac{1}{n}
\end{pmatrix}.
$$

By definition, $\mathbf{Y}_n$ is a normal vector, since it is a linear combination of the components of $\mathbf{X_n}$. By Theorem 2.5.2, and an easy computation, the covariance matrix of $\mathbf{Y}_n$ equals

$$
\mathrm{Cov}\,\mathbf{Y}_n = \mathbf{B}\mathbf{\Lambda}\mathbf{B}' = \mathbf{B}\mathbf{B}' = \begin{pmatrix} \frac{1}{n} & \mathbf{0} \\ \mathbf{0} & \mathbf{A} \end{pmatrix},
$$

where $\mathbf{A}$ is some matrix whose exact expression is of no importance. Namely, the essential point is the *structure* of the covariance matrix, which tells us that $\bar{X}_n$ and $(X_1-\bar{X}_n, X_2-\bar{X}_n, \ldots, X_n-\bar{X}_n)'$ are uncorrelated, and therefore, by Theorem 5.4, independent. The independence of $\bar{X}_n$ and $s_n^2$ finally follows by invoking Theorem 2.10.4, since $s_n^2$ is a function of the vector $(X_1-\bar{X}_n, X_2-\bar{X}_n, \ldots, X_n-\bar{X}_n)'$, or, equivalently, of the random variables $X_k-\bar{X}_n, 1 \le k \le n$.

This ends our discussion of the characteristic function, and we turn our attention to the other transforms mentioned in the introduction of this chapter: the cumulant generating function, the (probability) generating function, and the moment generating function.

## 6 The Cumulant Generating Function

This transform is closely related to the characteristic function.

**Definition 6.1.** *The* cumulant generating function *of a random variable $X$ is*

$$
\kappa_X(t) = \log \varphi_X(t),
$$

*where $\varphi_X(t)$ is the characteristic function of $X$.*  □

It turns out that this transform, or, rather, the coefficients in the Taylor expansion, at times are more easily accessible than those of the characteristic functions. For example, if $X \in N(0,1)$, then $\kappa_X(t) = -t^2/2$. Moreover, as we shall see in a minute, the cumulant generating function of a sum of independent random variables equals the *sum* (and not the product) of the individual cumulant generating functions.

Uniqueness is immediate, since characteristic functions uniquely determine the distribution, and since the logarithmic function is strictly increasing.

**Theorem 6.1.** (Uniqueness) *Let $X$ and $Y$ be random variables. If $\kappa_X = \kappa_Y$, then $X \overset{d}{=} Y$ (and conversely).*

Since the logarithm of a product equals the sum of the logarithms, the multiplication theorem for cumulant generating functions is, in fact, an addition theorem.

**Theorem 6.2.** ("Multiplication") *Let $X_1, X_2, \ldots, X_n$ be independent random variables, and set $S_n = X_1 + X_2 + \cdots + X_n$. Then*

$$\kappa_{S_n}(t) = \sum_{k=1}^{n} \kappa_{X_k}(t).$$

*If, in addition, $X_1, X_2, \ldots, X_n$ are equidistributed, then*

$$\kappa_{S_n}(t) = n\kappa_{X_1}(t).$$

By combining the Taylor expansion of the logarithm with Theorem 4.2, we obtain the following result; please check.

**Theorem 6.3.** *Let $X$ be a random variable with cumulant generating function $\kappa$. If $E\,|X|^n < \infty$ for some $n = 1, 2, \ldots$, then*

$$\kappa_X(t) = 1 + \sum_{k=1}^{n} \frac{(it)^k}{k!} \cdot \varkappa_k + o(|t|^n) \quad as \quad t \to 0.$$

The coefficients $\{\varkappa_k\}$ are called *cumulants* or *semi-invariants*, and were introduced in [242, 243], cf. also [56], p. 186.

Next, let $\mu_k = E\,X^k$, $k \geq 1$, denote the moments of the random variable $X$. In particular, $\mu = E\,X = \mu_1$, and $\sigma^2 = \operatorname{Var} X = \mu_2 - \mu_1^2$, all of this, provided the relevant quantities exist. By comparing the Taylor expansion of the characteristic function with that of the cumulant generating function, the following relations between moments and cumulants emerge.

**Theorem 6.4.** *Let $X$ be a random variable.*

(a)  *If $E|X| < \infty$,*  then  $\varkappa_1 = \mu_1$;
(b)  *If $E\,X^2 < \infty$,*  then  $\varkappa_2 = \mu_2 - \mu_1^2$;
(c)  *If $E|X|^3 < \infty$,*  then  $\varkappa_3 = \mu_3 - 3\mu_1\mu_2 + 2\mu_1^3$;
(d)  *If $E\,X^4 < \infty$,*  then  $\varkappa_4 = \mu_4 - 3\mu_2^2 - 4\mu_1\mu_3 + 12\mu_1^2\mu_2 - 6\mu_1^4$.

**Exercise 6.1.** If the cumulants are known one can derive the moments similarly. Do that for the first four moments.    □

In this connection we state the obvious fact that all odd moments are 0 for *all symmetric* distributions (provided the moments exist). A common measure of (a)symmetry is *skewness*, which, in the above notation, equals

$$\gamma_1 = \frac{E(X - \mu_1)^3}{\sigma^3}.$$

Another measure that relates different distributions, in particular to the normal one, is the *coefficient of excess*,

$$\gamma_2 = \frac{E(X - \mu_1)^4}{\sigma^4} - 3.$$

Note that both measures are normalized to be dimensionless. Moreover, $\gamma_1 = 0$ for all symmetric distributions, and $\gamma_2 = 0$ for the normal distribution.

**Exercise 6.2.** Prove that

$$\gamma_1 = \frac{\varkappa_3}{\varkappa_1^{3/2}} \quad \text{and that} \quad \gamma_2 = \frac{\varkappa_4}{\varkappa_2^2}.$$

Both measures are thus (somewhat) easier expressed in terms of cumulants than in terms of moments.                                                        □

# 7 The Probability Generating Function

Although, as has been mentioned, characteristic functions always exist for all random variables on the whole real axis, probability generating functions, which can be defined for non-negative random variables, are power series which have the advantage that they require less mathematics (no complex analysis for example) to be analyzed.

**Definition 7.1.** *Let $X$ be a non-negative, integer valued random variable. The (probability) generating function of $X$ is*

$$g_X(t) = E\, t^X = \sum_{n=0}^{\infty} t^n \cdot P(X = n).$$                                                        □

The generating function is defined at least for $|t| < 1$, since it is a power series with coefficients in $[0, 1]$. Moreover, $g_X(1) = \sum_{n=0}^{\infty} P(X = n) = 1$.

**Theorem 7.1.** (Uniqueness) *Let $X$ and $Y$ be non-negative, integer valued random variables. If $g_X = g_Y$, then $p_X = p_Y$ (and conversely).*

*Proof.* This follows, in fact, from the uniqueness theorem for power series. Namely, since power series can be differentiated term-wise strictly within their radius of convergence, we can do so with any generating function (at least) for $|t| < 1$, to obtain

$$g_X'(t) = \sum_{n=1}^{\infty} nt^{n-1} P(X = n), \tag{7.1}$$

$$g_X''(t) = \sum_{n=2}^{\infty} n(n-1)t^{n-2} P(X = n), \tag{7.2}$$

and, in general, for $k = 1, 2, \ldots$,

$$g_X^{(k)}(t) = \sum_{n=k}^{\infty} n(n-1)\cdots(n-k+1)t^{n-k}P(X=n). \qquad (7.3)$$

Putting $t = 0$ in the expressions for the derivatives yields

$$P(X=n) = \frac{g_X^{(n)}(0)}{n!}. \qquad (7.4)$$

Given a generating function we thus have an explicit, unique, way of computing the probabilities.  □

*Remark 7.1.* Formula (7.4) tells us that the probability generating function generates the probabilities, which makes the name of the transform most adequate.  □

The multiplication theorem is the natural analog of the corresponding one for characteristic functions and the proof is the natural analog of the corresponding proof.

**Theorem 7.2.** (Multiplication) *Let $X_1, X_2, \ldots, X_n$ be independent, non-negative, integer valued random variables, and set $S_n = X_1 + X_2 + \cdots + X_n$. Then*

$$g_{S_n}(t) = \prod_{k=1}^{n} g_{X_k}(t).$$

*If, in addition, $X_1, X_2, \ldots, X_n$ are equidistributed, then*

$$g_{S_n}(t) = (g_{X_1}(t))^n.$$

**Exercise 7.1.** Prove the theorem.

**Exercise 7.2.** Compute the generating function of some standard distributions, such as the binomial, geometric, and Poisson distributions.  □

Just as for characteristic functions one can also use generating functions to compute moments of random variables. These are obtained via the derivatives evaluated at $t = 1$. However, this requires a little more care as is seen by the following example.

*Example 7.1.* Let $X$ have probability function

$$p(n) = \frac{6}{\pi^2 n^2}, \quad n = 1, 2, 3, \ldots,$$

where the constant stems from the fact that $\sum_{n=1}^{\infty} 1/n^2 = \pi^2/6$. The divergence of the harmonic series tells us that the distribution does not have a finite mean.

The generating function is

$$g(t) = \frac{6}{\pi^2} \sum_{n=1}^{\infty} \frac{t^n}{n^2}, \quad \text{for} \quad |t| \leq 1.$$

The first derivative equals

$$g'(t) = \frac{6}{\pi^2} \sum_{n=1}^{\infty} \frac{t^{n-1}}{n} = -\frac{6}{\pi^2} \cdot \frac{\log(1-t)}{t} \nearrow +\infty \quad \text{as} \quad t \nearrow 1.$$

The point is that the generating function itself exists for $t = 1$ (they all do). The derivative, however, exists for all $t$ *strictly smaller* than 1. But not on the boundary, $t = 1$. ◻

Derivatives at $t = 1$ are therefore throughout to be interpreted as limits as $t \nearrow 1$. For simplicity, however, these derivatives will be denoted as $g'(1)$, $g''(1)$, and so on.

Here is now the connection between derivatives and moments.

**Theorem 7.3.** *Let $k \geq 1$, let $X$ be a non-negative, integer valued random variable, and suppose that $E|X|^k < \infty$. Then*

$$E\,X(X-1)\cdots(X-k+1) = g_X^{(k)}(1).$$

*In particular, if $E|X| < \infty$, then*

$$E\,X = g_X'(1),$$

*and if* $\mathrm{Var}\,X < \infty$, *then*

$$\mathrm{Var}\,X = g_X''(1) + g_X'(1) - \left(g_X'(1)\right)^2.$$

*Proof.* Letting $t \nearrow 1$ in (7.1)–(7.3) yields

$$g'(1) = E\,X,$$
$$g''(1) = E\,X(X-1),$$
$$\vdots$$
$$g^{(k)}(1) = E\,X(X-1)(X-2)\cdots(X-k+1).$$

This proves the first two assertions. The expression for the variance follows via

$$\mathrm{Var}\,X = E\,X^2 - (E\,X)^2 = E\,X(X-1) + E\,X - (E\,X)^2. \qquad ◻$$

## 7.1 Random Vectors

**Definition 7.2.** *Let* $\mathbf{X} = (X_1, X_2, \ldots, X_n)'$ *be a random vector. The (probability) generating function of* $\mathbf{X}$ *is*

$$g_{X_1,X_2,\ldots,X_n}(t_1, t_2, \ldots, t_n) = E\,t_1^{X_1} t_2^{X_2} \cdots t_n^{X_n}. \qquad ◻$$

We also note that, for example, $g_{X_1,X_2\ldots,X_n}(t, t, \ldots, t) = g_{X_1+X_2+\cdots+X_n}(t)$.

# 8 The Moment Generating Function

This is another real valued transform, which, in contrast to the probability generating function, can be defined for any distribution. The problem is, however, that it does not always exist. As we shall see in a minute, a necessary, but not sufficient, condition for existence is that all moments exist, which, immediately tells us that the Cauchy distribution does not possess a moment generating function. Nevertheless, once they exist they allow us to work in a real, less complex, world.

**Definition 8.1.** *Let $X$ be a random variable. The* moment generating function *of $X$ is*

$$\psi_X(t) = E\,e^{tX} = \int_{-\infty}^{\infty} e^{tx}\,dF_X(x),$$

provided *the expectation is finite for $|t| < h$, for some $h > 0$.*    □

*Remark 8.1.* Moment generating functions are also called two-sided Laplace transforms, the motivation being that, in analysis, Laplace transforms are defined for non-negative, real valued functions. Indeed, for a non-negative random variable $X$, one may define the Laplace transform

$$E\,e^{-sX}, \quad \text{for} \quad s \geq 0,$$

which, moreover, always exists (why?).

*Remark 8.2.* For non-negative, integer valued random variables with an existing moment generating function,

$$\psi(t) = g(e^t) \quad \text{for} \quad |t| < h.$$    □

Since we have a proviso "provided ..." let us first present an example (the Cauchy distribution, of course) for which the transform does not exist.

*Example 8.1.* Let $X$ be standard Cauchy, that is, let the density be

$$f_X(x) = \frac{1}{\pi(1+x^2)}, \quad -\infty < x < \infty.$$

The integral $\int_{-\infty}^{\infty} e^{tx}|f_X(x)|\,dx$ is clearly divergent for all $t \neq 0$. Hence, the moment generating function does not exist.    □

Next we prove the uniqueness theorem and the multiplicative property.

**Theorem 8.1.** (Uniqueness) *Let $X$ and $Y$ be random variables. If $\psi_X(t) = \psi_Y(t)$ when $|t| < h$ for some $h > 0$, then $X \overset{d}{=} Y$.*    □

*Proof.* A moment generating function which, as required, is finite in the interval $|t| < h$ can, according to (well-known) results from the theory of analytic functions, be extended to a complex function $E \exp\{zX\}$ for $|\mathfrak{Re}(z)| < h$. Putting $z = iy$, where $y$ is real, yields the characteristic function. Thus, if two moment generating functions are equal, then so are the corresponding characteristic functions, which, as we already know from Theorem 1.2, uniquely determines the distribution.    $\square$

**Theorem 8.2.** (Multiplication) *Let $X_1$, $X_2$, ..., $X_n$ be independent random variables, whose moment generating functions exist for $|t| < h$ for some $h > 0$, and set $S_n = X_1 + X_2 + \cdots + X_n$. Then*

$$\psi_{S_n}(t) = \prod_{k=1}^{n} \psi_{X_k}(t), \quad |t| < h.$$

*If, in addition, $X_1$, $X_2$, ..., $X_n$ are equidistributed, then*

$$\psi_{S_n}(t) = \big(\psi_{X_1}(t)\big)^n, \quad |t| < h.$$

**Exercise 8.1.** Prove the theorem.    $\square$

**Exercise 8.2.** Compute the moment generating function of some standard distributions, such as the binomial, geometric, Poisson, exponential, gamma, uniform, and normal distributions.    $\square$

Just as the derivatives at 0 of the probability generating function produce the probabilities (which motivated the name of the transform), the derivatives at 0 of the moment generating function produce the moments (motivating the name of the transform).

**Theorem 8.3.** *Let $X$ be a random variable whose moment generating function, $\psi_X(t)$, exists for $|t| < h$ for some $h > 0$. Then*

(i)    $E\,|X|^r < \infty \quad$ *for all* $\quad r > 0$;

(ii)    $E\,X^n = \psi_X^{(n)}(0) \quad$ *for* $\quad n = 1, 2, \ldots$;

(iii)    $\psi_X(t) = 1 + \sum_{n=1}^{\infty} \frac{t^n}{n!} E\,X^n \quad$ *fot* $\quad |t| < h$.

*Proof.* Let $r > 0$ and $|t| < h$ be given. Since $|x|^r / e^{|tx|} \to 0$ as $x \to \infty$ for all $r > 0$, we choose $A$ (depending on $r$) in such a way that

$$|x|^r \leq e^{|tx|} \quad \text{whenever} \quad |x| > A.$$

Then

$$E|X|^r = E|X|^r I\{|X| \leq A\} + E|X|^r I\{|X| > A\}$$
$$\leq A^r + Ee^{|tX|} I\{|X| > A\} \leq A^r + Ee^{|tX|} < \infty.$$

This proves (i), from which (ii) follows by differentiation (under the integral sign):

$$\psi_X^{(n)}(t) = \int_{-\infty}^{\infty} x^n e^{tx} \, dF(x),$$

which yields

$$\psi_X^{(n)}(0) = \int_{-\infty}^{\infty} x^n \, dF(x) = E\, X^n.$$

Finally, Taylor expansion and (ii) (which identifies the coefficients in the expansion) prove (iii). □

*Remark 8.3.* The idea in (i) is simply that the exponential function grows more rapidly than every polynomial. The proof amounts to translating this fact into mathematics. □

*Remark 8.4.* If we know the Taylor expansion of the moment generating function, then (iii) tells us that we may simply read off the moments; $E\, X^n$ is the coefficient of $\frac{t^n}{n!}$, $n = 1, 2, \ldots$, in the expansion. □

**Exercise 8.3.** Find the mean and variance of the binomial, Poisson, uniform, exponential, and standard normal distributions. □

## 8.1 Random Vectors

**Definition 8.2.** *Let* $\mathbf{X} = (X_1, X_2, \ldots, X_n)'$ *be a random vector. The mo-ment generating function of* $\mathbf{X}$ *is*

$$\psi_{X_1,X_2,\ldots,X_n}(t_1,\ldots,t_n) = E\, e^{t_1 X_1 + t_2 X_2 + \cdots + t_n X_n},$$

*provided there exist* $h_1, h_2, \ldots, h_n > 0$ *such that the expectation exists for* $|t_k| < h_k$, $k = 1, 2, \ldots, n$. □

*Remark 8.5.* Just as for characteristic functions one can rewrite the definition in vector notation:

$$\psi_{\mathbf{X}}(\mathbf{t}) = E\, e^{\mathbf{t}'\mathbf{X}},$$

provided there exists $\mathbf{h} > \mathbf{0}$, such that the expectation exists for $|\mathbf{t}| < \mathbf{h}$ (the inequalities being interpreted component-wise). □

## 8.2 Two Boundary Cases

The formulation of Theorem 8.3 suggests that there might exist distributions with moments of all orders and, yet, the moment generating function does not exist in any neighborhood of zero. The *log-normal distribution* is a famous such example: if $X \in \text{LN}(\mu, \sigma^2)$, then $X \stackrel{d}{=} e^Y$, where $Y \in N(\mu, \sigma^2)$.

Let $r > 0$. Then all moments exist, since

$$E\, X^r = E\, e^{rY} = \psi_Y(r) = \exp\{r\mu + \tfrac{1}{2}\sigma^2 r^2\}.$$

However, since $e^x \geq x^n/n!$ for any $n$, it follows, given any $t > 0$, that

$$
\begin{aligned}
E \exp\{tX\} = E \exp\{te^Y\} &\geq E \frac{(te^Y)^n}{n!} = \frac{t^n}{n!} E\, e^{nY} \\
&= \frac{t^n}{n!} \psi_Y(n) = \frac{t^n}{n!} \exp\{n\mu + \tfrac{1}{2}\sigma^2 n^2\} \\
&= \frac{1}{n!} \exp\{n(\log t + \mu + \tfrac{1}{2}\sigma^2 n)\} \to \infty \quad \text{as} \quad n \to \infty,
\end{aligned}
$$

since $\log t + \mu + \tfrac{1}{2}\sigma^2 n \geq \tfrac{1}{4}\sigma^2 n$ as $n \to \infty$, and $\exp\{cn^2\}/n! \to \infty$ as $n \to \infty$ for any positive constant $c$, and any $t$. The moment generating function thus does not exist.

Another example is provided by the generalized gamma distributions with density

$$
f(x) = \begin{cases} c \cdot x^{\beta-1} e^{-x^\alpha}, & \text{for} \quad x > 0, \\ 0, & \text{otherwise,} \end{cases}
$$

where $\beta > 0$, $0 < \alpha < 1$, and $c$ is a normalizing constant.

**Exercise 8.4.** Check that these distributions have the desired property (that is, of lacking a moment generating function). □

*Remark 8.6.* The moment generating function exists, however, when $\alpha \geq 1$. The case $\alpha = 1$ corresponds to the gamma distribution, and the case $\alpha = 2, \beta = 1$ corresponds to the absolute value of a normal distribution. □

We shall return to these two examples in Section 10 where we shall make some brief comments on the so-called moment problem, the problem whether or not a distribution is uniquely determined by its moment sequence.

## 9 Sums of a Random Number of Random Variables

This was the topic of Section 2.15. The model assumptions there were that $X, X_1, X_2, \ldots$ are independent, identically distributed random variables with partial sums $S_n$, $n \geq 1$, and $N$ a non-negative, integer valued random variable independent of $X_1, X_2, \ldots$. By the law of total probability, Proposition 1.4.1, we found that

$$
P(S_N \leq x) = \sum_{n=1}^{\infty} P(S_n \leq x) \cdot P(N = n), \quad -\infty < x < \infty,
$$

and, assuming absolute convergence, that

$$
E\, h(S_N) = \sum_{n=1}^{\infty} E\big(h(S_n)\big) \cdot P(N = n).
$$

By letting $h(x) = e^{itx}$ in the latter formula we obtain an expression for the characteristic function of $S_N$.

**Theorem 9.1.** *Under the above assumptions,*

$$\varphi_{S_N}(t) = g_N(\varphi_X(t)).$$

*Proof.* Recalling (2.15.4), and the fact that $E\,e^{itS_n} = \varphi_{S_n}(t) = (\varphi_X(t))^n$, we have

$$\varphi_{S_N}(t) = E\,e^{itS_N} = \sum_{n=1}^{\infty} E(e^{itS_n}) \cdot P(N = n)$$

$$= \sum_{n=1}^{\infty} (\varphi_X(t))^n \cdot P(N = n) = g_N(\varphi_X(t)). \qquad \square$$

*Remark 9.1.* In words the statement means that the characteristic function of $S_N$ is obtained by evaluating the (probability) generating function of $N$ at the point $\varphi_X(t)$. $\qquad \square$

*Example 9.1.* Let us illustrate Theorem 9.1 with our toy Example 2.15.2. The assumption there was that the number of customers in a store during one day was $Po(\lambda)$-distributed, and that the probability that a customer buys something was $p$. The number of customers that buy something was described by the random variable $S_N$, where $N \in Po(\lambda)$, and where $X_k = 1$ if customer $k$ shops and 0 otherwise. Under the independence assumptions of Theorem 9.1,

$$\varphi_{S_N}(t) = g_N(\varphi_X(t)) = \exp\{\lambda(q + pe^{it} - 1)\} = \exp\{\lambda p(e^{it} - 1)\},$$

which is the characteristic function of a $Po(\lambda p)$-distribution, so that $S_N \in Po(\lambda p)$ by the uniqueness theorem for characteristic functions, Theorem 1.2.

And if the amounts spent by the customers follow some distribution with characteristic function $\varphi$, then, by letting $\{Y_k,\ k \geq 1\}$ denote these amounts, the sum $S_N = \sum_{k=1}^{N} Y_k$ describes the total amount spent by the customers during one day, the characteristic function of which becomes

$$\varphi_{S_N}(t) = \exp\{\lambda(\varphi(t) - 1)\}. \qquad \square$$

In Theorem 4.2 we learned that mean and variance (if they exist) can be found by differentiation of the characteristic function. By differentiating both members in the relation $\varphi_{S_N}(t) = g_N(\varphi_X(t))$ we obtain

$$\varphi'_{S_N}(t) = g'_N(\varphi_X(t)) \cdot \varphi'_X(t),$$
$$\varphi''_{S_N}(t) = g''_N(\varphi_X(t)) \cdot (\varphi'_X(t))^2 + g'_N(\varphi_X(t)) \cdot \varphi''_X(t),$$

which, upon putting $t = 0$, and recalling Theorem 7.3, yields

$$iE\,S_N = \varphi'_{S_N}(0) = g'_N(\varphi_X(0)) \cdot \varphi'_X(0) = g'_N(1) \cdot \varphi'_X(0) = E\,N(iE\,X),$$

and

$$i^2 E(S_N)^2 = \varphi''_{S_N}(0) = g''_N(\varphi_X(0)) \cdot (\varphi'_X(0))^2 + g'_N(\varphi_X(0)) \cdot \varphi''_X(0)$$
$$= g''_N(1) \cdot (\varphi'_X(0))^2 + g'_N(1) \cdot \varphi''_X(0)$$
$$= E\,N(N-1)(iE\,X)^2 + E\,N(i^2 E\,X^2),$$

which, after cleaning up, reproves Theorem 2.15.1, where we found that, under appropriate assumptions,

$$E\,S_N = E\,N \cdot E\,X, \quad \text{and} \quad \operatorname{Var} S_N = E\,N \cdot \operatorname{Var} X + (E\,X)^2 \cdot \operatorname{Var} N.$$

The analogs of Theorem 9.1 for generating functions and moment generating functions run as follows.

**Theorem 9.2.** *Let* $X, X_1, X_2, \ldots$ *be independent, identically distributed, non-negative integer valued random variables with partial sums* $S_n$, $n \geq 1$, *and let* $N$ *be a non-negative, integer valued random variable independent of* $X_1, X_2, \ldots$. *Then*

$$g_{S_N}(t) = g_N(g_X(t)).$$

**Theorem 9.3.** *Let* $X, X_1, X_2, \ldots$ *be independent, identically distributed random variables whose moment generating function exists for* $|t| < h$ *for some* $h > 0$, *and let* $S_n$, $n \geq 1$, *denote their partial sums. Further, let* $N$ *be a non-negative, integer valued random variable independent of* $X_1, X_2, \ldots$. *Then*

$$\psi_{S_N}(t) = g_N(\psi_X(t)).$$

**Exercise 9.1.** Prove the two theorems.                                        □


## 10 The Moment Problem

The moment problem concerns the question whether or not a given sequence of moments, $\{m_n, n \geq 1\}$, uniquely determines the associated probability distribution or random variable. The case when the support of the summands is the whole real axis is called the *Hamburger* moment problem. When the support is the positive half-axis one talks about the *Stieltjes* moment problem. So far no conveniently applicable necessary *and* sufficient condition has been found.

A trivial *sufficient* condition is the existence of the moment generating function, recall Theorem 8.1. All standard distributions possessing moments of all orders, such as the normal, the Poisson, the binomial, the exponential, and the geometric distributions are uniquely determined by their moment sequences.

A more sophisticated sufficient condition is the Carleman condition, [37] (where the condition appears in the context of quasi-analytic functions), which states that a distribution is uniquely determined by its moment sequence if

$$\sum_{k=1}^{\infty} m_{2k}^{-1/2k} = \infty. \tag{10.1}$$

For non-negative random variables the condition becomes

$$\sum_{k=1}^{\infty} m_k^{-1/2k} = \infty. \tag{10.2}$$

A *necessary* (but not sufficient) condition in the absolutely continuous case is due to Krein, [164], who proved that, in order for a distribution to be uniquely determined by its moment sequence, it is necessary that

$$K_H = \int_{-\infty}^{\infty} \frac{-\log f(y)}{1 + y^2}\, dy = \infty,$$

in the Hamburger case. The analog for the Stieltjes problem, where the necessary condition is

$$K_S = \int_0^{\infty} \frac{-\log f(y^2)}{1 + y^2}\, dy = \infty,$$

was obtained by Slud; see [228]. In both cases $f$ is the corresponding density.

In order to find a moment sequence that does *not* uniquely determine the distribution we must look for distributions with moments of all orders and at the same time without an existing moment generating function.

Earlier in this chapter, in Subsection 4.8.2, we found that the log-normal distribution is an example of that kind. Heyde [133] proved that this distribution is not uniquely determined by the moment sequence by exhibiting a family of distributions having the same moments; see also [222]. Checking the Krein-Slud integral is an alternative way to prove this: The density of the log-normal distribution is

$$f(x) = \frac{1}{\sigma \log x \sqrt{2\pi}} \exp\left\{ -\frac{(\log x - \mu)^2}{2\sigma^2} \right\}, \quad x > 0,$$

so that

$$K_S = \int_0^{\infty} \frac{C + \log(2\log x) + \frac{(2\log x - \mu)^2}{2\sigma^2}}{1 + x^2}\, dx < \infty.$$

Another example was the generalized gamma distributions. In this case

$$K_S = \int_0^{\infty} \frac{C - 2(\beta - 1)\log x + x^{2\alpha}}{1 + x^2}\, dx \quad \begin{cases} < \infty, & \text{when} \quad \alpha < 1/2, \\ = \infty, & \text{when} \quad \alpha \geq 1/2. \end{cases}$$

It follows that the moments do not determine the distribution uniquely when $0 < \alpha < 1/2$. For $1/2 \leq \alpha < 1$ the integral diverges, but that is not enough for uniqueness. One possibility is to check the Carleman condition (10.2):

$$E\,X^k = C \int_0^\infty x^k x^{\beta-1} e^{-x^\alpha}\,dx = C \int_0^\infty y^{\frac{\beta+k}{\alpha}-1} e^{-y}\,dy = C\Gamma\left(\frac{\beta+k}{\alpha}\right)$$

$$\sim C\left(\frac{\beta+k}{e\alpha}\right)^{(\beta+k)/\alpha} \sqrt{2\pi\frac{\beta+k}{\alpha}} \quad \text{as} \quad k \to \infty,$$

which implies that

$$\sum_{k=1}^\infty m_k^{-1/2k} = C \sum_{n=1}^\infty \left(\frac{e\alpha}{\beta+k}\right)^{\frac{1}{2\alpha}+\frac{\beta}{2k\alpha}} \left(2\pi\frac{\alpha}{\beta+k}\right)^{1/4k}$$

$$\geq C \sum_{n=1}^\infty \left(\frac{e\alpha}{\beta+k}\right)^{\frac{1}{2\alpha}} \geq \sum_{n=1}^\infty \frac{1}{\beta+k} = \infty.$$

The moments thus determine the distribution uniquely when $1/2 \leq \alpha < 1$.

Just to complete the picture, when $\alpha \geq 1$ the moment generating function exists, and the moments (trivially) determine the distribution uniquely.

*Remark 10.1.* There exists a rather handy condition due to Lin [174], which, ensures uniqueness when the Krein-Slud integral diverges.                          □

## 10.1 The Moment Problem for Random Sums

Let, once again, $X$, $X_1$, $X_2$, ... be independent, identically distributed random variables with partial sums $S_n$, $n \geq 1$, and suppose that $N$ is a non-negative, integer valued random variable which is independent of $X_1$, $X_2$, .... (For $N = 0$ we set $S_N = S_0 = 0$.) All random variables are supposed to possess moments of all orders. The moment problem for geometrically compounded sums was investigated in [175], where also a number of conjectures concerning the determinacy of $S_N$ were raised, such as *if $X$ is uniquely determined by its moments and $N$ is not, then neither is $S_N$*. With the aid of Theorem 9.1 it is rather straightforward to verify (a little more than) the truth of this conjecture.

**Theorem 10.1.** (i) *If $N$ is not uniquely determined by its moment sequence, then neither is $S_N$.*
(ii) *If $X$ is not uniquely determined by its moment sequence, then neither is $S_N$.*

*Remark 10.2.* The content of the theorem thus is that if at least one of $X$ and $N$ is not uniquely determined by its moment sequence, then neither is $S_N$, or equivalently, if $S_N$ is uniquely determined by its moment sequence, then, necessarily, so are $X$ as well as $N$.                          □

*Proof.* (i): Let $N_1$ and $N_2$ be random variables with the same moment sequence but with different distributions. Theorem 9.1 then tells us that

$$\varphi_{S_{N_1}}(t) = g_{N_1}(\varphi_X(t)) \quad \text{and} \quad \varphi_{S_{N_2}}(t) = g_{N_2}(\varphi_X(t)).$$

By uniqueness of the transform, $S_{N_1}$ and $S_{N_2}$ have different distributions. However, by, for example, differentiation, we find that the moments of $S_{N_1}$ and $S_{N_2}$ coincide.
(ii): The proof is the same with $X_1$ and $X_2$ playing the role of $N_1$ and $N_2$, and $N$ the role of $X$. □

More on the conjectures can be found in [116].

# 11 Problems

1. Show, by using characteristic, or moment generating functions, that if

$$f_X(x) = \frac{1}{2}e^{-|x|}, \quad -\infty < x < \infty,$$

then $X \overset{d}{=} Y_1 - Y_2$, where $Y_1$ and $Y_2$ are independent, exponentially distributed random variables.

2. Let $X_1, X_2, \ldots, X_n$ be independent random variables with expectation 0 and finite third moments. Show, with the aid of characteristic functions, that

$$E(X_1 + X_2 + \cdots + X_n)^3 = E X_1^3 + E X_2^3 + \cdots + E X_n^3.$$

3. Let $X_1, X_2, \ldots, X_n$ be independent, Exp(1)-distributed random variables, and set

$$V_n = \max_{1 \le k \le n} X_k \quad \text{and} \quad W_n = X_1 + \tfrac{1}{2}X_2 + \tfrac{1}{3}X_3 + \cdots + \tfrac{1}{n}X_n.$$

Show that $V_n \overset{d}{=} W_n$.

4. Suppose that $P(X = 1) = P(X = -1) = 1/2$, that $Y \in U(-1,1)$ and that $X$ and $Y$ are independent.
   (a) Show, by direct computation, that $X + Y \in U(-2,2)$.
   (b) Translate the result to a statement about characteristic functions.
   (c) Which well-known trigonometric formula did you discover?

5. Let $X$ and $Y$ be independent random variables and let the superscript $s$ denote symmetrization (recall Section 3.6). Show that

$$(X + Y)^s \overset{d}{=} X^s + Y^s.$$

6. Show that one cannot find independent, identically distributed random variables $X$ and $Y$ such that $X - Y \in U(-1,1)$.

7. Consider the function

$$\varphi(t) = (1 + |t|)e^{-|t|}, \quad t \in \mathbb{R}.$$

   (a) Prove that $\varphi$ is a characteristic function.
   (b) Prove that the corresponding distribution is absolutely continuous.

(c) Prove, departing from $\varphi$ itself, that the distribution has finite mean and variance.

(d) Prove, without computation, that the mean equals 0.

(e) Compute the density.

8. Suppose that the moments of the random variable $X$ are constant, that is, suppose that $E\,X^n = c$ for all $n \geq 1$, for some constant $c$. Find the distribution of $X$.

9. Let $X$ be a random variable, such that

$$E\,X^n = \frac{2n!}{3\lambda^n}, \quad n = 1, 2, \ldots.$$

Find the distribution of $X$. Is it unique?

10. Prove that, if $\varphi(t) = 1 + o(t^2)$ as $t \to 0$ is a characteristic function, then $\varphi \equiv 1$.

11. Prove or disprove:

(a) If $\varphi$ is a characteristic function, then so is $\varphi^2$;

(b) If $\varphi$ is a non-negative characteristic function, then so is $\sqrt{\varphi}$.

12. Prove that

$$\frac{1}{\pi} \int_{-\infty}^{\infty} \frac{(\cos y)^2}{1 + y^2}\, dy = \frac{1 + e^{-2}}{2},$$

$$\int_{-\infty}^{\infty} \frac{1}{(1 + y^2)^2}\, dy = \frac{\pi}{2},$$

$$\frac{1}{\pi} \int_{-\infty}^{\infty} \frac{1 - \cos y}{y^2} e^{-|y|}\, dy = \frac{\pi}{2} - \log 2.$$

13. Invent and prove your own relation.

14. Let $\varphi$ be a characteristic function, and $F$ the distribution function of a non-negative random variable. Prove that

$$\int_0^1 \varphi(tu)\, du,$$

$$\int_{-\infty}^{\infty} \varphi(tu) e^{-|u|}\, du,$$

$$\int_{-\infty}^{\infty} \frac{\varphi(tu)}{1 + u^2}\, du,$$

$$\int_0^{\infty} \varphi(tu)\, dF(u),$$

are characteristic functions, and describe the corresponding random variables.

15. Recall formula $(2.3) - \frac{\sin t}{t} = \prod_{n=1}^{\infty} \cos\left(\frac{t}{2^n}\right)$. Derive *Vieta's formula*

$$\frac{2}{\pi} = \frac{\sqrt{2}}{2} \cdot \frac{\sqrt{2 + \sqrt{2}}}{2} \cdot \frac{\sqrt{2 + \sqrt{2 + \sqrt{2}}}}{2} \cdot \frac{\sqrt{2 + \sqrt{2 + \sqrt{2 + \sqrt{2}}}}}{2} \cdots\cdots.$$

16. Let $\varphi$ be a characteristic function. Prove that

$$|\varphi(t) - \varphi(t+h)|^2 \le 2(1 - \Re(\varphi(h))).$$

17. Let $X$ be a random variable with characteristic function $\varphi$. Show that

$$P(|X| > 1/h) \le \frac{7}{h} \int_0^h (1 - \Re(\varphi(t))) \, dt.$$

18. The characteristic function of $\mathbf{X}$ equals

$$\varphi(s,t,u) = \exp\left\{-\frac{s^2}{2} - t^2 - 2u^2 - \frac{st}{2} + \frac{su}{2} - \frac{tu}{2}\right\}.$$

Determine the distribution of $\mathbf{X}$.

19. Suppose that the mean vector and the covariance matrix of the three-dimensional normal random vector $\mathbf{X}$ are

$$\boldsymbol{\mu} = \begin{pmatrix} 3 \\ 4 \\ -3 \end{pmatrix} \quad \text{and} \quad \boldsymbol{\Lambda} = \begin{pmatrix} 2 & 1 & 3 \\ 1 & 4 & -2 \\ 3 & -2 & 8 \end{pmatrix},$$

respectively. Determine the distribution of $X_1 - 2X_2 + 3X_3$.

20. The random variables $X$ and $Y$ are independent and $N(0,1)$-distributed. Determine
    (a) $E(X \mid X > Y)$;
    (b) $E(X + Y \mid X > Y)$.

21. Suppose that $\varphi$ is a characteristic function, and let $p \in (0,1)$. Show that

$$\tilde{\varphi}(t) = \frac{p}{1 - (1-p)\varphi(t)}$$

is a characteristic function.

22. Let $N \in \mathrm{Bin}(n, 1 - e^{-m})$, let $X, X_1, X_2, \ldots$ have the same 0-truncated Poisson distribution,

$$P(X = x) = \frac{m^x}{x!}/(e^m - 1), \quad x = 1, 2, 3, \ldots,$$

and assume that $N, X_1, X_2, \ldots$ are independent.
    (a) Find the distribution of $Y = \sum_{k=1}^N X_k$, (where $Y = 0$ when $N = 0$).
    (b) Compute $E Y$ and $\mathrm{Var}\, Y$ without using (a).

23. Consider a branching process (Section 2.14), where, thus, $X(n)$ denotes the number of individuals in generation $n$, $n \ge 0$, with $X(0) = 1$, and where $g_n = g_{X(n)}$, with $g_1 = g$.
    (a) Prove that $g_2(t) = g(g(t))$, and, generally, that $g_n(t) = g(g_{n-1}(t))$.
    (b) Suppose that the mean number of children, $m = E X(1) < \infty$. Determine $E X(2)$, and, more generally, $E X(n)$.

(c) Let $T_n = \sum_{k=0}^{n} X(k)$ be the total progeny up to and including generation number $n$, and let $G_n(t)$ be the generating function of $T_n$. Show that

$$G_n(t) = t \cdot g\big(G_{n-1}(t)\big).$$

(d) Suppose that $m < 1$. Argue that the probability of extinction equals 1 (it can be shown that this is, indeed, the case), and find the expected value of the total progeny.

24. Suppose that the lifetimes, $X_1$, $X_2$, $\ldots$, of the individuals in a tribe are independent, $\mathrm{Exp}(\frac{1}{a})$-distributed random variables, that the number of individuals, $N$, in the tribe is $\mathrm{Fs}(p)$-distributed, viz.

$$P(N = n) = p(1 - p)^{n-1}, \quad \text{for} \quad n \in \mathbb{N},$$

and that $N$ and $X_1$, $X_2$, $\ldots$ are independent. Determine the distribution of the shortest lifetime.

25. Suppose that the random variables $X_1$, $X_2$, $\ldots$ are independent with common distribution function $F(x)$, and let $N$ be a non-negative, integer valued random variable with generating function $g(t)$. Finally, suppose that $N$ and $X_1$, $X_2$, $\ldots$ are independent. Set

$$Y = \max\{X_1, X_2, \ldots, X_N\}.$$

Show that

$$F_Y(y) = g\big(F(y)\big).$$

# 5

## Convergence

From the introductory chapter we remember that the basis of probability theory, the empirical basis upon which the modeling of random phenomena rests, is the stabilization of the relative frequencies. In statistics a rule of thumb is to base one's decisions or conclusions on *large* samples, if possible, because large samples have smoothing effects, the more wild randomness that is always there in small samples has been smeared out. The frequent use of the normal distribution (less nowadays, since computers can do a lot of numerical work within a reasonable time) is based on the fact that the arithmetic mean of some measurement in a sample is approximately normal when the sample is large. And so on. All of this triggers the notion of *convergence*. Let $X_1$, $X_2$, ... be random variables. What can be said about their sum, $S_n$, as the number of summands increases $(n \to \infty)$? What can be said about the largest of them, $\max\{X_1, X_2, \ldots, X_n\}$ as $n \to \infty$? What about the limit of sums of sequences? About functions of converging sequences? In mathematics one discusses point-wise convergence and convergence of integrals. When, if at all, can we assert that the integral of a limit equals the limit of the integrals? And what do such statements amount to in the context of random variables?

This and a lot more is what we are going to meet in the present chapter. The following three chapters are then devoted to the three most fundamental results in probability theory: the law of large numbers, the central limit theorem, and the law of the iterated logarithm, respectively.

We begin by defining various modes of convergence, prove uniqueness of the limits, and relate them into a hierarchical system. We then investigate what additional conditions might ensure implications between the concepts which are not there from the outset, such as when is the limit of an expectation equal to the expectation of the limit? Some of this is reminiscent of what we met in Section 2.5. Another section concerns transforms. In Chapter 4 we found results relating distributional equalities to equalities between transforms. In this chapter we relate results about "almost equality" to each other, that is, we prove convergence theorems. We also provide examples and applications.

# 1 Definitions

There are several convergence concepts in probability theory. We shall discuss five of them here. Let $X_1$, $X_2$, ... be random variables.

**Definition 1.1.** $X_n$ *converges* almost surely (a.s.) *to the random variable X as* $n \to \infty$ *iff*
$$P(\{\omega : X_n(\omega) \to X(\omega) \ as \ n \to \infty\}) = 1.$$
Notation: $X_n \overset{a.s.}{\to} X$ *as* $n \to \infty$.

**Definition 1.2.** $X_n$ *converges* in probability *to the random variable X as* $n \to \infty$ *iff, for every* $\varepsilon > 0$,
$$P(|X_n - X| > \varepsilon) \to 0 \quad as \quad n \to \infty.$$
Notation: $X_n \overset{p}{\to} X$ *as* $n \to \infty$.

**Definition 1.3.** $X_n$ *converges* in $r$-mean *to the random variable X as* $n \to \infty$ *iff*
$$E|X_n - X|^r \to 0 \quad as \quad n \to \infty.$$
Notation: $X_n \overset{r}{\to} X$ *as* $n \to \infty$.

**Definition 1.4.** *Let* $C(F_X) = \{x : F_X(x) \ is \ continuous \ at \ x\}$ = *the* continuity set *of* $F_X$. $X_n$ *converges* in distribution *to the random variable X as* $n \to \infty$ *iff*
$$F_{X_n}(x) \to F_X(x) \quad as \quad n \to \infty, \quad for \ all \quad x \in C(F_X).$$
Notation: $X_n \overset{d}{\to} X$ *as* $n \to \infty$.     □

These are the four traditional, most common, convergence concepts.

*Remark 1.1.* Almost sure convergence is also called convergence with probability 1 (w.p.1).

*Remark 1.2.* Definition 1.3 with $r = 2$ is called convergence in square mean to the (or mean-square convergence).

*Remark 1.3.* Since the random variables in Definition 1.4 are present only in terms of their distribution functions, they need not be defined on the same probability space.

*Remark 1.4.* We shall permit ourselves the convenient abuse of notation such as $X_n \overset{d}{\to} N(0, 1)$ instead of the formally more correct, but lengthier "$X_n \overset{d}{\to} X$ as $n \to \infty$, where $X \in N(0, 1)$".     □

The following more modern definition of distributional convergence avoids the mentioning of discontinuity points. The fact that the two definitions are equivalent will be established in Theorem 6.1.

**Definition 1.5.** $X_n \overset{d}{\to} X$ *as* $n \to \infty$ *iff, for every* $h \in C_B$,

$$E\, h(X_n) \to E\, h(X) \quad as \quad n \to \infty. \qquad \square$$

*Remark 1.5.* The definition in terms of expectations is most suitable for extensions to higher dimensions and to more general function spaces, such as $C[0, 1]$, the space of continuous functions on the interval $[0, 1]$ endowed with the uniform topology. Distributional convergence is often called weak convergence in these more general settings. We refer to [20] for an excellent treatment of this topic. $\qquad \square$

We shall also meet a somewhat less common, but very useful convergence concept, introduced by Hsu and Robbins in [140], which, as we immediately note from its nature, is closely related to the Borel-Cantelli lemmas.

**Definition 1.6.** $X_n$ converges completely *to the random variable* $X$ *as* $n \to \infty$ *iff*

$$\sum_{n=1}^{\infty} P(|X_n - X| > \varepsilon) < \infty, \quad for\ all \quad \varepsilon > 0.$$

Notation: $X_n \overset{c.c.}{\to} X$ *as* $n \to \infty$. $\qquad \square$

*Remark 1.6.* The limiting variable in [140] is supposed to be a constant. We have extended the definition here for convenience. $\qquad \square$

## 1.1 Continuity Points and Continuity Sets

In the traditional definition of convergence in distribution one has to check convergence at continuity points of the limiting distribution. To see that this makes sense, consider the following example.

*Example 1.1.* Suppose that $X_n \in \delta(\frac{1}{n})$, that is, $X_n$ is the one-point distribution with its mass concentrated at the point $1/n$. If Definition 1.4 makes sense one should have $X_n \overset{d}{\to} \delta(0)$ as $n \to \infty$. Checking the distribution function we have

$$F_{X_n}(x) = \begin{cases} 0, & \text{for } x < \frac{1}{n}, \\ 1, & \text{for } x \geq \frac{1}{n}, \end{cases} \quad \to \quad \begin{cases} 0, & \text{for } x \leq 0, \\ 1, & \text{for } x > 0. \end{cases}$$

Thus, $F_n(x) \to F_{\delta(0)}(x)$ as $n \to \infty$ for all $x \in C(F_{\delta(0)})$, but not for *every* $x$. If, on the other hand, $Y_n \in \delta(-\frac{1}{n})$, then

$$F_{Y_n}(x) = \begin{cases} 0, & \text{for } x < -\frac{1}{n}, \\ 1, & \text{for } x \geq -\frac{1}{n}, \end{cases} \quad \to \quad \begin{cases} 0, & \text{for } x < 0, \\ 1, & \text{for } x \geq 0, \end{cases}$$

so that, in this case we have convergence for *all* $x$. Now, since $X_n$ as well as $Y_n$ are "close" to 0 when $n$ is large, it would be awkward if only one of them would converge to 0. Luckily, the origin does not trouble us since it belongs to the discontinuity set, so that both sequences converge in distribution to the $\delta(0)$-distribution as $n \to \infty$ according to the definition.    □

In this example there was only one discontinuity point. However, since a distribution function has at most a countable number of discontinuities (Lemma 2.2.1), the set of discontinuity points has Lebesgue measure 0.

An additional way to illustrate the fact that only continuity points matter is via the second definition. Namely, one can show that Definition 1.5 is equivalent to

$$P(X_n \in A) \to P(X \in A) \quad \text{for every } P\text{-}continuity \text{ set,}$$

that is, for every set $A$, such that $P(\partial A) = 0$.

With respect to Definition 1.4 we observe that if $X$ is a random variable with distribution function $F$, then

$$x \in C(F) \quad \Longleftrightarrow \quad P(X = x) = P(X \in \partial((-\infty, x])) = 0,$$

that is $x$ is a continuity *point* precisely when $(-\infty, x]$ is a continuity *set*.

Another example is the number of successes in independent repetitions, which follows the binomial distributions. These are distributions with support on the non-negative integers. It is well known (and will be seen again in Chapter 7) that if $X_n$ counts the number of successes after $n$ repetitions, then $X_n$ suitably normalized, converges in distribution to a standard normal distribution, $N$, say. Now, $P(X_n \in \{0, 1, 2, \ldots, n\}) = 1$, so that, if $p$ is the success probability and we let

$$A = \left\{ \frac{j - np}{\sqrt{np(1-p)}} : j = 0, 1, \ldots, n, \ n = 1, 2, \ldots \right\},$$

then

$$P\left( \frac{X_n - np}{\sqrt{np(1-p)}} \in A \right) = 1, \quad \text{whereas} \quad P(N \in A) = 0.$$

In other words, we do not have convergence on the set $A$. This is, however, no problem, because $A$ is not an $N$-continuity set; $\partial A = \mathbb{R}$.

A more pathological example, typically of mathematics, cannot be resisted.

*Example 1.2.* Let $X_n$ be uniformly distributed on $\{\frac{1}{n}, \frac{2}{n}, \ldots, 1\}$. Then

$$F_n(x) = \begin{cases} 0, & \text{for } x < 0, \\ \frac{[nx]}{n} \to x, \quad \text{as} \quad n \to \infty, & \text{for } 0 \le x < 1, \\ 1, & \text{for } x \ge 1, \end{cases}$$

In other words, $X_n \overset{d}{\to} X \in U(0, 1)$ as $n \to \infty$ (which is the natural guess).

The pathology is explained via the rationals and irrationals as is common-place in mathematics; remember also Example 2.2.1. We have

$$P(X_n \in \mathbb{Q}) = 1, \quad \text{whereas} \quad P(X \in \mathbb{Q}) = 0.$$

The problem is, of course, no problem, since $P(X \in \partial\mathbb{Q}) = 1$, that is, $\mathbb{Q}$ is not an $X$-continuity set. $\qquad\qquad\square$

## 1.2 Measurability

Before we proceed we must check that the convergence concepts make mathematical sense, that is, that there are no measurability problems.

Inspecting the definitions we notice that the only one that needs to be taken care of is almost sure convergence in Definition 1.1, where we consider every $\omega \in \Omega$ and check whether or not the sequence of real numbers $X_n(\omega)$ converges to the real number $X(\omega)$ as $n \to \infty$. Almost sure convergence holds if the $\omega$-set for which there is convergence has probability 1 or, equivalently, if the $\omega$-set for which we do not have convergence has probability 0. So, we must prove that the convergence set

$$\{\omega : X_n(\omega) \to X(\omega) \quad \text{as} \quad n \to \infty\}$$

is measurable.

Now, this set can be rewritten as

$$A = \bigcap_{\varepsilon > 0} \bigcup_{m=1}^{\infty} \bigcap_{i=m}^{\infty} \{|X_i - X| \le \varepsilon\}, \tag{1.1}$$

which is *not* necessarily measurable, because an uncountable intersection is involved. However an equivalent way to express the convergence set is (why?)

$$A = \bigcap_{n=1}^{\infty} \bigcup_{m=1}^{\infty} \bigcap_{i=m}^{\infty} \left\{|X_i - X| \le \frac{1}{n}\right\}, \tag{1.2}$$

which is a measurable set. It thus makes sense to talk about $P(A)$. Moreover, the following convenient criteria for checking almost sure convergence emerge.

**Proposition 1.1.** *Let $X_1, X_2, \ldots$ be random variables. Then $X_n \overset{a.s.}{\to} X$ as $n \to \infty$ iff*

$$P(A) = P\left(\bigcap_{n=1}^{\infty} \bigcup_{m=1}^{\infty} \bigcap_{i=m}^{\infty} \left\{|X_i - X| \le \frac{1}{n}\right\}\right)$$

$$= \lim_{n \to \infty} P\left(\bigcup_{m=1}^{\infty} \bigcap_{i=m}^{\infty} \left\{|X_i - X| \le \frac{1}{n}\right\}\right)$$

$$= \lim_{n \to \infty} \lim_{m \to \infty} P\left(\bigcap_{i=m}^{\infty} \left\{|X_i - X| \le \frac{1}{n}\right\}\right) = 1,$$

*or, equivalently, iff*

$$P(A^c) = P\left( \bigcup_{n=1}^{\infty} \bigcap_{m=1}^{\infty} \bigcup_{i=m}^{\infty} \left\{ |X_i - X| > \frac{1}{n} \right\} \right)$$

$$= \lim_{n \to \infty} P\left( \bigcap_{m=1}^{\infty} \bigcup_{i=m}^{\infty} \left\{ |X_i - X| > \frac{1}{n} \right\} \right)$$

$$= \lim_{n \to \infty} \lim_{m \to \infty} P\left( \bigcup_{i=m}^{\infty} \left\{ |X_i - X| > \frac{1}{n} \right\} \right) = 0.$$

A minor rewriting of the conditions in Proposition 1.1, taking the equivalence between (1.1) and (1.2) into account, yields the following equivalent criteria for almost sure convergence.

**Proposition 1.2.** *Let $X_1$, $X_2$, ... be random variables. Then $X_n \overset{a.s.}{\to} X$ as $n \to \infty$ iff, for every $\varepsilon > 0$,*

$$P\left( \bigcup_{m=1}^{\infty} \bigcap_{i=m}^{\infty} \{ |X_i - X| \le \varepsilon \} \right) = \lim_{m \to \infty} P\left( \bigcap_{i=m}^{\infty} \{ |X_i - X| \le \varepsilon \} \right) = 1,$$

*or, equivalently, iff, for every $\varepsilon > 0$,*

$$P\left( \bigcap_{m=1}^{\infty} \bigcup_{i=m}^{\infty} \{ |X_i - X| > \varepsilon \} \right) = \lim_{m \to \infty} P\left( \bigcup_{i=m}^{\infty} \{ |X_i - X| > \varepsilon \} \right) = 0.$$

**Exercise 1.1.** Check the details.                                                    □

### 1.3 Some Examples

Before moving onto the theory we present some examples to illustrate the concepts. As for almost sure convergence we already encountered some cases in Section 2.18.

*Example 1.3.* Let $X_n \in \Gamma(n, \frac{1}{n})$. Show that $X_n \overset{p}{\to} 1$ as $n \to \infty$.

We first note that $E X_n = 1$ and that $\operatorname{Var} X_n = \frac{1}{n}$. An application of Chebyshev's inequality shows that, for all $\varepsilon > 0$,

$$P(|X_n - 1| > \varepsilon) \le \frac{1}{n\varepsilon^2} \to 0 \qquad \text{as} \quad n \to \infty.$$

*Example 1.4.* Let $X_1$, $X_2$, ... be independent with common density

$$f(x) = \begin{cases} \alpha x^{-\alpha-1}, & \text{for} \quad x > 1, \quad \alpha > 0, \\ 0, & \text{otherwise,} \end{cases}$$

and set $Y_n = n^{-1/\alpha} \cdot \max_{1 \le k \le n} X_k$, $n \ge 1$. Show that $Y_n$ converges in distribution as $n \to \infty$, and determine the limit distribution.

In order to solve this problem, we first compute the distribution function:

$$F(x) = \begin{cases} \int_1^x \alpha y^{-\alpha-1} \, dy = 1 - x^{-\alpha}, & \text{for} \quad x > 1, \\ 0, & \text{otherwise,} \end{cases}$$

from which it follows that, for any $x > 0$,

$$F_{Y_n}(x) = P(\max_{1 \le k \le n} X_k \le x n^{1/\alpha}) = \left(F(x n^{1/\alpha})\right)^n$$

$$= \left(1 - \frac{1}{n x^\alpha}\right)^n \to e^{-x^{-\alpha}} \quad \text{as} \quad n \to \infty.$$

*Example 1.5. The law of large numbers.* This is a very important result that will be proved in greater generality in Theorem 6.3.1. The current version is traditionally proved in a first course in probability.

Let $X_1, X_2, \ldots$ be a sequence of independent, identically distributed random variables with mean $\mu$ and finite variance $\sigma^2$, and set $S_n = X_1 + X_2 + \cdots + X_n$, $n \ge 1$. The law of large numbers states that

$$\frac{S_n}{n} \xrightarrow{p} \mu \quad \text{as} \quad n \to \infty.$$

To prove this statement we let $\varepsilon$ be arbitrary, and invoke Chebyshev's inequality:

$$P\left(\left|\frac{S_n}{n} - \mu\right| > \varepsilon\right) \le \frac{\sigma^2}{n\varepsilon^2} \to 0 \quad \text{as} \quad n \to \infty.$$

*Example 1.6.* This example concerns Poisson approximation of the binomial distribution. For the sake of illustration we assume, for simplicity, that $p = p_n = \lambda/n$.

Thus, suppose that $X_n \in \text{Bin}(n, \frac{\lambda}{n})$. Then

$$X_n \xrightarrow{d} \text{Po}(\lambda) \quad \text{as} \quad n \to \infty.$$

The elementary proof involves showing that, for fixed $k$,

$$\binom{n}{k}\left(\frac{\lambda}{n}\right)^k \left(1 - \frac{\lambda}{n}\right)^{n-k} \to e^{-\lambda}\frac{\lambda^k}{k!} \quad \text{as} \quad n \to \infty.$$

We omit the details. □

## 2 Uniqueness

We begin by proving that convergence is unique – in other words, that the limiting random variable is uniquely defined in the following sense: If $X_n \to X$ and $X_n \to Y$ completely, almost surely, in probability, or in $r$-mean, then $X = Y$ almost surely, that is, $P(X = Y) = 1$ (or, equivalently, $P(\{\omega : X(\omega) \ne Y(\omega)\}) = 0$). For distributional convergence, uniqueness means $F_X(x) = F_Y(x)$ for all $x$, that is, $X \overset{d}{=} Y$.

**Theorem 2.1.** *Let $X_1$, $X_2$, ... be a sequence of random variables. If $X_n$ converges completely, almost surely, in probability, in r-mean, or in distribution as $n \to \infty$, then the limiting random variable (distribution) is unique.*

*Proof.* Suppose first that $X_n \overset{a.s.}{\to} X$ and that $X_n \overset{a.s.}{\to} Y$ as $n \to \infty$. Let

$$N_X = \{\omega : X_n(\omega) \not\to X(\omega) \text{ as } n \to \infty\},$$

and

$$N_Y = \{\omega : X_n(\omega) \not\to Y(\omega) \text{ as } n \to \infty\}.$$

Since $N_X$ and $N_Y$ are null sets, so is their union. The conclusion follows via the triangle inequality, since, for $\omega \notin N_X \cup N_Y$,

$$|X(\omega) - Y(\omega)| \leq |X(\omega) - X_n(\omega)| + |X_n(\omega) - Y(\omega)| \to 0 \quad \text{as} \quad n \to \infty,$$

so that $X = Y$ a.s.

Next suppose that $X_n \overset{p}{\to} X$ and that $X_n \overset{p}{\to} Y$ as $n \to \infty$, and let $\varepsilon > 0$ be arbitrary. Then

$$P(|X - Y| > \varepsilon) \leq P\left(|X - X_n| > \frac{\varepsilon}{2}\right) + P\left(|X_n - Y| > \frac{\varepsilon}{2}\right) \to 0 \quad \text{as} \quad n \to \infty,$$

so that, once again, $P(X = Y) = 1$.

Uniqueness for complete convergence follows by summation;

$$\sum_{n=1}^{\infty} P(|X - Y| > \varepsilon) \leq \sum_{n=1}^{\infty} P\left(|X - X_n| > \frac{\varepsilon}{2}\right) + \sum_{n=1}^{\infty} P\left(|X_n - Y| > \frac{\varepsilon}{2}\right) < \infty,$$

for all $\varepsilon > 0$. (Note, in particular, that we have obtained an infinite, convergent, sum of identical terms!)

Now suppose that $X_n \overset{r}{\to} X$ and that $X_n \overset{r}{\to} Y$ as $n \to \infty$. The $c_r$-inequality (Theorem 3.2.2) yield

$$E|X - Y|^r \leq c_r(E|X - X_n|^r + E|X_n - Y|^r) \to 0 \quad \text{as} \quad n \to \infty,$$

so that $E|X - Y|^r = 0$, in particular $X - Y = 0$ a.s.

Finally, suppose that $X_n \overset{d}{\to} X$ and that $X_n \overset{d}{\to} Y$ as $n \to \infty$, and let $x \in C(F_X) \cap C(F_Y)$; remember that $(C(F_X) \cap C(F_Y))^c$ contains at most a countable number of points. Then, once again, by the triangle inequality,

$$|F_X(x) - F_Y(x)| \leq |F_X(x) - F_{X_n}(x)| + |F_{X_n}(x) - F_Y(x)| \to 0 \quad \text{as} \quad n \to \infty,$$

which shows that $F_X(x) = F_Y(x)$, $\forall x \in C(F_X) \cap C(F_Y)$. In view of the right continuity of distribution functions (Lemma 2.2.3), we finally conclude that $F_X(x) = F_Y(x)$ for *all* $x$.  □

# 3 Relations Between Convergence Concepts

A natural first problem is to determine the hierarchy between the convergence concepts.

**Theorem 3.1.** *Let $X$ and $X_1$, $X_2$, ... be random variables. The following implications hold as $n \to \infty$:*

$$X_n \overset{c.c.}{\to} X \quad \Longrightarrow \quad X_n \overset{a.s.}{\to} X \quad \Longrightarrow \quad X_n \overset{p}{\to} X \quad \Longrightarrow \quad X_n \overset{d}{\to} X$$

$$\Uparrow$$

$$X_n \overset{r}{\to} X$$

*All implications are strict.*

PROOF. We proceed via several steps.

**I.**  $X_n \overset{c.c.}{\to} X \Longrightarrow X_n \overset{a.s.}{\to} X$

Immediate from the Borel-Cantelli lemma; Theorem 2.18.1. Note also that $X_n \overset{c.c.}{\to} X \Longrightarrow X_n \overset{p}{\to} X$, since the terms of a convergent sum tend to 0.

**II.**  $X_n \overset{a.s.}{\to} X \Longrightarrow X_n \overset{p}{\to} X$

This follows from Proposition 1.2, since

$$\lim_{m \to \infty} P(|X_i - X| > \varepsilon) \leq \lim_{m \to \infty} P\left( \bigcup_{i=m}^{\infty} \{|X_i - X| > \varepsilon\} \right) = 0 \quad \text{for any} \quad \varepsilon > 0.$$

**III.**  $X_n \overset{r}{\to} X \Longrightarrow X_n \overset{p}{\to} X$

This is a consequence of Markov's inequality, Theorem 3.1.1, since

$$P(|X_n - X| > \varepsilon) \leq \frac{E|X_n - X|^r}{\varepsilon^r} \to 0 \quad \text{as} \quad n \to \infty \quad \text{for every} \quad \varepsilon > 0.$$

**IV.**  $X_n \overset{p}{\to} X \Longrightarrow X_n \overset{d}{\to} X$

Let $\varepsilon > 0$. Then

$$
\begin{aligned}
F_{X_n}(x) = P(X_n \leq x) &= P(\{X_n \leq x\} \cap \{|X_n - X| \leq \varepsilon\}) \\
&\quad + P(\{X_n \leq x\} \cap \{|X_n - X| > \varepsilon\}) \\
&\leq P(\{X \leq x + \varepsilon\} \cap \{|X_n - X| \leq \varepsilon\}) + P(|X_n - X| > \varepsilon) \\
&\leq P(X \leq x + \varepsilon) + P(|X_n - X| > \varepsilon),
\end{aligned}
$$

so that, by exploiting convergence in probability,

$$\limsup_{n\to\infty} F_{X_n}(x) \leq F_X(x+\varepsilon).$$

By switching $X_n$ to $X$, $x$ to $x - \varepsilon$, $X$ to $X_n$, and $x + \varepsilon$ to $x$, it follows, analogously, that

$$\liminf_{n\to\infty} F_{X_n} \geq F_X(x - \varepsilon).$$

The last two relations hold for all $x$ and for all $\varepsilon > 0$. To prove convergence in distribution, we finally suppose that $x \in C(F_X)$ and let $\varepsilon \to 0$, to conclude that

$$F_X(x) = F_X(x-) \leq \liminf_{n\to\infty} F_{X_n}(x) \leq \limsup_{n\to\infty} F_{X_n}(x) \leq F_X(x).$$

Since $x \in C(F_X)$ was arbitrary, we are done.

*Remark 3.1.* If $F_X$ has a jump at $x$, then we can only conclude that

$$F_X(x-) \leq \liminf_{n\to\infty} F_{X_n}(x) \leq \limsup_{n\to\infty} F_{X_n}(x) \leq F_X(x).$$

Since $F_X(x) - F_X(x-)$ equals the size of the jump we can never obtain convergence at a jump, which, however, is no problem.     □

Looking back we have justified all arrows, and it remains to show that they are strict, and that there are no other arrows.

The following example will be used for these purposes.

*Example 3.1.* This is a slight variation of Example 2.5.1.

Let, for $\alpha > 0$, $X_1$, $X_2$, ... be independent random variables such that

$$P(X_n = 0) = 1 - \frac{1}{n^\alpha} \quad \text{and} \quad P(X_n = n) = \frac{1}{n^\alpha}, \quad n \geq 1.$$

The following statements hold:

$$
\begin{aligned}
X_n &\xrightarrow{p} 0 &\text{as} \quad n \to \infty &\quad \text{even without independence,} \\
X_n &\xrightarrow{a.s.} 0 &\text{as} \quad n \to \infty &\quad \text{iff} \quad \alpha > 1, \\
X_n &\xrightarrow{c.c.} 0 &\text{as} \quad n \to \infty &\quad \text{iff} \quad \alpha > 1, \\
X_n &\xrightarrow{r} 0 &\text{as} \quad n \to \infty &\quad \text{iff} \quad \alpha > r.
\end{aligned}
$$

Convergence in probability is a consequence of the fact that

$$P(|X_n| > \varepsilon) = P(X_n = n) = \frac{1}{n^\alpha} \to 0 \quad \text{as} \quad n \to \infty.$$

The complete and almost sure facts follows from the Borel-Cantelli lemmas, since

$$\sum_{n=1}^{\infty} P(|X_n| > \varepsilon) \begin{cases} < +\infty & \text{when} \quad \alpha > 1, \\ = +\infty & \text{when} \quad \alpha \leq 1. \end{cases}$$

As for mean convergence,

$$E|X_n|^r = 0^r \cdot \left(1 - \frac{1}{n^\alpha}\right) + n^r \cdot \frac{1}{n^\alpha}$$

$$= n^{r-\alpha} \begin{cases} \to 0, & \text{for} \quad r < \alpha, \\ = 1, & \text{for} \quad r = \alpha, \quad \text{as} \quad n \to \infty. \\ \to +\infty, & \text{for} \quad r > \alpha. \end{cases}$$

Note that $E|X_n|^r$ neither converges to 0 nor diverges to infinity when $r = \alpha$, but equals (in fact, not even converges to) "the wrong number", 1.  □

We have now made the preparations we need in order to complete the proof of the theorem.

In order to show that the left-most arrow is strict we confine ourselves at present to referring to the already cited Example 2.18.2, and repeat that the proof of that statement will be given in Section 6.11 ahead.

As for the second arrow, we note that for $\alpha \leq 1$, we do not have almost sure convergence to 0. But we cannot have almost sure convergence to any other limit either. Namely, if $X_n \xrightarrow{a.s.} X$, say, then $X_n \xrightarrow{p} X$ as well, which is impossible because of the uniqueness. Consequently, $X_n \overset{a.s.}{\nrightarrow}$ as $n \to \infty$, which establishes the strictness of the second arrow.

That the vertical arrow is strict follows from the fact that $E|X_n - X|^r$ might not even exist. There are, however, cases when $X_n \xrightarrow{p} X$ as $n \to \infty$, whereas $E|X_n - X|^r \nrightarrow 0$ as $n \to \infty$. To see this we can exploit our favorite Example 3.1 with $\alpha < r$, because in that case we do not have $r$-mean convergence (to 0 and, by arguing as a few lines ago, not to any other limit either).

Next we note that for $1 < \alpha < r$, $X_n$ converges to 0 completely and almost surely but not in $r$-mean, and if $r < \alpha \leq 1$, then $X_n$ converges to 0 in $r$-mean but neither completely nor almost surely.

It remains to find an example where we have convergence in distribution, but not in probability. As we already have mentioned, distributional convergence does not require jointly distributed random variables, so it is necessarily the weakest convergence concept. However, the following example provides jointly distributed random variables that converge in distribution only.

*Example 3.2.* Toss a symmetric coin, set $X = 1$ for heads and $X = 0$ for tails, and let $X_{2n} = X$ and $X_{2n-1} = 1 - X$, $n \geq 1$. Since $X, X_1, X_2, \ldots$ all have the same distribution, it follows, in particular, that $X_n \xrightarrow{d} X$ as $n \to \infty$. However, $X_n \overset{p}{\nrightarrow} X$ as $n \to \infty$, because of the repeated sign change.  □

The proof of the theorem, finally, is complete.  □

## 3.1 Converses

Beyond Theorem 3.1 there exist, under certain additional assumptions, converses to some of the arrows. In the first two cases to follow we require the limit $X$ to be degenerate, that is, that $P(X = c) = 1$ for some constant $c$.

**Theorem 3.2.** *If $X_1$, $X_2$, ... are independent and $c$ a constant, then*

$$X_n \overset{c.c.}{\to} c \quad \Longleftrightarrow \quad X_n \overset{a.s.}{\to} c \quad as \quad n \to \infty.$$

*Proof.* Since both statements are equivalent to

$$\sum_{n=1}^{\infty} P(|X_n - c| > \varepsilon) < \infty \quad \text{for all} \quad \varepsilon > 0,$$

the conclusion follows from the Borel-Cantelli lemmas, Theorems 2.18.1 and 2.18.2.    □

*Remark 3.2.* It is vital that the limit is degenerate, because $\{X_n - X, n \geq 1\}$ are *not* independent random variables (in contrast to $\{X_n - c, n \geq 1\}$), so that the second Borel-Cantelli lemma is not applicable in the more general case.    □

**Theorem 3.3.** *Let $X_1$, $X_2$, ... be random variables and $c$ a constant. Then*

$$X_n \overset{d}{\to} \delta(c) \quad as \quad n \to \infty \quad \Longleftrightarrow \quad X_n \overset{p}{\to} c \quad as \quad n \to \infty.$$

*Proof.* In view of Theorem 3.1 we only have to prove the converse. Thus, assume that $X_n \overset{d}{\to} \delta(c)$ as $n \to \infty$, and let $\varepsilon > 0$. Then

$$\begin{aligned}
P(|X_n - c| > \varepsilon) &= 1 - P(c - \varepsilon \leq X_n \leq c + \varepsilon) \\
&= 1 - F_{X_n}(c + \varepsilon) + F_{X_n}(c - \varepsilon) - P(X_n = c - \varepsilon) \\
&\leq 1 - F_{X_n}(c + \varepsilon) + F_{X_n}(c - \varepsilon) \to 1 - 1 + 0 \\
&= 0 \quad \text{as} \quad n \to \infty,
\end{aligned}$$

since $F_{X_n}(c + \varepsilon) \to F_X(c + \varepsilon) = 1$, $F_{X_n}(c - \varepsilon) \to F_X(c - \varepsilon) = 0$, and $c + \varepsilon$ and $c - \varepsilon \in C(F_X) = \{x : x \neq c\}$.    □

Another kind of partial converse runs as follows.

**Theorem 3.4.** *Let $X_1$, $X_2$, ... be random variables such that $X_n \overset{p}{\to} X$ as $n \to \infty$. Then there exists a non-decreasing subsequence $\{n_k, k \geq 1\}$ of the positive integers, such that*

$$X_{n_k} \overset{c.c.}{\to} X \quad as \quad n \to \infty,$$

*in particular,*

$$X_{n_k} \overset{a.s.}{\to} X \quad as \quad n \to \infty.$$

*Proof.* By assumption there exists a non-decreasing subsequence, $\{n_k,\ k \geq 1\}$, such that

$$P\Big(|X_{n_k} - X| > \frac{1}{2^k}\Big) < \frac{1}{2^k}.$$

Consequently,

$$\sum_{k=1}^{\infty} P\Big(|X_{n_k} - X| > \frac{1}{2^k}\Big) < \infty.$$

Since $\frac{1}{2^k} < \varepsilon$ for any $\varepsilon > 0$ whenever $k > \log(1/\varepsilon)/\log 2$, it follows that

$$\sum_{k=1}^{\infty} P(|X_{n_k} - X| > \varepsilon) < \infty,$$

which proves complete convergence, from which almost sure convergence is immediate via Theorem 3.1. □

**Theorem 3.5.** *Let $X_1$, $X_2$, ... be a monotone sequence of random variables, and suppose that $X_n \xrightarrow{p} X$ as $n \to \infty$. Then*

$$X_n \xrightarrow{a.s.} X \qquad as \quad n \to \infty.$$

*Proof.* According to the previous result there exists an almost surely convergent subsequence $\{X_{n_k},\ k \geq 1\}$. Suppose w.l.o.g. that $X_1$, $X_2$, ... are increasing, and let $\omega$ outside the exceptional null set be given. Then, for any $\varepsilon > 0$, there exists $k_0(\omega)$, such that

$$X(\omega) - X_{n_k}(\omega) < \varepsilon \qquad \text{for all} \quad k \geq k_0(\omega).$$

The monotonicity now forces

$$X(\omega) - X_n(\omega) < \varepsilon \qquad \text{for all} \quad n \geq k_0(\omega),$$

which establishes almost sure convergence. □

There also exists a kind of converse to the implication $X_n \xrightarrow{a.s.} X \Longrightarrow X_n \xrightarrow{d} X$ as $n \to \infty$. The reason we write "a kind of converse" is that the converse statement (only) is a *representation*. The details of this result, which is known as Skorohod's representation theorem, will be given in Section 5.13.

It is also possible to turn the implication $X_n \xrightarrow{r} X \Longrightarrow X_n \xrightarrow{p} X$ as $n \to \infty$ around under some additional condition, namely, uniform integrability. The mathematical analog is to switch the order between taking limits and integrating;

$$\lim \int = \int \lim,$$

something we remember requires some additional condition such as uniformity or domination. This is the topic of the following two sections.

## 4 Uniform Integrability

Knowing that convergence in probability does not necessarily imply mean convergence, a natural question is whether there exist conditions that guarantee that a sequence that converges in probability (or almost surely or in distribution) also converges in $r$-mean. It turns out that *uniform integrability* is the adequate concept for this problem.

**Definition 4.1.** *A sequence $X_1$, $X_2$, ... is called* uniformly integrable *iff*

$$E|X_n|I\{|X_n| > a\} \to 0 \quad as \quad a \to \infty \quad uniformly \ in \ n. \qquad \square$$

Another, equivalent, way to express uniform integrability is via the distribution function; $X_1$, $X_2$, ... is uniformly integrable iff

$$\int_{|x|>a} |x| \, dF_{X_n}(x) \to 0 \quad as \quad a \to \infty \quad uniformly \ in \ n.$$

*Remark 4.1.* The assumption that $X_1$, $X_2$, ... have finite mean, implies that $E|X_n|I\{|X_n| > a\} \to 0$ as $a \to \infty$ for every $n$; the tails of convergent integrals converge to 0. The requirement that the sequence is uniformly integrable means that the contributions in the tails of the integrals tend to 0 *uniformly* for all members of the sequence. $\qquad \square$

Since for a uniformly integrable sequence of random variables,

$$E|X_n| = E|X_n|I\{|X_n| \le a\} + E|X_n|I\{|X_n| > a\} \le a + 1,$$

for $a$ large enough, it follows immediately that the moments are uniformly bounded. However, uniform integrability is more, which is illustrated by the following theorem.

**Theorem 4.1.** *The random variables $X_1$, $X_2$, ... are uniformly integrable iff*
(i) $\sup_n E|X_n| < \infty$;
(ii) *for any $\varepsilon > 0$ there exists $\delta > 0$, such that for any set $A$ with $P(A) < \delta$,*

$$E|X_n|I\{A\} < \varepsilon \quad uniformly \ in \ n.$$

*Proof.* Suppose first that $X_1$, $X_2$, ... is uniformly integrable, let $\varepsilon$ be given and let $A$ be such that $P(A) < \delta$. The uniform boundedness of the moments has already been verified a few lines ago. As for (ii),

$$E|X_n|I\{A\} = E|X_n|I\{A \cap \{|X_n| \le a\}\} + E|X_n|I\{A \cap \{|X_n| > a\}\}$$
$$\le aP(A) + E|X_n|I\{|X_n| > a\} \le a\delta + \varepsilon/2 < \varepsilon,$$

if we first choose $a$ large enough to make the second term small enough and then $\delta$ small enough in order to ensure that $a\delta < \varepsilon/2$.

If, on the other hand, the conditions of the theorem are fulfilled, we set $A_n = \{|X_n| > a\}$, and apply Markov's inequality and (i) to obtain

$$P(A_n) \leq \frac{E|X_n|}{a} \leq \frac{\sup_n E|X_n|}{a} < \delta \quad \text{uniformly in } n,$$

for $a$ sufficiently large, which, by (ii), shows that

$$E|X_n|I\{|X_n| > a\} = E|X_n|I\{A_n\} < \varepsilon \quad \text{uniformly in } n,$$

thus establishing uniform integrability. $\quad\square$

It may be difficult at times to verify uniform integrability directly. Following are some convenient sufficient criteria.

**Theorem 4.2.** *Let $X_1$, $X_2$, ... be random variables, and suppose that*

$$\sup_n E|X_n|^p < \infty \quad \text{for some} \quad p > 1.$$

*Then $\{X_n, n \geq 1\}$ is uniformly integrable. In particular this is the case if $\{|X_n|^p, n \geq 1\}$ is uniformly integrable for some $p > 1$.*

*Proof.* We have

$$E|X_n|I\{|X_n| > a\} \leq a^{1-p}E|X_n|^p I\{|X_n| > a\} \leq a^{1-p}E|X_n|^p$$
$$\leq a^{1-p}\sup_n E|X_n|^p \to 0 \quad \text{as} \quad a \to \infty,$$

independently, hence uniformly, in $n$.

The particular case is immediate since more is assumed. $\quad\square$

With a little bit more effort one can prove the following generalization.

**Theorem 4.3.** *Let $X_1$, $X_2$, ... be random variables and $g$ a non-negative increasing function such that $g(x)/x \to \infty$ as $x \to \infty$. If*

$$\sup_n E\,g(X_n) < \infty,$$

*then $\{X_n, n \geq 1\}$ is uniformly integrable.*

*Proof.* By assumption,

$$\frac{g(x)}{x} > b \quad \text{for all } x > \text{some } a(b) > 0.$$

Hence, given $\varepsilon > 0$,

$$E|X_n|I\{|X_n| > a\} \leq \frac{1}{b}E\,g(X_n)I\{|X_n| > a\} \leq \frac{1}{b}\sup_n E\,g(X_n) < \varepsilon,$$

independently of $n$ if $b$ is large enough, that is, for all $a > a(b)$. $\quad\square$

Theorem 4.2 corresponds, of course, to the case $g(x) = |x|^p$. However, Theorem 4.3 also covers situations when no polynomial moment of order strictly greater than 1 exists. One such example is $g(x) = |x|\log(1+|x|)$. The following kind of "in between" result only presupposes first moments, but in terms of a dominating random variable. The proof is short, but the assumption about domination is rather strong. On the other hand, no higher-order moments are required.

**Theorem 4.4.** *Suppose that $X_1$, $X_2$, ... are random variables such that*

$$|X_n| \leq Y \quad a.s. \quad \text{for all} \quad n,$$

*where $Y$ is a positive integrable random variable. Then $\{X_n, n \geq 1\}$ is uniformly integrable.*

*Proof.* This follows by observing that

$$E|X_n|I\{|X_n| > a\} \leq EYI\{Y > a\} \to 0 \quad \text{as} \quad a \to \infty,$$

independently of (hence uniformly in) $n$. □

**Corollary 4.1.** *Let $X_1$, $X_2$, ... be a sequence of random variables. If*

$$E\sup_n |X_n| < \infty,$$

*then $\{X_n, n \geq 1\}$ is uniformly integrable.*

*Remark 4.2.* The converse is not true in general. Namely, let $Y_1$, $Y_2$, ... are independent, identically distributed random variables, and let $X_n$, $n \geq 1$, denote the arithmetic means; $X_n = \frac{1}{n}\sum_{k=1}^n Y_k$, $n \geq 1$. Then (note that $X_1 = Y_1$)

$$\{X_n, n \geq 1\} \text{ is uniformly integrable} \quad \Longleftrightarrow \quad E|X_1| < \infty,$$
$$E\sup_n |X_n| < \infty \quad \Longleftrightarrow \quad E|X_1|\log^+ |X_1| < \infty.$$

Proofs of these fact and a bit more will be given in Subsection 10.16.1. □

An analysis of the proof of Theorem 4.4 reveals that the following stregthening holds.

**Theorem 4.5.** *Let $X_1$, $X_2$, ... be random variables such that*

$$|X_n| \leq Y_n \quad a.s. \quad \text{for all} \quad n,$$

*where $Y_1$, $Y_2$, ... are positive integrable random variables. If $\{Y_n, n \geq 1\}$ is uniformly integrable, then so is $\{X_n, n \geq 1\}$.*

**Exercise 4.1.** Prove Theorem 4.5. □

In order to prove uniform integrability it may happen that one succeeds in doing so for all elements in a sequence with an index beyond some number. The following lemma shows that this is enough for the whole sequence to be uniformly integrable. This may seem obvious, however, the conclusion is not necessarily true for *families* of random variables, the reason being that there is an uncountable number of variables with an index prior to any given fixed index in that case.

**Lemma 4.1.** *Let $X_1$, $X_2$, ... be random variables. If $\{X_n, n \geq N > 1\}$ is uniformly integrable, then so is $\{X_n, n \geq 1\}$.*

*Proof.* By assumption,

$$\sup_{n \geq N} E|X_n|I\{|X_n| > a\} < \varepsilon \quad \text{for} \quad a > a_0.$$

Moreover, for $1 \leq n < N$,

$$E|X_n|I\{|X_n| > a\} < \varepsilon \quad \text{for} \quad a > a_n.$$

Combining these facts we find that

$$\sup_{n \geq 1} E|X_n|I\{|X_n| > a\} < \varepsilon \quad \text{for} \quad a > \max\{a_0, a_1, a_2, \ldots, a_{N-1}\}. \qquad \square$$

Another useful fact is that the sum of two uniformly integrable sequences is uniformly integrable.

**Theorem 4.6.** *If $\{X_n, n \geq 1\}$ and $\{Y_n, n \geq 1\}$ are uniformly integrable, then so is $\{X_n + Y_n, n \geq 1\}$.*

*Proof.* Let $a > 0$. Then,

$$|X_n + Y_n|I\{|X_n + Y_n| > a\} \leq 2\max\{|X_n|, |Y_n|\}I\{2\max\{|X_n|, |Y_n|\} > a\}$$
$$\leq 2|X_n|I\{|X_n| > a/2\} + 2|Y_n|I\{|Y_n| > a/2\}.$$

Taking expectations, and letting $a \to \infty$ proves the claim. $\qquad \square$

Note that this result is *not* transferable to products, since the product of two integrable random variables need not even be integrable. Handling products is a different story; recall the Hölder inequality, Theorem 3.2.4. However, the following holds.

**Theorem 4.7.** *Let $p, q > 1$, with $p^{-1} + q^{-1} = 1$. If $\{|X_n|^p, n \geq 1\}$ and $\{|Y_n|^q, n \geq 1\}$ are uniformly integrable, then so is $\{X_n \cdot Y_n, n \geq 1\}$.*

*Proof.* Let $a > 0$. We have

$$|X_n \cdot Y_n|I\{|X_n \cdot Y_n| > a\} \leq |X_n| \cdot |Y_n|I\{|X_n| > \sqrt{a}\} + |X_n| \cdot |Y_n|I\{|Y_n| > \sqrt{a}\}.$$

Taking expectations, noticing that $(I\{\cdot\})^\alpha = I\{\cdot\}$ for any $\alpha > 0$, and applying the Hölder inequality yields

$$
\begin{aligned}
E|X_n \cdot Y_n| I\{|X_n \cdot Y_n| > a\} &\leq \|X_n I\{|X_n| > \sqrt{a}\}\|_p \cdot \|Y_n\|_q \\
&\quad + \|X_n\|_p \cdot \|Y_n I\{|Y_n| > \sqrt{a}\}\|_q \\
&\leq \|X_n I\{|X_n| > \sqrt{a}\}\|_p \cdot \sup_n \|Y_n\|_q + \sup_n \|X_n\|_p \cdot \|Y_n I\{|Y_n| > \sqrt{a}\}\|_q \\
&\to 0 \quad \text{as} \quad a \to \infty \quad \text{uniformly in } n.
\end{aligned}
$$

The argument for the uniformity is that the individual sequences being uniformly integrable implies that the moments of orders $p$ and $q$, respectively, are uniformly bounded by Theorem 4.1(i). The other factor in each term converges uniformly to 0 as $n \to \infty$ due to the same uniform integrability. $\qquad\square$

# 5 Convergence of Moments

We are now in the position to show that uniform integrability is the "correct" concept, that is, that a sequence that converges almost surely, in probability, or in distribution, and is uniformly integrable, converges in the mean, that moments converge and that uniform integrability is the minimal additional assumption for this to happen.

## 5.1 Almost Sure Convergence

The easiest case is when $X_n \overset{a.s.}{\to} X$ as $n \to \infty$. A result that "always holds" is Fatou's lemma. Although we have already seen the lemma in connection with the development of the Lebesgue integral in Chapter 2, we recall it here for easy reference.

**Theorem 5.1.** (Fatou's lemma) *Let $X$ and $X_1$, $X_2$, ... be random variables, and suppose that $X_n \overset{a.s.}{\to} X$ as $n \to \infty$. Then*

$$
E|X| \leq \liminf_{n\to\infty} E|X_n|.
$$

Here is now a convergence theorem that shows the intimate connection between uniform integrability and moment convergence.

**Theorem 5.2.** *Let $X$ and $X_1$, $X_2$, ... be random variables, and suppose that $X_n \overset{a.s.}{\to} X$ as $n \to \infty$. Let $r > 0$. The following are equivalent:*

(a)  $\{|X_n|^r, n \geq 1\}$ *is uniformly integrable;*

(b)  $X_n \overset{r}{\to} X$ *as $n \to \infty$;*

(c)  $E|X_n|^r \to E|X|^r$ *as $n \to \infty$.*

*Moreover, if $r \geq 1$ and one of the above holds, then $E X_n \to E X$ as $n \to \infty$.*

*Proof.* We proceed via (a) $\implies$ (b) $\implies$ (c) $\implies$ (a).

*(a) $\implies$ (b).* First of all, by Fatou's lemma and Theorem 4.1(i),

$$E|X|^r \leq \liminf_{n\to\infty} E|X_n|^r \leq \sup_n E|X_n|^r < \infty.$$

Secondly, we note that, since $|X_n - X|^r \leq 2^r(|X_n|^r + |X|^r)$, it follows from Theorem 4.6 that $\{|X_n - X|^r, \ n \geq 1\}$ is uniformly integrable.

Next, let $\varepsilon > 0$. Then

$$E|X_n - X|^r = E|X_n - X|^r I\{|X_n - X| \leq \varepsilon\} + E|X_n - X|^r I\{|X_n - X| > \varepsilon\}$$
$$\leq \varepsilon^r + E|X_n - X|^r I\{|X_n - X| > \varepsilon\},$$

so that

$$\limsup_{n\to\infty} E|X_n - X|^r \leq \varepsilon^r,$$

which, due to the arbitrariness of $\varepsilon$ proves that implication.

*(b) $\implies$ (c).* Suppose first that $0 < r \leq 1$. By the $c_r$-inequality, Theorem 3.2.2,

$$\left|E|X_n|^r - E|X|^r\right| \leq E|X_n - X|^r \to 0 \quad \text{as} \quad n \to \infty.$$

For $r \geq 1$ we use the Minkowski inequality, Theorem 3.2.6 to obtain

$$\left|\|X_n\|_r - \|X\|_r\right| \leq \|X_n - X\|_r \to 0 \quad \text{as} \quad n \to \infty.$$

*(c) $\implies$ (a).* Set $A_n = \{|X_n - X| > 1\}$. Then

$$E|X_n|^r I\{|X_n| > a+1\} = E|X_n|^r I\{\{|X_n| > a+1\} \cap A_n\}$$
$$+E|X_n|^r I\{\{|X_n| > a+1\} \cap A_n^c\}$$
$$\leq E|X_n|^r I\{A_n\} + E|X_n|^r I\{|X| > a\}$$
$$\leq \left|E\big((|X_n|^r - |X|^r)I\{A_n\}\big)\right| + E|X|^r I\{A_n\} + E|X_n|^r I\{|X| > a\}$$
$$\leq \left|E|X_n|^r - E|X|^r\right| + E|X|^r I\{A_n\} + E|X_n|^r I\{|X| > a\}.$$

Let $\varepsilon > 0$ be given. The first term converges to 0 as $n \to \infty$ by assumption, which means that,

$$\sup_{n\geq n_0} \left|E|X_n|^r - E|X|^r\right| < \varepsilon \quad \text{for some} \quad n_0. \tag{5.1}$$

The second term converges to 0 as $a \to \infty$ (independently of $n$) by Proposition 2.6.3(ii), since, by assumption, $P(A_n) \to 0$ as $n \to \infty$. As for the third one,

$$E|X|^r I\{|X| \leq a\} \leq \liminf_{n\to\infty} E|X_n|^r I\{|X| \leq a\},$$

by Fatou's lemma, (since $|X_n|I\{|X| \leq a\} \overset{a.s.}{\to} |X|I\{|X| \leq a\}$ as $n \to \infty$), so that

$$\limsup_{n\to\infty} E|X_n|^r I\{|X| > a\} = \limsup_{n\to\infty} \left( E|X_n|^r - E|X_n|^r I\{|X| \le a\} \right)$$
$$\le E|X|^r - \liminf_{n\to\infty} E|X_n|^r I\{|X| \le a\}$$
$$\le E|X|^r - E|X|^r I\{|X| \le a\} = E|X|^r I\{|X| > a\}.$$

Thus,
$$\sup_{n \ge n_1} E|X_n|^r I\{|X| > a\} < \varepsilon \quad \text{for some} \quad n_1,$$

provided $a$ is sufficiently large, since $E|X|^r I\{|X| > a\} \to 0$ as $a \to \infty$.

Combining the above estimates for the three terms, it follows that

$$\sup_{n \ge \max\{n_0, n_1\}} E|X_n|^r I\{|X| > a\} < \varepsilon,$$

for $a$ sufficiently large, which means that the sequence

$$\{|X_n|^r, \, n \ge \max\{n_0, n_1\}\} \quad \text{is uniformly integrable.}$$

Uniform integrability of the whole sequence follows from Lemma 4.1.

Convergence of the expected values, finally, follows via the triangle inequality (and Lyapounov's inequality, Theorem 3.2.5). $\qquad \square$

## 5.2 Convergence in Probability

We begin with Fatou's lemma assuming only convergence in probability. But first an auxiliary result.

**Lemma 5.1.** *Let $\{c_n, \, n \ge 1\}$ be a sequence of real numbers. If for every subsequence there exists a further subsequence with a limit superior that does not exceed some number, $a$, say, then $\limsup_{n\to\infty} c_n \le a$. Similarly, if for every subsequence there exists a further subsequence with a limit inferior that exceeds some number, $b$, say, then $\liminf_{n\to\infty} c_n \ge b$.*

This kind of result is well known for *convergence* of real numbers. The present variation is maybe less known.

**Exercise 5.1.** Prove the lemma. $\qquad \square$

**Theorem 5.3.** *Let $X$ and $X_1, X_2, \ldots$ be random variables, and suppose that $X_n \xrightarrow{p} X$ as $n \to \infty$. Then*

$$E|X| \le \liminf_{n\to\infty} E|X_n|.$$

*Proof.* By Theorem 3.4 we know that there exists an almost surely convergent subsequence. Exploiting this fact, we can, for every subsequence $\{X_{n_k}, k \ge 1\}$ (which also converges in probability to $X$ as $n \to \infty$) find an almost surely convergent subsequence $\{X_{n_{k_j}}, j \ge 1\}$ to which we can apply Theorem 5.1, to obtain

$$E|X| \le \liminf_{j\to\infty} E|X_{n_{k_j}}|.$$

An application of Lemma 5.1 finishes the proof. $\qquad \square$

**Theorem 5.4.** *Let $X$ and $X_1, X_2, \ldots$ be random variables, suppose that $X_n \xrightarrow{p} X$ as $n \to \infty$, and let $r > 0$. The following are equivalent:*

(a)   $\{|X_n|^r, n \geq 1\}$ *is uniformly integrable;*

(b)   $X_n \xrightarrow{r} X$ *as $n \to \infty$;*

(c)   $E|X_n|^r \to E|X|^r$ *as $n \to \infty$.*

*Moreover, if $r \geq 1$ and one of the above holds, then $E\,X_n \to E\,X$ as $n \to \infty$.*

*Proof.* With Theorem 5.3 replacing Theorem 5.1, the proof of Theorem 5.2 goes through unchanged.                                                        $\square$

**Exercise 5.2.** Check that this is true.                                $\square$

In the last part of the proof of Theorem 5.2 we showed, in a somewhat lengthy way, that

$$\limsup_{n\to\infty} E|X_n|^r I\{|X| > a\} \leq E|X|^r I\{|X| > a\}.$$

If we had applied Fatou's lemma directly to the upper tail we would have obtained

$$E|X|^r I\{|X| > a\} \leq \liminf_{n\to\infty} E|X_n|^r I\{|X| > a\}.$$

By combining the two we have shown that if $X_n \xrightarrow{a.s.} X$ and $E|X_n|^r \to E|X|^r$ as $n \to \infty$, then $E|X_n|^r I\{|X| > a\} \to E|X|^r I\{|X| > a\}$ as $n \to \infty$ (and likewise, by subtraction, for $E|X_n|^r I\{|X| \leq a\}$).

The conclusion is intuitively reasonable, since $|X_n|^r I\{|X| > a\}$ converges almost surely and is dominated by $|X_n|^r$, the expectations of which are convergent. However, the domination is *not* by a single, integrable random variable, but by a *sequence* whose moments converge. The legitimacy of this procedure is the following (version of the) strengthening of the Lebesgue dominated convergence theorem, called *Pratt's lemma*; see [202].

**Theorem 5.5.** *Let $X$ and $X_1, X_2, \ldots$ be random variables. Suppose that $X_n \xrightarrow{a.s.} X$ as $n \to \infty$, and that*

$$|X_n| \leq Y_n \text{ for all } n, \quad Y_n \xrightarrow{a.s.} Y, \quad E\,Y_n \to E\,Y \text{ as } n \to \infty.$$

*Then*

$$X_n \to X \quad \text{in } L^1 \qquad \text{and} \quad E\,X_n \to E\,X \quad \text{as} \quad n \to \infty.$$

*The theorem remains true with almost sure convergence replaced by convergence in probability.*

*Remark 5.1.* Note that the special case $Y_n = Y$ for all $n$ is the dominated convergence theorem.                                                        $\square$

**Exercise 5.3.** Prove Pratt's lemma by suitably modifying the arguments preceding the statement of the theorem.

**Exercise 5.4.** Use Pratt's lemma to check that if $X_n \overset{a.s.}{\to} X$ and $EX_n \to EX$ as $n \to \infty$, then, for any measurable set $A$, $X_nI\{A\} \overset{a.s.}{\to} XI\{A\}$, and $EX_nI\{A\} \to EXI\{A\}$ as $n \to \infty$. This is then, for example, true for the set $A = I\{|X| > a\}$, where $a$ is some positive number.                                                $\square$

Next we return for a moment to Example 3.1 (with $\alpha = 1$), in which $P(X_n = 0) = 1 - \frac{1}{n}$, and $P(X_n = n) = \frac{1}{n}$, $n \geq 1$, so that $X_n \overset{p}{\to} 0$ and $E|X_n| = EX_n = 1 \neq 0$ as $n \to \infty$. In other words,

$$\lim_{n\to\infty} EX_n \neq E \lim_{n\to\infty} X_n.$$

This implies that $X_1, X_2, \ldots$ cannot be uniformly integrable. Indeed, it follows from the definition that (for $a > 1$)

$$E|X_n|I\{|X_n| > a\} = \begin{cases} n \cdot \frac{1}{n} = 1 & \text{for} \quad a \leq n, \\ 0 & \text{for} \quad a > n, \end{cases}$$

so that $\sup_n E|X_n|I\{|X_n| > a\} = 1$ for all $a$.

## 5.3 Convergence in Distribution

As the assumptions get weaker we need more and more preparations before we can really take off. We begin with two limit theorems that, in addition, will be used to prove equivalence between the two definitions of distributional convergence that were given in Section 5.1, after which we provide a Fatou lemma adapted to the setup of this section.

**Theorem 5.6.** *Let $X_1, X_2, \ldots$ be a sequence of random variables and suppose that $X_n \overset{d}{\to} X$ as $n \to \infty$. If $h$ is a real valued or complex valued, continuous function defined on the bounded interval $[a, b]$, where $a, b \in C(F_X)$, then*

$$E\,h(X_n) \to E\,h(X) \quad as \quad n \to \infty.$$

*Proof.* The complex case follows from the real valued case by considering real and imaginary parts separately and adding them, so we only have to prove the latter one.

We shall exploit Lemma A.9.3. Let $A \subset C(F_X) \subset \mathbb{R}$ be a countable dense set (such a set exists; why?). If $h(x) = I_{\{(c,d]\}}(x)$, the indicator of the interval $(c, d]$, for some $c, d \in A$, $a \leq c < d \leq b$, the statement of the theorem reduces to $P(c < X_n \leq d) \to P(c < X \leq d)$ which holds by assumption. By linearity, the conclusion holds for simple functions whose "steps" have their endpoints in $A$. Now, let $h \in C[a, b]$, and let $g$ be an approximating simple function as provided by Lemma A.9.3. Then,

$$|E\,h(X_n) - E\,h(X)| \le |E\,h(X_n) - E\,g(X_n)| + |E\,g(X_n) - E\,g(X)|$$
$$+|E\,g(X) - E\,h(X)|$$
$$\le E|h(X_n) - g(X_n)| + |E\,g(X_n) - E\,g(X)|$$
$$+E|g(X) - h(X)|$$
$$\le \varepsilon + |E\,g(X_n) - E\,g(X)| + \varepsilon.$$

Since, by what has already been shown for simple functions, the middle term converges to 0 as $n \to \infty$, we obtain

$$\limsup_{n\to\infty} |E\,h(X_n) - E\,h(X)| < 2\varepsilon,$$

which, due to the arbitrariness of $\varepsilon$, proves the theorem.    $\square$

In Theorem 5.6 the interval was bounded but not necessarily the function. In the following theorem it is the other way around; the function is bounded but not the interval.

**Theorem 5.7.** *Let $X_1$, $X_2$, ... be a sequence of random variables and suppose that $X_n \overset{d}{\to} X$ as $n \to \infty$. If $h$ is a real valued or complex valued, bounded, continuous function, then*

$$E\,h(X_n) \to E\,h(X) \quad as \quad n \to \infty.$$

*Proof.* With $|h| \le M$, say,

$$|E\,h(X_n) - E\,h(X)| \le |E\,h(X_n)I\{|X_n| \le A\} - E\,h(X)I\{|X| \le A\}|$$
$$+|Eh(X_n)I\{|X_n| > A\}| + |E\,h(X)I\{|X| > A\}|$$
$$\le |E\,h(X_n)I\{|X_n| \le A\} - E\,h(X)I\{|X| \le A\}|$$
$$+E|h(X_n)|I\{|X_n| > A\} + E|h(X)|I\{|X| > A\}$$
$$\le |E\,h(X_n)I\{|X_n| \le A\} - E\,h(X)I\{|X| \le A\}|$$
$$+MP(|X_n| > A) + MP(|X| > A).$$

Let $\varepsilon > 0$, and choose $A \in C(F_X)$ so large that $2MP(|X| > A) < \varepsilon$. Exploiting Theorem 5.6 on the first term and distributional convergence on the two other ones, we find that

$$\limsup_{n\to\infty} |E\,h(X_n) - E\,h(X)| < 2MP(|X| > A) < \varepsilon.$$    $\square$

**Theorem 5.8.** *Let $X$ and $X_1$, $X_2$, ... be random variables, and suppose that $X_n \overset{d}{\to} X$ as $n \to \infty$. Then*

$$E|X| \le \liminf_{n\to\infty} E|X_n|.$$

*Proof.* Let $A \in C(F_X)$ be a positive number. Theorem 5.6 then tells us that

$$\liminf_{n\to\infty} E|X_n| \geq \lim_{n\to\infty} E|X_n|I\{|X_n| \leq A\} = E|X|I\{|X| \leq A\}.$$

The conclusion now follows by letting $A$ tend to infinity through continuity points of $F_X$ (do not forget Theorem 2.2.1(iii)).    □

*Remark 5.2.* Notice that, since convergence in probability implies convergence in distribution the proof of this version of Fatou's lemma also works for Theorem 5.3 where convergence in probability was assumed.    □

Here is, finally, the promised result. However, we do not have equivalences as in Theorems 5.2 and 5.4.

**Theorem 5.9.** *Let $X$ and $X_1$, $X_2$, ... be random variables, and suppose that $X_n \overset{d}{\to} X$ as $n \to \infty$.*
(i) *If, for some $r > 0$, $\{|X_n|^r, n \geq 1\}$ is uniformly integrable, then*

$$E|X_n|^r \to E|X|^r \quad as \quad n \to \infty.$$

(ii) *For $r \geq 1$ we also have $E\,X_n \to E\,X$ as $n \to \infty$.*

*Proof.* By uniform integrability and the distributional version of Fatou's lemma, we first note that

$$E|X|^r \leq \liminf_{n\to\infty} E|X_n|^r \leq \sup_n E|X_n|^r < \infty.$$

In order to apply Theorem 5.7 we must separate the tails from the center again:

$$\begin{aligned}
\big|E|X_n|^r - E|X|^r\big| &\leq \big|E|X_n|^r I\{|X_n| \leq A\} - E|X|^r I\{|X| \leq A\}\big| \\
&\quad + E|X_n|^r I\{|X_n| > A\} + E|X|^r I\{|X| > A\} \\
&\leq \big|E|X_n|^r I\{|X_n| \leq A\} - E|X|^r I\{|X| \leq A\}\big| \\
&\quad + \sup_n E|X_n|^r I\{|X_n| > A\} + E|X|^r I\{|X| > A\}.
\end{aligned}$$

Let $\varepsilon > 0$. Since $E|X|^r < \infty$ we may choose $A_1 \in C(F_X)$ so large that $E|X|^r I\{|X| > A_1\} < \varepsilon$. The uniform integrability assumption implies that $\sup_n E|X_n|^r I\{|X_n| > A_2\} < \varepsilon$ whenever $A_2$ is sufficiently large. This, together with Theorem 5.7 applied to the first term, proves that, for $A \in C(F_X)$, $A > \max\{A_1, A_2\}$, we have

$$\limsup_{n\to\infty} \big|E|X_n|^r - E|X|^r\big| < 2\varepsilon,$$

which concludes the proof of (i), from which (ii) follows as in Theorems 5.2 and 5.4.    □

*Remark 5.3.* Note how very nicely the proof illustrates the role of uniform integrability in that it is precisely what is needed in order for the proof to go through. Namely, the first term is taken care of by Theorem 5.7, and the last one is a consequence of the integrability of $X$. It remains to take care of $\sup_n E|X_n|^r I\{|X_n| > A\}$, and this is exactly what uniform integrability does for us.

*Remark 5.4.* We also mention, without proof, that if, in addition, $X$ and $X_1, X_2, \ldots$ are non-negative and integrable then convergence of the expectations implies uniform integrability; see [20], Section 3.                    □

## 6 Distributional Convergence Revisited

Let, again, $\{X_n, n \geq 1\}$ be random variables with distribution functions $\{F_n, n \geq 1\}$. In Theorem 5.7 we proved that, if $X_n \overset{d}{\to} X$ as $n \to \infty$ and $h \in C_B$, then

$$E\,h(X_n) \to E\,h(X) \quad \text{as} \quad n \to \infty.$$

However, this coincides with the alternative Definition 1.5 of distributional convergence. So, as a logical next task, we prove the converse, thereby establishing the equivalence of the two definitions.

**Theorem 6.1.** *Definitions 1.4 and 1.5 are equivalent.*

*Proof.* Let $a, b \in C(F)$, $-\infty < a < b < \infty$, set

$$h_k(x) = \begin{cases} 0, & \text{for} \quad x < a - \delta_k, \\ \frac{x - (a - \delta_k)}{\delta_k}, & \text{for} \quad x \in [a - \delta_k, a], \\ 1, & \text{for} \quad x \in [a, b], \\ \frac{b + \delta_k - x}{\delta_k}, & \text{for} \quad x \in [b, b + \delta_k], \\ 0, & \text{for} \quad x > b + \delta_k, \end{cases}$$

with $\delta_k \downarrow 0$ as $k \to \infty$, and suppose that $X_n \overset{d}{\to} X$ as $n \to \infty$ in the sense of Definition 1.5. Theorem 5.7 and the monotonicity of distribution functions then tell us that

$$F_n((a, b]) = \int_a^b dF_n(x) \leq E\,h_k(X_n) \to E\,h_k(X) \quad \text{as} \quad n \to \infty.$$

In other words,

$$\limsup_{n \to \infty} F_n((a, b]) \leq E\,h_k(X) = F([a - \delta_k, b + \delta_k]).$$

Since $k$ was arbitrary, $\delta_k \to 0$ as $k \to \infty$, $h_k(x) \downarrow I[a, b]$ as $k \to \infty$, and $a, b$ are continuity points of $F$, we further conclude that

$$\limsup_{n\to\infty} F_n((a,b]) \leq F((a,b]).$$

If, instead

$$h_k(x) = \begin{cases} 0, & \text{for } x < a, \\ \frac{x-a}{\delta_k}, & \text{for } x \in [a, a + \delta_k], \\ 1, & \text{for } x \in [a + \delta_k, b - \delta_k], \\ \frac{b-x}{\delta_k}, & \text{for } x \in [b - \delta_k, b], \\ 0, & \text{for } x > b + \delta_k, \end{cases}$$

the same arguments yield

$$\liminf_{n\to\infty} F_n((a,b]) \geq F([a + \delta_k, b - \delta_k]),$$

and, since, this time, $h_k(x) \uparrow I(a,b)$, finally,

$$\liminf_{n\to\infty} F_n((a,b]) \geq F((a,b]).$$

By joining the inequalities for limsup and liminf we have shown that $X_n \xrightarrow{d} X$ as $n \to \infty$ in the sense of Definition 1.4.                                   □

As a corollary, or special case, we obtain the following uniqueness theorem.

**Theorem 6.2.** *Let $X$ and $Y$ be random variables. Then*

$$X \stackrel{d}{=} Y \quad \Longleftrightarrow \quad E\, h(X) = E\, h(Y) \quad \text{for every} \quad h \in C_B.$$

*Proof.* The conclusion is immediate upon setting $X_n \equiv Y$ for all $n$.          □

*Remark 6.1.* The logical treatment if one departs from the alternative definition of distributional convergence is of course to start with uniqueness and move on to convergence. The converse order here is due to our entering the alternative viewpoint through the back door.                                   □

**Exercise 6.1.** Use the corollary to reprove the uniqueness theorem for distributional convergence.

**Exercise 6.2.** Prove that a sequence that convergences in probability also converges in distribution by showing that Definition 1.5 is satisfied.          □

## 6.1 Scheffé's Lemma

So far we have studied the problem of whether or not convergence implies convergence of moments, that is, of integrals. A related question is to what extent, if at all, convergence in distribution and convergence of densities, that is, of derivatives in the absolutely continuous case, imply each other.

*Example 6.1.* Let $X_1$, $X_2$, ... be random variables with densities

$$f_{X_n}(x) = \begin{cases} 1 - \cos(2\pi n x), & \text{for } 0 \le x \le 1, \\ 0, & \text{otherwise.} \end{cases}$$

Clearly,

$$F_{X_n}(x) = \begin{cases} 0, & \text{for } x \le 0, \\ x - \frac{\sin(2\pi n x)}{2\pi n} \to x, & \text{for } 0 \le x \le 1, \\ 1, & \text{for } x \ge 1, \end{cases}$$

which tells us that $X_n \overset{d}{\to} U(0,1)$ as $n \to \infty$. However, the density is oscillating and, hence, cannot converge. □

The following result provides a criterion for when densities do converge, but first a definition and an auxiliary result.

**Definition 6.1.** (i)    *The variational distance between the distribution functions $F$ and $G$ is*

$$d(F,G) = \sup_{A \in \mathcal{R}} |F(A) - G(A)|.$$

*Associating the random variables $X$ and $Y$ with $F$ and $G$, respectively, the definition is equivalent to*

$$d(X,Y) = \sup_{A \in \mathcal{R}} |P(X \in A) - P(Y \in A)|.$$

(ii)    *If $X$, $X_1$, $X_2$, ... are random variables such that*

$$d(X_n, X) \to 0 \quad \text{as} \quad n \to \infty,$$

*we say that $X_n$ converges to $X$ in total variation as $n \to \infty$.* □

The second part of the definition introduces another convergence concept. How does it relate to those already defined?
    Since the sets $(-\infty, x]$ are Borel sets for any given $x$, it is clear that

$$|P(X_n \le x) - P(X \le x)| \le \sup_{A \in \mathcal{R}} |P(X_n \in A) - P(X \in A)|,$$

which establishes the following hierarchy.

**Lemma 6.1.** *Let $X_1$, $X_2$, ... be random variables. If $X_n \to X$ in total variation as $n \to \infty$, then $X_n \overset{d}{\to} X$ as $n \to \infty$.*

Here is now Scheffé's result.

**Theorem 6.3.** (Scheffé's lemma) *Suppose that $X$, $X_1$, $X_2$, ... are absolutely continuous random variables. Then*

$$\sup_{A \in \mathcal{R}} |P(X_n \in A) - P(X \in A)| \le \int_{\mathbb{R}} |f_{X_n}(x) - f_X(x)| \, dx, \qquad (6.1)$$

and if $f_{X_n}(x) \to f_X(x)$ almost everywhere as $n \to \infty$, then

$$d(X_n, X) \to 0 \quad as \quad n \to \infty,$$

in particular, $X_n \xrightarrow{d} X$ as $n \to \infty$.

*Proof.* Let $A \in \mathcal{R}$. By the triangle inequality,

$$|P(X_n \in A) - P(X \in A)| = \left| \int_A f_{X_n}(x) \, dx - \int_A f_X(x) \, dx \right|$$
$$\le \int_A |f_{X_n}(x) - f_X(x)| \, dx \le \int_{\mathbb{R}} |f_{X_n}(x) - f_X(x)| \, dx,$$

which proves (6.1).

To prove convergence, we first observe that

$$\int_{\mathbb{R}} (f_X(x) - f_{X_n}(x)) \, dx = 1 - 1 = 0,$$

so that

$$\int_{\mathbb{R}} (f_X(x) - f_{X_n}(x))^+ dx = \int_{\mathbb{R}} (f_X(x) - f_{X_n}(x))^- dx.$$

Moreover, $f_X(x) - f_{X_n}(x) \to 0$ a.e. as $n \to \infty$, and $0 \le (f_X(x) - f_{X_n}(x))^+ \le f_X(x)$, which, via dominated convergence, yields

$$\int_{\mathbb{R}} |f_X(x) - f_{X_n}(x)| \, dx = 2 \int_{\mathbb{R}} (f_X(x) - f_{X_n}(x))^+ dx \to 0 \quad as \quad n \to \infty.$$

This proves convergence in total variation, from which distributional convergence follows from Lemma 6.1. $\qquad \square$

A discrete version runs as follows.

**Theorem 6.4.** *Suppose that $X, X_1, X_2, \ldots$ are integer valued random variables. Then*

$$\sup_{A \in \mathcal{R}} |P(X_n \in A) - P(X \in A)| \le \sum_k |P(X_n = k) - P(X = k)|,$$

and if $P(X_n = k) \to P(X = k)$ as $n \to \infty$ for all $k$, then

$$d(X_n, X) \to 0 \quad as \quad n \to \infty,$$

in particular, $X_n \xrightarrow{d} X$ as $n \to \infty$.

The proof amounts, essentially, to a translation of the previous one. We leave the details as an exercise.

*Remark 6.2.* With some additional effort one can show that

$$d(X_n, X) = \frac{1}{2} \int_{\mathbb{R}} |f_{X_n}(x) - f_X(x)| \, dx,$$

and

$$d(X_n, X) = \frac{1}{2} \sum_k |P(X_n = k) - P(X = k)|. \qquad \square$$

# 7 A Subsequence Principle

It is well known that a sequence of reals converges if and only if every subsequence contains a convergent subsequence. It is natural to ask whether or not such a principle holds for the various modes of convergence. More precisely, given a sequence $X_1$, $X_2$, ... of random variables and a mode of convergence, is it true that $X_n \to X$ as $n \to \infty$ in that mode if and only if every subsequence contains a subsequence that converges to $X$ as $n \to \infty$ in that very mode?

## Complete Convergence

The answer is negative since the subsequence principle does not hold for sums of reals; consider, for example, the harmonic series.

## Almost Sure Convergence

The answer is negative as is shown with the aid of the following modification of our favorite Example 3.1. Namely, let

$$P(X_n = 0) = 1 - \frac{1}{n} \quad \text{and} \quad P(X_n = 1) = \frac{1}{n}, \quad n \geq 1,$$

be independent random variables. Once again, $X_n \overset{p}{\to} 0$ but $X_n \overset{a.s.}{\not\to} 0$ as $n \to \infty$. Nevertheless, for any subsequence it is easy to construct a further subsequence that is almost surely convergent.

## Convergence in Probability

The negative answer for almost sure convergence shows that this concept cannot be defined via a metric in a metric space (since the subsequence principle holds in such spaces). Convergence in probability, on the other hand, is a metrizable concept. This is the content of the following proposition; recall, by the way, Section 3.4.

**Proposition 7.1.** *Let $X$ and $X_1$, $X_2$, ... be random variables. Then*

$$X_n \overset{p}{\to} X \quad as \quad n \to \infty \quad \Longleftrightarrow \quad E\left(\frac{|X_n - X|}{1 + |X_n - X|}\right) \to 0 \quad as \quad n \to \infty.$$

*Proof.* Since the function $\frac{|x|}{1+|x|}$ is bounded (by 1), Theorem 5.4 tells us that

$$\frac{|X_n - X|}{1 + |X_n - X|} \overset{p}{\to} 0 \text{ as } n \to \infty \quad \Longleftrightarrow \quad E\left(\frac{|X_n - X|}{1 + |X_n - X|}\right) \to 0 \text{ as } n \to \infty,$$

and since

$$P\left(\frac{|X_n - X|}{1 + |X_n - X|} > \varepsilon\right) \le P(|X_n - X| > \varepsilon) \le P\left(\frac{|X_n - X|}{1 + |X_n - X|} > \frac{\varepsilon}{1+\varepsilon}\right),$$

it follows that

$$\frac{|X_n - X|}{1 + |X_n - X|} \overset{p}{\to} 0 \text{ as } n \to \infty \quad \Longleftrightarrow \quad X_n - X \overset{p}{\to} 0 \text{ as } n \to \infty,$$

and we are done (since the latter is equivalent to $X_n \overset{p}{\to} X$ as $n \to \infty$).    □

The content of the proposition thus is that $X_n \overset{p}{\to} X$ as $n \to \infty$ if and only if $E\frac{|X_n-X|}{1+|X_n-X|} \to 0$ as $n \to \infty$, which, in turn, holds iff every subsequence of the expectations contains a further subsequence that is convergent. And, in view of the proposition, this holds iff every subsequence of $X_n$ contains a subsequence that converges in probability.

## Mean Convergence

There is nothing to prove, since the definition is in terms of sequences of real numbers; the answer is positive.

## Distributional Convergence

Once again the definition is in terms of real numbers that belong to the continuity set of the limit distribution, the complement of which is a Lebesgue null set, so that the answer is positive here as well.

# 8 Vague Convergence; Helly's Theorem

We have seen in Theorem 3.4 that if a sequence of random variables converges in a weak sense, then a stronger kind of convergence may hold along subsequences. In this section we first present a famous theorem of that kind, *Helly's selection theorem*. However, in order to describe this result we introduce a mode of convergence which is slightly weaker than convergence in distribution.

## 8.1 Vague Convergence

The distinction between vague convergence and distributional convergence is that the limiting random variable is not (need not be) *proper* in the former case, that is, probability mass may escape to infinity. We therefore state the definition in terms of distribution functions. Although a limiting *proper* random variable does not necessarily exist, it is, however, sometimes convenient to call the limit distribution a *pseudo-distribution*, or *sub-probability distribution*, and such "random variables" *improper*.

**Definition 8.1.** *A sequence of distribution functions* $\{F_n, n \geq 1\}$ *converges* vaguely *to the pseudo-distribution function* $H$ *if, for every finite interval* $I = (a, b] \subset \mathbb{R}$, *where* $a, b \in C(H)$,

$$F_n(I) \to H(I) \quad as \quad n \to \infty.$$

Notation: $F_n \overset{v}{\to} H$ *as* $n \to \infty$.    □

Once again, the distinction to distributional convergence is that the total mass of $H$ is *at most* equal to 1.

*Example 8.1.* Suppose that $P(X_n = n) = P(X_n = -n) = 1/2$. Then

$$F_n(x) = \begin{cases} 0, & \text{for} \quad x < -n, \\ \frac{1}{2}, & \text{for} \quad -n \leq x < n, \\ 1, & \text{for} \quad x \geq n, \end{cases}$$

so that, for all $x \in \mathbb{R}$,

$$F_n(x) \to H(x) = \frac{1}{2} \quad \text{as} \quad n \to \infty.$$

Clearly $H$ has all properties of a distribution function except that the limits at $-\infty$ and $+\infty$ are not 0 and 1, respectively.

*Example 8.2.* Let $X_n \in U(-n, n)$, $n \geq 1$. The distribution function equals

$$F_{X_n}(x) = \begin{cases} 0, & \text{for} \quad x < -n, \\ \frac{x+n}{2n}, & \text{for} \quad -n \leq x < n, \\ 1, & \text{for} \quad x \geq n, \end{cases}$$

and converges to

$$H(x) = \frac{1}{2} \quad \text{as} \quad n \to \infty \quad \text{for all} \quad x \in \mathbb{R}.$$

Once again we have an improper distribution in the limit.    □

If $X_n \overset{d}{\to} X$ as $n \to \infty$, then, by definition, $F_{X_n}(x) \to F_X(x)$ as $n \to \infty$ for $x \in C(F_X)$, from which it is immediate that, for any bounded interval $I = (a, b]$, such that $a, b \in C(F_X)$,

$$F_{X_n}(I) = F_{X_n}(b) - F_{X_n}(a) \to F_X(b) - F_X(a) = F_X(I) \quad \text{as} \quad n \to \infty,$$

so that convergence in distribution always implies vague convergence; distributional convergence is a stronger mode of convergence than vague convergence.

A necessary and sufficient condition for the converse is that the tails of the distribution functions are uniformly small. We shall soon prove a theorem to that effect. However, we have to wait a moment until we have some necessary tools at our disposal.

## 8.2 Helly's Selection Principle

The first one is Helly's theorem that tells us that vague convergence always holds along *some* subsequence.

**Theorem 8.1.** (Helly's selection principle)
*Let $\{F_n, n \geq 1\}$ be be a sequence of distribution functions. Then there exists a non-decreasing subsequence, $\{n_k, k \geq 1\}$, such that*

$$F_{n_k} \overset{v}{\to} H \quad as \quad k \to \infty,$$

*for some pseudo-distribution function $H$.*

*Proof.* The proof uses the diagonal method. Let $\{r_k, k \geq 1\}$ be an enumeration of $\mathbb{Q}$. Since the sequence $\{F_n(r_1), n \geq 1\}$ is bounded, it follows from the Bolzano-Weierstrass theorem that there exists an accumulation point $j_1$ and a subsequence $\{F_{k,1}(r_1), k \geq 1\}$, such that

$$F_{k,1}(r_1) \to j_1 \quad as \quad k \to \infty,$$

where, of course, $0 \leq j_1 \leq 1$. The same argument applied to the subsequence $\{F_{k,1}(r_2), k \geq 1\}$ produces an accumulation point $j_2$, and a subsubsequence $\{F_{k,2}(r_2), k \geq 1\}$, such that

$$F_{k,2}(r_2) \to j_2 \quad as \quad k \to \infty,$$

where $0 \leq j_2 \leq 1$, and so on. Moreover the subsubsequence also converges for $x = r_1$, since it is a subsequence of the subsequence. Continuing this procedure leads to the following as $k \to \infty$:

$$F_{k,1}(r_1) \to j_1,$$
$$F_{k,2}(r_i) \to j_i \quad \text{for} \quad i = 1, 2,$$
$$F_{k,3}(r_i) \to j_i \quad \text{for} \quad i = 1, 2, 3,$$

$$\cdots\cdots\cdots\cdots\cdots$$

$$F_{k,m}(r_i) \to j_i \quad \text{for} \quad i = 1, 2, \ldots, m,$$

$$\cdots\cdots\cdots\cdots\cdots$$

The trick now is to consider the infinite diagonal, $\{F_{k,k},\ k \geq 1\}$, which, apart from a finite number of terms in the beginning, is a subsequence of *every* horizontal subsequence, and therefore converges on all of $\mathbb{Q}$.

We have thus exhibited a countable, dense set of values for which the sequence converges.

Next we define a function $H$ as follows: Set

$$H(r_i) = j_i, \quad \text{for} \quad r_i \in \mathbb{Q}.$$

Recalling that $0 \leq j_i \leq 1$ for all $i$, it follows automatically that

$$0 \leq H(r_i) \leq 1 \quad \text{for} \quad r_i \in \mathbb{Q}.$$

Moreover, if $r < s$, $r, s \in \mathbb{Q}$, then

$$0 \leq F_{k,k}(s) - F_{k,k}(r) \to H(s) - H(r) \quad \text{as} \quad k \to \infty,$$

which shows that $H$ is non-decreasing on $\mathbb{Q}$.

Next, let $x \in \mathbb{R}$ be arbitrary, and choose $s, r \in \mathbb{Q}$, such that $r < x < s$. Then, since $F_{k,k}(r) \leq F_{k,k}(x) \leq F_{k,k}(s)$, it follows, upon taking limits, that

$$H(r) \leq \liminf_{k \to \infty} F_{k,k}(x) \leq \limsup_{k \to \infty} F_{k,k}(x) \leq H(s),$$

and since $r$ as well as $s$ may be chosen arbitrarily close to $x$, that

$$H(x - 0) \leq \liminf_{k \to \infty} F_{k,k}(x) \leq \limsup_{k \to \infty} F_{k,k}(x) \leq H(x + 0).$$

This proves that there always exists a subsequence that converges to a non-decreasing function $H$ taking its values in $[0, 1]$ at the continuity points of $H$. In order to "make" $H$ right-continuous we complete the definition of $H$ by setting

$$H(x) = \lim_{r_i \downarrow x} H(r_i) \quad \text{for} \quad x \in \mathbb{R}.$$

This means that we "choose" the right-continuous version of $H$.

The proof of the theorem is complete.                                    □

**Exercise 8.1.** Prove that $H$ as defined is, indeed, right-continuous.      □

*Remark 8.1.* In books where distribution functions are defined to be left-continuous one would extend $H$ by defining it as the limit as $r_i \uparrow x$.       □

The second auxiliary result is an extension of Theorem 5.6.

**Theorem 8.2.** *Let $F_n$, $n \geq 1$, be distribution functions, and suppose that $F_n \overset{v}{\to} H$ for some pseudo-distribution function $H$ as $n \to \infty$. If $h$ is a real valued or complex valued, continuous function, defined on the bounded interval $[a, b]$, where $a, b \in C(H)$, then*

$$\int_a^b h(x)\, dF_n(x) \to \int_a^b h(x)\, dH(x) \quad \text{as} \quad n \to \infty.$$

*Proof.* The proof of Theorem 5.6 carries over without changes, because we only deal with finite intervals (please check).  □

Theorem 5.7 on the other hand, does not carry over to vague convergence, as the function $h(x) \equiv 1$ easily illustrates. However the theorem remains true if we add the assumption that $h$ vanishes at the infinities.

**Theorem 8.3.** *Let $\{F_n, n \geq 1\}$ be a sequence of distribution functions, and suppose that $F_n \overset{v}{\to} H$ as $n \to \infty$. If $h$ is a real valued or complex valued, continuous function, such that $|h(x)| \to 0$ as $x \to \pm\infty$, then*

$$\int_{-\infty}^{\infty} h(x)\, \mathrm{d}F_n(x) \to \int_{-\infty}^{\infty} h(x)\, \mathrm{d}H(x) \quad as \quad n \to \infty.$$

*Proof.* The proof is almost the same as that of Theorem 5.7. The problem is that the terms involving the tails no longer have to be small. In order to "make" them small in the present case, we use the assumption about the vanishing at infinity to conclude that, for any $\varepsilon > 0$ there exists $A_0$, such that $|h(x)| \leq \varepsilon$, whenever $A > A_0$. By selecting $A \in C(H)$ exceeding $A_0$, the proof of Theorem 5.7 modifies into

$$\left| \int_{-\infty}^{\infty} h(x)\, \mathrm{d}F_n(x) - \int_{-\infty}^{\infty} h(x)\, \mathrm{d}H(x) \right|$$

$$\leq \int_{|x|>A} |h(x)|\, \mathrm{d}F_n(x) + \left| \int_{|x|\leq A} h(x)\, \mathrm{d}F_n(x) - \int_{|x|\leq A} h(x)\, \mathrm{d}H(x) \right|$$

$$+ \int_{|x|>A} |h(x)|\, \mathrm{d}H(x)$$

$$\leq \varepsilon \cdot 1 + \left| \int_{|x|\leq A} h(x)\, \mathrm{d}F_{X_n}(x) - \int_{|x|\leq A} h(x)\, \mathrm{d}H(x) \right| + \varepsilon \cdot 1,$$

where 1 is the upper bound of the probability mass in the tails, which is at most equal to 1 also for $H$ whether it is a pseudo-distribution function or not.

Since the central term converges to 0 as $n \to \infty$ by Theorem 8.2 – note that $h$, being continuous, is bounded for $|x| \leq A$ – we have shown that

$$\limsup_{n\to\infty} \left| \int_{-\infty}^{\infty} h(x)\, \mathrm{d}F_n(x) - \int_{-\infty}^{\infty} h(x)\, \mathrm{d}H(x) \right| < 2\varepsilon,$$

and the conclusion follows.  □

*Remark 8.2.* The last two results have been formulated in terms of integrals rather than expectations, the reason being that the limit $H$ need not be a *proper* distribution function.  □

## 8.3 Vague Convergence and Tightness

We are now ready to provide a connection between vague and distributional convergence.

**Theorem 8.4.** *Let* $\{F_n,\, n \geq 1\}$ *be distribution functions. In order for every vaguely convergent subsequence to converge in distribution, that is, in order for the limit to be a proper distribution function, it is necessary and sufficient that*

$$\int_{|x|>a} dF_n(x) \to 0 \quad as \quad a \to \infty \quad uniformly\ in\ n. \tag{8.1}$$

*Proof of the sufficiency.* Suppose that $F_{n_k} \overset{v}{\to} H$ for some pseudo-distribution function $H$ as $n \to \infty$, and that (8.1) is satisfied. Since $H$ has finite total mass it follows that, for any $\varepsilon > 0$,

$$\int_{|x|>a} dH(x) < \varepsilon \quad for \quad a > a_H.$$

By assumption we also have

$$\sup_n \int_{|x|>a} dF_n(x) < \varepsilon \quad for \quad a > a_F.$$

Thus, for $a > \max\{a_H, a_F\}$, it therefore follows that, for every $k$,

$$1 - \int_{-\infty}^{\infty} dH(x) = \int_{-\infty}^{\infty} dF_{n_k}(x) - \int_{-\infty}^{\infty} dH(x)$$

$$\leq \int_{|x|>a} dF_{n_k}(x) + \left| \int_{-a}^{a} dF_{n_k}(x) - \int_{-a}^{a} dH(x) \right| + \int_{|x|>a} dH(x)$$

$$\leq \varepsilon + \left| \int_{-a}^{a} dF_n(x) - \int_{-a}^{a} dH(x) \right| + \varepsilon$$

so that, by Theorem 8.2,

$$1 - \int_{-\infty}^{\infty} dH(x) < 2\varepsilon,$$

which, due to the arbitrariness of $\varepsilon$, forces $H$ to have total mass equal to 1, and thus to be a proper distribution function.

*Proof of the necessity.* We now know that $F_n \overset{v}{\to} H$ and that $F_n \overset{d}{\to} H$ as $n \to \infty$, so that $H$ is proper, and we wish to show that (8.1) is satisfied.

Suppose that (8.1) is violated. Then, given $\varepsilon$, there exist increasing subsequences $\{n_k,\, k \geq 1\}$ and $\{a_k,\, k \geq 1\}$, such that

$$\int_{|x|>a_k} dF_{n_k}(x) > 2\varepsilon.$$

Moreover,

$$\int_{|x|>a} dH(x) < \varepsilon \quad for \quad a > a_H,$$

as before. In addition, since a countable number of distributions are involved, and a distribution function has at most a countable number of discontinuities (Lemma 2.2.1), we may, without restriction, assume that $a$ and $\{a_k\}$ are continuity points of every distribution function involved.

Now, for any $a > a_H$, there exists $k_0$, such that $a_k > a$ for $k > k_0$. Let $a$ and $k > k_0$ be given. Combining our facts yields

$$
1 - \int_{-\infty}^{\infty} dH(x) = \int_{-\infty}^{\infty} dF_{n_k}(x) - \int_{-\infty}^{\infty} dH(x)
$$

$$
= \int_{|x|>a} dF_{n_k}(x) - \int_{|x|>a} dH(x) + \left( \int_{-a}^{a} dF_{n_k}(x) - \int_{-a}^{a} dH(x) \right)
$$

$$
> 2\varepsilon - \varepsilon + \left( \int_{-a}^{a} dF_{n_k}(x) - \int_{-a}^{a} dH(x) \right),
$$

so that, by Theorem 8.2,

$$
1 - \int_{-\infty}^{\infty} dH(x) > \varepsilon,
$$

that is, $H$ is *not* a proper distribution function. The assumption that (8.1) is violated thus leads to a contradiction, and the proof of the theorem is complete.    □

A sequence of distributions satisfying (8.1) does not allow any mass to escape to infinity because of the uniformity in $n$; all distributions have uniformly small masses outside sufficiently large intervals. A sequence of distributions satisfying this condition is called *tight*. If $\{X_n, n \geq 1\}$ are random variables associated with a tight sequence of distribution functions we say that $\{X_n, n \geq 1\}$ is tight. Notice also that if, for a moment, we call a sequence of random variables $\{X_n, n \geq 1\}$, such that $\{|X_n|^r, n \geq 1\}$ is uniformly integrable for some $r > 0$ "uniformly integrable of order $r$", then, by rewriting Condition (8.1) as

$$
E\, I\{|X_n| > a\} \to 0 \quad \text{as} \quad a \to \infty \quad \text{uniformly in } n,
$$

we may say that tightness corresponds to the sequence being "uniformly integrable of order 0".

With this interpretation, and Theorem 4.2 in mind, we notice that, for any $p > 0$,

$$
P(|X_n| > a) \leq \frac{1}{a^p} E|X_n|^p I\{|X| > a\} \leq \frac{1}{a^p} \sup_n E|X_n|^p,
$$

from which we obtain the first part of the following corollary. The proof of the second half is similar; please check this.

**Corollary 8.1.** (i) *Let $p > 0$. An $L^p$-bounded sequence is tight. In particular, uniformly integrable sequences are tight.*
(ii) *Let $g$ be a non-negative function increasing to $+\infty$, and suppose that $\sup_n E\, g(X_n) < \infty$. Then $\{X_n, n \geq 1\}$ is tight.*

**Exercise 8.2.** Show that (a) if $\sup_n E\log^+|X_n| < \infty$, then $\{X_n, n \geq 1\}$ is tight. (b): if $X_n \in U(-n, n)$, $n \geq 1$, then $\{X_n, n \geq 1\}$ is *not* tight.          □

By an obvious modification of the proof of Lemma 4.1 it follows that a sequence of random variables that is tight remains so if a finite number of random variables are added to the sequence.

**Lemma 8.1.** *Let* $X_1$, $X_2$, ... *be random variables. If* $\{X_n, n \geq N > 1\}$ *is tight, then so is* $\{X_n, n \geq 1\}$.

**Exercise 8.3.** Prove the lemma.          □

With the aid of Theorem 8.4 we can prove the following, useful, subsequence principle for distributional convergence.

**Theorem 8.5.** *Suppose that* $X_1$, $X_2$, ... *is tight. Then* $X_n \overset{d}{\to} X$ *as* $n \to \infty$ *iff every subsequence contains a subsequence that converges vaguely to* $X$.

*Proof.* The assumptions imply, in view of Theorem 8.4, that the vaguely convergent subsequences, in fact, converge in distribution. The conclusion therefore follows via the subsequence principle for distributional convergence; recall Section 5.7.          □

## 8.4 The Method of Moments

In Section 4.10 we discussed criteria for when a sequence of moments uniquely determines the corresponding distribution, the so-called moment problem. A reasonable extension of this problem would be a result that tells us that if the moments of a sequence of random variables converge to those of some random variable $X$, then one has convergence in distribution. Provided the limiting sequence uniquely determines the distribution of $X$. The following result provides a positive answer to this suggestion.

**Theorem 8.6.** *Let* $X$ *and* $X_1$, $X_2$, ... *be random variables with finite moments of all orders, and suppose that*

$$E|X_n|^k \to E|X|^k \quad as \quad n \to \infty, \quad for \quad k = 1, 2, \ldots.$$

*If the moments of* $X$ *determine the distribution of* $X$ *uniquely, then*

$$X_n \overset{d}{\to} X \quad as \quad n \to \infty.$$

*Proof.* Since the moments converge they are uniformly bounded, which shows that $\{X_n, n \geq 1\}$ is tight (Corollary 8.1) and, moreover, that $\{|X_n|^p, n \geq 1\}$ is uniformly integrable for every $p > 0$ (Theorem 4.2). Therefore, every vaguely convergent sequence is distributionally convergent by Theorem 8.4; note that such sequences do exist due to Helly's theorem, Theorem 8.1.

Now, suppose that $\{X_{n_j}, j \geq 1\}$ is such a sequence and that $Y$ is the limiting variable, that is, suppose that $X_{n_j} \xrightarrow{v}$, and, hence, $\xrightarrow{d} Y$ as $j \to \infty$. It then follows from Theorem 5.9 that the moments converge, that is, that

$$E|X_{n_j}|^p \to E|Y|^p \quad \text{as} \quad j \to \infty, \quad \text{for every} \quad p > 0.$$

However, since, by assumption, the moments converge to those of $X$, the distribution of which was uniquely determined by its moments, we must have $X \overset{d}{=} Y$. An appeal to Theorem 8.5 finishes the proof. $\qquad \square$

# 9 Continuity Theorems

Transforms are, as we have seen, very useful for determining distributions of new random variables, particularly for sums of independent random variables. One essential feature that made this successful was the uniqueness theorems, which stated that if two transforms are equal then so are the associated distributions. It is then not far-fetched to guess that if two transforms are almost equal, then so are the associated distributions. Mathematically, one would translate this idea into the statement that if a sequence of transforms converges, then so do the associated distributions or random variables. In this section we shall see that this is indeed the case. Theorems of this kind are called *continuity theorems*.

## 9.1 The Characteristic Function

This is the most useful transform, since it exists for *all* random variables.

**Theorem 9.1.** *Let* $X$, $X_1$, $X_2$, ... *be random variables. Then*

$$\varphi_{X_n}(t) \to \varphi_X(t) \quad \text{as} \quad n \to \infty \quad \text{for all} \quad t,$$

*if and only if*
$$X_n \xrightarrow{d} X \quad \text{as} \quad n \to \infty.$$

*Proof of the necessity.* This is the easy half. Namely, if $X_n \xrightarrow{d} X$ as $n \to \infty$, then, since $|e^{itx}| = 1$, Theorem 5.7 (the complex version) tells us that

$$E\,e^{itX_n} \to E\,e^{itX} \quad \text{as} \quad n \to \infty,$$

that is, that the sequence of characteristic functions converges.
*Proof of the sufficiency.* The first step is to prove that the sequence is tight. We state this fact as a separate lemma.

**Lemma 9.1.** *Let* $X$, $X_1$, $X_2$, ... *be random variables. If* $\varphi_{X_n}(t) \to \varphi_X(t)$ *as* $n \to \infty$ *for all t, then* $\{X_n, n \geq 1\}$ *is tight.*

*Proof of Lemma 9.1.* Let $\varepsilon > 0$. Since $\varphi_X$ is continuous and equals 1 at the origin, it follows that there exists $h_0 > 0$, such that

$$\frac{1}{h} \int_{|t|<h} (1 - \varphi_X(t))\, dt < \varepsilon \quad \text{for} \quad h < h_0.$$

Moreover, by assumption, $1 - \varphi_{X_n}(t) \to 1 - \varphi_X(t)$ as $n \to \infty$, so that, by Lemma A.6.3,

$$\frac{1}{h} \int_{|t|<h} (1 - \varphi_{X_n}(t))\, dt \to \frac{1}{h} \int_{|t|<h} (1 - \varphi_X(t))\, dt \quad \text{as} \quad n \to \infty,$$

which means that there exists $n_0$, such that

$$\sup_{n>n_0} \frac{1}{h} \int_{|t|<h} (1 - \varphi_{X_n}(t))\, dt < 2\varepsilon.$$

Recalling from Lemma 4.4.1 that

$$P(|X_n| > 2/h) \leq \frac{1}{h} \int_{|t|<h} (1 - \varphi_{X_n}(t))\, dt,$$

we finally obtain, upon replacing $2/h$ by $a$, that

$$\sup_{n>n_0} P(|X_n| > a) < 2\varepsilon \quad \text{for all} \quad a > a_0.$$

Invoking Corollary 8.1 this establishes the tightness of $\{X_n, n > n_0\}$, which, in view of Lemma 8.1 proves tightness of the whole sequence. $\qquad\square$

*Proof of the sufficiency (continued).* By the necessity part of the theorem and the uniqueness theorem for characteristic functions, Theorem 4.1.2, it follows that every vaguely convergent subsequence (remember that Helly's theorem guarantees their existence) must converge to $X$, so that an application of Theorem 8.5 tells us that $X_n \xrightarrow{d} X$ as $n \to \infty$. $\qquad\square$

In case the characteristic functions converge to "some" function one should be able to draw the conclusion that $X_n$ converges in distribution to "some" random variable.

**Theorem 9.2.** *Let $X_1, X_2, \ldots$ be random variables, and suppose that*

$$\varphi_{X_n}(t) \to \varphi(t) \quad \text{as} \quad n \to \infty, \quad \text{for} \quad -\infty < t < \infty,$$

*where $\varphi(t)$ is continuous at $t = 0$. Then there exists a random variable $X$ with characteristic function $\varphi$, such that*

$$X_n \xrightarrow{d} X \quad \text{as} \quad n \to \infty.$$

*Proof.* The proof is the same as the proof of Theorem 9.1 with one exception. There we *knew* that the characteristic functions converges to the characteristic function of some random variable, that is, we *knew* that the limiting characteristic function indeed was continuous at 0. Here we only know that the characteristic functions converge, which forces us to *assume* continuity at 0. But once that has been done the previous proof works fine. □

*Example 9.1.* Remember Example 8.2, which was used to illustrate the difference between vague convergence and distributional convergence.

Let $X_n \in U(-n, n)$ for $n \geq 1$. Then

$$\varphi_{X_n}(t) = \frac{\sin nt}{nt} \to \begin{cases} 1, & \text{when} \quad t = 0, \\ 0, & \text{when} \quad t \neq 0. \end{cases}$$

The limit is *not* continuous at $t = 0$. □

A particular case of interest is when the limiting random variable $X$ is degenerate, since in that case convergence in probability is equivalent to distributional convergence; cf. Theorem 5.3.3. The following result is a useful consequence of this fact.

**Corollary 9.1.** *Let* $X_1, X_2, \ldots$ *be random variables, and suppose that, for some real number* $c$,

$$\varphi_{X_n}(t) \to e^{itc} \quad \text{as} \quad n \to \infty, \quad \text{for} \quad -\infty < t < \infty.$$

*Then*

$$X_n \xrightarrow{p} c \quad \text{as} \quad n \to \infty.$$

**Exercise 9.1.** Prove the corollary. □

*Example 9.2.* Here is a solution of Example 1.3 based on characteristic functions. Admittedly the proof is more clumsy than the elementary one. The only reason for presenting it here is to illustrate a method.

The characteristic function of the $\Gamma(p, a)$-distribution equals $(1 - ait)^{-p}$, so that with $p = n$ and $a = 1/n$ we obtain,

$$\varphi_{X_n}(t) = \left(1 - \frac{it}{n}\right)^{-n} \to \frac{1}{e^{-it}} = e^{it} = \varphi_{\delta(1)}(t) \quad \text{as} \quad n \to \infty,$$

which, in view of Corollary 9.1, shows that $X_n \xrightarrow{p} 1$ as $n \to \infty$. □

## 9.2 The Cumulant Generating Function

Since the logarithmic function is strictly increasing, this one is easy.

**Theorem 9.3.** *Let* $X, X_1, X_2, \ldots$ *be random variables, and suppose that*

$$\kappa_{X_n}(t) \to \kappa_X(t) \quad \text{as} \quad n \to \infty.$$

*Then*

$$X_n \xrightarrow{d} X \quad \text{as} \quad n \to \infty.$$

**Exercise 9.2.** Isn't it? □

## 9.3 The (Probability) Generating Function

**Theorem 9.4.** *Let* $X, X_1, X_2, \ldots$ *be non-negative, integer valued random variables, and suppose that*

$$g_{X_n}(t) \to g_X(t) \qquad as \quad n \to \infty.$$

*Then*

$$X_n \xrightarrow{d} X \qquad as \quad n \to \infty.$$

*Proof.* This should be (is?) a well-known fact for a power series. Nevertheless, we thus know that

$$\sum_{k=0}^{\infty} P(X_n = k)t^k \to \sum_{k=0}^{\infty} P(X = k)t^k \qquad as \quad n \to \infty,$$

which, putting $a_{n,k} = P(X_n = k) - P(X = k)$ is the same as

$$h_n(t) := \sum_{k=0}^{\infty} a_{n,k}t^k \to 0 \qquad as \quad n \to \infty,$$

where, in addition, we know that $|a_{n,k}| \le 1$ for all $k$ and $n$ and that the series is absolutely convergent for $|t| < 1$. We wish to show that $a_{n,k} \to 0$ as $n \to \infty$ for every $k$.

Let $0 < t < \varepsilon < 1/2$. Then

$$|h_n(t) - a_{n,0}| \le \sum_{k=1}^{\infty} |a_{n,k}|t^k \le \sum_{k=1}^{\infty} t^k = \frac{t}{1-t} < 2\varepsilon,$$

so that, by the triangle inequality,

$$|a_{n,0}| < 2\varepsilon + |h_n(t)|,$$

which tells us that

$$\limsup_{n\to\infty} |a_{n,0}| < 2\varepsilon,$$

which, due to our assumption and the arbitrariness of $\varepsilon$ proves that $a_{n,0} \to 0$ as $n \to \infty$.

We now repeat the argument:

$$t^{-1}|h_n(t) - a_{n,0} - a_{n,1}t| \le \sum_{k=2}^{\infty} |a_{n,k}|t^{k-1} < 2\varepsilon,$$

which proves that

$$|a_{n,1}| < 2\varepsilon + t^{-1}|h_n(t) - a_{n,0}|,$$

so that

$$\limsup_{n\to\infty} |a_{n,1}| < 2\varepsilon.$$

Iterating this procedure (formally, by induction), we find that $a_{n,k} \to 0$ as $n \to \infty$ for every $k$. $\square$

*Remark 9.1.* Note that this approach can be used to provide a (slightly) different proof of the uniqueness theorem for generating functions, Theorem 4.7.1. Namely, the modification amounts to showing that if

$$\sum_{k=0}^{\infty} a_k t^k = 0,$$

for $t$ in a neighborhood of 0, then $a_k = 0$ for all $k$. □

**Exercise 9.3.** Write down the details. □

Let us illustrate the theorem by proving the Poisson approximation of the binomial distribution in the simplest case (this can, of course, also be done directly via the probability function as was pointed out in Example 1.6).

*Example 9.3.* Let $X_n \in \text{Bin}(n, \frac{\lambda}{n})$. We wish to show that $X_n \overset{d}{\to} \text{Po}(\lambda)$ as $n \to \infty$.

To achieve this, we compute the generating function of $X_n$:

$$g_{X_n}(t) = \left(1 - \frac{\lambda}{n} + \frac{\lambda}{n}t\right)^n = \left(1 + \frac{\lambda(t-1)}{n}\right)^n \to e^{\lambda(t-1)} = g_{\text{Po}(\lambda)}(t),$$

as $n \to \infty$, and the conclusion follows from Theorem 9.4. □

**Exercise 9.4.** Prove, more generally, that, if $X_n \in \text{Bin}(n, p_n)$, and $n \to \infty$ in such a way that $p_n \to 0$ and $np_n \to \lambda$ as $n \to \infty$, then $X_n \overset{d}{\to} \text{Po}(\lambda)$. □

## 9.4 The Moment Generating Function

**Theorem 9.5.** *Let $X_1, X_2, \ldots$ be random variables such that $\psi_{X_n}(t)$ exists for $|t| < h$, for some $h > 0$, and for all $n$. Suppose further that $X$ is a random variable whose moment generating function, $\psi_X(t)$, exists for $|t| \leq h_1 < h$ for some $h_1 > 0$, and that*

$$\psi_{X_n}(t) \to \psi_X(t) \quad as \quad n \to \infty, \quad for \quad |t| < h_1.$$

*Then*

$$X_n \overset{d}{\to} X \quad as \quad n \to \infty.$$

*Moreover, $\{|X_n|^r \, n \geq 1\}$ is uniformly integrable, and*

$$E|X_n|^r \to E|X|^r \quad as \quad n \to \infty, \quad for \, all \quad r > 0.$$

*Proof.* The convergence of the moment generating functions implies their uniform boundedness, which, in particular, implies that $\{X_n, n \geq 1\}$ is tight; recall Corollary 8.1(ii) (or note that uniform boundedness of the moment generating function implies the same for the moments and apply the first part of the same corollary).

Moreover, by the uniqueness theorem, the limit $X$ is unique, so that every vaguely convergent subsequence must converge to $X$, from which the first conclusion follows by an appeal to Theorem 8.5.

For the remaining statements we recall that convergence of the moment generating functions implies their uniform boundedness, so that Theorem 4.3 implies uniform integrability of $\{|X_n|^r,\ n \geq 1\}$ for all $r > 0$, after which Theorem 5.9 yields moment convergence.

Alternatively, one may appeal to the theory of analytic functions, as in the proof of Theorem 4.8.1, and extend the moment generating function analytically to the strip $|\mathfrak{Re}(z)| < h_1 < h$. Putting $z = iy$, where $y$ is real, then takes us to the realm of characteristic functions, so that the conclusion follows from Theorem 9.1.                                                                          □

**Exercise 9.5.** Solve Example 1.3 using moment generating functions.

**Exercise 9.6.** Re-prove the results of Example 9.3 and the accompanying exercise with the aid of moment generating functions.                                         □

Let us pause for a moment and summarize. In the proofs of Theorems 8.6, 9.5, and 9.1 the common denominator was to establish (i) tightness and (ii) that every vague limit must be the unique limiting random variable, after which an application of Theorem 8.5 finished the proof.

In the first case the assumption was that all moments converge to those of a random variable that was uniquely determined by its moments. So, we had to prove tightness. In the second case we used characteristic functions to prove tightness, and the necessity part for uniqueness. Note that there is no necessity part for moment generating functions since they do not always exist. In the third case, convergence of the moment generating functions implied tightness as well as uniqueness of the limit.

We close this section by pointing out, once again, that, surely, certain results are more easily accessible with transforms than with a pedestrian approach. However, one must not forget that *behind* these more elegant solutions or derivations there are sophisticated continuity theorems. So, starting out at a higher level there is obviously a shorter route to go than if we start at a low level.

# 10 Convergence of Functions of Random Variables

Suppose that $X_1$, $X_2$, ... is a sequence of random variables that converges in some sense to the random variable $X$ and that $h$ is a real valued function. Is it true that the sequence $h(X_1)$, $h(X_2)$, ... converges (in the same sense)? If so, does the limiting random variable equal $h(X)$?

We begin with the simplest case, almost sure convergence.

**Theorem 10.1.** *Let* $X$, $X_1$, $X_2$, ... *be random variables such that*

$$X_n \overset{a.s.}{\to} X \quad as \quad n \to \infty,$$

*and suppose that* $h$ *is a continuous function. Then*

$$h(X_n) \overset{a.s.}{\to} h(X) \quad as \quad n \to \infty.$$

*Proof.* The definition of continuity implies that,

$$\{\omega : X_n(\omega) \to X(\omega)\} \subset \{\omega : h(X_n(\omega)) \to h(X(\omega))\},$$

and since the former set has probability 1, so has the latter.     □

Next we turn to convergence in probability.

**Theorem 10.2.** *Let* $X_1$, $X_2$, ... *be random variables such that*

$$X_n \overset{p}{\to} a \quad as \quad n \to \infty,$$

*and suppose that* $h$ *is a function that is continuous at* $a$. *Then*

$$h(X_n) \overset{p}{\to} h(a) \quad as \quad n \to \infty.$$

*Proof.* The assumption is that

$$P(|X_n - a| > \delta) \to 0 \quad as \quad n \to \infty \quad for \ all \quad \delta > 0,$$

and we wish to prove that

$$P(|h(X_n) - h(a)| > \varepsilon) \to 0 \quad as \quad n \to \infty \quad for \ all \quad \varepsilon > 0.$$

The continuity of $h$ at $a$ implies that

$$\forall \varepsilon > 0 \ \exists \delta > 0, \quad such \ that \quad |x - a| < \delta \implies |h(x) - h(a)| < \varepsilon,$$

or, equivalently, that

$$\forall \varepsilon > 0 \ \exists \delta > 0, \quad such \ that \quad |h(x) - h(a)| > \varepsilon \implies |x - a| > \delta,$$

which implies that

$$\{\omega : |h(X_n(\omega)) - h(a)| > \varepsilon\} \subset \{\omega : |X_n(\omega) - a| > \delta\},$$

that is, $\forall \varepsilon > 0 \ \exists \delta > 0$ such that

$$P(|h(X_n) - h(a)| > \varepsilon) \leq P(|X_n - a| > \delta).$$

Since the latter probability tends to 0 for *all* $\delta$, this is particularly true for the $\delta$ corresponding to the arbitrary $\varepsilon > 0$ of our choice.     □

What if $X_n \overset{p}{\to} X$ as $n \to \infty$, where $X$ is some (non-degenerate) random variable?

**Exercise 10.1.** Why does the above proof not carry over as it is?    □

*Answer:* The inclusion

$$\{\omega : |h(X_n(\omega)) - h(X(\omega))| > \varepsilon\} \subset \{\omega : |X_n(\omega) - X(\omega)| > \delta\}, \quad (10.1)$$

requires that $h$ be *uniformly* continuous; the point $X(\omega)$ varies as $\omega$ varies. This means that the theorem remains true as is provided we strengthen $h$ to be uniformly continuous. However, the result remains true also without this stronger assumption. Namely, let $\varepsilon > 0$ and $\eta > 0$ be given, and choose $A$ so large that

$$P(|X| > A) < \eta/2.$$

Since $h$ is uniformly continuous on $[-A, A]$, and we can apply (10.1) precisely as in the proof of Theorem 10.2, and it follows that

$$\begin{aligned}
P(|h(X_n) - h(X)| > \varepsilon) &= P(\{|h(X_n) - h(X)| > \varepsilon\} \cap \{|X| \leq A\}) \\
&\quad + P(\{|h(X_n) - h(X)| > \varepsilon\} \cap \{|X| > A\}) \\
&\leq P(|X_n - X| > \delta) + P(|X| > A) \leq \frac{\eta}{2} + \frac{\eta}{2} = \eta,
\end{aligned}$$

for *all* $\delta$ as soon as $n$ is large enough, that is, in particular for the $\delta$ corresponding to our $\varepsilon > 0$.

This establishes the following extension of Theorem 10.2

**Theorem 10.3.** *Let* $X, X_1, X_2, \ldots$ *be random variables such that*

$$X_n \overset{p}{\to} X \quad as \quad n \to \infty,$$

*and suppose that $h$ is a continuous function. Then*

$$h(X_n) \overset{p}{\to} h(X) \quad as \quad n \to \infty.$$

*Remark 10.1.* The proof above is interesting because it stresses the difference between continuity and uniform continuity. Another proof of the theorem is obtained by an appeal to Theorems 3.4 and 10.1, and Proposition 7.1.    □

**Exercise 10.2.** Carry out the proof as indicated in the remark.    □

## 10.1 The Continuous Mapping Theorem

The analog for distributional convergence has this special name. We shall provide two proofs here. One via characteristic functions, and one via Definition 1.5. A third proof with the aid of the so-called Skorohod representation theorem will be given at the end of Section 5.13.

For notational reasons we prefer to use the letter $g$ rather than $h$ in the present section.

**Theorem 10.4.** (The continuous mapping theorem)
*Let $X$, $X_1$, $X_2$, ... be random variables, and suppose that*

$$X_n \overset{d}{\to} X \quad as \quad n \to \infty.$$

*If $g$ is continuous, then*

$$g(X_n) \overset{d}{\to} g(X) \quad as \quad n \to \infty.$$

*First proof.* An application of Theorem 5.7 (the complex version) shows that

$$\varphi_{g(X_n)}(t) = E \exp\{itg(X_n)\} \to E \exp\{itg(X)\} = \varphi_{g(X)}(t) \quad as \quad n \to \infty,$$

after which the continuity theorem for characteristic functions, Theorem 9.1, does the rest.

*Second proof.* Let $h \in C_B$. Since $h(g(x)) \in C_B$ it follows from Theorem 5.7 that $E\,h(g(X_n)) \to E\,h(g(X))$ as $n \to \infty$, and since this is true for *any* $h \in C_B$, the conclusion follows via an appeal to Definition 1.5.    □

*Remark 10.2.* With a little additional work one can show that the theorem remains true also if $g$ is measurable and the discontinuity set of $g$ has Lebesgue measure 0; see [20], Theorem 2.7. Alternatively, try Problem 8.24 first.    □

A very useful application of this result is the so-called *Cramér-Wold device*. The content is that a sequence of random vectors converges in distribution if and only if every linear combination of its components does so. For the proof we rely on some facts about random vectors that only have been demonstrated here for real valued random variables. We apologize for this.

**Theorem 10.5.** *Let* $\mathbf{X}$ *and* $\mathbf{X}_n$, $n \geq 1$, *be $k$-dimensional random vectors. Then*

$$\mathbf{X}_n \overset{d}{\to} \mathbf{X} \text{ as } n \to \infty \quad \Longleftrightarrow \quad \mathbf{t}' \cdot \mathbf{X}_n \overset{d}{\to} \mathbf{t}' \cdot \mathbf{X} \text{ as } n \to \infty \quad for \text{ all } \mathbf{t}.$$

*Proof.* Suppose that all linear combinations converge. Then, by Theorem 5.7,

$$\varphi_{\mathbf{t}' \cdot \mathbf{X}_n}(u) \to \varphi_{\mathbf{t}' \cdot \mathbf{X}}(u) \quad as \quad n \to \infty.$$

In particular, rewriting this for $u = 1$, we have

$$E \exp\{i\mathbf{t}' \cdot \mathbf{X}_n\} \to E \exp\{i\mathbf{t}' \cdot \mathbf{X}\} \quad as \quad n \to \infty,$$

which is the same as

$$\varphi_{\mathbf{X}_n}(\mathbf{t}) \to \varphi_{\mathbf{X}}(\mathbf{t}) \quad as \quad n \to \infty.$$

An application of the continuity theorem for characteristic functions (the vector valued version that we have not proved here) shows that $\mathbf{X}_n \overset{d}{\to} \mathbf{X}$ as $n \to \infty$.

The converse follows from (a vector valued version of) the Theorem 10.4, since the function $g(\mathbf{x}) = \mathbf{t}' \cdot \mathbf{x}$ is continuous.    □

*Remark 10.3.* Note that the convergence criterion of the theorem is an "almost equal version" of the definition of the multivariate distribution in Chapter 2: a random vector is normal iff every linear combination of its components is (one-dimensional) normal.                                                              $\square$

# 11 Convergence of Sums of Sequences

Let $X_1$, $X_2$, ... and $Y_1$, $Y_2$, ... be sequences of random variables. Suppose that $X_n \to X$ and that $Y_n \to Y$ as $n \to \infty$ in some sense. To what extent can we conclude that $X_n + Y_n \to X + Y$ as $n \to \infty$?

**Theorem 11.1.** *Let $X_1$, $X_2$, ... and $Y_1$, $Y_2$, ... be sequences of random variables such that*

$$X_n \to X \quad and \quad Y_n \to Y \quad as \quad n \to \infty,$$

*completely, almost surely, in probability or in r-mean, respectively. Then*

$$X_n + Y_n \to X + Y \quad as \quad n \to \infty,$$

*completely, almost surely, in probability or in r-mean, respectively.*

*Proof.* With $N_X$ and $N_Y$ being the null sets in Theorem 2.1 it follows that $X_n(\omega) \to X(\omega)$ and $Y_n(\omega) \to Y(\omega)$ as $n \to \infty$, and, hence, that

$$X_n(\omega) + Y_n(\omega) \to X(\omega) + Y_n(\omega) \quad as \quad n \to \infty,$$

for all $\omega \in (N_X \cup N_Y)^c$.

For complete convergence and convergence in probability it suffices to observe that

$$\{|X_n + Y_n - (X + Y)| > \varepsilon\} \subset \{|X_n - X)| > \varepsilon/2\} \cup \{|Y_n - Y)| > \varepsilon/2\},$$

and for convergence in $r$-mean the conclusion follows by an application of the $c_r$-inequality; Theorem 3.2.2.                                                              $\square$

As for convergence in distribution, a little more care is needed, in that some additional assumption is required.

**Theorem 11.2.** *Let $X_1$, $X_2$, ... and $Y_1$, $Y_2$, ... be sequences of random variables such that*

$$X_n \overset{d}{\to} X \quad and \quad Y_n \overset{d}{\to} Y \quad as \quad n \to \infty.$$

*Suppose that $X_n$ and $Y_n$ are independent for all $n$ and that $X$ and $Y$ are independent. Then*

$$X_n + Y_n \overset{d}{\to} X + Y \quad as \quad n \to \infty.$$

*Proof.* The independence assumption suggests the use of transforms. In view of Theorem 4.1.8 we have

$$\varphi_{X_n+Y_n}(t) = \varphi_{X_n}(t) \cdot \varphi_{Y_n}(t) \to \varphi_X(t) \cdot \varphi_Y(t) = \varphi_{X+Y}(t) \quad \text{as} \quad n \to \infty,$$

which, via the continuity theorem, Theorem 9.1, proves the conclusion.     □

If one does not like the independence assumption one can avoid it, provided it is compensated by the assumption that one of the limits is degenerate.

**Theorem 11.3.** *Let $X_1$, $X_2$, ... and $Y_1$, $Y_2$, ... be sequences of random variables such that*

$$X_n \overset{d}{\to} X \quad \text{and} \quad Y_n \overset{p}{\to} a \quad \text{as} \quad n \to \infty,$$

*where $a$ is some constant. Then*

$$X_n + Y_n \overset{d}{\to} X + a \quad \text{as} \quad n \to \infty.$$

*Proof.* The proof is similar to that of Step IV in the proof of Theorem 3.1.
Let $\varepsilon > 0$ be given. Then

$$\begin{aligned}
F_{X_n+Y_n}(x) &= P\big(\{X_n + Y_n \le x\} \cap \{|Y_n - a| \le \varepsilon\}\big) \\
&\quad + P\big(\{X_n + Y_n \le x\} \cap \{|Y_n - a| > \varepsilon\}\big) \\
&\le P\big(\{X_n \le x - a + \varepsilon\} \cap \{|Y_n - a| \le \varepsilon\}\big) + P(|Y_n - a| > \varepsilon) \\
&\le P(X_n \le x - a + \varepsilon) + P(|Y_n - a| > \varepsilon) \\
&= F_{X_n}(x - a + \varepsilon) + P(|Y_n - a| > \varepsilon),
\end{aligned}$$

from which it follows that

$$\limsup_{n\to\infty} F_{X_n+Y_n}(x) \le F_X(x - a + \varepsilon) \quad \text{for} \quad x - a + \varepsilon \in C(F_X).$$

A similar argument shows that

$$\liminf_{n\to\infty} F_{X_n+Y_n}(x) \ge F_X(x - a - \varepsilon) \quad \text{for} \quad x - a - \varepsilon \in C(F_X).$$

Since $\varepsilon > 0$ may be arbitrarily small (and since $F_X$ has only at most a countable number of discontinuity points – Lemma 2.2.1), we finally conclude that

$$F_{X_n+Y_n}(x) \to F_X(x - a) = F_{X+a}(x) \quad \text{as} \quad n \to \infty,$$

for $x - a \in C(F_X)$, that is, for $x \in C(F_{X+a})$.     □

Theorems 11.1 and 11.3 also hold for differences, products, and ratios. We confine ourselves to formulating the counterpart of Theorem 11.3.

**Theorem 11.4.** *Let $X_1, X_2, \ldots$ and $Y_1, Y_2, \ldots$ be sequences of random variables. Suppose that*

$$X_n \xrightarrow{d} X \quad and \quad Y_n \xrightarrow{p} a \quad as \quad n \to \infty,$$

*where $a$ is some constant. Then*

$$X_n + Y_n \xrightarrow{d} X + a,$$
$$X_n - Y_n \xrightarrow{d} X - a,$$
$$X_n \cdot Y_n \xrightarrow{d} X \cdot a,$$
$$\frac{X_n}{Y_n} \xrightarrow{d} \frac{X}{a}, \quad for \quad a \neq 0,$$

*as $n \to \infty$.*

*Remark 11.1.* Theorem 11.4 is called *Cramér's theorem* or *Slutsky's theorem* (depending on from which part of the world one originates). □

**Exercise 11.1.** Prove Theorem 11.4. □

## 11.1 Applications

To illustrate the usefulness and efficiency of Cramér's theorem we first review the use of the normal quantiles instead of the more exact quantiles in the upper and lower bounds for confidence intervals based on large samples, after which we re-prove some results for characteristic functions that were earlier proved in a more purely analytical fashion.

### Confidence Intervals

From statistics we know that in order to obtain confidence intervals for $\mu$ when $\sigma$ is unknown in the normal distribution one uses the $t$-statistic, and also that one approximates the $t$-quantile with the corresponding quantile of the standard normal distribution when the number of degrees of freedom is large. As a nice application of Cramér's theorem we show that this is reasonable in the sense that for the $t(n)$-distribution, that is, the $t$-distribution with $n$ degrees of freedom, we have $t(n) \xrightarrow{d} N(0,1)$ as $n \to \infty$. This is not exactly the same as replacing quantiles by quantiles, but since the normal distribution is absolutely continuous, this is no problem.

*Example 11.1.* Let $Z_n \in N(0,1)$ and $V_n \in \chi^2(n)$ be independent random variables, and set

$$T_n = \frac{Z_n}{\sqrt{\frac{V_n}{n}}}, \quad n = 1, 2, \ldots.$$

The numerator, $Z_n$, is always standard normal, in particular, asymptotically so. As for the denominator, $E\,V_n = n$ and $\mathrm{Var}\,V_n = 2n$, so that

$$\frac{V_n}{n} \xrightarrow{p} 1 \quad \text{as} \quad n \to \infty,$$

by Chebyshev's inequality. Since the function $h(x) = \sqrt{x}$ is continuous at $x = 1$, it further follows, from Theorem 10.2, that

$$\sqrt{\frac{V_n}{n}} \xrightarrow{p} 1 \quad \text{as} \quad n \to \infty.$$

An application of Cramér's theorem finishes the proof.                    □

*Remark 11.2.* The result as such does not require independence of $Z_n$ and $V_n$. We assumed independence here only because this is part of the definition of the $t$-distribution.                    □

In the following exercise the reader is asked to prove the analog for Bernoulli trials or coin-tossing experiments when the success probability is unknown.

**Exercise 11.2.** Let $X_1, X_2, \ldots, X_n$ be independent, $\mathrm{Be}(p)$-distributed random variables, $0 < p < 1$, and set $Y_n = \frac{1}{n}\sum_{k=1}^{n} X_k$. The interval

$$Y_n \pm \lambda_{\alpha/2}\sqrt{Y_n(1 - Y_n)/n},$$

where $\lambda_\alpha$ is the $\alpha$-quantile of the standard normal distribution ($\Phi(\lambda_\alpha) = 1 - \alpha$), is commonly used as an approximative confidence interval for $p$ on the confidence level $1 - \alpha$ for $n$ large. Show that this is acceptable in the sense that

$$\frac{Y_n - p}{\sqrt{Y_n(1 - Y_n)/n}} \xrightarrow{d} N(0, 1) \quad \text{as} \quad n \to \infty.$$                    □

## The Uniqueness Theorem for Characteristic Functions

One can also use Cramér's theorem to provide another proof of the uniqueness theorem for characteristic functions, Theorem 4.1.2, as follows:

We thus wish to prove that the characteristic function uniquely determines the distribution. The proof is based on the smoothing technique discussed in Section 4.3, which amounted to adding an independent normal random variable, thereby producing a sum which is absolutely continuous; recall Section 2.19. In the final step one lets some suitable parameter converge to 0 or infinity to obtain the desired conclusion.

Thus, let $X$ be our random variable under investigation and let $Y_a$ be normal with mean 0 and variance $1/a^2$. The point is that the density and the characteristic function of the standard normal distribution look the same (except for the factor $1/\sqrt{2\pi}$), so that the switching from density to characteristic function which is the essence of Parseval's relation (Subsection 4.2.5) can be interpreted more freely if one of the distributions is normal.

With our choice we thus have

$$f_{Y_a}(y) = \frac{a}{\sqrt{2\pi}} \exp\left\{-\frac{1}{2}a^2 y^2\right\} \quad \text{and} \quad \gamma(s) = \exp\left\{-\frac{s^2}{2a^2}\right\}.$$

Inserting this into Parseval's relation yields

$$\int_{-\infty}^{\infty} e^{-iuy}\varphi(y)\frac{a}{\sqrt{2\pi}}\exp\left\{-\frac{1}{2}a^2y^2\right\}dy = \int_{-\infty}^{\infty} \exp\left\{-\frac{(x-u)^2}{2a^2}\right\}dF(x). (11.1)$$

Recalling the convolution formula, this can be reformulated as

$$f_{X+Y_a}(u) = \frac{1}{2\pi}\int_{-\infty}^{\infty} e^{-iuy}\varphi(y)e^{-a^2y^2/2}\,dy. \tag{11.2}$$

Integrating from $-\infty$ to $x$ produces an expression for the distribution function of $X + Y_a$:

$$F_{X+Y_a}(x) = \frac{1}{2\pi}\int_{-\infty}^{x}\int_{-\infty}^{\infty} e^{-iuy}\varphi(y)e^{-a^2y^2/2}\,dy\,du.$$

By Cramér's theorem we know that $X + Y_a \xrightarrow{d} X$ as $a \to \infty$, since $Y_a \xrightarrow{p} 0$ as $a \to \infty$. The right-hand side therefore provides a *unique* way of computing the distribution function of $X$, namely, for all $x \in C(F_X)$, we have

$$F_X(x) = \lim_{a\to\infty}\frac{1}{2\pi}\int_{-\infty}^{x}\int_{-\infty}^{\infty} e^{-ity}\varphi(y)e^{-a^2y^2/2}\,dy\,du. \tag{11.3}$$

This (re)establishes the uniqueness theorem for characteristic functions.

## The Continuity Theorem for Characteristic Functions

Suppose that $X_1, X_2, \ldots$ is a sequence of random variables, and that we know that $\varphi_{X_n}(t) \to \varphi(t)$ as $n \to \infty$, where $\varphi$ is continuous at 0.

By Helly's selection theorem, Theorem 8.1, there exists a vaguely convergent subsequence, $\{X_{n_k}, k \geq 1\}$; $F_{n_k} \xrightarrow{v} F$ as $k \to \infty$, where $F$ is a, possibly degenerate, distribution function. Applying (11.1) to the subsequence yields

$$\int_{-\infty}^{\infty} e^{-iuy}\varphi_{X_{n_k}}(y)\frac{a}{\sqrt{2\pi}}\exp\left\{-\frac{1}{2}a^2y^2\right\}dy = \int_{-\infty}^{\infty}\exp\left\{-\frac{(x-u)^2}{2a^2}\right\}dF_{X_{n_k}}(x).$$

Let $k \to \infty$. The right-hand side then converges to the integral with respect to $F$ by Theorem 8.3, and the left-hand side converges to the same integral with respect to $\varphi$ instead of $\varphi_{X_{n_k}}$, by bounded convergence, since $\varphi_{X_{n_k}} \to \varphi$ as $k \to \infty$. Thus,

$$\int_{-\infty}^{\infty} e^{-iuy}\varphi(y)\frac{a}{\sqrt{2\pi}}\exp\left\{-\frac{1}{2}a^2y^2\right\}dy = \int_{-\infty}^{\infty}\exp\left\{-\frac{(x-u)^2}{2a^2}\right\}dF(x).$$

Next, let $a \to \infty$. A rewriting into probabilistic language of the integral in the left-hand side is

$$E\big(\exp\{-iuY_a\}\varphi(Y_a)\big), \quad \text{where} \quad Y_a \in N(0, 1/a^2).$$

Now, since $Y_a \overset{p}{\to} 0$ as $a \to \infty$, the left-hand side converges to $\exp\{-iu0\}\varphi(0) = 1$ as $a \to \infty$ by Theorem 10.2. The right-hand side converges to the total mass of $F$ as $a \to \infty$ by Theorem 8.3, since the exponential converges to 1. This forces $F$ to have total mass 1. We have thus shown that $F_{n_k} \overset{d}{\to} F$ as $k \to \infty$.

Finally: This procedure can be applied to any vaguely convergent subsequence of any subsequence. Since the limit law is unique, recall (11.3), it follows from the subsequence principle (Section 5.7) that $F_n \overset{d}{\to} F$ as $n \to \infty$, that is, that $X_n$ converges in distribution as $n \to \infty$, to a random variable $X$ whose distribution function equals $F$.

This concludes the proof of the continuity theorem.

**The Inversion Formula for Densities**

It is also possible to (re)prove the inversion formula for densities given in Theorem 4.1.4 with the present approach. Namely, suppose that $\varphi \in L^1$ and consider the density $f_{X+Y_a}$ in (11.2). By bounded convergence,

$$f_{X+Y_a}(u) \to f(u) = \frac{1}{2\pi} \int_{-\infty}^{\infty} e^{-iuy}\varphi(y)\, dy \quad \text{as} \quad a \to \infty.$$

The right-hand side thus is the density we are aiming at for the limit $X$. It remains to show that the density is, indeed, the density we are looking for. To see this, let $c, d \in C(F_X)$, $-\infty < c < d < \infty$. Then

$$\int_c^d f_{X+Y_a}(u)\, du \to \begin{cases} \int_c^d f(u)\, du, \\ F_X(d) - F_X(c), \end{cases} \quad \text{as} \quad a \to \infty.$$

Since $c, d$ are arbitrary it follows that $f$ must be the density of $X$.

**11.2 Converses**

In Chapter 3 we proved results such as "if $X$ and $Y$ are integrable, then so is their sum". We also found that a converse need not be true. In Section 5.10 we studied to what extent sums of sequences converge if the individual ones do. In this section we shall look for converses, that is, if $X_1, X_2, \ldots$ and $Y_1, Y_2, \ldots$ are sequences of random variables such that $X_n + Y_n$ converges in some mode of convergence, is it then true that $X_n$ and $Y_n$ converge?

The guess is that it cannot always be true, and that a degenerate example illustrates this.

*Example 11.2.* Suppose that $P(X_n = n) = P(Y_n = -n) = 1$ for $n \geq 1$. Then $X_n + Y_n = 0$ with probability 1, in particular, the sum converges in all convergence modes. However, the individual sequences diverge.

Note also that being degenerate, $X_n$ and $Y_n$ are independent for all $n$, so that an independence assumption only, will not be of any help.   □

The problem is that even though the sum is small, the individual "centers", that is, the medians, drift off to infinity. And this happens in such a way that the drifts "happen to" cancel when the sequences are added. This suggests two solutions, namely, that assuming that

- one of the sequences is well behaved, or
- symmetry

might make things work out all right.

**Theorem 11.5.** *Let $a, b \in \mathbb{R}$, let $X_1, X_2, \ldots$ and $Y_1, Y_2, \ldots$ be two sequences of random variables.*
*(i) Suppose that $X_n$ and $Y_n$ are independent for all $n \geq 1$. If $X_n + Y_n \overset{c.c.}{\to} a+b$ and $Y_n \overset{p}{\to} b$ as $n \to \infty$, then*

$$X_n \overset{c.c.}{\to} a \quad and \quad Y_n \overset{c.c.}{\to} b \quad as \quad n \to \infty.$$

*(ii) Suppose that $X_1, X_2, \ldots, X_n$ and $Y_n$ are independent for all $n \geq 1$. If $X_n + Y_n \overset{a.s.}{\to} a+b$ and $Y_n \overset{p}{\to} b$ as $n \to \infty$, then*

$$X_n \overset{a.s.}{\to} a \quad and \quad Y_n \overset{a.s.}{\to} b \quad as \quad n \to \infty.$$

*(iii) Suppose that $X_n$ and $Y_n$ are independent for all $n \geq 1$. If $X_n + Y_n \overset{p}{\to} a+b$ and $Y_n \overset{p}{\to} b$ as $n \to \infty$, then*

$$X_n \overset{p}{\to} a \quad as \quad n \to \infty.$$

*Proof.* We assume, without restriction, that $a = b = 0$ during the proof.

Our first observation is that

$$\{|X_n + Y_n| > \varepsilon\} \supset \{|X_n| > 2\varepsilon\} \cap \{|Y_n| < \varepsilon\}. \tag{11.4}$$

For all $\varepsilon > 0$, there exists, by assumption, $n_0(\varepsilon)$, such that

$$P(|Y_n| > \varepsilon) < \frac{1}{2} \quad for \quad n > n_0(\varepsilon), \tag{11.5}$$

which, in view (11.4) and independence, shows that, given any such $\varepsilon$,

$$P(|X_n + Y_n| > \varepsilon) \geq P(|X_n| > 2\varepsilon) \cdot P(|Y_n| < \varepsilon) \geq \frac{1}{2}P(|X_n| > 2\varepsilon),$$

for $n > n_0(\varepsilon)$. Therefore, under the assumption of (i), we have

$$\infty > \sum_{n>n_0(\varepsilon)} P(|X_n + Y_n| > \varepsilon) \geq \frac{1}{2} \sum_{n>n_0(\varepsilon)} P(|X_n| > 2\varepsilon),$$

which proves that $X_n \overset{c.c.}{\to} 0$ as $n \to \infty$. The same then follows for $Y_n$ via Theorem 11.1.

In order to prove (ii), we first note that, by Theorem 1.3.1,

$$0 = P(|X_n + Y_n| > \varepsilon \text{ i.o.}) = P\left( \bigcap_{m=1}^{\infty} \bigcup_{n=m}^{\infty} \{|X_n + Y_n| > \varepsilon\} \right)$$

$$= \lim_{m \to \infty} P\left( \bigcup_{n=m}^{\infty} \{|X_n + Y_n| > \varepsilon\} \right). \tag{11.6}$$

Next, let $\varepsilon$ be given, set, for $n \geq m \geq n_0(\varepsilon)$,

$$A_n^{(m)} = \{ \max_{m \leq k \leq n-1} |X_k| \leq 2\varepsilon, |X_n| > 2\varepsilon \},$$

and note that, for a fixed $m$, the sets $A_n^{(m)}$, are disjoint for $n \geq m$, (11.4).

This fact, independence, and (11.5), together yield

$$P\left( \bigcup_{n=m}^{\infty} \{|X_n + Y_n| > \varepsilon\} \right) \geq P\left( \bigcup_{n=m}^{\infty} A_n^{(m)} \cap \{|Y_n| < \varepsilon\} \right)$$

$$= \sum_{n=m}^{\infty} P(A_n^{(m)} \cap \{|Y_n| < \varepsilon\}) = \sum_{n=m}^{\infty} P(A_n^{(m)}) \cdot P(|Y_n| < \varepsilon)$$

$$\geq \frac{1}{2} \sum_{n=m}^{\infty} P(A_n^{(m)}) = \frac{1}{2} P\left( \bigcup_{n=m}^{\infty} A_n^{(m)} \right)$$

Joining this with (11.6) shows that

$$P\left( \bigcap_{n=1}^{\infty} \bigcup_{n=m}^{\infty} A_n^{(m)} \right) = \lim_{m \to \infty} P\left( \bigcup_{m=1}^{\infty} A_n^{(m)} \right) = 0,$$

which, together with the fact that at most one of the events $A_n^{(m)}$, $n \geq m$, can occur for every fixed $m$, shows that $P(|X_n| > 2\varepsilon \text{ i.o.}) = 0$, and, hence, that $X_n \overset{a.s.}{\to} 0$ as $n \to \infty$. The same then follows for $Y_n$ by Theorem 11.1.

The final assertion is contained in Theorem 11.1.     □

The conclusion of the theorem is that the first suggestion, that one of the sequences is well behaved, is in order (under additional independence assumptions). Let us check the second one. Note that the symmetry assumption forces the limits $a$ and $b$ to be equal to 0.

**Exercise 11.3.** Check the last claim, that is, prove that, if $X_1, X_2, \ldots$ are symmetric random variables such that $X_n \to a$ for some constant $a$, completely, almost surely, or in probability as $n \to \infty$, then $a = 0$.     □

**Theorem 11.6.** *Let $X_1$, $X_2$, ... and $Y_1$, $Y_2$, ... be sequences of symmetric random variables.*
(i) *If $X_n$ and $Y_n$ are independent for all $n \geq 1$, then*

$$X_n + Y_n \overset{c.c.}{\to} 0 \text{ as } n \to \infty \quad \Longrightarrow \quad X_n \overset{c.c.}{\to} 0 \text{ and } Y_n \overset{c.c.}{\to} 0 \text{ as } n \to \infty.$$

(ii) *If $X_1$, $X_2$, ..., $X_n$ and $Y_n$ are independent for all $n \geq 1$, then*

$$X_n + Y_n \overset{a.s.}{\to} 0 \text{ as } n \to \infty \quad \Longrightarrow \quad X_n \overset{a.s.}{\to} 0 \text{ and } Y_n \overset{a.s.}{\to} 0 \text{ as } n \to \infty.$$

(iii) *If $X_n$ and $Y_n$ are independent for all $n \geq 1$, then*

$$X_n + Y_n \overset{p}{\to} 0 \text{ as } n \to \infty \quad \Longrightarrow \quad X_n \overset{p}{\to} 0 \text{ and } Y_n \overset{p}{\to} 0 \text{ as } n \to \infty.$$

*Proof.* In this setup the proof follows (almost) immediately via Corollary 3.7.1, since
$$P(|X_n| > \varepsilon) \leq 2P(|X_n + Y_n| > \varepsilon).$$

Complete convergence for the sum therefore implies complete convergence for each of the individual sequences. Similarly for convergence in probability.

Finally, if the sum converges almost surely it also converges in probability, so that each of the sequences does so too. An application of Theorem 11.5(ii) completes the proof. □

## 11.3 Symmetrization and Desymmetrization

Let us make the statement concerning the "centers" prior to Theorem 11.5 a little more precise. As we have already mentioned (and shall see in Chapter 6), a common way to prove limit theorems is to consider the symmetric case first, and then to "desymmetrize". The symmetrization inequalities, Propositions 3.6.2 and 3.6.3, then tell us that the asymptotics of the medians are the crucial issue. The following two propositions shed some light on that. Recall that med $(X)$ is a median of the random variable $X$.

**Proposition 11.1.** *Let $X_1$, $X_2$, ... be random variables, let $\{X_k^s, k \geq 1\}$ be the symmetrized sequence, and let $a_n \in \mathbb{R}$, $n \geq 1$. Then,*

$$X_n - a_n \overset{p}{\to} 0 \implies X_n^s \overset{p}{\to} 0 \implies X_n - \text{med}(X_n) \overset{p}{\to} 0$$
$$\implies \text{med}(X_n) - a_n \to 0 \quad as \quad n \to \infty.$$

*In particular,*

$$X_n \overset{p}{\to} 0 \implies \text{med}(X_n) \to 0 \quad as \quad n \to \infty.$$

**Proposition 11.2.** *Let $X_1$, $X_2$, ... be random variables, let $\{X_k^s, k \geq 1\}$ be the symmetrized sequence, and let $a_n \in \mathbb{R}$, $n \geq 1$. Then,*

$$X_n - a_n \overset{a.s.}{\to} 0 \implies X_n^s \overset{a.s.}{\to} 0 \implies X_n - \operatorname{med}(X_n) \overset{a.s.}{\to} 0$$
$$\implies \operatorname{med}(X_n) - a_n \to 0 \quad as \quad n \to \infty.$$

*In particular,*

$$X_n \overset{a.s.}{\to} 0 \implies \operatorname{med}(X_n) \to 0 \quad as \quad n \to \infty.$$

**Exercise 11.4.** Review the symmetrization inequalities and prove the propositions. □

*Remark 11.3.* The converses are obvious, that is, if $X_n \to X$ almost surely or in probability as $n \to \infty$, then $X_n^s$ converge too (Theorem 11.1). □

## 12 Cauchy Convergence

For a sequence of real numbers it is well known that convergence is equivalent to Cauchy convergence, viz., for a sequence $\{a_n, n \geq 1\}$ of reals,

$$a_n \to a \quad as \quad n \to \infty \iff a_n - a_m \to 0 \quad as \quad n, m \to \infty.$$

The fancy terminology is that $(\mathbb{R}, \mathcal{R})$, that is, the space of reals, together with its $\sigma$-algebra of Borel sets is *complete*.

In this subsection we check possible completeness for (most of) our modes of convergence.

We begin by noticing that convergence always implies Cauchy convergence via some kind of triangle inequality, together with the results in Section 5.11. For example, if $X_n \overset{p}{\to} X$ as $n \to \infty$, then

$$|X_n - X_m| \leq |X_n - X| + |X - X_m| \overset{p}{\to} 0 \quad as \quad n, m \to \infty,$$

by Theorem 11.1, and for distributional convergence,

$$|F_{X_n}(x) - F_{X_m}(x)| \leq |F_{X_n}(x) - F_X(x)| + |F_X(x) - F_{X_m}(x)|$$
$$\to 0 \quad as \quad n, m \to \infty, \quad \text{for} \quad x \in C(F_X).$$

Now, let us turn to the converses. We thus assume that our sequence is Cauchy convergent.

### Almost Sure Convergence

For a.s. convergence this follows from the corresponding result for real numbers, since the assumption is that $\{X_n(\omega), n \geq 1\}$ is Cauchy convergent for almost all $\omega$.

## Convergence in Probability

By modifying the proof of Theorem 3.4 one shows that there exists a subsequence, $\{X_{n_k}, k \geq 1\}$, that is almost surely Cauchy convergent, from which we conclude that there exists a limiting random variable $X$. This tells us that $X_{n_k}$ converges almost surely, and, hence, in probability to $X$ as $n \to \infty$, from which the conclusion follows via the triangle inequality,

$$|X_n - X| \leq |X_n - X_{n_k}| + |X_{n_k} - X|,$$

and Theorem 11.1.

**Exercise 12.1.** Prove the modification of Theorem 3.4 referred to. □

**Exercise 12.2.** An alternative proof would be to exploit Proposition 7.1. Verify this. □

## Mean Convergence

Given $\{X_n, n \geq 1\}$, we thus assume that the sequence $\{E|X_n|^r, n \geq 1\}$ is Cauchy convergent for some $r > 0$. By Markov's inequality, Theorem 3.1.1,

$$P(|X_m - X_n| > \varepsilon) \leq \frac{E|X_n - X_m|^r}{\varepsilon^r} \to 0 \quad \text{as} \quad n \to \infty,$$

that is, $\{X_n, n \geq 1\}$ is Cauchy convergent in probability, so that there exists a limiting random variable, $X$, such that $X_n \overset{p}{\to} X$ as $n \to \infty$.

Moreover, by Theorem 3.4, there exists a subsequence, $\{X_{n_k}, k \geq 1\}$, such that

$$X_{n_k} \overset{a.s.}{\to} X \quad \text{as} \quad n \to \infty,$$

so that, for $m$ fixed,

$$X_m - X_{n_k} \overset{a.s.}{\to} 0 \quad \text{as} \quad k \to \infty.$$

An application of Fatou's lemma then shows that

$$E|X_m - X|^r \leq \liminf_{k\to\infty} E|X_m - X_{n_k}|^r,$$

from which it follows that

$$\lim_{m\to\infty} E|X_m - X|^r \leq \lim_{m\to\infty} \liminf_{k\to\infty} E|X_m - X_{n_k}|^r = 0.$$

This proves that $X_n \overset{r}{\to} X$ as $n \to \infty$.

Note that we thereby have fulfilled our promise in Section 3.4 to show that the $L^p$-spaces equipped with the norm $\| \cdot \|_p$ are complete for $p \geq 1$.

**Distributional Convergence**

This case is immediate, since distribution functions are real valued; if

$$F_n(x) - F_m(x) \to 0 \quad \text{as} \quad n \to \infty,$$

then there exists a limiting value, $F(x)$, say, such that $F_n(x) \to F(x)$ as $n \to \infty$.

# 13 Skorohod's Representation Theorem

In Theorem 3.1 we established a hierarchy between the various modes of convergence. Later we have established some converses, and encountered some cases where additional assumptions ensured a converse. In this section we shall prove a theorem due to Skorohod [224], according to which, given that $X_n \overset{d}{\to} X$ as $n \to \infty$ there exists a "parallel" sequence which converges almost surely. The usefulness, and beauty, of the result is that it provides easy proofs of theorems such as the continuous mapping theorem, Theorem 10.4.

**Theorem 13.1.** (Skorohod's representation theorem)
Let $\{X_n, n \geq 1\}$ be *random variables such that*

$$X_n \overset{d}{\to} X \quad \text{as} \quad n \to \infty.$$

*Then there exist random variables* $X'$ *and* $\{X'_n, \geq 1\}$ *defined on the Lebesgue probability space, such that*

$$X'_n \overset{d}{=} X_n \quad \text{for} \quad n \geq 1, \quad X' \overset{d}{=} X, \quad \text{and} \quad X'_n \overset{a.s.}{\to} X' \quad \text{as} \quad n \to \infty.$$

*Remark 13.1.* In probabilistic language the Lebesgue probability space corresponds to a $U(0,1)$-distributed random variable.    □

*Proof.* Let $F_n$, be the distribution function of $X_n$, $n \geq 1$, $F$ that of $X$, and set

$$X'_n(\omega) = \inf\{\omega : F_n(x) \geq \omega\}, \quad n \geq 1, \quad \text{and} \quad X'(\omega) = \inf\{\omega : F(x) \geq \omega\},$$

so that

$$\{F_n(x) \geq \omega\} = \{X'_n(\omega) \leq x\}, \quad n \geq 1, \quad \text{and} \quad \{F(x) \geq \omega\} = \{X'(\omega) \leq x\},$$

which implies that

$$F'_n(x) = P'(X'_n \leq x) = P'(\{\omega : \omega \in (0, F_n(x)]\}) = F(x), \quad n \geq 1,$$
$$F'(x) = P'(X' \leq x) = P'(\{\omega : \omega \in (0, F(x)]\}) = F(x).$$

So far we have found random variables with the correct distributions. In order to prove convergence, let $\omega \in (0,1)$ and $\varepsilon > 0$ be given and select $x \in C(F)$, such that

$$F(x) < \omega \leq F(x+\varepsilon).$$

Since, by assumption, $F_n(x) \to F(x)$ as $n \to \infty$, there exists $n_0$, such that

$$F_n(x) < \omega \leq F(x+\varepsilon) \qquad \text{for all} \quad n \geq n_0.$$

In view of the inverse relationships this can, equivalently, be rewritten as

$$X'(\omega) - \varepsilon \leq x < X'_n(\omega) \qquad \text{for all} \quad n \geq n_0.$$

Due to the arbitrary choices of $\omega$ and $\varepsilon$ it follows that

$$\liminf_{n \to \infty} X'_n(\omega) \geq X'(\omega) \qquad \text{for all} \quad \omega. \tag{13.1}$$

Next, let $\omega^* \in (\omega, 1)$, $\varepsilon > 0$ and choose $x \in C(F)$, such that

$$F(x - \varepsilon) < \omega^* \leq F(x).$$

By arguing as in the previous portion of the proof we first have

$$F(x - \varepsilon) < \omega^* < F_n(x) \qquad \text{for all} \quad n \geq n_0,$$

which is the same as

$$X'_n(\omega^*) < x < X'(\omega^*) + \varepsilon \qquad \text{for all} \quad n \geq n_0.$$

Moreover, the fact that $\omega < \omega^*$, forces

$$X'_n(\omega) < x < X'(\omega^*) + \varepsilon \qquad \text{for all} \quad n \geq n_0,$$

so that

$$\limsup_{n \to \infty} X'_n(\omega) \leq X'(\omega^*) \qquad \text{for all} \quad \omega^* > \omega. \tag{13.2}$$

Joining (13.1) with (13.2), and noticing that the choice of $\omega^*$ was arbitrary (as long as it exceeded $\omega$), shows that $X_n(\omega) \to X(\omega)$ as $n \to \infty$ for all $\omega \in C(X)$. However, $X$ as a function of $\omega$ belongs to $D^+$, so that the complement of $C(X)$ is at most countable by Lemma A.9.1(i). Thus,

$$X_n(\omega) \to X(\omega) \qquad \text{as} \quad n \to \infty, \qquad \text{for all} \quad \omega \text{ outside a null set,}$$

which is exactly what was to be demonstrated.    $\square$

Having the representation theorem at our disposal, let us illustrate its usefulness by re-proving the continuous mapping theorem, Theorem 10.4.

We are thus given a sequence $X_1, X_2, \ldots$, where $X_n \overset{d}{\to} X$ as $n \to \infty$, and a function $g \in C$. The Skorohod representation theorem provides us with a sequence $X_1', X_2', \ldots$, and a random variable $X'$, such that $X_n' \overset{a.s.}{\to} X'$ as $n \to \infty$. Theorem 10.1 then tells us that

$$g(X_n') \overset{a.s.}{\to} g(X') \quad \text{as} \quad n \to \infty.$$

Since almost sure convergence implies distributional convergence, it follows that

$$g(X_n') \overset{d}{\to} g(X') \quad \text{as} \quad n \to \infty,$$

and, finally, since every primed random variable has the same distribution as the original, unprimed, one, we also have

$$g(X_n) \overset{d}{\to} g(X) \quad \text{as} \quad n \to \infty,$$

which is the conclusion of the continuous mapping theorem.

Once again, this is a pretty and easy proof, BUT it relies on a very sophisticated prerequisite. In other words, we do not get anything for free.

**Exercise 13.1.** Prove the Fatou lemmas, i.e., Theorems 5.3 and 5.8, via Skorohod representation.

**Exercise 13.2.** Prove Theorem 5.7 with the aid of Skorohod's result.

# 14 Problems

1. Let $\{X_n, n \geq 1\}$ be a sequence of independent random variables with common density

$$f(x) = \begin{cases} e^{-(x-a)}, & \text{for} \quad x \geq a, \\ 0, & \text{for} \quad x < a. \end{cases}$$

Set $Z_n = \min\{X_1, X_2, \ldots, X_n\}$. Show that

$$Z_n \overset{p}{\to} a \quad \text{as} \quad n \to \infty.$$

2. Suppose that $X_k \in \Gamma(3, k)$, that is, that

$$f_{X_k}(x) = \begin{cases} \frac{x^2}{2k^3} e^{-x/k}, & \text{for} \quad x > 0, \\ 0, & \text{otherwise.} \end{cases}$$

Show that

$$\sum_{k=1}^{n} \frac{1}{X_k} - \frac{1}{2} \log n \quad \text{converges in probability as } n \to \infty.$$

♠ Remember that $\sum_{k=1}^{n} \frac{1}{k} - \log n \to \gamma$ as $n \to \infty$, where $\gamma = 0.5772\ldots$ is Euler's constant.

3. Suppose that $X_n \in \text{Ge}(\frac{\lambda}{n+\lambda})$, $n = 1, 2, \ldots$, where $\lambda$ is a positive constant. Show that $X_n/n$ converges in distribution to an exponential distribution as $n \to \infty$, and determine the parameter of the limit distribution.

4. Suppose that $X_n \in \text{Ge}(p_n)$, $n \geq 1$. Show that

(a) if $p_n \to p > 0$, then $X_n \overset{d}{\to} \text{Ge}(p)$ as $n \to \infty$.

(b) if $p_n \to 0$ and $np_n \to \lambda > 0$, then $X_n/n \overset{d}{\to} \text{Exp}(1/\lambda)$ as $n \to \infty$.

5. Let $X, X_1, X_2, \ldots$ be independent, identically distributed random variables, such that $\sup\{x : F(x) < 1\} = +\infty$, and let

$$\tau(t) = \min\{n : X_n > t\}, \quad t > 0,$$

that is, $\tau(t)$ is the index of the first $X$-variable that exceeds the level $t$. Show that

$$p_t \tau(t) \overset{d}{\to} \text{Exp}(1) \quad \text{as} \quad t \to \infty,$$

where $p_t = P(X > t)$.

♣ In a typical application we would consider $\tau(t)$ as the number of "shocks" until the failure of a system and $X_{\tau(t)}$ as the size of the fatal "shock".

6. Suppose that $X \in N(\mu, \sigma^2)$, and that $X_n \in N(\mu_n, \sigma_n^2)$, $n \geq 1$. Prove that

$$X_n \overset{d}{\to} X \text{ as } n \to \infty \quad \Longleftrightarrow \quad \mu_n \to \mu \text{ and } \sigma_n \to \sigma \quad \text{as} \quad n \to \infty.$$

7. Suppose that $X \in \text{Po}(\lambda)$, and that $X_n \in \text{Po}(\lambda_n)$, $n \geq 1$. Prove that

$$X_n \overset{d}{\to} X \text{ as } n \to \infty \quad \Longleftrightarrow \quad \lambda_n \to \lambda \quad \text{as} \quad n \to \infty.$$

8. Let $X_1, X_2, \ldots$ be independent random variables with common characteristic function

$$\varphi(t) = \begin{cases} 1 - \sqrt{|t|(2 - |t|)}, & \text{for } |t| \leq 1, \\ 0, & \text{otherwise,} \end{cases}$$

and let $\{S_n, n \geq 1\}$ denote the partial sums. Show that $\frac{S_n}{n^2}$ converges in distribution as $n \to \infty$ and determine the limit.

9. Can you find a sequence of absolutely continuous random variables that converges distribution to a discrete random variable? Can you find a sequence of discrete random variables with an absolutely continuous limit?

10. Let $X_1, X_2, \ldots$ be random variables, and suppose that

$$P(X_n < a \text{ i.o.} \quad \text{and} \quad x_n > b \text{ i.o.}) = 0 \quad \text{for all} \quad a < b.$$

Prove that $X_n$ converges a.s. as $n \to \infty$.

♠ Do not forget the rationals, $\mathbb{Q}$.

11. Show that, if $X_1, X_2, \ldots$ are independent, identically distributed random variables, then

$$P(X_n \text{ converges}) = 0.$$

12. Let $X_{n1}$, $X_{n2}$, $\ldots$, $X_{nn}$ be independent random variables, with a common distribution given as follows:

$$P(X_{nk} = 0) = 1 - \tfrac{1}{n} - \tfrac{1}{n^2}, \quad P(X_{nk} = 1) = \tfrac{1}{n}, \quad P(X_{nk} = 2) = \tfrac{1}{n^2},$$

where $k = 1, 2, \ldots, n$ and $n = 1, 2, \ldots$. Set

$$S_n = X_{n1} + X_{n2} + \cdots + X_{nn}, \quad n \geq 1.$$

Show that

$$S_n \xrightarrow{d} \mathrm{Po}(1) \quad \text{as} \quad n \to \infty.$$

13. Let $X_1$, $X_2$, $\ldots$ be independent, standard Cauchy-distributed random variables, and $a_k$, $k \geq 1$, real numbers. Prove that $\sum_{k=1}^{n} a_k X_k$ converges in distribution if and only if

$$\sum_{n=1}^{\infty} |c_n| < \infty.$$

14. Suppose that $X_n \in N(\mu_n, \sigma_n^2)$, $n \geq 1$. When is $\{X_n, n \geq 1\}$ uniformly integrable?

15. Suppose that $X_n \in \mathrm{Exp}(\lambda_n)$, $n \geq 1$. When is $\{X_n, n \geq 1\}$ uniformly integrable?

16. *Pólya's theorem.* The aim of this problem is to prove the following result:

**Theorem 14.1.** *Suppose that $\varphi$ is a continuous, even function on $\mathbb{R}$, that $\varphi(0) = 0$, that $\varphi(t) \to 0$ as $t \to \pm\infty$, and that $\varphi$ is convex on $\mathbb{R}^+$. Then $\varphi$ is a characteristic function.*

To prove this we begin by recalling, from Section 4.3, that the function

$$\varphi(t) = \begin{cases} 1 - |t|, & \text{for} \quad |t| < 1, \\ 0, & \text{otherwise}, \end{cases}$$

is a characteristic function (corresponding to a random variable with density $(1 - \cos x)/(\pi x^2)$).

(a) Let, for $1 \leq k \leq n$, $n \geq 1$, $p_{k,n}$, be positive numbers that add to 1, and $\{a_{k,n}, 1 \leq k \leq n\}$, a sequence of positive, strictly increasing reals. Prove that

$$\varphi_n(t) = \sum_{k=1}^{n} p_{k,n} \varphi(t/a_{k,n})$$

is a characteristic function.

♣ Note that $\tilde{\varphi}$ in Section 4.3 is of this kind.

(b) Let $n \to \infty$ and use a suitable approximation result relating convex functions and polygons to finish the proof.

17. Suppose that $\varphi(t, u)$ is a characteristic function as a function of $t$ for every fixed $u$, and continuous as a function of $u$ for every fixed $t$, and that $G$ is the distribution function of some random variable. Prove that

$$\tilde{\varphi}(t) = \exp\left\{\int_{\mathbb{R}} (\varphi(t, u) \, \mathrm{d}G(u)\right\}$$

is a characteristic function.

18. Let $a > 0$, suppose that $X_1, X_2, \ldots$ are independent random variables with common density

$$f_X(x) = \frac{1}{2a} e^{-|x|/a}, \quad -\infty < x < \infty,$$

and let $N \in \mathrm{Po}(m)$ be independent of $X_1, X_2, \ldots$. Determine the limit distribution of $S_N = X_1 + X_2 + \cdots + X_N$ (where $S_0 = 0$) as $m \to \infty$ and $a \to 0$ in such a way that $m \cdot a^2 \to 1$.

19. Let $X_k \in \mathrm{Exp}(k!)$, $k = 1, 2, \ldots$, and suppose that $X_1, X_2, \ldots$ are independent. Set $S_n = \sum_{k=1}^{n} X_k$, $n \geq 1$. Show that

$$\frac{S_n}{n!} \xrightarrow{d} \mathrm{Exp}(1) \quad \text{as} \quad n \to \infty.$$

♠ What is the distribution of $X_n/n!$?

20. Let $X_1, X_2, \ldots$ be $U(-1, 1)$-distributed random variables, and set

$$Y_n = \begin{cases} X_n, & \text{for} \quad |X_n| \leq 1 - \frac{1}{n}, \\ n, & \text{otherwise}. \end{cases}$$

(a) Show that $Y_n$ converges in distribution as $n \to \infty$, and determine the limit distribution.

(b) Let $Y$ denote the limiting random variable. Consider the statements $EY_n \to EY$ and $\mathrm{Var}\, Y_n \to \mathrm{Var}\, Y$ as $n \to \infty$. Are they true or false?

21. Let $X \in N(0, 1)$, let $X_1, X_2, \ldots$ be random variables defined by

$$P(X_n = 1) = 1 - \frac{1}{n} \quad \text{and} \quad P(X_n = n) = \frac{1}{n}, \quad n \geq 1,$$

and suppose that all random variables are independent. Set

$$Y_n = X \cdot X_n, \quad n \geq 1.$$

Show that

$$Y_n \xrightarrow{d} N(0, 1) \quad \text{as} \quad n \to \infty,$$
$$EY_n = 0,$$
$$\mathrm{Var}\, Y_n \to +\infty \quad \text{as} \quad n \to \infty.$$

22. Let $d \geq 2$, and let $\mathbf{X}_n = (X_n^{(1)}, X_n^{(2)}, \ldots, X_n^{(d)})$, $n \geq 1$, be random vectors. We say that $\mathbf{X}_n \xrightarrow{p} \mathbf{X}$ as $n \to \infty$ iff

$$\|\mathbf{X}_n - \mathbf{X}\| \xrightarrow{p} 0 \quad \text{as} \quad n \to \infty,$$

where $\| \cdot \|$ is the Euclidean norm; $\|x\| = \sqrt{\sum_{k=1}^{d} x_k^2}$ for $\mathbf{x} \in \mathbb{R}^d$.
(a) Prove that

$$\mathbf{X}_n \xrightarrow{p} \mathbf{X} \quad \Longleftrightarrow \quad X_n^{(k)} \to X^{(k)} \text{ for } k = 1, 2, \ldots, d \quad \text{as} \quad n \to \infty.$$

(b) Suppose that $h : \mathbb{R}^d \to \mathbb{R}^{d'}$ is continuous. Prove that

$$\mathbf{X}_n \xrightarrow{p} \mathbf{X} \text{ as } n \to \infty \quad \Longrightarrow \quad h(\mathbf{X}_n) \to h(\mathbf{X}) \text{ as } n \to \infty.$$

♣ This generalizes Theorem 10.3 to $d$ dimensions.

23. Suppose that $\{U_n, n \geq 1\}$ and $\{V_n, n \geq 1\}$ are sequences of random variables, such that

$$U_n \xrightarrow{d} U \quad \text{and} \quad V_n \xrightarrow{p} a \quad \text{as} \quad n \to \infty,$$

for some random variable $U$, and finite constant $a$. Prove that

$$\max\{U_n, V_n\} \xrightarrow{d} \max\{U, a\} \quad \text{as} \quad n \to \infty,$$
$$\min\{U_n, V_n\} \xrightarrow{d} \min\{U, a\} \quad \text{as} \quad n \to \infty.$$

♣ This is a kind of Cramér theorem for the maximum and the minimum.

24. Let $X_1, X_2, \ldots$ be random variables, such that $X_n \xrightarrow{d} X$ as $n \to \infty$, let $g : \mathbb{R} \to \mathbb{R}$, and set $E = \{x : g \text{ is discontinuous at } x\}$. Show that

$$P(X \in E) = 0 \quad \Longrightarrow \quad g(X_n) \xrightarrow{d} g(X) \quad \text{as} \quad n \to \infty.$$

♣ This extends Theorem 10.4.

25. Let $X$ and $Y$ be random variables. The *Lévy distance* between $X$ and $Y$ is

$$d_L(X, Y) = \inf\{\varepsilon > 0 : F_Y(x - \varepsilon) - \varepsilon \leq F_X(x) \leq F_Y(x + \varepsilon) + \varepsilon \text{ for all } x\}.$$

Show that, if $X, X_1, X_2, \ldots$ are random variables, then

$$X_n \xrightarrow{d} X \text{ as } n \to \infty \quad \Longleftrightarrow \quad d_L(X_n, X) \to 0 \text{ as } n \to \infty.$$

# 6

## The Law of Large Numbers

We have mentioned (more than once) that the basis for probabilistic modeling is the stabilization of the relative frequencies. Mathematically this phenomenon can be formulated as follows: Suppose that we perform independent repetitions of an experiment, and let $X_k = 1$ if round $k$ is successful and $0$ otherwise, $k \geq 1$. The relative frequency of successes is described by the arithmetic mean, $\frac{1}{n} \sum_{k=1}^{n} X_k$, and the stabilization of the relative frequencies corresponds to

$$\frac{1}{n} \sum_{k=1}^{n} X_k \to p \quad \text{as} \quad n \to \infty,$$

where $p = P(X_1 = 1)$ is the success probability.

Note that the convergence arrow is a *plain* arrow! The reason for this is that the first thing to wonder about is: *Convergence in what sense?*

The basis for the probability model was the observation that whenever such a random experiment is performed the relative frequencies stabilize. The word "whenever" indicates that the interpretation must be "almost sure convergence". The stabilization thus is translated into

$$\frac{1}{n} \sum_{k=1}^{n} X_k \overset{a.s.}{\to} p \quad \text{as} \quad n \to \infty.$$

This, in turn, means that the validation of the Ansatz is that a theorem to that effect must be contained in our theory. The *strong law of large numbers*, which is due to Kolmogorov, is a more general statement to the effect that if $X_1, X_2, \ldots$ are arbitrary independent, identically distributed random variables with finite mean, $\mu$, then the arithmetic mean converges almost surely to $\mu$. Moreover, finite mean is necessary for the conclusion to hold. If, in particular, the summands are indicators, the result reduces to the *almost sure* formulation of the stabilization of the relative frequencies.

There also exist weak laws, which means convergence in probability.

Other questions concern (i) moment convergence, which means questions related to uniform integrability (recall Section 5.4), (ii) whether other limit theorems can be obtained by other normalizations under suitable conditions, and (iii) laws of large numbers for randomly indexed sequences. We shall also prove a fact that has been announced before, namely that complete convergence requires more than almost sure convergence.

Applications to normal numbers, the Glivenko-Cantelli theorem, renewal theory, and records will be given, which for the latter two means a continuation of earlier visits to those topics.

The final section, entitled "Some Additional Results and Remarks", preceding the problem section, contains different aspects of convergence rates. This section may be considered as less "hot" (for the non-specialist) and can therefore be skipped, or skimmed through, at a first reading.

# 1 Preliminaries

A common technique is to use what is called *truncation*, which means that one creates a new sequence of random variables which is asymptotically equivalent to the sequence of interest, and easier to deal with than the original one.

## 1.1 Convergence Equivalence

The first thing thus is to find criteria for two sequences of random variables to be equivalent.

**Definition 1.1.** *The sequences* $X_1, X_2, \ldots$ *and* $Y_1, Y_2, \ldots$ *of random variables are said to be* convergence equivalent *if*

$$\sum_{n=1}^{\infty} P(X_n \neq Y_n) < \infty.$$

□

The first Borel-Cantelli lemma, Theorem 2.18.1, immediately tells us the following.

**Theorem 1.1.** *If* $X_1, X_2, \ldots$ *and* $Y_1, Y_2, \ldots$ *are convergence equivalent, then*

(i)    $P(X_n \neq Y_n \text{ i.o.}) = 0$;

(ii)   $\sum_{n=1}^{\infty}(X_n - Y_n)$ *converges a.s.;*

(iii)  *if* $b_n \in \mathbb{R}$, $n \geq 1$, $b_n \uparrow \infty$ *as* $n \to \infty$, *then*

$$\frac{1}{b_n} \sum_{k=1}^{n}(X_k - Y_k) \overset{a.s.}{\to} 0 \quad as \quad n \to \infty.$$

*Proof.* The first statement follows from Theorem 2.18.1. The significance is that $X_n$ and $Y_n$ differ only a *finite* number of times, that is, there exists a random index $n(\omega)$ after which $X_n$ and $Y_n$ are equal, so that the sum in (ii) contains only a finite (but random) number of terms for almost all $\omega$. The last claim follows by the same argument, together with the fact that $b_n \uparrow \infty$ as $n \to \infty$. $\qquad\square$

*Example 1.1.* If $X_1, X_2, \ldots$ is such that $\sum_{n=1}^{\infty} P(|X_n| > a) < \infty$, then we obtain a convergence equivalent sequence by introducing the random variables $Y_n = X_n I\{|X_n| \le a\}$. $\qquad\square$

Another example follows after the following useful tool.

**Proposition 1.1.** *Suppose that $X, X_1, X_2, \ldots$ are independent, identically distributed random variables, and let $r > 0$. The following are equivalent:*

(i)    $E|X|^r < \infty$;

(ii)   $\sum_{n=1}^{\infty} P(|X_n| > n^{1/r}\varepsilon) < \infty$ for all $\varepsilon > 0$;

(iii)  $P(|X_n| > n^{1/r}\varepsilon$ i.o.$) = 0$ for all $\varepsilon > 0$;

(iv)   $\frac{X_n}{n^{1/r}} \xrightarrow{a.s.} 0$ as $n \to \infty$.

*Proof.* We know from Theorem 2.12.1, scaling, and equidistribution that

$$\text{(i)} \iff \sum_{n=1}^{\infty} P(|X|^r > n) < \infty \iff \sum_{n=1}^{\infty} P(|X_n| > n^{1/r}\varepsilon) < \infty.$$

This proves that (i) and (ii) are equivalent. The equivalence with (iii) and (iv) is a consequence of the Borel-Cantelli lemmas and the definition of almost sure convergence. As for (iv) one may also review Corollary 2.18.1. $\qquad\square$

Proposition 1.1 will be frequently used in the sequel as follows.

**Proposition 1.2.** *Suppose that $X, X_1, X_2, \ldots$ are independent, identically distributed random variables, such that $E|X|^r < \infty$ for some $r > 0$, and set*

$$Y_n = X_n I\{|X_n| \le n^{1/r}\}, \quad n \ge 1.$$

*Then $X_1, X_2, \ldots$ and $Y_1, Y_2, \ldots$ are convergence equivalent.*

**Exercise 1.1.** Although the proof is fairly immediate it is a healthy exercise to put it on paper. $\qquad\square$

## 1.2 Distributional Equivalence

This is a weaker equivalence concept, in the same way as convergence in distribution is weaker than almost sure convergence.

**Definition 1.2.** *The sequences* $X_1, X_2, \ldots$ *and* $Y_1, Y_2, \ldots$ *of random variables are said to be* distributionally equivalent *if*

$$P(X_n \neq Y_n) \to 0 \quad as \quad n \to \infty. \qquad \square$$

Since the terms of a convergent series tend to 0 it follows immediately that:

**Proposition 1.3.** *If* $X_1, X_2, \ldots$ *and* $Y_1, Y_2, \ldots$ *are convergence equivalent then they are also distributionally equivalent.*

Following are some elementary facts concerning distributional equivalence, the second of which explains the name of the equivalence concept.

**Theorem 1.2.** *If* $X_1, X_2, \ldots$ *and* $Y_1, Y_2, \ldots$ *are distributionally equivalent, then*

(i)   $X_n - Y_n \xrightarrow{p} 0 \quad as \quad n \to \infty;$

(ii)   *if* $Y_n \xrightarrow{d} Y$ *as* $n \to \infty$, *then* $X_n \xrightarrow{d} Y$ *as* $n \to \infty$.

*Proof.* To prove (i) we simply note that, for any $\varepsilon > 0$,

$$P(|X_n - Y_n| > \varepsilon) \leq P(X_n \neq Y_n) \to 0 \quad as \quad n \to \infty,$$

after which (ii) follows from (i) and Cramér's theorem, Theorem 5.11.3.   $\square$

## 1.3 Sums and Maxima

As a further preparation we point out a relation between sums of *symmetric* random variables and their partial maxima.

**Proposition 1.4.** *Let* $X_1, X_2, \ldots$ *be independent,* symmetric *random variables, and set* $Y_n = \max_{1 \leq k \leq n} |X_k|$, *and* $S_n = \sum_{k=1}^{n} X_k$, $n \geq 1$. *Then,*

$$P(Y_n > 2x) \leq P(\max_{1 \leq k \leq n} |S_k| > x) \leq 2P(|S_n| > x), \quad x > 0.$$

*Proof.* Since $|X_n| \leq |S_n| + |S_{n-1}|$ it follows that $Y_n \leq 2\max_{1 \leq k \leq n} |S_k|$. Now apply the Lévy inequalities, Theorem 3.7.1.   $\square$

## 1.4 Moments and Tails

In this subsection we collect some facts that will be used several times later. It is, maybe, a bit pathetic to collect them in a proposition, but (a) some of the facts are easily overlooked, and (b) it is convenient for reference.

**Proposition 1.5.** (i) *Let* $r > 0$. *Suppose that* $X$ *is a non-negative random variable. Then*

$$E\,X^r < \infty \quad \implies \quad x^r P(X > x) \to 0 \quad as \quad x \to \infty,$$

*but not necessarily conversely.*

(ii) *Suppose that $X$, $X_1$, $X_2$, ... are independent, identically distributed random variables with mean 0. Then, for any $a > 0$,*

$$E\,XI\{|X| \le a\} = -E\,XI\{|X| > a\},$$

*and*

$$\left| E \sum_{k=1}^{n} X_k I\{|X_k| \le a\} \right| \le nE|X|I\{|X| > a\}.$$

(iii) *Let $a > 0$. If $X$ is a random variable with mean 0, then $Y = XI\{|X| \le a\}$ does* not *in general have mean 0. However, if $X$ is symmetric, then $E\,Y = 0$.*

*Proof.* (i): We have

$$x^r P(X > x) = x^r \int_x^\infty \mathrm{d}F(y) \le \int_x^\infty y^r\,\mathrm{d}F(y) \to 0 \quad \text{as} \quad x \to \infty,$$

being the tail of a convergent integral.

If, on the other hand, $X$ is a random variable with density

$$f(x) = \begin{cases} \frac{c}{x^{r+1} \log x}, & \text{for} \quad x > e, \\ 0, & \text{otherwise,} \end{cases}$$

say, where $c$ is a normalizing constant, then

$$x^r P(X > x) \sim Cx^r \frac{1}{x^r \log x} = \frac{C}{\log x} \to 0 \quad \text{as} \quad x \to \infty$$

via partial integration, but

$$E\,X^r = c \int_e^\infty \frac{\mathrm{d}x}{x \log x} = +\infty.$$

(ii): The first result follows from the fact that the mean is zero;

$$0 = E\,X = E\,XI\{|X| \le a\} + E\,XI\{|X| > a\}.$$

The second result follows from the first one and the triangle inequality.
(iii): Once the statement has been made (but not always before) the conclusion is "trivial". After all, a skew random variable with mean 0 that is truncated symmetrically certainly does not retain mean 0.

If, on the other hand, we truncate a symmetric random variable symmetrically, the resulting variable is symmetric again, and has mean 0. Notice that any truncated symmetric random variable $X$ has mean 0, whether the original mean exists or not.    □


## 2 A Weak Law for Partial Maxima

Before we enter the discussion of weak laws of large numbers we squeeze a weak law for partial *maxima* into the text. Partly because the law is of interest in its own right, but also for later use.

**Theorem 2.1.** *Let $X, X_1, X_2, \ldots$ be independent, identically distributed random variables, and set $Y_n = \max_{1 \le k \le n} |X_k|$, $n \ge 1$. If $\{b_n, n \ge 1\}$ is a sequence of non-decreasing positive reals, then*

$$\frac{Y_n}{b_n} \xrightarrow{p} 0 \text{ as } n \to \infty \quad \Longleftrightarrow \quad nP(|X| > b_n\varepsilon) \to 0 \quad \text{for all } \varepsilon \quad \text{as} \quad n \to \infty.$$

*In particular, for $r > 0$,*

$$\frac{Y_n}{n^{1/r}} \xrightarrow{p} 0 \text{ as } n \to \infty \quad \Longleftrightarrow \quad nP(|X| > n^{1/r}) \to 0 \text{ as } n \to \infty.$$

*Proof.* The proof is based on the double inequality

$$\frac{1}{2} nP(|X| > b_n\varepsilon) \le P(Y_n > b_n\varepsilon) \le nP(|X| > b_n\varepsilon), \quad \text{for } n \text{ large.} \quad (2.1)$$

The upper inequality is immediate, since $\{Y_n > b_n\varepsilon\} \subset \bigcup_{k=1}^{n}\{|X_k| > b_n\varepsilon\}$.

As for the lower bound, we have

$$P(Y_n > b_n\varepsilon) = 1 - \left(P(|X| \le b_n\varepsilon)\right)^n = 1 - \left(1 - P(|X| > b_n\varepsilon)\right)^n.$$

so that, for $n$ sufficiently large, the inequality follows via an application of Lemma A.4.2 with $\delta = 1/2$. □

*Remark 2.1.* A sufficient condition in the particular case is $E|X|^r < \infty$ (cf. Proposition 1.5). □

**Exercise 2.1.** Check that the "obvious" one-sided analog holds similarly, that is, show that

$$\frac{\max_{1 \le k \le n} X_k}{b_n} \xrightarrow{p} 0 \text{ as } n \to \infty \quad \Longleftrightarrow \quad nP(X > b_n\varepsilon) \to 0 \quad \text{for all } \varepsilon \text{ as } n \to \infty.$$

**Exercise 2.2.** Suppose that $X_1, X_2, \ldots$ are independent random variables, with $Y_n = \max_{1 \le k \le n} |X_k|$, $n \ge 1$, and let $\{b_n, n \ge 1\}$ be a sequence of non-decreasing positive reals as in the theorem. Prove (e.g., by extending Lemma A.4.2) that

$$\frac{Y_n}{b_n} \xrightarrow{p} 0 \quad \text{as} \quad n \to \infty \quad \Longleftrightarrow \quad \sum_{k=1}^{n} P(|X_k| > b_n\varepsilon) \to 0 \quad \text{for all } \varepsilon \quad \text{as} \quad n \to \infty,$$

and correspondingly for the one-sided analog. □

## 3 The Weak Law of Large Numbers

Here is a first, easily accessible, result.

**Theorem 3.1.** *Let $X, X_1, X_2, \ldots$ be independent, identically distributed random variables with finite mean, $\mu$, and let $S_n, n \ge 1$, denote their partial sums. Then*

$$\frac{S_n}{n} \xrightarrow{p} \mu \quad \text{as} \quad n \to \infty.$$

*Proof.* The essential tool to prove this is the truncated Chebyshev inequality, Theorem 3.1.5. By centering it is no restriction to assume that $\mu = 0$ in the proof.

Let $\varepsilon > 0$, and set, for $k = 1, 2, \ldots, n$, $n \geq 1$,

$$Y_{k,n} = X_k I\{|X_k| \leq n\varepsilon^3\}, \quad \text{and} \quad S_n' = \sum_{k=1}^{n} Y_{k,n}.$$

Using the cited inequality, we have

$$
\begin{aligned}
P(|S_n - E S_n'| > n\varepsilon) &\leq \frac{1}{n\varepsilon^2} \operatorname{Var} Y_{1,n} + nP(|X| > n\varepsilon^3) \\
&\leq \frac{1}{n\varepsilon^2} E Y_{1,n}^2 + nP(|X| > n\varepsilon^3) \\
&= \frac{1}{n\varepsilon^2} E(X^2 I\{|X| \leq n\varepsilon^3\}) + nP(|X| > n\varepsilon^3) \\
&\leq \varepsilon E|X| I\{|X| \leq n\varepsilon^3\} + nP(|X| > n\varepsilon^3) \\
&\leq \varepsilon E|X| + nP(|X| > n\varepsilon^3).
\end{aligned}
$$

Thus, by Proposition 1.5(i),

$$\limsup_{n \to \infty} P(|S_n - E S_n'| > n\varepsilon) \leq \varepsilon E|X|,$$

so that, $\varepsilon$ being arbitrary, we have asserted that

$$\frac{S_n - E S_n'}{n} \xrightarrow{p} 0 \quad \text{as} \quad n \to \infty.$$

To finish off we note that, the mean being 0, it follows from Proposition 1.5(ii) that

$$|E S_n'| = |nE X I\{|X| \leq n\varepsilon^3\}| \leq nE|X| I\{|X| > n\varepsilon^3\},$$

so that

$$\frac{E S_n'}{n} \to 0 \quad \text{as} \quad n \to \infty,$$

and the conclusion follows (via Theorem 5.11.1).

As an alternative we may use characteristic functions and Corollary 5.9.1, according to which the conclusion follows if we can show that

$$\varphi_{\bar{X}_n}(t) \to e^{it\mu} \quad \text{as} \quad n \to \infty, \quad \text{for} \quad -\infty < t < \infty.$$

Now, by Theorems 4.1.10 and 4.1.8,

$$\varphi_{\bar{X}_n}(t) = \varphi_{S_n}(t/n) = (\varphi_X(t/n))^n,$$

which, together with Theorem 4.4.2, yields

$$\varphi_{\bar{X}_n}(t) = \left(1 + i\frac{t}{n}\mu + o\left(\frac{t}{n}\right)\right)^n \to e^{it\mu} \quad \text{as} \quad n \to \infty, \quad \text{for all} \quad t.$$

Admittedly, this proof is shorter. However, we must keep in mind that we rest on Theorem 3.3 and a deep continuity theorem!     □

The following example pertains to a sequence of random variables whose mean does not exist.

*Example 3.1.* Let $X_1$, $X_2$, ... be independent, standard Cauchy-distributed random variables. The characteristic function is $\varphi(t) = e^{-|t|}$, and computations as above yield

$$\varphi_{\bar{X}_n}(t) = \left(\varphi(t/n)\right)^n = \left(e^{-|t/n|}\right)^n = e^{-|t|} = \varphi(t),$$

which, in view of the uniqueness theorem for characteristic functions shows that

$$\bar{X}_n \stackrel{d}{=} X_1, \quad \text{for all} \quad n.$$

The arithmetic mean of any number of observations thus has the same distribution as a single observation. This is most counterintuitive, since any "reasonable" model implies that arithmetic means have a "better precision" than individual observations. The counter-intuitiveness is explained by the fact that the law of large numbers does not hold, which, in turn is no contradiction, since the mean of the Cauchy distribution does not exist.    $\square$

The weak law of large numbers, assuming finite mean and a normalization by $n$, can be generalized to distributions with a finite moment of any order between 0 and 2, and with a normalizing sequence that is a suitable power of $n$ as follows.

**Theorem 3.2.** (The Marcinkiewicz-Zygmund weak law)
*Let $0 < r < 2$. Suppose that $X$, $X_1$, $X_2$, ... are independent, identically distributed random variables, such that $E|X|^r < \infty$, and let $S_n$, $n \geq 1$, denote their partial sums. We also suppose, without restriction, that $E\,X = 0$ when $1 \leq r < 2$. Then*

$$\frac{S_n}{n^{1/r}} \stackrel{p}{\to} 0 \quad as \quad n \to \infty.$$

*Proof.* The main difference compared to the previous proof is that we are facing a bit more technical trouble at the end in order to take care of the truncated means. In addition, since the mean does not exist when $0 < r < 1$, we shall solve the two cases $0 < r < 1$ and $1 < r < 2$ differently (the case $r = 1$ is Theorem 3.1).

We begin by considering the case $1 < r < 2$. Let $\varepsilon > 0$, and set, for $k = 1, 2, \ldots, n$, $n \geq 1$,

$$Y_{k,n} = X_k I\{|X_k| \leq n^{1/r}\varepsilon^{\frac{3}{2-r}}\}, \quad \text{and} \quad S'_n = \sum_{k=1}^{n} Y_{k,n}.$$

By proceeding exactly as in the proof of Theorem 3.1, we arrive at

$$P(|S_n - E\,S'_n| > n^{1/r}\varepsilon) \leq \frac{1}{n^{(2/r)-1}\varepsilon^2}\operatorname{Var} Y_{1,n} + nP(|X| > n^{1/r}\varepsilon^{\frac{3}{2-r}})$$

$$\leq \varepsilon E|X|^r + nP(|X| > n^{1/r}\varepsilon^{\frac{3}{2-r}}),$$

which, via Proposition 1.5(i), shows that

$$\limsup_{n\to\infty} P(|S_n - E\,S_n'| > n^{1/r}\varepsilon) \leq \varepsilon E|X|^r.$$

It remains to show that $E\,S_n' = o(n^{1/r})$ as $n \to \infty$, after which the conclusion follows (via Theorem 5.11.1). We proceed basically as in the previous proof:

$$|E\,S_n'| = |nE\,XI\{|X| \leq n^{1/r}\varepsilon^{\frac{3}{2-r}}\}| \leq nE|X|I\{|X| > n^{1/r}\varepsilon^{\frac{3}{2-r}}\}$$
$$\leq n(n^{1/r}\varepsilon^{\frac{3}{2-r}})^{1-r}E|X|^r I\{|X| > n^{1/r}\varepsilon^{\frac{3}{2-r}}\}$$
$$= \varepsilon^{\frac{3(1-r)}{2-r}} n^{1/r}E|X|^r I\{|X| > n^{1/r}\varepsilon^{\frac{3}{2-r}}\} = o(n^{1/r}) \quad \text{as} \quad n \to \infty,$$

since $E|X|^r < \infty$.

Now let $0 < r < 1$. In this case we shall, in fact, show that convergence holds in $L^r$, from which it follows that it also holds in probability (Theorem 5.3.1).

Toward this end, let $M > 0$ be so large that $E|X|^r I\{|X| > M\} < \varepsilon$. This is possible for any $\varepsilon > 0$, since $E|X|^r < \infty$. Set

$$Y_k = X_k I\{|X_k| \leq M\} \quad \text{and} \quad Z_k = X_k I\{|X_k| > M\}, \quad k = 1, 2, \ldots.$$

Then, by the $c_r$-inequality,

$$E|S_n|^r \leq E\left|\sum_{k=1}^{n} Y_k\right|^r + E\left|\sum_{k=1}^{n} Z_k\right|^r \leq (nM)^r + nE|Z_1|^r$$
$$= (nM)^r + nE|X|^r I\{|X| > M\} \leq (nM)^r + n\varepsilon,$$

so that

$$\limsup_{n\to\infty} \frac{E|S_n|^r}{n} \leq \varepsilon.$$

This proves that $\frac{S_n}{n^{1/r}} \to 0$ in $r$-mean as $n \to \infty$. $\qquad\square$

*Remark 3.1.* Convergence in $L^r$ also holds when $1 \leq r < 2$, but that is a bit harder to prove. We shall return to this problem in Section 6.10. $\qquad\square$

*Example 3.2.* Let $X_1, X_2, \ldots$ have a symmetric stable distribution with index $\alpha \in (0, 2)$, that is, suppose that, for some $c > 0$,

$$\varphi_{X_n}(t) = e^{-c|t|^\alpha}, \quad -\infty < t < \infty,$$

and let, as always, $S_n$, $n \geq 1$, denote the partial sums. By arguing as in Example 3.1 – please check – we find that

$$\frac{S_n}{n^{1/\alpha}} \overset{d}{=} X_1,$$

so that, with $r = \alpha$ we find that the Marcinkiewicz-Zygmund weak law does not hold. Which is no contradiction, since stable distributions with index in $(0, 2)$ have finite moments of order strictly smaller than $\alpha$, whereas moments of higher order do not exist (as we shall find out in Chapter 9). $\qquad\square$

By inspecting the proof of Theorem 3.1 more closely, and by modifying it appropriately the following result emerges.

**Theorem 3.3.** *Let $X_1$, $X_2$, ... be independent random variables with partial sums $\{S_n, n \geq 1\}$, and let $\{b_n, n \geq 1\}$ be a sequence of positive reals, increasing to $+\infty$. Further, set, for $k = 1, 2, \ldots, n$, $n \geq 1$,*

$$Y_{k,n} = X_k I\{|X_k| \leq b_n\}, \quad S'_n = \sum_{k=1}^{n} Y_{k,n}, \quad and \quad \mu_n = E\,S'_n.$$

*If*

$$\sum_{k=1}^{n} P(|X_k| > b_n) \to 0, \quad as \quad n \to \infty \tag{3.1}$$

*and*

$$\frac{1}{b_n^2} \sum_{k=1}^{n} \operatorname{Var} Y_{k,n} \to 0 \quad as \quad n \to \infty, \tag{3.2}$$

*then*

$$\frac{S_n - \mu_n}{b_n} \xrightarrow{p} 0 \quad as \quad n \to \infty. \tag{3.3}$$

*If, in addition,*

$$\frac{\mu_n}{b_n} \to 0 \quad as \quad n \to \infty,$$

*then*

$$\frac{S_n}{b_n} \xrightarrow{p} 0 \quad as \quad n \to \infty. \tag{3.4}$$

*Conversely, if the weak law (3.3) holds, then so do (3.1) and (3.2).*

*Proof of the sufficiency.* An application of the truncated Chebyshev inequality, Theorem 3.1.5, tells us that

$$P(|S_n - \mu_n| > b_n\varepsilon) \leq \frac{1}{b_n^2 \varepsilon^2} \sum_{k=1}^{n} \operatorname{Var} Y_{k,n} + \sum_{k=1}^{n} P(|X_k| > b_n),$$

upon which we observe that the assumptions are tailor made for (3.3) to hold. The rest is immediate.

*Proof of the necessity.* Suppose that (3.3) holds. Due to Proposition 5.11.1 we know that

$$\frac{S_n^s}{b_n} \xrightarrow{p} 0 \quad as \quad n \to \infty.$$

Joining this with Proposition 1.4 we obtain a weak law for the partial maxima;

$$\frac{\max_{1 \leq k \leq n} |X_k^s|}{b_n} \xrightarrow{p} 0 \quad as \quad n \to \infty,$$

which, by invoking (the exercise following) Theorem 2.1 shows that (3.1) is satisfied in the symmetric case.

To verify (3.2), likewise in the symmetric case, we wish to show that

$$\sum_{k=1}^{n} \text{Var}\left(\frac{Y_{k,n}^s}{b_n}\right) \to 0 \quad \text{as} \quad n \to \infty.$$

First of all, since we now know that (3.1) holds, it follows that

$$P\left(\sum_{k=1}^{n} Y_{k,n}^s \neq \sum_{k=1}^{n} X_k^s\right) \leq \sum_{k=1}^{n} P(|X_k^s| > b_n) \to 0 \quad \text{as} \quad n \to \infty, \quad (3.5)$$

which means that, by distributional equivalence, the weak law also holds for the truncated sequence; recall, e.g., Theorem 1.2.

In order to return to the sum of variances, let $n_0$ be so large that

$$P\left(\left|\sum_{k=1}^{n} \frac{Y_{k,n}^s}{b_n}\right| > \varepsilon\right) < \delta < 1/2 \quad \text{for} \quad n > n_0,$$

let $n > n_0$, and note that $\{Y_{k,n}^s/b_n, 1 \leq k \leq n\}$ are uniformly bounded (by 2) random variables. Applying the corollary to "the other Kolmogorov inequality", Corollary 3.1.1, and the Lévy inequality (and the monotonicity of the function $x/(1-x)$ for $0 < x < 1$), yields

$$\sum_{k=1}^{n} \text{Var}\left(\frac{Y_{k,n}^s}{b_n}\right) \leq \varepsilon^2 + (\varepsilon + 2)^2 \cdot \frac{P\left(\max_{1 \leq k \leq n}\left|\sum_{j=1}^{k} \frac{Y_{j,n}^s}{b_n}\right| > \varepsilon\right)}{1 - P\left(\max_{1 \leq k \leq n}\left|\sum_{j=1}^{k} \frac{Y_{j,n}^s}{b_n}\right| > \varepsilon\right)}$$

$$\leq \varepsilon^2 + (\varepsilon + 2)^2 \cdot \frac{2P\left(\left|\sum_{k=1}^{n} \frac{Y_{k,n}^s}{b_n}\right| > \varepsilon\right)}{1 - 2P\left(\left|\sum_{k=1}^{n} \frac{Y_{k,n}^s}{b_n}\right| > \varepsilon\right)},$$

so that, remembering that the weak law (also) holds for the truncated random variables, we conclude that

$$\limsup_{n \to \infty} \sum_{k=1}^{n} \text{Var}\left(\frac{Y_{k,n}^s}{b_n}\right) \leq \varepsilon^2.$$

The arbitrariness of $\varepsilon$ closes the case, and it remains to desymmetrize.

The weak symmetrization inequalities, Proposition 3.6.2, applied to (3.5) show that

$$\sum_{k=1}^{n} P(|X_k - \text{med}(X_k)| > b_n) \to 0 \quad \text{as} \quad n \to \infty,$$

and, since $X_n/b_n \xrightarrow{p} 0$ as $n \to \infty$, an application of Proposition 5.11.1 shows that $\text{med}(X_n/b_n) \to 0$ as $n \to \infty$, so that, finally, (3.1) is also satisfied in the general case.

The desymmetrization of the truncated variances is an easier job, since

$$\sum_{k=1}^{n} \operatorname{Var}\left(\frac{Y_{k,n}}{b_n}\right) = \frac{1}{2} \sum_{k=1}^{n} \operatorname{Var}\left(\frac{Y_{k,n}^s}{b_n}\right).$$

The proof is complete.                                                    □

*Remark 3.2.* Since the truncated Chebyshev inequality remains valid under the assumption that $X_1$, $X_2$, ... are (only) pair-wise independent (Remark 3.1.2), the same holds true for the sufficiency.                    □

### 3.1 Two Applications

Before we proceed to some weak laws in the i.i.d. case, in which the moment assumptions can be slightly relaxed, we pause for two pleasant illustrations.

### Empirical Distributions

Let $X_1$, $X_2$, ..., $X_n$ be a sample from the (typically unknown) distribution $F$. We define the *empirical distribution function* $F_n(x)$ as follows:

$$F_n(x) = \frac{1}{n} \sum_{k=1}^{n} I\{X_k \leq x\},$$

that is, $nF_n(x)$ equals the number of observations among the first $n$ that are at most equal to $x$. Suppose we wish to estimate the distribution function (at $x$). Since the indicator functions are independent, identically distributed random variables with mean $F(x)$, the weak law of large numbers tells us that

$$F_n(x) \xrightarrow{p} F(x) \quad \text{as} \quad n \to \infty.$$

This means that if, in a "large" sample, we check the relative number of observations with a value at most equal to $x$, we should obtain a number "close" to the true value, $F(x)$.

But, how large is "large"? How close is "close"? And how much do the answers to these questions depend on $x$?

Here are some answers. Since $nF_n(x) \in \operatorname{Bin}(n, F(x))$, we know that

$$E(F_n(x)) = F(x) \quad \text{and that} \quad \operatorname{Var}(F_n(x)) = \frac{F(x)(1 - F(x))}{n} \leq \frac{1}{4n},$$

(since $y(1 - y) \leq 1/4$ for $0 < y < 1$) so that by Chebyshev's inequality,

$$P(|F_n(x) - F(x)| > \varepsilon) \leq \frac{1}{4n\varepsilon^2} \to 0 \quad \text{as} \quad n \to \infty \quad \text{uniformly in } n.$$

This implies, in particular, that if we desire, or need, a given precision, Chebyshev's inequality provides a lower bound for the sample size, which, although crude, is independent of the value $x$.

## The Weierstrass Approximation Theorem

The message of the Weierstrass approximation theorem is that every continuous function on the unit interval (on a finite interval) can be approximated
by a polynomial with a uniform precision. The closeness of the approximation
and the degree of the polynomial are of course linked together. This is an important theorem, but it lacks one thing. Namely, we are informed about the
*existence* of a polynomial. For somebody in the real world it is more important
to *know* the polynomial.

The weak law together with Chebyshev's inequality provide an answer.
Suppose that $u$ is a continuous function on $[0, 1]$ – and, thus, uniformly continuous and bounded, by $M$, say. Our approximating polynomial of degree $n$
is the Bernstein polynomial

$$u_n(x) = \sum_{k=0}^{n} u\left(\frac{k}{n}\right)\binom{n}{k}x^k(1-x)^{n-k}.$$

In order to see this, let $X, X_1, X_2, \ldots$ be independent, identically distributed,
Bernoulli variables; $P(X = 1) = 1 - P(X = 0) = x$, and let $Y_n = \frac{1}{n}\sum_{k=1}^{n} X_k$,
$n \geq 1$. Since $E\,u(Y_n) = u_n(x)$, our task is to show that $E\,u(Y_n) \approx u(x)$ for $n$
large (in a mathematically more precise manner).

By the weak law of large numbers, $Y_n \xrightarrow{p} x$ as $n \to \infty$, and due to the
continuity and boundedness of the function $u$ it follows, by Theorems 5.10.2
and 5.5.4, that

$$u(Y_n) \xrightarrow{p} u(x) \quad \text{and} \quad u_n(x) = E\,u(Y_n) \to u(x) \quad \text{as} \quad n \to \infty.$$

Thus, $u_n(x) \approx u(x)$ for every $x$ when $n$ is large. To prove uniformity we
compute the mean and split it into the regions where $k/n$ is "close" to $x$ and
where not.

By the uniform continuity we can, for any given $\varepsilon > 0$ choose $\delta$ such that
$|u(x) - u(y)| < \varepsilon$ whenever $|x - y| < \delta$. Let one such $\varepsilon$ be given with its fellow
$\delta$. Then

$$|u_n(x) - u(x)| = |E\,u(Y_n) - u(x)| = \left|\sum_{k=0}^{n}\left(u\left(\frac{k}{n}\right) - u(x)\right)\binom{n}{k}x^k(1-x)^{n-k}\right|$$

$$\leq \sum_{\{k:|\frac{k}{n}-x|\leq\delta\}}\left|u\left(\frac{k}{n}\right) - u(x)\right|\binom{n}{k}x^k(1-x)^{n-k}$$

$$+ \sum_{\{k:|\frac{k}{n}-x|>\delta\}}\left|u\left(\frac{k}{n}\right) - u(x)\right|\binom{n}{k}x^k(1-x)^{n-k}$$

$$\leq \sum_{\{k:|\frac{k}{n}-x|\leq\delta\}}\varepsilon\binom{n}{k}x^k(1-x)^{n-k} + \sum_{\{k:|\frac{k}{n}-x|>\delta\}}2M\binom{n}{k}x^k(1-x)^{n-k}$$

$$\leq \varepsilon\cdot 1 + 2MP(|Y_n - x| > \delta) \leq \varepsilon + \frac{M}{2n\delta^2},$$

where we used Chebyshev's inequality (and the fact that $\operatorname{Var} Y_n \leq 1/4n$) in the last step. Note that the bound is uniform in $x$, since it does not involve $x$.

We have thus shown that if we know the values of a function at the points $k/n$, $k = 0, 1, 2, \ldots, n$, then we can approximate it throughout the whole interval uniformly well. Suppose, for example, that we are interested in obtaining a given precision $\eta$. In order to achieve this we may (for example) choose $\varepsilon < \eta/2$, and then $n > M/(\delta^2 \eta)$.

## 4 A Weak Law Without Finite Mean

Theorem 3.1 concerned an independent sequence with finite mean. The proof was a fairly immediate consequence of the truncated Chebyshev inequality, and was also the basis for the general weak law of large numbers. A natural question is whether or not finite mean is necessary for a weak law.

Typical counter-examples in the area involve Pareto-like densities, that is, densities where the tails drop off like a negative power of $x$, possibly multiplied by (a power of) a logarithm. It is also frequently easier to consider a symmetric distribution, mainly since truncated means are (remain, if they exist) equal to 0. So, let's try:

*Example 4.1.* Suppose that $X$, $X_1$, $X_2$, ... are independent random variables with common density

$$f(x) = \begin{cases} \frac{c}{x^2 \log |x|}, & \text{for } |x| > 2, \\ 0, & \text{otherwise,} \end{cases}$$

where $c$ is a normalizing constant (without importance).

We begin by verifying that the mean does not exist;

$$\int_{|x|>2} \frac{c|x|}{x^2 \log |x|} \, dx = 2c \int_2^\infty \frac{1}{x \log x} \, dx = +\infty.$$

With $b_n = n$ in Theorem 3.3 – the natural guess for a law of large numbers – the first condition becomes

$$nP(|X| > n) = 2n \int_n^\infty \frac{c}{x^2 \log x} \, dx \sim n \frac{C}{n \log n} = \frac{C}{\log n} \to 0 \quad \text{as} \quad n \to \infty,$$

and the second one becomes

$$\frac{1}{n^2} n E |X|^2 I\{|X| \leq n\} = \frac{2}{n} \int_2^n x^2 \frac{c}{x^2 \log x} \, dx = \frac{2c}{n} \int_2^n \frac{1}{\log x} \, dx \sim \frac{C}{n \log n} \frac{n}{}$$

$$= \frac{C}{\log n} \to 0 \quad \text{as} \quad n \to \infty,$$

so that both conditions are satisfied, and, hence, the weak law holds.    □

We have thus exhibited an example where the weak law holds, and where the mean does not exist. Note also that the estimate for (3.2) coincides with that of (3.1).

So, is this the general story? In the i.i.d. case? The *Kolmogorov-Feller law of large numbers*, see [160, 162] and [88], Section VII.7, tells us that, yes, this is, indeed, the general case. The proof amounts to showing precisely that if we assume (3.1), then (3.2) holds automatically.

**Theorem 4.1.** *Suppose that $X$, $X_1$, $X_2$, ... are independent, identically distributed random variables with partial sums $S_n$, $n \geq 1$. Then*

$$\frac{S_n - nE\,XI\{|X| \leq n\}}{n} \xrightarrow{P} 0 \quad as \quad n \to \infty,$$

*if and only if*

$$nP(|X| > n) \to 0 \quad as \quad n \to \infty. \tag{4.1}$$

*Proof.* As explained above, the proof of the sufficiency amounts to verifying (3.1) and (3.2).

The first one is precisely (4.1), and the second one reduces (as in the example) to $\frac{1}{n}E\,X^2I\{|X| \leq n\}$, which is estimated with the aid of a standard "slicing" device. Roughly speaking, we thereby improve the trivial $O(1)$-estimate $E|X|$ to an $o(1)$-estimate.

$$\frac{1}{n^2}nE\,X^2I\{|X| \leq n\} = \frac{1}{n}E\,X^2I\{|X| \leq n\}$$

$$= \frac{1}{n}\sum_{k=1}^{n}E\,X^2I\{k-1 < |X| \leq k\} \leq \frac{1}{n}\sum_{k=1}^{n}k^2P(k-1 < |X| \leq k)$$

$$\leq \frac{1}{n}\sum_{k=1}^{n}\left(\sum_{j=1}^{k}2j\right)P(k-1 < |X| \leq k) = \frac{1}{n}\sum_{j=1}^{n}2j\sum_{k=j}^{n}P(k-1 < |X| \leq k)$$

$$= \frac{2}{n}\sum_{j=1}^{n}jP(j-1 < |X| \leq n) \leq \frac{2}{n}\sum_{j=1}^{n}jP(j-1 < |X|)$$

$$= \frac{2}{n}\sum_{j=0}^{n-1}(j+1)P(j < |X|) \leq \frac{4}{n}\sum_{j=0}^{n-1}jP(|X| > j) \to 0 \quad as \quad n \to \infty.$$

The convergence to 0 is justified by Lemma A.6.1, since we are faced with (four times) an arithmetic mean of objects that converge to 0.

As for the necessity, we can copy the arguments from the proof of the necessity of (3.1) in the proof of Theorem 3.3.

**Exercise 4.1.** Check the details of the proof of the necessity.          □

*Example 4.2.* Consider, as in Example 3.1, standard Cauchy-distributed random variables, $X$, $X_1$, $X_2$, .... We already know from there that the weak law does not hold. It follows from Theorem 4.1 that condition (4.1) does not (should not) hold.

Now, the density equals

$$f(x) = \frac{1}{\pi} \cdot \frac{1}{1+x^2}, \quad -\infty < x < \infty,$$

so that

$$xP(|X| > x) = x\frac{2}{\pi}\left(\frac{\pi}{2} - \arctan x\right) = \frac{2x}{\pi}\arctan\frac{1}{x} \to \frac{2}{\pi} \quad \text{as} \quad x \to \infty,$$

and condition (4.1) does not hold – as expected.    □

Consider the following variation:

*Example 4.3.* Suppose that $X, X_1, X_2, \ldots$ are independent random variables with common density

$$f(x) = \begin{cases} \frac{1}{2x^2}, & \text{for } |x| > 1, \\ 0, & \text{otherwise.} \end{cases}$$

One readily checks that the mean does not exist.

The first condition with $b_n = n$ becomes

$$nP(|X| > n) = n\int_n^\infty \frac{1}{x^2}\,dx = 1,$$

so that the Kolmogorov-Feller law does not hold. However, inspecting the conditions in Theorem 3.3 with $b_n = n\log n$ instead, we find, for the first one, that

$$nP(|X| > n\log n) = n\int_{n\log n}^\infty \frac{1}{x^2}\,dx = \frac{1}{\log n} \to 0 \quad \text{as} \quad n \to \infty,$$

and for the second one that

$$\frac{1}{(n\log n)^2}nE\,X^2I\{|X| \le n\log n\} = \frac{1}{n(\log n)^2}\int_1^{n\log n} 1\,dx$$

$$\le \frac{1}{\log n} \to 0 \quad \text{as} \quad n \to \infty.$$

By Theorem 3.3 we therefore conclude that

$$\frac{S_n}{n\log n} \xrightarrow{p} 0 \quad \text{as} \quad n \to \infty.$$

We have thus obtained a weak law, albeit with a different normalization.    □

A second look at the example reveals that, once again, the two conditions in the theorem coincide, and one can ask again, was this an exceptional case or is there a general result around?

The following result, taken from [117], provides an extension of Theorem 4.1 to more general normalizing sequences. It involves regularly varying functions for which we refer to Section A.7 for a short background and some references.

**Theorem 4.2.** *Suppose that $X$, $X_1$, $X_2$, ... are independent, identically distributed random variables with partial sums $S_n$, $n \geq 1$. Further, let, for $x > 0$, $b \in \mathcal{RV}(1/\rho)$ for some $\rho \in (0,1]$, that is, let $b(x) = x^{1/\rho}\ell(x)$, where $\ell \in \mathcal{SV}$. Finally, set $b_n = b(n)$, $n \geq 1$. Then*

$$\frac{S_n - nE\,XI\{|X| \leq b_n\}}{b_n} \xrightarrow{p} 0 \quad as \quad n \to \infty,$$

*if and only if*

$$nP(|X| > b_n) \to 0 \quad as \quad n \to \infty. \tag{4.2}$$

*In particular, for $0 < r \leq 1$, we have*

$$\frac{S_n - nE\,XI\{|X| \leq n^{1/r}\}}{n^{1/r}} \xrightarrow{p} 0 \quad as \quad n \to \infty,$$

*if and only if*

$$nP(|X| > n^{1/r}) \to 0 \quad as \quad n \to \infty. \tag{4.3}$$

*Proof.* The proof amounts to minor modifications of the proof of Theorem 4.1.

The first condition, (3.1), is precisely our assumption (4.2). As for the second one,

$$\frac{n}{b_n^2}E\,X^2I\{|X| \leq b_n\} = \frac{n}{b_n^2}\sum_{k=1}^{n}E\,X^2I\{b_{k-1} < |X| \leq b_k\}$$

$$\leq \frac{n}{b_n^2}\sum_{k=1}^{n}b_k^2 P(b_{k-1} < |X| \leq b_k)$$

$$= \frac{n}{b_n^2}\sum_{k=1}^{n}k^{2/\rho}(\ell(k))^2 P(b_{k-1} < |X| \leq b_k)$$

$$\leq C\frac{n}{b_n^2}\sum_{k=1}^{n}\left(\sum_{j=1}^{k}j^{(2/\rho)-1}(\ell(j))^2\right)P(b_{k-1} < |X| \leq b_k)$$

$$= C\frac{n}{b_n^2}\sum_{j=1}^{n}j^{(2/\rho)-1}(\ell(j))^2\sum_{k=j}^{n}P(b_{k-1} < |X| \leq b_k)$$

$$= C\frac{n}{b_n^2}\sum_{j=1}^{n}j^{(2/\rho)-1}(\ell(j))^2 P(b_{j-1} < |X| \leq b_n)$$

$$\leq C\frac{n}{b_n^2}\sum_{j=1}^{n}j^{(2/\rho)-1}(\ell(j))^2 P(|X| > b_{j-1})$$

$$\leq C\frac{1}{n^{(2/\rho)-1}(\ell(n))^2}\sum_{j=0}^{n-1}\{j^{(2/\rho)-2}(\ell(j))^2\}jP(|X| > b_j),$$

which converges to 0 as $n \to \infty$. Once again, the convergence to 0 is justified by Lemma A.6.1, but by the second half. Namely, we are faced with a weighted average of quantities that converge to 0. The weights are $j^{(2/\rho)-2}(\ell(j))^2$, the sum of which (Lemma A.7.3) behaves like

$$\sum_{j=1}^{n} j^{(2/\rho)-2}(\ell(j))^2 \sim \frac{\rho}{2-\rho} n^{(2/\rho)-1}(\ell(n))^2 \quad \text{as} \quad n \to \infty.$$

This finishes the proof of the sufficiency.

The necessity, follows, once again, by the arguments leading to (3.1) in the proof of Theorem 3.3: symmetrize to obtain a weak law for the partial maxima, apply Theorem 2.1, and desymmetrize.

The particular case is immediate upon putting $r = \rho$ and $\ell(x) \equiv 1$.    □

A pleasant application of Theorem 4.2 is the classical *St. Petersburg game* that we present in a separate subsection following the model examples based on (two-sided) Pareto distributions.

*Example 4.4.* Let $0 < \rho \leq 1$, and suppose that $X, X_1, X_2, \ldots$ are independent random variables with common density

$$f(x) = \begin{cases} \frac{\rho}{2|x|^{1+\rho}}, & \text{for} \quad |x| > 1, \\ 0, & \text{otherwise.} \end{cases}$$

An analysis of condition (4.2) yields

$$\frac{S_n}{(n \log n)^{1/\rho}} \xrightarrow{p} 0 \quad \text{as} \quad n \to \infty.$$

The normalizing sequence is $b_n = (n \log n)^{1/\rho} \in \mathcal{RV}(1/\rho)$ as should be.    □

In the same way as Theorem 3.2 with $r = 1$ coincides with the classical weak law, Theorem 4.2 with $\rho = 1$ and $\ell(x) \equiv 1$ reduces to Theorem 4.1. For $0 < r = \rho < 1$ and $\ell(x) \equiv 1$ the theorem relates to the Marcinkiewicz-Zygmund weak law, Theorem 3.2, in the same way as the Kolmogorov-Feller law relates to the classical weak law. Let us, accordingly, extend Example 3.2 as follows.

*Example 4.5.* Let, again, $X_1, X_2, \ldots$ have a symmetric stable distribution with index $\alpha \in (0,1)$. We found, in Example 3.2, that $\frac{S_n}{n^{1/\alpha}} \stackrel{d}{=} X_1$, so that, with $r = \alpha$, the Marcinkiewicz-Zygmund weak law does not hold.

Unfortunately, since there only exist explicit expressions for stable densities for $\alpha = 1/2$ and $\alpha = 1$ (the Cauchy distribution), we have to resort to the fact that one can show that, if $X$ is symmetric stable with index $\alpha$, then, for some constant $C$,

$$x^{\alpha} P(|X| > x) \to C \frac{2-\alpha}{\alpha} \quad \text{as} \quad x \to \infty;$$

see [88], Theorem XVII.5.1. With $r = \alpha$ as before, this is equivalent to the statement

$$xP(|X| > x^{1/r}) \to C\frac{2-r}{r} \quad \text{as} \quad x \to \infty,$$

that is, condition (4.3) is not satisfied (as expected).    □

We shall soon see that finiteness of the adequate moment is necessary as well as sufficient in the corresponding *strong* law. The results of this section thus tell us that a slightly weaker requirement suffices for weak laws of large numbers in the i.i.d. case.

**Exercise 4.2.** Let $0 < \rho \leq 1$, and suppose that $X, X_1, X_2, \ldots$ are independent random variables with common density

$$f(x) = \begin{cases} \frac{c}{|x|^{1+\rho}(\log |x|)^\beta}, & \text{for} \quad |x| > 2, \\ 0, & \text{otherwise}, \end{cases}$$

where $-\infty < \beta < \infty$, and $c$ is a normalizing constant. State and prove a weak law for the partial maxima and a weak law of large numbers.

**Exercise 4.3.** Let $0 < \rho \leq 1$, and suppose that $X, X_1, X_2, \ldots$ are independent, positive, random variables with common probability function

$$p(k) = \begin{cases} \frac{c}{k^{1+\rho}(\log k)^\beta}, & \text{for} \quad k = 2, 3, \ldots, \\ 0, & \text{otherwise}, \end{cases}$$

where $-\infty < \beta < \infty$, and $c$ is a normalizing constant. State and prove a weak law for the partial maxima and a weak law of large numbers.    □

## 4.1 The St. Petersburg Game

This game was called a paradox, because when it was "invented" the formalism to handle random variables with infinite expectation seemed paradoxical. The game is defined as follows: I toss a fair coin repeatedly until heads appears. If this happens at trial number $n$ you receive $2^n$ Euros. The problem is what a *fair price* would be for you to participate in this game.

A natural fair price would be the expected value, since this would imply that, in the long run, you would neither win nor lose. However, the random variable $X$ behind the game is

$$P(X = 2^n) = \frac{1}{2^n}, \quad n = 1, 2, \ldots,$$

which has infinite mean – $E\,X = \sum_{n=1}^{\infty} 2^n \frac{1}{2^n} = \sum_{n=1}^{\infty} 1 = +\infty$. A fair price thus seems impossible.

One variant of the game is to set a maximal number of trials, and you win nothing if head never appears. This is obviously less favorable to you. The solution is to set the fee as a function of the number of games, as $b_n$, where

$b_n$ is defined so that a weak law of large numbers holds, in the sense that if $X, X_1, X_2, \ldots$ are independent, identically distributed random variables, and $S_n = \sum_{k=1}^n X_k$, $n \geq 1$, then

$$\frac{S_n}{b_n} \xrightarrow{p} 1 \quad \text{as} \quad n \to \infty. \tag{4.4}$$

As it turns out, such a weak law follows from Theorem 4.2. Namely, let $\log_2$ denote the logarithm relative to base 2. Noticing that

$$P(X > 2^n) = \sum_{k=n+1}^{\infty} \frac{1}{2^k} = \frac{1}{2^n},$$

and that $x \log_2 x = 2^{\log_2(x \log_2 x)}$, we obtain

$$xP(X > x\log_2 x) = x\left(\frac{1}{2}\right)^{[\log_2(x\log_2 x)]} \leq 2x\frac{1}{x\log_2 x}$$

$$= \frac{2}{\log_2 x} \to 0 \quad \text{as} \quad x \to \infty.$$

Theorem 4.2 therefore tells us that

$$\frac{S_n - nE\,I\{|X| \leq n\log_2 n\}}{n\log_2 n} \xrightarrow{p} 0 \quad \text{as} \quad n \to \infty.$$

This, and the fact that

$$nE\,I\{|X| \leq n\log_2 n\} = n \sum_{k=1}^{[\log_2(n\log_2 n)]} 2^k\frac{1}{2^k} = n[\log_2(n\log_2 n)] \sim n\log_2 n,$$

as $n \to \infty$, shows that the desired weak law, (4.4), holds.

For more on this, see [87], Chapter X, and [88], Chapter VII, and further references given there.


# 5 Convergence of Series

Apart from having an interest as such, the convergence of sums of independent random variables has important connections with almost sure convergence, in particular, with strong laws. More precisely, the convergence of $\sum_{k=1}^{\infty} \frac{X_k}{k}$, together with the Kronecker lemma, Lemma A.6.2, will imply a strong law. The following, what might be called a *random Kronecker lemma*, provides the link.

**Lemma 5.1.** *Suppose that $\{X_n, n \geq 1\}$ are random variables, set $a_0 = 0$, and let $\{a_n, n \geq 1\}$ be positive numbers increasing to $+\infty$. Then*

$$\sum_{k=1}^{\infty} \frac{X_k}{a_k} \quad \text{converges a.s.} \quad \Longrightarrow \quad \frac{1}{a_n}\sum_{k=1}^n X_k \xrightarrow{a.s.} 0 \quad \text{as} \quad n \to \infty.$$

*Proof.* Let

$$A = \left\{ \omega : \sum_{k=1}^{\infty} \frac{X_k(\omega)}{a_k} \text{ converges} \right\}, \quad \text{and}$$

$$B = \left\{ \omega : \frac{1}{a_n} \sum_{k=1}^{n} X_k(\omega) \to 0 \quad \text{as} \quad n \to \infty \right\}.$$

Then $P(A) = 1$ by assumption, and the Kronecker lemma tells us that $A \subset B$, so that $P(B) = 1$. $\qquad\square$

Thus: if we wish to prove a strong law, $\frac{S_n}{n} \overset{a.s.}{\to} 0$, then this can be achieved by finding conditions under which $\sum_{k=1}^{\infty} \frac{X_k}{k}$ converges almost surely. We therefore first turn our attention to facts about the convergence of sums of independent random variables.

One can obviously talk about the probability that a given sum of independent random variables converges. As a consequence of the Kolmogorov zero-one law, Theorem 2.10.6, this probability can only assume the two values 0 and 1. However, in the particular case of the present section this can, as we shall see, be proved directly.

As a first rather elementary result we prove the following lemma on $L^2$-convergence.

**Lemma 5.2.** *($L^2$-lemma) Let $X_1, X_2, \ldots$ be independent random variables with partial sums $S_n$, $n \geq 1$. Then*

$$\sum_{n=1}^{\infty} \text{Var} \, X_n < \infty \quad \Longleftrightarrow \quad \sum_{n=1}^{\infty} (X_n - E \, X_n) \text{ converges in } L^2.$$

*Proof.* The sum of the variances converges if and only if it is Cauchy-convergent, in which case

$$\sum_{k=n+1}^{m} \text{Var} \, X_k \to 0 \quad \text{as} \quad m, n \to \infty,$$

which, due to independence, is the same as

$$\text{Var} \, (S_m - S_n) = E\big((S_m - E \, S_m) - (S_n - E \, S_n)\big)^2 \to 0 \quad \text{as} \quad m, n \to \infty,$$

which is the same as $\{S_n - E \, S_n, n \geq 1\}$ being $L^2$-Cauchy-convergent, and, hence, $L^2$-convergent. $\qquad\square$

As an immediate corollary it follows that, if $X_1, X_2, \ldots$ are independent and identically distributed random variables, then the sum cannot converge in square mean.

But a weighted sum can:

**Theorem 5.1.** *Let $X_1$, $X_2$, ... be independent, identically distributed random variables with mean 0 and finite variance, and let $\{a_n, n \geq 1\}$ be real numbers. Then*

$$\sum_{n=1}^{\infty} a_n X_n \text{ converges in } L^2 \quad \Longleftrightarrow \quad \sum_{n=1}^{\infty} a_n^2 < \infty.$$

**Exercise 5.1.** Prove the theorem.    □

*Remark 5.1.* The essence of the $L^2$-lemma is, in fact, not really the independence, rather a kind of orthogonality; note that uncorrelatedness – which is a kind of orthogonality – suffices for the lemma to hold.    □

## 5.1 The Kolmogorov Convergence Criterion

This is a most useful criterion for determining whether or not a series is convergent. As a first step toward that result we prove

**Theorem 5.2.** (The Kolmogorov convergence criterion)
*Let $X_1$, $X_2$, ... be independent random variables with partial sums $S_n$, $n \geq 1$. Then*

$$\sum_{n=1}^{\infty} \operatorname{Var} X_n < \infty \quad \Longrightarrow \quad \sum_{n=1}^{\infty} (X_n - E\,X_n) \text{ converges a.s.}$$

*If, in addition,*

$$\sum_{n=1}^{\infty} E\,X_n \text{ converges,}$$

*then*

$$\sum_{n=1}^{\infty} X_n \text{ converges a.s.}$$

*Proof.* Let $\varepsilon > 0$, suppose that the sum of the variances converges, let $n < m$, and consider a Cauchy sequence. By the Kolmogorov inequality, Theorem 3.1.6, we then obtain

$$P(\max_{n \leq k \leq m} |(S_k - E\,S_k) - (S_n - E\,S_n)| > \varepsilon)$$

$$= P\left(\max_{n \leq k \leq m} \left| \sum_{j=n+1}^{k} (X_j - E\,X_j) \right| > \varepsilon \right) \leq \frac{\sum_{k=n+1}^{m} \operatorname{Var} X_k}{\varepsilon^2}$$

$$\leq \frac{\sum_{k=n+1}^{\infty} \operatorname{Var} X_k}{\varepsilon^2}.$$

Since the left-hand side does not depend on $m$, it follows that, for any $\varepsilon > 0$,

$$P(\sup_{k \geq n} |(S_k - E\,S_k) - (S_n - E\,S_n)| > \varepsilon) \leq \frac{\sum_{k=n+1}^{\infty} \operatorname{Var} X_k}{\varepsilon^2} \to 0 \quad \text{as} \quad n \to \infty,$$

which, in view of Proposition 5.1.2, proves almost sure convergence of the centered sum.

The second statement follows, of course, from the fact that

$$X_n = (X_n - E\,X_n) + E\,X_n.$$ □

For independent, *uniformly bounded*, random variables there is also a converse, thanks to "the other Kolmogorov inequality", Theorem 3.1.7.

**Theorem 5.3.** *Let $X_1, X_2, \ldots$ be independent, uniformly bounded random variables. Then*

$$\sum_{n=1}^{\infty}(X_n - E\,X_n) \text{ converges a.s.} \quad \Longleftrightarrow \quad \sum_{n=1}^{\infty} \operatorname{Var} X_n < \infty.$$

*Indeed,*

$$P\left(\sum_{n=1}^{\infty}(X_n - E\,X_n) \text{ converges}\right) = \begin{cases} 1, & \text{if } \sum_{n=1}^{\infty} \operatorname{Var} X_n < \infty \\ 0, & \text{if } \sum_{n=1}^{\infty} \operatorname{Var} X_n = \infty. \end{cases}$$

*If, in addition,*

$$\sum_{n=1}^{\infty} E\,X_n \text{ converges},$$

*then*

$$P\left(\sum_{n=1}^{\infty} X_n \text{ converges}\right) = \begin{cases} 1, & \text{if } \sum_{n=1}^{\infty} \operatorname{Var} X_n < \infty \\ 0, & \text{if } \sum_{n=1}^{\infty} \operatorname{Var} X_n = \infty. \end{cases}$$

*Proof.* Suppose that $\sup_n |X_n| \leq A$ for some $A > 0$, so that, consequently, $\sup_n |X_n - E\,X_n| \leq 2A$, and that the sum of the variances *diverges*. Then

$$P(\max_{n \leq k \leq m} |(S_k - E\,S_k) - (S_n - E\,S_n)| > \varepsilon) \geq 1 - \frac{(\varepsilon + 2A)^2}{\sum_{k=n+1}^{m} \operatorname{Var} X_k},$$

which, upon letting $m \to \infty$ in the left-hand side, and then in the right-hand side, shows that

$$P\left(\sup_{k \geq n} |(S_k - E\,S_k) - (S_n - E\,S_n)| > \varepsilon\right) \geq 1,$$

so that almost sure convergence fails.

But it more than fails. Namely, we have just shown that, in fact, the convergence set is a null set.

The last statement follows as in Theorem 5.2. □

A fascinating and somewhat frustrating example is the harmonic series with random signs.

*Example 5.1.* Consider the series

$$\sum_{n=1}^{\infty} \pm \frac{1}{n},$$

more formally, let $X$, $X_1$, $X_2$, ... be independent coin-tossing random variables, i.e., $P(X = 1) = P(X = -1) = 1/2$, and consider the series

$$\sum_{n=1}^{\infty} \frac{X_n}{n}.$$

Since $EX = 0$ and $\text{Var } X = 1$, an application of the Kolmogorov convergence criterion shows that *the sum is almost surely convergent.* And, yet, the harmonic series, which causes mathematicians so much trouble and pain, but also excitement and pleasure, is divergent!

Note also that the harmonic series corresponds to all $X$'s being equal to $+1$, which occurs with probability 0 in view of the Borel-Cantelli lemma (as we have already seen in Section 2.18). As a contrast, the alternating series

$$\sum_{n=1}^{\infty} \frac{(-1)^n}{n}$$

is well known to be convergent. According to the law of large numbers one should expect equally many positive and negative terms, so, in this sense, the convergence of the alternating series is as expected.                                  □

## 5.2 A Preliminary Strong Law

Although the ultimate goal is to prove a strong law under the necessary and sufficient condition of finite mean, it is hard to resist the temptation to provide the following weaker variant which, given the Kolmogorov convergence criterion, comes for free.

**Theorem 5.4.** (The Kolmogorov sufficient condition)
*Let $X_1$, $X_2$, ... be independent random variables with mean 0 and finite variances, $\sigma_n^2$, $n \geq 1$, and set $S_n = \sum_{k=1}^{n} X_k$, $n \geq 1$. Then*

$$\sum_{n=1}^{\infty} \frac{\sigma_n^2}{n^2} < \infty \implies \frac{S_n}{n} \xrightarrow{a.s.} 0 \quad as \quad n \to \infty.$$

*Proof.* The Kolmogorov convergence criterion shows that $\sum_{k=1}^{\infty} \frac{X_k}{k}$ is a.s. convergent. The random Kronecker lemma, Lemma 5.1 does the rest.          □

**Corollary 5.1.** *Suppose that $X_1$, $X_2$, ... are independent, identically distributed random variables with mean $\mu$ and finite variance. Then*

$$\frac{1}{n} \sum_{k=1}^{n} X_k \xrightarrow{a.s.} \mu \quad as \quad n \to \infty.$$

## 5.3 The Kolmogorov Three-series Theorem

So far we have shown, based on the Kolmogorov convergence criterion, that a strong law in the i.i.d. case holds if the variance of the summands is finite. In order to remove the assumption of finite variance, we shall, via truncation, exhibit a more well behaved convergence equivalent sequence for which the Kolmogorov convergence criterion applies. Recall that $X_1, X_2, \ldots$ and $Y_1, Y_2, \ldots$ are convergence equivalent if

$$\sum_{n=1}^{\infty} P(X_n \neq Y_n) < \infty.$$

With this as a motivating background, here is now a celebrated result, that provides necessary and sufficient conditions for a series to converge.

**Theorem 5.5.** (The Kolmogorov three-series theorem)
*Let $A > 0$. Suppose that $X_1, X_2, \ldots$ are independent random variables and set, for $k \geq 1$,*

$$Y_k = \begin{cases} X_k, & \text{when } |X_k| \leq A, \\ 0, & \text{otherwise.} \end{cases}$$

*Then*

$$\sum_{k=1}^{\infty} X_k \quad \text{converges almost surely as } n \to \infty,$$

*if and only if*

(i)  $\sum_{k=1}^{\infty} P(X_k \neq Y_k) = \sum_{k=1}^{\infty} P(|X_k| > A) < \infty;$
(ii)  $\sum_{k=1}^{\infty} E Y_k$ *converges;*
(iii)  $\sum_{k=1}^{\infty} \text{Var } Y_k < \infty.$

*Proof.* If the variables are uniformly bounded, then, by choosing $A$ larger than the bound, the first two sums vanish, and we know, from Theorem 5.3, that a.s. convergence holds if and only if the third sum converges.

Next, suppose that $X_1, X_2, \ldots$ are symmetric. Then $Y_1, Y_2, \ldots$ are symmetric too, and the second sum vanishes. The convergence of the first sum implies that $X_1, X_2, \ldots$ and $Y_1, Y_2, \ldots$ are convergence equivalent, and since the latter sum is a.s. convergent, so is the former.

Conversely, if $\sum_{n=1}^{\infty} X_n$ is a.s. convergent, then $X_n \overset{a.s.}{\to} 0$ as $n \to \infty$, so that $P(|X_n| > A \text{ i.o.}) = 0$, which, by the second Borel-Cantelli lemma, Theorem 2.18.2, implies that the first sum converges. By convergence equivalence it therefore follows that $\sum_{n=1}^{\infty} Y_n$ is a.s. convergent, and, hence, by Theorem 5.3, that the third sum converges.

It remains to consider the general case. Convergence of the last two sums implies, by Theorem 5.3, that $\sum_{n=1}^{\infty} Y_n$ is a.s. convergent, which, in view of (i), proves that $\sum_{n=1}^{\infty} X_n$ is a.s. convergent.

Conversely, if $\sum_{n=1}^{\infty} X_n$ is a.s. convergent, then so is the symmetrized series $\sum_{n=1}^{\infty} X_n^s$. Mimicking the arguments a few lines back we conclude that $X_n^s \overset{a.s.}{\to} 0$ as $n \to \infty$, that $\sum_{n=1}^{\infty} P(|X_n^s| > A) < \infty$, that $\sum_{n=1}^{\infty} Y_n^s$ is a.s. convergent, and that $\sum_{n=1}^{\infty} \operatorname{Var} Y_n^s < \infty$.

The latter fact first implies that $\sum_{n=1}^{\infty} \operatorname{Var} Y_n = \frac{1}{2} \sum_{n=1}^{\infty} \operatorname{Var} Y_n^s < \infty$, and secondly, by Theorem 5.3, that $\sum_{n=1}^{\infty} (Y_n - E Y_n)$ is a.s. convergent. Finally, (ii) follows from the fact that

$$E Y_n = Y_n - (Y_n - E Y_n). \qquad \square$$

As the reader has, hopefully, noticed, the difficulty in the proof lies in keeping track of the different steps, in how one successively approaches the final result. One way to memorize this is that for bounded random variables the first and second sum vanish, and the rest is taken care of by Theorem 5.3. For symmetric random variables the second sum vanishes, and one truncates to reduce the problem to the bounded case. The general case is partly resolved by symmetrization.

Note also that the final step was to take care of the second sum. And that's the general story: the second sum is usually the hardest one to verify. Recall that even if the original summands have mean 0, this is not so in general for the truncated ones.

Let us, in this connection, mention that the final step in the proof of the strong law of large numbers in the following section, to take care of the truncated means, appears *after* Lemma 5.1 has been applied, and that this does *not* amount to verifying (ii).

In the bounded case, Theorem 5.3, we found that the probability that the (centered) series is convergent is 0 or 1. This remains true also in the general case.

**Corollary 5.2.** *Let* $X_1, X_2, \ldots$ *be independent random variables. Then*

$$P\left( \sum_{k=1}^{\infty} X_k \ converges \right) = 0 \quad or \quad 1.$$

*Proof.* If Theorem 5.5(i) is satisfied, the conclusion holds by convergence equivalence and Theorem 5.3. If the first sum is divergent, then $X_n \overset{a.s.}{\not\to} 0$ as $n \to \infty$, so that the probability that the first sum converges is 0. $\qquad \square$

As has been mentioned before, and just seen again, symmetrization plays a central role in many proofs. For sequences it was easy to see (as it is for sums) that convergence implies convergence of the symmetrized sequence. The problem was the converse. The following result is a sum analog to Proposition 5.11.1, and shows that if the sum of the symmetrized random variables converges, then the original sum *properly centered* converges too.

**Theorem 5.6.** *Let $A > 0$. Suppose that $X_1$, $X_2$, ... are independent random variables. If $\sum_{n=1}^{\infty} X_n^s$ converges almost surely as $n \to \infty$, then*

$$\sum_{n=1}^{\infty} (X_n - \mathrm{med}\,(X_n) - E[(X_n - \mathrm{med}\,(X_n))I\{|X_n - \mathrm{med}\,(X_n)| \le A\}])$$

*converges almost surely as $n \to \infty$.*

*Proof.* Since the symmetrized sum converges almost surely, we know from the three-series theorem that the sums (i) and (iii) converge for the symmetrized random variables (and that the sum in (ii) vanishes).

Set $Y_n = X_n - \mathrm{med}\,(X_n)$, $n \ge 1$. By the weak symmetrization inequalities, Proposition 3.6.2,

$$\sum_{n=1}^{\infty} P(|Y_n| > A) \le 2 \sum_{n=1}^{\infty} P(|X_n^s| > A) < \infty,$$

so that the sum in (i) converges for $\{Y_n, n \ge 1\}$.

By partial integration and the weak symmetrization inequalities, cf. also the proof of Theorem 2.12.1,

$$\mathrm{Var}\,(Y_n I\{|Y_n| \le A\}) \le E Y_n^2 I\{|Y_n| \le A\}$$

$$= -A^2 P(|Y_n| > A) + 2 \int_0^A x P(|Y_n| > x)\,dx \le 4 \int_0^A x P(|X_n^s| > x)\,dx$$

$$= 2\mathrm{Var}\,(X_n^s I\{|X_n^s| \le A\}) + 2A^2 P(|X_n^s| > A),$$

which shows that

$$\sum_{n=1}^{\infty} \mathrm{Var}\,(Y_n I\{|Y_n| \le A\}) < \infty.$$

This proves the finiteness of the third sum for the $Y$-sequence.

Since the second one is a troublemaker, we center to make it 0, and since centering does not change the variance and only doubles the first sum (at most) we have shown that

$$\sum_{n=1}^{\infty} (Y_n I\{|Y_n| \le A\} - E(Y_n I\{|Y_n| \le A\}))$$

converges almost surely as $n \to \infty$.

Finally, since, once again, the sum in (i) converges we know, by convergence equivalence, that also

$$\sum_{n=1}^{\infty} (Y_n - E(Y_n I\{|Y_n| \le A\})$$

converges almost surely as $n \to \infty$, which, recalling that $Y_n = X_n - \mathrm{med}\,(X_n)$, is precisely the claim. $\qquad\square$

## 5.4 Lévy's Theorem on the Convergence of Series

The results of this subsection on the convergence of series are not immediately connected with the law of large numbers. The first one is a famous theorem due to Lévy.

**Theorem 5.7.** *Let $X_1$, $X_2$, ... be independent random variables. Then*

$$\sum_{n=1}^{\infty} X_n \text{ converges in probability} \quad \Longleftrightarrow \quad \sum_{n=1}^{\infty} X_n \text{ converges almost surely.}$$

*Proof.* Since almost sure convergence always implies convergence in probability, only the opposite implication has to be proved.

The first observation is that there is no assumption about the existence of moments, which means that probabilities of maxima, which are bound to enter, have to be estimated by "individual" probabilities. The inequality that does it for us is Skorohod's, or Ottaviani's, inequality, Theorem 3.7.7.

Since the sum converges in probability it is also Cauchy-convergent in probability, which implies that, given $0 < \varepsilon < 1/2$, there exists $n_0$, such that, for all $n, m$, with $n_0 < n < m$, $P(|S_m - S_n| > \varepsilon) < \varepsilon$, in particular,

$$\beta = \max_{n \le k \le m} P(|S_m - S_k| > \varepsilon) < \varepsilon, \quad \text{for} \quad n_0 < n < m,$$

which, inserted into Theorem 3.7.7 yields

$$P(\max_{n \le k \le m} |S_k - S_n| > 2\varepsilon) \le \frac{1}{1-\beta} P(|S_m - S_n| > \varepsilon) \le \frac{1}{1-\varepsilon} \cdot \varepsilon < 2\varepsilon.$$

An appeal to Proposition 5.1.1 establishes almost sure convergence.    □

With the aid of characteristic functions it is, in fact, possible to prove that the weaker assumption of convergence in distribution implies almost sure convergence.

**Theorem 5.8.** *Let $X_1$, $X_2$, ... be independent random variables. Then*

$$\sum_{n=1}^{\infty} X_n \text{ converges in distribution} \quad \Longleftrightarrow \quad \sum_{n=1}^{\infty} X_n \text{ converges almost surely.}$$

*Proof.* Since almost sure convergence always implies convergence in distribution, we only have to prove the opposite implication.

We thus assume that $\sum_{n=1}^{\infty} X_n$ converges in distribution as $n \to \infty$. By Theorem 5.9.1 we then know that

$$\prod_{k=1}^{n} \varphi_{X_k}(t) \to \varphi(t) \quad \text{as} \quad n \to \infty,$$

for all $t$, where $\varphi$ is some characteristic function. By considering the symmetrized random variables we also have

$$\prod_{k=1}^{n}|\varphi_{X_k^s}(t)|^2 \to |\varphi(t)|^2 \quad \text{as} \quad n \to \infty, \tag{5.1}$$

for all $t$; note, in particular, that $|\varphi(t)|^2$ is real valued, and, moreover, positive in some neighborhood of 0, $|t| < h$, say, for some $h > 0$.

Our next task is to prove that the Kolmogorov three-series theorem, Theorem 5.5, applies.

Lemma 4.4.1, the fact that $1 - x < e^{-x}$ for $x > 0$, and (5.1), together show that

$$\sum_{n=1}^{\infty} P(|X_n^s| > 2/h) \le \sum_{n=1}^{\infty}\frac{1}{h}\int_{|t|<h}(1-|\varphi_{X_n^s}(t)|^2)\,dt$$

$$= \frac{1}{h}\int_{|t|<h}\left(\sum_{n=1}^{\infty}(1-|\varphi_{X_n^s}(t)|^2)\right)dt \le \frac{1}{h}\int_{|t|<h}\sum_{n=1}^{\infty}\exp\{-|\varphi_{X_n^s}(t)|^2\}\,dt$$

$$= \frac{1}{h}\int_{|t|<h}\exp\left\{-\prod_{n=1}^{\infty}|\varphi_{X_n^s}(t)|^2\right\}dt = \frac{1}{h}\int_{|t|<h}\exp\{-|\varphi(t)|^2\}\,dt < \infty.$$

This proves that the first sum in Theorem 5.5 converges.

The second sum vanishes since we consider symmetric random variables.

For the third sum we exploit (the second half of) Lemma 4.4.2 to obtain

$$\sum_{n=1}^{\infty}E|X_n^s|^2I\{|X_n^s|<2/h\} \le \sum_{n=1}^{\infty}3(1-\varphi_{X_n^s}(h)) \le 3\sum_{n=1}^{\infty}\exp\{-|\varphi_{X_n^s}(h)|^2\}$$

$$\le 3\exp\left\{-\prod_{n=1}^{\infty}|\varphi_{X_n^s}(h)|^2\right\} = 3\exp\{-|\varphi(h)|^2\} < \infty.$$

All sums being convergent, the three-series theorem now tells us that

$$\sum_{n=1}^{\infty}X_n^s \quad \text{converges almost surely as} \quad n \to \infty,$$

and the proof is complete for symmetric random variables.

For the general case, we set

$$a_n = \text{med}(X_n) + E((X_n - \text{med}(X_n))I\{|X_n - \text{med}(X_n)| \le 2/h\}).$$

Theorem 5.6 (with $A = 2/h$) then tells us that

$$\sum_{n=1}^{\infty}(X_n - a_n) \quad \text{converges almost surely as} \quad n \to \infty,$$

so that, by Theorem 5.9.1,

$$\prod_{k=1}^{n} \varphi_{(X_k - a_k)}(t) = \exp\left\{-it\sum_{k=1}^{n} a_k\right\}\prod_{k=1}^{n} \varphi_{X_k}(t) \to \tilde{\varphi}(t), \quad \text{say,} \quad \text{as} \quad n \to \infty.$$

However, for $t$ sufficiently close to 0 the characteristic functions $\varphi$ and $\tilde{\varphi}$ both are close to 1 by continuity; in particular, they are non-zero. This shows that, for $t$ sufficiently small,

$$\exp\left\{-it\sum_{k=1}^{n} a_k\right\} \to \frac{\varphi(t)}{\tilde{\varphi}(t)} \neq 0 \quad \text{as} \quad n \to \infty.$$

It follows that $\sum_{k=1}^{n} a_k$ converges as $n \to \infty$ or moves around along periods of multiples of $2\pi/t$. However, since the arguments are valid for all $t$ in a neighborhood of 0, only convergence is possible. This, finally, establishes, via Theorem 5.11.1, that

$$\sum_{k=1}^{n} X_k = \sum_{k=1}^{n}(X_k - a_k) + \sum_{k=1}^{n} a_k \quad \text{converges almost surely as} \quad n \to \infty. \quad \square$$

**Exercise 5.2.** In the notation of Theorem 5.1, extend the $L^2$-convergence result there to

$$\sum_{n=1}^{\infty} a_n X_n < \infty \text{ converges a.s.} \quad \Longleftrightarrow \quad \sum_{n=1}^{\infty} a_n^2 < \infty. \quad \square$$

## 6 The Strong Law of Large Numbers

There are many important and profound results in probability theory. One of them is the strong law of large numbers that we are finally ready to meet.

However, in order not to disturb the flow of the proof we shall prove the following technical lemma as a prelude.

**Lemma 6.1.** *Let $0 < r < 2$. Suppose that $X, X_1, X_2, \ldots$ are independent, identically distributed random variables, and set*

$$Y_n = X_n I\{|X_n| \leq n^{1/r}\}, \quad n \geq 1.$$

*If $E|X|^r < \infty$, then*

$$\sum_{n=1}^{\infty} \text{Var}\left(\frac{Y_n}{n^{1/r}}\right) = \sum_{n=1}^{\infty} \frac{\text{Var}\,Y_n}{n^{2/r}} < \infty.$$

*Proof.* We shall provide two proofs. The first, more elegant one, is based on the interchange of expectation and summation (and Lemma A.3.1).

$$\sum_{n=1}^{\infty} \mathrm{Var}\left(\frac{Y_n}{n^{1/r}}\right) \leq \sum_{n=1}^{\infty} \frac{EY_n^2}{n^{2/r}} = \sum_{n=1}^{\infty} \frac{E(X^2 I\{|X| \leq n^{1/r}\})}{n^{2/r}}$$

$$= EX^2\left(\sum_{n=1}^{\infty} \frac{I\{|X| \leq n^{1/r}\}}{n^{2/r}}\right) = EX^2\left(\sum_{n \geq |X|^r \vee 1} \frac{1}{n^{2/r}}\right)$$

$$= EX^2\left(\sum_{n \geq |X|^r} \frac{1}{n^{2/r}}\right)\left(I\{1 \leq |X| < 2^{1/r}\} + I\{|X| \geq 2^{1/r}\}\right)$$

$$\leq 2^{2/r}\sum_{n=1}^{\infty} \frac{1}{n^{2/r}} + \frac{2^{(2/r)-1}}{(2/r)-1}EX^2\frac{1}{(|X|^r)^{(2/r)-1}} = C + CE|X|^r.$$

The second, traditional proof, is based on the slicing technique. Via Lemma A.3.1 we obtain

$$\sum_{n=1}^{\infty} \mathrm{Var}\left(\frac{Y_n}{n^{1/r}}\right) \leq \sum_{n=1}^{\infty} \frac{EY_n^2}{n^{2/r}} = \sum_{n=1}^{\infty} \frac{E(X^2 I\{|X| \leq n^{1/r}\})}{n^{2/r}}$$

$$= \sum_{n=1}^{\infty} \frac{1}{n^{2/r}}\sum_{k=1}^{n} E\big(X^2 I\{(k-1)^{1/r} < |X| \leq k^{1/r}\}\big)$$

$$= \sum_{n=1}^{\infty} \frac{1}{n^{2/r}}E(X^2 I\{|X| \leq 1\})$$

$$+ \sum_{k=2}^{\infty}\left(\sum_{n=k}^{\infty} \frac{1}{n^{2/r}}\right)E\big(X^2 I\{(k-1)^{1/r} < |X| \leq k^{1/r}\}\big)$$

$$\leq C + \frac{2^{(2/r)-1}}{(2/r)-1}\sum_{k=2}^{\infty} \frac{1}{k^{(2/r)-1}}E\big(X^2 I\{(k-1)^{1/r} < |X| \leq k^{1/r}\}\big)$$

$$\leq C + C\sum_{k=1}^{\infty} \frac{1}{k^{(2/r)-1}} \cdot (k^{1/r})^{2-r}E\big(|X|^r I\{(k-1)^{1/r} < |X| \leq k^{1/r}\}\big)$$

$$= C + C\sum_{k=1}^{\infty} E|X|^r I\{k-1 < |X|^r \leq k\} = C + CE|X|^r < \infty. \qquad \square$$

Here is now the strong law.

**Theorem 6.1.** (The Kolmogorov strong law)

(a) *If $E|X| < \infty$ and $EX = \mu$, then*

$$\frac{S_n}{n} \xrightarrow{a.s.} \mu \qquad as \quad n \to \infty.$$

(b) *If $\frac{S_n}{n} \xrightarrow{a.s.} c$ for some constant $c$, as $n \to \infty$, then*

$$E|X| < \infty \qquad and \quad c = EX.$$

(c) *If $E|X| = \infty$, then*

$$\limsup_{n\to\infty} \frac{S_n}{n} = +\infty.$$

*Remark 6.1.* Strictly speaking, we presuppose in (b) that the limit can only be a constant. That this is indeed the case follows from the Kolmogorov zero-one law. Considering this, (c) is somewhat more general than (b).    □

*Proof of (a).* Set

$$Y_n = X_n I\{|X_n| \le n\} = \begin{cases} X_n, & \text{if } |X_n| \le n, \\ 0, & \text{otherwise.} \end{cases}$$

Then $X_1, X_2, \ldots$ and $Y_1, Y_2, \ldots$ are convergence equivalent by Proposition 1.2. Spelled out this means that

$$\sum_{n=1}^{\infty} P\left(\left|\frac{X_n}{n}\right| > 1\right) < \infty,$$

so that the first sum in the Kolmogorov three-series theorem, Theorem 5.5, is convergent (with $A = 1$).

Next, Lemma 6.1 (with $r = 1$) tells us that

$$\sum_{n=1}^{\infty} \operatorname{Var}\left(\frac{Y_n}{n}\right) < \infty,$$

so that the third sum in the three-series theorem is convergent. An application of the Kolmogorov convergence criterion, Theorem 5.2, therefore yields

$$\sum_{n=1}^{\infty} \frac{Y_n - EY_n}{n} \quad \text{converges a.s.,}$$

so that, by the random Kronecker lemma, Lemma 5.1,

$$\frac{1}{n}\sum_{k=1}^{n}(Y_k - EY_k) \overset{a.s.}{\to} 0 \quad \text{as} \quad n \to \infty.$$

Finally,

$$EY_n = EX_n I\{|X_n| \le n\} = EXI\{|X| \le n\} \to EX = \mu \quad \text{as} \quad n \to \infty,$$

so that, by Lemma A.6.1,

$$\frac{1}{n}\sum_{k=1}^{n} EY_k \to \mu \quad \text{as} \quad n \to \infty,$$

which implies that

$$\frac{1}{n}\sum_{k=1}^{n} Y_k \overset{a.s.}{\to} \mu \quad \text{as} \quad n \to \infty,$$

which, due to the convergence equivalence noted above, proves the strong law.
*Proof of (b)*. Given the strong law, we have (Theorem 5.11.1),

$$\frac{X_n}{n} = \frac{S_n}{n} - \frac{n-1}{n}\cdot\frac{S_n}{n} \overset{a.s.}{\to} c - 1\cdot c = 0 \quad \text{as} \quad n \to \infty,$$

and the mean is finite (Proposition 1.1). An application of the sufficiency part shows that $c = E\,X$.
*Proof of (c)*. If $E|X| = \infty$, Proposition 1.1 and the second Borel-Cantelli lemma, Theorem 2.18.2, tell us that

$$P(|X_n| > nc \text{ i.o.}) = 1 \quad \text{for } every \quad c > 0,$$

and, hence, since $|X_n| \le |S_n| + |S_{n-1}|$, that

$$P(|S_n| > nc/2 \text{ i.o.}) = 1 \quad \text{for } every \quad c > 0,$$

which is equivalent to our claim. $\qquad\square$

*Remark 6.2.* Note that in the proof of the weak law we studied $S_n$ for a fixed $n$ and truncated $X_k$ at $n$ (or $n\varepsilon^3$) for every $k = 1, 2, \ldots, n$, in other words, for every fixed $n$ the truncation level was the same for all summands with index less than or equal to $n$. Here we treat all summands and partial sums simultaneously, and each $X$ has its own truncation level.

*Remark 6.3.* Etemadi [82] has shown that pair-wise independence is enough for the strong law of large numbers to hold. $\qquad\square$

Since the strong law requires finite mean, Example 4.1 illustrates that there are cases when the weak law holds but the strong law does not – the laws are *not* equivalent.
We complete the discussion by exploiting Example 4.1 a bit further and show that, in fact, the two non-trivial sums related to the Kolmogorov three-series theorem do not converge for any truncation level (due to the symmetry the sum of the expectations of the truncated variables vanishes). As a corollary, $\sum_{k=1}^{\infty} \frac{X_k}{k}$ diverges almost surely (the divergence of one of the two sums is of course enough for that conclusion).
Let us recall the situation: $X_1, X_2, \ldots$ are independent random variables with common density

$$f(x) = \begin{cases} \frac{c}{x^2 \log |x|}, & \text{for} \quad |x| > 2, \\ 0, & \text{otherwise}, \end{cases}$$

where $c$ is some normalizing constant.

Let $A > 0$, set $S_n = \sum_{k=1}^{n} X_k$, $n \geq 1$, and

$$
Y_k = \begin{cases} \frac{X_k}{k}, & \text{when } |\frac{X_k}{k}| \leq A, \\ 0, & \text{otherwise,} \end{cases} \qquad k = 1, 2, \dots .
$$

Then (stealing from Example 4.1),

$$
P(|Y_k| > A) = P(|X_k| > kA) = 2c \int_{kA}^{\infty} \frac{dx}{x^2 \log x} = \mathcal{O}\left(\frac{1}{k \log k}\right) \qquad \text{as} \quad k \to \infty,
$$

and

$$
\operatorname{Var} Y_k = E\left(\frac{X}{k}\right)^2 I\{|X| \leq kA\} = \frac{2c}{k^2} \int_{2}^{kA} \frac{dx}{\log x} = \mathcal{O}\left(\frac{1}{k \log k}\right) \qquad \text{as} \quad k \to \infty,
$$

so that both sums diverge for all $A$ as claimed.

# 7 The Marcinkiewicz-Zygmund Strong Law

Just as we met a generalization of the weak law of large numbers in Theorem 3.2, there exists a strong law generalizing Theorem 3.2 – the Marcinkiewicz-Zygmund strong law – which first appeared in their 1937 paper [180].

**Theorem 7.1.** (The Marcinkiewicz-Zygmund strong law)
*Let $0 < r < 2$. Suppose that $X, X_1, X_2, \dots$ are independent, identically distributed random variables. If $E|X|^r < \infty$, and $E X = 0$ when $1 \leq r < 2$, then*

$$
\frac{S_n}{n^{1/r}} \xrightarrow{a.s.} 0 \qquad \text{as} \quad n \to \infty.
$$

*Conversely, if almost sure convergence holds as stated, then $E|X|^r < \infty$, and $E X = 0$ when $1 \leq r < 2$.*

*Proof of the sufficiency 1.* The first part of the proof is essentially the same as that for the strong law. The difference enters in the getting rid of the truncated means. We therefore reduce the details somewhat in the first part.
    Set

$$
Y_n = X_n I\{|X_n| \leq n^{1/r}\} = \begin{cases} X_n, & \text{if } |X_n| \leq n^{1/r}, \\ 0, & \text{otherwise.} \end{cases}
$$

Then $X_1, X_2, \dots$ and $Y_1, Y_2, \dots$ are convergence equivalent, spelled out,

$$
\sum_{n=1}^{\infty} P\left(\left|\frac{X_n}{n^{1/r}}\right| > 1\right) < \infty.
$$

Lemma 6.1 now tells us that

$$\sum_{n=1}^{\infty} \text{Var}\left(\frac{Y_n}{n^{1/r}}\right) < \infty,$$

the Kolmogorov convergence criterion yields

$$\sum_{n=1}^{\infty} \frac{Y_n}{n^{1/r}} \quad \text{converges a.s.,}$$

and the random Kronecker lemma,

$$\frac{1}{n^{1/r}} \sum_{k=1}^{n} (Y_k - E Y_k) \overset{a.s.}{\to} 0 \quad \text{as} \quad n \to \infty.$$

Next we wish to show that

$$\frac{1}{n^{1/r}} \sum_{k=1}^{n} E Y_k \to \mu \quad \text{as} \quad n \to \infty, \tag{7.1}$$

in order to conclude that

$$\frac{1}{n} \sum_{k=1}^{n} Y_k \overset{a.s.}{\to} \mu \quad \text{as} \quad n \to \infty,$$

after which the convergence equivalence finishes the proof.

Since the truncation in this proof differs from that of the corresponding weak law, Theorem 3.2, we have to argue somewhat differently.

First, let $0 < r < 1$. Then

$$|E S_n'| \leq \sum_{k=1}^{n} E|X_k| I\{|X_k| \leq k^{1/r}\}$$

$$= \sum_{k=1}^{n} E|X_k| I\{|X_k| \leq k^{1/(2r)}\} + \sum_{k=1}^{n} E|X_k| I\{k^{1/(2r)} < |X_k| \leq k^{1/r}\}$$

$$\leq \sum_{k=1}^{n} (k^{1/(2r)})^{1-r} E|X|^r I\{|X| \leq k^{1/(2r)}\}$$

$$\quad + \sum_{k=1}^{n} (k^{1/r})^{1-r} E|X|^r I\{k^{1/(2r)} < |X| \leq k^{1/r}\}$$

$$\leq \sum_{k=1}^{n} k^{(1-r)/(2r)} E|X|^r + \sum_{k=1}^{n} k^{(1/r)-1} E|X|^r I\{|X| > k^{1/(2r)}\}$$

$$\leq C n^{(1+r)/(2r)} E|X|^r + \sum_{k=1}^{n} k^{(1/r)-1} E|X|^r I\{|X| > k^{1/(2r)}\},$$

where, in the last inequality, we glanced at Lemma A.3.1 without specifying the constant. Thus,

$$\frac{|E\,S_n'|}{n^{1/r}} \le n^{(1/2)-(1/r)} \cdot E|X|^r + \frac{1}{n^{1/r}} \sum_{k=1}^{n} k^{(1/r)-1} E|X|^r I\{|X| > k^{1/(2r)}\},$$

which converges to 0 as $n \to \infty$, since the first term in the right-hand side converges to 0 (the exponent is negative), and the second term is a weighted average of terms tending to 0 (Lemma A.6.1).

Next, let $1 < r < 2$. Then

$$|E\,S_n'| = \left| \sum_{k=1}^{n} E\,X_k I\{|X_k| \le k^{1/r}\} \right| \le \sum_{k=1}^{n} E|X_k| I\{|X| > k^{1/r}\}$$

$$\le \sum_{k=1}^{n} (k^{1/r})^{1-r} E|X|^r I\{|X| > k^{1/r}\}$$

$$= \sum_{k=1}^{n} k^{(1/r)-1} E|X|^r I\{|X| > k^{1/r}\},$$

which means that

$$\frac{|E\,S_n'|}{n^{1/r}} \le \frac{1}{n^{1/r}} \sum_{k=1}^{n} k^{(1/r)-1} E|X|^r I\{|X| > k^{1/r}\} \to 0 \quad \text{as} \quad n \to \infty,$$

since the right-hand side is a weighted average of objects converging to 0; Lemma A.6.1 again. This proves (7.1) for that case.

*Proof of the sufficiency 2.* Another way to prove this is to proceed as in the first half of the proof, but to start with the symmetric case. This procedure has the advantage that (7.1) is void.

Assuming that $E|X|^r < \infty$, it follows, by the $c_r$-inequality, Theorem 3.2.2, that also $E|X^s|^r < \infty$, so that, by arguing as in the first proof via convergence equivalence, the Kolmogorov convergence criterion, the random Kronecker lemma, we conclude that

$$\frac{S_n^s}{n^{1/r}} \overset{a.s.}{\to} 0 \quad \text{as} \quad n \to \infty.$$

In order to desymmetrize, we apply the *strong* symmetrization inequalities, Theorem 3.6.3, to obtain

$$\frac{S_n - \operatorname{med}(S_n)}{n^{1/r}} \overset{a.s.}{\to} 0 \quad \text{as} \quad n \to \infty.$$

To get rid of the median we combine this with the *weak* Marcinkiewicz-Zygmund law, Theorems 3.2, to obtain

$$\frac{\operatorname{med}(S_n)}{n^{1/r}} \to 0 \quad \text{as} \quad n \to \infty,$$

and the conclusion follows.

*Proof of the necessity.* We argue as in the proof of the strong law, Theorem 6.1(b):

$$\frac{X_n}{n^{1/r}} = \frac{S_n}{n^{1/r}} - \left(\frac{n-1}{n}\right)^{1/r} \cdot \frac{S_n}{n^{1/r}} \stackrel{a.s.}{\to} 0 - 1 \cdot 0 = 0 \quad \text{as} \quad n \to \infty,$$

so that $E|X|^r < \infty$ by Proposition 1.1.

If $1 < r < 2$ the strong law, Theorem 6.1, holds. In particular, the mean is finite. But

$$\frac{S_n}{n} = n^{(1/r)-1} \frac{S_n}{n^{1/r}} \stackrel{a.s.}{\to} 0 \quad \text{as} \quad n \to \infty,$$

so that $EX = 0$ by the converse of the strong law.

If $0 < r < 1$ there is nothing more to prove.

This finishes the proof of the necessity and, hence, of the theorem.    □

*Remark 7.1.* In the first proof the last part amounted to "get rid of" the truncated mean. In the second proof the final part was to "get rid of" the medians. These are the standard patterns in these two approaches. Moreover, means are truncated means, whereas medians are medians.

*Remark 7.2.* Note that in order to take care of the truncated mean in the sufficiency part for $0 < r < 1$, we had to do some extra work, since we could not turn around the inequality in the indicator, which, in turn, was due to the fact that the mean does not exist in that case. Going straight along the lines of the first part would have resulted in a weighted average of objects tending to $E|X|^r$ and not to 0.

*Remark 7.3.* Another alternative for the truncated means is the slicing technique.    □

**Exercise 7.1.** Check the statements in the last two remarks.

**Exercise 7.2.** Formulate and prove a Marcinkiewicz-Zygmund analog to Theorem 6.1(c).

**Exercise 7.3.** Modify the computations at the end of Section 6.6 concerning Example 4.1 in order to exhibit an example where the weak Marcinkiewicz-Zygmund law holds but the strong one does not.    □

# 8 Randomly Indexed Sequences

Traditional limit theorems concern asymptotics "as $n \to \infty$", that is, when a *fixed* index tends to infinity. In the law of large numbers we have met above we considered the arithmetic mean as the *fixed* number of terms increased. In many applications, however, one studies some process during a *fixed period of time*, which means that the number of observations is random. A law of large numbers in such a situation would involve the arithmetic mean of a *random number of terms* as this random number increases, which it does if time increases.

To make things a bit more precise, and at the same time more general, suppose that $Y_1, Y_2, \ldots$ are random variables, such that

$$Y_n \to Y \quad \text{in some sense as } n \to \infty,$$

and let $\{N(t),\, t \geq 0\}$ be a family of positive, integer valued random variables, such that

$$N(t) \to \infty \quad \text{in some sense as } t \to \infty.$$

Can we, or when can we, or can we not, conclude that

$$Y_{N(t)} \to Y \quad \text{in some sense as } t \to \infty\,?$$

A reasonable guess is that the easiest case should be almost sure convergence.

**Theorem 8.1.** *Suppose that $Y_1, Y_2, \ldots$ are random variables, such that*

$$Y_n \overset{a.s.}{\to} Y \quad \text{as} \quad n \to \infty,$$

*and that $\{N(t),\, t \geq 0\}$ is a family of positive, integer valued random variables, such that*

$$N(t) \overset{a.s.}{\to} \infty \quad \text{as} \quad t \to \infty.$$

*Then*

$$Y_{N(t)} \overset{a.s.}{\to} Y \quad \text{as} \quad t \to \infty.$$

*Proof.* Let $A = \{\omega : Y_n(\omega) \nrightarrow Y\}$ as $n \to \infty$, $B = \{\omega : N(t,\omega) \nrightarrow \infty\}$ as $t \to \infty$, and $C = \{\omega : Y_{N(t,\omega)}(\omega) \nrightarrow Y\}$ as $t \to \infty$.

Then $C \subset A \cup B$, so that

$$P(C) \leq P(A \cup B) \leq P(A) + P(B) = 0. \qquad \square$$

With the aid of this result the following law of large numbers is within easy reach.

**Theorem 8.2.** *Let $X, X_1, X_2, \ldots$ be independent, identically distributed random variables, set $S_n = \sum_{k=1}^{n} X_k$, $n \geq 1$, and suppose that $\{N(t),\, t \geq 0\}$ is a family of positive, integer valued random variables, such that*

$$N(t) \overset{a.s.}{\to} \infty \quad \text{as} \quad t \to \infty.$$

(i)    *Let $r > 0$. If $E|X|^r < \infty$, then*

$$\frac{X_{N(t)}}{(N(t))^{1/r}} \overset{a.s.}{\to} 0 \quad \text{as} \quad t \to \infty.$$

(ii)   *Let $0 < r < 2$. If $E|X|^r < \infty$, and $EX = 0$ when $1 \leq r < 2$, then*

$$\frac{S_{N(t)}}{(N(t))^{1/r}} \stackrel{a.s.}{\to} 0 \quad as \quad t \to \infty.$$

(iii)  *If $EX = \mu$, then*

$$\frac{S_{N(t)}}{N(t)} \stackrel{a.s.}{\to} \mu \quad as \quad t \to \infty.$$

(iv)   *If, in addition, $\frac{N(t)}{t} \stackrel{a.s.}{\to} \theta$, for some $\theta \in (0, \infty)$, then, in addition,*

$$\frac{X_{N(t)}}{t^{1/r}} \stackrel{a.s.}{\to} 0, \qquad \frac{S_{N(t)}}{t^{1/r}} \stackrel{a.s.}{\to} 0, \qquad and \qquad \frac{S_{N(t)}}{t} \stackrel{a.s.}{\to} \mu\theta, \qquad respectively.$$

*Proof.* Combining Theorem 8.1 with Proposition 1.1 yields (i), combining it with Theorems 7.1 and 6.1 proves (ii) and (iii), respectively. Finally, (iv) follows from the fact that the product of two almost surely convergent sequences (or families) converges almost surely, recall Theorem 5.11.1, or note, as in the previous proof, that the union of two null sets is a null set. □

The first variation of these results is when one of the convergences is almost sure, and the other one is in probability. There are two possible setups; $Y_n \stackrel{a.s.}{\to} Y$ and $N(t) \stackrel{p}{\to} \infty$, and $Y_n \stackrel{p}{\to} Y$ and $N(t) \stackrel{a.s.}{\to} \infty$, respectively. One is true, the other one is false.

**Exercise 8.1.** Which is true, which is not? □

**Theorem 8.3.** *Suppose that $Y_1, Y_2, \ldots$ are random variables, such that*

$$Y_n \stackrel{a.s.}{\to} Y \quad as \quad n \to \infty,$$

*and that $\{N(t), t \geq 0\}$ is a family of positive, integer valued random variables, such that*

$$N(t) \stackrel{p}{\to} \infty \quad as \quad t \to \infty.$$

*Then*

$$Y_{N(t)} \stackrel{p}{\to} Y \quad as \quad t \to \infty.$$

*Proof.* We apply the subsequence principle (Section 5.7) according to which we must prove that every subsequence of $\{Y_{N(t)}\}$ contains a further subsequence that converges almost surely as $t \to \infty$, and, thus, also in probability.

Since $N(t) \stackrel{p}{\to} \infty$ as $t \to \infty$, this is also true for every subsequence, from which we can select a further subsequence that converges almost surely, according to Theorem 5.3.4. In other words, for every subsequence $\{t_k, k \geq 1\}$ there exists a subsubsequence $\{t_{k_j}, j \geq 1\}$, such that $N(t_{k_j}) \stackrel{a.s.}{\to} \infty$ as $j \to \infty$. An application of Theorem 8.1 then shows that $Y_{N(t_{k_j})} \to \infty$ almost surely and, hence, also in probability as $j \to \infty$. □

The conclusion here is convergence in probability. Intuitively it is not to expect that the convergence mode in the conclusion should be stronger than the weakest of those in the assumptions. A simple example, given to me by Svante Janson, shows that this is, indeed, the case.

*Example 8.1.* Let $Y_1$, $Y_2$, ... be random variables such that $P(Y_n = 1/n) = 1$. Obviously, $Y_n \overset{a.s.}{\to} 0$ as $n \to \infty$. But no matter which family $\{N(t), t \geq 0\}$ we pick we always have $P(Y_{N(t)} = 1/N(t)) = 1$, which converges (if at all) in the same convergence mode as $N(t)$ does. □

We also present an example from [209] because of its connection with the following one, which, in turn, is a counter-example to "the other variation" mentioned above.

*Example 8.2.* Let $\Omega = [0,1]$, with $\mathcal{F}$ the corresponding $\sigma$-algebra, and $P$ Lebesgue measure. Set

$$Y_n(\omega) = \begin{cases} \frac{1}{m+1}, & \text{for } \frac{j}{2^m} \leq \omega < \frac{j+1}{2^m}, \\ 0, & \text{otherwise}, \end{cases}$$

where $n = 2^m + j$, $0 \leq j \leq 2^m - 1$, and let

$$N(t,\omega) = \begin{cases} 1, & \text{for } \frac{s}{2^r} \leq \omega < \frac{s+1}{2^r}, \\ \min\{k \geq 2^t : Y_k(\omega) > 0\}, & \text{otherwise}, \end{cases}$$

where $t = 2^r + s$, $0 \leq s \leq 2^r - 1$.

In this setup, $Y_n \overset{a.s.}{\to} 0$ as $n \to \infty$, and $N(t) \overset{p}{\to} \infty$ as $t \to \infty$. However,

$$Y_{N(t,\omega)} = \begin{cases} 1, & \text{for } \frac{s}{2^r} \leq \omega < \frac{s+1}{2^r}, \\ \frac{1}{t+1}, & \text{otherwise}, \end{cases}$$

where, again, $t = 2^r + s$, $0 \leq s \leq 2^r - 1$.

Clearly, $Y_{N(t)} \overset{p}{\to} 0$ as $t \to \infty$, but, since $P(Y_{N(t)} = 1 \text{ i.o.}) = 1$, $Y_{N(t)}$ cannot converge almost surely to 1. □

Next we provide the counter-example, also from [209], that we promised half a page ago.

*Example 8.3.* Let the probability space be the same as in Example 8.2, set

$$Y_n(\omega) = \begin{cases} 1, & \text{for } \frac{j}{2^m} \leq \omega < \frac{j+1}{2^m}, \\ 0, & \text{otherwise}, \end{cases}$$

where $n = 2^m + j$, $0 \leq j \leq 2^m - 1$, and let

$$N(t) = \min\{k \geq 2^t : Y_k(\omega) > 0\}.$$

In this setup $Y_n \overset{p}{\to} 0$ as $n \to \infty$, but *not* almost surely. Moreover, $N(t) \overset{a.s.}{\to} \infty$ as $t \to \infty$. However,

$$Y_{N(t)} = 1 \quad \text{a.s.} \quad \text{for all } t. \qquad\qquad □$$

*Remark 8.1.* The crux in the counter-example is that $Y_n$ does not converge along sample paths to 0 (only in probability), and that $N(t)$ is designed to pick the "evil" points.                                                                      □

# 9 Applications

## 9.1 Normal Numbers

In his celebrated paper [26] Borel proved that almost all real numbers are normal. This result, which is a purely mathematical statement, can be derived as a corollary of the strong law as follows.

A number is normal with respect to base 10 if the relative frequency of every decimal converges to $1/10$. Analogously for any base. For reasons of uniqueness we identify any non-terminating expansion with its terminating equivalent. For example, the number $0, 39999\ldots$ is identified with 0.4. It suffices to consider the interval $[0, 1]$, since the number of intervals of unit length is countable, and a countable union of sets of measure 0 has measure 0.

**Theorem 9.1.** *Almost all numbers in $[0, 1]$ are normal with respect to all bases.*

*Proof.* Consider a given base $k \geq 1$ with decimals $j = 0, 1, \ldots, k - 1$. Pick a number according to the $U(0, 1)$-distribution. One readily checks that the decimals in the $k$-ary expansion are equidistributed and independent. Thus, if $X_i$ is the $i$th decimal of the given number, then

$$P(X_i = j) = \frac{1}{k} \quad \text{for } j = 0, 1, \ldots, k - 1 \text{ and all } i,$$

and $\{X_i, i \geq 1\}$ are independent. An application of the strong law of large numbers thus tells us that

$$\frac{1}{n} \sum_{i=1}^{n} X_i \overset{a.s.}{\to} \frac{1}{k} \quad \text{as} \quad n \to \infty,$$

which shows that the number is normal relative to base $k$.

Now, let

$$B_k = \{x \in [0, 1] : x \text{ is } not \text{ normal with respect to base } k\}.$$

We have just shown that $P(B_k) = 0$ for any given $k$. Consequently,

$$P\left( \bigcup_{k=1}^{\infty} B_k \right) \leq \sum_{k=1}^{\infty} P(B_k) = 0.$$                                                                    □

**Exercise 9.1.** Write down a normal number.

**Exercise 9.2.** Write down another one.                                         □

*Remark 9.1.* It is not known whether such famous numbers as $\pi$ or $e$ are normal or not (as far as I know).                                                 □

## 9.2 The Glivenko-Cantelli Theorem

Let $X_1, X_2, \ldots, X_n$ be a sample from the distribution $F$. In Subsection 6.3.1 we proved that the empirical distribution function

$$F_n(x) = \frac{1}{n} \sum_{k=1}^{n} I\{X_k \leq x\}$$

was uniformly close to the true distribution function in probability. In this subsection we prove the Glivenko-Cantelli theorem which states that the almost sure closeness is uniform in $x$.

The strong law immediately tells us that, for all $x$

$$F_n(x) \overset{a.s.}{\to} F(x) \quad \text{as} \quad n \to \infty. \tag{9.1}$$

**Theorem 9.2.** *Under the above setup,*

$$\sup_x |F_n(x) - F(x)| \overset{a.s.}{\to} 0 \quad \text{as} \quad n \to \infty.$$

*Proof.* The proof of the theorem thus amounts to showing uniformity in $x$.

Let $\mathbb{J}_F$ be the countable (Proposition 2.2.1) set of jumps of $F$, and let $J_n(x)$ and $J(x)$ denote the jumps of $F_n$ and $F$, respectively, at $x$. In addition to (9.1), which we exploit at continuity points of $F$, another application of the strong law tells us that

$$J_n(x) \overset{a.s.}{\to} J(x) \quad \text{for all} \quad x \in \mathbb{J}_F.$$

The conclusion now follows via Lemma A.9.2(iii).    □

*Remark 9.2.* If, in particular, $F$ is continuous, the proof is a little easier (cf. Lemma A.9.2).    □

For more about empirical distributions, cf., e.g., [200, 221, 248].

## 9.3 Renewal Theory for Random Walks

We recall the setup from Subsection 2.16.3, that is, $X, X_1, X_2, \ldots$ are independent, identically distributed random variables, with positive, finite, mean $EX = \mu$, partial sums $S_n = \sum_{k=1}^{n} X_k$, $n \geq 1$, and first passage times

$$\tau(t) = \min\{n : S_n > t\}, \quad t \geq 0.$$

Here is a strong law for first passage times.

**Theorem 9.3.** *In the above setup,*

$$\frac{\tau(t)}{t} \overset{a.s.}{\to} \frac{1}{\mu} \quad \text{as} \quad t \to \infty.$$

*Proof.* First of all, $\tau(t) \xrightarrow{p} \infty$ as $t \to \infty$ (why?), and since $\tau(t)$ is non-decreasing, an application of Theorem 5.3.5 shows that $\tau(t) \xrightarrow{a.s.} \infty$ as $t \to \infty$, which, by the "random index strong law", Theorem 8.2, tells us that

$$\frac{S_{\tau(t)}}{\tau(t)} \xrightarrow{a.s.} \mu \quad \text{and that} \quad \frac{X_{\tau(t)}}{\tau(t)} \xrightarrow{a.s.} 0 \quad \text{as} \quad t \to \infty. \tag{9.2}$$

Moreover, by construction,

$$t < S_{\tau(t)} = S_{\tau(t)-1} + X_{\tau(t)} \le t + X_{\tau(t)}, \tag{9.3}$$

so that

$$\frac{t}{\tau(t)} < \frac{S_{\tau(t)}}{\tau(t)} \le \frac{t}{\tau(t)} + \frac{X_{\tau(t)}}{\tau(t)}.$$

Joining the two numbered formulas (and Theorem 5.11.1), finally, yields

$$\limsup_{t \to \infty} \frac{t}{\tau(t)} < \mu \le \liminf_{t \to \infty} \frac{t}{\tau(t)} + 0,$$

which implies the statement of the theorem. □

**Exercise 9.3.** Modify the proof in order to prove the analogous Marcinkiewicz-Zygmund law: If $E|X|^r < \infty$, $1 < r < 2$, then

$$\frac{\tau(t) - \frac{t}{\mu}}{t^{1/r}} \xrightarrow{a.s.} 0 \quad \text{as} \quad t \to \infty,$$

([103], Theorem 2.8; also [110], Theorem III.4.4). □

## 9.4 Records

The model as described in Subsection 2.17.2 was a sequence $X_1, X_2, \ldots$ of independent, identically distributed, continuous random variables, with record times $L(1) = 1$ and, recursively,

$$L(n) = \min\{k : X_k > X_{L(n-1)}\}, \quad n \ge 2.$$

The associated counting process, $\{\mu(n), n \ge 1\}$, was defined by

$$\mu(n) = \#\text{records among } X_1, X_2, \ldots, X_n = \max\{k : L(k) \le n\} = \sum_{k=1}^{n} I_k,$$

where $I_k = 1$ when $X_k$ is a record, and $I_k = 0$ otherwise. Moreover, the indicators are independent, and $I_k \in \text{Be}(1/k)$, $k \ge 1$.

In the present subsection we prove strong laws of large numbers for the record times and the counting process.

We have seen that the counting process is described by a sum of independent, but not identically distributed random variables. This is on the one hand a complication compared to the i.i.d. case. On the other hand, the summands are indicators, and, hence, bounded, which is a simplification; all moments exist, and so on.

**Theorem 9.4.** *We have*

$$\frac{\mu(n)}{\log n} \overset{a.s.}{\to} 1 \quad as \quad n \to \infty;$$

$$\frac{\log L(n)}{n} \overset{a.s.}{\to} 1 \quad as \quad n \to \infty.$$

*Proof.* The conclusion for the counting process follows from the Kolmogorov three-series theorem, Theorem 5.3, or rather, from the Kolmogorov convergence criterion, Theorem 5.2, because of the boundedness of the summands.

Namely, $I_k - 1/k$, $k \geq 1$, are independent, uniformly bounded random variables with mean 0 and variance $\frac{1}{k}(1 - \frac{1}{k})$. Therefore, since

$$\sum_{k=1}^{\infty} \text{Var}\left(\frac{I_k - \frac{1}{k}}{\log k}\right) = \sum_{k=1}^{\infty} \frac{\frac{1}{k}(1 - \frac{1}{k})}{(\log k)^2} \leq \sum_{k=1}^{\infty} \frac{1}{k(\log k)^2} < \infty,$$

it follows that

$$\sum_{k=1}^{\infty} \frac{I_k - \frac{1}{k}}{\log k} \quad \text{converges a.s.,}$$

so that, by the random Kronecker lemma, Lemma 5.1,

$$\frac{1}{\log n} \sum_{k=1}^{n} \left(I_k - \frac{1}{k}\right) \overset{a.s.}{\to} 0 \quad as \quad n \to \infty.$$

The desired limiting result now follows from Lemma A.3.1(iii) (and Theorem 5.11.1).

Next we turn our attention to record times. Rényi's proof in [205] was based on inversion;

$$\{L(n) \geq k\} = \{\mu(k) \leq n\}.$$

Inspired by the corresponding proof for the first passage times for random walks it is a natural hope that a similar approach would work here too. And, indeed, it works ([111]), in fact, more easily, since, in the present context, the boundary is hit exactly, by which we mean that

$$\mu(L(n)) = n.$$

After all, the number of records obtained at the moment of occurrence of record number $n$ clearly must be equal to $n$.

Now, since $L(n) \overset{a.s.}{\to} \infty$ as $n \to \infty$ (please check), we obtain

$$\frac{\mu(L(n))}{\log L(n)} \overset{a.s.}{\to} 1 \quad as \quad n \to \infty,$$

via Theorem 8.2, so that

$$\frac{n}{\log L(n)} \overset{a.s.}{\to} 1 \quad as \quad n \to \infty,$$

which is our claim turned upside down. □

# 10 Uniform Integrability; Moment Convergence

In order to prove moment convergence we first have to prove uniform integrability, and then join this with Theorem 5.5.2.

We begin with the Kolmogorov strong law, Theorem 6.1.

**Theorem 10.1.** *Suppose that* $X$, $X_1$, $X_2$, ... *are independent, identically distributed random variables with finite mean,* $\mu$, *and set* $S_n = \sum_{k=1}^{n} X_k$, $n \geq 1$. *Then*

$$\frac{S_n}{n} \to \mu \quad in \quad L^1 \quad as \quad n \to \infty.$$

*Proof.* In view of the strong law and Theorem 5.5.2 we have to prove that the sequence of arithmetic means is uniformly integrable, and in order to achieve this we lean on Theorem 4.1. We thus wish to show that

$$\sup_n E\left|\frac{S_n}{n}\right| < \infty,$$

and that for any $\varepsilon > 0$ there exists $\delta > 0$, such that for any set $A$ with $P(A) < \delta$,

$$E\left|\frac{S_n}{n}\right| I\{A\} < \varepsilon.$$

The first relation follows via the triangle inequality:

$$E\left|\frac{S_n}{n}\right| \leq \frac{1}{n} \sum_{k=1}^{n} E|X_k| = E|X|.$$

Similarly for the second one. Namely, for every $\varepsilon > 0$ there exists $\delta > 0$ such that $E|X|I\{A\} < \varepsilon$ whenever $P(A) < \delta$; cf. Proposition 2.6.3. Thus, if $A$ is one such set, then

$$E\left|\frac{S_n}{n}\right| I\{A\} \leq \frac{1}{n} \sum_{k=1}^{n} E|X_k|I\{A\} = E|X|I\{A\} < \varepsilon.$$

Alternatively, we may verify that the definition of uniform integrability is satisfied directly. Set $B_n = \{|\frac{S_n}{n}| > \varepsilon\}$. We then know that $P(B_n) \to 0$ as $n \to \infty$, so that, for $n_0$ sufficiently large,

$$E|X|I\{B_n\} < \varepsilon \quad \text{for} \quad n > n_0.$$

As above it then follows that, for $n > n_0$

$$E\left|\frac{S_n}{n}\right| I\{B_n\} \leq \frac{1}{n} \sum_{k=1}^{n} E|X_k|I\{B_n\} = E|X|I\{B_n\} < \varepsilon,$$

which proves that $\{\frac{S_n}{n}, n \geq n_0\}$ is uniformly integrable. Since adding a finite number of random variables to the sequence preserves uniform integrability (Lemma 5.4.1) we have completed a second proof.

For a third proof, let $a > 0$. A direct computation, a small trick, and Markov's inequality yield

$$E\left|\frac{S_n}{n}\right| I\left\{\left|\frac{S_n}{n}\right| > a\right\} \leq \frac{1}{n}\sum_{k=1}^{n} E|X_k| I\left\{\left|\frac{S_n}{n}\right| > a\right\} = E|X| I\left\{\left|\frac{S_n}{n}\right| > a\right\}$$

$$= E|X| I\left\{\left\{\left|\frac{S_n}{n}\right| > a\right\} \cap \{|X| \leq \sqrt{a}\}\right\}$$

$$+ E|X| I\left\{\left\{\left|\frac{S_n}{n}\right| > a\right\} \cap \{|X| > \sqrt{a}\}\right\}$$

$$\leq \sqrt{a}P\left(\left|\frac{S_n}{n}\right| > a\right) + E|X| I\{|X| > \sqrt{a}\}$$

$$\leq \sqrt{a}\frac{E|X|}{a} + E|X| I\{|X| > \sqrt{a}\} \to 0 \quad \text{as} \quad a \to \infty,$$

independently of (and hence uniformly in) $n$.     □

*Remark 10.1.* A fourth proof will be given in Subsection 10.16.1.     □

If higher-order moments exist can prove convergence in $r$-mean by exploiting the convexity of the function $|x|^r$ for $r \geq 1$ as follows:

**Theorem 10.2.** *Let* $r \geq 1$. *Suppose that* $X, X_1, X_2, \ldots$ *are independent, identically distributed random variables with finite mean,* $\mu$, *and set* $S_n = \sum_{k=1}^{n} X_k$, $n \geq 1$. *If* $E|X|^r < \infty$, *then*

$$\frac{S_n}{n} \to \mu \quad \text{in} \quad L^r \quad \text{and} \quad E\left|\frac{S_n}{n}\right|^r \to \mu^r \quad \text{as} \quad n \to \infty.$$

*Proof.* We wish to prove that $\{|\frac{S_n}{n}|^r, n \geq 1\}$ is uniformly integrable. To check the conditions of Theorem 5.4.1, let $x_1, x_2, \ldots, x_n$ are positive reals. Then, by convexity,

$$\left(\frac{1}{n}\sum_{k=1}^{n} x_k\right)^r \leq \frac{1}{n}\sum_{k=1}^{n}(x_k)^r,$$

so that, (via the triangle inequality),

$$E\left|\frac{S_n}{n}\right|^r \leq E\left(\frac{1}{n}\sum_{k=1}^{n}|X_k|\right)^r \leq \frac{1}{n}\sum_{k=1}^{n} E|X_k|^r = E|X|^r < \infty.$$

Similarly, given $\varepsilon$, $\delta$, and $A$ as required,

$$E\left|\frac{S_n}{n}\right|^r I\{A\} \leq E|X|^r I\{A\} < \varepsilon,$$

since now $E|X|^r I\{A\} < \varepsilon$ whenever $P(A) < \delta$.     □

For an analogous result for the Marcinkiewicz-Zygmund strong law we profit from the following result due to Pyke and Root [203]. Their proof, however, differs from ours.

**Theorem 10.3.** *Let $0 < r < 2$. Suppose that $X$, $X_1$, $X_2$, ... are independent, identically distributed random variables, and set $S_n = \sum_{k=1}^{n} X_k$, $n \geq 1$. If $E|X|^r < \infty$, and $EX = 0$ when $1 \leq r < 2$, then*

$$E\left|\frac{S_n}{n^{1/r}}\right|^r = E\frac{|S_n|^r}{n} \to 0 \quad as \quad n \to \infty.$$

*Proof.* Let $1 \leq r < 2$. As in the proof of the weak Marcinkiewicz-Zygmund law, Theorem 3.2, we let, for a given, arbitrary $\varepsilon > 0$, $M > 0$ be so large that $E|X|^r I\{|X| > M\} < \varepsilon$, and set

$$Y_k = X_k I\{|X_k| \leq M\} \quad \text{and} \quad Z_k = X_k I\{|X_k| > M\}, \quad k = 1, 2, \ldots.$$

By the Marcinkiewicz-Zygmund inequalities, Theorem 3.8.1, followed by two applications of the $c_r$-inequality (note that $r/2 < 1$), we obtain

$$E|S_n|^r \leq B_r E\left|\sum_{k=1}^{n} X_k^2\right|^{r/2} \leq B_r E\left|\sum_{k=1}^{n} Y_k^2\right|^{r/2} + B_r E\left|\sum_{k=1}^{n} Z_k^2\right|^{r/2}$$

$$\leq B_r(nM^2)^{r/2} + B_r nE\big((Z_1^2)^{r/2}\big) = B_r n^{r/2} M^r + B_r nE|Z_1|^r$$

$$\leq B_r n^{r/2} M^r + B_r n\varepsilon,$$

so that

$$\limsup_{n \to \infty} E\frac{|S_n|^r}{n} \leq B_r \varepsilon.$$

The conclusion follows.

The case $0 < r < 1$ was actually already proved as part of the proof of the weak Marcinkiewicz-Zygmund law. Let us, however, for convenience, quickly recall that, using the $c_r$-inequality directly, one obtains

$$E|S_n|^r \leq (nM)^r + nE|X|^r I\{|X| > M\} \leq (nM)^r + n\varepsilon,$$

after which the conclusion follows as for the case $1 < r < 2$.  $\square$

**Exercise 10.1.** Had we known the Pyke-Root theorem before the Marcinkiewicz strong law, an alternative for the latter would have been to provide a proof for symmetric random variables, to desymmetrize via the *strong* symmetrization inequalities and, finally, to use Theorem 10.3 to take care of the median. Check the details of this plan.  $\square$

# 11 Complete Convergence

As mentioned in Chapter 5, the concept complete convergence was introduced in 1947 by Hsu and Robbins [140], who proved that the sequence of arithmetic means of independent, identically distributed random variables converges completely to the expected value of the variables, provided their variance is finite. The necessity was proved somewhat later by Erdős [74, 75].

We begin, however, with a result on the maximum of a sequence of independent, identically distributed random variables paralleling Theorem 2.1.

**Theorem 11.1.** *Let $\beta > 0$, and suppose that $X$, $X_1$, $X_2$, ... are independent, identically distributed random variables. Further, let $b_n$, $n \geq 1$, be a sequence of positive reals, increasing to $+\infty$, and set $Y_n = \max_{1 \leq k \leq n} |X_k|$, $n \geq 1$. Then*

$$\sum_{n=1}^{\infty} n^{\beta-2} P(Y_n > b_n) < \infty \quad \Longleftrightarrow \quad \sum_{n=1}^{\infty} n^{\beta-1} P(|X| > b_n) < \infty.$$

*In particular, for $r > 0$,*

$$\sum_{n=1}^{\infty} n^{\beta-2} P(Y_n > n^{1/r}) < \infty \quad \Longleftrightarrow \quad E|X|^{r\beta} < \infty,$$

*and*

$$\frac{Y_n}{n^{1/r}} \xrightarrow{c.c.} 0 \quad \Longleftrightarrow \quad E|X|^{2r} < \infty.$$

*Proof.* The first part is immediate from relation (2.1), viz.,

$$\frac{1}{2} n P(|X| > b_n \varepsilon) \leq P(Y_n > b_n \varepsilon) \leq n P(|X| > b_n \varepsilon),$$

for $n$ sufficiently large (and $\varepsilon > 0$).

To prove the second statement, set $b_n = n^{1/r}$. The first part of the theorem then tells us that

$$\sum_{n=1}^{\infty} n^{\beta-2} P(Y_n > n^{1/r}) < \infty \quad \Longleftrightarrow \quad \sum_{n=1}^{\infty} n^{\beta-1} P(|X| > n^{1/r}) < \infty,$$

which, in turn, holds if and only if $E\big((|X|^r)^{\beta}\big) < \infty$ (Theorem 2.12.1), that is, if and only if $E|X|^{r\beta} < \infty$.

Finally, let $\varepsilon > 0$. Putting $b_n = n^{1/r}\varepsilon$ and $\beta = 2$, we conclude from our findings so far that if $E|X|^{2r} < \infty$, then $E|X/\varepsilon|^{2r} < \infty$, so that

$$\sum_{n=1}^{\infty} P(Y_n > n^{1/r}\varepsilon) < \infty \quad \text{for all} \quad \varepsilon > 0.$$

Conversely, if the sum converges for some $\varepsilon > 0$, then $E|X/\varepsilon|^{2r} < \infty$, and $E|X|^{2r} < \infty$. □

## 11.1 The Hsu-Robbins-Erdős Strong Law

**Theorem 11.2.** *Let $X$, $X_1$, $X_2$, ... be independent, identically distributed random variables, and set $S_n = \sum_{k=1}^{n} X_k$, $n \geq 1$. If $EX = 0$ and $EX^2 < \infty$, then*

$$\sum_{n=1}^{\infty} P(|S_n| > n\varepsilon) < \infty \quad \text{for all} \quad \varepsilon > 0.$$

*Conversely, if the sum is finite for some $\varepsilon > 0$, then $EX = 0$, $EX^2 < \infty$, and the sum is finite for all $\varepsilon > 0$.*

*Proof of the sufficiency.* The first obvious attempt is to try Chebyshev's inequality. However

$$\sum_{n=1}^{\infty} P(|S_n| > n\varepsilon) \leq \varepsilon^{-2}\sigma^2 \sum_{n=1}^{\infty} \frac{1}{n} = +\infty.$$

So that attempt fails.

The classical proof of the sufficiency runs, essentially, as follows. Let $2^i \leq n < 2^{i+1}$ and set

$$A_n^{(1)} = \{|X_k| > 2^{i-2} \text{ for at least one } k \leq n\},$$
$$A_n^{(2)} = \{|X_{k_1}| > n^\gamma, |X_{k_2}| > n^\gamma \text{ for at least two } k_i \leq n\},$$
$$A_n^{(3)} = \{|\sum\nolimits' X_k| > 2^{i-2}\},$$

where $\gamma$ is "suitably" chosen at a later time, and where $\sum'$ denotes summation over those indices which are not among the first two sets. After this one observes that

$$\{|S_n| > n\varepsilon\} \subset A_n^{(1)} \cup A_n^{(2)} \cup A_n^{(3)}.$$

A fair amount of computations yield the desired result.

A more efficient way to prove the result is to symmetrize and desymmetrize as described in Subsection 5.11.3, and to exploit the Kahane-Hoffmann-Jørgensen inequality, Theorem 3.7.5.

Therefore, suppose first that the random variables are *symmetric*. Then, by the KHJ-inequality, Chebyshev's inequality, and Theorem 2.12.1,

$$\sum_{n=1}^{\infty} P(|S_n| > 3n\varepsilon) \leq \sum_{n=1}^{\infty} nP(|X| > n\varepsilon) + 4\big(P(|S_n| > n\varepsilon)\big)^2$$

$$\leq E(X/\varepsilon)^2 + 4\sum_{n=1}^{\infty} \left(\frac{\sigma^2}{n\varepsilon^2}\right)^2 = \frac{EX^2}{\varepsilon^2} + 4\frac{\sigma^4}{\varepsilon^4}\frac{\pi^2}{6}.$$

The beauty of this attack is the squaring of the original probability, in that $\sum_{n=1}^{\infty} n^{-1}$ (which is divergent) is replaced by $\sum_{n=1}^{\infty} n^{-2}$ (which is convergent).

To desymmetrize, let $X, X_1, X_2, \ldots$ be the original sequence. Then, $\operatorname{Var} X^s = 2\operatorname{Var} X$, so that we know that the theorem holds for the symmetrized sequence. By the weak symmetrization inequalities (Proposition 3.6.2) we therefore also know that

$$\sum_{n=1}^{\infty} P(|S_n - \operatorname{med}(S_n)| > n\varepsilon) < \infty \quad \text{for all} \quad \varepsilon > 0.$$

This implies, in particular, that

$$\frac{S_n - \operatorname{med}(S_n)}{n} \xrightarrow{p} 0 \quad \text{as} \quad n \to \infty,$$

which, in view of the weak law of large numbers (and Theorem 5.11.1) shows that

$$\frac{\text{med}\,(S_n)}{n} = \frac{S_n}{n} - \frac{S_n - \text{med}\,(S_n)}{n} \xrightarrow{p} 0 \quad \text{as} \quad n \to \infty,$$

that is $\frac{\text{med}\,(S_n)}{n} \to 0$ as $n \to \infty$. Thus, given $\varepsilon > 0$, there exists $n_0$, such that

$$\frac{|\text{med}\,(S_n)|}{n} < \frac{\varepsilon}{2} \quad \text{for} \quad n > n_0,$$

so that

$$\sum_{n=n_0+1}^{\infty} P(|S_n| > n\varepsilon) \le \sum_{n=n_0+1}^{\infty} P(|S_n - \text{med}\,(S_n)| > n\varepsilon/2) < \infty.$$

*Proof of the necessity.* Once again we begin with the symmetric case.

Set, for $n \ge 1$, $Y_n = \max_{1 \le k \le n} |X_k|$. Recalling Proposition 1.4 we know that

$$P(Y_n > 2n\varepsilon) \le P(\max_{1 \le k \le n} |S_k| > n\varepsilon) \le 2P(|S_n| > n\varepsilon).$$

Combining this with the fact that the Borel-Cantelli sum is finite (for *some* $\varepsilon$), we conclude that

$$\sum_{n=1}^{\infty} P(|Y_n| > 2n\varepsilon) < \infty \quad \text{for that very} \quad \varepsilon > 0.$$

An application of Theorem 11.1 then tells us that

$$\sum_{n=1}^{\infty} nP(|X/2| > n\varepsilon) < \infty,$$

which proves that $E(X/(2\varepsilon))^2 < \infty$, and therefore also that $E X^2 < \infty$.

To desymmetrize we use the weak symmetrization inequalities to conclude that if the sum is finite for the original random variables, then it is also finite for the symmetrized ones, so that $E(X^s)^2 < \infty$. But

$$\text{Var}\,X = \frac{1}{2}\text{Var}\,X^s = \frac{1}{2}E(X^s)^2 < \infty.$$

Moreover, since the variance is finite, the mean must be finite, so that the strong law of large numbers holds, which, in turn, forces the mean to equal 0.

Finally, knowing this we can apply the sufficiency part of the theorem, according to which the Borel-Cantelli sum converges *for all* $\varepsilon > 0$.

The proof of the theorem is complete.    □

## 11.2 Complete Convergence and the Strong Law

Now is the time to return to Example 2.18.2 and our promise made in connection with the second Borel-Cantelli lemma, of a Borel-Cantelli sum that diverges, and, yet, the probability of infinitely many events occurring is 0.

Namely, if the mean is finite and the variance is infinite, then Theorem 11.2 states that the Borel-Cantelli sum is divergent. But the strong law of large numbers, Theorem 6.1, holds, so that

$$\sum_{n=1}^{\infty} P(|S_n| > n\varepsilon) = \infty \quad \text{and (but)} \quad P(|S_n| > n\varepsilon \text{ i.o.}) = 0 \quad \text{for all} \quad \varepsilon > 0.$$

A simple examples is provided by the two-sided Pareto distribution with density

$$f(x) = \begin{cases} \frac{\beta}{2|x|^{\beta+1}}, & \text{for} \quad |x| > 1, \\ 0, & \text{otherwise.} \end{cases}$$

with $\beta \in (1,2]$, for which $E|X|^r < \infty$ for $r < \beta$, $EX = 0$, and $E|X|^r = \infty$ for $r \geq \beta$, in particular, the variance is infinite.

Another way to illustrate this discrepancy is via the counting variable: Let, for $\varepsilon > 0$,

$$N(\varepsilon) = \sum_{n=1}^{\infty} I\{|S_n| > n\varepsilon\} = \text{Card}\,\{n : |S_n| > n\varepsilon\},$$

that is, we count the number of times the value of the arithmetic mean falls outside the strip $\pm\varepsilon$. The strong law of large numbers states that this number is (a.s.) finite if and only if the mean of the summands is 0;

$$P(N(\varepsilon) < \infty) = 1 \quad \Longleftrightarrow \quad EX = 0.$$

The Hsu-Robbins-Erdős law tells us that the *expected number* of exceedances is finite if and only if the variance is finite and the mean is 0;

$$E\,N(\varepsilon) = \sum_{n=1}^{\infty} P(|S_n| > n\varepsilon) < \infty \quad \Longleftrightarrow \quad EX = 0 \quad \text{and} \quad EX^2 < \infty.$$

This means that if the mean is finite, but the variance is infinite, then the number of times the arithmetic mean falls outside any strip is a.s. finite, but the *expected number* of times this happens is infinite.

# 12 Some Additional Results and Remarks

Much more can be said about the law of large numbers, but one has to stop somewhere. In this section we collect a few additional pieces that are interesting and illuminating, but less central.

## 12.1 Convergence Rates

Whereas convergence is a qualitative result in the sense that it tells us that convergence holds, a rate result is a quantitative result in the sense that it tells us *how fast* convergence is obtained, and how large a sample must be for a certain precision.

In addition to being a result on another kind of convergence, Theorem 11.2 can be viewed as a result on *the rate of convergence* in the law of large numbers. Namely, not only do the terms $P(|S_n| > \varepsilon n)$ have to tend to 0, the sum of them has to converge, which is a little more.

This idea can be pursued further. Following is a more general result linking integrability of the summands to the rate of convergence in the law of large numbers. However, the proof is more technical, since dealing with moment inequalities for sums of an arbitrary order is much more messy than adding variances.

**Theorem 12.1.** *Let $p, r > 0$, $r \geq p$, $p < 2$. Suppose that $X, X_1, X_2, \ldots$ are independent, identically distributed random variables with partial sums $S_n = \sum_{i=1}^{n} X_k$, $n \geq 1$. If*

$$E|X|^r < \infty \quad \text{and, if } r \geq 1, \quad EX = 0, \tag{12.1}$$

*then*

$$\sum_{n=1}^{\infty} n^{(r/p)-2} P(|S_n| > n^{1/p}\varepsilon) < \infty \quad \text{for all } \varepsilon > 0; \tag{12.2}$$

$$\sum_{n=1}^{\infty} n^{(r/p)-2} P(\max_{1 \leq k \leq n} |S_k| > n^{1/p}\varepsilon) < \infty \quad \text{for all } \varepsilon > 0. \tag{12.3}$$

*If $r > p$ we also have*

$$\sum_{n=1}^{\infty} n^{(r/p)-2} P(\sup_{k \geq n} |S_k/k^{1/p}| > \varepsilon) < \infty \quad \text{for all } \varepsilon > 0. \tag{12.4}$$

*Conversely, if one of the sums is finite for some $\varepsilon > 0$, then so are the others (for appropriate values of $r$ and $p$), $E|X|^r < \infty$ and, if $r \geq 1$, $EX = 0$.*

*Remark 12.1.* Theorem 12.1 has a long history. For $r = 1$, $p = 1$ the equivalence between (12.1) and (12.2) is a famous result due to Spitzer [232]; see also [234]. For $r = 2$, $p = 1$ the equivalence between (12.1) and (12.2) is precisely Theorem 11.2. Departing from these, Katz, and later Baum and Katz, proved the equivalence between (12.1), (12.2), and (12.4) [153, 13]; and Chow established equivalence between (12.1) and (12.3) [44]. For a generalization to random variables with multi-dimensional indices we refer to [104], which we mention because several parts of the proof below are taken from there. For an extension to Banach space valued random variables, see [146].          □

*Proof.* The proof proceeds in steps.

Until further notice we consider symmetric random variables.
*Proof of (12.1) $\Longrightarrow$ (12.2), $r = p$.*
We first note that, necessarily, $r < 2$. Set, for $n \geq 1$,

$$Y_{k,n} = X_k I\{|X_k| \le n^{1/r}\}, \quad k = 1, 2, \ldots, n, \quad \text{and} \quad S'_n = \sum_{k=1}^{n} Y_{k,n}.$$

By the truncated Chebyshev inequality, Theorem 3.1.5,

$$P(|S_n| > n^{1/r}\varepsilon) \le \frac{\operatorname{Var} Y_{1,n}}{n^{(2/r)-1}\varepsilon^2} + nP(|X| > n^{1/r}),$$

so that (remember that $r = p$)

$$\sum_{n=1}^{\infty} n^{(r/p)-2} P(|S_n| > n^{1/p}\varepsilon) = \sum_{n=1}^{\infty} \frac{1}{n} P(|S_n| > n^{1/r}\varepsilon)$$

$$\le \frac{1}{\varepsilon^2} \sum_{n=1}^{\infty} \frac{\operatorname{Var} Y_{1,n}}{n^{2/r}} + \sum_{n=1}^{\infty} P(|X| > n^{1/r}) < \infty,$$

by Lemma 6.1, and Proposition 1.1, respectively.

*Proof of (12.1) $\implies$ (12.2), $r > p$, $r \le 1$.*

Applying, successively, the KHJ-inequality, Theorem 3.7.5, Markov's inequality, Theorem 3.1.1, and the $c_r$-inequality, Theorem 3.2.2, yields

$$\sum_{n=1}^{\infty} n^{(r/p)-2} P(|S_n| > 3n^{1/p}\varepsilon)$$

$$\le \sum_{n=1}^{\infty} n^{(r/p)-1} P(|X| > n^{1/p}\varepsilon) + 4 \sum_{n=1}^{\infty} n^{(r/p)-2} \left( \frac{E|S_n|^r}{(n^{1/p}\varepsilon)^r} \right)^2$$

$$\le \sum_{n=1}^{\infty} n^{(r/p)-1} P(|X| > n^{1/p}\varepsilon) + 4 \sum_{n=1}^{\infty} n^{(r/p)-2} \left( \frac{nE|X|^r}{(n^{1/p}\varepsilon)^r} \right)^2$$

$$= \sum_{n=1}^{\infty} n^{(r/p)-1} P(|X| > n^{1/p}\varepsilon) + \frac{4(E|X|^r)^2}{\varepsilon^{2r}} \sum_{n=1}^{\infty} \frac{1}{n^{r/p}} < \infty.$$

The finiteness of the first term follows from Theorem 2.12.1(iv), and that of the last one because $r > p$.

*Proof of (12.1) $\implies$ (12.2), $r > p$, $1 < r < 2$.*

The same procedure with the $c_r$-inequality replaced by Corollary 3.8.2 yields

$$\sum_{n=1}^{\infty} n^{(r/p)-2} P(|S_n| > 3n^{1/p}\varepsilon)$$

$$\le \sum_{n=1}^{\infty} n^{(r/p)-1} P(|X| > n^{1/p}\varepsilon) + 4 \sum_{n=1}^{\infty} n^{(r/p)-2} \left( \frac{E|S_n|^r}{(n^{1/p}\varepsilon)^r} \right)^2$$

$$\le \sum_{n=1}^{\infty} n^{(r/p)-1} P(|X| > n^{1/p}\varepsilon) + 4 \sum_{n=1}^{\infty} n^{(r/p)-2} \left( \frac{B_r n E|X|^r}{(n^{1/p}\varepsilon)^r} \right)^2$$

$$= \sum_{n=1}^{\infty} n^{(r/p)-1} P(|X| > n^{1/p}\varepsilon) + \frac{4(B_r E|X|^r)^2}{\varepsilon^{2r}} \sum_{n=1}^{\infty} \frac{1}{n^{r/p}} < \infty.$$

*Proof of (12.1) $\Longrightarrow$ (12.2), $r \geq 2$.*
This time we use the iterated Kahane-Hoffman-Jørgensen inequality. Let $j > 1$ to be specified later.

$$\sum_{n=1}^{\infty} n^{(r/p)-2} P(|S_n| > 3^j n^{1/p}\varepsilon)$$

$$\leq C_j \sum_{n=1}^{\infty} n^{(r/p)-1} P(|X| > n^{1/p}\varepsilon) + D_j \sum_{n=1}^{\infty} n^{(r/p)-2} \left(\frac{E|S_n|^r}{(n^{1/p}\varepsilon)^r}\right)^{2^j}$$

$$\leq C_j \sum_{n=1}^{\infty} n^{(r/p)-1} P(|X| > n^{1/p}\varepsilon) + D_j \sum_{n=1}^{\infty} n^{(r/p)-2} \left(\frac{B_r n^{r/2} E|X|^r}{(n^{1/p}\varepsilon)^r}\right)^{2^j}$$

$$= C_j \sum_{n=1}^{\infty} n^{(r/p)-1} P(|X| > n^{1/p}\varepsilon) + D_j \left(\frac{B_r E|X|^r}{\varepsilon^r}\right)^{2^j} \sum_{n=1}^{\infty} n^{\beta},$$

where

$$\beta = (r/p) - 2 + (r/2)2^j - (r/p)2^j = (r/p) - 2 + \frac{r(p-2)}{p}2^{j-1}.$$

The first sum converges as in the previous steps. The second sum converges because $\beta < -1$, provided $j$ is chosen sufficiently large.

*Proof of (12.2) $\Longleftrightarrow$ (12.3).*
This one is easy for a change. Namely,

$$P(|S_n| > n^{1/p}\varepsilon) \leq P(\max_{1 \leq k \leq n} |S_k| > n^{1/p}\varepsilon) \leq 2P(|S_n| > n^{1/p}\varepsilon).$$

The first inequality is trivial, and the second one is a consequence of the Lévy inequalities, Theorem 3.7.1.

*Proof of (12.2) $\Longrightarrow$ (12.4).*
The idea is to divide the supremum into slices along powers of 2. Let $j \geq 1$, and let $2^{j-1} \leq k < 2^j$.

Via an application of the Lévy inequality at the end, we have

$$P(\max_{2^{j-1} \leq k < 2^j} |S_k/k^{1/p}| > \varepsilon) \leq P(\max_{2^{j-1} \leq k < 2^j} |S_k| > 2^{(j-1)/p}\varepsilon)$$

$$\leq P(\max_{1 \leq k \leq 2^j} |S_k| > 2^{(j-1)/p}\varepsilon) \leq 2P(|S_{2^j}| > 2^{(j-1)/p}\varepsilon).$$

Next, set $C(r,p) = 2^{(r/p)-2}$ if $r > 2p$ and $= 1$ otherwise. Inserting the previous estimate at third inequality below, and applying Corollary 3.7.1 at the last one, yields

$$\sum_{n=1}^{\infty} n^{(r/p)-2} P(\sup_{k \geq n} |S_k/k^{1/p}| > 2^{2/p}\varepsilon)$$

$$= \sum_{i=0}^{\infty} \sum_{j=2^i}^{2^{i+1}-1} j^{(r/p)-2} P(\sup_{k \geq j} |S_k/k^{1/p}| > 2^{2/p}\varepsilon)$$

$$\leq \sum_{i=0}^{\infty} 2^i C(r,p)\, 2^{i((r/p)-2)} P\big(\sup_{k\geq 2^i} |S_k/k^{1/p}| > 2^{2/p}\varepsilon\big)$$

$$\leq C(r,p) \sum_{i=0}^{\infty} 2^{i((r/p)-1)} \sum_{j=i+1}^{\infty} P\big(\max_{2^{j-1}\leq k<2^j} |S_k/k^{1/p}| > 2^{2/p}\varepsilon\big)$$

$$\leq C(r,p) \sum_{i=0}^{\infty} 2^{i((r/p)-1)} \sum_{j=i+1}^{\infty} 2P(|S_{2^j}| > 2^{(j+1)/p}\varepsilon)$$

$$= 2C(r,p) \sum_{j=0}^{\infty} \Big(\sum_{i=0}^{j-1} 2^{i((r/p)-1)}\Big) P(|S_{2^j}| > 2^{(j+1)/p}\varepsilon)$$

$$\leq 2C(r,p) \sum_{j=0}^{\infty} 2^{j((r/p)-1)} P(|S_{2^j}| > 2^{(j+1)/p}\varepsilon)$$

$$= 2C(r,p) \sum_{j=0}^{\infty} \sum_{i=2^j}^{2^{j+1}-1} 2^{j((r/p)-2)} P(|S_{2^j}| > 2^{(j+1)/p}\varepsilon)$$

$$\leq 2C(r,p) \sum_{j=0}^{\infty} \sum_{i=2^j}^{2^{j+1}-1} C(r,p)\, i^{(r/p)-2} 2P(|S_i| > i^{1/p}\varepsilon)$$

$$= 4(C(r,p))^2 \sum_{j=0}^{\infty} j^{(r/p)-2} P(|S_j| > j^{1/p}\varepsilon) < \infty.$$

So far we have shown that (12.1) $\Longrightarrow$ (12.2) $\Longleftrightarrow$ (12.3), and that (12.2) $\Longrightarrow$ (12.4) for symmetric random variables.

*Desymmetrization*

This is achieved the usual way. If $E|X|^r < \infty$, then $E|X^s|^r < \infty$ as well, so that (12.2) holds for the symmetrized random variables. By the weak symmetrization inequalities, Proposition 3.6.2, this implies that

$$\sum_{n=1}^{\infty} n^{(r/p)-2} P(|S_n - \mathrm{med}\,(S_n)| > n^{1/p}\varepsilon) < \infty \quad \text{for all} \quad \varepsilon > 0,$$

from which (12.2) follows in the general case, since $\mathrm{med}\,(S_n)/n^{1/r} \to 0$ as $n \to \infty$.

The procedure for (12.3) and (12.4) are the same, however, with the strong symmetrization inequalities, Proposition 3.6.3, instead of the weak ones. We leave the details to the reader.

*Proof of (12.4) $\Longrightarrow$ (12.2).*

Trivial.

*Proof of (12.2) $\Longrightarrow$ (12.1).*

This is the last step. We follow the proof of the necessity in Theorem 11.2.

Suppose first that (12.2) holds in the symmetric case, that is, that the sum is convergent for *some* $\varepsilon > 0$. Leaning onto Proposition 1.4 we conclude that

$$\sum_{n=1}^{\infty} n^{(r/p)-2} P(\max_{1 \le k \le n} |X_k| > n^{1/p}\varepsilon) < \infty,$$

and, via Theorem 11.1, that

$$\sum_{n=1}^{\infty} n^{(r/p)-1} P(|X| > n^{1/p}\varepsilon) < \infty,$$

which is equivalent to $E|(X/\varepsilon)^p|^{r/p} < \infty$, and therefore to $E|X|^r < \infty$.

Now, suppose that (12.2) holds for some $\varepsilon > 0$ in the general case. Then it also does so for the symmetrized variables (with $\varepsilon/2$). Hence, $E|X^s|^r < \infty$, from which we conclude that $E|X|^r < \infty$ too (Theorem 3.2.3). Moreover, for $r \ge 1$ we know that the strong law holds, which necessitates $E X = 0$.

And now there is nothing more around to prove.  $\square$

The Pareto distributions are often suitable for illustrations as we have already seen. Partly, of course, since they are easy to handle, more importantly because by choosing the parameter carefully one can exhibit a distribution that has moments of every order strictly less than some prescribed level, and none above that level.

*Example 12.1.* Let $\alpha > 0$, and suppose that $X, X_1, X_2, \ldots$ are independent random variables with common density

$$f(x) = \begin{cases} \frac{\alpha}{2|x|^{\alpha+1}}, & \text{for } |x| > 1, \\ 0, & \text{otherwise.} \end{cases}$$

By joining (2.1) and Proposition 1.4 (note that the distribution is symmetric), we have, for $0 < \delta < 1$ and $n$ large,

$$(1-\delta)n \frac{1}{(2x)^\alpha} \le 2P(|S_n| > x), \quad x > 0, \tag{12.5}$$

where $S_n$, $n \ge 1$, as always, denotes partial sums.

For the specific parameter choice $\alpha = p$ we have $E|X|^r < \infty$ for $r < p$ and $E|X|^r = \infty$ for $r \ge p$. Moreover, setting $x = n^{1/p}\varepsilon$ in (12.5) yields

$$2P(|S_n| > n^{1/p}\varepsilon) \ge (1-\delta)n \frac{1}{n(2\varepsilon)^p} = \frac{1-\delta}{(2\varepsilon)^p},$$

so that the sum in (12.2) diverges for every $r \ge p$ (as expected).  $\square$

## 12.2 Counting Variables

In connection with Theorem 11.2 we introduced the counting variable

$$N(\varepsilon) = \sum_{n=1}^{\infty} I\{|S_n| > \varepsilon n^{1/p}\} = \text{Card}\{n : |S_n| > \varepsilon n^{1/p}\},$$

for $p = 1$, and showed that finite mean is equivalent to $N(\varepsilon)$ being a.s. finite, and finite variance is equivalent to $N(\varepsilon)$ having finite mean (throughout $E X = 0$).

In this subsection we shall connect the counting variable to Theorem 12.1, but, more so, we shall connect the theorem to

$$L(\varepsilon) = \sup\{n : |S_n| > \varepsilon n^{1/p}\},$$

that is to the *last time* the sequence of normalized normalized partial sums leaves the strip $\pm\varepsilon$.

We note immediately that $N(\varepsilon) \leq L(\varepsilon)$, but an intuitive guess is that these quantities should not differ all that much.

Since

$$\{L(\varepsilon) \geq n\} = \{\sup_{k \geq n} |S_k/k^{1/p}| > \varepsilon),$$

the following corollary emerges immediately by combining Theorems 12.1 and 3.12.1. The result for the counting variable was proved in [226] for the case $p = 1$.

**Corollary 12.1.** *Let $p, r > 0$, $r > p$, $p < 2$. Suppose that $X, X_1, X_2, \ldots$ are independent, identically distributed random variables with partial sums $S_n = \sum_{k=1}^n X_k$, $n \geq 1$, and let the counting variable $N(\varepsilon)$ and last exit time $L(\varepsilon)$ be defined as above. Then, for all $\varepsilon > 0$,*

$$E|X|^r < \infty \quad \text{and, if } r \geq 1, \quad EX = 0 \quad \Longrightarrow \quad E\,L(\varepsilon)^{(r/p)-1} < \infty,$$
$$E|X|^r < \infty \quad \text{and, if } r \geq 1, \quad EX = 0 \quad \Longrightarrow \quad E\,N(\varepsilon)^{(r/p)-1} < \infty.$$

*Conversely, if $E\,L(\varepsilon)^{(r/p)-1} < \infty$ for some $\varepsilon > 0$, then $E|X|^r < \infty$ and, if $r \geq 1$, $EX = 0$.*

*Remark 12.2.* We also remark that, although there is no equivalence in the second statement, the Pareto example 12.1, shows that there is no general improvement available.     □

## 12.3 The Case $r = p$ Revisited

The case $r = p$ was not allowed in (12.4) in Theorem 12.1 (because of the divergence of the harmonic series). There exists a variant of the theorem when $r = p$, however, with slight alterations in the first three conditions. We confine ourselves to formulating the result and refer to the various sources that were collected in Remark 12.1.

**Theorem 12.2.** *Let $0 < r < 2$. Suppose that $X, X_1, X_2, \ldots$ are independent, identically distributed random variables with partial sums $S_n = \sum_{k=1}^n X_k$, $n \geq 1$, and let $L(\varepsilon)$ be the last exit time. If*

$$E|X|^r \log^+ |X| < \infty \quad \text{and, if } r \geq 1, \quad EX = 0, \tag{12.6}$$

*then*

$$\sum_{n=1}^{\infty} \frac{\log n}{n} P(|S_n| > n^{1/r}\varepsilon) < \infty \quad \textit{for all } \varepsilon > 0; \tag{12.7}$$

$$\sum_{n=1}^{\infty} \frac{\log n}{n} P(\max_{1 \le k \le n} |S_k| > n^{1/r}\varepsilon) < \infty \quad \textit{for all } \varepsilon > 0; \tag{12.8}$$

$$\sum_{n=1}^{\infty} \frac{1}{n} P(\sup_{k \ge n} |S_k/k^{1/r}| > \varepsilon) < \infty \quad \textit{for all } \varepsilon > 0; \tag{12.9}$$

$$E \log L(\varepsilon) < \infty \quad \textit{for all } \varepsilon > 0. \tag{12.10}$$

*Conversely, if one of (12.7)–(12.10) holds for some $\varepsilon > 0$, then so do the others, $E|X|^r < \infty$ and, if $r \ge 1$, $EX = 0$.*

## 12.4 Random Indices

In this subsection we present some random index extensions of convergence rate results. We confine ourselves to statements, and refer to the relevant literature for proofs.

### Randomly Indexed Sums

Here is an extension of Theorem 12.1.

**Theorem 12.3.** *Let $p, r > 0$, $r \ge p$, $p < 2$, let $X, X_1, X_2, \ldots$ be independent, identically distributed random variables with partial sums $S_n = \sum_{k=1}^{n} X_k$, $n \ge 1$, and let $\{N_n, n \ge 1\}$ be non-negative, integer valued random variables. Suppose that $E|X|^r < \infty$, and that $EX = 0$ when $r \ge 1$. If, for some $\theta \in (0, \infty)$ and some $\delta \in (0, \infty)$,*

$$\sum_{n=1}^{\infty} n^{(r/p)-2} P(|N_n - n\theta| > n\delta) < \infty,$$

*then*

$$\sum_{n=1}^{\infty} n^{(r/p)-2} P(|S_{N_n}| > N_n^{1/p}\varepsilon) < \infty \quad \textit{for all } \varepsilon > 0,$$

*and*

$$\sum_{n=1}^{\infty} n^{(r/p)-2} P(|S_{N_n}| > n^{1/p}\varepsilon) < \infty \quad \textit{for all } \varepsilon > 0.$$

The proof of the first claim is based on the decomposition

$$P(|S_{N_n}| > N_n^{1/p}\varepsilon) = P(\{|S_{N_n}| > N_n^{1/p}\varepsilon\} \cap \{|N_n - n\theta| \le n\delta\})$$
$$+ P(\{|S_{N_n}| > N_n^{1/p}\varepsilon\} \cap \{|N_n - n\theta| > n\delta\}),$$

after which the first term in the decomposition is taken care of by Theorem 12.1, and the second one by the rate assumption on the index sequence. For the details we refer to [107].

**First Passage Times For Random Walks**

For a proof of the following result we refer to [107].

**Theorem 12.4.** *Let* $r \geq 1$, $1 \leq p < 2$, *and* $r \geq p$. *Let* $X, X_1, X_2, \ldots$ *be independent, identically distributed random variables with mean* $\mu > 0$, *and partial sums* $S_n = \sum_{k=1}^{n} X_k$, $n \geq 1$, *and let* $\tau(t) = \min\{n : S_n > t\}$, $t \geq 0$. *If* $E|X|^r < \infty$, *then*

$$\sum_{n=1}^{\infty} n^{(r/p)-2} P\left( \left| \tau(n) - \frac{n}{\mu} \right| > n^{1/p}\delta \right) < \infty \quad \text{for all} \quad \delta > 0.$$

Results of this kind have been proved for record times $\{L(n), n \geq 1\}$ and the counting process $\{\mu(n), n \geq 1\}$ in [112].

We finally mention that one can also investigate at which rate probabilities, such as $P(|S_n| > \varepsilon n)$ tend to 1 as $\varepsilon \searrow 0$. For example, $\varepsilon^2$ multiplied by the Borel-Cantelli sum of Theorem 11.2 converges to the second moment of the summands; see [136]. We shall (briefly) discuss such results in Subsection 7.7.8.

# 13 Problems

1. Let $X_1, X_2, \ldots$ be independent, identically distributed random variables. Suppose that $x_\infty = \sup\{x : F(x) < 1\} < \infty$, and set $Y_n = \max_{1 \leq k \leq n} X_k$, $n \geq 1$.
   (a) Prove that
   $$Y_n \xrightarrow{p} x_\infty \quad \text{as} \quad n \to \infty.$$
   (b) Prove that $Y_n \xrightarrow{a.s.} x_\infty$ as $n \to \infty$.

   Suppose, in addition, that $x_{-\infty} = \inf\{x : F(x) > 0\} > -\infty$, and set $Z_n = \min_{1 \leq k \leq n} X_k$, $n \geq 1$.
   (c) Prove that
   $$\frac{Y_n}{Z_n} \xrightarrow{p} \frac{x_\infty}{x_{-\infty}} \quad \text{as} \quad n \to \infty.$$
   (d) Prove that
   $$\frac{Y_n}{Z_n} \xrightarrow{a.s.} \frac{x_\infty}{x_{-\infty}} \quad \text{as} \quad n \to \infty.$$

2. Let $X_1, X_2, \ldots$ be independent, identically distributed random variables with mean $\mu$ and finite variance $\sigma^2$. Show that
   $$\frac{X_1 + X_2 + \cdots + X_n}{X_1^2 + X_2^2 + \cdots + X_n^2} \xrightarrow{a.s.} \frac{\mu}{\sigma^2 + \mu^2} \quad \text{as} \quad n \to \infty.$$

3. Let $(X_k, Y_k)'$, $1 \leq k \leq n$ be a sample from a two-dimensional distribution with mean vector and covariance matrix

$$\boldsymbol{\mu} = \begin{pmatrix} \mu_x \\ \mu_y \end{pmatrix}, \qquad \boldsymbol{\Lambda} = \begin{pmatrix} \sigma_x^2 & \rho \\ \rho & \sigma_y^2 \end{pmatrix},$$

respectively, and let

$$\bar{X}_n = \frac{1}{n} \sum_{k=1}^n X_k, \qquad s_{n,x}^2 = \frac{1}{n-1} \sum_{k=1}^n (X_k - \bar{X}_n)^2,$$

$$\bar{Y}_n = \frac{1}{n} \sum_{k=1}^n Y_k, \qquad s_{n,y}^2 = \frac{1}{n-1} \sum_{k=1}^n (Y_k - \bar{Y}_n)^2,$$

denote arithmetic means and sample variances.

(a) Prove that

$$s_{n,x}^2 \overset{a.s.}{\to} \sigma_x^2 \quad \text{and} \quad s_{n,y}^2 \overset{a.s.}{\to} \sigma_y^2 \quad \text{as} \quad n \to \infty.$$

(b) The *empirical correlation coefficient* is defined as

$$r_n = \frac{\sum_{k=1}^n (X_k - \bar{X}_n)(Y_k - \bar{Y}_n)}{\sqrt{\sum_{k=1}^n (X_k - \bar{X}_n)^2 \sum_{k=1}^n (Y_k - \bar{Y}_n)^2}}.$$

Prove that

$$r_n \overset{a.s.}{\to} \rho \quad \text{as} \quad n \to \infty.$$

4. Let $X_1$, $X_2$, ... be independent $U(0,1)$-distributed random variables, and set $Y_n = \min_{1 \leq k \leq n} X_k$, $n \geq 1$. Show that

$$V_n = \frac{\sum_{k=1}^n Y_k}{\log n} \overset{p}{\to} 1 \quad \text{as} \quad n \to \infty.$$

♠ Compute $E V_n$ and $\mathrm{Var}\, V_n$.

5. We are given $m$ boxes into which balls are thrown independently, and uniformly, that is, the probability of a ball falling into a given box is $1/m$ for all boxes. Let $N_n$ denote the number of empty boxes after $n$ balls have been distributed.

(a) Compute $E N_n$ and $\mathrm{Var}\, N_n$.

(b) Let $n, m \to \infty$ in such a way that $n/m \to \lambda$. Prove that

$$\frac{N_n}{n} \overset{p}{\to} c,$$

and determine the constant $c$.

6. Suppose that $X$ and $\{X_k,\ k \geq 1\}$ are independent, identically distributed random variables, such that

$$P(X = n) = \frac{1}{n(n-1)}, \quad n = 2, 3, \ldots.$$

Can you prove a strong law of large numbers? A weak law?

7. Suppose that $\{X_k,\ k \geq 1\}$ are independent random variables, such that

$$P(X_k = -k^2) = \frac{1}{k^2}, \quad P(X_k = -k^3) = \frac{1}{k^3}, \quad P(X_k = 2) = 1 - \frac{1}{k^2} - \frac{1}{k^3}.$$

Prove that $\sum_{k=1}^n X_k \overset{a.s.}{\to} +\infty$ as $n \to \infty$.

8. Let $\{a_n \in \mathbb{R}^+,\ n \geq 1\}$, and suppose that $\{X_k,\ k \geq 1\}$ are independent random variables, such that

$$P(X_n = -3) = 1 - p_n, \quad P(X_n = a_n) = p_n, \ n \geq 1.$$

Prove that

$$\sum_{n=1}^{\infty} p_n < \infty \quad \Longrightarrow \quad \sum_{k=1}^n X_k \overset{a.s.}{\to} -\infty \quad \text{as} \quad n \to \infty.$$

What about a converse?

9. Let $X_1, X_2, \ldots$ be independent, identically distributed random variables. Show that

(a) $\sum_{n=1}^{\infty} X_n$ does not converge almost surely;

(b) $P(\sum_{n=1}^{\infty} X_n \text{ converges}) = 0$.

10. Let $X_1, X_2, \ldots$ be independent, identically distributed random variables with mean $\mu > 0$. Show that

$$\sum_{k=1}^n X_k \overset{a.s.}{\to} +\infty \quad \text{as} \quad n \to \infty.$$

♣ If the mean is negative the sum converges almost surely to $-\infty$. If the mean equals 0 one can show that the sum, almost surely, oscillates between the infinities.

11. Let $X_1, X_2, \ldots$ be random variables with partial sums $S_n, n \geq 1$. Prove that

(a) $X_n \overset{a.s.}{\to} 0 \quad \Longrightarrow \quad \frac{S_n}{n} \overset{a.s.}{\to} 0$ as $n \to \infty$;

(b) $X_n \overset{a.s.}{\to} 0 \quad \Longrightarrow \quad \frac{1}{\log n} \sum_{k=1}^n \frac{X_k}{k} \overset{a.s.}{\to} 0$ as $n \to \infty$;

(c) $X_n \overset{a.s.}{\to} 0 \quad \Longrightarrow \quad \frac{1}{\log \log n} \sum_{k=1}^n \frac{X_k}{k \log k} \overset{a.s.}{\to} 0$ as $n \to \infty$;

(d) $X_n \overset{r}{\to} 0 \quad \Longrightarrow \quad \frac{S_n}{n} \overset{r}{\to} 0$ as $n \to \infty$, $(r \geq 1)$;

(e) $X_n \overset{p}{\to} 0 \quad \nRightarrow \quad \frac{S_n}{n} \overset{p}{\to} 0$ as $n \to \infty$.

♦ Try $P(X_n = 0) = 1 - \frac{1}{n}$ and $P(X_n = 2^n) = \frac{1}{n}, n \geq 1$.

12. Suppose that $X_1, X_2, \ldots$ are independent random variables, such that $X_n \in N(\mu_n, \sigma_n^2)$, $n \geq 1$.

(a) Prove that

$$\sum_{n=1}^{\infty} X_n \text{ converges a.s} \iff \sum_{n=1}^{\infty} \mu_n \text{ converges} \quad \text{and} \quad \sum_{n=1}^{\infty} \sigma_n^2 < \infty.$$

(b) Which of the following three statements are true?

$$\sum_{n=1}^{\infty} X_n \in N\left(\sum_{n=1}^{\infty} \mu_n, \sum_{n=1}^{\infty} \sigma_n^2\right) ?$$

$$E\left(\sum_{n=1}^{\infty} X_n\right) = \sum_{n=1}^{\infty} \mu_n ?$$

$$\text{Var}\left(\sum_{n=1}^{\infty} X_n\right) = \sum_{n=1}^{\infty} \sigma_n^2 ?$$

13. Suppose that $X_n \in \text{Ge}(p_n)$, $n \geq 1$, and that $X_1, X_2, \ldots$ are independent. State and prove the analog of the previous problem.

14. Let $X_1, X_2, \ldots$ be independent, standard Cauchy-distributed random variables, and $a_k$, $k \geq 1$, be real numbers. Prove that

$$\sum_{n=1}^{\infty} a_n X_n \text{ converges a.s.} \iff \sum_{n=1}^{\infty} |a_n| < \infty,$$

and determine the distribution of the limiting random variable.

15. Prove the following weaker strong law: Let $X$, $X_1, X_2, \ldots$ be independent, identically distributed random with partial sums $S_n$, $n \geq 1$. Suppose that $E X = 0$, and that $E X^4 < \infty$.

(a) Prove that $E(S_n)^4 \leq An^2 + Bn$, and determine $A$ and $B$.

(b) Prove that

$$\frac{S_{n^2}}{n^2} \xrightarrow{a.s.} 0 \quad \text{as} \quad n \to \infty.$$

(c) Set $T_n = \max_{n^2 \leq k \leq (n+1)^2} |S_k - S_{n^2}|$. Prove that

$$P(T_n > n^2 \varepsilon \text{ i.o.}) = 0 \quad \text{for all} \quad \varepsilon > 0.$$

(d) Alternatively, set $V_n = \max_{n^2 \leq k \leq (n+1)^2} \frac{|S_k|}{k}$, and prove that

$$P(V_n > \varepsilon \text{ i.o.}) = 0 \quad \text{for all} \quad \varepsilon > 0.$$

(e) Prove that

$$\frac{S_n}{n} \xrightarrow{a.s.} 0 \quad \text{as} \quad n \to \infty.$$

♣ It obviously suffices to solve (c) *or* (d) in order to solve the problem; it is, however, not forbidden to solve both.

16. Let $X_1$, $X_2$, ... be *uncorrelated* random variables with partial sums $S_n$, $n \geq 1$. Suppose that $E X = 0$, and that $\sup_n E(X_n)^4 = M < \infty$. Prove that

$$\frac{S_n}{n} \stackrel{a.s.}{\to} 0 \quad \text{as} \quad n \to \infty.$$

♣ Review your proof from the previous problem and modify where necessary.

17. *Numerical integration.* Suppose that $g$ is a non-negative, continuous function on the unit interval, and that $\sup_x g(x) \leq 1$. The following procedure aims at approximating the integral of $g$ over the unit interval. Pick $n$ points uniformly in the unit square, and let $U_n$ equal the number of points falling below the curve $y = g(x)$. Prove that

$$\frac{U_n}{n} \stackrel{a.s.}{\to} \int_0^1 g(x)\, dx \quad \text{as} \quad n \to \infty.$$

18. A stick of length 1 is randomly broken, which means that the remaining piece is $U(0,1)$-distributed. The remaining piece is broken similarly, and so on.
    (a) Let $X_n$ be the length of the piece that remains after the stick has been broken $n$ times. Describe $X_n$ as a product.
    (b) Prove that

$$\frac{\log X_n}{n} \stackrel{a.s.}{\to} \quad \text{as} \quad n \to \infty,$$

and determine the limit.
    (c) Find the distribution of the length of the first piece that was *thrown away* – don't compute, think! Then compute.

19. Let $X$ be a random variable with mean 0 and variance 1, and suppose that $X_1$, $X_2$, ... are independent random variables, such that $X_k \stackrel{d}{=} kX$ for all $k$, and set $S_n = \sum_{k=1}^n X_k$, $n \geq 1$. Show that
    (a) $\frac{S_n}{n^2} \stackrel{P}{\to} 0$ as $n \to \infty$;
    (b) $\frac{S_n}{n^2} \stackrel{a.s.}{\to} 0$ as $n \to \infty$.

20. Find the appropriate moment condition and normalization if we assume that, for some $\beta > 0$, $X_k \stackrel{d}{=} k^\beta X$ for all $k$ in the previous problem, and prove the corresponding laws of large numbers.

21. Let $X_1$, $X_2$, ... be independent, identically distributed random variables with mean 0, and $\{a_n, n \geq 1\}$ a sequence of bounded reals; $\sup_n |a_n| \leq A < \infty$. Prove that
    (a) $\frac{1}{n} \sum_{k=1}^n a_k X_k \stackrel{P}{\to} 0$ as $n \to \infty$;
    (b) $\frac{1}{n} \sum_{k=1}^n a_k X_k \stackrel{a.s.}{\to} 0$ as $n \to \infty$.

22. Let $X_1$, $X_2$, ... be independent, identically distributed random variables, such that $E X \neq 0$, set $S_n = \sum_{k=1}^n X_k$ and $Y_k = \max_{1 \leq k \leq n} X_k$. Prove that

$$\frac{Y_n}{S_n} \stackrel{a.s.}{\to} 0 \quad \text{as} \quad n \to \infty.$$

# 7

# The Central Limit Theorem

The law of large numbers states that the arithmetic mean of independent, identically distributed random variables converges to the expected value. One interpretation of the central limit theorem is as a (distributional) rate result. Technically, let $X, X_1, X_2, \ldots$ be independent, identically distributed random variables with mean $\mu$. The weak and strong laws of large numbers state that $\frac{1}{n}\sum_{k=1}^{n} X_k \to \mu$ in probability and almost surely, respectively, as $n \to \infty$. A distributional rate result deals with the question of how one should properly "blow up" the difference $\frac{1}{n}\sum_{k=1}^{n} X_k - \mu$ in order for the limit to have a non-trivial limit as $n$ tends to infinity. The corresponding theorem was first stated by Laplace. The first general version with a rigorous proof is due to Lyapounov [178, 179].

It turns out that if, in addition, the variance exists, then a multiplication by $\sqrt{n}$ yields a normal distribution in the limit. Our first result is a proof of this fact. We also prove the Lindeberg-Lévy-Feller theorem which deals with the same problem under the assumption that the summands are independent, but not identically distributed, and Lyapounov's version. Another variant is Anscombe's theorem, a special case of which is the central limit theorem for randomly indexed sums of random variables.

After this we turn our attention to the celebrated Berry-Esseen theorem, which is a convergence rate result for the central limit theorem, in that it provides an upper bound for the difference between the distribution functions of the standardized arithmetic mean and the normal distribution, under the additional assumption of a finite third moment.

The remaining part of the chapter (except for the problems) might be considered as somewhat more peripheral for the non-specialist. It contains various rate results for tail probabilities, applications to our companions renewal theory and records, some remarks on so-called *local limit theorems* for discrete random variables, and a mention of the concept of *large deviations*.

There also exist limit theorems when the variance does not exist and/or when the summands are not independent. An introduction to these topics will be given in Chapter 9.

# 1 The i.i.d. Case

In order to illustrate the procedure, we begin with the following warm-up; the
i.i.d. case.

**Theorem 1.1.** *Let $X$, $X_1$, $X_2$, ... be independent, identically distributed random variables with finite expectation $\mu$, and positive, finite variance $\sigma^2$, and
set $S_n = X_1 + X_2 + \cdots + X_n$, $n \geq 1$. Then*

$$\frac{S_n - n\mu}{\sigma\sqrt{n}} \xrightarrow{d} N(0,1) \quad as \quad n \to \infty.$$

*Proof.* In view of the continuity theorem for characteristic functions (Theorem
5.9.1), it suffices to prove that

$$\varphi_{\frac{S_n-n\mu}{\sigma\sqrt{n}}}(t) \to e^{-t^2/2} \quad as \quad n \to \infty, \quad for \quad -\infty < t < \infty.$$

Since $(S_n - n\mu)/\sigma\sqrt{n} = \left(\sum_{k=1}^{n}(X_k - \mu)/\sigma\right)/\sqrt{n}$, we may assume w.l.o.g.
throughout the proof, that $\mu = 0$ and $\sigma = 1$. With the aid of Theorems
4.1.10, 4.1.8, and 4.4.2, we then obtain

$$\varphi_{\frac{S_n-n\mu}{\sigma\sqrt{n}}}(t) = \varphi_{\frac{S_n}{\sqrt{n}}}(t) = \varphi_{S_n}\left(\frac{t}{\sqrt{n}}\right) = \left(\varphi_X\left(\frac{t}{\sqrt{n}}\right)\right)^n = \left(1 - \frac{t^2}{2n} + o\left(\frac{t^2}{n}\right)\right)^n$$

$$\to e^{-t^2/2} \quad as \quad n \to \infty. \qquad \square$$

# 2 The Lindeberg-Lévy-Feller Theorem

Let $X_1$, $X_2$, ... be independent random variables with finite variances, and
set, for $k \geq 1$, $E X_k = \mu_k$, $\operatorname{Var} X_k = \sigma_k^2$, and, for $n \geq 1$, $S_n = \sum_{k=1}^{n} X_k$,
and $s_n^2 = \sum_{k=1}^{n} \sigma_k^2$. We disregard throughout the degenerate case that all
variances are equal to zero.

The two fundamental conditions involved in the general form of the central
limit theorem are the so-called *Lindeberg conditions*

$$L_1(n) = \max_{1 \leq k \leq n} \frac{\sigma_k^2}{s_n^2} \to 0 \quad as \quad n \to \infty, \tag{2.1}$$

and

$$L_2(n) = \frac{1}{s_n^2} \sum_{k=1}^{n} E|X_k - \mu_k|^2 I\{|X_k - \mu_k| > \varepsilon s_n\} \to 0 \quad as \quad n \to \infty. \tag{2.2}$$

Here is now the legendary result.

**Theorem 2.1.** (Lindeberg-Lévy-Feller) *Let* $X_1$, $X_2$, ... *be given as above.*
(i) *If (2.2) is satisfied, then so is (2.1), and*

$$\frac{1}{s_n}\sum_{k=1}^{n}(X_k - \mu_k) \xrightarrow{d} N(0,1) \quad as \quad n \to \infty. \tag{2.3}$$

(ii) *If (2.1) and (2.3) are satisfied, then so is (2.2).*

*Remark 2.1.* Lindeberg proved the sufficiency by replacing one summand after the other with a normal one; see [176]. Lévy gave a proof based on characteristic functions; cf. his book [170], pp. 242. Feller [83] proved the necessity; see also [171], Théorème VI, and [172], Théorème 35.     □

We first prove that if (2.1) holds, then

$$s_n^2 \to \infty \quad as \quad n \to \infty, \tag{2.4}$$

and that $(2.2) \Longrightarrow (2.1)$, after which we provide two proofs of the sufficiency via characteristic functions – the more classical one and a more recent variation, and then another proof by the replacement method. We conclude with a proof of the necessity.
    *We assume, without restriction, that $\mu_k = 0$ throughout the proof.*

**Proof of the Implication (2.1) $\Longrightarrow$ (2.4)**

Since we have excluded the case when all variances are equal to zero, there exists $m$, such that $\sigma_m^2 > 0$. For $n > m$ it then follows that

$$\frac{\sigma_m^2}{s_n^2} \leq L_1(n) \to 0 \quad as \quad n \to \infty,$$

from which the conclusion follows.

**Proof of the Implication (2.2) $\Longrightarrow$ (2.1)**

For any $\varepsilon > 0$,

$$L_1(n) \leq \max_{1\leq k\leq n}\frac{1}{s_n^2}EX_k^2 I\{|X_k| \leq \varepsilon s_n\} + \max_{1\leq k\leq n}\frac{1}{s_n^2}EX_k^2 I\{|X_k| > \varepsilon s_n\}$$
$$\leq \varepsilon^2 + L_2(n),$$

so that

$$\limsup_{n\to\infty} L_1(n) \leq \varepsilon^2,$$

which proves the assertion.     □

**Proof of the Sufficiency; Characteristic Functions 1**

The traditional method is to write a product as an exponential of sums of logarithms, use Taylor expansion, and then the refined estimate for characteristic functions given in Theorem 4.4.1(i), where the cruder estimate is used outside the interval $[-\varepsilon s_n, \varepsilon s_n]$ and the finer estimate inside the same interval. In order to indicate the procedure we begin with a rougher preliminary version – note that $1 - \varphi_{X_k}(t/s_n)$ is "small" when $n$ is "large":

$$\varphi_{S_n/s_n}(t) = \varphi_{S_n}(t/s_n) = \prod_{k=1}^{n} \varphi_{X_k}(t/s_n) = \exp\left\{ \sum_{k=1}^{n} \log \varphi_{X_k}(t/s_n) \right\}$$

$$\approx \exp\left\{ -\sum_{k=1}^{n} \left( 1 - \varphi_{X_k}(t/s_n) \right) \right\}$$

$$\approx \exp\left\{ -\sum_{k=1}^{n} \left( 1 - \left( 1 + \frac{it}{s_n} \cdot 0 + \frac{(it)^2}{2s_n^2} \cdot \sigma_k^2 \right) \right) \right\}$$

$$= \exp\left\{ -\sum_{k=1}^{n} \left( 1 - \left( 1 - \frac{t^2}{2s_n^2} \cdot \sigma_k^2 \right) \right) \right\}$$

$$= \exp\left\{ -\frac{t^2}{2s_n^2} \sum_{k=1}^{n} \sigma_k^2 \right\} = \exp\{-t^2/2\} = \varphi_{N(0,1)}(t),$$

after which an application of the continuity theorem for characteristic functions, Theorem 5.9.1, would finish the proof.

Let us now straighten out the $\approx$ into $=$ in order to make this rigorous. This amounts to proving that

$$\left| \sum_{k=1}^{n} \left( \log \varphi_{X_k}(t/s_n) + \left( 1 - \varphi_{X_k}(t/s_n) \right) \right) \right| \to 0 \quad \text{as} \quad n \to \infty, \quad (2.5)$$

and that

$$\left| \sum_{k=1}^{n} \left( \varphi_{X_k}(t/s_n) - \left( 1 - \frac{\sigma_k^2 t^2}{2s_n^2} \right) \right) \right| \to 0 \quad \text{as} \quad n \to \infty. \quad (2.6)$$

The key estimates from Theorem 4.4.1(i) are:

$$\left| \varphi_{X_k}(t/s_n) - 1 \right| \le E \min\left\{ \frac{2|tX|}{s_n}, \frac{t^2 X_k^2}{2s_n^2} \right\}, \quad (2.7)$$

$$\left| \varphi_{X_k}(t/s_n) - \left( 1 - \frac{t^2 \sigma_k^2}{2s_n^2} \right) \right| \le E \min\left\{ \frac{t^2 X_k^2}{2s_n^2}, \frac{|t|^3 |X_k|^3}{6s_n^3} \right\}. \quad (2.8)$$

Since

$$|1 - \varphi_{X_k}(t/s_n)| \le E \frac{t^2 X_k^2}{2s_n^2} \le \frac{t^2}{2} L_1(n) \to 0 \quad \text{as} \quad n \to \infty, \qquad (2.9)$$

uniformly for $k = 1, 2, \ldots, n$, formula (A.A.5) with $z = 1 - \varphi_{X_k}(t/s_n)$ tells us that, for $n$ sufficiently large,

$$\left| \sum_{k=1}^{n} \left( \log \varphi_{X_k}(t/s_n) + \left(1 - \varphi_{X_k}(t/s_n)\right) \right) \right|$$

$$\le \sum_{k=1}^{n} \left| \log \left(1 - (1 - \varphi_{X_k}(t/s_n))\right) + \left(1 - \varphi_{X_k}(t/s_n)\right) \right|$$

$$\le \sum_{k=1}^{n} \left| 1 - \varphi_{X_k}(t/s_n) \right|^2 \le \max_{1 \le k \le n} |1 - \varphi_{X_k}(t/s_n)| \sum_{k=1}^{n} |1 - \varphi_{X_k}(t/s_n)|$$

$$\le \frac{t^2}{2} L_1(n) \sum_{k=1}^{n} E \frac{t^2 X_k^2}{2s_n^2} = \frac{t^4}{4} L_1(n) \to 0 \quad \text{as} \quad n \to \infty.$$

This justifies the first approximation, (2.5).

As for the second one, applying (2.8) and splitting at $\pm \varepsilon s_n$, yield

$$\left| \sum_{k=1}^{n} \left( \varphi_{X_k}(t/s_n) - \left(1 - \frac{\sigma_k^2 t^2}{2s_n^2}\right) \right) \right| \le \sum_{k=1}^{n} \left| \varphi_{X_k}(t/s_n) - \left(1 - \frac{\sigma_k^2 t^2}{2s_n^2}\right) \right|$$

$$\le \sum_{k=1}^{n} E \min \left\{ \frac{t^2 X_k^2}{s_n^2}, \frac{|t|^3 |X_k|^3}{6 s_n^3} \right\}$$

$$\le \sum_{k=1}^{n} E \frac{|t|^3 |X_k|^3}{6 s_n^3} I\{|X_k| \le \varepsilon s_n\} + \sum_{k=1}^{n} E \frac{t^2 X_k^2}{s_n^2} I\{|X_k| > \varepsilon s_n\}$$

$$\le \sum_{k=1}^{n} \frac{|t|^3 \varepsilon s_n}{6 s_n^3} E|X_k|^2 I\{|X_k| \le \varepsilon s_n\} + t^2 L_2(n) \le \frac{|t|^3 \varepsilon}{6} + t^2 L_2(n).$$

Consequently,

$$\limsup_{n \to \infty} \left| \sum_{k=1}^{n} \varphi_{X_k}(t/s_n) - \left(1 - \frac{\sigma_k^2 t^2}{2s_n^2}\right) \right|$$

$$\le \limsup_{n \to \infty} \sum_{k=1}^{n} \left| \varphi_{X_k}(t/s_n) - \left(1 - \frac{\sigma_k^2 t^2}{2s_n^2}\right) \right| \le \frac{|t|^3 \varepsilon}{6}, \qquad (2.10)$$

which, due to the arbitrariness of $\varepsilon$, proves (2.6).  $\square$

## Proof of the Sufficiency; Characteristic Functions 2

This proof starts off a little differently. Instead of the $\varphi = \exp\{\log \varphi\}$ device, one uses the following lemma, which leads to a slightly slicker, more elegant, proof.

**Lemma 2.1.** *Suppose that $\{z_k, 1 \le k \le n\}$ and $\{w_k, 1 \le k \le n\}$ are complex numbers, such that $|z_k| \le 1$ and $|w_k| \le 1$ for all $k$. Then*

$$\left| \prod_{j=1}^{n} z_j - \prod_{j=1}^{n} w_j \right| \le \sum_{k=1}^{n} |z_k - w_k|.$$

*Proof.* The proof proceeds by induction. The claim is obviously true for $n = 1$. For the induction step we set $z_n^* = \prod_{j=1}^{n} z_j$, and $w_n^* = \prod_{j=1}^{n} w_j$. Then

$$z_n^* - w_n^* = (z_n - w_n)z_{n-1}^* + w_n(z_{n-1}^* - w_{n-1}^*),$$

so that

$$|z_n^* - w_n^*| \le |z_n - w_n| \cdot |z_{n-1}^*| + |w_n| \cdot |z_{n-1}^* - w_{n-1}^*|$$
$$\le |z_n - w_n| + |z_{n-1}^* - w_{n-1}^*|. \qquad \square$$

The plan now is to exploit the lemma in proving that

$$|\varphi_{S_n/s_n}(t) - e^{-t^2/2}| \to 0 \quad \text{as} \quad n \to \infty.$$

Via a glance at the beginning of the "approximative proof" above, and the identity

$$e^{-t^2/2} = \prod_{k=1}^{n} \exp\left\{ -\frac{\sigma_k^2 t^2}{2s_n^2} \right\},$$

this is the same as proving that

$$\left| \prod_{k=1}^{n} \varphi_{X_k}(t/s_n) - \prod_{k=1}^{n} \exp\left\{ -\frac{\sigma_k^2 t^2}{2s_n^2} \right\} \right| \to 0 \quad \text{as} \quad n \to \infty.$$

However, for this it suffices, in view Lemma 2.1, to show that

$$\sum_{k=1}^{n} \left| \varphi_{X_k}(t/s_n) - \exp\left\{ -\frac{\sigma_k^2 t^2}{2s_n^2} \right\} \right| \to 0 \quad \text{as} \quad n \to \infty.$$

Now, since the first two terms in the Taylor expansions of the characteristic function and the exponential coincide, we can add and subtract them, so that, by the triangle inequality, we are done if we can show that

$$\sum_{k=1}^{n} \left| \varphi_{X_k}(t/s_n) - \left(1 - \frac{\sigma_k^2 t^2}{2s_n^2}\right) \right| \to 0 \quad \text{as} \quad n \to \infty, \qquad (2.11)$$

$$\sum_{k=1}^{n} \left| \exp\left\{ -\frac{\sigma_k^2 t^2}{2s_n^2} \right\} - \left(1 - \frac{\sigma_k^2 t^2}{2s_n^2}\right) \right| \to 0 \quad \text{as} \quad n \to \infty. \qquad (2.12)$$

However, since (2.12) actually is a particular case of (2.11), namely, the case when $X_k \in N(0, \sigma_k^2)$ for all $k$, our task finally reduces to proving (2.11) only. This, however, coincides with (2.10) from the preceding proof. $\qquad \square$

**Proof of the Sufficiency; Replacement**

In addition to our original sequence $X_1, X_2, \ldots$ we introduce a sequence of independent random variables $Y_k \in N(0, \sigma_k^2)$, $k \geq 1$, with partial sums $Z_n$, $n \geq 1$. We also assume that the two sequences are independent. The idea is to estimate the difference between the partial sums successively by exchanging one $X_k$ for one $Y_k$ at a time. For this we shall exploit the following slight variation of Definition 5.1.5:

$$X \overset{d}{=} Y \iff E\,h(X) = E\,h(Y) \quad \text{for every} \quad h \in C_B, \tag{2.13}$$

*with three bounded continuous derivatives.*

Accepting this, our object is to show that, for any such $h$,

$$E\,h(S_n/s_n) \to E\,h(N(0,1)) \quad \text{as} \quad n \to \infty. \tag{2.14}$$

By Taylor expansion and the mean value theorem,

$$h(x+u) = h(x) + u\,h'(x) + \frac{u^2}{2}h''(x) + r_0(u), \tag{2.15}$$

where $|r_0(u)| \leq C\min\{u^2, |u|^3\}$ for some fixed constant $C$.

Set, for $1 \leq j \leq n$, $n \geq 1$,

$$S_n^{(j)} = Y_1 + Y_2 + \cdots + Y_{j-1} + X_{j+1} + X_{j+2} + \cdots + X_n,$$

and note, in particular, that $X_1 + S_n^{(1)} = S_n$, that $S_n^{(n)} + Y_n = Z_n$, and that $S_n^{(j)}$, $X_j$, and $Y_j$ are independent random variables. Inserting this into (2.15) with $x = S_n^{(j)}/s_n$, for $j = 1, 2, \ldots, n$, and noticing that the terms $h(S_n^{(j)}/s_n)$ cancel for all $j$ as we pass the second inequality sign, yields

$$|E\,h(S_n/s_n) - E\,h(N(0,1))| = |E\,h(S_n/s_n) - E\,h(Z_n/s_n)|$$

$$\leq \sum_{j=1}^{n} \left| E\Big( h((S_n^{(j)} + X_k)/s_n) - h((S_n^{(j)} + Y_j)/s_n) \Big) \right|$$

$$\leq \sum_{j=1}^{n} \left| E\Big( (X_j - Y_j)/s_n \Big) \cdot h'(S_n^{(j)}/s_n) \right|$$

$$+ \sum_{j=1}^{n} \left| E\Big( (X_j^2 - Y_j^2)/2s_n^2 \Big) \cdot h''(S_n^{(j)}/s_n) \right| + \sum_{j=1}^{n} E\big(r_0(X_j) + r_0(Y_j)\big)$$

$$= \sum_{j=1}^{n} \left| E\Big( (X_j - Y_j)/s_n \Big) \cdot E\,h'(S_n^{(j)}/s_n) \right|$$

$$+ \sum_{j=1}^{n} \left| E\Big( (X_j^2 - Y_j^2)/2s_n^2 \Big) E\,h''(S_n^{(j)}/s_n) \right| + \sum_{j=1}^{n} E\big(r_0(X_j) + r_0(Y_j)\big)$$

$$= \sum_{j=1}^{n} E\big(r_0(X_j) + r_0(Y_j)\big),$$

where the last equality is a consquence of the fact that all variables have mean 0 and that the variances of $X_j$ and $Y_j$ coincide for all $j$.

It remains to investigate the remainders. Since

$$r_0(X_j) + r_0(Y_j) \leq C \min\{(X_j/s_n)^2, |X_j/s_n|^3\} + C \min\{(Y_j/s_n)^2, |Y_j/s_n|^3\},$$

our task is to show that

$$\sum_{j=1}^{n} E \min\{(X_j/s_n)^2, |X_j/s_n|^3\} \to 0 \quad \text{as} \quad n \to \infty, \tag{2.16}$$

$$\sum_{j=1}^{n} E \min\{(Y_j/s_n)^2, |Y_j/s_n|^3\} \to 0 \quad \text{as} \quad n \to \infty. \tag{2.17}$$

Before doing so we note that (2.17) is a particular case of (2.16), namely, the case when we consider normal distributions. However, remember that we assume that $X_1, X_2, \ldots$ satisfy (2.2), so that what we really have to prove is that (2.16) holds and that $Y_1, Y_2, \ldots$ satisfy (2.2).

To prove (2.16) we split, as before, at $\pm \varepsilon s_n$.

$$\sum_{j=1}^{n} E \min\{(X_j/s_n)^2, |X_j/s_n|^3\}$$

$$\leq \frac{1}{s_n^3} \sum_{j=1}^{n} E|X_j|^3 I\{|X_j| \leq \varepsilon s_n\} + \frac{1}{s_n^2} \sum_{j=1}^{n} E X_j^2 I\{|X_j| > \varepsilon s_n\}$$

$$\leq \varepsilon \frac{1}{s_n^2} \sum_{j=1}^{n} E|X_j|^2 I\{|X_j| \leq \varepsilon s_n\} + L_2(n) \leq \varepsilon + L_2(n),$$

which proves that

$$\limsup_{n \to \infty} \sum_{j=1}^{n} E \min\{(X_j/s_n)^2, |X_j/s_n|^3\} \leq \varepsilon,$$

and, hence, (2.16), in view of the arbitrariness of $\varepsilon$.

Finally, to verify (2.2) for the normal random variables we can exploit the fact that higher moments exist:

$$\frac{1}{s_n^2} \sum_{j=1}^{n} E Y_j^2 I\{|Y_j| > \varepsilon s_n\} \leq \frac{1}{\varepsilon^2 s_n^4} \sum_{j=1}^{n} E Y_j^4 I\{|Y_j| > \varepsilon s_n\} \leq \frac{1}{\varepsilon^2 s_n^4} \sum_{j=1}^{n} E Y_j^4$$

$$= \frac{1}{\varepsilon^2 s_n^4} \sum_{j=1}^{n} 3\sigma_j^4 \leq \frac{3}{\varepsilon^2 s_n^2} \max_{1 \leq j \leq n} \frac{\sigma_j^2}{s_n^2} \sum_{j=1}^{n} \sigma_j^2 = \frac{3}{\varepsilon^2} L_1(n) \to 0 \quad \text{as} \quad n \to \infty,$$

by assumption. $\square$

**Proof of the Necessity**

We thus assume that (2.1) is satisfied and that $S_n/s_n \overset{d}{\to} N(0,1)$ as $n \to \infty$, which, by Theorem 5.9.1, implies that

$$\varphi_{S_n/s_n}(t) = \prod_{k=1}^{n} \varphi_{X_k}(t/s_n) \to e^{-t^2/2} \quad \text{as} \quad n \to \infty. \qquad (2.18)$$

From the justification of (2.5), which only exploited (2.1), we can recycle

$$\sum_{k=1}^{n} |1 - \varphi_{X_k}(t/s_n)|^2 \leq \frac{t^4}{4} L_1(n) \to 0 \quad \text{as} \quad n \to \infty. \qquad (2.19)$$

Set $z = \varphi_{X_k}(t/s_n) - 1$. Then $|z| \leq 1/2$ for $n$ large, which, by (A.A.4) tells us that $|e^z - 1 - z| \leq |z|^2$, so that

$$\begin{aligned}
\big| \exp\{\varphi_{X_k}(t/s_n) - 1\} - 1 \big| &\leq \big| \exp\{\varphi_{X_k}(t/s_n) - 1\} - \varphi_{X_k}(t/s_n) \big| \\
&\quad + |\varphi_{X_k}(t/s_n) - 1| \\
&\leq |\varphi_{X_k}(t/s_n) - 1|^2 + |\varphi_{X_k}(t/s_n) - 1| \leq \frac{3}{4},
\end{aligned}$$

so that, by Lemma 2.1 with $z = \varphi_{X_k}(t/s_n)$ and $w = \exp\{\varphi_{X_k}(t/s_n) - 1\}$, we obtain, recalling (2.19),

$$\begin{aligned}
\bigg| \prod_{k=1}^{n} \varphi_{X_k}(t/s_n) &- \prod_{k=1}^{n} \exp\{\varphi_{X_k}(t/s_n) - 1\} \bigg| \\
&\leq \sum_{k=1}^{n} \big| \varphi_{X_k}(t/s_n) - \exp\{\varphi_{X_k}(t/s_n) - 1\} \big| \\
&= \sum_{k=1}^{n} \big| \exp\{\varphi_{X_k}(t/s_n) - 1\} - 1 - (\varphi_{X_k}(t/s_n) - 1) \big| \\
&\leq \sum_{k=1}^{n} |1 - \varphi_{X_k}(t/s_n)|^2 \to 0 \quad \text{as} \quad n \to \infty.
\end{aligned}$$

By the triangle inequality and (2.18), this entails that

$$\prod_{k=1}^{n} \exp\{\varphi_{X_k}(t/s_n) - 1\} \to e^{-t^2/2} \to 0 \quad \text{as} \quad n \to \infty,$$

which is the same as

$$\exp\left\{ \sum_{k=1}^{n} \left( \varphi_{X_k}(t/s_n) - 1 + \frac{\sigma_k^2 t^2}{2s_n^2} \right) \right\} \to 1 \quad \text{as} \quad n \to \infty.$$

However, the limit being real, the real part of the left-hand side must also converge to 1. Therefore, the real part of the exponent must converge to 0, viz.,

$$\sum_{k=1}^{n}\left(E\cos(X_k t/s_n) - 1 + \frac{\sigma_k^2 t^2}{2s_n^2}\right) \to 0 \quad \text{as} \quad n \to \infty.$$

Since the integrand is non-negative $(\cos x - 1 + x^2/2 \geq 0)$, the expectation restricted to any subset also converges to 0 as $n \to \infty$, so that, i.a.,

$$E\sum_{k=1}^{n}\left(\cos(X_k t/s_n) - 1 + \frac{\sigma_k^2 t^2}{2s_n^2}\right)I\{|X_k| > \varepsilon s_n\} \to 0 \quad \text{as} \quad n \to \infty,$$

which is the same as

$$\frac{t^2}{2}L_2(n) - \sum_{k=1}^{n}E\big(1 - \cos(X_k t/s_n)\big)I\{|X_k| > \varepsilon s_n\} \to 0 \quad \text{as} \quad n \to \infty.$$

Finally, the fact that $|1 - \cos y| \leq 2$ for all $y$, and Markov's inequality, together yield

$$\sum_{k=1}^{n}E\big(1 - \cos(X_k t/s_n)\big)I\{|X_k| > \varepsilon s_n\} \leq \sum_{k=1}^{n}2P(|X_k| > \varepsilon s_n)$$

$$\leq 2\sum_{k=1}^{n}\frac{\sigma_k^2}{(\varepsilon s_n)^2} = \frac{2}{\varepsilon^2},$$

so that

$$0 \leq \limsup_{n\to\infty} L_2(n) \leq \frac{4}{t^2\varepsilon^2},$$

which can be made arbitrarily small by choosing $t$ sufficiently large, and (2.2) follows. □

**Exercise 2.1.** Check the Lindeberg conditions for the i.i.d. case, and reconfirm the validity of Theorem 1.1.

**Exercise 2.2.** Reprove Theorem 1.1 with the second proof, that is, with the aid of Lemma 2.1. □

*Remark 2.2.* For the converse it is, in fact, possible to replace the assumption (2.1) with the weaker

$$\max_{1\leq k\leq n} P(|X_k/s_n| > \varepsilon) \to 0 \quad \text{as} \quad n \to \infty.$$

Namely, one then uses the inequality $\big|\varphi_{X_k}(t/s_n) - 1\big| \leq E\min\{2, |tX|\}\}$ from Theorem 5.4.1, and splits the right-hand side at $\pm\varepsilon s_n$ to prove that

$$\max_{1\leq k\leq n} |1 - \varphi_{X_k}(t/S_n)| \to 0 \quad \text{as} \quad n \to \infty,$$

and then (2.8) to prove that

$$\sum_{k=1}^{n} \left| 1 - \varphi_{X_k}(t/S_n) \right|^2 \to 0 \quad \text{as} \quad n \to \infty,$$

after which the proof proceeds as above.                                      □

**Exercise 2.3.** Write out the details.                                      □

## 2.1 Lyapounov's Condition

The Lindeberg condition may be a bit unpleasant to verify. A slightly stronger sufficient condition is the *Lyapounov condition* [178, 179].

**Theorem 2.2.** *Let* $X_1$, $X_2$, ... *be given as before, and assume, in addition, that* $E|X_k|^r < \infty$ *for all* $k$. *If, for some* $r > 2$,

$$\beta(n,r) = \frac{\sum_{k=1}^{n} E|X_k - \mu_k|^r}{s_n^r} \to 0 \quad \text{as} \quad n \to \infty, \tag{2.20}$$

*then (2.3) – the central limit theorem – holds.*

*Proof.* In view of Theorem 2.1 it suffices to show that (2.20) $\implies$ (2.2).

Toward this end, let $\varepsilon > 0$. Assuming, once again, w.l.o.g., that the means are equal to 0, we obtain

$$\frac{1}{s_n^2} \sum_{k=1}^{n} E\, X_k^2 I\{|X_k| > \varepsilon s_n\} \leq \frac{1}{s_n^2} \sum_{k=1}^{n} \frac{1}{(\varepsilon s_n)^{r-2}} E|X_k|^r I\{|X_k| > \varepsilon s_n\}$$

$$\leq \frac{1}{\varepsilon^{r-2} s_n^r} \sum_{k=1}^{n} E|X_k|^r = \frac{1}{\varepsilon^{r-2}} \beta(n,r),$$

which converges to 0 as $n \to \infty$ for any $\varepsilon > 0$.                                      □

From the formulation it is clear that there should exist cases where Lyapounov's condition is not applicable, that is, when the variance is finite, but higher-order moments are infinite.

*Example 2.1.* The "natural" example is, as often before, a Pareto-type distribution, in this case, given by the density

$$f(x) = \begin{cases} \frac{c}{|x|^3 (\log|x|)^2}, & \text{for} \quad |x| > 2, \\ 0, & \text{otherwise}, \end{cases}$$

where $c$ is a normalizing constant.

If $X$, $X_1$, $X_2$, ... are independent random variables with this density, then $\sigma^2 = \text{Var}\, X = 2c \int_2^{\infty} \frac{dx}{x(\log x)^2} < \infty$, whereas $E|X|^r = \infty$ for all $r > 2$, so that there is no Lyapounov condition.

In this case we know, of course, already from Theorem 1.1 that the central limit theorem holds. Nevertheless, condition (2.2) becomes

$$\frac{2c}{\sigma^2} \int_{\varepsilon\sigma\sqrt{n}}^{\infty} \frac{dx}{x(\log x)^2} \sim C\frac{1}{\log(\varepsilon\sqrt{n})} \to 0 \quad \text{as} \quad n \to \infty,$$

so that (2.2) is satisfied. This is no news, since the condition is always satisfied in the i.i.d. case (cf. the exercise 2 pages ago). Alternatively, since (2.1) and the central limit theorem hold, we know from Feller's converse that (2.2) must hold too.    □

**Exercise 2.4.** The i.i.d. case is the least exciting case in this context. Suppose therefore that $X_1$, $X_2$, ... are independent random variables, such that $X_k \overset{d}{=} a_k X$ for all $k$, where $X$ has the above density, and $a_k \in \mathbb{R}$. Provide conditions on the coefficients in order to exhibit a more general example than the previous one.    □

### 2.2 Remarks and Complements

In this subsection we collect some remarks and exercises around the central limit theorem.

*Remark 2.3.* There also exists an operator method due to Trotter [246], which is somewhat reminiscent of the exchange method above. See also [88], Chapters VIII and IX.

*Remark 2.4.* A consequence of the Lindeberg condition (2.2) is that

$$P(\max_{1\leq k\leq n} |X_k - \mu_k| > \varepsilon s_n) \leq P\left(\bigcup_{k=1}^{n} \{|X_k - \mu_k| > \varepsilon s_n\}\right)$$

$$\leq \sum_{k=1}^{n} P(|X_k - \mu_k| > \varepsilon s_n) \leq \frac{1}{(\varepsilon s_n)^2} \sum_{k=1}^{n} E|X_k - \mu_k|^2 I\{|X_k - \mu_k| > \varepsilon s_n\}$$

$$= \frac{1}{\varepsilon^2} L_2(n).$$

The second Lindeberg condition therefore tells us that no individual summand may dominate any other one. The summands are, what is called *uniformly asymptotically negligible*.

*Remark 2.5.* The first analysis of the assumptions lead to the conclusion (2.4), that the sum of the variances diverges. Suppose therefore, on the contrary, that

$$s_n^2 \nearrow s^2 < \infty \quad \text{as} \quad n \to \infty.$$

Then, for $n > m$

$$E(S_n - S_m)^2 = \sum_{k=m+1}^{n} \sigma_k^2 \to 0 \quad \text{as} \quad n, m \to \infty,$$

which means that $S_n$ converges in quadratic mean, and, hence, in distribution as $n \to \infty$, in addition, *without normalization*.

This implies, by the following famous theorem due to Cramér [54] that every summand must be normal, that is, every partial sum *is* normal from the beginning, which implies that there is nothing to prove.

**Theorem 2.3.** *If $X$ and $Y$ are independent, non-degenerate random variables and $X + Y$ is normal, then $X$ and $Y$ are both normal.*

We do not prove the theorem here, merely provide an outline of the arguments.

Supposing that the sum is standard normal, which is no restriction, the assumption implies that

$$\varphi_X(t) \cdot \varphi_Y(t) = \mathrm{e}^{-t^2/2}.$$

One then concludes that $\varphi_X$ and $\varphi_Y$ are entire functions of order at most 2, which, according to a theorem of Hadamard, implies that they are of the form $\exp\{g(t)\}$, where $g$ is a polynomial of degree at most equal to 2. Comparing means and variances finally proves that $\varphi_X$ and $\varphi_Y$ must be of the desired form.

Returning to our problem, we have shown that $S_n$ converges in distribution to $S$, say. The same argument applied to $S_n(j) = \sum_{k=1, k \neq j}^n X_k$ shows that $S_n(j)$ converges in distribution to $S(j)$, say, which means that

$$S(j) + X_j \overset{d}{=} S.$$

An application of Cramér's result then shows that if the limit $S$ is normal, then $X_j$ must be normal, and since $j$ is arbitrary, every summand must be normal.

The conclusion is that $s_n^2 \nearrow \infty$ is necessary in order to obtain a non-trivial result.

*Remark 2.6.* The normal distribution can be used for further comments. If, once again, $s_n^2 \nearrow s^2 < \infty$ as $n \to \infty$ and the summands are normal with mean 0, then, as we have seen, $S_n/s_n \in N(0,1)$ for all $n$, in particular, in the limit. But

$$L_2(n) = \frac{1}{s_n^2} \sum_{k=1}^n E\, X_k^2 I\{|X_k| > \varepsilon s_n\} \geq \frac{1}{s_n^2} E\, X_1^2 I\{|X_1| > \varepsilon s_n\}$$

$$\geq \frac{1}{s^2} E\, X_1^2 I\{|X_1| > \varepsilon s\} > 0,$$

that is, (2.2) is not satisfied.

Moreover, if, say, $\sigma_k^2 = \frac{1}{2^k}$, for all $k$, then $s_n^2 = 1 - \frac{1}{2^n}$, so that

$$L_1(n) = \max_{1 \leq k \leq n} \frac{\sigma_k^2}{s_n^2} = \frac{1}{1 - (1/2)^n} \to 1 \neq 0 \quad \text{as} \quad n \to \infty,$$

that is, (2.1) does not hold either.

The problem is (of course) that the summands are *not* uniformly asymptotically negligible.

Alternatively, if $X_1, X_2, \ldots$ are independent random variables, where $X_n \in \text{Po}(1/2^n)$ for $n \geq 1$, then $\sum_{n=1}^{\infty} X_k \in \text{Po}(1)$, that is, the sum converges without normalization. And not to a normal distribution.

For another example, suppose that $X_n$ is the $n$th decimal of the dyadic expansion of $Y \in U(0, 1)$. Then $P(Y \leq 1/2) = P(X_1 = 1) = 1/2$, which tells us that $X_1$, *the first decimal alone*, decides in which half of the unit interval $Y$ will end up.

*Remark 2.7.* If, instead of decreasing rapidly, the variances increase very rapidly one obtains another kind of exception. Namely, suppose that the variances increase rapidly enough to ensure that

$$\frac{\sigma_n^2}{s_n^2} \to 1 \quad \text{as} \quad n \to \infty,$$

or, equivalently, that

$$\frac{s_{n-1}^2}{s_n^2} \to 0 \quad \text{as} \quad n \to \infty.$$

Such examples appear if the variances grow faster than exponentially, for example, $\sigma_k^2 = k^k$ or $\sigma_k^2 = 2^{2^k}$. Then, for any $\varepsilon > 0$ (assuming zero means),

$$P\left(\left|\frac{S_{n-1}}{s_n}\right| > \varepsilon\right) \leq \frac{s_{n-1}^2}{\varepsilon^2 s_n^2} \to 0 \quad \text{as} \quad n \to \infty,$$

so that

$$\frac{S_{n-1}}{s_n} \xrightarrow{p} 0 \quad \text{as} \quad n \to \infty,$$

which, by Cramér's theorem, Theorem 5.11.3, shows that $X_n/s_n$ and $S_n/s_n$ have the same limit distribution (if any). Thus, assuming that the sum is asymptotically standard normal, this must also be the case for $X_n/s_n$, which, in turn, implies that

$$\frac{X_n}{\sigma_n} \xrightarrow{d} N(0, 1) \quad \text{as} \quad n \to \infty.$$

One specific example is obtained by letting $X_n = Z \cdot Y_n$ for all $n$, where

$$Z \in N(0, 1), \quad \text{and} \quad P(Y_n = 1) = 1 - 1/n, \quad P(Y_n = c_n) = 1/n, \quad n \geq 1,$$

with $c_n$ growing rapidly, and by assuming that all variables are independent – please, check the details for $c_n = n^n$ or $2^{2^n}$.

This shows that asymptotic normality is possible without (2.1). And since (2.1) does not hold neither can (2.2).

*Remark 2.8.* Another anomaly is when asymptotic normality holds and (2.1) is satisfied, but (2.2) is not. This may appear as a contradiction, but the catch is that the asymptotic distribution is not the standard normal one; the variance is not "the correct" one.

To see that this may happen, let $Y_1, Y_2, \ldots$ be independent, identically distributed random variables with mean 0 and variance 1, and let $Z_1, Z_2, \ldots$ be independent random variables, defined by

$$P(Z_k = k^2) = P(Z_k = -k^2) = \frac{1}{2k^2}, \quad P(Z_k = 0) = 1 - \frac{1}{2k^2}, \quad k \geq 1.$$

Then $Z_k$ also has mean 0 and variance 1 for each $k$. Finally, assume that the two sequences are independent and set, for $k \geq 1$, $X_k = Y_k + Z_k$, and $S_n = \sum_{k=1}^{n} X_k$, $n \geq 1$.

Then,

$$\frac{\sum_{k=1}^{n} Y_k}{\sqrt{n}} \xrightarrow{d} N(0,1) \quad \text{as} \quad n \to \infty,$$

and

$$\sum_{n=1}^{\infty} P(|Z_n| > \varepsilon) = \sum_{n=1}^{\infty} \frac{1}{n^2} < \infty,$$

so that, by the Borel-Cantelli lemma, Theorem 2.18.1, $P(Z_n \neq 0 \text{ i.o}) = 0$, which (i.a.) implies that

$$\frac{\sum_{k=1}^{n} Z_k}{\sqrt{n}} \xrightarrow{a.s.} 0 \quad \text{as} \quad n \to \infty.$$

Combining the limits of the individual sequences with Cramér's theorem, Theorem 5.11.4, we find that

$$\frac{S_n}{\sqrt{n}} \xrightarrow{d} N(0,1) \quad \text{as} \quad n \to \infty.$$

On the other hand, $X_k$ all have mean 0 and variance 2 (two!), so

$$\frac{S_n}{s_n} = \frac{S_n}{\sqrt{2n}} \xrightarrow{d} N(0, 1/2) \quad \text{as} \quad n \to \infty.$$

In this case $L_1(n) = \frac{2}{2n}$, so that (2.1) holds trivially. Since Theorem 2.1 does not hold as stated, it follows from Feller's converse that (2.2) must be violated, which shows that asymptotic normality may hold without (2.2).  □

## 2.3 Pair-wise Independence

In Chapter 1 we presented an example showing that pair-wise independence was a weaker concept than independence. In Chapter 2 we proved the same for random variables. We have also found that the variance of the sum of random variables equals the sum of the variances of the summands also for pair-wise independent random variables, in fact, even for uncorrelated ones, and (Remark 6.3.2) that the law of large numbers remains valid if the independence assumption is replaced by pair-wise independence.

The corresponding weakening of the assumption of independence is, however, *not* possible in the central limit theorem! In fact, Janson [149] provides examples of sequences of equidistributed, pair-wise independent random variables with finite (non-zero) variance that do not "properly" obey the central limit theorem.

More precisely, he presents a sequence $X_1$, $X_2$, ... of equidistributed, pair-wise independent random variables with mean 0 and variance $\sigma^2 \in (0, \infty)$, such that

$$S_n = \sum_{k=1}^{n} X_k \xrightarrow{d} S \quad \text{as} \quad n \to \infty,$$

where $S$ is a non-normal, non-degenerate distribution. This implies, in particular, that

$$\frac{S_n}{\sigma \sqrt{n}} \xrightarrow{p} 0 \quad \text{as} \quad n \to \infty.$$

In other words, the non-normalized sequence converges in distribution, and the properly normalized one is asymptotically degenerate.

One can also find examples in [149], such that

$$\frac{S_n}{\sigma \sqrt{n}} \xrightarrow{d} V \quad \text{as} \quad n \to \infty,$$

where $V$ is neither normal nor degenerate.

## 2.4 The Central Limit Theorem for Arrays

A useful extension is to consider *(triangular) arrays* of random variables instead of sequences.

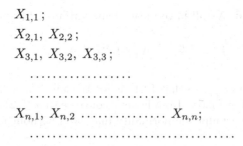

$X_{1,1}$;

$X_{2,1}$, $X_{2,2}$;

$X_{3,1}$, $X_{3,2}$, $X_{3,3}$;

. . . . . . . . . . . . . . . . .

. . . . . . . . . . . . . . . . . . . . . .

$X_{n,1}$, $X_{n,2}$ . . . . . . . . . . . . . . . $X_{n,n}$;

. . . . . . . . . . . . . . . . . . . . . . . . . . . . . . .

**Figure 7.1.** A triangular array of random variables

Thus, instead of investigating $X_1$, $X_2$, ... we define, for each $n$, sequences of random variables $X_{n,j}$, $1 \le j \le n$, and consider the sums $S_n = \sum_{k=1}^{n} X_{n,j}$, $n \ge 1$.

In the standard case the random variables in each row are independent (and identically) distributed. The distributions in different rows typically are not the same and the rows are not independent. Moreover, nothing prevents the size of the rows to be some function of the row number, $n$.

The simplest example in which the distributions in different rows differ, and where there is dependence between the rows, is obtained by considering a sequence $X_1, X_2, \ldots$ of independent, identically distributed random variables with mean 0 and finite variance, $\sigma^2$, and by putting

$$X_{n,j} = \frac{X_j}{\sigma\sqrt{n}}, \quad 1 \le j \le n, \quad \text{and} \quad S_n = \sum_{j=1}^n X_{n,j}, \quad n \ge 1.$$

Following is a restatement, or extension, of the sufficiency part of Theorem 2.1. The proof amounts to reviewing the former proof and modifying it appropriately.

**Theorem 2.4.** *Let $\{(X_{n,j}, 1 \le j \le n), n \ge 1\}$ be a triangular array of row-wise independent random variables, set $S_n = \sum_{j=1}^n X_{n,j}$, $s_n^2 = \sum_{j=1}^n \sigma_{n,j}^2$, $n \ge 1$, where $\sigma_{n,j}^2 = \operatorname{Var} X_{n,j}$, $1 \le j \le n$, and suppose, without restriction, that*

$$E\,X_{n,j} = 0 \text{ for } 1 \le j \le n, \, n \ge 1, \quad \text{and that} \quad s_n^2 = 1 \text{ for all } n. \quad (2.21)$$

*If every row satisfies the Lindeberg condition (2.2), then*

$$\frac{S_n}{s_n} \xrightarrow{d} N(0,1) \quad \text{as} \quad n \to \infty.$$

# 3 Anscombe's Theorem

We have already pointed out several times that, in practice, it is frequently more natural to study random processes during fixed time intervals, which means that the number of observations is random. Following is a central limit theorem for randomly indexed partial sums of independent, identically distributed random variables. But first, the more general result, due to Anscombe [5], which was only established in 1952.

**Theorem 3.1.** *Suppose that $Y_1, Y_2, \ldots$ are random variables, such that*

$$Y_n \xrightarrow{d} Y \quad \text{as} \quad n \to \infty,$$

*and that $\{N(t), t \ge 0\}$ is a family of positive, integer valued random variables, such that, for some family of positive reals $\{b(t), t \ge 0\}$, where $b(t) \nearrow \infty$ as $t \to \infty$,*

$$\frac{N(t)}{b(t)} \xrightarrow{p} 1 \quad \text{as} \quad t \to \infty. \quad (3.1)$$

*Finally, suppose that, given $\varepsilon > 0$, there exist $\eta > 0$ and $n_0$, such that, for all $n > n_0$,*

$$P\left(\max_{\{k:|k-n|<n\delta\}} |Y_k - Y_n| > \varepsilon\right) < \eta. \tag{3.2}$$

*Then*

$$Y_{N(t)} \xrightarrow{d} Y \quad as \ t \to \infty.$$

*Remark 3.1.* Condition (3.2) is called the *Anscombe condition*. In his paper Anscombe labels the condition as *uniform continuity in probability*.

*Remark 3.2.* The important feature of Anscombe's theorem is that *nothing is assumed about independence between the summands and the index family*.

*Remark 3.3.* Note also that the limit 1 in (3.1) is no restriction, since any other constant can be absorbed in the normalizing sequence.

*Remark 3.4.* The constant 1 can be replaced by a positive random variable; see, e.g., [24] and [253].                    □

For a proof we refer to the original paper. The special case that concerns us here is the following version, which was first given with a direct proof by Rényi [204]. The essence is that, instead of verifying the Anscombe condition, Rényi provides a direct proof (which essentially amounts to the same work).

**Theorem 3.2.** *Let $X, X_1, X_2, \ldots$ be independent, identically distributed random variables with mean 0 and positive, finite, variance $\sigma^2$, and set $S_n = \sum_{k=1}^{n} X_k$, $n \geq 1$. Suppose that $\{N(t), t \geq 0\}$ is a family of positive, integer valued random variables, such that, for some $0 < \theta < \infty$,*

$$\frac{N(t)}{t} \xrightarrow{p} \theta \quad as \ t \to \infty. \tag{3.3}$$

*Then*

$$\frac{S_{N(t)}}{\sigma\sqrt{N(t)}} \xrightarrow{d} N(0,1) \quad as \ t \to \infty,$$

$$\frac{S_{N(t)}}{\sigma\sqrt{\theta t}} \xrightarrow{d} N(0,1) \quad as \ t \to \infty.$$

*Proof.* We assume w.l.o.g. that $\sigma^2 = \theta = 1$ throughout the proof. With $n_0 = [nt]$ we then obtain

$$\frac{S_{N(t)}}{\sqrt{N(t)}} = \left(\frac{S_{n_0}}{\sqrt{n_0}} + \frac{S_{N(t)} - S_{n_0}}{\sqrt{n_0}}\right)\sqrt{\frac{n_0}{N(t)}}.$$

Now, by assumption,

$$\frac{S_{n_0}}{\sqrt{n_0}} \overset{d}{\to} N(0,1) \quad \text{and} \quad \frac{N(t)}{n_0} \overset{p}{\to} 1 \quad \text{as} \quad t \to \infty,$$

so that, in view of Theorem 5.10.2, and Cramér's theorem 5.11.4 (twice), it remains to prove that

$$\frac{S_{N(t)} - S_{n_0}}{\sqrt{n_0}} \overset{p}{\to} 0 \quad \text{as} \quad t \to \infty$$

for the first claim, after which another application of Cramér's theorem yields the second one.

Let $\varepsilon \in (0, 1/3)$ be given and set $n_1 = [n_0(1-\varepsilon^3)]+1$ and $n_2 = [n_0(1+\varepsilon^3)]$. Then, by the Kolmogorov inequality, Theorem 3.1.6,

$$
\begin{aligned}
P(|S_{N(t)} - S_{n_0}| > \varepsilon\sqrt{n_0}) &= P(\{|S_{N(t)} - S_{n_0}| > \varepsilon\sqrt{n_0}\} \cap \{N(t) \in [n_1, n_2]\}) \\
&\quad + P(\{|S_{N(t)} - S_{n_0}| > \varepsilon\sqrt{n_0}\} \cap \{N(t) \notin [n_1, n_2]\}) \\
&\leq P(\max_{n_1 \leq k \leq n_0} |S_k - S_{n_0}| > \varepsilon\sqrt{n_0}) + P(\max_{n_0 \leq k \leq n_2} |S_k - S_{n_0}| > \varepsilon\sqrt{n_0}) \\
&\quad + P(N(t) \notin [n_1, n_2]) \\
&\leq \frac{n_0 - n_1}{\varepsilon^2 n_0} + \frac{n_2 - n_0}{\varepsilon^2 n_0} + P(N(t) \notin [n_1, n_2]) \\
&= \frac{n_0 - [n_0(1-\varepsilon^3)] - 1}{\varepsilon^2 n_0} + \frac{[n_0[1+\varepsilon^3)] - n_0}{\varepsilon^2 n_0} + P(N(t) \notin [n_1, n_2]),
\end{aligned}
$$

so that, recalling (3.3),

$$\limsup_{t \to \infty} P(|S_{N(t)} - S_{n_0}| > \varepsilon\sqrt{n_0}) < 2\varepsilon,$$

which, due to the arbitrariness of $\varepsilon$ proves the conclusion. □

**Exercise 3.1.** Prove the theorem via a direct verification of the Anscombe condition (3.2). □

For the law of large numbers it was sufficient that $N(t) \overset{a.s.}{\to} +\infty$ as $t \to \infty$. That this is not enough for a "random-sum central limit theorem" can be seen as follows.

*Example 3.1.* Let $X, X_1, X_2, \ldots$ be independent coin-tossing random variables; $P(X = 1) = P(X = -1) = 1/2$. Let $S_n$, $n \geq 1$, be the partial sums and let $N(n)$, $n \geq 1$, be the index of $S_n$ at the time of the $n$th visit to 0. It follows from random walk theory (recall Subsection 2.18.4) that $P(S_n = 0 \text{ i.o.}) = 1$, so that $N(n) \overset{a.s.}{\to} \infty$ as $n \to \infty$. However

$$\frac{S_{N(n)}}{\sqrt{N(n)}} = 0 \quad \text{for all} \quad n,$$

which is far from asymptotic normality. Thus, something more than $N(t) \overset{a.s.}{\to} +\infty$ as $t \to \infty$ is necessary. □

There also exists a version for sums of non-identically distributed random variables based on the Lindeberg conditions. For this we need the following *generalized Anscombe condition*: A sequence $Y_1, Y_2, \ldots$ satisfies the generalized Anscombe condition with norming sequence $\{k_n, n \geq 1\}$ if, for every $\varepsilon > 0$, there exists $\delta > 0$, such that

$$\limsup_{n \to \infty} P\left( \max_{\{j : |k_j^2 - k_n^2| \leq \delta k_n^2\}} |Y_j - Y_n| > \varepsilon \right) < \varepsilon). \tag{3.4}$$

**Theorem 3.3.** *Let $X_1, X_2, \ldots$ be independent random variables with finite variances, and set, for $k \geq 1$, $E\,X_k = \mu_k$, $\mathrm{Var}\,X_k = \sigma_k^2$, and, for $n \geq 1$, $S_n = \sum_{k=1}^n X_k$, and $s_n^2 = \sum_{k=1}^n \sigma_k^2$. Suppose that the Lindeberg conditions (2.1) and (2.2) are satisfied, that $\{(S_n - \sum_{k=1}^n \mu_k)/s_n, n \geq 1\}$ satisfies the generalized Anscombe condition for some normalizing sequence $\{k_n, n \geq 1\}$, and that $\{N_n, n \geq 1\}$ is a sequence of positive, integer valued random variables, such that*

$$\frac{k_{N(n)}}{k_{a_n}} \xrightarrow{p} 1 \quad as \quad n \to \infty, \tag{3.5}$$

*for some sequence $\{a_n, n \geq 1\}$ of positive integers increasing to $+\infty$. Then,*

$$\frac{S_{N_n} - \sum_{k=1}^{N_n} \mu_k}{s_{N_n}} \xrightarrow{d} N(0, 1) \quad as \quad n \to \infty.$$

The formulation looks more technical and involved than it really is. We first assume that enough is satisfied in order for Theorem 2.1 to apply. Moreover, the set of indices $\{j : |j - n| < \delta n\}$ must be modified into some kind of non-uniformity, due to fact that the summands are non-identically distributed (in the same way as $n\sigma^2$ is modified into $s_n^2$). Note that in the i.i.d.-case, we may choose $k_n = n$ and $a_n = n$ or $\theta n$ (apart from the fact that it should be an integer).

For more, see [57, 58] and references given there.

**Exercise 3.2.** Prove the theorem.                                  □

*Remark 3.5.* For an application to record times, see Subsection 7.4.4.     □


# 4 Applications

In this subsection we provide proofs of the asymptotic normality of the first passage times of a random walk across a horizontal boundary, and of the record times and the associated counting process. In both cases the random index technique that was used in connection with the law of large numbers in the previous chapter will be exploited. Another application is a nice proof of Stirling's formula, due to Khan [157]. But first, a presentation of the so-called Delta method.

## 4.1 The Delta Method

This is a method to prove convergence of functions of arithmetic means of independent, identically distributed random variables.

Suppose that $X_1, X_2, \ldots$ are independent, identically distributed random variables with mean $\mu$ and variance $\sigma^2$, and set $S_n = \sum_{k=1}^n X_k$, $n \geq 1$. Further, suppose that $g$ is a real valued, continuously differentiable function, such that

$$g'(\mu) \neq 0.$$

Then, by Taylor expansion and the mean value theorem,

$$g\left(\frac{S_n}{n}\right) = g(\mu) + \left(\frac{S_n}{n} - \mu\right)g'(\theta_n)$$

where $|\theta_n - \frac{S_n}{n}| \leq |\mu - \frac{S_n}{n}|$. This can be rewritten as

$$\sqrt{n}\left(g\left(\frac{S_n}{n}\right) - g(\mu)\right) = \sqrt{n}\left(\frac{S_n}{n} - \mu\right)g'(\theta_n).$$

Since $\sqrt{n}(\frac{S_n}{n} - \mu) \overset{d}{\to} N(0, \sigma^2)$, and $g'(\theta_n) \overset{a.s.}{\to} g'(\mu)$ as $n \to \infty$, we conclude, via Cramér's theorem, Theorem 5.11.3, that

$$\sqrt{n}\left(g\left(\frac{S_n}{n}\right) - g(\mu)\right) \overset{d}{\to} N\left(0, \sigma^2(g'(\mu))^2\right) \quad \text{as} \quad n \to \infty. \tag{4.1}$$

If, on the other hand, $g'(\mu) = 0$, $g$ is twice continuously differentiable, and

$$g''(\mu) \neq 0,$$

then, adding one term in the Taylor expansion, yields

$$g\left(\frac{S_n}{n}\right) = g(\mu) + \left(\frac{S_n}{n} - \mu\right)g'(\mu) + \frac{1}{2}\left(\frac{S_n}{n} - \mu\right)^2 g''(\theta_n)$$
$$= g(\mu) + \frac{1}{2}\left(\frac{S_n}{n} - \mu\right)^2 g''(\theta_n),$$

where, again, $|\theta_n - \frac{S_n}{n}| \leq |\mu - \frac{S_n}{n}|$. In this case this is the same as

$$n\left(g\left(\frac{S_n}{n}\right) - g(\mu)\right) = \left(\frac{S_n - n\mu}{\sigma\sqrt{n}}\right)^2 \cdot \sigma^2 g''(\theta_n).$$

Since $g(x) = x^2$ is continuous it follows that $\left(\frac{S_n - n\mu}{\sigma\sqrt{n}}\right)^2 \overset{d}{\to} \chi^2(1)$ as $n \to \infty$, so that

$$n\left(g\left(\frac{S_n}{n}\right) - g(\mu)\right) \overset{d}{\to} \sigma^2 g''(\mu)\chi^2(1) \quad \text{as} \quad n \to \infty. \tag{4.2}$$

**Exercise 4.1.** These proofs lean on the central limit theorem and Cramér's theorem. An alternative would be to exploit the deeper Skorohod representation from Section 5.13. Re-prove the results with this approach.

**Exercise 4.2.** Let $X_1, X_2, \ldots$ be independent, identically distributed random variables with finite variance, and set $S_n = \sum_{k=1}^n X_k$, $n \geq 1$. Prove that $S_n^2$, appropriately normalized, converges in distribution as $n \to \infty$. □

## 4.2 Stirling's Formula

There exist many more-or-less complicated integrals, sums, or other expressions that a mathematician has to encounter. As probabilists we often recognize such objects as being equal to 1 or to the expected value of some random variable, and so on. A beautiful example was the relation $\frac{\sin t}{t} = \prod_{k=1}^{\infty} \cos\left(\frac{t}{2^k}\right)$ in Subsection 4.2.2.

The following lines provide an outline of a probabilistic proof of Stirling's formula via convergence in distribution and moment convergence. Namely, let $X_1, X_2, \ldots$ be independent standard exponential random variables, and set $S_n = \sum_{k=1}^{n}, n \geq 1$. The central limit theorem tells us that

$$U_n = \frac{S_n - n}{\sqrt{n}} \overset{d}{\to} N(0,1) \quad \text{as} \quad n \to \infty,$$

and, since, for example, $E\,U_n^2 = 1$ for all $n$ (so that the second moments are uniformly bounded), it follows from Theorem 5.4.2 that all lower powers are uniformly integrable, and (Theorem 5.5.9) that moments of lower order converge to those of the standard normal distribution. This implies, in particular, that

$$\lim_{n\to\infty} E|U_n| = E|N(0,1)| = \sqrt{\frac{2}{\pi}}.$$

Since $U_n$ is a scaled gamma distribution we can spell out the limit relation exactly:

$$\int_0^\infty \left|\frac{x-n}{\sqrt{n}}\right| \frac{1}{\Gamma(n)} x^{n-1} e^{-x}\,dx \to \sqrt{\frac{2}{\pi}} \quad \text{as} \quad n \to \infty.$$

Splitting the integral into two at $x = n$, and a change of variable, $u = x/n$, followed by some additional computations finally lead to Stirling's formula

$$\lim_{n\to\infty} \frac{\left(\frac{n}{e}\right)^n \sqrt{2n\pi}}{n!} = 1.$$

**Exercise 4.3.** For details, see [157] or, better still, carry out the program.    □

## 4.3 Renewal Theory for Random Walks

This topic was introduced in Subsection 2.16.3: Let $X, X_1, X_2, \ldots$ be independent, identically distributed random variables with positive, finite, mean $\mu$, partial sums $S_n, n \geq 1$, and the associated first passage process, $\{\tau(t), t \geq 0\}$, defined by

$$\tau(t) = \min\{n : S_n > t\}, \quad t \geq 0.$$

In Subsection 6.9.3 we proved a strong law – $\frac{\tau(t)}{t} \overset{a.s.}{\to} \frac{1}{\mu}$ as $t \to \infty$. Here is the corresponding central limit theorem.

**Theorem 4.1.** *If, in addition* $\operatorname{Var} X = \sigma^2 < \infty$, *then*

$$\frac{\tau(t) - t/\mu}{\sqrt{\frac{\sigma^2 t}{\mu^3}}} \xrightarrow{d} N(0,1) \quad as \quad t \to \infty.$$

*Proof.* The central limit theorem and Anscombe's theorem together yield

$$\frac{S_{\tau(t)} - \mu\tau(t)}{\sqrt{\sigma^2 \tau(t)}} \xrightarrow{d} N(0,1) \quad as \quad t \to \infty.$$

By Theorem 6.8.2(i), the sandwich formula $t < S_{\tau(t)} \leq t + X_{\tau(t)}$, (recall (6.9.3)), and Cramér's theorem, Theorem 5.11.3, we next obtain

$$\frac{t - \mu\tau(t)}{\sqrt{\sigma^2 \tau(t)}} \xrightarrow{d} N(0,1) \quad as \quad t \to \infty.$$

The strong law, Theorem 6.9.3, another application of Cramér's theorem, and the symmetry of the normal distribution finish the proof.                           □

**Exercise 4.4.** Spell out the details.                                         □

If the summands are positive, we remember from (2.16.1) that $\{S_n, n \geq 1\}$ and the counting process $\{N(t), t \geq 0\}$ are inverses of each other, and in this case, asymptotic normality was originally proved in [85, 241] in the lattice case and the continuous case, respectively, by exploiting this relationship. The advantage with the above proof is that it works, not only for both cases simultaneously, but also, as we have just seen, for random walks.

## 4.4 Records

Let $X_1, X_2, \ldots$ be independent, identically distributed, continuous random variables. From Subsection 2.18.3 we recall that the record times are $L(1) = 1$ and, recursively,

$$L(n) = \min\{k : X_k > X_{L(n-1)}\}, \quad n \geq 2,$$

and that the associated counting process $\{\mu(n), n \geq 1\}$ was defined by

$$\mu(n) = \# \text{records among } X_1, X_2, \ldots, X_n = \max\{k : L(k) \leq n\}.$$

In the previous chapter we proved a strong law for the record times and the counting processes. Now we are ready for the central limit theorem.

**Theorem 4.2.** *We have*

$$\frac{\mu(n) - \log n}{\sqrt{\log n}} \xrightarrow{d} N(0,1) \quad as \quad n \to \infty.$$

*Proof.* One way to prove the theorem is to check the Lyapounov condition, another proof can be obtained via characteristic functions or moment generating functions; note, in particular, that the random variables $|I_k - \frac{1}{k}|$, $k \geq 1$, are uniformly bounded.     □

**Exercise 4.5.** Carry out the details.     □

The corresponding result for the sequence of record times, originally due to Rényi [205], and proved via inversion, runs as follows.

**Theorem 4.3.** *We have*

$$\frac{\log L(n) - n}{\sqrt{n}} \xrightarrow{d} N(0,1) \quad as \quad n \to \infty.$$

*Proof.* The proof follows the same pattern as the proof of Theorem 4.1. The main difference is that here we need the more general Anscombe theorem, Theorem 3.3.

In view of Theorems 4.2 and 3.3, it follows that

$$\frac{\mu(L(n)) - \log L(n)}{\sqrt{\log L(n)}} \xrightarrow{d} N(0,1) \quad as \quad n \to \infty,$$

(where we leave the verification of the generalized Anscombe condition (3.4) to the reader).

From Subsection 6.9.4 we then remember that $\mu(L(n)) = n$, which, implies that

$$\frac{n - \log L(n)}{\sqrt{\log L(n)}} \xrightarrow{d} N(0,1) \quad as \quad n \to \infty.$$

The strong law for record times, Theorem 6.9.4, the symmetry of the normal distribution, and an application of Cramér's theorem, Theorem 5.11.3 finish the proof.     □

# 5 Uniform Integrability; Moment Convergence

In Section 5.5 we presented some results on moment convergence, and showed that uniformly integrable sequences that converge almost surely, in probability, or in distribution also converge in $L^1$. In Subsection 7.4.2 we made use of such results in order to provide a probabilistic proof of Stirling's formula. In this subsection we prove the optimal result in the i.i.d. case, namely that if the summands of a sequence of independent, identically distributed random variables have a finite moment of order at least 2, then we also have moment convergence of the same order in the central limit theorem. Since uniformly *bounded* moments of some order always imply convergence of moments of *lower* order, the optimality lies in the fact that existence of moments of a given order implies uniform integrability of *the same* order.

**Theorem 5.1.** *Let $X, X_1, X_2, \ldots$ be independent identically distributed random variables, and set $S_n = X_1 + X_2 + \cdots + X_n$, $n \geq 1$. Suppose that $E|X|^r < \infty$ for some $r \geq 2$, and set $EX = \mu$, and $\operatorname{Var} X = \sigma^2$. Then*

$$\frac{S_n - n\mu}{\sigma\sqrt{n}} \xrightarrow{r} N(0,1) \quad as \quad n \to \infty,$$

$$E\left|\frac{S_n - n\mu}{\sigma\sqrt{n}}\right|^r \to E|N(0,1)|^r \quad as \quad n \to \infty.$$

*Remark 5.1.* Before turning our attention to the proof, we remark that if $E|X|^r < \infty$, then $\sup_n E|\frac{S_n - n\mu}{\sqrt{n}}|^r < \infty$ by the corollary to the Marcinkiewicz-Zygmund inequalities, Corollary 3.8.2. The proof thus amounts to improving our knowledge from uniform boundedness to uniform integrability. □

*Proof.* Once again we assume w.l.o.g. that $\mu = 0$ and that $\sigma^2 = 1$. In view of Theorem 5.5.9 we must (and it suffices to) show that

$$\left\{\left|\frac{S_n}{\sqrt{n}}\right|^r, n \geq 1\right\} \quad \text{is uniformly integrable.}$$

Let $\varepsilon > 0$, choose $A$ large enough to ensure that

$$E|X|^r I\{|X| > A\} < \varepsilon, \tag{5.1}$$

set

$$X_k' = X_k I\{|X| \leq A\} - E(X_k I\{|X| \leq A\}), \quad \text{and} \quad S_n' = \sum_{k=1}^n X_k',$$

$$X_k'' = X_k I\{|X| > A\} - E(X_k I\{|X| > A\}), \quad \text{and} \quad S_n'' = \sum_{k=1}^n X_k'',$$

and note that $E X_k' = E X_k'' = 0$, that $X_k' + X_k'' = X_k$, and that $S_n' + S_n'' = S_n$. Let $a > 0$. To take care of $S_n'$ we note that

$$E\left|\frac{S_n'}{\sqrt{n}}\right|^r I\left\{\left|\frac{S_n'}{\sqrt{n}}\right| > a\right\} \leq \frac{1}{a^r} E\left|\frac{S_n'}{\sqrt{n}}\right|^{2r} I\left\{\left|\frac{S_n'}{\sqrt{n}}\right| > a\right\} \leq \frac{1}{a^r} E\left|\frac{S_n'}{\sqrt{n}}\right|^{2r}$$

$$\leq \frac{1}{(na)^r} B_{2r} n^r E|X_1'|^{2r} \leq \frac{B_{2r}(2A)^{2r}}{a^r}, \tag{5.2}$$

where we have used Corollary 3.8.2 in the second-last inequality, and the uniform boundedness of the primed summands in the last one.

As for $S_n''$, the same Marcinkiewicz-Zygmund inequality and (5.1) yield

$$E\left|\frac{S_n''}{\sqrt{n}}\right|^r I\left\{\left|\frac{S_n''}{\sqrt{n}}\right| > a\right\} \leq E\left|\frac{S_n''}{\sqrt{n}}\right|^r \leq \frac{B_r n^{r/2} E|X_1''|^r}{n^{r/2}} \leq B_r 2^r \varepsilon, \tag{5.3}$$

where the factor $2^r$ follows from an application of the $c_r$-inequalities, Theorem 3.2.2:

$$E|X_1''|^r = E|X_1I\{|X_1| > A\} - E(X_1I\{|X_1| > A\})|^r$$
$$\leq 2^{r-1}2E|X_1I\{|X_1| > A\}|^r < 2^r\varepsilon.$$

By joining (5.2) and (5.3) we obtain, via a glance at the proof of Theorem 5.4.6, that

$$E\left|\frac{S_n}{\sqrt{n}}\right|^r I\left\{\left|\frac{S_n}{\sqrt{n}}\right| > 2a\right\} \leq E\left(\left|\frac{S_n'}{\sqrt{n}}\right| + \left|\frac{S_n'}{\sqrt{n}}\right|\right)^r I\left\{\left|\frac{S_n'}{\sqrt{n}}\right| + \left|\frac{S_n''}{\sqrt{n}}\right| > 2a\right\}$$

$$\leq 2^r E\left|\frac{S_n'}{\sqrt{n}}\right|^r I\left\{\left|\frac{S_n'}{\sqrt{n}}\right| > a\right\} + 2^r E\left|\frac{S_n''}{\sqrt{n}}\right|^r I\left\{\left|\frac{S_n''}{\sqrt{n}}\right| > a\right\}$$

$$\leq \frac{2^r B_{2r} M^{2r}}{a^r} + 2^{2r} B_r \varepsilon.$$

It follows that

$$\limsup_{a \to \infty} E\left|\frac{S_n}{\sqrt{n}}\right|^r I\left\{\left|\frac{S_n}{\sqrt{n}}\right| > 2a\right\} \leq 2^{2r} B_r \varepsilon,$$

*independently of*, and hence, *uniformly in*, n, which, due to the arbitrariness of $\varepsilon$, proves the desired uniform integrability, and the proof is complete.  □

Returning for a second to the applications to renewal theory and records, we mention that there exist uniform integrability results and theorems on moment convergence.

In the case of renewal theory for random walks the results were first proved in [46]; see also [110], Chapter III.

For the record counting process the result is immediate, since the moment generating function of $\frac{\mu(n) - \log n}{\sqrt{\log n}}$ converges – Theorem 5.9.5 – and for the record times the prettiest way is to apply a beautiful result due to Williams [251], according to which the record times can be approximated by $\Gamma$-distributed random variables, and then apply Theorem 5.1. For details, see [111]. Williams' theorem can also be used to prove many other results for record times, such as the strong law and the asymptotic normality we have proved via "$S_N$-sums".

# 6 Remainder Term Estimates

Let $X_1, X_2, \ldots$ be independent, identically distributed random variables with finite expectation $\mu$ and finite variance $\sigma^2$, and set $S_n = X_1 + X_2 + \cdots + X_n$, $n \geq 1$. The message of the (weak) law of large numbers is that $\bar{X}_n \overset{p}{\to} \mu$ as $n \to \infty$ for any $\varepsilon > 0$, which means that $\bar{X}_n - \mu$ is "small" (with high probability) when $n$ is "large". The central limit theorem tells us that $\sigma^{-1}\sqrt{n}(\bar{X}_n - \mu) \overset{d}{\to} N$

as $n \to \infty$, where $N \in N(0,1)$, which means that "$\bar{X}_n - \mu \approx N\sigma/\sqrt{n}$" when $n$ is "large". The central limit theorem thus provides information on the *rate of convergence* in the law of large numbers.

A natural next step concerns the closeness between the distribution of $\bar{X}_n - \mu$ and the appropriate normal distribution, which means asking for the *rate of convergence* in the central limit theorem. This is the topic of the present section. Some references are the original works by Berry [17] and Esseen [76, 77], and the books [88, 98, 194, 195].

## 6.1 The Berry-Esseen Theorem

The following result was proved independently by Berry [17] and Esseen [76]. The reason that the authors were unaware of each other is that this was during the Second World War when journals were not free to travel.

**Theorem 6.1.** *Let $X$, $X_1$, $X_2$, ... be independent, identically distributed random variables with partial sums $\{S_n, n \geq 1\}$, set $\mu = EX$, $\sigma^2 = \mathrm{Var}\,X$, and suppose that $\gamma^3 = E|X|^3 < \infty$. Then*

$$\sup_x |F_{\frac{S_n - n\mu}{\sigma\sqrt{n}}}(x) - \Phi(x)| \leq C \cdot \frac{\gamma^3}{\sigma^3\sqrt{n}},$$

*where $C$ is a purely numerical constant.*

Much work has been devoted to the search for an exact value of the constant. The current best upper bounds are 0.7975 [14] and 0.7655 [220]. Recent bounds depending on $n$ have been proved in [40]. For $n \geq 65$ the bound is smaller that 0.7655, and approaches 0.7164 as $n \to \infty$.

The other half of the competition is devoted to lower bounds. Toward that end, Esseen [78] first noticed that if $X$ puts mass $1/2$ at $\pm 1/2$, and $X$, $X_1$, $X_2$, ... are independent, identically distributed random variables, then

$$\lim_{n\to\infty} \sqrt{n} \sup_x |F_{\frac{S_n - n\mu}{\sigma\sqrt{n}}}(x) - \Phi(x)| = \lim_{n\to\infty} \sqrt{n} \sup_x |F_{\frac{S_n}{\frac{1}{2}\sqrt{n}}}(x) - \Phi(x)| = \frac{1}{\sqrt{2\pi}},$$

and, after a further analysis, that the modified two-point distribution

$$P(X = -[(4-\sqrt{10})/2]h) = \frac{\sqrt{10}-2}{2}, \quad P(X = [(\sqrt{10}-2)/2]h) = \frac{4-\sqrt{10}}{2},$$

where $h > 0$ is some parameter, yields

$$\lim_{n\to\infty} \sqrt{n} \sup_x |F_{\frac{S_n - n\mu}{\sigma\sqrt{n}}}(x) - \Phi(x)| = \frac{\sqrt{10}+3}{6\sqrt{2\pi}} \approx 0.4097.$$

This means that currently (June 2004)

$$0.4097 \leq C \leq 0.7655.$$

Since the proof of the theorem is very much the same as the analog for independent, not necessarily identically distributed summands, we only prove the latter, which is stated next. The proof will be given in Subsection 7.6.2, after some additional remarks.

**Theorem 6.2.** *Let $X_1$, $X_2$, ... be independent random variables with zero mean and with partial sums $\{S_n, n \geq 1\}$. Suppose that $\gamma_k^3 = E|X_k|^3 < \infty$ for all $k$, and set $\sigma_k^2 = \operatorname{Var} X_k$, $s_n^2 = \sum_{k=1}^n \sigma_k^2$, and $\beta_n^3 = \sum_{k=1}^n \gamma_k^3$. Then*

$$\sup_x |F_{\frac{S_n}{s_n}}(x) - \Phi(x)| \leq C \cdot \frac{\beta_n^3}{s_n^3},$$

*where $C$ is a purely numerical constant.*    □

*Remark 6.1.* We shall not aim at finding best constants, merely establish their existence. The proof below produces a constant slightly smaller than 36. The upper bound 0.7915 in Theorem 6.2 is given in [220].    □

Before proving anything we quote some extensions to moments of order $2 + \delta$ for some $\delta \in (0, 1)$; [154, 193] and [195], Theorem 5.6, p. 151.

**Theorem 6.3.** *Let $X_1$, $X_2$, ... be independent, with zero mean, and set $S_n = \sum_{k=1}^n X_k$, and $s_n^2 = \sum_{k=1}^n \operatorname{Var} X_k^2$, $n \geq 1$. Further, let $\mathfrak{G}$ denote the class of functions on $\mathbb{R}$, which are non-negative, even, and such that $x/g(x)$ and $g(x)$ are non-decreasing on $\mathbb{R}^+$. If, for some function $g \in \mathfrak{G}$,*

$$E X_k^2 g(X_k) < \infty \quad \text{for all} \quad k,$$

*then*

$$\sup_x |F_{\frac{S_n}{s_n}}(x) - \Phi(x)| \leq C \cdot \frac{\sum_{k=1}^n E X_k^2 g(X_k)}{s_n^2 g(s_n)},$$

*where $C$ is a purely numerical constant.*

*In particular, if $X$, $X_1$, $X_2$, ..., in addition, are identically distributed with common variance 1, then*

$$\sup_x |F_{\frac{S_n}{\sqrt{n}}}(x) - \Phi(x)| \leq C \cdot \frac{E X^2 g(X)}{g(\sqrt{n})}.$$

The ultimate goal would be to move down to nothing more than second moments, but this is not feasible, since finite variances only yield *existence* of the limit. More has to be assumed in order to obtain a rate result. However, in [81] existence of higher-order moments is replaced by assumptions on *truncated* moments.

**Theorem 6.4.** *Let $X_1$, $X_2$, ... be independent random variables with zero mean and partial sums $\{S_n, n \geq 1\}$. Set $\sigma_k^2 = \operatorname{Var} X_k$, $s_n^2 = \sum_{k=1}^n \sigma_k^2$, and*

$$\rho_k = \sup_{x>0} \left( E|X_k|^3 I\{|X| \leq x\} + x E|X_k|^2 I\{|X| \geq x\} \right), \quad k = 1, 2, \ldots, n.$$

*Then*

$$\sup_x \left| F_{\frac{S_n}{s_n}}(x) - \Phi(x) \right| \le C \cdot \frac{\sum_{k=1}^n \rho_k}{s_n^3},$$

*where $C$ is a purely numerical constant.*

## 6.2 Proof of the Berry-Esseen Theorem 6.2

We are thus given independent random variables $X_1, X_2, \ldots$ with partial sums $\{S_n, n \ge 1\}$. As for notation, recall that

$$\sigma_k^2 = \operatorname{Var} X_k, \quad s_n^2 = \sum_{k=1}^n \sigma_k^2, \quad \gamma_k^3 = E|X_k|^3 < \infty, \quad \beta_n^3 = \sum_{k=1}^n \gamma_k^3.$$

Moreover, by Lyapounov's inequality, Theorem 3.2.5,

$$\sigma_k \le \gamma_k \quad \text{for all} \quad k.$$

The proof is based on characteristic functions. Just as a continuity theorem translates or inverts convergence to distributional convergence, the essential tool here is a result that translates (inverts) distance between characteristic functions to distance between distributions. The following *Esseen's lemma*, [76], Theorem 1, and [77], Theorem 2a, does that for us.

**Lemma 6.1.** *Let $U$ and $V$ be random variables, and suppose that*

$$\sup_{x \in \mathbb{R}} F_V'(x) \le A. \tag{6.1}$$

*Then*

$$\sup_x |F_U(x) - F_V(x)| \le \frac{1}{\pi} \int_{-T}^T \left| \frac{\varphi_U(t) - \varphi_V(t)}{t} \right| \left( 1 - \frac{|t|}{T} \right) dt + \frac{24A}{\pi T}$$

$$\le \frac{1}{\pi} \int_{-T}^T \left| \frac{\varphi_U(t) - \varphi_V(t)}{t} \right| dt + \frac{24A}{\pi T}.$$

*Proof.* From the early examples in Chapter 4 and scaling (Theorem 4.1.10) it follows that a $U(-T, T)$-distributed random variable has density $(1-|x/T|)/T$ for $|x| \le T$, and 0 otherwise, and characteristic function $\left( \sin(tT/2)/(tT/2) \right)^2$. The inversion formula, Theorem 4.1.4, therefore tells us that

$$\frac{1}{T}\left(1 - \frac{|x|}{T}\right) = \frac{1}{2\pi} \int_{-\infty}^\infty e^{-itx} \frac{4\sin^2(tT/2)}{(tT)^2} dt,$$

so that, by symmetry, and by letting $x$ and $t$ switch roles,

$$1 - \frac{|t|}{T} = \int_{-\infty}^\infty e^{ixt} \frac{2\sin^2(xT/2)}{\pi(xT)^2} dx = \int_{-\infty}^\infty e^{ixt} \frac{1 - \cos xT}{\pi T x^2} dx.$$

Let $Z_T$ be a random variable with density $\frac{1-\cos xT}{\pi T x^2}$ and characteristic function $1 - \frac{|t|}{T}$, $|t| \leq T$.

We are now all set for the details of the proof. First of all we may assume that

$$\int_{-T}^{T} \left| \frac{\varphi_U(t) - \varphi_V(t)}{t} \right| dt < \infty,$$

since otherwise there is nothing to prove.

Set

$$\Delta(x) = F_U(x) - F_V(x),$$

$$\Delta_T(x) = \int_{-\infty}^{\infty} \Delta(x-y) f_{Z_T}(y) \, dy = F_{U+Z_T}(x) - F_{V+Z_T}(x),$$

$$\Delta^* = \sup_{x \in \mathbb{R}} |\Delta(x)| \quad \text{and} \quad \Delta_T^* = \sup_{x \in \mathbb{R}} |\Delta_T(x)|.$$

Since $U + Z_T$ and $V + Z_T$ are absolutely continuous random variables (recall Subsection 4.2.2) it follows that the "characteristic function" corresponding to $\Delta_T(x)$ equals

$$\varphi_U(t)\varphi_{Z_T}(t) - \varphi_V(t)\varphi_{Z_T}(t) = (\varphi_U(t) - \varphi_V(t))\varphi_{Z_T}(t), \quad |t| \leq T,$$

and 0 otherwise. The quotation marks are there because the transform is a difference between the characteristic functions of $U + Z_T$ and $V + Z_T$. Note also that $\varphi_{Z_T}(t) \to 1$ as $T \to \infty$, so that $Z_T \xrightarrow{p} 0$ as $T \to \infty$, which, by Cramér's theorem, implies that $U + Z_T \xrightarrow{p} U$ and $V + Z_T \xrightarrow{p} V$ as $T \to \infty$, and, hence, that

$$\Delta_T(x) \to \Delta(x) \quad \text{as} \quad T \to \infty \quad \text{for all} \quad x, \tag{6.2}$$

since $\Delta_T(x)$ is continuous.

The plan now is to estimate $\Delta_T^*$, and, from there, $\Delta^*$.

Being absolutely integrable, the inversion theorem for densities implies that

$$f_{U+Z_T}(x) - f_{V+Z_T}(x) = \frac{1}{2\pi} \int_{-T}^{T} e^{-itx} (\varphi_U(t) - \varphi_V(t))\varphi_{Z_T}(t) \, dt,$$

which, after integration with respect to $x$, yields

$$\Delta_T(x) = \frac{1}{2\pi} \int_{-T}^{T} e^{-itx} \frac{\varphi_U(t) - \varphi_V(t)}{-it} \varphi_{Z_T}(t) \, dt.$$

Note that we have integrated the difference between two densities. However, since all distribution functions agree at $-\infty$ and $+\infty$, no additional constant appears. This provides the following estimate(s) for $\Delta_T^*$, namely,

$$\Delta_T^* \leq \frac{1}{2\pi} \int_{-T}^{T} \left| \frac{\varphi_U(t) - \varphi_V(t)}{t} \right| \left( 1 - \frac{|t|}{T} \right) dt \leq \frac{1}{2\pi} \int_{-T}^{T} \left| \frac{\varphi_U(t) - \varphi_V(t)}{t} \right| dt.$$

The proof of the lemma thus will be completed by showing that

$$\Delta^* \le 2\Delta_T^* + \frac{24A}{\pi T}. \tag{6.3}$$

In order to show this, we note that $\Delta(-\infty) = \Delta(+\infty) = 0$, and, being the difference of two distribution functions, it is right-continuous with left-hand limits. This implies that

$$\Delta^* = |\Delta(x_0)| \quad \text{or} \quad \Delta^* = |\Delta(x_0-)| \quad \text{for some} \quad x_0,$$

which produces four possibilities. We choose to treat the case when $\Delta^* = \Delta(x_0)$, the others being similar.

By (6.1),

$$\Delta(x_0 + s) \ge \Delta^* - As \quad \text{for} \quad s > 0,$$

in particular,

$$\Delta\left(x_0 + \frac{\Delta^*}{2A} + y\right) \ge \Delta^* - A\left(\frac{\Delta^*}{2A} + y\right) = \frac{\Delta^*}{2} - Ay, \quad \text{for} \quad |y| \le \frac{\Delta^*}{2A}.$$

With this estimate when $|y| \le \Delta^*/2A$, and the estimate $\Delta(x_0 + \frac{\Delta^*}{2A} + y) \ge -\Delta^*$ for $|y| > \Delta^*/2A$, we obtain, recalling the definition, and noticing that the distribution of $Z_T$ is symmetric,

$$\Delta_T\left(x_0 + \frac{\Delta^*}{2A}\right) = \int_{-\infty}^{\infty} \Delta\left(x_0 + \frac{\Delta^*}{2A} - y\right) f_{Z_T}(y)\,\mathrm{d}y$$

$$= \int_{|y| \le \Delta^*/2A} \Delta\left(x_0 + \frac{\Delta^*}{2A} - y\right) f_{Z_T}(y)\,\mathrm{d}y$$

$$+ \int_{|y| > \Delta^*/2A} \Delta\left(x_0 + \frac{\Delta^*}{2A} - y\right) f_{Z_T}(y)\,\mathrm{d}y$$

$$\ge \int_{|y| \le \Delta^*/2A} \left(\frac{\Delta^*}{2} - Ay\right) f_{Z_T}(y)\,\mathrm{d}y - \Delta^* \int_{|y| > \Delta^*/2A} f_{Z_T}(y)\,\mathrm{d}y$$

$$= \frac{\Delta^*}{2} P\left(|Z_T| \le \frac{\Delta^*}{2A}\right) - \Delta^* P\left(|Z_T| > \frac{\Delta^*}{2A}\right)$$

$$= \frac{\Delta^*}{2}\left(1 - 3P\left(|Z_T| > \frac{\Delta^*}{2A}\right)\right),$$

which shows that

$$\Delta^* \le 2\Delta_T^* + 3\Delta^* P\left(|Z_T| > \frac{\Delta^*}{2A}\right).$$

It thus remains to prove that

$$P\left(|Z_T| > \frac{\Delta^*}{2A}\right) \le \frac{8A}{\pi \Delta^* T}.$$

Now, by symmetry and a change of variables,

$$P\left(|Z_T| > \frac{\Delta^*}{2A}\right) = \int_{|x|>\Delta^*/2A} \frac{1-\cos xT}{\pi T x^2}\,dx = 2\int_{\Delta^*/2A}^{\infty} \frac{1-\cos xT}{\pi T x^2}\,dx$$

$$= \int_{\Delta^*T/4A}^{\infty} \frac{1-\cos 2y}{\pi y^2}\,dy = \frac{2}{\pi}\int_{\Delta^*T/4A}^{\infty} \frac{\sin^2 y}{y^2}\,dy$$

$$\leq \frac{2}{\pi}\int_{\Delta^*T/4A}^{\infty} \frac{1}{y^2}\,dy = \frac{8A}{\pi\Delta^*T}. \qquad \square$$

**Exercise 6.1.** Check the three other cases connected with the proof of (6.3).   $\square$

In order to apply the lemma to our setting we need an estimate between the difference of the characteristic functions of

$$U = U_n = \frac{S_n}{s_n} \quad \text{and} \quad V \in N(0,1).$$

**Lemma 6.2.** *With the above notation,*

$$\left|\varphi_{S_n/s_n}(t) - e^{-t^2/2}\right| \leq 16\frac{\beta_n^3}{s_n^3}|t|^3 e^{-t^2/3} \quad \text{for} \quad |t| \leq \frac{s_n^3}{4\beta_n^3}.$$

*Proof.* The proof proceeds in steps. Set $\varphi_n(t) = \varphi_{S_n/s_n}(t)$.

**I.**  $|\varphi_n(t)| \leq e^{-t^2/3}$ for $|t| \leq \frac{s_n^3}{4\beta_n^3}$.
To prove this it is tempting to use (4.4.1) and the triangle inequality. However, this is not convenient since $\varphi_n$ is complex valued. We therefore resort to the symmetrized random variables $\{X_k^s, k \geq 1\}$, which are independent, identically distributed random variables, such that, for all $k$,

$$E\,X_k^s = 0, \quad \text{Var } X_k^s = 2\sigma_k^2, \quad E|X_k^s|^3 \leq 8\gamma_k^3,$$

where the inequality is due to the $c_r$-inequalities, Theorem 3.2.2.

The main feature is that $\varphi_{X_k^s}(t) = |\varphi_{X_k}(t)|^2$ is real valued. Applying (4.4.1) to the symmetrized random variables yields

$$\left|\varphi_{X_k^s}(t) - (1 - t^2\sigma_k^2)\right| \leq E\min\left\{t^2(X_k^s)^2, \frac{|tX_k^s|^3}{6}\right\} \leq \frac{|t|^3 8\gamma_k^3}{6},$$

so that, by the triangle inequality,

$$|\varphi_{X_k/s_n}(t)|^2 = \varphi_{X_k^s/s_n}(t) \leq 1 - \frac{t^2\sigma_k^2}{s_n^2} + \frac{4|t|^3\gamma_k^3}{3s_n^3} \leq \exp\left\{-\frac{t^2\sigma_k^2}{s_n^2} + \frac{4|t|^3\gamma_k^3}{3s_n^3}\right\}.$$

It follows, taking the upper bound on $t$ into account, that

$$|\varphi_n(t)|^2 = \varphi_{S_n^s/s_n}(t) = \prod_{k=1}^{n}\varphi_{X_k^s/s_n}(t) \leq \prod_{k=1}^{n}\exp\left\{-\frac{t^2\sigma_k^2}{s_n^2} + \frac{4|t|^3\gamma_k^3}{3s_n^3}\right\}$$

$$= \exp\left\{-\sum_{k=1}^{n}\left(\frac{t^2\sigma_k^2}{s_n^2} + \frac{4|t|^3\gamma_k^3}{3s_n^3}\right)\right\} \leq \exp\left\{-t^2 + 4t^2\frac{s_n^3}{4\beta_n^3}\sum_{k=1}^{n}\frac{\gamma_k^3}{3s_n^3}\right\}$$

$$= \exp\left\{-t^2 + \frac{t^2}{3}\right\} = \exp\left\{-\frac{2t^2}{3}\right\},$$

which, upon taking the square root, proves the first assertion.

**II.**   The lemma holds for $\frac{s_n}{2\beta_n} \leq |t| \leq \frac{s_n^3}{4\beta_n^3}$.

The lower bound for $t$ implies that

$$1 \leq \frac{2|t|\beta_n}{s_n} \leq \frac{8|t|^3\beta_n^3}{s_n^3}.$$

This, together with the inequality established in the first step and the triangle inequality, therefore yields

$$\left|\varphi_n(t) - e^{-t^2/2}\right| \leq |\varphi_n(t)| + e^{-t^2/2} \leq e^{-t^2/3} + e^{-t^2/2} \leq 2e^{-t^2/3}$$

$$\leq \frac{8|t|^3\beta_n^3}{s_n^3} 2e^{-t^2/3} = \frac{16|t|^3\beta_n^3}{s_n^3} e^{-t^2/3}.$$

**III.**   The lemma holds for $|t| < \frac{s_n}{2\beta_n}$, in fact (which is more than we need),

$$\left|\varphi_n(t) - e^{-t^2/2}\right| \leq 0.5\frac{|t|^3\beta_n^3}{s_n^3} e^{-t^2/2} \leq 0.5\frac{|t|^3\beta_n^3}{s_n^3} e^{-t^2/3} \quad \text{for} \quad |t| < \frac{s_n}{2\beta_n}.$$

As a preparation for Taylor expansion of the characteristic function (Theorem 4.4.1) and then of the logarithm (Lemma A.1.1), we note that, by Lyapounov's inequality and the bound on $t$,

$$\frac{\sigma_k}{s_n}|t| \leq \frac{\gamma_k}{s_n}|t| \leq \frac{\beta_n}{s_n}|t| < \frac{1}{2}, \tag{6.4}$$

uniformly in $k$, and, hence, also that

$$\left|-\frac{\sigma_k^2 t^2}{2s_n^2} + \frac{\gamma_k^3|t|^3}{6s_n^3}\right| \leq \frac{\sigma_k^2 t^2}{2s_n^2} + \frac{\gamma_k^3|t|^3}{6s_n^3} \leq \frac{1}{2}\cdot\frac{1}{4} + \frac{1}{6}\cdot\frac{1}{8} = \frac{7}{48} < \frac{1}{2}.$$

The announced expansions therefore yield

$$\varphi_{X_k/s_n}(t) = \varphi_{X_k}\left(\frac{t}{s_n}\right) = 1 - \frac{t^2\sigma_k^2}{2s_n^2} + r_n',$$

$$\log\varphi_{X_k/s_n}(t) = -\frac{t^2\sigma_k^2}{2s_n^2} + r_n' + r_n'',$$

where, for some $|\theta_n'| \leq 1$,

$$|r_k'| = \left|\theta_n'\frac{\gamma_k^3|t|^3}{6s_n^3}\right| \leq \frac{\gamma_k^3|t|^3}{6s_n^3} \leq \frac{1}{48},$$

and where, by the $c_r$-inequalities and (6.4),

$$|r_k''| \leq \left|-\frac{t^2\sigma_k^2}{2s_n^2} + r_k'\right|^2 \leq 2\left|\frac{t^2\sigma_k^2}{2s_n^2}\right|^2 + 2|r_k'|^2 \leq 2\left|\frac{t^2\sigma_k^2}{2s_n^2}\right|^2 + 2\left|\frac{\gamma_k^3|t|^3}{6s_n^3}\right|^2$$

$$= \frac{1}{2}\frac{\sigma_k^3|t|^3}{s_n^3}\frac{\sigma_k|t|}{s_n} + \frac{1}{18}\left(\frac{\gamma_k^3|t|^3}{s_n^3}\right)^2 \leq \frac{1}{2}\frac{\gamma_k^3|t|^3}{s_n^3}\frac{1}{2} + \frac{1}{18}\frac{\gamma_k^3|t|^3}{s_n^3}\frac{1}{8} = \frac{37}{144}\frac{\gamma_k^3|t|^3}{s_n^3}.$$

Adding the individual logarithms we find that

$$\log \varphi_n(t) = \sum_{k=1}^{n} \log \varphi_{X_k/s_n}(t) = -\sum_{k=1}^{n} \frac{t^2 \sigma_k^2}{2s_n^2} + r_n = -\frac{t^2}{2} + r_n,$$

where

$$|r_n| \le \sum_{k=1}^{n}(r_k' + r_k'') \le \sum_{k=1}^{n}\left(\frac{\gamma_k^3 |t|^3}{6s_n^3} + \frac{37}{144}\frac{\gamma_k^3 |t|^3}{s_n^3}\right) \le \frac{61}{144} \cdot \frac{\beta_n^3 |t|^3}{s_n^3} \le \frac{61}{144} \cdot \frac{1}{8} < \frac{1}{18},$$

the second to last inequality being a consequence of (6.4).

Finally, since $|e^z - 1| \le |z|e^{|z|}$ for $z \in \mathbb{C}$ (Lemma A.1.1) we find, using the estimates on $r_n$, that

$$|\varphi_n(t) - e^{-t^2/2}| = |e^{-t^2/2 + r_n} - e^{-t^2/2}| = e^{-t^2/2}|e^{r_n} - 1|$$

$$\le e^{-t^2/2}|r_n|e^{|r_n|} \le e^{-t^2/2} \frac{61}{144} \cdot \frac{\beta_n^3 |t|^3}{s_n^3} e^{5/144} < 0.5 \frac{\beta_n^3 |t|^3}{s_n^3} e^{-t^2/2}. \qquad \square$$

To prove the theorem we finally join the two lemmas with $A = 1/\sqrt{2\pi}$ (the upper bound of the standard normal density), $T = T_n = s_n^3/4\beta_n^3$, and the fact that $\int_{-\infty}^{\infty} t^2 e^{-t^2/3}\, dt = \frac{3}{2}\sqrt{3\pi}$, to obtain,

$$\sup_{x} |F_{\frac{S_n - n\mu}{\sigma\sqrt{n}}}(x) - \Phi(x)| \le \frac{16}{\pi} \frac{\beta_n^3}{s_n^3} \int_{-\beta_n^3/s_n^3}^{\beta_n/s_n^3} t^2 e^{-t^2/3}\, dt + \frac{96\beta_n^3}{\pi\sqrt{2\pi}s_n^3}$$

$$\le \left(\frac{24\sqrt{3}}{\sqrt{\pi}} + \frac{96}{\pi\sqrt{2\pi}}\right)\frac{\beta_n^3}{s_n^3} \le 36\frac{\beta_n^3}{s_n^3}.$$

The proof of the theorem is complete. $\qquad \square$

# 7 Some Additional Results and Remarks

In this supplementary section we collect some results and applications related to the central limit theorem, which are somewhat less central to the mainstream.

## 7.1 Rates of Rates

We have interpreted the central limit theorem as a convergence rate result with respect to the law of large numbers. Similarly, the Berry-Esseen theorem is a rate result with respect to the central limit theorem. This search for more detailed precision can be continued. The general solution to this kind of problems is that, under the assumption of the existence of successively higher moments, one can continue to replace approximations with rates, that is, results about "closeness" with results about "how close". Such additional results are grouped under the heading of *Edgeworth expansions*. Here we shall state *one* such result, namely [76], Theorem 3; see also [77], Theorem 2, p. 49, and refer to the specialized literature, such as the books [88, 98, 194, 195] for more.

**Theorem 7.1.** *Let $X$, $X_1$, $X_2$, ... be a sequence of independent, identically, non-lattice distributed random variables with partial sums $S_n$, $n \geq 1$. Suppose that $E|X|^3 < \infty$, that $EX = 0$, and set $\sigma^2 = \operatorname{Var} X$ $(> 0)$, and $\alpha_3 = EX^3$. Then*

$$F_{\frac{S_n}{\sigma\sqrt{n}}}(x) = \Phi(x) + \frac{\alpha_3}{6\sigma^3\sqrt{2\pi n}}(1 - x^2)e^{-x^2/2} + o\left(\frac{1}{\sqrt{n}}\right) \quad as \quad n \to \infty.$$

The series can be further developed under the assumption of higher moments. The polynomials that appear as coefficients are *Hermite polynomials*.

## 7.2 Non-uniform Estimates

Another aspect is that the remainder term estimates so far have been *uniform* in $x$. As a glimpse into the area of estimates that depend on $x$ we state two results, and refer to the literature for proofs and more. The first one is [77], Theorem 1, p. 70.

**Theorem 7.2.** *Let $X$, $X_1$, $X_2$, ... be a sequence of independent, random variables with partial sums $S_n$, $n \geq 1$, $EX_k = 0$, and $\sigma_k^2 = \operatorname{Var} X_k < \infty$. Set $s_n^2 = \sum_{k=1}^{n} \sigma_k^2$, and*

$$\Delta_n = \sup_x |F_{\frac{S_n}{s_n}}(x) - \Phi(x)|, \quad n \geq 1.$$

*If $\Delta_n \leq 1/2$ for $n > n_0$, then there exists a constant $C$, such that, for $n > n_0$,*

$$|F_{\frac{S_n}{s_n}}(x) - \Phi(x)| \leq \min\left\{\Delta_n, C\frac{\Delta_n \log(1/\Delta_n)}{1 + x^2}\right\} \quad for \ all \quad x.$$

The first part of the following result, for which we refer to (e.g.) [194], Theorem 14, [195], Theorem 5.16, provides a non-uniform estimate which is valid for all $x$. The other two parts, which are reduced versions of [77], Theorem 2, p. 73, and Theorem 3, p. 75, respectively, provide estimates for $x$ above and below $\sqrt{(1 + \delta) \log n}$, respectively. The $\sqrt{\log n}$ boundary is natural since the function $e^{-x^2/2}$ reduces to a negative power of $\sqrt{n}$ for $x = \sqrt{\log n}$.

**Theorem 7.3.** *Under the assumptions of Theorem 7.1,*

$$|F_{\frac{S_n}{\sigma\sqrt{n}}}(x) - \Phi(x)| \leq C\frac{\gamma^3}{\sigma^3} \frac{1}{(1 + |x|^3)\sqrt{n}} \quad for \ all \ x \in \mathbb{R},$$

$$|F_{\frac{S_n}{\sigma\sqrt{n}}}(x) - \Phi(x)| \leq \frac{C(\delta, \gamma)}{(1 + |x|^3)\sqrt{n}} \quad for \ |x| \geq \sqrt{(1 + \delta) \log n},$$

$$|F_{\frac{S_n}{\sigma\sqrt{n}}}(x) - \Phi(x)| \leq \frac{C(\delta, \gamma)}{\sqrt{n}}\left((1 + |x|^3)e^{-x^2/2} + 1\right) \quad for \ |x| \leq \sqrt{(1 + \delta) \log n},$$

*where $C(\delta, \gamma)$ is a constant depending only on $\delta \in (0, 1)$ and $\gamma^3$.*

For analogs when higher-order moments are assumed to exist we refer to the above sources.

## 7.3 Renewal Theory

Using inversion, Englund [72] has proved the following Berry-Esseen theorem for the counting process of a renewal process.

**Theorem 7.4.** *Suppose that $X$, $X_1$, $X_2$, ... are independent, non-negative, identically distributed random variables, set $\mu = EX$, $\sigma^2 = \text{Var}\, X$, and let $\{S_n, n \geq 1\}$ and $\{N(t), t \geq 0\}$ be the associated renewal and renewal counting processes, respectively. If $\gamma^3 = E|X|^3 < \infty$, then*

$$\sup_n \left| P(N(t) < n) - \Phi\left(\frac{(n\mu - t)\sqrt{\mu}}{\sigma\sqrt{t}}\right) \right| \leq 4\frac{\gamma^3 \sqrt{\mu}}{\sigma^3 \sqrt{t}}.$$

## 7.4 Records

Suppose that $X_1$, $X_2$, ... are independent, identically distributed, continuous random variables, and let $\{L_n, n \geq 1\}$ and $\{\mu(n), n \geq 1\}$ be the record times and the counting process, respectively. The following Berry-Esseen theorem for these processes was obtained in [111] via Englund's technique from [72].

**Theorem 7.5.** *For $k \geq 2$,*

$$\sup_n \left| P(\mu(k) \leq n) - \Phi\left(\frac{n - \log k}{\sqrt{\log k}}\right) \right| \leq \frac{1.9}{\sqrt{\log k}},$$

$$\sup_k \left| P(L(n) \geq k) - \Phi\left(\frac{n - \log k}{\sqrt{n}}\right) \right| \leq \frac{4.3}{\sqrt{n}}.$$

*Remark 7.1.* Since records are rather rare events one might imagine that a Poisson approximation is more natural, as well as sharper, for the counting process. That this is indeed the case will be hinted at in Chapter 9, where we shall provide material enough to conclude that, if $V_k \in \text{Po}(E(\mu_k))$, then

$$\sup_{A \subset \mathbb{Z}^+} |P(\mu(k) \in A) - P(V_k \in A)| \leq \frac{\pi^2}{6 \log k}.$$

*Remark 7.2.* The rate in the Poisson approximation should be compared with the rate $\mathcal{O}(\frac{1}{\sqrt{\log k}})$ in the normal approximation. The discrepancy stems from the fact that the approximation from Poisson to normal is of the order $\mathcal{O}(\frac{1}{\sqrt{\log k}})$ (due to the central limit theorem; recall the Berry-Esseen theorem Theorem 6.2).                                                  $\square$

## 7.5 Local Limit Theorems

In a first course in probability theory one learns that the binomial distribution may be approximated by the normal distribution with the aid of a "half-correction" which, in particular, permits one to approximate the *probability function* at some point $k$ by integrating the normal density from $k - 1/2$ to $k + 1/2$. More precisely, if $Y_n \in \text{Bin}(n, p)$ and $n$ is "large", then one learns that

$$P(Y_n = k) \approx \int_{k-1/2}^{k+1/2} \frac{1}{\sqrt{2\pi np(1-p)}} \exp\left\{-\frac{(x - np)^2}{2np(1-p)}\right\} dx.$$

This kind of approximation holds more generally for *lattice distributions*, that is, distributions whose support is concentrated on a set of the form $\{kd + \lambda : k = 0, \pm 1, \pm 2, \ldots\}$, for some $d > 0$ and $\lambda \in \mathbb{R}$; cf. Chapter 4.

The following result, which we state without proof, is due to Gnedenko [96]; see also [88], p. 517. For more on this topic we refer to [97, 98, 142, 194].

**Theorem 7.6.** *Suppose that* $X$, $X_1$, $X_2$, ... *are independent, identically distributed lattice random variables with span* $d$, *set* $\mu = E X$, $\sigma^2 = \text{Var} X$, *and* $S_n = \sum_{k=1}^n X_k$, $k \geq 1$. *Then*

$$\lim_{n\to\infty} \frac{\sigma \sqrt{n}}{d} P(S_n = kd + n\lambda) - \frac{1}{\sqrt{2\pi}} \exp\left\{-\frac{(kd + n\lambda - n\mu)^2}{2n\sigma^2}\right\} = 0,$$

*uniformly in* $k \in \mathbb{Z}$.

Two references for corresponding results in the renewal theoretic context, which thus approximate the first passage time probabilities $P(\nu(t) = k)$ by a normal density, are [4, 165]. In the formulation of the latter,

$$\frac{\mu^{3/2}\sqrt{t}}{\sigma} P\big(\nu(t) = [\mu^{-1}t + u\sqrt{\sigma^2\mu^{-3}t}]\big) \sim \frac{1}{\sqrt{2\pi}} e^{-u^2/2} \qquad \text{as} \quad t \to \infty.$$

## 7.6 Large Deviations

Let $X$, $X_1$, $X_2$, ... be a sequence of independent, identically distributed random variables, with partial sums $S_n$, $n \geq 1$. Suppose that $E X = 0$, that $\sigma^2 = \text{Var} X < \infty$, and let $F_n$ denote the distribution function of $S_n/\sigma\sqrt{n}$. One consequence of the central limit theorem is that

$$\frac{1 - F_n(x)}{1 - \Phi(x)} \to 1 \quad \text{and} \quad \frac{F_n(-x)}{\Phi(-x)} \to 1 \qquad \text{as} \quad n \to \infty. \tag{7.1}$$

The problem of large deviations concerns the behavior of these ratios for $x = x_n$, where $x_n \to \infty$ as $n \to \infty$. It turns out that the behavior depends on the actual growth rate of $x_n$.

The general problem of when and how (7.1) holds when $x$ is allowed to depend on $n$ is called the theory of large deviations. Some authors use the term "moderate deviations" for $x = x_n$ growing at some kind of moderate rate, such as a logarithmic one.

The basic standard assumption, usually called the Cramér condition, probably since the first general limit theorem in the area was presented in [55], is the existence of the moment generating function $\psi$. Following, for the right-hand tails, are two results without proofs. The obvious analogs for the left-hand tails hold too.

Relevant introductory literature is [88], Section XVI.7; [142], Chapter VI; [194], Chapter VIII; and [195], Section 5.8.

**Theorem 7.7.** *Suppose that $\psi_X(t)$ exists in a neighborhood of $t = 0$.*
*(i) If $|x| = o(n^{1/6})$ as $n \to \infty$, then (7.1) holds.*
*(ii) If $x = o(\sqrt{n})$, then*

$$\frac{1 - F_n(x)}{1 - \Phi(x)} = \exp\left\{ x^3 \lambda\left(\frac{x}{\sqrt{n}}\right)\left(1 + \mathcal{O}\left(\frac{x}{\sqrt{n}}\right)\right)\right\} \quad as \quad n \to \infty,$$

*where $\lambda(x)$ is a power series, called the Cramér series, that depends on the cumulants of $X$, and which converges for sufficiently small arguments.*

If, in particular, $x = o(n^{1/4})$, then it turns out that

$$\frac{1 - F_n(x)}{1 - \Phi(x)} \sim \exp\left\{ \frac{E\,X^3}{6\sigma^3\sqrt{n}} x^3\right\}, \quad as \quad n \to \infty,$$

which turns the mind somewhat back to Theorem 7.1.

## 7.7 Convergence Rates

There exist two kinds of rate results. One kind pertains to tail probabilities, the other, more directly, to the random variables and their distributions. This subsection is devoted to the former kind, more precisely to a central limit theorem analog of the results in Subsection 6.12.1.

**Theorem 7.8.** *Let $X$, $X_1$, $X_2$, ... be independent, identically distributed random variables, such that $E\,X = 0$, and $\operatorname{Var} X = 1$, and let $\{S_n, n \geq 1\}$ denote their partial sums. Further, set $\sigma_n^2 = \operatorname{Var}\left( X\,I\{|X| \leq \sqrt{n}\}\right)$.*
*(i) We have*

$$\sum_{n=1}^{\infty} \frac{1}{n} \sup_x \left| P\left(\frac{S_n}{\sqrt{n}} \leq x\right) - \Phi\left(\frac{x}{\sigma_n}\right)\right| < \infty. \tag{7.2}$$

*Moreover,*

$$\sum_{n=1}^{\infty} \frac{1}{n} \sup_x \left| P\left(\frac{S_n}{\sqrt{n}} \leq x\right) - \Phi(x)\right| < \infty, \tag{7.3}$$

*if and only if*

$$\sum_{n=1}^{\infty} \frac{1}{n}(1 - \sigma_n^2) < \infty, \tag{7.4}$$

*or, equivalently, if and only if* $E X^2 \log^+ |X| < \infty.$
(ii) *Let* $2 < r < 3.$ *If* $E|X|^r < \infty,$ *then*

$$\sum_{n=1}^{\infty} n^{\frac{r}{2}-2} \sup_x \left| P\left(\frac{S_n}{\sqrt{n}} \leq x\right) - \Phi\left(\frac{x}{\sigma_n}\right) \right| < \infty, \tag{7.5}$$

$$\sum_{n=1}^{\infty} n^{\frac{r}{2}-2} \sup_x \left| P\left(\frac{S_n}{\sqrt{n}} \leq x\right) - \Phi(x) \right| < \infty. \tag{7.6}$$

The first result in this direction is due to Spitzer [233], who proved that the sum

$$\sum_{n=1}^{\infty} \frac{1}{n}\left(P(S_n \leq 0) - \frac{1}{2}\right) \tag{7.7}$$

is convergent; see also [234], p. 199. By Fourier methods Rosén [213] proved that the sum is *absolutely* convergent, which amounts to proving (i) for $x = 0$. Roséns method was extended in [12], where the sufficiency in (ii) was proved for $x = 0$. The paper closes by pointing out that, by letting $X$ be a coin-tossing random variable, one can show that an analogous theorem cannot hold for $r \geq 3$, (and thus, even less so in the general case).

For general $x \in \mathbb{R}$, it is shown in [90] that (7.6) holds if and only if

$$\sum_{n=1}^{\infty} n^{\frac{r}{2}-2}(1 - \sigma_n^2) < \infty. \tag{7.8}$$

With the aid of Fourier methods Heyde, [134], proves that, if $E X = 0$ and $\text{Var}\, X < \infty$, then (7.3) holds if and only if $E X^2 \log^+ |X| < \infty$, and that (7.6) holds if and only if $E|X|^r < \infty$; cf. also [114], where a slightly stronger version of the necessity is proved. Namely, that if (7.3) holds, then $E X^2 \log^+ |X| < \infty$ and $E X = 0$. We also remark that whereas Rosén's result is used in [90] for their proof, we do not need that result, so that, as a corollary, we obtain the absolute convergence of (7.7) by probabilistic methods.

We precede the proof of the theorem with some auxiliary matter.

**Lemma 7.1.** *Let* $2 \leq r < 3.$ *In the above notation,*

$$\sum_{n=1}^{\infty} \frac{1}{n}(1 - \sigma_n^2) < \infty \quad \Longleftrightarrow \quad E X^2 \log^+ |X| < \infty,$$

$$\sum_{n=1}^{\infty} n^{\frac{r}{2}-2}(1 - \sigma_n^2) < \infty \quad \Longleftrightarrow \quad E|X|^r < \infty.$$

*Proof.* We first note that

$$
\begin{aligned}
E\,X^2 I\{|X| > \sqrt{n}\} &= 1 - E\,X^2 I\{|X| \le \sqrt{n}\} \le 1 - \sigma_n^2 \\
&= 1 - E\,X^2 I\{|X| \le \sqrt{n}\} + (E\,XI\{|X| \le \sqrt{n}\})^2 \\
&= E\,X^2 I\{|X| > \sqrt{n}\} + (-E\,XI\{|X| > \sqrt{n}\})^2 \\
&\le 2E\,X^2 I\{|X| > \sqrt{n}\},
\end{aligned}
$$

where we used the fact that $E\,X = 0$ and Lyapounov's inequality, respectively, in the last two steps. Consequently,

$$
\sum_{n=1}^{\infty} n^{\frac{r}{2}-2}(1 - \sigma_n^2) < \infty \quad \Longleftrightarrow \quad \sum_{n=1}^{\infty} n^{\frac{r}{2}-2} E\,X^2 I\{|X| > \sqrt{n}\} < \infty.
$$

By Fubini's theorem, the latter sum is finite if and only if

$$
E\left(\left(\sum_{n \le X^2} n^{\frac{r}{2}-2}\right)X^2\right) < \infty \quad \Longleftrightarrow \quad
\begin{cases}
E(\log^+(X^2)X^2), & \text{when} \quad r = 2, \\
E((X^2)^{\frac{r}{2}-1}X^2), & \text{when} \quad 2 < r < 3,
\end{cases}
$$

which proves the desired statements. $\qquad\square$

We also need [73], Lemma 2.9 (which is an exercise in Taylor expansion; note that one may assume that $\left|\frac{x}{y} - 1\right| \le 0.8$ in the proof).

**Lemma 7.2.** *For* $x, y > 0$, *and* $z \in \mathbb{R}$,

$$
\left|\Phi\left(\frac{z}{x}\right) - \left(\frac{z}{y}\right)\right| \le 1.25\left|\frac{x}{y} - 1\right|.
$$

**Proof of the Theorem**

(i): We first wish to prove that the sum in (7.2) converges. Toward this end, set, for $1 \le k \le n$, and $n \ge 1$,

$$
Y_{n,k} = \begin{cases} X_k, & \text{when} \quad |X_k| \le \sqrt{n}, \\ 0, & \text{otherwise}, \end{cases}
$$

$S_n' = \sum_{k=1}^{n} Y_{n,k}$, $\mu_n = E\,Y_{n,1}$, and recall that $\sigma_n^2 = \operatorname{Var} Y_{n,1}$.

Since $\sigma_n \to 1$ as $n \to \infty$, there exists $n_0$ such that $\sigma_n > 1/2$ for $n \ge n_0$. Now, since

$$
\left|P\left(\frac{S_n}{\sqrt{n}} \le x\right) - P\left(\frac{S_n'}{\sqrt{n}} \le x\right)\right| \le n P(|X| > \sqrt{n}),
$$

it follows, via the triangle inequality, that

$$\sum_{n=1}^{\infty} \frac{1}{n} \sup_x \left| P\left(\frac{S_n}{\sqrt{n}} \le x\right) - \Phi\left(\frac{x}{\sigma_n}\right) \right|$$

$$\le \sum_{n=1}^{\infty} P(|X| > \sqrt{n}) + \sum_{n=1}^{\infty} \frac{1}{n} \sup_x \left| P\left(\frac{S_n'}{\sqrt{n}} \le x\right) - \Phi\left(\frac{x - \mu_n \sqrt{n}}{\sigma_n}\right) \right|$$

$$+ \sum_{n=1}^{\infty} \frac{1}{n} \sup_x \left| \Phi\left(\frac{x - \mu_n \sqrt{n}}{\sigma_n}\right) - \Phi\left(\frac{x}{\sigma_n}\right) \right|$$

$$= \Sigma_1 + \Sigma_2 + \Sigma_3. \tag{7.9}$$

Next we examine each of the sums separately.

By Theorem 2.12.1 we know that

$$\Sigma_1 < \infty \quad \Longleftrightarrow \quad E\,X^2 < \infty.$$

As for $\Sigma_2$, Theorem 6.1, and the $c_r$-inequalities together yield

$$\sup_x \left| P\left(\frac{S_n'}{\sqrt{n}} \le x\right) - \Phi\left(\frac{x - \mu_n \sqrt{n}}{\sigma_n}\right) \right| \le C \frac{E|Y_{n,1} - \mu_n|^3}{\sqrt{n}\sigma_n^3}$$

$$\le C \frac{4(E|Y_{n,1}|^3 + |\mu_n|^3)}{\sqrt{n}\sigma_n^3} \le C \frac{8E|Y_{n,1}|^3}{\sqrt{n}\sigma_n^3},$$

so that, by Lemma A.3.1 and Fubini's theorem,

$$\Sigma_2 \le C \sum_{n=1}^{\infty} \frac{1}{n} \frac{E|Y_{n,1}|^3}{\sqrt{n}\sigma_n^3} \le C + C \sum_{n \ge n_0} \frac{1}{n^{3/2}} E|Y_{n,1}|^3$$

$$= C + C \sum_{n \ge n_0} \frac{1}{n^{3/2}} E|X|^3 I\{|X| \le \sqrt{n}\}$$

$$\le C + CE\left(\left(\sum_{n \ge X^2} \frac{1}{n^{3/2}}\right)|X|^3\right) \le C + CE\left(\frac{1}{\sqrt{X^2}}|X|^3\right)$$

$$= C + CE\,X^2.$$

For the third sum we apply the mean value theorem, together with the facts that the standard normal density never exceeds $1/\sqrt{2\pi}$, that $\sigma_n > 1/2$ when $n \ge n_0$, that $E\,X = 0$, and Fubini:

$$\Sigma_3 \le \sum_{n=1}^{\infty} \frac{1}{n} \frac{1}{\sqrt{2\pi}} \frac{|\mu_n|\sqrt{n}}{\sigma_n} \le \frac{1}{\sqrt{2\pi}} \sum_{n=1}^{\infty} \frac{1}{\sigma_n \sqrt{n}} E|X| I\{|X| > \sqrt{n}\}$$

$$\le C + \sqrt{\frac{2}{\pi}} \sum_{n \ge n_0} \frac{1}{\sqrt{n}} E|X| I\{|X| > \sqrt{n}\}$$

$$\le C + CE\left(\left(\sum_{n < X^2} \frac{1}{\sqrt{n}}\right)|X|\right) \le C + CE\sqrt{X^2}|X| = C + CE\,X^2.$$

This proves (7.2).

Next we turn our attention to the second sum. Let $\Sigma_{(i)}$ and $\Sigma_{(ii)}$ denote the sums in (7.2) and (7.3), respectively. Then

$$\left|\Sigma_{(ii)} - \Sigma_{(i)}\right| \le \sum_{n=1}^{\infty} \frac{1}{n} \sup_x \left|\Phi(x) - \Phi\left(\frac{x}{\sigma_n}\right)\right| \le \Sigma_{(ii)} + \Sigma_{(i)}. \qquad (7.10)$$

Since $\sigma_n^2 \le E\,X^2 I\{|X| \le \sqrt{n}\} \le E\,X^2 = 1$, it follows that

$$1 - \sigma_n \le 1 - \sigma_n^2 = (1 - \sigma_n)(1 + \sigma_n) \le 2(1 - \sigma_n), \qquad (7.11)$$

which, together with Lemma 7.2, shows that

$$\sup_x \left|\Phi(x) - \Phi\left(\frac{x}{\sigma_n}\right)\right| \le 1.25(1 - \sigma_n) \le 1.25(1 - \sigma_n^2). \qquad (7.12)$$

On the other hand, by the mean value theorem and (7.11),

$$\sup_x \left|\Phi(x) - \Phi\left(\frac{x}{\sigma_n}\right)\right| \ge (1 - \sigma_n)\frac{1}{2\sqrt{2\pi}} e^{-\frac{1}{2}} \ge C(1 - \sigma_n^2) \quad \text{for} \quad n \ge n_0. \text{(7.13)}$$

Combining (7.10), (7.12), and (7.13) therefore shows that, if $\Sigma_{(i)}$ converges, then so does $\Sigma_{(ii)}$ if and only if $\sum_{n=1}^{\infty} \frac{1}{n}(1 - \sigma_n^2) < \infty$, that is (Lemma 7.1), if and only if $E\,X^2 \log^+ |X| < \infty$.

(ii): The proof of this half follows the same path as that of the first half. We therefore confine ourselves to a sketch, leaving it to the reader to complete the details.

The first goal is to prove that

$$\sum_{n=1}^{\infty} n^{r/2-2} \sup_x \left|P\left(\frac{S_n}{\sqrt{n}} \le x\right) - \Phi\left(\frac{x}{\sigma_n}\right)\right| < \infty.$$

The analogs of the three sums, $\Sigma_1, \Sigma_2,$ and $\Sigma_3$ are estimated as follows.

- By Lemma 2.12.1,

$$\Sigma_1' = \sum_{n=1}^{\infty} n^{r/2-1} P(|X| > \sqrt{n}) < \infty \quad \Longleftrightarrow \quad E|X|^r < \infty.$$

- Secondly, (note that $(r-5)/2 < -1$ since $r < 3$),

$$\Sigma_2' \le C \sum_{n=1}^{\infty} n^{r/2-2-1/2} \frac{E|Y_{n,1}|^3}{\sigma_n^3} \le C + CE\left(\left(\sum_{n \ge X^2} n^{(r-5)/2}\right)|X|^3\right)$$

$$\le C + CE\left((X^2)^{(r-3)/2}|X|^3\right).$$

- Finally,

$$\Sigma'_3 \leq \sum_{n=1}^{\infty} n^{r/2-2} \frac{1}{\sqrt{2\pi}} \frac{|\mu_n|\sqrt{n}}{\sigma_n} \leq \frac{1}{\sqrt{2\pi}} \sum_{n=1}^{\infty} n^{(r-3)/2} \frac{1}{\sigma_n} E|X|I\{|X| \geq \sqrt{n}\}$$

$$\leq C + CE\left(\left(\sum_{n \leq X^2} n^{(r-3)/2}\right)|X|\right) \leq C + CE\left((X^2)^{(r-1)/2}\right).$$

This takes care of the first sum.

The conclusion for the second sum follows as in the first part of the theorem with the second half of Lemma 7.1 replacing the first half.

This finishes the proof of (ii).　　　　　　□

*Remark 7.3.* The finiteness of the second sum was derived by the successive approximations

$$P\left(\frac{S_n}{\sqrt{n}} \leq x\right) \longrightarrow P\left(\frac{S'_n}{\sqrt{n}} \leq x\right) \longrightarrow \Phi\left(\frac{x - \mu_n\sqrt{n}}{\sigma_n}\right) \longrightarrow \Phi\left(\frac{x}{\sigma_n}\right) \longrightarrow \Phi(x).$$

An inspection of the proof shows that the first three arrows require that $E\,X^2 < \infty$, whereas the last one requires $E\,X^2 \log^+ |X| < \infty$. This means that the convergence rate of the standardized sum to $\Phi(x/\sigma_n)$ is faster than the convergence rate to $\Phi(x)$ in the sense that additional integrability is required in the latter case. For a related remark, see [127], p. 1038.　　　　　□

## 7.8 Precise Asymptotics

In our discussion of convergence rates in Chapter 6 one kind concerned the rate at which probabilities such as $P(|S_n| > n\varepsilon)$ tend to 0 as $n \to \infty$, thus providing rates in terms of $n$, the number of summands. Another approach is to investigate at which rate such probabilities tend to 1 as $\varepsilon \searrow 0$, in other words, in terms of the closeness of $\varepsilon$ to 0. Although this is a law of large numbers problem it belongs to this chapter, since normal approximation is a fundamental tool for the proof. The first result of this kind is due to Heyde [136].

**Theorem 7.9.** *Let $X$, $X_1$, $X_2$, ... be independent, identically distributed random variables with mean 0 and finite variance, and set $S_n = \sum_{k=1}^{n} X_k$, $n \geq 1$. Then*

$$\lim_{\varepsilon \searrow 0} \varepsilon^2 \sum_{n=1}^{\infty} P(|S_n| \geq n\varepsilon) = E\,X^2.$$

Note the relation to Theorem 11.2.

The next step was taken by Chen [41], whose result reduces to Heyde's theorem for $r = p = 2$.

**Theorem 7.10.** *Let* $1 \leq p < 2 \leq r$. *Suppose that* $X$, $X_1$, $X_2$, ... *are independent, identically distributed random variables with mean* $0$, *and (for simplicity) variance* $1$, *and suppose that* $E|X|^r < \infty$. *Set* $S_n = \sum_{k=1}^n X_k$, $n \geq 1$. *Then*

$$\lim_{\varepsilon \searrow 0} \varepsilon^{\frac{2p}{2-p}(\frac{r}{p}-1)} \sum_{n=1}^{\infty} n^{\frac{r}{p}-2} P(|S_n| \geq n^{1/p}\varepsilon) = E|N(0,1)|^{\frac{2p}{2-p}(\frac{r}{p}-1)}$$

$$= 2^{\frac{p}{2-p}(\frac{r}{p}-1)} \frac{\Gamma(\frac{1}{2} + \frac{p}{2-p}(\frac{r}{p}-1))}{(\frac{r}{p}-1)\Gamma(\frac{1}{2})} .$$

The strategy of the proof is to begin by proving the conclusion for the normal distribution, then to approximate the sum for "lower" indices by normal approximation and, finally, to take care of the tails separately. We confine ourselves to providing an outline of the proof, and refer to the cited articles for details.

*Proof of Step I.* The first step thus is to prove the result for $X \in N(0,1)$. Putting $\Psi(x) = P(|X| > x) = 1 - \Phi(x) + \Phi(-x)$, and noticing that $S_n/\sqrt{n} \overset{d}{=} X$, we have

$$\int_1^{\infty} x^{r/p-2}\Psi(x^{1/p-1/2}\varepsilon)\,dx \leq \sum_{n=1}^{\infty} n^{r/p-2} P(|S_n| \geq n^{1/p}\varepsilon)$$

$$\leq \int_0^{\infty} x^{r/p-2}\Psi(x^{1/p-1/2}\varepsilon)\,dx,$$

which, after the change of variable $y = \varepsilon x^{1/p-1/2}$, turns into

$$\frac{2p}{2-p} \int_\varepsilon^{\infty} y^{\frac{2p}{2-p}(\frac{r}{p}-1)-1}\Psi(y)\,dy \leq \varepsilon^{\frac{2p}{2-p}(\frac{r}{p}-1)} \sum_{n=1}^{\infty} n^{r/p-2} P(|S_n| \geq n^{1/p}\varepsilon)$$

$$\leq \frac{2p}{2-p} \int_0^{\infty} y^{\frac{2p}{2-p}(\frac{r}{p}-1)-1}\Psi(y)\,dy ,$$

which, in view of Theorem 2.12.1, proves the conclusion in the standard normal case.

*Proof of Step II.* The second step is to take care of "lower" indices by the normal approximation.

Toward this end, set

$$\Delta_n = \sup_{x \in \mathbb{R}} \left| P(|S_n| \geq n^{1/2}x) - \Psi(x) \right|,$$

and let $a(\varepsilon) = M\varepsilon^{-2p/(2-p)}$, where $M$ is some fixed, large number.

Since $\Delta_n \to 0$ as $n \to \infty$, Lemma A.6.1 tells us that the same is true for a weighted average, which, in our case, yields

$$\frac{1}{(a(\varepsilon))^{(r/p)-1}} \sum_{n \leq a(\varepsilon)} n^{\frac{r}{p}-2} \Delta_n \to 0 \quad \text{as} \quad \varepsilon \to 0,$$

which, by properly spelling this out, yields

$$\lim_{\varepsilon \searrow 0} \varepsilon^{\frac{2p}{2-p}(\frac{r}{p}-1)} \sum_{n \leq a(\varepsilon)} n^{\frac{r}{p}-2} \left| P(|S_n| > n^{1/p}\varepsilon) - \Psi(n^{1/p-1/2}\varepsilon) \right| = 0.$$

*Proof of Step III.* This step is devoted to the normal tail. Since the summands are monotonically decreasing for $n$ large enough, letting $\varepsilon$ be sufficiently small, an integral comparison, followed by a change of variable $y = x^{1/p-1/2}\varepsilon$, yields

$$\varepsilon^{\frac{2p}{2-p}(\frac{r}{p}-1)} \sum_{n > a(\varepsilon)} n^{\frac{r}{p}-2} \Psi(n^{1/p-1/2}\varepsilon) \leq \varepsilon^{\frac{2p}{2-p}(\frac{r}{p}-1)} \int_{a(\varepsilon)}^{\infty} x^{\frac{r}{p}-2} \Psi(x^{1/p-1/2}\varepsilon)\,\mathrm{d}x$$

$$= \int_{M^{(2-p)/2p}y^{\frac{2r}{2-p}-1}} \Psi(y)\,\mathrm{d}y \to 0, \qquad \text{as } \varepsilon \to 0, \text{ and then } M \to \infty.$$

*Proof of Step IV.* The last, and most difficult, step is devoted to the analog for general summands, that is, to show that

$$\varepsilon^{\frac{2p}{2-p}(\frac{r}{p}-1)} \sum_{n > a(\varepsilon)} n^{\frac{r}{p}-2} P(|S_n| > n^{1/p}\varepsilon) \to 0, \qquad \text{as } \varepsilon \to 0, \text{ and then } M \to \infty.$$

For this, some sharp inequalities for tail probabilities and some additional estimates are needed. Once again, we refer to the original sources.

Note also that, in fact, Step III is a special case of Step IV.    □

*Remark 7.4.* Most proofs follow this pattern. Sometimes, however, the introduction of the additional constant $M$ is not needed.    □

In the remainder of this subsection we present a sample of extensions and generalizations, together with some references.

Analogs for the counting process of a renewal process and first passage times of random walks across horizontal boundaries have been proved in [122]. Here is one example.

**Theorem 7.11.** *Let $X, X_1, X_2, \ldots$ be independent, identically distributed random variables with positive mean $\mu$ and finite variance $\sigma^2$. Set $S_n = \sum_{k=1}^{n} X_k$, $n \geq 1$, and let $\{\tau(t), t \geq 0\}$ be the usual first passage time process defined by*

$$\tau(t) = \min\{n : S_n > t\}, \quad t \geq 0.$$

*Let $0 < p < 2$ and $r \geq 2$. If $E|X|^r < \infty$, then*

$$\lim_{\varepsilon \searrow 0} \varepsilon^{\frac{2p}{2-p}(\frac{r}{p}-1)} \int_1^{\infty} t^{r/p-2} P\left(|\tau(t) - \frac{t}{\mu}| > t^{1/p}\varepsilon\right) \mathrm{d}t$$

$$= \left(\frac{\sigma^2}{\mu^3}\right)^{\frac{r-p}{2-p}} \frac{p}{r-p} E|N|^{\frac{2(r-p)}{2-p}},$$

*where $N$ is a standard normal random variable.*

*Remark 7.5.* For $r = 2$ and $p = 1$ the conclusion reduces to

$$\lim_{\varepsilon \searrow 0} \varepsilon^2 \int_1^\infty P\left(\left|\tau(t) - \frac{t}{\mu}\right| > t\varepsilon\right) dt = \frac{\sigma^2}{\mu^3}.$$

Observe the relation to Theorem 7.9, in that the normalization is the same and that $\sigma^2/\mu^3$ here plays the role of $\sigma^2$ there. □

The corresponding result for record times and the associated counting process has been proved in [115].

**Theorem 7.12.** *Suppose that $X_1$, $X_2$, ... are independent, identically distributed, absolutely continuous random variables, let $\{L_n, n \geq 1\}$ be the record times, and $\{\mu(n), n \geq 1\}$ the counting process. Then, for $1 \leq p < 2$, and $\delta > -1$,*

$$\lim_{\varepsilon \searrow 0} \varepsilon^{\frac{2p(1+\delta)}{2-p}} \sum_{n \geq 3} \frac{(\log n)^\delta}{n} P(|\mu(n) - \log n| > (\log n)^{1/p}\varepsilon) = \frac{1}{1+\delta} E|N|^{\frac{2p(1+\delta)}{2-p}},$$

*and for $0 < p < 2$,*

$$\lim_{\varepsilon \searrow 0} \varepsilon^{\frac{2p}{2-p}(\frac{r}{p}-1)} \sum_{n=1}^\infty n^{\frac{r}{p}-2} P(|\log L(n) - n| \geq n^{1/p}\varepsilon) = \frac{p}{r-p} E|N|^{\frac{2p}{2-p}(\frac{r}{p}-1)}.$$

*where, again, $N$ is a standard normal random variable.*

## 7.9 A Short Outlook on Extensions

The central limit theorem, the remainder term estimates, all of this chapter, is devoted to limit distributions for sums of independent random variables with finite variances. Having reached this point, at the end of the chapter, there are two natural questions popping up.

The first one is "what can be said if the variance does not exist?"

The answer to this question leads to the theory of stable distributions, domains of attraction, and as an extension, to the theory of infinitely divisible distributions. A short(er) introduction and overview of these topics will be given in Chapter 9.

The second question is "what happens if the summands are no longer independent?"

There is an answer to this one too. There exist different dependence concepts. One is the martingale concept, that we shall meet in Chapter 10. Another one, which is more oriented toward extensions of the classical limit theorems, concerns different notions of mixing, to which we shall be introduced in Section 9.5.

Another generalization, that pertains in a different manner to the present chapter, is the extension to the space $C[0, 1]$ of continuous functions on the unit interval, which was briefly introduced in Subsection 2.3.2. Namely, in addition to considering the limit of $S_n$ as $n \to \infty$, we may study the whole path simultaneously.

Let $\xi_1, \xi_2, \ldots$ be independent, identically distributed random variables with mean 0 and variance $\sigma^2 \in (0, \infty)$, and consider the linearly interpolated, normalized partial sums, $S_n$, $n \geq 1$. For $0 \leq t \leq 1$ we thus construct a random element $X_n \in C[0, 1]$ via

$$X_n(t, \omega) = \frac{S_{k-1}(\omega)}{\sigma \sqrt{n}} + \frac{t - (k/n)}{1/n} \frac{\xi_k(\omega)}{\sigma \sqrt{n}} \quad \text{for} \quad t \in \left[\frac{k-1}{n}, \frac{k}{n}\right], \ k = 1, 2, \ldots, n.$$

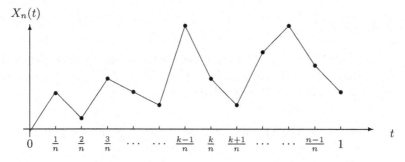

**Figure 7.2.** Linear interpolation of partial sums

By construction, $X_n(t, \omega)$ is a continuous function for every $\omega$, for every $n$. This means that $X_n$ is, indeed, a random element in the space $C[0, 1]$, which we endow with the uniform topology. The point is that one can show that $X_n$ behaves asymptotically like standard Brownian motion, also called the Wiener process, on the unit interval, that is, as a stochastic process with independent, stationary normally distributed increments, and with continuous sample paths. Formally, if $W = \{W(t), 0 \leq t \leq 1\}$ is such a process, then, for $0 \leq s < t \leq 1$,

$$W(t) - W(s) \in N(0, t - s).$$

To prove this one shows that all finite-dimensional distributions converge to the corresponding multivariate normal distribution, and that the sequence $\{X_n, n \geq 1\}$ is tight in the space $C[0, 1]$ (for random variables the concept of tightness was introduced in Section 5.8). The theorem thus obtained is called *Donsker's theorem*.

**Theorem 7.13.** *In the above setting,*

$$X_n \overset{d}{\to} W \quad as \quad n \to \infty,$$

*where $W$ is the Wiener measure on $[0, 1]$.*

*Remark 7.6.* The arrow $\overset{d}{\to}$ is to be interpreted as convergence of measures in the space $C[0, 1]$, which is defined in analogy with the second definition of distributional convergence, Definition 5.1.5. Recall also the remark immediately following that definition.                                                                              □

The reference for Donsker's theorem is [63]. For this and a lot more in the area we refer to [20].

Theorems of this kind are called (weak) *invariance principles* or *functional central limit theorems*. The reason for the first name is that, by applying a continuous mapping theorem (cf. Subsection 5.10.1), one can prove the analog of Theorem 5.10.4, and, as a consequence, find distributions of functionals of the Wiener measure by computing limits of the corresponding functional of $X_n$ for a suitable distribution, which typically may be the simple, symmetric random walk. Once again, consult [20]!

# 8 Problems

1. Suppose that $X_1, X_2, \ldots$ are independent random variables, such that $X_k \in \text{Be}(p_k)$, $k \geq 1$, and set $S_n = \sum_{k=1}^n X_k$, $m_n = \sum_{k=1}^n p_k$, and $s_n^2 = \sum_{k=1}^n p_k(1 - p_k)$, $n \geq 1$. Show that

$$\frac{S_n - m_n}{s_n} \xrightarrow{d} N(0,1) \quad \text{as} \quad n \to \infty \quad \Longleftrightarrow \quad \sum_{n=1}^\infty p_k(1 - p_k) = +\infty.$$

2. Suppose that $X_1, X_2, \ldots$ are independent $U(-1,1)$-distributed random variables, let $m_k$, $k \geq 1$, be a sequence of non-decreasing positive integers, and set $S_n = \sum_{k=1}^n X_k^{m_k}$, $n \geq 1$. Prove that

$$\frac{S_n - E\, S_n}{\sqrt{\text{Var}\, S_n}} \xrightarrow{d} N(0,1) \quad \text{as} \quad n \to \infty \quad \Longleftrightarrow \quad \sum_{n=1}^\infty \frac{1}{m_k} = +\infty.$$

3. Let $X_1, X_2, \ldots$ be independent random variables, such that

$$P(X_k = k^\alpha) = P(X_k = -k^\alpha) = 1/2, \quad k \geq 1,$$

where $\alpha \in \mathbb{R}$. Prove that the central limit theorem applies iff $\alpha \geq -1/2$.

4. Suppose that $X_k \in U(-k^\alpha, k^\alpha)$, where $\alpha \in \mathbb{R}$, and that $X_1, X_2, \ldots$ are independent. Determine the limit distribution (whenever it exists) of $\sum_{k=1}^n X_k$, suitably normalized, as $n \to \infty$.

5. Let $X_1, X_2, \ldots$ be independent, identically distributed random variables with mean 0 and variance $\sigma^2$, and set $S_n = \sum_{k=1}^n X_k$, $n \geq 1$. Determine $\gamma > 0$ so that

$$\frac{\sum_{k=1}^n S_k}{n^\gamma} \xrightarrow{d} N(0, b^2) \quad \text{as} \quad n \to \infty,$$

and determine $b^2 > 0$.

6. Prove that

$$\lim_{n \to \infty} e^{-n} \sum_{k=0}^n \frac{n^k}{k!} = \frac{1}{2}.$$

7. Let $X_1$, $X_2$, ... be independent random variables, such that

$$P(X_k = \sqrt{k}) = P(X_k = -\sqrt{k}) = 1/2, \quad k \geq 1,$$

and set

$$Y_n = \prod_{k=1}^{n} (1 + X_k)^{1/\sqrt{\log n}}, \quad n \geq 1.$$

Prove that $Y_n$, suitably normalized, converges in distribution as $n \to \infty$, and determine the limit distribution.

8. There exist central limit theorems in spite of the fact that the variance is infinite: Consider the two-sided Pareto distribution with density

$$f(x) = \begin{cases} \frac{1}{|x|^3}, & \text{for } |x| \geq 1, \\ 0, & \text{otherwise.} \end{cases}$$

(a) Prove that the characteristic function

$$\varphi(t) = 1 - t^2 \left( \log \frac{1}{|t|} + \mathcal{O}(1) \right) \quad \text{as } t \to 0.$$

(b) Let $X_1$, $X_2$, ... be independent, identically distributed random variables with common density as above, and set $S_n = \sum_{k=1}^{n} X_k$. Prove that

$$\frac{S_n}{\sqrt{n \log n}} \xrightarrow{d} N(0,1) \quad \text{as } n \to \infty.$$

♣ Note that the normalization is *not* the usual $\sqrt{n}$.

9. Consider the setup of the previous problem. For an alternative proof, set $Y_n = X_n I\{|X_n| \leq b_n\}$, for some "convenient" choice of $b_n$.
   (a) Compute $E Y_n$ and $\operatorname{Var} Y_n$.
   (b) Show that $Y_1$, $Y_2$, ..., $Y_n$ satisfies Lyapounov's condition, and state the corresponding limit theorem for $\sum_{k=1}^{n} Y_k$.
   (c) Use convergence equivalence or distributional equivalence, depending on the choice of truncation, to prove the asymptotic normality of $S_n$ as stated in the previous problem.

10. Show that the central limit theorem cannot be extended to convergence in probability, that is, if $X_1$, $X_2$, ... are independent, identically distributed random variables with finite variance, then the sum, suitably normalized, converges *in distribution* to the standard normal distribution but *not in probability*.

11. (a) Prove the central limit theorem for arrays, Theorem 2.4.
    (b) Check that assumption (2.21) does not restrict generality.
    (c) Specialize to rediscover Theorem 2.1.

12. The central limit theorem for uniformly bounded random variables is rather immediate. Here is an analog for arrays: Let $\{(X_{n,k}, k \geq 1), n \geq 1\}$

be an array of row-wise independent random variables, such that, for every $n \geq 1$,

$$|X_{n,k}| \leq A_n \quad \text{for} \quad k = 1, 2, \ldots, n,$$

where $A_n$, $n \geq 1$, are positive numbers, such that

$$\frac{A_n}{s_n} \to 0 \quad \text{as} \quad n \to \infty.$$

Show, by verifying the Lyapounov and/or Lindeberg conditions, that the central limit theorem holds.

♣ Note that the random variables are uniformly bounded *in each row*, but that the bounds (may) tend to infinity.

13. An extension of Lyapounov's theorem, Theorem 2.2: Let $X_1$, $X_2$, ... be independent random variables with mean 0. Suppose that $g$ is a non-negative function, such that $g(x)/x^2 \nearrow +\infty$ as $x \to \infty$, and such that $E\,g(X_k) < \infty$ for all $k$. Finally, set $S_n = \sum_{k=1}^n X_k$ and $s_n^2 = \sum_{k=1}^n \operatorname{Var} X_k$, $n \geq 1$. Show that, if, for every $\varepsilon > 0$,

$$\frac{\sum_{k=1}^n E\,g(X_k)}{g(\varepsilon s_n)} \to 0 \quad \text{as} \quad n \to \infty,$$

then

$$\frac{S_n}{s_n} \xrightarrow{d} N(0, 1) \quad \text{as} \quad n \to \infty.$$

♣ For $g(x) = |x|^{2+\delta}$ the result reduces to Theorem 2.2. The case $g(x) = x^2 \log^+ |x|$ provides a true extension, and, moreover, does not necessitate $E|X|^r < \infty$ for any $r > 2$.

♣ We have tacitly used the fact that the variances are finite. Why is this in order?

14. Suppose that $X$ and $Y$ are independent, identically distributed random variables with mean 0 and variance 1, such that

$$\frac{X + Y}{\sqrt{2}} \stackrel{d}{=} X.$$

Show that $X$ and $Y$ are standard normal.

♣ If we know that $X$ and $Y$ are standard normal, then it is easy to verify the relation. The idea thus is to prove the converse.

15. We know, or else, it is not hard to prove, that if $X$ and $Y$ are independent standard normal random variables, then $X + Y$ and $X - Y$ are independent. This problem is devoted to a converse. Suppose that $X$ and $Y$ are independent *symmetric* random variables with mean 0 and variance 1, such that $X + Y$ and $X - Y$ are independent. Let $\varphi$ denote the common characteristic function of $X$ and $Y$. Prove, along the following path, that $X$ and $Y$ must be standard normal random variables:
    (a) Prove that $\varphi(2t) = (\varphi(t))^4$.
    (b) Iterate.

16. Consider the previous problem without assuming symmetry.

    (a) Prove that $\varphi(2t) = (\varphi(t))^3 \overline{\varphi(t)}$.

    (b) Prove that $\varphi(t) \neq 0$ for all $t$.

    (c) Set $\gamma(t) = \varphi(t)/\overline{\varphi(t)}$ and show that $\gamma(2t) = (\gamma(t))^2$.

    (d) Prove that $\gamma(t) \equiv 1$.

    (e) Prove that $\varphi(2t) = (\varphi(t))^4$.

    (f) Iterate.

    ♣ Note that the distribution must be symmetric as a consequence of (d). Why?

17. *The alternating renewal process.* Consider a machine that alternates be-
    tween being busy and idle, or a person at a cash register who alternates
    between serving customers and reading a novel. Let $Y_1, Y_2, \ldots$ denote
    the duration of the busy periods, and $Z_1, Z_2, \ldots$ the durations of the idle
    periods, and suppose that a busy period starts at time 0. This means, i.a.,
    that $T_n = \sum_{k=1}^n (Y_k + Z_k)$ equals the time that has elapsed after $n$ visits
    to each of the two states, and that $S_n = \sum_{k=1}^n Y_k$ equals the amount of
    time spent in the busy period at time $T_n$. Let $L(t)$ denote the *availability*
    during the time span $(0, t]$, that is, the relative amount of time spent in
    the busy period during that time span.

    (a) Prove that, if $\mu_y = EY < \infty$ and $\mu_z = EZ < \infty$, then

    $$\frac{L(t)}{t} \overset{a.s.}{\to} \frac{\mu_y}{\mu_y + \mu_z} \quad \text{as} \quad t \to \infty,$$

    and interpret why this is a reasonable answer.

    (b) Prove that if, in addition, $\sigma_y^2 = \text{Var}\, Y < \infty$ and $\sigma_z^2 = \text{Var}\, Z < \infty$, then

    $$\frac{L(t) - \frac{\mu_y}{\mu_y+\mu_z}t}{\sqrt{t}} \overset{d}{\to} N\left(0, \frac{\mu_y^2\sigma_z^2 + \mu_z^2\sigma_y^2}{(\mu_y+\mu_z)^3}\right) \quad \text{as} \quad t \to \infty.$$

    ♠ Note that $\mu_y Z_k - \mu_z Y_k$, $k \geq 1$, are independent, identically distributed
    random variables with mean 0.

18. Let $X_1, X_2, \ldots$ be independent, identically distributed, non-negative,
    random variables with finite variance. Find the limit distribution of $(S_n)^p$
    as $n \to \infty$ $(p > 0)$.

    ♣ The case $p = 1/2$ is special, because the conjugate rule can be exploited.

19. Consider an experiment with success probability $p$. The *entropy* is

    $$H(p) = -p\log p - (1-p)\log(1-p).$$

    If $p$ is unknown a natural (and best) estimate after $n$ repetitions is the
    relative frequency of successes, $p_n^* = X_n/n$, where $X_n$ equals the number
    of successes. In order to find an estimate of $H(p)$ it seems reasonable to
    try $H(p_n^*)$. Determine the asymptotic distribution of $H(p_n^*)$ as $n \to \infty$.

    ♠ Do not forget to distinguish between the cases $p = 1/2$ and $p \neq 1/2$.

    ♣ We have thus replaced the estimate of the function with the function of the
    estimate.

20. Let $X_n \in \Gamma(n,1)$, $n \geq 1$. Prove that

$$\frac{X_n - n}{\sqrt{X_n}} \xrightarrow{d} N(0,1) \quad \text{as} \quad n \to \infty.$$

21. *Problem 6.13.3(a), cont'd.* Let $X$, $X_1$, $X_2$, $\ldots$, $X_n$ be independent, identically distributed random variables with $E\,X^4 < \infty$, and set $\mu = E\,X$, $\sigma^2 = \operatorname{Var} X$, and $\mu_4 = E(X - \mu)^4$. Furthermore, let $\bar{X}_n$ and $s_n^2$ denote the sample mean and sample variances, respectively. Show that

$$\frac{s_n^2 - \sigma^2}{\sqrt{n}} \xrightarrow{d} N(0, \mu_4 - \sigma^4) \quad \text{as} \quad n \to \infty.$$

22. *Problem 6.13.3(b), cont'd.* Let $(X_k, Y_k)'$, $1 \leq k \leq n$, be a sample from a two-dimensional distribution with component-wise arithmetic means $\bar{X}_n = \frac{1}{n} \sum_{k=1}^{n} X_k$, and $\bar{Y}_n = \frac{1}{n} \sum_{k=1}^{n} Y_k$, respectively, and set $\mu_x = E\,X_1$, $\sigma_x^2 = \operatorname{Var} X_1$ $\mu_y = E\,Y_1$, and $\sigma_y^2 = \operatorname{Var} Y_1$. Suppose, in addition, that the correlation coefficient $\rho = 0$, and that

$$\mu_{xy}^2 = E(X_1 - \mu_x)^2 (Y_1 - \mu_y)^2 < \infty.$$

Prove the following limit theorem for the empirical correlation coefficient:

$$\sqrt{n} r_n \xrightarrow{d} N(0, b^2) \quad \text{as} \quad n \to \infty,$$

and determine $b^2$.

23. *Self-normalized sums.* Let $X_1$, $X_2$, $\ldots$ be independent, identically distributed random variables with mean 0, and set $S_n = \sum_{k=1}^{n} X_k$, $n \geq 1$. Consider the statistic

$$T_n(p) = \frac{S_n}{\left( \sum_{k=1}^{n} |X_k|^p \right)^{1/p}},$$

where $0 < p < \infty$. For "$p = \infty$" the statistic is interpreted as $S_n/Y_n$, where $Y_n = \max_{1 \leq k \leq n} |X_k|$, and suppose that $\operatorname{Var} X < \infty$. Show that $T_n(2)$, suitably normalized, converges to a standard normal distribution as $n \to \infty$.
   ♣ In terms of Euclidean norms the denominator equals $\|\mathbf{X}_n\|_p$, where $\mathbf{X}_n = (X_1, X_2, \ldots, X_n)$, and if we let $\mathbf{1}_n$ denote the vector $(1, 1, \ldots, 1)'$, we may rewrite the statistic, using scalar products as

$$T_n(p) = \frac{\mathbf{X}_n \cdot \mathbf{1}_n}{\|\mathbf{X}_n\|_p}.$$

For $p = 2$ the statistic is reminiscent of the $t$-statistic if the sample is from a normal distribution.

24. Extend Theorem 5.1 to the Lindeberg-Lévy case.

25. We found in Section 4.3 that if two characteristic functions agree on an interval around 0, then this was not enough for the distributions to coincide. However, a certain closeness can be estimated. Namely, let $U$ and $V$ be random variables with characteristic functions $\varphi_U$ and $\varphi_V$, respectively, and suppose that

$$\varphi_U(t) = \varphi_V(t) \quad \text{for} \quad |t| \leq T.$$

(a) Prove that, if $F_V$ is differentiable, then

$$\sup_{x \in \mathbb{R}} |F_U(x) - F_V(x)| \leq C \frac{\sup_x F_V'(x)}{T}.$$

(b) Show that

$$\int_{-\infty}^{\infty} |F_U(x) - F_V(x)| \, dx \leq \frac{C}{T}.$$

# 8

# The Law of the Iterated Logarithm

The central limit theorem tells us that suitably normalized sums can be approximated by a normal distribution. Although arbitrarily large values may occur, and will occur, one might try to bound the magnitude in some manner. This is what the law of the iterated logarithm (LIL) does, in that it provides a parabolic bound on how large the oscillations of the partial sums may be as a function of the number of summands.

In Theorem 6.9.1 we presented Borel's theorem [26], stating that almost all numbers are normal. Later Khintchine [159], proved that, if $N_n$ equals the number of ones among the first $n$ decimals in the binary expansion of a number (in the unit interval), then

$$\limsup_{n \to \infty} \frac{N_n - n/2}{\sqrt{\frac{1}{2}n \log \log n}} = 1 \quad \text{a.s.}$$

By symmetry, the liminf equals $-1$ almost surely. We also observe that $n/2$ is the expected number of ones. The conclusion thus tells us that the fluctuations around the expected value stay within a precisely given parabola except, possibly, for a finite number of visits outside. So:

- To what extent can this be generalized?
- The result provides the extreme limit points. Are there any more?

The answer to these questions belong to the realm of *the law of the iterated logarithm*. If the law of large numbers and the central limit theorem are the two most central and fundamental limit theorems, the law of the iterated logarithm is the hottest candidate for the third position.

In this chapter the main focus is on the Hartman-Wintner-Strassen theorem, which deals with the i.i.d. case; Hartman and Wintner [130] proved the sufficiency, and Strassen [238] the necessity. The theorem exhibits the extreme limit points. We shall also prove an extension, due to de Acosta [2], to the effect that the interval whose endpoints are the extreme limit points is a cluster set, meaning that, almost surely, the set of limit points is equal to the interval whose endpoints are the extreme limit points. During the course of the proof we find results for subsequences from [109] and [245], respectively, where it, for example, is shown that the cluster set shrinks for rapidly increasing subsequences.

The "Some Additional Results and Remarks" section contains a proof of how one can derive the general law from the result for normal random variables via the Berry-Essen theorem (under the assumption of an additional third moment), more on rates, and additional examples and complements.

# 1 The Kolmogorov and Hartman-Wintner LILs

In a seminal paper, Kolmogorov [161] proved the following result for independent, not necessarily identically distributed, random variables.

**Theorem 1.1.** *Suppose that $X_1$, $X_2$, ... are independent random variables with mean 0 and finite variances $\sigma_k^2$, $k \geq 1$, set $S_n = \sum_{k=1}^{n} X_k$, and $s_n^2 = \sum_{k=1}^{n} \sigma_k^2$, $n \geq 1$. If*

$$|X_n| \leq o\left(\frac{s_n}{\sqrt{\log \log s_n}}\right) \quad for \ all \quad n, \tag{1.1}$$

*then*

$$\limsup_{n \to \infty} (\liminf_{n \to \infty}) \frac{S_n}{\sqrt{2s_n^2 \log \log s_n^2}} = +1 \, (-1) \quad a.s.$$

*Remark 1.1.* Marcinkiewicz and Zygmund [181] provide an example which shows that the growth condition (1.1) cannot in general be weakened.    □

Condition (1.1) looks strange if one also assumes that the random variables are identically distributed; the condition should disappear.

The sufficiency part of the following result is due to Hartman and Wintner [130]. The necessity is due to Strassen [238].

**Theorem 1.2.** *Suppose that $X$, $X_1$, $X_2$, ... are independent random variables with mean 0 and finite variance $\sigma^2$, and set $S_n = \sum_{k=1}^{n} X_k$, $n \geq 1$. Then*

$$\limsup_{n \to \infty} (\liminf_{n \to \infty}) \frac{S_n}{\sqrt{2\sigma^2 n \log \log n}} = +1 \, (-1) \quad a.s. \tag{1.2}$$

*Conversely, if*

$$P\left(\limsup_{n \to \infty} \frac{|S_n|}{\sqrt{n \log \log n}} < \infty\right) > 0,$$

*then $E\,X^2 < \infty$, $E\,X = 0$, and (1.2) holds.*

**Exercise 1.1.** Check that Khintchine's result is a special case.    □

*Remark 1.2.* As was mentioned in connection with the Kolmogorov zero-one law, Theorem 2.10.6, the limsup is finite with probability 0 or 1, and, if finite, the limit equals a constant almost surely. Thus, assuming in the converse that the probability is positive is in reality assuming that it is equal to 1.

*Remark 1.3.* By symmetry it suffices to consider the limit superior (replace $X$ by $-X$). □

As a distinction between convergence in probability and point-wise behavior we observe that, by Chebyshev's inequality,

$$P\left(\left|\frac{S_n}{\sqrt{2\sigma^2 n \log\log n}}\right| > \varepsilon\right) \leq \frac{1}{2\varepsilon^2 \log\log n} \to 0 \quad \text{as} \quad n \to \infty,$$

so that

$$\frac{S_n}{\sqrt{2\sigma^2 n \log\log n}} \xrightarrow{p} 0 \quad \text{as} \quad n \to \infty,$$

under the assumptions of Theorem 1.2. In other words in $\frac{S_n}{\sqrt{2\sigma^2 n \log\log n}}$ is close to 0 with a probability close to 1 for large $n$, but sample-wise, or path-wise, one observes oscillations between $-1$ and $+1$.

## 1.1 Outline of Proof

Proofs of laws of the iterated logarithm are rather technical. The first, and a central, tool are the Kolmogorov upper and lower exponential bounds for tail probabilities. The proof of the upper bound is manageable and will be presented, but we omit the proof of the lower bound.

The second step is to apply these bounds to Borel-Cantelli sums for a geometrically increasing subsequence of the partial sums. The convergence part and the first Borel-Cantelli lemma then provide an upper bound for the limit superior along that subsequence, after which an application of the Lévy inequalities takes care of the behavior between subsequence points.

As for the divergence part we must establish a divergence part for increments, since the partial sums themselves are not independent. An application of the second Borel-Cantelli lemma and some additional arguments then provide a lower bound – which coincides with the upper bound.

## 2 Exponential Bounds

Let $Y_1, Y_2, \ldots$ be independent random variables with mean 0, set $\sigma_k^2 = \operatorname{Var} Y_k$ $k \geq 1$, and, for $n \geq 1$, $s_n^2 = \sum_{k=1}^n \sigma_k^2$. Finally, suppose that, for $c_n > 0$,

$$|Y_k| \leq c_n s_n \text{ a.s.} \quad \text{for} \quad k = 1, 2, \ldots, n, \ n \geq 1.$$

**Lemma 2.1.** (The upper exponential bound)
*For $0 < x < 1/c_n$,*

$$P\left(\sum_{k=1}^{n} Y_k > x s_n\right) \leq \exp\left\{-\frac{x^2}{2}\left(1 - \frac{x c_n}{2}\right)\right\}.$$

*Proof.* In order to apply Markov's inequality, Theorem 3.1.1, we first derive the following refinement of Theorem 3.1.2:

$$\psi_n(t) = E \exp\left\{t\sum_{k=1}^{n} Y_k\right\} \leq \exp\left\{\frac{t^2 s_n^2}{2}\left(1 + \frac{t c_n s_n}{2}\right)\right\}. \qquad (2.1)$$

Let $n \geq 1$, and $0 < t < 1/(c_n s_n)$. For $1 \leq k \leq n$, Taylor expansion, and the fact that the mean equals 0, yield

$$\psi_{Y_k}(t) = E \exp\{t Y_k\} = 1 + \sum_{j=2}^{\infty} \frac{t^j}{j!} E(Y_k^j)$$

$$\leq 1 + \frac{t^2}{2!} E(Y_k^2)\left(1 + 2\sum_{j=3}^{\infty} \frac{(t c_n s_n)^{j-2}}{j!}\right)$$

$$\leq 1 + \frac{t^2 \sigma_k^2}{2}\left(1 + 2 t c_n s_n \sum_{j=3}^{\infty} \frac{1}{j!}\right)$$

$$\leq 1 + \frac{t^2 \sigma_k^2}{2}\left(1 + 2 t c_n s_n (e - 2.5)\right)$$

$$\leq 1 + \frac{t^2 \sigma_k^2}{2}\left(1 + \frac{t c_n s_n}{2}\right) \leq \exp\left\{\frac{t^2 \sigma_k^2}{2}\left(1 + \frac{t c_n s_n}{2}\right)\right\},$$

the last inequality being a consequence of the inequality $1 + x < e^x$.
By independence we then obtain

$$\psi_n(t) = \prod_{k=1}^{n} \psi_{Y_k}(t) \leq \exp\left\{\frac{t^2}{2}\sum_{k=1}^{n} \sigma_k^2\left(1 + \frac{t c_n s_n}{2}\right)\right\} = \exp\left\{\frac{t^2}{2} s_n^2\left(1 + \frac{t c_n s_n}{2}\right)\right\},$$

which completes the proof of (2.1), after which an application of Markov's inequality with $t = x/s_n$ yields

$$P\left(\sum_{k=1}^{n} Y_k > x s_n\right) \leq \frac{\psi_n(t)}{e^{t x s_n}} \leq \exp\left\{-t x s_n + \frac{t^2 s_n^2}{2}\left(1 + \frac{t c_n s_n}{2}\right)\right\}$$

$$= \exp\left\{-x^2 + \frac{x^2}{2}\left(1 + \frac{x c_n}{2}\right)\right\}. \qquad \square$$

**Lemma 2.2.** (The lower exponential bound)
*Suppose that $\gamma > 0$. There exist constants $x(\gamma)$ and $\kappa(\gamma)$, such that, for $x(\gamma) \leq x \leq \kappa(\gamma)/c_n$,*

$$P\left(\sum_{k=1}^{n} Y_k > x s_n\right) \geq \exp\left\{-\frac{x^2}{2}(1 + \gamma)\right\}.$$

As has already been mentioned, this one is no pleasure to prove. We refer to [239], Theorem 5.2.2, for details.

In order to make believe that the exponential bounds are in order, suppose, for a moment, that $Y_1$, $Y_2$, ... are independent $N(0, \sigma^2)$-distributed random variables. The exponential bounds

$$P\left(\sum_{k=1}^{n} Y_k > x\sigma\sqrt{n}\right) \begin{cases} \leq \frac{1}{x} \cdot \exp\{-\frac{1}{2}x^2\}, & \text{for } x > 0, \\ \geq \frac{1}{2x} \cdot \exp\{-\frac{1}{2}x^2\}, & \text{for } x > \sqrt{2}, \end{cases} \quad (2.2)$$

are immediate consequences of the fact that $\sum_{k=1}^{n} Y_k \overset{d}{=} \sqrt{n}Y_1$, and Mill's ratio, Lemma A.2.1.

*Remark 2.1.* Note that the bounds are somewhat sharper than in the general case in that the exponent in the right-hand side is exactly equal to $-x^2/2$. On the other hand, here we are dealing with an exact normal distribution, not an approximate one.                                                    □

# 3 Proof of the Hartman-Wintner Theorem

The first step is to investigate Borel-Cantelli sums for geometrically increasing subsequences.

**Lemma 3.1.** *Let $X$, $X_1$, $X_2$, ... be independent, identically distributed random variables with mean 0, finite variance, $\sigma^2$, and partial sums $S_n = \sum_{k=1}^{n} X_k$, $n \geq 1$. Furthermore, let $\lambda > 1$, and set $n_k = [\lambda^k]$. Then*

$$\sum_{k=1}^{\infty} P(|S_{n_k}| > \varepsilon\sqrt{n_k \log\log n_k}) < \infty \quad \text{for } \varepsilon > \sigma\sqrt{2}, \quad (3.1)$$

$$\sum_{k=1}^{\infty} P(S_{n_k} > \varepsilon\sqrt{n_k \log\log n_k}) = \infty \quad \text{for } \varepsilon < \sigma\sqrt{2}. \quad (3.2)$$

Since the proof is rather technical we begin by checking the normal case.

**Borel-Cantelli Sums; Normal Random Variables**

Let $X_1$, $X_2$, ... be independent $N(0, \sigma^2)$-distributed random variables with partial sums $S_n$, $n \geq 1$. Since, in view of the exponential bounds,

$$\sum_{n=1}^{\infty} P(S_n > \varepsilon\sqrt{n \log\log n}) \sim \sum_{n=1}^{\infty} \frac{\sigma}{\varepsilon\sqrt{\log\log n}} \cdot (\log n)^{-\frac{\varepsilon^2}{2\sigma^2}} = \infty$$

for *any* $\varepsilon > 0$, there is no hope for an application of the Borel-Cantelli lemma to the whole sequence. We must resort to subsequences.

Toward this end, let $\lambda > 1$, set $n_k = [\lambda^k]$, and let $k_0$ be so large that

$$n_{k+1} > n_k + 1, \quad 1 < \frac{n_{k+1}}{n_k} < \lambda^2 \quad \text{and} \quad 1 < \frac{n_{k+1}\log\log n_{k+1}}{n_k \log\log n_k} < \lambda^3; \quad (3.3)$$

note that these requirements can always be met, since the ratio $n_{k+1}/n_k \to \lambda > 1$ as $k \to \infty$. Actually, one should throughout think of $\lambda$ as being "close" to 1.

In order to prove (3.1) we use the upper exponential bound from (2.2) (once for each tail) and (3.3):

$$\sum_{k=k_0}^{\infty} P(|S_{n_k}| > \varepsilon\sqrt{n_k \log\log n_k}) \leq 2 \sum_{k=k_0}^{\infty} \frac{\sigma}{\varepsilon\sqrt{\log\log n_k}} \cdot (\log n_k)^{-\frac{\varepsilon^2}{2\sigma^2}}$$

$$\leq \frac{2\sigma}{\varepsilon} \sum_{k=k_0}^{\infty} \left(\log\left(\lambda^{k-1}\right)\right)^{-\frac{\varepsilon^2}{2\sigma^2}} = \frac{2\sigma}{\varepsilon} \sum_{k=k_0}^{\infty} \left((k-1)\log\lambda\right)^{-\frac{\varepsilon^2}{2\sigma^2}}.$$

For the lower bound we use the lower bound:

$$\sum_{k=k_0}^{\infty} P(S_{n_k} > \varepsilon\sqrt{n_k \log\log n_k}) \geq \sum_{k=k_0}^{\infty} \frac{\sigma}{2\varepsilon\sqrt{\log\log n_k}} \cdot (\log n_k)^{-\frac{\varepsilon^2}{2\sigma^2}}$$

$$\geq \frac{\sigma}{2\varepsilon} \sum_{k=k_0}^{\infty} \frac{1}{\sqrt{\log(\lambda\log k)}} (k\log\lambda)^{-\frac{\varepsilon^2}{2\sigma^2}}.$$

This establishes the lemma for normal random variables.

## Borel-Cantelli Sums; The General Case – Convergence

We proceed by truncation. The Kolmogorov LIL forces a truncation at $o(s_n/\sqrt{\log\log s_n}) = o(\sqrt{n/\log\log n})$. The typical upper truncation when the variance is finite is $\mathcal{O}(\sqrt{n})$. Unfortunately, this leaves a gap between the two levels, which forces us to truncate at two levels. And, although narrow, it turns out that the middle part is the hardest one to deal with.

Thus, let $X, X_1, X_2, \ldots$ be independent, identically distributed random variables with mean 0 and finite variance $\sigma^2$, and set $S_n = \sum_{k=1}^{n} X_k$, $n \geq 1$. Let $0 < \delta < 1/2$, and set, for $\varepsilon > 0$,

$$c_n = \frac{2\delta\sigma}{\varepsilon\sqrt{\log\log n}} \quad \text{and} \quad b_n = c_n\sigma\sqrt{n} = \frac{2\delta\sigma^2}{\varepsilon}\sqrt{\frac{n}{\log\log n}}, \quad n \geq 1. \quad (3.4)$$

For each $n \geq 1$ we define, for $j = 1, 2, \ldots, n$, the truncated random variables

$$X'_{n,j} = X_j I\{|X_j| < \tfrac{1}{2}b_n\}, \quad X''_{n,j} = X_j I\{|X_j| > \sqrt{n}\},$$
$$X'''_{n,j} = X_j I\{\tfrac{1}{2}b_n \leq |X_j| \leq \sqrt{n}\},$$

and note that $X'_{n,j} + X''_{n,j} + X'''_{n,j} = X_j$ for all $j$. We also tacitly suppose that $\frac{1}{2}b_n < \sqrt{n}$.

In addition we define the respective partial sums

$$S'_n = \sum_{k=1}^{n} X'_{n,j}, \quad S''_n = \sum_{k=1}^{n} X''_{n,j}, \quad S'''_n = \sum_{k=1}^{n} X'''_{n,j},$$

so that $S'_n + S''_n + S'''_n = S_n$.

The three sums will now be taken care of separately. Remember that the goal was to prove convergence of Borel-Cantelli sums for the subsequence $\{S_{n_k}, k \geq 1\}$, where $n_k = [\lambda^k]$ for some $\lambda > 1$. We also remind the reader of $k_0$ defined in (3.3).

## $S'_n$

Since the truncated random variables need not have mean 0 we first have to estimate their means.

$$|E X'_{n,j}| = |E X_j I\{|X_j| < \tfrac{1}{2}b_n\}| = |-E X_j I\{|X_j| \geq \tfrac{1}{2}b_n\}|$$
$$\leq E|X_j| I\{|X_j| \geq \tfrac{1}{2}b_n\} \leq \frac{2}{b_n} E X^2 I\{|X| \geq \tfrac{1}{2}b_n\},$$

so that

$$|E S'_n| \leq n|E X'_{n,j}| \leq \frac{2n}{b_n} E X^2 I\{|X| \geq \tfrac{1}{2}b_n\} = o(n/b_n) = o(\sqrt{n \log\log n}),$$

as $n \to \infty$, that is, for our given $\delta$ above, there exists $n_0$, such that

$$|E S'_n| \leq \varepsilon\delta\sqrt{n \log\log n} \quad \text{for} \quad n > n_0. \tag{3.5}$$

Moreover,

$$\text{Var } X'_{n,1} \leq E\big((X'_{n,1})^2\big) \leq E X^2 = \sigma^2.$$

By centering $S'_{n_k}$ via (3.5), and applying the upper exponential bound, Lemma 2.1, with $x = \frac{\varepsilon(1-\delta)}{\sigma}\sqrt{\log\log n_k}$, and $t = 2\delta/c_{n_k}$ (once for each tail), we then obtain

$$P(|S'_{n_k}| > \varepsilon\sqrt{n_k \log\log n_k}) \leq P(|S'_{n_k} - E S'_{n_k}| > \varepsilon(1-\delta)\sqrt{n_k \log\log n_k})$$
$$\leq 2\exp\left\{-\frac{1}{2}\frac{\varepsilon^2(1-\delta)^2}{\sigma^2}\log\log n_k(1-\delta)\right\}$$
$$= 2\exp\left\{-\frac{\varepsilon^2}{2\sigma^2}(1-\delta)^3\log\log n_k\right\} = 2(\log n_k)^{-\frac{\varepsilon^2}{2\sigma^2}(1-\delta)^3}, \tag{3.6}$$

for $k$ large, more precisely, for $k \geq k_0$ and $n_{k_0} \geq n_0$. Redefining, $k_0$, if necessary, so that these conditions are met, we have shown that

$$\sum_{k=k_0}^{\infty} P(|S'_{n_k}| > \varepsilon\sqrt{n_k \log\log n_k}) \leq 2\sum_{k=k_0}^{\infty} (\log n_k)^{-\frac{\varepsilon^2}{2\sigma^2}(1-\delta)^3}$$

$$\leq 2\sum_{k=k_0}^{\infty} \left((k-1)\log\lambda\right)^{-\frac{\varepsilon^2}{2\sigma^2}(1-\delta)^3} < \infty \quad \text{for} \quad \varepsilon > \frac{\sigma\sqrt{2}}{(1-\delta)^{3/2}}. \quad (3.7)$$

## $S''_n$

Let $\eta > 0$. Since

$$\{|S''_n| > \eta\sqrt{n\log\log n}\} \subset \bigcup_{j=1}^{n}\{X''_{j,n} \neq 0\} = \bigcup_{j=1}^{n}\{|X_j| > \sqrt{n}\},$$

it follows that

$$\sum_{k=k_0}^{\infty} P(|S''_{n_k}| > \eta\sqrt{n_k \log\log n_k}) \leq \sum_{k=k_0}^{\infty} P\left(\bigcup_{j=1}^{n_k}\{|X_j| > \sqrt{n_k}\}\right)$$

$$\leq \sum_{k=k_0}^{\infty} n_k P(|X| > \sqrt{n_k}) \leq \sum_{k=k_0}^{\infty} \lambda^k P(|X| > \sqrt{\lambda^{k-1}})$$

$$= \lambda \sum_{k=k_0-1}^{\infty} \lambda^k P(X^2 > \lambda^k) < C E X^2 < \infty. \quad (3.8)$$

Here we used Theorem 2.12.4 to bound the last sum by the second moment.

## $S'''_n$

Let $\eta > 0$. We first estimate the truncated means.

$$|E X'''_{n,j}| = |E X_j I\{\tfrac{1}{2}b_n < |X_j| < \sqrt{n}\}| \leq E|X_j| I\{\tfrac{1}{2}b_n < |X_j| < \sqrt{n}\}$$

$$\leq E|X_j| I\{|X_j| \geq \tfrac{1}{2}b_n\} \leq \frac{2}{b_n} E X^2 I\{|X| \geq \tfrac{1}{2}b_n\},$$

which is the same upper bound as for $|E X'_{j,n}|$, so that (this time)

$$|E S'''_n| \leq \frac{\eta}{2}\sqrt{n\log\log n} \quad \text{for} \quad n > n_\eta.$$

By Chebyshev's inequality we thus have

$$P(|S'''_{n_k}| > \eta\sqrt{n_k \log\log n_k}) \leq P(|S'''_{n_k} - E S'''_{n_k}| > \frac{\eta}{2}\sqrt{n_k \log\log n_k})$$

$$\leq \frac{4\text{Var } S'''_{n_k}}{\eta^2 n_k \log\log n_k} \leq \frac{4E(X'''_{1,n_k})^2}{\eta^2 \log\log n_k} = \frac{4E X_k^2 I\{\tfrac{1}{2}b_{n_k} \leq |X_k| \leq \sqrt{n_k}\}}{\eta^2 \log\log n_k},$$

which (by redefining $k_0$ if necessary, so that also $n_{k_0} > n_\eta$) implies that

$$\frac{\eta^2}{4} \sum_{k=k_0}^{\infty} P(|S_{n_k}'''| > \eta\sqrt{n_k \log\log n_k}) \leq \sum_{k=k_0}^{\infty} \frac{E\, X_k^2 I\{\frac{1}{2}b_{n_k} \leq |X_k| \leq \sqrt{n_k}\}}{\log\log n_k}$$

$$= \sum_{k=k_0}^{\infty} \frac{1}{\log\log n_k} \int_{\frac{1}{2}b_{n_k}}^{\sqrt{n_k}} x^2 \,dF(x) = \int_{k_*}^{\infty} \Big( \sum_{A(k,x)} \frac{1}{\log\log n_k} \Big) x^2 \,dF(x),$$

where $k_*$ is some lower irrelevant limit, and

$$A(k,x) = \{k : \tfrac{1}{2}b_{n_k} \leq |x| \leq \sqrt{n_k}\}.$$

In order to invert the double inequality we first observe that, since the inverse of $\sqrt{y/\log\log y}$ is of the order $y^2 \log\log y$ as $y \to \infty$, we have

$$\{k : b_{n_k} \leq |x|\} \subset \Big\{k : n_k \leq 2\Big(\frac{\delta\sigma^2}{\varepsilon}\Big)^2 x^2 \log\log x\Big\} \quad \text{for} \quad k \geq k_1,$$

so that, for $k = k_2 \geq \max\{k_0, k_*, k_1\}$,

$$A(k,x) \subset \Big\{k : x^2 \leq n_k \leq 2\Big(\frac{\varepsilon}{\delta\sigma^2}\Big)^2 x^2 \log\log |x|\Big\},$$

which, via (3.3), yields

$$A(k,x) \subset A^*(k,x)$$
$$= \Big\{k : \frac{2\log|x|}{\log\lambda} + 1 \leq k \leq \frac{\log 2 + 2\log\frac{\varepsilon}{\delta\sigma^2} + 2\log|x| + \log\log\log|x|}{\log\lambda}\Big\}.$$

This tells us that, for any $x$,

$$\sum_{A(k,x)} \frac{1}{\log\log n_k} \leq \sum_{A^*(k,x)} \frac{1}{\log\log(x^2)} = \frac{1}{\log(2\log|x|)} \cdot |A^*(k,x)|$$

$$\leq \frac{\log 2 + 2\log\frac{\varepsilon}{\delta\sigma^2} + \log\log\log|x|}{\log\log|x|} = \frac{C + \log\log\log|x|}{\log\log|x|},$$

which, inserted into the integral, proves that, for all $\eta > 0$,

$$\sum_{k=k_0}^{\infty} P(|S_{n_k}'''| > \eta\sqrt{n_k \log\log n_k}) \leq \frac{4}{\eta^2} \int_{k_*}^{\infty} \Big( \sum_{A(k,x)} \frac{1}{\log\log n_k} \Big) x^2 \,dF(x)$$

$$\leq \frac{4}{\eta^2} \int_{k_*}^{\infty} \frac{C + \log\log\log|x|}{\log\log|x|} x^2 \,dF(x)$$

$$\leq C E\, X^2 \frac{\log\log\log|X|}{\log\log|X|} < \infty. \tag{3.9}$$

Putting the estimates for the three sums together via the inclusion

$$\{|S_{n_k}| > (\varepsilon + 2\eta)\sqrt{n_k \log\log n_k}\} \subset \{|S_{n_k}'| > \varepsilon\sqrt{n_k \log\log n_k}\}$$
$$\cup \{|S_{n_k}''| > \eta\sqrt{n_k \log\log n_k}\} \bigcup \{|S_{n_k}'''| > \eta\sqrt{n_k \log\log n_k}\},$$

shows that, for $\eta > 0$ and $\varepsilon + 2\eta > \frac{\sigma\sqrt{2}}{(1-3\delta)^{3/2}}$,

$$\sum_{k=1}^{\infty} P(|S_{n_k}| > (\varepsilon + 2\eta)\sqrt{n_k \log\log n_k}) < \infty,$$

which, $\delta$ and $\eta$ being arbitrary, proves (3.1).

*Remark 3.1.* Under the slightly stronger assumption that

$$EX^2 \log^+ \log^+ |X| < \infty,$$

it suffices to truncate at $\frac{1}{2}b_n$. Namely, in this case one can join $X_{n,j}''$ and $X_{n,j}'''$ into $X_j I\{|X_j| > \frac{1}{2}b_n\}$. By proceeding as above, the computation for $S_n''$ and $S_n'''$ then simplifies into

$$\sum_{k=k_0}^{\infty} P\left(\sum_{j=1}^{k} X_j I\{|X_j| > \tfrac{1}{2}b_{n_k}\}\right) \le \lambda \sum_{k=k_0-1}^{\infty} \lambda^k P(|X| > c\sqrt{\lambda^k / \log\log(\lambda^k)})$$
$$< CE\, X^2 \log^+ \log^+ |X| < \infty,$$

since, asymptotically, $x^2 \log\log x$ is the inverse of $\sqrt{x/\log\log x}$.    □

**Exercise 3.1.** Check the details.    □

### Borel-Cantelli Sums; The General Case – Divergence

For this half of the proof we use the lower exponential bound applied to $S_{n_k}'$. For the other two sums we use estimates (3.8) and (3.9) from above, since they are valid for *all* $\eta > 0$.

In addition to the truncated expectation, $E\,S_{n_k}'$, that has been estimated in (3.5), we need a lower bound for the truncated variance. Remembering that

$$(E\,X_{1,n})^2 = (-E\,XI\{|X| > \tfrac{1}{2}b_n\})^2 \le E\,X^2 I\{|X| > \tfrac{1}{2}b_n\},$$

via Lyapounov's inequality, Theorem 3.2.5, we obtain

$$\begin{aligned}
\text{Var } X_{1,n}' &= E\,X_{1,n}^2 - (E\,X_{1,n})^2 = E\,X^2 - E\,X^2 I\{|X| > \tfrac{1}{2}b_n\} - (E\,X_{1,n})^2 \\
&\ge \sigma^2 - 2E\,X^2 I\{|X| > \tfrac{1}{2}b_n\} \ge \sigma^2(1 - \delta),
\end{aligned}$$

for $n \ge n_1$, so that

$$\text{Var } S_{n_k}' \ge n_k \sigma^2(1 - \delta) \quad \text{for} \quad k > k_3.$$

Thus, via centering and the lower exponential bound (3.2), we conclude that, for $\delta > 0$ and $k > k_4 = \max\{k_2, k_3\}$,

$$P(S'_{n_k} > \varepsilon \sqrt{n_k \log \log n_k})$$

$$\geq P\left(S'_{n_k} - E\, S'_{n_k} > \varepsilon \frac{(1+\delta)}{\sigma\sqrt{1-\delta}} \mathrm{Var}\,(S'_{n_k}) \sqrt{\log \log n_k}\right)$$

$$\geq \exp\left\{-\frac{1}{2}\left(\frac{\varepsilon(1+\delta)}{\sigma\sqrt{1-\delta}}\sqrt{\log \log n_k}\right)^2 (1+\gamma)\right\}$$

$$= \left(\log n_k\right)^{-\frac{\varepsilon^2(1+\delta)^2(1+\gamma)}{2\sigma^2(1-\delta)}} \geq \left(k \log \lambda\right)^{-\frac{\varepsilon^2(1+\delta)^2(1+\gamma)}{2\sigma^2(1-\delta)}}, \qquad (3.10)$$

and, hence, that

$$\sum_{k=k_4}^{\infty} P(S'_{n_k} > \varepsilon \sqrt{n_k \log \log n_k}) \geq \sum_{k=k_4}^{\infty} \left(\log n_k\right)^{-\frac{\varepsilon^2(1+\delta)^2(1+\gamma)}{2\sigma^2(1-\delta)}}$$

$$\geq \sum_{k=k_4}^{\infty} \left(k \log \lambda\right)^{-\frac{\varepsilon^2(1+\delta)^2(1+\gamma)}{2\sigma^2(1-\delta)}} = \infty \quad \text{for} \quad \varepsilon < \frac{\sigma\sqrt{2}}{1+\delta}\sqrt{\frac{1-\delta}{1+\gamma}}.$$

By combining this estimate with those in (3.8), (3.9), and the inclusion

$$\left\{S'_{n_k} > \varepsilon\sqrt{n_k \log \log n_k}\right)\right\} \subset \left\{S_{n_k} > (\varepsilon - 2\eta)\sqrt{n_k \log \log n_k}\right)\right\}$$
$$\bigcup \left\{|S''_{n_k}| > \eta\sqrt{n_k \log \log n_k}\right)\right\} \bigcup \left\{|S'''_{n_k}| > \eta\sqrt{n_k \log \log n_k})\right\},$$

it follows that, for $\eta > 0$ and $\varepsilon - 2\eta < \frac{\sigma\sqrt{2}}{1+\delta}\sqrt{\frac{1-\delta}{1+\gamma}}$,

$$\sum_{k=1}^{\infty} P(S_{n_k} > (\varepsilon - 2\eta)\sqrt{n_k \log \log n_k}) = \infty,$$

which, in view of the arbitrariness of $\delta$, $\gamma$, and $\eta$, finishes the proof of (3.2), and thereby the proof of Lemma 3.1. □

## Applying the Borel-Cantelli Lemmas

In this part of the proof we apply the Borel-Cantelli lemmas to the convergence and divergence parts, respectively, of Lemma 3.1. For the convergence part this is straightforward, in that the first Borel-Cantelli lemma applied to (3.1) tells us that $P(S_{n_k} > \varepsilon\sqrt{n_k \log \log n_k} \text{ i.o.}) = 0$ almost surely for $\varepsilon > \sigma\sqrt{2}$, and, hence that

$$\limsup_{k \to \infty} \frac{S_{n_k}}{\sqrt{n_k \log \log n_k}} \leq \sigma\sqrt{2} \quad \text{a.s.}$$

In other words, we have exhibited a *subsequence* that behaves as desired. In order to prove that the whole sequence behaves properly we must prove that nothing terrible occurs between the subsequence points.

Now, for $k \geq k_0$,

$$P\left(\max_{n_k \leq j \leq n_{k+1}} \left|\frac{S_j}{\sqrt{j \log \log j}}\right| > \varepsilon\right) \leq P(\max_{n_k \leq j \leq n_{k+1}} |S_j| > \varepsilon \sqrt{n_k \log \log n_k})$$

$$\leq P(\max_{1 \leq j \leq n_{k+1}} |S_j| > \varepsilon \sqrt{n_k \log \log n_k})$$

$$\leq 2P(|S_{n_{k+1}}| > \varepsilon \sqrt{n_k \log \log n_k} - \sigma \sqrt{2n_{k+1}})$$

$$\leq 2P(|S_{n_{k+1}}| > \lambda^{-3/2}\varepsilon \sqrt{n_{k+1} \log \log n_{k+1}} - \sigma \sqrt{2n_{k+1}})$$

$$\leq 2P(|S_{n_{k+1}}| > \lambda^{-2}\varepsilon \sqrt{n_{k+1} \log \log n_{k+1}}).$$

Here we have applied the Lévy inequalities, Theorem 3.7.2, in the third inequality, (3.3) in the fourth, and the fact that $\lambda^{-3/2}\varepsilon - \sigma\sqrt{2/\log\log n_{k+1}} > \lambda^{-2}\varepsilon$ for $k > k_0$ in the last one (possibly an increased $k_0$).

Combining this with the first part of Lemma 3.1, leads to the following upper class result.

**Lemma 3.2.** *In the above setting,*

$$\sum_{k=1}^{\infty} P(\max_{n_k \leq j \leq n_{k+1}} |S_j| > \varepsilon \sqrt{n_k \log \log n_k}) < \infty \ \text{for } \varepsilon > \lambda^2 \sigma \sqrt{2},$$

$$P(|S_n| > \varepsilon \sqrt{n \log \log n} \ \text{i.o.}) = 0 \ \text{for } \varepsilon > \lambda^2 \sigma \sqrt{2},$$

$$\limsup_{n \to \infty} \frac{S_n}{\sqrt{n \log \log n}} \leq \sigma \sqrt{2} \ \text{a.s.}$$

*Proof.* The first relation follows from Lemma 3.1 and the computations preceding the present lemma. The convergence of these sums, together with the first Borel-Cantelli lemma, Theorem 2.18.1, implies that

$$P(|S_{n_k}| > \varepsilon \sqrt{n_k \log \log n_k} \ \text{i.o.}) = 0 \ \text{for } \varepsilon > \sigma\sqrt{2},$$

$$P(\max_{n_k \leq j \leq n_{k+1}} |S_j| > \varepsilon \sqrt{n_k \log \log n_k}) \ \text{for } \varepsilon > \lambda^2 \sigma \sqrt{2},$$

that is, only finitely many sums from the subsequence surpass the given level, and only finitely many increments between the subsequence points do so for $\varepsilon > \lambda^2 \sigma \sqrt{2}$. Putting these facts together we find that only finitely many partial sums surpass the given level. This proves the second statement. The third one follows from the fact that $\lambda$ may be chosen arbitrarily close to 1. $\square$

It remains to prove that

$$\limsup_{n \to \infty} \frac{S_n}{\sqrt{n \log \log n}} \geq \sigma \sqrt{2} \quad \text{a.s.,} \tag{3.11}$$

which is something we shall achieve via the second Borel-Cantelli lemma. However, we have to transfer to increments in order to obtain a Borel-Cantelli sum for independent random variables.

**Lemma 3.3.** *Let $0 < \delta < 1/3$, and set, for $k \geq 1$ and some fixed positive integer $j$, $m_k = [\lambda^{kj}] (= [(\lambda^j)^k])$. Then, for $\varepsilon < \sigma\sqrt{2} < \varepsilon\delta\lambda^{j/2}$,*

$$\sum_{k=1}^{\infty} P(|S_{m_{k-1}}| > \varepsilon\delta\sqrt{m_k \log\log m_k}) < \infty, \tag{3.12}$$

$$\sum_{k=1}^{\infty} P(S_{m_k} - S_{m_{k-1}} > \varepsilon(1-\delta)\sqrt{m_k \log\log m_k}) = \infty. \tag{3.13}$$

*Proof.* We have

$$\{S_{m_k} > \varepsilon\sqrt{m_k \log\log m_k}\} \subset \{S_{m_k} - S_{m_{k-1}} > \varepsilon(1-\delta)\sqrt{m_k \log\log m_k}\}$$
$$\cup \{S_{m_{k-1}} > \varepsilon\delta\sqrt{m_k \log\log m_k}\}. \tag{3.14}$$

Now, let $k_5$ be large enough to ensure that $\frac{m_k}{m_{k-1}} > \lambda^{j+1}$ for $k > k_5$, which (cf. (3.3)) is possible in view of the fact that the ratio converges to $\lambda^j$ as $k \to \infty$. Then,

$$\sum_{k=k_5}^{\infty} P(S_{m_{k-1}} > \varepsilon\delta\sqrt{m_k \log\log m_k})$$

$$\leq \sum_{k=k_5}^{\infty} P(|S_{m_{k-1}}| > \varepsilon\delta\sqrt{\frac{m_k}{m_{k-1}}}\sqrt{m_{k-1} \log\log m_{k-1}})$$

$$\leq \sum_{k=k_5}^{\infty} P(|S_{m_{k-1}}| > \varepsilon\delta\lambda^{(j+1)/2}\sqrt{m_{k-1} \log\log m_{k-1}}) < \infty.$$

This proves (3.12), which, together with (3.2) and (3.14), proves (3.13). □

To conclude the proof of the theorem it only remains to exploit Lemma 3.3 to prove (3.11), which amounts to proving that

$$P(S_n > \varepsilon\sqrt{n \log\log n} \text{ i.o.}) = 1 \quad \text{for} \quad \varepsilon < \sigma\sqrt{2}. \tag{3.15}$$

In order to achieve this, we need the following variation of (3.14):

$$\{|S_{m_{k-1}}| \leq 2\varepsilon\delta\sqrt{m_k \log\log m_k}\} \cap \{S_{m_k} - S_{m_{k-1}} > \varepsilon(1-\delta)\sqrt{m_k \log\log m_k}\}$$
$$\subset \{S_{m_k} > \varepsilon(1-3\delta)\sqrt{m_k \log\log m_k}\}. \tag{3.16}$$

Let $\varepsilon < \sigma\sqrt{2}$. By (3.12) and the first Borel-Cantelli lemma we know, in particular, that

$$P(|S_{m_{k-1}}| > 2\varepsilon\delta\sqrt{m_k \log\log m_k} \text{ i.o.}) = 0,$$

and, remembering that the *increments* involved in the sum in (3.13) are independent, by (3.13) and the second Borel-Cantelli lemma, that

$$P(S_{m_k} - S_{m_{k-1}} > \varepsilon(1-\delta)\sqrt{m_k \log\log m_k} \text{ i.o.}) = 1.$$

These facts, together with (3.16), then show that

$$P(S_{m_k} > \varepsilon(1 - 3\delta)\sqrt{m_k \log\log m_k} \text{ i.o.}) = 1,$$

which, due to $\delta$ being arbitrary (and $\{m_k, \, k \geq 1\}$ being a subsequence), proves
(3.15).

The proof of the Hartman-Wintner law of the iterated logarithm has,
finally, reached its end.                                                  □

### The Kolmogorov Law of the Iterated Logarithm

Theorem 1.1 may be proved along the same lines with some modifications.
(i) One relief is that no truncation is necessary.
(ii) In order to select the ideal subsequence one should observe that $n_k$ as
defined in the proof of Theorem 1.2 does not refer to the *number* of summands
but to the *variance*. The natural subsequence to try for Theorem 1.1 therefore
is $n_k = \min\{j : s_n^2 > \lambda^k\}$ (which reduces to $[\lambda^k]$ in the previous setting (if
$\sigma^2 = 1$)).

**Exercise 3.2.** Prove Theorem 1.1.                                        □

## 4 Proof of the Converse

Instead of Strassen's original proof we present (a variation of) a shorter one
due to Feller [86]; see also [239], Theorem 5.3.5. For other proofs, see [135, 236].

We thus assume that

$$P\left( \limsup_{n\to\infty} \frac{|S_n|}{\sqrt{n\log\log n}} < \infty \right) > 0, \qquad (4.1)$$

and we wish to show that $E\,X^2 < \infty$, $E\,X = 0$, and that (1.2) holds.

First of all, once the conditions on the first two moments have been established the validity of the Hartman-Wintner law is immediate from what has
already been shown.

Following Feller's proof as referred to, we begin by noticing that the probability that the limsup is finite is 0 or 1 by the Kolmogorov zero-one law,
Theorem 2.10.6. This means that

$$P\left( \limsup_{n\to\infty} \frac{|S_n|}{\sqrt{n\log\log n}} < \infty \right) = 1,$$

which, in turn, implies that

$$P\left( \limsup_{n\to\infty} \frac{|S_n|}{n} = 0 \right) = 1,$$

so that the strong law of large numbers holds. From its converse, Theorem 6.6.1(b), we therefore conclude that $E\,X = 0$.

Now, suppose that we know that the result holds for symmetric random variables. In order to desymmetrize we use the usual technique from Chapter 6. Namely, if (4.1) holds, then, by the triangle inequality, (4.1) also holds for the symmetrized sequence, from which it follows that $\operatorname{Var} X^s < \infty$, and thus, that $\operatorname{Var} X = \frac{1}{2}\operatorname{Var} X^s < \infty$. We may therefore restrict ourselves to the symmetric case.

Suppose, on the contrary, that $\operatorname{Var} X = +\infty$. Let $M > 0$, and set $Y_k = X_k I\{|X_k| \le M\}$, $k \ge 1$, and $S'_n = \sum_{k=1}^{n} Y_k$, $n \ge 1$. Since the truncated random variables are bounded, their variance $\sigma_M^2$ must be finite, in fact, at most equal to $M^2$. It therefore follows from the Hartman-Wintner law, Theorem 1.2, that (for example)

$$P\left(\frac{S'_n}{\sqrt{n\log\log n}} > \sigma_M \text{ i.o.}\right) = 1. \tag{4.2}$$

Since, by symmetry,

$$S_n - S'_n = \sum_{k=1}^{n} X_k I\{|X| > M\} \stackrel{d}{=} \sum_{k=1}^{n} -X_k I\{|X| > M\} = S_n - S'_n,$$

it follows that the two events

$$\left\{\left\{\frac{S'_n}{\sqrt{n\log\log n}} > \sigma_M\right\} \bigcap \{S_n - S'_n \ge 0\} \text{ i.o.}\right\} \quad \text{and}$$

$$\left\{\left\{\frac{S'_n}{\sqrt{n\log\log n}} > \sigma_M\right\} \bigcap \{S_n - S'_n \le 0\} \text{ i.o.}\right\}$$

are equiprobable, and since the union of them equals $\left\{\frac{S'_n}{\sqrt{n\log\log n}} > \sigma_M \text{ i.o.}\right\}$, the two probabilities must be at least equal to $1/2$, because of (4.2). This implies, in particular, that

$$P\left(\frac{S_n}{\sqrt{n\log\log n}} > \sigma_M \text{ i.o.}\right)$$
$$\ge P\left(\left\{\frac{S'_n}{\sqrt{n\log\log n}} > \sigma_M\right\} \bigcap \{S_n - S'_n \ge 0\} \text{ i.o.}\right) \ge \frac{1}{2}. \tag{4.3}$$

Now, the choice of $M$ is arbitrary, and the variance is infinite, so that we may choose a non-decreasing sequence $\{M_k, k \ge 1\}$, such that $\sigma_{M_k}^2 \ge k^2$ for all $k$. Moreover, since the events

$$\left\{\frac{S_n}{\sqrt{n\log\log n}} > k \text{ i.o.}\right\} \quad \text{are decreasing in } k,$$

and

$$P\left(\frac{S_n}{\sqrt{n \log \log n}} > k \text{ i.o.}\right) \geq \frac{1}{2} \quad \text{for all} \quad k,$$

it follows that, for all $k$,

$$P\left(\bigcap_{j=i}^{k}\left\{\frac{S_n}{\sqrt{n \log \log n}} > j \text{ i.o.}\right\}\right) = P\left(\frac{S_n}{\sqrt{n \log \log n}} > k \text{ i.o.}\right) \geq \frac{1}{2}.$$

Since the lower bound, $1/2$, is independent of $k$ we may let $k \to \infty$, and conclude that

$$P\left(\left\{\frac{S_n}{\sqrt{n \log \log n}} > k\right\} \text{ i.o. for all } k\right) \geq \frac{1}{2}.$$

Finally, since this is a statement about the probability of a tail event, the Kolmogorov zero-one law, Theorem 2.10.6, forces the probability to be equal to 1, which, differently expressed, tells us that

$$P\left(\limsup_{n \to \infty} \frac{S_n}{\sqrt{n \log \log n}} = \infty\right) = 1,$$

which contradicts (4.1). □

*Remark 4.1.* The stronger Hewitt-Savage zero-one law referred to in [86] and [239], p. 297, implies more directly that the probability in (4.3) equals 1. After having established this fact one argues that, by choosing $M$ appropriately, one may achieve an arbitrarily large $\sigma_M^2$, from which the contradiction follows. The original reference for the Hewitt-Savage zero-one law is [132]. Proofs can also be found in, e.g., [19, 48]. The above detour via the Kolmogorov zero-one law was given to me by Svante Janson. □

*Remark 4.2.* Assumption (4.1) and the Kolmogorov zero-one law together imply that $\limsup_{n \to \infty} \frac{|X_n|}{\sqrt{n \log \log n}} < \infty$ almost surely, which, in view of the second Borel-Cantelli lemma and the equidistribution, shows that

$$\sum_{n=1}^{\infty} P(|X| > \sqrt{n \log \log n}) < \infty,$$

which, in turn, shows that $E\,X^2 / \log^+ \log^+ |X| < \infty$ via inversion and Theorem 2.12.3. The hard part thus is to move from $E\,X^2 / \log^+ \log^+ |X| < \infty$ to finite variance. □

## 5 The LIL for Subsequences

It is frequently important, interesting, and developing to lean back and review a proof. It is, in particular, advisable to find out where a given condition has been used (let alone *if*), and how. Sometimes one can thereby discover more general results. It may even happen that a more general assumption can simplify a proof, because some more specialized condition or assumption does not have to be checked.

The Hartman-Wintner law is, as it turns out, such a case. Namely, the geometrically increasing sequence $\{n_k\}$ forces condition (3.3) upon the proof for the passage from $n_k$ to $\lambda \log(k-1)$. By reviewing the proof of the Borel-Cantelli sums we observe that the steps preceding the application of (3.3) in both the upper and lower bounds for $S'_{n_k}$ coincide, except for details in the exponent. More precisely, they are

$$\sum_{k=k_0}^{\infty} (\log n_k)^{-\frac{\varepsilon^2}{2\sigma^2}(1-\delta)^3} \quad \text{and} \quad \sum_{k=k_2}^{\infty} (\log n_k)^{-\frac{1}{2}\frac{\varepsilon^2(1+\delta)^2(1+\gamma)}{(1-\delta)\sigma^2}}.$$

Since the sums concerning $S''_{n_k}$ and $S'''_{n,k}$ were always convergent, the crucial object seems to be

$$\sum_{k}^{\infty} (\log n_k)^{-\frac{1}{2}\frac{\varepsilon^2}{\sigma^2}(1\pm\delta)}, \tag{5.1}$$

where the boundary for convergence or divergence is at $\delta = 0$.

Once we have realized this, the natural question pops up, namely, "what conclusions can be drawn for other subsequences, such as powers (which increase more slowly than geometrically), or for very rapidly increasing subsequences, such as $n_k = 2^{2^k}$ ?"

The answer is that the conclusion is the same as before for sequences that grow at most geometrically, but differs for more rapidly increasing subsequences. This is the topic of the present section, which is, essentially, borrowed from [109].

We begin with the easy one, namely, the *dense* subsequences, that is, subsequences that increase at *most* geometrically.

**Theorem 5.1.** *Let* $X, X_1, X_2, \ldots$ *be independent random variables with mean* 0 *and finite variance* $\sigma^2$, *and set* $S_n = \sum_{k=1}^{n} X_k$, $n \geq 1$. *Further, let* $\{n_k, k \geq 1\}$ *be s strictly increasing subsequence of the integers, such that*

$$\liminf_{k\to\infty} \frac{n_k}{n_{k+1}} > 0. \tag{5.2}$$

*Then*

$$\limsup_{k\to\infty} (\liminf_{k\to\infty}) \frac{S_{n_k}}{\sqrt{n_k \log\log n_k}} = +\sigma\sqrt{2} \, (-\sigma\sqrt{2}) \quad a.s.$$

*Proof.* The conclusion is a consequence of the fact that the limsup of the subsequence is at most equal to the limsup of the whole sequence which is $\sigma\sqrt{2}$, and at least equal to the limsup of a geometric subsubsequence, which is also $\sigma\sqrt{2}$.

For the upper bound this is translated into

$$\limsup_{k\to\infty} \frac{S_{n_k}}{\sqrt{n_k \log\log n_k}} \leq \limsup_{n\to\infty} \frac{S_n}{\sqrt{n \log\log n}} = \sigma\sqrt{2} \quad \text{a.s.}$$

For the lower bound, let $M > 2$ be an integer and define

$$m_j = \min\{k : n_k > M^j\} \quad \text{for} \quad j = 1, 2, \ldots.$$

Then, because of (5.2), there exists $L > 1$, such that $n_{k+1} \leq Ln_k$, so that, for all $j \geq 1$,

$$M^j \leq n_{m_j} \leq LM^j \quad \text{and} \quad \frac{1}{LM} \leq \frac{n_{m_j}}{n_{m_{j+1}}} \leq \frac{L}{M}.$$

In other words, we have exhibited a geometrically growing subsubsequence with the aid of which we are in the position to conclude that

$$\limsup_{k\to\infty} \frac{S_{n_k}}{\sqrt{n_k \log\log n_k}} \geq \limsup_{j\to\infty} \frac{S_{n_{m_j}}}{\sqrt{n_{m_j} \log\log n_{m_j}}} = \sigma\sqrt{2} \quad \text{a.s.} \qquad \square$$

*Remark 5.1.* With the aid of the same subsequence the theorem can alternatively, and maybe more sophisticatedly, be deduced from the next one.     $\square$

The next one that was just alluded to concerns *sparse* subsequences, that is, subsequences that increase *at least* geometrically. The crucial sum in (5.1) suggests the definition of

$$\varepsilon^* = \inf\left\{\varepsilon > 0 : \sum_{k=3}^{\infty}(\log n_k)^{-\varepsilon^2/2} < \infty\right\}. \tag{5.3}$$

The reason is that $\varepsilon^*$, being the boundary between convergence and divergence, is likely to play the role of $\sqrt{2}$ in the present context. This is confirmed by the following result.

**Theorem 5.2.** *Let $X, X_1, X_2, \ldots$ be independent random variables with mean 0 and finite variance $\sigma^2$, and set $S_n = \sum_{k=1}^{n} X_k$, $n \geq 1$. Further, let $\{n_k, k \geq 1\}$ be s strictly increasing subsequence of the integers, such that*

$$\limsup_{k\to\infty} \frac{n_k}{n_{k+1}} < 1. \tag{5.4}$$

*Then*

$$\limsup_{k\to\infty} (\liminf_{k\to\infty}) \frac{S_{n_k}}{\sqrt{n_k \log\log n_k}} = +\sigma\varepsilon^* \ (-\sigma\varepsilon^*) \quad \text{a.s.}$$

*Remark 5.2.* Theorems 5.1 and 5.2 meet at subsequences that increase geometrically (and hence, at most as well as at least, geometrically). Such sequences are of the form $n_k = \lambda^k$ for some $\lambda > 1$, and $\varepsilon^* = \sqrt{2}$ in that case.     $\square$

The decisive tool for the proof of Theorem 5.2 is the following result on Borel-Cantelli sums for sparse sequences.

## 5.1 A Borel-Cantelli Sum for Subsequences

**Lemma 5.1.** *Under the assumptions of the theorem we have*

$$\sum_{k=3}^{\infty} P(|S_{n_k}| > \sqrt{n_k \log\log n_k}) < \infty \quad \text{for all} \quad \varepsilon > \sigma\varepsilon^*, \tag{5.5}$$

$$\sum_{k=3}^{\infty} P(|S_{n_k}| > \sqrt{n_k \log\log n_k}) = \infty \quad \text{for all} \quad \varepsilon < \sigma\varepsilon^*. \tag{5.6}$$

*Proof.* Most of the proof of Lemma 3.1 carries over to this, more general, case. Using the same notation and truncations as there we find that (3.6) remains unchanged, so that

$$\sum_{k=k_0}^{\infty} P(|S'_{n_k}| > \varepsilon\sqrt{n_k \log\log n_k}) \leq 2 \sum_{k=k_0}^{\infty} (\log n_k)^{-\frac{\varepsilon^2}{2\sigma^2}(1-\delta)^3} < \infty \tag{5.7}$$

for $\varepsilon > \frac{\sigma\varepsilon^*}{(1-\delta)^{3/2}}$.

Instead of (3.8) we obtain

$$\sum_{k=k_0}^{\infty} P(|S''_{n_k}| > \eta\sqrt{n_k \log\log n_k}) \leq \sum_{k=k_0}^{\infty} n_k P(|X| > \sqrt{n_k})$$

$$= \sum_{k=k_0}^{\infty} n_k P(X^2 > n_k) \leq C E X^2,$$

where the finiteness is due to Theorem 2.12.6.

As for (3.9), we must find an upper bound for $|A^*(k,x)|$ in order to estimate $\sum_{A(k,x)} \frac{1}{\log\log n_k}$.

Since the subsequence increases at least geometrically, there exists (as we have noted before) $\lambda > 1$, such that

$$n_{k+1} \geq \lambda n_k \quad \text{for all} \quad k. \tag{5.8}$$

Moreover, since $n_k \geq x^2$ for $k \in A(k,x)$, it follows that

$$|A^*(k,x)| \leq \min\{k : x^2\lambda^k \geq 2\left(\frac{\varepsilon}{\delta\sigma^2}\right)^2 x^2 \log\log|x|\}$$

$$= \min\{k : \lambda^k \geq 2\left(\frac{\varepsilon}{\delta\sigma^2}\right)^2 \log\log|x|\}$$

$$= \min\left\{k : k \geq \frac{\log 2 + 2\log\left(\frac{\varepsilon}{\delta\sigma^2}\right) + \log\log\log|x|)}{\log\lambda}\right\}$$

$$\leq C + C\log\log\log|x|.$$

Having this estimate at our disposal we are now able to compute the third sum:

$$\sum_{k=k_0}^{\infty} P(|S_{n_k}'''| > \eta\sqrt{n_k \log\log n_k}) \leq \frac{4}{\eta^2} \int_{k_*}^{\infty} \Big( \sum_{A(k,x)} \frac{1}{\log\log n_k} \Big) x^2 \, dF(x)$$

$$\leq \frac{4}{\eta^2} \int_{k_*}^{\infty} \frac{1}{\log\log(x^2)} |A^*(k,x)| x^2 \, dF(x)$$

$$\leq \frac{4}{\eta^2} \int_{k_*}^{\infty} \frac{C + C \log\log\log x}{\log\log x} x^2 \, dF(x) < \infty.$$

This proves the finiteness of the three sums. From here on the convergence part follows as before, that is, the arguments following (3.9) carry over without change.

To prove the divergence part we only have to revise the computations for the first sum, since two other ones always converge. We obtain (cf. (3.10))

$$\sum_{k=k_2}^{\infty} P(S_{n_k}' > \varepsilon\sqrt{n_k \log\log n_k}) \geq \sum_{k=k_2}^{\infty} \big( \log n_k \big)^{-\frac{\varepsilon^2(1+\delta)^2(1+\gamma)}{2(1-\delta)\sigma^2}} = \infty$$

for $\varepsilon < \frac{\sigma\varepsilon^*}{1+\delta}\sqrt{\frac{1-\delta}{1+\gamma}}$.

From here on the arguments remain, once again, the same as before, which completes the proof of the lemma.    □

### 5.2  Proof of Theorem 5.2

Having established the behavior of the two Borel-Cantelli sums, the next step amounts to applying the Borel-Cantelli lemmas.

For the convergence part this is easy, since the first Borel-Cantelli lemma applied to (5.5) immediately tells us that

$$\limsup_{k\to\infty} \frac{S_{n_k}}{\sqrt{n_k \log\log n_k}} \leq \sigma\varepsilon^*.$$

Note also that we do not need any Lévy inequality here since we are only interested in subsequences.

To prove the reverse inequality we must, as always, begin by proving a divergence result for increments via a suitable subsubsequence. And, once again, it turns out that geometric thinning works.

Thus, let $j \geq 1$, set $m_k = n_{jk}$, $k \geq 1$, and define an $\varepsilon^*$ for that sequence;

$$\varepsilon_j^* = \inf\Big\{ \varepsilon : \sum_{k=1}^{\infty} (\log m_k)^{-\varepsilon^2/2} = \sum_{k=1}^{\infty} (\log n_{jk})^{-\varepsilon^2/2} = \infty \Big\}.$$

We first claim that

$$\varepsilon_j^* = \varepsilon^*. \tag{5.9}$$

To see this, we first note that $\varepsilon_j^* \leq \varepsilon^*$, since $\{m_k\}$ is a subsequence of $\{n_k\}$. For the converse, let $\varepsilon < \varepsilon^*$, so that

$$\sum_{k=1}^{\infty} (\log n_k)^{-\varepsilon^2/2} = \infty.$$

Then, necessarily,

$$\sum_{k=1}^{\infty} (\log n_{jk+i})^{-\varepsilon^2/2} = \infty \quad \text{for at least one} \quad i = 0, 1, 2, \ldots j-1,$$

so that, in particular,

$$\sum_{k=1}^{\infty} (\log m_k)^{-\varepsilon^2/2} = \sum_{k=1}^{\infty} (\log n_{jk})^{-\varepsilon^2/2} = \infty,$$

since $\{n_k\}$ is strictly increasing. The reverse inequality follows.

The next thing is to prove that

$$\sum_{k=1}^{\infty} P(S_{m_k} > \varepsilon \sqrt{m_k \log \log m_k}) = \infty \quad \text{for} \quad \varepsilon < \varepsilon^*. \qquad (5.10)$$

Now, by (5.4) it follows that

$$\limsup_{k \to \infty} \frac{m_k}{m_{k+1}} = \limsup_{k \to \infty} \frac{n_{kj}}{n_{(k+1)j}} = \limsup_{k \to \infty} \prod_{i=0}^{j-1} \frac{n_{kj+i}}{n_{kj+i+1}} < 1,$$

since there are always exactly $j$ factors involved (or, else, directly, by (5.8), since $\frac{m_k}{m_{k+1}} \leq \lambda^{-j} < 1$). The subsubsequence therefore satisfies the sparseness condition, so that an application of Lemma 5.1 tells us that (5.10) holds.

From here on on can argue as in the proof of Lemma 3.3, upon noticing that $\sqrt{\frac{m_k}{m_{k-1}}} \geq \sqrt{\lambda}$, and by replacing $\sqrt{2}$ by $\varepsilon^*$.

This completes the proof of the theorem. $\qquad \square$

**Exercise 5.1.** Complete the details that we have omitted in the proof. $\qquad \square$

A particular case is when $\varepsilon^* = 0$, corresponding to "very rapidly" increasing subsequences, that is, to "very sparse" subsequences.

**Corollary 5.1.** *Let* $X, X_1, X_2, \ldots$ *be independent random variables with mean 0 and finite variance* $\sigma^2$, *and set* $S_n = \sum_{k=1}^{n} X_k$, $n \geq 1$. *Further, let* $\{n_k, k \geq 1\}$ *be a strictly increasing subsequence of the integers, such that*

$$\limsup_{k \to \infty} \frac{n_k}{n_{k+1}} < 1 \quad \text{and} \quad \varepsilon^* = 0.$$

*Then*

$$\frac{S_{n_k}}{\sqrt{n_k \log \log n_k}} \xrightarrow{a.s.} 0 \quad \text{as} \quad n \to \infty.$$

The corollary is, in fact, contained in Theorem 5.2, and does not require any additional proof. The interesting fact is that the fluctuations disappear, so that almost sure convergence holds. We shall return to this discussion in a moment.

## 5.3 Examples

To make the results a bit more concrete we present some examples of sequences that are sparse, some that are dense, and some that are not covered by the theorems.

*Example 5.1.* The first example (although we have briefly met it) is $n_k = 2^k$, for which $n_k/n_{k+1} = 1/2$, so that the subsequence is sparse as well as dense, and both theorems apply. And, luckily, $\varepsilon^* = \sqrt{2}$.

*Example 5.2.* A sparser one is $n_k = 2^{2^k}$, for which $n_k/n_{k+1} = 2^{-2^k} \to 0$ as $n \to \infty$, $\log n_k = 2^k \log 2$, and $\varepsilon^* = 0$, in particular, the corollary applies.

*Example 5.3.* Subsequences that grow like powers are dense. For $n_k = k^d$, for some $d = 2, 3, \ldots$, we have $n_k/n_{k+1} = k/(k+1) \to 1$ as $n \to \infty$, so that Theorem 5.1 is the appropriate one. More generally, $n_k = [ck^\alpha]$, $k \geq 1$, for some $c > 0$, $\alpha > 1$ is also dense.

*Example 5.4.* The example one longs for at this point is one where the extreme limit points are strictly between 0 and $\pm\sigma\sqrt{2}$. One family of examples satisfying this desire is $n_k = [2^{k^\alpha}]$, $k \geq 1$, for some $\alpha > 1$, which increases a bit faster than geometrically, but not as fast as in the second example. Namely, in this case, $n_k/n_{k+1} \sim 2^{-\alpha k^{\alpha-1}} \to 0$ as $n \to \infty$; it is a sparse subsequence. Since $\log n_k \sim k^\alpha \log 2$ we find that $\varepsilon^* = \sqrt{2/\alpha}$, so that, by Theorem 5.2,

$$\limsup_{k\to\infty} (\liminf_{k\to\infty}) \frac{S_{n_k}}{\sqrt{n_k \log\log n_k}} = +\sigma\sqrt{\frac{2}{\alpha}} \quad \left(-\sigma\sqrt{\frac{2}{\alpha}}\right) \quad \text{a.s.}$$

*Example 5.5.* For $0 < \alpha < 1$ the previous subsequence is dense.

*Example 5.6.* The subsequence $n_k = k!$, $k \geq 1$, is dense; one readily checks that $n_k/n_{k+1} = 1/(k+1) \to 0$ as $n \to \infty$. □

## 6 Cluster Sets

In this section we present the following more general result, namely that, in fact, all points between the extreme ones provided by the Hartman-Wintner law are limit points (almost surely). Not only do we thereby obtain a more general result, as it turns out, we also replace the use of the lower exponential bound by some more easily accessible tool. In addition, we obtain cluster analogs of Theorems 5.2 and 5.1.

The proofs follow a blend of [2], [3], [109], and [245]. However, before we can state the result we need to define the concept "cluster set of a sequence of random variables".

**Definition 6.1.** *For a sequence* $Y_1$, $Y_2$, ... *of random variables we define*

$$C(\{Y_n, \ n \geq 1\}) = \text{the cluster set} = \text{the set of limit points of } \{Y_n, \ n \geq 1\}. \quad \Box$$

**Theorem 6.1.** *Let* $X$, $X_1$, $X_2$, ... *be independent random variables with mean* 0 *and finite variance* $\sigma^2$, *and set* $S_n = \sum_{k=1}^{n} X_k$, $n \geq 1$. *Then*

$$C\left(\left\{\frac{S_n}{\sqrt{2n\sigma^2 \log \log n}}, \ n \geq 9\right\}\right) = [-1, +1] \quad a.s.,$$

*in particular,*

$$\limsup_{n \to \infty} (\liminf_{n \to \infty}) \frac{S_n}{\sqrt{n \log \log n}} = +\sigma\sqrt{2} \ (-\sigma\sqrt{2}) \quad a.s.$$

Just as in the previous proofs we proceed via subsequences. *For notational convenience we assume throughout that for any increasing subsequence* $\{n_k, \ k \geq 1\}$ *we always have* $n_1 \geq 9$ *in order for* $\log \log n_k$ *to be larger than* 1 *for all* $k$.

The following result, which extends Theorem 5.2 is due to Torrång [245]; see Theorem 2.1 there.

**Theorem 6.2.** *Let* $X$, $X_1$, $X_2$, ... *be independent random variables with mean* 0 *and finite variance* $\sigma^2$, *and set* $S_n = \sum_{k=1}^{n} X_k$, $n \geq 1$. *Further, let* $\{n_k, \ k \geq 1\}$ *be a sequence of positive integers, strictly increasing to* $+\infty$, *such that*

$$\limsup_{k \to \infty} \frac{n_k}{n_{k+1}} < 1,$$

*and let* $\varepsilon^* = \inf\{\varepsilon > 0 : \sum_{k=1}^{\infty} (\log n_k)^{-\varepsilon^2/2} < \infty\}$ *as before. Then*

$$C\left(\left\{\frac{S_{n_k}}{\sqrt{n_k \log \log n_k}}, \ k \geq 1\right\}\right) = [-\sigma\varepsilon^*, +\sigma\varepsilon^*] \quad a.s.,$$

*in particular,*

$$\limsup_{k \to \infty} (\liminf_{k \to \infty}) \frac{S_{n_k}}{\sqrt{n_k \log \log n_k}} = \sigma\varepsilon^* \ (-\sigma\varepsilon^*) \quad a.s.$$

As pointed out earlier, this verifies, in particular, Theorem 5.2 without using the lower exponential bounds, Lemma 2.2. The same goes for the following result, which follows as a corollary in the same way as Theorem 5.1 was obtained from Theorem 5.2; cf. [245].

**Theorem 6.3.** *Let* $X, X_1, X_2, \ldots$ *be independent random variables with mean 0 and finite variance* $\sigma^2$, *and set* $S_n = \sum_{k=1}^{n} X_k$, $n \geq 1$. *Further, let* $\{n_k, k \geq 1\}$ *be a sequence of positive integers, strictly increasing to* $+\infty$, *such that*

$$\liminf_{k \to \infty} \frac{n_k}{n_{k+1}} > 0.$$

*Then*

$$C\left(\left\{\frac{S_{n_k}}{\sqrt{n_k \log \log n_k}}, k \geq 1\right\}\right) = [-\sigma\sqrt{2}, +\sigma\sqrt{2}] \quad a.s.,$$

*in particular,*

$$\limsup_{k \to \infty} (\liminf_{k \to \infty}) \frac{S_{n_k}}{\sqrt{n_k \log \log n_k}} = \sigma\sqrt{2} \, (-\sigma\sqrt{2}) \quad a.s.$$

## 6.1 Proofs

Having stated the theorems it remains to prove them. We begin by proving Theorem 6.2, after which Theorems 6.1 and 6.3 will be taken care of in that order.

### Proof of Theorem 6.2

The first step is to show that

$$\limsup_{k \to \infty} \frac{S_{n_k}}{\sqrt{n_k \log \log n_k}} \leq \sigma\varepsilon^*. \tag{6.1}$$

But this follows from the first half of Lemma 5.1 and the first Borel-Cantelli lemma as before.

We have thus found an upper (lower) bound for the extreme limit points, which means that we have shown that

$$C\left(\left\{\frac{S_{n_k}}{\sqrt{n_k \log \log n_k}}\right\}\right) \subset [-\sigma\varepsilon^*, +\sigma\varepsilon^*] \quad a.s. \tag{6.2}$$

It thus remains to show that the cluster set consists of all points in between. In order to fulfill this task we follow closely the method of [2], cf. also [3], a method that was later pursued in [245].

The procedure involves two steps. The first one is [2], Lemma 2.4; see also [3], Lemma 3.1.

**Lemma 6.1.** *Let* $X, X_1, X_2, \ldots$ *be independent, identically distributed random variables with partial sums* $\{S_n, n \geq 1\}$. *Suppose that* $E X = 0$, *and set* $\sigma^2 = \operatorname{Var} X < \infty$. *Further, let* $m_k \in \mathbb{N}$, *and* $\alpha_k \in \mathbb{R}^+$, $k \geq 1$, *be such that*

$$\frac{\alpha_k}{m_k} \to 0 \quad and \quad \frac{\alpha_k^2}{m_k} \to \infty \quad as \quad k \to \infty.$$

*Then, for every $\varepsilon > 0$ and $b \in \mathbb{R}$,*

$$\liminf_{k \to \infty} \frac{m_k}{\alpha_k^2} \log P\left(\left|\frac{S_{m_k}}{\alpha_k} - b\right| < \varepsilon\right) \geq -\frac{b^2}{2\sigma^2}.$$

*Remark 6.1.* For future use, think $m_k = n_k$ and $\alpha_k = \sqrt{n_k \log \log n_k}$.     $\square$

*Proof.* Let $N \in N(0, \sigma^2)$, and set, for $k \geq 1$,

$$p_k = \left[\frac{m_k^2 t^2}{\alpha_k^2}\right], \quad q_k = \left[\frac{\alpha_k^2}{m_k t^2}\right], \quad r_k = \frac{\alpha_k}{t q_k}.$$

Note that, asymptotically,

$$p_k \sim r_k^2, \quad p_k q_k \sim m_k, \quad \frac{q_k m_k}{\alpha_k^2} \sim \frac{1}{t^2}. \tag{6.3}$$

For $-\infty < c + \delta < d - \delta < \infty$,

$$\left(P\left(t(c+\delta) < \frac{S_{p_k}}{r_k} < t(d-\delta)\right)\right)^{q_k}$$

$$= \prod_{j=0}^{q_k-1} P\left(t(c+\delta) < \frac{S_{(j+1)p_k} - S_{jp_k}}{r_k} < t(d-\delta)\right)$$

$$= P\left(\bigcap_{j=0}^{q_k-1} \left\{t(c+\delta) < \frac{S_{(j+1)p_k} - S_{jp_k}}{r_k} < t(d-\delta)\right\}\right)$$

$$\leq P\left(tq_k(c+\delta) < \frac{S_{p_k q_k}}{r_k} < tq_k(d-\delta)\right) = P\left(c+\delta < \frac{S_{p_k q_k}}{\alpha_k} < d-\delta\right),$$

and

$$P\left(c+\delta < \frac{S_{p_k q_k}}{\alpha_k} < d-\delta\right) \cdot P\left(\left|\frac{S_{m_k} - S_{p_k q_k}}{\alpha_k}\right| < \delta\right)$$

$$= P\left(\left\{c+\delta < \frac{S_{p_k q_k}}{\alpha_k} < d-\delta\right\} \cap \left\{\left|\frac{S_{m_k} - S_{p_k q_k}}{\alpha_k}\right| < \delta\right\}\right)$$

$$\leq P\left(c < \frac{S_{m_k}}{\alpha_k} < d\right),$$

which, upon setting $\lambda_k = P(|\frac{S_{m_k} - S_{p_k q_k}}{\alpha_k}| \geq \delta)$, combined with the previous computation and (6.3), tells us that

$$\log P(c\alpha_k < S_{m_k} < d\alpha^k) \geq \log(1 - \lambda_k)$$

$$+ q_k \log P\left(t(c+\delta) < \frac{S_{p_k}}{r_k} < t(d-\delta)\right). \tag{6.4}$$

Next we note that, by Chebyshev's inequality,

$$\lambda_k \le \frac{(m_k - p_k q_k)\sigma^2}{\alpha^2 \delta^2} \to 0 \quad \text{as} \quad k \to \infty,$$

and, by the central limit theorem (and, strictly speaking, by Cramér's theorem 5.11.3, since the relations in (6.3) are $\sim$-relations),

$$\frac{S_{p_k}}{r_k} \xrightarrow{d} N(0, \sigma^2) \quad \text{as} \quad k \to \infty,$$

which, inserted into (6.4) and recalling (6.3), yields

$$\liminf_{k \to \infty} \frac{m_k}{\alpha_k^2} P\Big(c < \frac{S_{m_k}}{\alpha_k} < d\Big)$$

$$\ge \liminf_{k \to \infty} \Big(\frac{m_k}{\alpha_k^2} \log(1 - \lambda_k) + \frac{q_k m_k}{\alpha_k^2} P(t(c + \delta) < \frac{S_{p_k}}{r_k} < t(d - \delta)))\Big)$$

$$\ge \liminf_{k \to \infty} \frac{m_k}{\alpha_k^2} \log(1 - \lambda_k) + \liminf_{k \to \infty} \frac{q_k m_k}{\alpha_k^2} P\Big(t(c + \delta) < \frac{S_{p_k}}{r_k} < t(d - \delta)\Big)$$

$$= 0 + \liminf_{k \to \infty} \frac{1}{t^2} P\Big(t(c + \delta) < \frac{S_{p_k}}{r_k} < t(d - \delta)\Big)$$

$$= \frac{1}{t^2} P(t(c + \delta) < N < t(d - \delta)).$$

Since $\delta > 0$ may be arbitrarily small this establishes that

$$\liminf_{k \to \infty} \frac{m_k}{\alpha_k^2} P\Big(c < \frac{S_{m_k}}{\alpha_k} < d\Big) \ge \frac{1}{t^2} P(tc < N < td). \tag{6.5}$$

Next, let $c = b - \varepsilon$ and $d = b + \varepsilon$. Then, since $N - tb \in N(-tb, \sigma^2)$, it follows that

$$P(|N - tb| < t\varepsilon) = \int_{-t\varepsilon}^{t\varepsilon} \frac{1}{\sqrt{2\pi}} e^{-(x+tb)^2/2\sigma^2} \, dx$$

$$\ge \exp\Big\{-\frac{t^2 b^2}{2\sigma^2}\Big\} \int_{-t\varepsilon}^{t\varepsilon} \frac{1}{\sqrt{2\pi}} e^{-x^2/2\sigma^2} \, dx$$

$$= \exp\Big\{-\frac{t^2 b^2}{2\sigma^2}\Big\} P(|N| < t\varepsilon),$$

and, hence, that

$$\frac{1}{t^2} \log P(|N - tb| < t\varepsilon) \ge -\frac{b^2}{2\sigma^2} + \frac{1}{t^2} \log P(|N| < t\varepsilon),$$

so that

$$\liminf_{t \to \infty} \frac{1}{t^2} \log P(|N - tb| < t\varepsilon) \ge -\frac{b^2}{2\sigma^2}.$$

This means that, for any given $\eta > 0$, we may choose $t_0$ so large that

$$\frac{1}{t^2} \log P(|N - tb| < t\varepsilon) \geq -\frac{b^2}{2\sigma^2} - \eta \quad \text{for} \quad t > t_0,$$

which, in turn, inserted into (6.5), implies that

$$\liminf_{k \to \infty} \frac{m_k}{\alpha_k^2} P\left(\left|\frac{S_{m_k}}{\alpha_k} - b\right| < \varepsilon\right) \geq -\frac{b^2}{2\sigma^2} - \eta,$$

which establishes the lemma.  □

In the second step of the proof we wish to verify the opposite of (6.2), which amounts to proving that

$$P\left(\liminf_{k \to \infty} \left|\frac{S_{n_k}}{\sqrt{n_k \log \log n_k}} - b\right| = 0\right) = 1 \quad \text{for all} \quad b \in [-\sigma\varepsilon^*, +\sigma\varepsilon^*], \quad (6.6)$$

because once this has been achieved it follows that, for any countable, dense set $D \subset [-\sigma\varepsilon^*, +\sigma\varepsilon^*]$ (for example $\mathbb{Q} \cap [-\sigma\varepsilon^*, +\sigma\varepsilon^*]$),

$$P\left(D \subset C\left(\left\{\frac{S_{n_k}}{\sqrt{n_k \log \log n_k}}\right\}\right)\right) = 1,$$

which, together with the fact that the cluster set is closed, establishes that

$$P\left([-\sigma\varepsilon^*, +\sigma\varepsilon^*] \subset C\left(\left\{\frac{S_{n_k}}{\sqrt{n_k \log \log n_k}}\right\}\right)\right) = 1,$$

which is precisely the opposite of (6.2).

*Proof of 6.6.* The proof contains two main ingredients from the corresponding half of the proof of Theorem 1.2, namely, thinning and transition to increments, in order to apply the second Borel-Cantelli lemma; cf. Lemma 3.3.

Thus, set, as in the proof of Lemma 5.1, $m_k = n_{jk}$, $k \geq 1$, and

$$\varepsilon_j^* = \inf\left\{\varepsilon : \sum_{k=1}^{\infty}(\log m_k)^{-\varepsilon^2/2} = \sum_{k=1}^{\infty}(\log n_{jk})^{-\varepsilon^2/2} = \infty\right\},$$

and recall from (5.9) that $\varepsilon_j^* = \varepsilon^*$. We also recall that the sparseness implies that there exists $\lambda > 1$, such that $n_k/n_{k+1} < \lambda < 1$, which, for the thinned subsequence implies that

$$\frac{m_k}{m_{k+1}} < \lambda^j < 1. \quad (6.7)$$

Let $\varepsilon > 0$. In order to take care of the increments, we apply Lemma 6.1 with $m_k - m_{k-1}$ and $\sqrt{m_k \log \log m_k}$ here playing the roles of $m_k$ and $\alpha_k$ there. The lemma then tells us that there exists $k_0$, such that for any $b \in \mathbb{R}$ and $\delta > 0$,

$$\frac{m_k - m_{k-1}}{m_k \log \log m_k} \log P\left(\left|\frac{S_{m_k} - S_{m_{k-1}}}{\sqrt{m_k \log \log m_k}} - b\right| < \varepsilon\right) \geq -\frac{b^2 + \delta}{2\sigma^2} \quad \text{for} \quad k > k_0,$$

which, rewritten, taking (6.7) into account, tells us that, for $k > k_0$,

$$P\left(\left|\frac{S_{m_k} - S_{m_{k-1}}}{\sqrt{m_k \log \log m_k}} - b\right| < \varepsilon\right) \geq \exp\left\{-\frac{b^2 + \delta}{2\sigma^2} \cdot \frac{m_k \log \log m_k}{m_k - m_{k-1}}\right\}$$

$$\geq \exp\left\{-\frac{(b^2 + \delta) \log \log m_k}{2\sigma^2(1 - \lambda^j)}\right\} = \left(\log m_k\right)^{-\frac{b^2 + \delta}{2\sigma^2(1 - \lambda^j)}}.$$

For $|b| < \sigma\varepsilon^*\sqrt{1 - \lambda^j}$ and $\delta$ so small that $\frac{b^2 + \delta}{\sigma^2(1 - \lambda^j)} < (\varepsilon^*)^2$, it therefore follows that

$$\sum_{k=1}^{\infty} P\left(\left|\frac{S_{m_k} - S_{m_{k-1}}}{\sqrt{m_k \log \log m_k}} - b\right| < \varepsilon\right) = +\infty,$$

which, by the second Borel-Cantelli lemma, Theorem 2.18.2, and the arbitrariness of $\varepsilon$ and $\delta$, shows that

$$P\left(\liminf_{k\to\infty}\left|\frac{S_{m_k} - S_{m_{k-1}}}{\sqrt{m_k \log \log m_k}} - b\right| = 0\right) = 1 \quad \text{for all} \quad |b| < \sigma\varepsilon^*\sqrt{1 - \lambda^j}. \tag{6.8}$$

The next step is to estimate the sum $S_{m_{k-1}}$ that precedes this increment. Toward this end, let $|b| < \sigma\varepsilon^*$, and let $j$ be so large that

$$|b| < \sigma\varepsilon^*\sqrt{1 - \lambda^j}. \tag{6.9}$$

From (6.2) we know that

$$P\left(\limsup_{k\to\infty}\left|\frac{S_{m_{k-1}}}{\sqrt{m_{k-1} \log \log m_{k-1}}}\right| \leq \sigma\varepsilon^*\right) = 1,$$

which, via the bound (6.7), implies that, almost surely,

$$\limsup_{k\to\infty}\left|\frac{S_{m_{k-1}}}{\sqrt{m_k \log \log m_k}}\right| = \limsup_{k\to\infty}\left|\frac{S_{m_{k-1}}}{\sqrt{m_{k-1} \log \log m_{k-1}}}\right.$$

$$\left.\times\sqrt{\frac{m_{k-1} \log \log m_{k-1}}{m_k \log \log m_k}}\right|$$

$$\leq \limsup_{k\to\infty}\left|\frac{S_{m_{k-1}}}{\sqrt{m_{k-1} \log \log m_{k-1}}}\right|\lambda^{j/2} \leq \sigma\varepsilon^*\lambda^{j/2}.$$

Now, by the triangle inequality,

$$\left|\frac{S_{m_k}}{\sqrt{m_k \log \log m_k}} - b\right| \leq \left|\frac{S_{m_{k-1}}}{\sqrt{m_k \log \log m_k}}\right| + \left|\frac{S_{m_k} - S_{m_{k-1}}}{\sqrt{m_k \log \log m_k}} - b\right|,$$

so that, by joining this with our knowledge about the two terms in the right-hand side, we obtain,

$$\liminf_{k\to\infty}\left|\frac{S_{m_k}}{\sqrt{m_k \log \log m_k}} - b\right| \leq \lambda^{j/2}\sigma\varepsilon^* + \liminf_{k\to\infty}\left|\frac{S_{m_k} - S_{m_{k-1}}}{\sqrt{m_k \log \log m_k}} - b\right|$$

$$\leq \lambda^{j/2}\sigma\varepsilon^* + 0 = \lambda^{j/2}\sigma\varepsilon^* \quad \text{a.s.}$$

Finally, since

$$\liminf_{k\to\infty}\left|\frac{S_{n_k}}{\sqrt{n_k\log\log n_k}}-b\right|\leq\liminf_{k\to\infty}\left|\frac{S_{m_k}}{\sqrt{m_k\log\log m_k}}-b\right|,$$

or, equivalently,

$$C\left(\left\{\frac{S_{m_k}}{\sqrt{m_k\log\log m_k}}\right\}\right)\subset C\left(\left\{\frac{S_{n_k}}{\sqrt{n_k\log\log n_k}}\right\}\right),$$

and since $j$ may be chosen arbitrarily large, we conclude, with (6.9) in mind, that

$$P\left(\liminf_{k\to\infty}\left|\frac{S_{n_k}}{\sqrt{n_k\log\log n_k}}-b\right|=0\right)=1\quad\text{for}\quad|b|\leq\sigma\varepsilon^*,$$

which is precisely (6.6).

This completes the proof of Theorem 6.2

## Proof of Theorem 6.1

From the proof of the upper half of the Hartman-Wintner theorem, Lemma 3.2, we know that

$$\limsup_{n\to\infty}\frac{S_n}{\sqrt{n\log\log n}}\leq\sigma\sqrt{2},$$

which, in particular, implies that

$$C\left(\left\{\frac{S_n}{\sqrt{n\log\log n}}\right\}\right)\subset[-\sigma\sqrt{2},+\sigma\sqrt{2}]\quad\text{a.s.}$$

In order to prove the opposite inclusion we choose a geometrically increasing subsequence, for example, $n_k=2^k$, $k\geq 1$, for which $\varepsilon^*=\sigma\sqrt{2}$. The proof is completed by observing that, almost surely,

$$[-\sigma\sqrt{2},+\sigma\sqrt{2}]=C\left(\left\{\frac{S_{n_k}}{\sqrt{n_k\log\log n_k}}\right\}\right)\subset C\left(\left\{\frac{S_n}{\sqrt{n\log\log n}}\right\}\right),$$

where the (almost sure) equality emanates from Theorem 6.2.     □

## Proof of Theorem 6.3

For the upper bound,

$$C\left(\left\{\frac{S_{n_k}}{\sqrt{n_k\log\log n_k}}\right\}\right)\subset[-\sigma\sqrt{2},+\sigma\sqrt{2}],$$

we invoke Theorem 5.1, and for the lower bound we choose the subsequence from the proof of Theorem 5.1, namely

$$m_j = \min\{k : n_k > M^j\} \quad \text{for} \quad j = 1, 2, \ldots, \quad \text{and some} \quad M = 2, 3, \ldots,$$

which is sparse (as well as dense), so that, on the one hand,

$$C\left(\left\{\frac{S_{m_k}}{\sqrt{m_k \log \log m_k}}\right\}\right) = [-\sigma\sqrt{2}, +\sigma\sqrt{2}],$$

and on the other,

$$C\left(\left\{\frac{S_{m_k}}{\sqrt{m_k \log \log m_k}}\right\}\right) \subset C\left(\left\{\frac{S_{n_k}}{\sqrt{n_k \log \log n_k}}\right\}\right). \qquad \square$$

**Exercise 6.1.** Check the details. $\qquad \square$

# 7 Some Additional Results and Remarks

In this section we collect some complements pertaining to the law of the iterated logarithm.

## 7.1 Hartman-Wintner via Berry-Esseen

Since, as we have seen earlier in this chapter, the steps corresponding to the derivation of the results for the Borel-Cantelli sums were more easily derived for normal random variables, it is tempting to derive the general result from the normal case with the aid of the Berry-Esseen theorem, Theorem 7.6.1; albeit, under the additional assumption of a finite third moment.

**Theorem 7.1.** *Suppose that $X$, $X_1$, $X_2$, ... are independent, identically distributed random variables with mean 0 and finite variance $\sigma^2$, and set $S_n = \sum_{k=1}^{n} X_k$, $n \geq 1$. Suppose, in addition, that $\gamma^3 = E|X|^3 < \infty$. Then*

$$\sum_{k=1}^{\infty} P(|S_{n_k}| > \varepsilon\sqrt{n_k \log \log n_k}) < \infty \quad \text{for} \quad \varepsilon > \sigma\sqrt{2}, \qquad (7.1)$$

$$\sum_{k=1}^{\infty} P(S_{n_k} > \varepsilon\sqrt{n_k \log \log n_k}) = \infty \quad \text{for} \quad \varepsilon < \sigma\sqrt{2}, \qquad (7.2)$$

*and (thus)*

$$\limsup_{n \to \infty} \frac{S_n}{\sqrt{2\sigma^2 n \log \log n}} = +1 \quad \text{a.s.} \qquad (7.3)$$

*Proof.* Let $k_0$ be defined as in (3.3). By the triangle inequality and the Berry-Esseen theorem,

$$\left| \sum_{k=k_0}^{\infty} P(|S_{n_k}| > \varepsilon\sqrt{n_k \log\log n_k}) - \sum_{k=k_0}^{\infty} 2\left(1 - \Phi\left(\frac{\varepsilon\sqrt{\log\log n_k}}{\sigma}\right)\right) \right|$$

$$\leq \sum_{k=k_0}^{\infty} \left| P(|S_{n_k}| > \varepsilon\sqrt{n_k \log\log n_k}) - 2\left(1 - \Phi\left(\frac{\varepsilon\sqrt{\log\log n_k}}{\sigma}\right)\right) \right|$$

$$\leq \sum_{k=k_0}^{\infty} 2\sup_{x\in\mathbb{R}} \left| P\left(\frac{S_{n_k}}{\sigma\sqrt{n_k}} \leq x\right) - \Phi(x) \right| \leq \sum_{k=k_0}^{\infty} \frac{C\gamma^3}{\sigma^3} \cdot \frac{1}{\sqrt{n_k}}$$

$$\leq \frac{C\gamma^3}{\sigma^3} \sum_{k=k_0}^{\infty} \frac{1}{\lambda^{(k-1)/2}} \leq \frac{C\gamma}{\sigma^3} \cdot \frac{\sqrt{\lambda}}{\sqrt{\lambda}-1} < \infty.$$

This proves the claims about the Borel-Cantelli sums, after which (7.3) follows as before.    □

## 7.2 Examples Not Covered by Theorems 5.2 and 5.1

Having met examples that are covered by the theorems it is reasonable to ask for sequences that are not. Such sequences necessarily have the property that

$$\liminf_{k\to\infty} \frac{n_k}{n_{k+1}} = 0 \quad \text{and} \quad \limsup_{k\to\infty} \frac{n_k}{n_{k+1}} = 1.$$

We shall provide two examples, both building on the sequence $n_k = 2^{2^k}$, $k \geq 1$. For one of them we shall find that $\varepsilon^* = 0$, whereas the sum $\sum_{n=1}^{\infty}(\log n_k)^{-\varepsilon^2/2}$ diverges for all $\varepsilon > 0$ in the other example.

*Example 7.1.* Let $n_{2k} = 2^{2^k}$ and $n_{2k+1} = 2^{2^k} + 1$, $k \geq 1$. Since $\{n_{2k}, k \geq 1\}$ and $\{n_{2k+1}, k \geq 1\}$ both are sparse sequences with $\varepsilon^* = 0$, it follows from Corollary 5.1 that the normalized limit for both sequences equals 0, and since the sum of two sequences that converge almost surely converges almost surely (Theorem 5.11.1), we conclude that

$$\frac{S_{n_k}}{\sqrt{n_k \log\log n_k}} \overset{a.s.}{\to} 0 \quad \text{as} \quad n \to \infty.$$

*Example 7.2.* This time, let $A_k = \{2^{2^k} + 1, 2^{2^k} + 2, \ldots, 2^{2^k+1}\}$, for $k \geq 1$, and set

$$B_1 = \bigcup_{k=1}^{\infty} A_{2k} \quad \text{and} \quad B_2 = \bigcup_{k=0}^{\infty} A_{2k+1}.$$

Then

$$\sum_{k\in B_1} (\log n_k)^{-\varepsilon^2/2} = \sum_{k\in B_2} (\log n_k)^{-\varepsilon^2/2} = \infty \quad \text{for all} \quad \varepsilon > 0.$$

Moreover, since $P(S_n > \varepsilon\sqrt{n \log \log n}$ i.o.$) = 1$ when $\varepsilon < \sigma\sqrt{2}$, we must have

$$P(S_n > \varepsilon\sqrt{n \log \log n} \text{ i.o.}, \; n \in B_j) = 1 \quad \text{for} \quad \varepsilon < \sigma\sqrt{2},$$

for at least one of $j = 1, 2$; it follows, in fact, that both probabilities are equal to 1.

To summarize, this means that

$$\limsup_{k \to \infty} \frac{S_{n_k}}{\sqrt{n_k \log \log n_k}} = \sigma\sqrt{2} \quad \text{a.s.,}$$

for (at least one of) the sequences $\{n_k, \; k \in B_j\}$, $j = 1, 2$, and at the same time there is no finite $\varepsilon^*$. □

The last two examples show that one has to check each individual subsequence that is not covered by Theorems 5.1 or 5.2 separately. For some extensions, see [225], where also Banach space valued random elements are considered.

## 7.3 Further Remarks on Sparse Subsequences

The law of the iterated logarithm provides one description on the fluctuations or oscillations of the random walk $\{S_n, \; n \geq 1\}$. We have seen, in Theorem 5.2, that the extreme limit points are smaller than $\sigma\sqrt{2}$ for subsequences that increase faster than geometrically, that is, when $\varepsilon^* < \sqrt{2}$. This means that the fluctuations for those subsequences are smaller. In particular, for $\varepsilon^* = 0$ they are invisible on the $\sqrt{n \log \log n}$ scale. In order to make them visible, we therefore have to turn to another scale.

**Theorem 7.2.** ([109]) *Let* $X, X_1, X_2, \ldots$ *be independent random variables with mean 0 and finite variance* $\sigma^2$, *and set* $S_n = \sum_{k=1}^{n} X_k$, $n \geq 1$. *Further, let* $\{n_k, \; k \geq 1\}$ *be a strictly increasing subsequence of the integers, such that*

$$\limsup_{k \to \infty} \frac{n_k}{n_{k+1}} < 1.$$

*Then*

$$\limsup_{k \to \infty} (\liminf_{k \to \infty}) \frac{S_{n_k}}{\sqrt{n_k \log k}} = +\sigma\sqrt{2} \, (-\sigma\sqrt{2}) \quad a.s.$$

The proof runs along the same lines as before. However, the truncation in (3.4) has to be modified; in the present case we truncate at

$$b_n = \frac{2\delta\sigma^2}{\varepsilon} \sqrt{\frac{n}{\log^+ m(n)}} \quad \text{and} \quad \sqrt{n},$$

where $m(x) = \#\{k : n_k \leq x\}$, $x > 0$. Moreover, $\log \log n$ has to be replaced by $\log m(n)$ at relevant places; note that $\log \log n_k$ then will be replaced by $\log m(n_k) = \log k$ in the proof.

**Exercise 7.1.** Carry out the details of the proof of the theorem.     □

Theorem 7.2 does not provide any essential novelty compared to Theorem 5.2 when $0 < \varepsilon^* < \sqrt{2}$, since then $\log \log n_k \sim \alpha \log k$, that is, the normalizations only differ by a non-zero scaling factor, which, obviously, reappears in the right-hand side. However, for $\varepsilon^* = 0$ the situation is different in that we have obtained a scaling that makes the fluctuations visible, the new scaling is of a different order of magnitude.

The correctness of the logarithmic order can somewhat heuristically be seen as follows. If the subsequence increases "very, very" rapidly, then $S_{n_{k-1}}$ should be essentially negligible compared to $S_{n_k}$, since the former is the sum of "just the first few terms" of the latter. Moreover, $S_{n_{k-1}}$ and $S_{n_k}$ should be close to independent. In view of the central limit theorem, the sequence $\{S_{n_k}/\sqrt{n_k \log k}, \; k \geq 3\}$ can therefore be expected to behave (asymptotically) essentially as the sequence $\{Z_k/\log k, \; k \geq 3\}$, where $\{Z_k, \; k \geq 3\}$ are independent $N(0, \sigma^2)$-distributed random variables.

Now, Mill's ratio, Lemma A.2.1, tells us that, as $k \to \infty$,

$$P(|Z_3| > \varepsilon \sqrt{\log k}) \sim \frac{\sigma}{\sqrt{2\pi} \varepsilon \sqrt{\log k}} \exp\left\{ -\frac{\varepsilon^2 \log k}{2\sigma^2} \right\} = \frac{C}{\sqrt{\log k}} \cdot \frac{1}{k^{\varepsilon^2/2\sigma^2}},$$

so that

$$\sum_{k=3}^{\infty} P(|Z_k| > \varepsilon \sqrt{\log k}) < \infty \quad \Longleftrightarrow \quad \varepsilon > \sigma\sqrt{2},$$

and since $\{Z_k, \; k \geq 3\}$ are independent, both Borel-Cantelli lemmas are applicable, from which we conclude (using the symmetry of the normal distribution) that

$$\limsup_{k\to\infty} (\liminf_{k\to\infty}) \frac{Z_k}{\sqrt{\log k}} = +\sigma\sqrt{2} \, (-\sigma\sqrt{2}) \quad \text{a.s.}$$

The cluster analog of this theorem runs as follows.

**Theorem 7.3.** *Let $X, X_1, X_2, \ldots$ be independent random variables with mean 0 and finite variance $\sigma^2$, and set $S_n = \sum_{k=1}^{n} X_k$, $n \geq 1$. Further, let $\{n_k, \; k \geq 1\}$ be a sequence of positive integers, strictly increasing to $+\infty$, such that*

$$\limsup_{k\to\infty} \frac{n_k}{n_{k+1}} < 1.$$

*Then*

$$C\left(\left\{ \frac{S_{n_k}}{\sqrt{n_k \log k}}, \; k \geq 3 \right\}\right) = [-\sigma\sqrt{2}, +\sigma\sqrt{2}] \quad \text{a.s.}$$

**Exercise 7.2.** Prove the theorem.     □

We have seen that the fluctuations for rapidly increasing subsequences are smaller than for the full sequence. One might also be tempted to ask whether, possibly, less than finite variance would suffice for some positive result in such cases. And, in fact, in [109], Section 7, it is shown that if, in addition, $\log k / \log \log n_k \downarrow 0$ as $k \to \infty$, then $S_{n_k} / \sqrt{n_k \log \log n_k} \overset{a.s.}{\to} 0$ as $k \to \infty$ under a moment condition which is (slightly) weaker than finite variance. For example, for our favorite subsequence $\{2^{2^k}, k \geq 1\}$, the conclusion holds under the assumption that $E\,X^2 \frac{\log^+ \log^+ \log^+ |X|}{\log^+ \log^+ |X|} < \infty$.

## 7.4 An Anscombe LIL

Since the law of the iterated logarithm is a path-wise result it is natural that the assumption of "asymptotic linear growth in probability" for the sequence of random indices in the original Anscombe theorem is exchanged for "almost sure asymptotic linear growth".

**Theorem 7.4.** *Let $X, X_1, X_2, \ldots$ be independent, identically distributed random variables with mean $0$ and variance $\sigma^2$, and set $S_n, n \geq 1$. Further, suppose that $\{N(t), t \geq 0\}$ is a family of positive, integer valued random variables, such that, for some $\theta \in (0, \infty)$,*

$$\frac{N(t)}{t} \overset{a.s.}{\to} \theta \quad as \quad t \to \infty.$$

*Then*

$$C\left(\left\{\frac{S_{N(t)}}{\sqrt{N(t) \log^+ \log^+ N(t)}}\right\}\right) = [-\sigma\sqrt{2}, +\sigma\sqrt{2}],$$

$$C\left(\left\{\frac{S_{N(t)}}{\sqrt{t \log \log t}}\right\}\right) = [-\sigma\sqrt{2\theta}, +\sigma\sqrt{2\theta}].$$

*In particular,*

$$\limsup_{t \to \infty} (\liminf_{t \to \infty}) \frac{S_{N(t)}}{\sqrt{N(t) \log^+ \log^+ N(t)}} = \sigma\sqrt{2} \, (-\sigma\sqrt{2}) \quad a.s.,$$

$$\limsup_{t \to \infty} (\liminf_{t \to \infty}) \frac{S_{N(t)}}{\sqrt{t \log \log t}} = \sigma\sqrt{2\theta} \, (-\sigma\sqrt{2\theta}) \quad a.s.$$

For an "elementary" proof (in discrete time) that proceeds via deterministic subsequences we refer to [245].

*Remark 7.1.* Elementary means that basic methods are used, that one does not appeal to some abstract, esoteric theorem from somewhere else. Elementary should not be confused with easy or simple. Elementary proofs can be very complicated – and sophisticated. □

## 7.5 Renewal Theory for Random Walks

The setup was first described in Subsection 2.16.3. In subsequent chapters we have established the law of large numbers and the central limit theorem. Here is now the LIL:

**Theorem 7.5.** ([45]) *If $X_1$, $X_2$, ... are independent, identically distributed random variables with mean $\mu > 0$, finite variance, $\sigma^2$, partial sums $S_n$, $n \geq 1$, and first passage times $\tau(t) = \min\{n : S_n > t\}$ for $t \geq 0$, then*

$$C\left(\left\{\frac{\tau(t) - t/\mu}{\sqrt{t \log \log t}}, t \geq e^e\right\}\right) = \left[-\frac{\sigma\sqrt{2}}{\mu^{3/2}}, \frac{\sigma\sqrt{2}}{\mu^{3/2}}\right] \quad a.s.$$

*In particular,*

$$\limsup_{t\to\infty} (\liminf_{t\to\infty}) \frac{\tau(t) - t/\mu}{\sqrt{t \log \log t}} = \frac{\sigma\sqrt{2}}{\mu^{3/2}} \left(-\frac{\sigma\sqrt{2}}{\mu^{3/2}}\right) \quad a.s.$$

For a proof using the random index method with the aid of Theorem 7.4, that is, the analog of the proof of Theorem 7.4.1, see [108], Section 4 (for a sketch, see [110], Section III.11).

## 7.6 Record Times

For a background to this problem we refer to Subsection 2.18.3, and for the law of large numbers and the central limit theorem we refer the corresponding chapters. We are thus given a sequence $X_1$, $X_2$, ... of independent, identically distributed, continuous random variables, with record times $L(n)$, $n \geq 1$, and counting variables $\mu(n)$, $n \geq 1$.

The law of the iterated logarithm for these two sequences, that is, the second half of the following theorem, is due to Rényi [205].

**Theorem 7.6.** *We have*

$$C\left(\left\{\frac{\mu(n) - \log n}{\sqrt{\log n \log \log \log n}}, n \geq e^e\right\}\right) = [-\sqrt{2}, \sqrt{2}] \quad a.s.,$$

$$C\left(\left\{\frac{\log L(n) - n}{\sqrt{n \log \log n}}, n \geq e^{e^e}\right\}\right) = [-\sqrt{2}, \sqrt{2}] \quad a.s.$$

*In particular,*

$$\limsup_{n\to\infty} (\liminf_{n\to\infty}) \frac{\mu(n) - \log n}{\sqrt{\log n \log \log \log n}} = \sqrt{2} \, (-\sqrt{2}) \quad a.s.,$$

$$\limsup_{n\to\infty} (\liminf_{n\to\infty}) \frac{\log L(n) - n}{\sqrt{n \log \log n}} = \sqrt{2} \, (-\sqrt{2}) \quad a.s.$$

The proof of the law of the iterated logarithm for the counting process follows from the Kolmogorov law of the iterated logaritm; Theorem 1.1. Rényi's proof for the record times then proceeds via inversion. The result for record times, alternatively, follows via William's representation [251].

However, more is true. Namely, William's representation, together with Theorem 6.1, may be exploited to provide the cluster set LIL for the record times, after which Theorem 7.4 (and a little more) proves the cluster LIL for the counting process. For details, see [111].

## 7.7 Convergence Rates

Let, again $X, X_1, X_2, \ldots$ be independent, identically distributed random variables with mean 0, variance $\sigma^2$, and partial sums $S_n$, $n \geq 1$. The problem of rate results in connection with the law of the iterated logarithm is more subtle than for the law of large numbers; Subsection 6.12.1. Namely, if the summands are standard normal then an application of Mill's ratio, Lemma A.2.1, tells us that

$$\sum_{n=3}^{\infty} n^\alpha P(|S_n| > \varepsilon \sqrt{n \log \log n}) = \infty \quad \text{for all} \quad \alpha > -1,$$

so that there is no hope for convergence.

The first result to that effect is due to Slivka [226], who proved that the counting variable

$$N(\varepsilon) = \#\{n : |S_n| > \varepsilon \sqrt{n \log \log n}\},$$

has no finite moments of any order.

Introducing the last exit times

$$L(\varepsilon) = \sup\{n : |S_n| > \varepsilon \sqrt{n \log \log n}\},$$

we have

$$E\, L(\varepsilon)^r \geq E\, N(\varepsilon)^r = \infty \quad \text{for all} \quad r > 0.$$

It is, however, possible to prove a positive result for logarithmic moments.

**Theorem 7.7.** *Suppose that $X, X_1, X_2, \ldots$ are independent, identically distributed random variables with mean 0 and finite variance $\sigma^2$. Further, let $S_n$, $n \geq 1$, denote the partial sums, and let $L(\varepsilon)$ be the last exit time. If*

$$E\, X^2 \frac{\log^+ |X|}{\log^+ \log^+ |X|} < \infty \quad \text{and} \quad E\, X = 0, \tag{7.4}$$

*then*

$$\sum_{n=1}^{\infty} \frac{\log n}{n} P(|S_n| > \varepsilon \sqrt{n \log \log n}) < \infty \quad \textit{for all} \quad \varepsilon > 2\sigma; \tag{7.5}$$

$$\sum_{n=3}^{\infty} \frac{\log n}{n} P(\max_{1 \le k \le n} |S_k| > \varepsilon \sqrt{n \log \log n}) < \infty \quad \textit{for all} \quad \varepsilon > 2\sigma; \tag{7.6}$$

$$\sum_{n=3}^{\infty} \frac{1}{n} P(\sup_{k \ge n} |S_k / \sqrt{k \log \log k}| > \varepsilon) < \infty \quad \textit{for all} \quad \varepsilon > 2\sigma; \tag{7.7}$$

$$E \log L(\varepsilon) < \infty \quad \textit{for all} \quad \varepsilon > 2\sigma. \tag{7.8}$$

*Conversely, if one of (7.5)–(7.8) holds for some $\varepsilon > 0$, then so do the others as well as (7.4).*

For proofs and related material we refer to [61, 62, 105].

The law of the iterated logarithm tells us that $L(\varepsilon) < \infty$ almost surely for $\varepsilon > \sigma\sqrt{2}$. Slivka's result implies that no moments exist. Relation (7.8) states that *logarithmic* moments exist, but not all the way down to $\sigma\sqrt{2}$, only to $2\sigma$. However, in [215] it is shown that if $\operatorname{Var} X = \sigma^2 < \infty$, then $E \log \log L(\varepsilon) < \infty$ for $\varepsilon > \sigma\sqrt{2}$.

Another way to express this in prose is to say that the law of the iterated logarithm "barely" holds.

## 7.8 Precise Asymptotics

There also exist analogs to those from Subsection 7.7.8. We confine ourselves to presenting one example, namely [121], Theorem 2.

**Theorem 7.8.** *Let $X$, $X_1$, $X_2$, ... be independent, identically distributed random variables, and suppose that $E X = 0$ and $\operatorname{Var} X = \sigma^2 < \infty$. Then*

$$\lim_{\varepsilon \searrow 0} \varepsilon^2 \sum_{n=3}^{\infty} \frac{1}{n \log n} P(|S_n| \ge \varepsilon \sqrt{n \log \log n}) = \sigma^2.$$

A further reference in the area is [173].

## 7.9 The Other LIL

Let $X_1$, $X_2$, ... be independent, (identically distributed) random variables with mean 0 and finite variances. One can show that the extreme limit points of the maximal partial sums are the same as those for the sums themselves. This is maybe no surprise in view of the Lévy inequalities, Theorem 3.7.1. However, Chung [47] proves that

$$\liminf_{n \to \infty} \frac{\max_{1 \le k \le n} |S_k|}{\sqrt{\frac{s_n^2}{\log \log s_n^2}}} = \frac{\pi}{\sqrt{8}} \quad \text{a.s.,}$$

(where as always, $S_n$ and $s_n^2$, $n \ge 1$, are the partial sums and the sums of the variances, respectively) under the assumption of finite third moments and Lyapounov type conditions.

## 8 Problems

1. Let $X_1$, $X_2$, ... be independent, identically distributed random variables with mean 0 and variance $\sigma^2$, and set $S_n = \sum_{k=1}^{n} X_k$, $n \geq 1$. Show that

$$\frac{S_n}{\sqrt{n}(\log n)^\beta} \overset{a.s.}{\to} 0 \quad \text{as} \quad n \to \infty, \quad \text{for any} \quad \beta > 0.$$

2. Let $X, X_1, X_2, \ldots$ be independent, identically distributed random variables with $EX = 0$ and $\operatorname{Var} X = \sigma^2$, and set $S_n = \sum_{k=1}^{n} X_k$, $n \geq 1$. Prove the following two LILs for subsequences:

$$\limsup_{k\to\infty} (\liminf_{k\to\infty}) \frac{S_{k^k}}{\sqrt{k^k \log k}} = \sigma\sqrt{2}\, (-\sigma\sqrt{2}) \quad \text{a.s.,}$$

$$\limsup_{k\to\infty} (\liminf_{k\to\infty}) \frac{S_{[k^\alpha]}}{\sqrt{k^\alpha \log\log k}} = \sigma\sqrt{2}\, (-\sigma\sqrt{2}) \quad \text{a.s.}$$

3. Let $E \subset \mathbb{R}$, and suppose that $\{Y(t), t \geq 1\}$ is a family of random variables, such that

$$C\{Y(t), t \geq 1\} = E \quad \text{a.s.}$$

Further, let $\{\xi(t), t \geq 1\}$ and $\{\eta(t), t \geq 1\}$ be families of random variables, such that

$$\xi(t) \overset{a.s.}{\to} 1 \quad \text{and} \quad \eta(t) \overset{a.s.}{\to} 0 \quad \text{as} \quad t \to \infty.$$

Prove that

$$C\{\xi(t)Y(t) + \eta(t), t \geq 1\} = E \quad \text{a.s.}$$

♣ This is a kind of Cramér theorem; cf. Theorems 5.11.3 and 5.11.4.

4. The purpose of this problem is to prove Theorem 7.5. Let $X_1$, $X_2$, ... be independent, identically distributed random variables with mean $\mu > 0$ and finite variance, $\sigma^2$, set $S_n = \sum_{k=1}^{n} X_k$, $n \geq 1$, and let

$$\tau(t) = \min\{n : S_n > t\}, \quad t \geq 0.$$

(a) Prove that

$$C\left(\left\{\frac{S_{\tau(t)} - \mu\tau(t)}{\sqrt{\tau(t)\log^+ \log^+ \tau(t)}}\right\}\right) = [-\sigma\sqrt{2}, \sigma\sqrt{2}] \quad \text{a.s.}$$

(b) Prove that

$$C\left(\left\{\frac{S_{\tau(t)} - \mu\tau(t)}{\sqrt{t\log\log t}}, t \geq e^e\right\}\right) = [-\sigma\sqrt{2/\mu}, \sigma\sqrt{2/\mu}] \quad \text{a.s.}$$

(c) Prove that

$$\frac{X_{\tau(t)}}{\sqrt{t\log\log t}} \overset{a.s.}{\to} 0 \quad \text{as} \quad n \to \infty.$$

(d) Prove the theorem.

5. Prove the LIL for the counting process in Theorem 7.6.
6. Let $X_1, X_2, \ldots$ be independent, identically distributed random variables with mean 0 and finite variance, and set $S_n = \sum_{k=1}^n X_k$, $M_n^{(1)} = \max_{1 \le k \le n} S_k$, and $M_n^{(2)} = \max_{1 \le k \le n} |S_k|$, $n \ge 1$. State and prove LILs for $M_n^{(1)}$ and $M_n^{(2)}$.

# 9

## Limit Theorems; Extensions and Generalizations

Let us recapitulate what we have learned so far. After an introductory chapter containing some set theory and measure theory, we met a chapter on random variables and expectations, the probabilistic equivalent of Lebesgue integration on finite measure spaces. This was then followed by a number of probabilistic tools and methods, and one chapter each on the three cornerstone results, the law of large numbers, the central limit theorem, and the law of the iterated logarithm – LLN,CLT, and LIL. So, what's up next?

The main emphasis in the last three chapters has been on sums of independent, identically distributed random variables. In the last two chapters finite variance was an additional assumption. Some natural questions that arise are the following:

- What happens if the variance does not exist?
- What happens if the summands are no longer independent?
- Are there interesting quantities aside from sums?

These and other problems certainly deserve a chapter of their own in a book like this. However, one has to make choices. Such choices necessarily are made via a blend of "importance of the various topics" and "personal taste" (which are not completely disjoint). In this chapter we provide an introduction, sometimes a little more than that, to some more general limit theorems, with the hope that the reader will be tempted to look into the literature for more. However, there is one exception – the theory of martingales, which is given a chapter of its own; the next, and final one.

The first three sections are devoted to an extension of the central limit problem. The problem was "what can be said if the variance does not exist?" By departing from the normal analog we remember from Subsection 4.2.4 that if $X_1$, $X_2$, ... have a symmetric stable distribution with index $\alpha$, where $0 < \alpha \leq 2$, then so has $\sum_{k=1}^{n} X_k / n^{1/\alpha}$.

Is there a connection? Is it possible that if the variance does not exist, then the sum $\sum_{k=1}^{n} X_k$, suitably normalized – by $n^{1/\alpha}$ for some convenient $\alpha$? – is asymptotically stable?

The answer to this question is essentially "yes" and the limits are the stable distributions, but the details are technically more sophisticated than when the variance is finite. So, before turning to asymptotics we have to investigate the stable distribution themselves, and to prove the so-called *convergence to types theorem*, which states that if a sequence of random variables converges in distribution and a linearly transformed sequence does too, then the limits must be linear transformations of each other.

A further extension is to consider *arrays* of random variables and the corresponding classes of limit distributions. The class of limit distributions for arrays is the class of *infinitely divisible distributions*. This will be the topic of Section 9.4.

Another generalization is to suppress the independence assumption. In Section 9.5 we provide an introduction to the area, where we present some of the dependence concepts and connections.

As a parallel to the limit theory for sums there exists a limit theory for *extremes*. In Section 9.6 we define max-stable distributions and extremal distributions, which are the ones that may occur as limit distributions; that they coincide is a consequence of the convergence to types theorem. We also prove a celebrated theorem due to Gnedenko [95], which states that there exist *three* possible classes of limit distributions for partial maxima, and present some criteria for convergence to the respective classes. The section concludes with the corresponding limit theorem for record values.

The *Stein-Chen method* is a device to prove distributional convergence to the Poisson distribution via total variation distance. A short introduction to the method will be given in Section 9.7.

# 1 Stable Distributions

The name of the class of distributions, the stability, pertains to the fact that a sum of independent, identically distributed random variables has the same distribution as a linearly transformed summand.

**Definition 1.1.** *The distribution of a random variable $X$ is stable in the broad sense if $X$, $X_1$, $X_2$, ... are independent, identically distributed random variables with partial sums $S_n$, $n \geq 1$, and there exist constants, $c_n > 0$, and $d_n \in \mathbb{R}$, $n \geq 1$, such that*

$$S_n \stackrel{d}{=} c_n X + d_n \quad \text{for all} \quad n.$$

*The distribution is strictly stable if $d_n = 0$ for all $n$.*    □

According to history, this class was discovered by Paul Lévy after a lecture in 1919 by Kolmogorov, when someone told him that the normal distribution was the only stable one. He went home and discovered the symmetric stable

distributions the same day. Lévy's early work on stable laws is summarized in
his two books [170, 172]. An excellent recent source is [216].

We observe immediately that the normal distribution with mean 0 is
strictly stable with $c_n = \sqrt{n}$, and that the standard Cauchy distribution
is strictly stable with $c_n = n$.

Generally, let $X, X_1, X_2, \ldots$ be independent random variables with a stable distribution, and *suppose, for simplicity, throughout the discussion that the
distribution is symmetric.* Then, by interpreting $S_{2n}$ in two different ways,
namely,

$$S_{2n} = \sum_{k=1}^{n} X_k + \sum_{k=n+1}^{2n} X_k \quad \text{and} \quad S_{2n} = \sum_{k=1}^{2n} X_k,$$

respectively, and at the same time remembering that $S_n \overset{d}{=} c_n X$, we find that

$$c_2 S_n \overset{d}{=} S_{2n} \overset{d}{=} c_{2n} X,$$

which tells us that

$$c_{2n} = c_2 \cdot c_n.$$

Similarly, by splitting the sum $S_{mn}$ into blocks of equal lengths, one finds that
the sequence $\{c_n, n \geq 1\}$ is multiplicative;

$$c_{mn} = c_m \cdot c_n \quad \text{for all} \quad n, m \in \mathbb{N},$$
$$c_{n^k} = (c_n)^k \quad \text{for all} \quad n, k \in \mathbb{N}.$$

The first of these equations is the *Hamel equation.* Moreover, the second
equation and Corollary 3.7.1, together imply that

$$P(X_1 > x) = P(c_{n^k} X_1 > c_{n^k} x) = P(S_{c_{n^k}} > c_{n^k} x) \leq 2P(S_{c_{(n+1)^k}} > c_{n^k} x)$$
$$= P(c_{(n+1)^k} X_1 > c_{n^k} x) = P\left(X_1 > \left(\frac{c_n}{c_{n+1}}\right)^k x\right).$$

Since the left-most member is some fixed quantity between 0 and 1, $(c_n/c_{n+1})^k$
must remain bounded for all $n$ and $k$ in order to avoid that the right-hand
side converges to 0 as $k \to \infty$, from which it follows that $c_n/c_{n+1} < 1$ for all
$n$, that is, $\{c_n\}$ is increasing. Lemma A.8.2(b) therefore tells us that,

$$c_n = n^{1/\alpha} \quad \text{for} \quad \alpha > 0. \tag{1.1}$$

Having thus extablished the nature of the normalizing sequence, what does
the characteristic function look like?

Since $S_n \overset{d}{=} n^{1/\alpha} X$, the characteristic function of a symmetric stable distribution must satisfy

$$\varphi(t) = \left(\varphi(t/n^{1/\alpha})\right)^n \quad \text{for all} \quad n \in \mathbb{N}.$$

Since the characteristic functions of the normal and Cauchy distributions are exponentials with the argument raised to 2 and 1, respectively, it is tempting to try the Ansatz

$$\varphi(t) = e^{-c|t|^{\alpha}}, \tag{1.2}$$

where $c > 0$ is some constant. And, indeed, it works:

$$\exp\{-c|t/n^{1/\alpha}|^{\alpha}\}^n = e^{-c|t|^{\alpha}}.$$

For $\alpha > 2$ this means that $\varphi(t) = 1 + o(t)$ for $t$ in the neighborhood of the origin, which, by Theorem 4.4.3, implies that the variance equals 0, and, hence, by uniqueness that $\varphi$ cannot be a characteristic function (this accidentally solves Exercise 4.4.2). Thus, *the only possible case is $0 < \alpha \leq 2$.*

What about moments? The normal distribution has moments of all orders. The Cauchy distribution possesses moments of all orders strictly less than 1. More generally, *if $X$ has a moment of order $r$,* then, by the scaling property on the one hand, and Theorem 3.6.1 on the other,

$$E|S_n|^r \begin{cases} = n^{r/\alpha}E|X|^r, \\ \leq nE|X|^r, \end{cases}$$

which only makes sense for $r \leq \alpha$. For the full story we exploit Lemma 4.4.4, which tells us that, if $E|X|^r < \infty$ for some $r \in (0, 2)$, then

$$E|X|^r = C(r) \int_{-\infty}^{\infty} \frac{1 - \Re(\varphi(t))}{|t|^{r+1}}\, dt,$$

and conversely, where $C(r) = \frac{1}{\pi}\Gamma(r+1)\sin(r\pi/2)$.

For a symmetric stable distribution this means that, for $E|X|^r$ to be finite, it is necessary that

$$\int_{-\infty}^{\infty} \frac{1 - e^{-c|t|^{\alpha}}}{|t|^{r+1}}\, dt < \infty.$$

Since the integrand $\sim c|t|^{\alpha - r - 1}$ as $t \to 0$, it follows that $E|X|^r = \infty$ for $r \geq \alpha$, $0 < \alpha < 2$. Moreover, since the integral is, indeed, finite for $r < \alpha$ we also conclude that $E|X|^r < \infty$ for $r < \alpha < 2$. Summarizing the above yields

**Proposition 1.1.** *If $X$ has a symmetric stable distribution with index $\alpha$, and characteristic function $\varphi$, then*

- $0 < \alpha \leq 2$;
- $\varphi(t) = \exp\{-c|t|^{\alpha}\}, \quad c > 0$;
- $E|X|^r < \infty \quad$ *for* $\quad r < \alpha < 2$;
- $E|X|^r = \infty \quad$ *for* $\quad r \geq \alpha$, $0 < \alpha < 2$.

We now turn to the general case. As for moments, it follows from the inequalities in Chapter 3 that

$$E|X|^r \leq E|X^s|^r \leq 2E|X|^r \quad \text{for} \quad 0 < r \leq 2,$$

(where a superscript $s$ denotes symmetrized random variables as usual), so that the conclusion of the proposition remains unchanged.

Furthermore, it can be shown – see, e.g., [88, 98, 142, 177, 216] – that the characteristic function is of the form

$$\varphi(t) = \exp\left\{ i\mu t - c|t|^\alpha \left( 1 + i\beta \frac{t}{|t|} \tan \frac{\pi\alpha}{2} \right) \right\} \text{ for } \begin{cases} 0 < \alpha \leq 2, \ \alpha \neq 1, \ |\beta| \leq 1, \\ \mu \in \mathbb{R}, \ c > 0, \end{cases}$$

and

$$\varphi(t) = \exp\left\{ i\mu t - c|t|^\alpha \left( 1 + i\beta \frac{t}{|t|} \frac{2}{\pi} \log |t| \right) \right\} \text{ for } \begin{cases} \alpha = 1, \ |\beta| \leq 1, \\ \mu \in \mathbb{R}, \ c > 0, \end{cases}$$

respectively. The parameter $\alpha$ is the index, and the parameter $\beta$ is a skewness parameter; note that the distribution is symmetric around $\mu$ when $\beta = 0$.

Moreover, since the characteristic function is integrable, it follows from Theorem 4.1.4 that stable distributions are absolutely continuous, and that the density can be obtained by the inversion formula there. However, only the three cases $\alpha = 1/2, 1$, and $2$ produce closed expressions; for other values of $\alpha$ the densities are expressed as infinite sums. For details we refer to the literature mentioned above.

## 2 The Convergence to Types Theorem

One consequence of the continuous mapping theorem, Theorem 5.10.4, is that if $X_1, X_2, \ldots$ is a sequence of random variables, such that $X_n \overset{d}{\to} X$ as $n \to \infty$ for some random variable $X$, then, for $a_n \in \mathbb{R}^+$ and $b_n \in \mathbb{R}$, such that $a_n \to a$ and $b_n \to b$ as $n \to \infty$, we have $a_n X_n + b_n \overset{d}{\to} aX + b$. Since general questions about convergence should have answers which in some sense do not depend on linear transformations, it is natural to introduce the concept of *types* of random variables or distributions in the following sense: To each random variable $X$ we assign a *type* of laws, namely,

$$\{aX + b : a \in \mathbb{R}^+, \ b \in \mathbb{R}\},$$

that is, the collection of distributions obtained by linear transformations of the original random variable. The particular case when $a = 0$, that is, the constants, is called the *degenerate type*.

For example, the central limit problem is independent under linear transformations; if the central limit theorem holds, then, since linear transformations of normal random variables remain normal, we can say, not only that the original sequence of distributions obeys the central limit theorem, but that the *type* does. This also includes the degenerate case; a one-point distribution may be interpreted as a normal distribution with variance 0.

Following is the main theorem in this direction.

**Theorem 2.1.** (The convergence to types theorem)
Let $X_1, X_2, \ldots$ be a sequence of random variables, $a_n, \alpha_n \in \mathbb{R}^+$, $b_n, \beta_n \in \mathbb{R}$, $n \geq 1$, be such that

$$\frac{X_n - \beta_n}{\alpha_n} \xrightarrow{d} U \quad and \quad \frac{X_n - b_n}{a_n} \xrightarrow{d} V \quad as \quad n \to \infty,$$

where $U$ and $V$ are non-degenerate random variables. Then there exist $A \in \mathbb{R}^+$ and $B \in \mathbb{R}$, such that

$$\frac{a_n}{\alpha_n} \to A \quad and \quad \frac{b_n - \beta_n}{\alpha_n} \to B \quad as \quad n \to \infty.$$

In particular, if

$$X_n \xrightarrow{d} U \quad and \quad V_n = \frac{X_n - b_n}{a_n} \xrightarrow{d} V \quad as \quad n \to \infty,$$

where $U$ and $V$ are as before, then there exist $A \in \mathbb{R}^+$ and $B \in \mathbb{R}$, such that

$$a_n \to A \quad and \quad b_n \to B \quad as \quad n \to \infty,$$

and, hence,

$$V \overset{d}{=} \frac{U - B}{A} \quad or, \; equivalently, \quad U \overset{d}{=} AV + B.$$

*Proof.* Since the general case follows from the particular one by rescaling, it suffices to prove the latter.

The proof proceeds via the following steps:

(a)    $X_n \xrightarrow{d} U$, $a_n \to a > 0$, $b_n \to b$    $\implies$    $V_n \xrightarrow{d} \frac{X-b}{a}$    as    $n \to \infty$;

(b)    $X_n \xrightarrow{d} U$, $a_n \to +\infty$    $\implies$    $\frac{X_n}{a_n} \xrightarrow{P} 0$    as    $n \to \infty$;

(c)    $X_n \xrightarrow{d} U$, $\sup_n |b_n| = \infty$    $\implies$    $X_n - b_n \overset{d}{\nrightarrow}$    as    $n \to \infty$;

(d)    $X_n \xrightarrow{d} U$, $V_n \xrightarrow{d} V$    $\implies$    $0 < \inf_n a_n \leq \sup_n a_n < \infty$, $\sup_n |b_n| < \infty$;

(e)    $U \overset{d}{=} \frac{U-b}{a}$    $\implies$    $a = 1, b = 0$.

We shall repeatedly use the continuous mapping theorem, Theorem 5.10.4, without explicit mentioning.

(a) and (b): Immediate (by Theorem 5.10.4).

(c): The unboundedness of the normalizing sequence implies that there exist subsequences $b'_n \to +\infty$ and/or $b''_n \to -\infty$ as $n', n'' \to \infty$, respectively, which implies that, for all $x$,

$$P(X'_n - b'_n \leq x) = P(X'_n \leq x + b'_n) \to 1 \quad \text{as} \quad n' \to \infty,$$
$$P(X''_n - b''_n \leq x) = P(X''_n \leq x + b''_n) \to 0 \quad \text{as} \quad n'' \to \infty.$$

(d): If $\sup_n a_n = +\infty$, there exists a subsequence $\{n_k, \ k \geq 1\}$, such that $a_{n_k} \to \infty$ as $k \to \infty$, so that, by (b), $X_{n_k}/a_{n_k} \overset{P}{\to} 0$ as $k \to \infty$. But then, remembering that $V_{n_k} \overset{d}{\to} V$ as $k \to \infty$, (c) tells us that $\sup_k |b_{n_k}|/a_{n_k} < +\infty$, which, in turn, implies that there exists a further subsequence $n_{k_j}, \ j \geq 1$, such that $b_{n_{k_j}}/a_{n_{k_j}} \to c$ for some $c \in \mathbb{R}$ as $j \to \infty$. Joining this with (a) we conclude that $V_{n_{k_j}} \overset{P}{\to} -c$ as $j \to \infty$, which contradicts the non-degeneracy of $V$, and establishes that $\sup_n a_n < \infty$. Reversing the roles of $X_n$ and $V_n$ shows that $\inf_n a_n > 0$:

$$X_n \overset{d}{\to} U \text{ and } V_n \overset{d}{\to} V \iff V_n \overset{d}{\to} V \text{ and } a_n V_n + b_n \overset{d}{\to} U.$$

It remains to show that $\sup_n |b_n| < \infty$. Toward this end, suppose the contrary, and choose a subsequence $\{n_k, \ k \geq 1\}$ (in two steps), such that $a_{n_k} \to a$ and $b_{n_k} \to \infty$ as $k \to \infty$. Then $X_{n_k}/a_{n_k}$ converges in distribution as $k \to \infty$, but $V_{n_k}$ does not in view of (c). The same argument (with the same conclusion) can be applied to a subsequence $\{m_j, \ j \geq 1\}$, such that $b_{m_j} \to -\infty$ as $j \to \infty$. This completes the proof of (d).

(e): Iterating the assumption yields

$$U \overset{d}{=} \frac{\frac{U-b}{a} - b}{a} = \frac{U}{a^2} - \frac{b}{a^2} - \frac{b}{a} \overset{d}{=} \cdots \overset{d}{=} \frac{U}{a^n} - b \sum_{k=1}^{n} \frac{1}{a^k} = \frac{U - a^n b \sum_{k=0}^{n-1} a^k}{a^n}.$$

An application of (d) therefore tells us that

$$0 < \inf_n a_n \leq \sup_n a_n < \infty \quad \text{and} \quad \sup_n a^n |b| \sum_{k=0}^{n-1} a^k | < \infty,$$

which forces $a = 1$ and $b = 0$.

We are now set for the final argument. Namely, the assumptions of the theorem and the boundedness implied by (d) allow us to select a subsequence $\{n_k, \ k \geq 1\}$, such that $a_{n_k} \to A > 0$ and $b_{n_k} \to B \in \mathbb{R}$ as $k \to \infty$, which, implies that

$$V_{n_k} \overset{d}{\to} \frac{U - B}{A} \quad \text{as} \quad k \to \infty,$$

so that

$$V \overset{d}{=} \frac{U - B}{A}.$$

Now, if, along another subsequence, $\{m_j, \ j \geq 1\}$, we have $a_{m_j} \to A^* > 0$ and $b_{m_j} \to B^* \in \mathbb{R}$ as $j \to \infty$, and, hence,

$$V_{m_j} \overset{d}{\to} \frac{U - B^*}{A^*} \quad \text{as} \quad k \to \infty,$$

so that

$$V \overset{d}{=} \frac{U - B^*}{A^*},$$

it follows, by (e), that $B = B^*$ and $A = A^*$. This shows that every convergent pair of normalizations yields the same limit, and, hence, that the whole sequence does too.

The proof the theorem is complete.    □

**Exercise 2.1.** Deduce the first, more general, statement in Theorem 2.1 from the particular one.    □

**Exercise 2.2.** Prove Theorem 2.1 using characteristic functions. Theorem 4.1.7 may be helpful with regard to certain (non)-degeneracy aspects.    □

An immediate consequence of the theorem is that the normal distribution is the only one that can appear as a limit distribution for normalized sums of independent, identically distributed random variables with finite variance, which means that the central limit theorem is "the only one" in that context. In the same vein the next section is devoted to the question of finding all limit distributions when the variance is no longer assumed to be finite.

It should also be mentioned that the convergence to types theorem does not prevent us from producing a normalizing sequence that yields a degenerate limit.

**Theorem 2.2.** *For any sequence of random variables, $X_1, X_2, \ldots$, there always exists $\{a_n, n \geq 1\}$, such that*

$$\frac{X_n}{a_n} \overset{p}{\to} 0 \quad as \quad n \to \infty.$$

*Proof.* Since $P(|X_n| > x) \to 0$ as $x \to \infty$ for all $n$, there exists $c_n$, such that

$$P(|X_n| > c_n) < \frac{1}{n} \quad \text{for all} \quad n.$$

By putting $a_n = nc_n$ we obtain, for any $\varepsilon > 0$ and $n > \varepsilon^{-1}$, that

$$P\left(\frac{|X_n|}{a_n} > \varepsilon\right) = P(|X_n| > nc_n\varepsilon) \leq P(|X_n| > c_n) < \frac{1}{n} \to 0 \quad as \quad n \to \infty. \ \square$$

# 3 Domains of Attraction

In this section we shall provide the answer to the question raised in the opening of this chapter, namely: What are the possible limit distributions of a normalized sequence of independent, identically distributed random variables when the variance does not exist? In order to investigate this problem we introduce the notion of *domains of attraction*.

**Definition 3.1.** *Let $X$, $X_1$, $X_2$, ... be independent, identically distributed random variables with partial sums $S_n$, $n \geq 1$. We say that $X$, or, equivalently, the distribution $F_X$, belongs to the domain of attraction of the (non-degenerate) distribution $G$ if there exist normalizing sequences $\{a_n > 0, n \geq 1\}$, and $\{b_n, n \geq 1\}$, such that*

$$\frac{S_n - b_n}{a_n} \xrightarrow{d} G \quad as \quad n \to \infty.$$

*The notation is $F_X \in \mathcal{D}(G)$, alternatively, $X \in \mathcal{D}(Z)$ if $Z \in G$.* □

The first observation is that, if $\operatorname{Var} X < \infty$, the central limit theorem tells us that $X$ belongs to the domain of attraction of the normal distribution; choose $b_n = nE\,X$, and $a_n = \sqrt{n \operatorname{Var} X}$. In particular, the normal distribution belongs to its own domain of attraction. Our discussion in the first section above shows that the stable distributions belong to their own domain of attraction. However, more is true.

**Theorem 3.1.** *Only the stable distributions or random variables possess a domain of attraction.*

*Proof.* That they do possess a domain of attraction has already been established. That they are the only ones follows from the convergence to types theorem, Theorem 2.1. Namely, suppose that $X_1$, $X_2$, ... are independent, identically distributed random variables with partial sums $S_n$, $n \geq 1$, such that

$$Z_n = \frac{S_n - b_n}{a_n} \xrightarrow{d} G \quad as \quad n \to \infty,$$

and consider blocks as in the previous section: Let $n = km$ with $k$ fixed, set

$$V_k = \sum_{j=1}^{k} \left( \frac{\sum_{i=(j-1)m+1}^{jm} X_i - b_m}{a_m} \right),$$

and observe that

$$V_k \stackrel{d}{=} \sum_{j=1}^{k} Z_j^{(m)},$$

where $Z_1^{(m)}, Z_2^{(m)}, \ldots, Z_k^{(m)}$ are independent copies of $Z_m$.

Now, if $Z_n \in \mathcal{D}(Y)$ for some random variable $Y$, then, by Theorem 5.11.2,

$$V_k \xrightarrow{d} \sum_{j=1}^{k} Y_j \quad as \quad n \to \infty,$$

where $Y_1, Y_2, \ldots, Y_k$ are independent copies of $Y$. Since $Z_n$ and $V_k$ only differ by scale and location, the same holds true for the limits $Y$ and $\sum_{j=1}^{k} Y_j$, which, by the convergence to types theorem, forces the distribution of $Y$ to be stable. □

So, the stable distributions are the only ones with a domain of attraction. The next problem is to find criteria for a distribution to belong to the domain of attraction of some given (stable) distribution for which we need some facts about regular and slow variation from Section A.7.

The main theorem runs as follows.

**Theorem 3.2.** *A random variable X with distribution function F belongs to the domain of attraction of a stable distribution iff there exists $L \in SV$, such that*

$$U(x) = E\, X^2 I\{|X| \le x\} \sim x^{2-\alpha} L(x) \quad as \quad x \to \infty, \qquad (3.1)$$

*and, moreover, for $\alpha \in (0, 2)$,*

$$\frac{P(X > x)}{P(|X| > x)} \to p \quad and \quad \frac{P(X < -x)}{P(|X| > x)} \to 1 - p \quad as \quad x \to \infty. \qquad (3.2)$$

*Remark 3.1.* Requirement (3.1) provides information on the growth rate of the truncated second moment. The second condition describes the proportion of the left-hand and right-hand tails relative to the total mass in the tails. □

By partial integration and Lemma A.7.1(i) one can show that (3.1) is equivalent to

$$\frac{x^2 P(|X| > x)}{U(x)} \to \frac{2 - \alpha}{\alpha} \quad as \quad x \to \infty, \quad for \quad 0 < \alpha \le 2, \qquad (3.3)$$

$$P(|X| > x) \sim \frac{2 - \alpha}{\alpha} \cdot \frac{L(x)}{x^\alpha} \quad as \quad x \to \infty, \quad for \quad 0 < \alpha < 2. \qquad (3.4)$$

Joining this with Theorem 3.1 yields the following alternative formulation of Theorem 3.2.

**Theorem 3.3.** *A random variable X with distribution function F belongs to the domain of attraction of*
(a) *the normal distribution iff $U \in SV$.*
(b) *a stable distribution with index $\alpha \in (0, 2)$ iff (3.4) and (3.2) hold.*

*Remark 3.2.* Here we have used the fact that we already know that *any possible limit is stable*. However, a full proof starting from scratch, which, i.a., reproves this fact without resorting to the convergence to types theorem is also possible. One approach is based on characteristic functions; see [98], Chapter 4. An elegant development of the theory based on convolution semi-groups can be found in [88], Chapter IX. □

The proof proceeds in two main steps. The first one consists of finding criteria under which the normalized partial sums converge. A detailed proof of this part would take us far beyond the scope of this book and will therefore only be sketched. In the second step the criteria are applied to prove Theorems 3.2 and 3.3. That part will be given greater attention.

## 3.1 Sketch of Preliminary Steps

A necessary condition for convergence is that $\{(S_n - b_n)/a_n, \, n \geq 1\}$ is *stochastically bounded*, that is, that, for every $\varepsilon > 0$, there exists $M$, such that

$$P\left(\left|\frac{S_n - b_n}{a_n}\right| > M\right) < \varepsilon. \tag{3.5}$$

For this to happen we must trivially assume that $a_n \to \infty$ as $n \to \infty$.

Throughout this subsection we reformulate, for convenience, the problem in the language of arrays; recall Subsection 7.2.4. For each $n \geq 1$ we thus set

$$X_{n,k} = \frac{X_k}{a_n} \quad \text{for} \quad 1 \leq k \leq n, \tag{3.6}$$

and consider $S_n = \sum_{k=1}^n X_{n,k}$.

To find conditions for stochastic boundedness, one uses standard truncation techniques. Let $u > 0$, and set, for $1 \leq k \leq n$, $n \geq 1$,

$$X'_{n,k} = \begin{cases} -u, & \text{when} \quad X_k \leq -u, \\ X_{n,k}, & \text{when} \quad |X_{n,k}| \leq u, \\ u, & \text{when} \quad X_k \geq u, \end{cases} \tag{3.7}$$

and $S'_n = \sum_{k=1}^n X'_{n,k}$.

The first step is to apply the truncated Chebyshev inequality, Theorem 3.1.5. In order to simplify matters one starts by assuming that the distribution of the summands is symmetric, which implies that $E\,X'_{n,1} = 0$, so that

$$P(|S_n| > M) \leq \frac{n\,\mathrm{Var}\,X'_{n,1}}{M^2} + nP(|X| > u). \tag{3.8}$$

An analysis of the two terms in the right-hand side provides necessary and sufficient criteria for stochastic boundedness:

**Lemma 3.1.** (i) *Suppose that the distribution is* symmetric. *For the sequence* $\{S_n, \, n \geq 1\}$ *to be stochastically bounded it is necessary and sufficient that, for every* $\varepsilon > 0$, *there exist* $M$ *and* $u_0$, *such that*

$$n\,\mathrm{Var}\,X'_{n,1} < M\varepsilon, \tag{3.9}$$
$$nP(|X_{n,1}| > u) < \varepsilon \quad \text{for} \quad u > u_0. \tag{3.10}$$

(ii) *In the general, non-symmetric, case the conditions are necessary and sufficient for the stochastic boundedness of*

$$\{S_n - \mathrm{med}\,(S_n), \, n \geq 1\}.$$

We also have the following corollary.

**Corollary 3.1.** *In the general case the sequence*

$$\{S_n - E\,S_n',\ n \geq 1\} \quad \text{is stochastically bounded}$$

*under the conditions of the lemma.*

*Proof.* The conclusion is immediate from the truncated Chebyshev inequality, since, in the general case, $S_n$ in (3.8) is replaced by $S_n - E\,S_n'$.    □

In order to postpone the problem of centering, we note that, in the symmetric case, (3.9) is the same as

$$nE\big((X_{n,1}')^2\big) = nE\,X_{n,1}^2 I\{|X_{n,1}| \leq u\} < M\varepsilon. \tag{3.11}$$

In the general case, they are equivalent if

$$(E\,X_{n,1}')^2 = o\big(E(X_{n,1}')^2\big) \quad \text{as} \quad n \to \infty. \tag{3.12}$$

One such example is when the mean exists and equals 0, because $E\,X_{n,1}' \to 0$ as $n \to \infty$ in this case.

The next point is to prove existence of a so-called *canonical measure*, $\Pi$, that is finite for finite intervals, and such that the integrals

$$\gamma^+(x) = \int_x^\infty \frac{d\Pi(y)}{y^2} \quad \text{and} \quad \gamma^-(x) = \int_{-\infty}^{-x} \frac{d\Pi(y)}{y^2} \tag{3.13}$$

are convergent for all $x > 0$. This is achieved via a limiting procedure applied to the measures $\Pi_n$, defined by

$$\Pi_n((c,d]) = nE\,X_{n,1}^2 I\{c < X_{n,1} \leq d\} \quad \text{for} \quad -\infty < c < d < \infty. \tag{3.14}$$

**Lemma 3.2.** *Suppose that (3.12) holds. The measure $\Pi_n$ is bounded on bounded intervals iff (3.9) holds.*

*Proof.* The conclusion follows from the relation

$$\text{Var}\,S_n' + \big(E\,S_n'\big)^2 = E\big((S_n')^2\big) = \Pi_n(-u,u) + u^2 P(|X_{n,1}| > u).    □$$

*Remark 3.3.* Condition (3.12) is automatic in the symmetric case, and in the general case with $E\,X = 0$.    □

Next one applies Helly's selection principle, Theorem 5.8.1, to conclude that

$$nP(c < X_{n,1} \leq d) = \int_c^d \frac{d\Pi_n(y)}{y^2} \to \int_c^d \frac{d\Pi(y)}{y^2},$$

along a subsequence, after which one lets $d \to \infty$ to obtain

$$nP(X_{n,1} > c) \to \int_c^\infty \frac{d\Pi(y)}{y^2} = \gamma^+(c),$$

along the same subsequence, and similarly for the negative tail. Some additional arguments then show that the limit is unique, after which one arrives at

**Lemma 3.3.** (i) *If (3.10) and (3.12) are satisfied, and the measure* $\Pi_n$ *defined in (3.14) is bounded on bounded intervals, then*

$$\{S_n - E\,S_n',\, n \geq 1\} \qquad and \qquad \{S_n - \mathrm{med}\,(S_n),\, n \geq 1\}$$

*are stochastically bounded, and*

- $\Pi_n \to \Pi$ *as* $n \to \infty$;
- $\Pi$ *is bounded on bounded intervals;*
- $nP(X_{n,1} > c) \to \gamma^+(c)$ *as* $n \to \infty$;
- $nP(X_{n,1} < -c) \to \gamma^-(c)$ *as* $n \to \infty$.

(ii) *If the distribution of the summands is* symmetric, *the conclusions remain the same without assumption (3.12).*

The final step amounts to proving that a suitably centered sum is convergent under the conditions of Lemma 3.3.

**Theorem 3.4.** *If (3.10) and (3.12) are satisfied, and the measure* $\Pi_n$ *defined in (3.14) is bounded on bounded intervals, then*

$$S_n - nE\,X_{n,1}' \overset{d}{\to} Y_\alpha,$$

*where* $Y_\alpha$ *is a non-degenerate stable distribution.*

*If the distribution is* symmetric, *convergence holds without assumption (3.12) and without the centering of* $S_n$.

*Remark 3.4.* (a) That the limit distribution necessarily is stable is a consequence of Theorem 3.1.
(b) One can show that the sufficient conditions are also necessary.     □

After this *sketch* we shall see how our main result follows from this, more general, setting.

## 3.2 Proof of Theorems 3.2 and 3.3

We only prove the sufficiency, and refer to the already cited literature for additional details.

We modify the original notation slightly in that we consider the sequence

$$\frac{S_n}{a_n} - b_n, \quad n \geq 1,$$

because it facilitates the desymmetrization procedure.

Before we take off we make some preliminary observations.
(i): If $U(\infty) < \infty$, then the variance exists, and the central limit theorem applies. This case is therefore no longer of interest (here).

(ii): In the more general scenario of the previous subsection we introduced condition (3.12) to facilitate certain operations. In the present context, upon noticing, for example, that $|x| = |x|^{\alpha/4}|x|^{1-\alpha/4}$, we have

$$
\begin{aligned}
\left(E|X|I\{|X| \le x\}\right)^2 &= \left(E|X|^{\alpha/4} \cdot E|X|^{1-\alpha/4}I\{|X| \le x\}\right)^2 \\
&\le EX^{\alpha/2}E|X|^{2-\alpha/2}I\{|X| \le x\} \\
&\le C\left(U(x)\right)^{1-\alpha/4} = o(U(x)) \quad \text{as} \quad x \to \infty,
\end{aligned}
$$

by Proposition 1.1, Cauchy's inequality, and Lyapounov's inequality, which shows that Condition (3.12) is always satisfied whenever $U(\infty) = \infty$.

(iii): In Proposition 1.1 we established, i.a., that stable distributions only possess moments of order strictly less than $\alpha$. The same is true for distributions belonging to the domain of attraction of stable distributions. To see this one combines assumption (3.4) on the decay of the tails and Theorem 2.12.1.

**Theorem 3.5.** *If $X$ belongs to the domain of attraction of a stable distribution with index $\alpha \in (0,2)$, then*

$$
E|X|^r < \infty \quad \text{for} \quad r < \alpha, \quad \text{and} \quad E|X|^r = \infty \quad \text{for} \quad r \ge \alpha. \qquad \square
$$

**Exercise 3.1.** Check the details. $\qquad \square$

We finally turn our attention to Theorems 3.2 and 3.3. Toward this end, we define a normalizing sequence $\{a_n, n \ge 1\}$ as follows:

$$
a_n = \sup\left\{x : \frac{nU(x)}{x^2} \le 1\right\}, \quad n \ge 1.
$$

In view of assumption (3.1), this produces a sequence of positive numbers, such that

$$
\lim_{n \to \infty} \frac{nU(a_n)}{a_n^2} = 1. \tag{3.15}
$$

Because of the regular variation of $U$ this implies (Section A.7) that

$$
\begin{aligned}
\lim_{n \to \infty} \Pi_n(-x, x) = \lim_{n \to \infty} \frac{nU(a_n x)}{a_n^2} &= \lim_{n \to \infty} \frac{U(a_n x)}{U(a_n)} \cdot \frac{nU(a_n)}{a_n^2} \\
&= x^{2-\alpha} \cdot 1 = x^{2-\alpha} = \Pi(-x, x),
\end{aligned}
$$

which shows that $\Pi$ is bounded on bounded intervals of the form $(-x, x)$.

For $\alpha = 2$ we obtain $\Pi = \delta(0)$, after which the rest is automatic, so that there is nothing more to prove.

For $0 < \alpha < 2$ an application of (3.2) and (3.4) shows that

$$
\begin{aligned}
\frac{x^\alpha P(X > x)}{L(x)} &= \frac{P(X > x)}{P(|X| > x)} \cdot \frac{x^2 P(|X| > x)}{U(x)} \cdot \frac{U(x)}{x^{2-\alpha}L(x)} \\
&\sim p \cdot \frac{2-\alpha}{\alpha} \quad \text{as} \quad x \to \infty,
\end{aligned}
$$

so that, by (3.15) and the slow variation of $L$,

$$nP(X > a_n x) \sim \frac{p(2-\alpha)}{\alpha} \cdot \frac{nL(a_n x)}{(a_n x)^\alpha} = \frac{p(2-\alpha)}{\alpha} \cdot \frac{1}{x^\alpha} \cdot \frac{L(a_n x)}{L(a_n)} \cdot \frac{nU(a_n)}{a_n^2}$$

$$\sim \frac{p(2-\alpha)}{\alpha} \cdot \frac{1}{x^\alpha} \quad ( = \gamma^+(x) ) \quad \text{as} \quad x \to \infty.$$

The mirrored argument proves the analog for the other tail.

Combining the two shows that $\Pi$ is bounded for arbitrary intervals that stay away from 0, and, hence, that $\Pi$ is the desired canonical measure. Next, set

$$X'_{n,k} = \begin{cases} -u, & \text{for} \quad X_k \leq -a_n u, \\ \frac{X_k}{a_n}, & \text{for} \quad |X_{n,k}| \leq a_n u, \\ u, & \text{for} \quad X_k \geq a_n u. \end{cases}$$

An appeal to Theorem 3.4 then establishes that

$$\frac{S_n - E\, S'_n}{a_n} = \frac{S_n}{a_n} - nE\, X'_{n,1} \xrightarrow{d} Y_\alpha \quad \text{as} \quad n \to \infty,$$

where $Y_\alpha$ is stable with index $\alpha$.

It remains to investigate the centering.

The tail behavior (3.4), (3.15), the slow variation of $L$, and Theorem 2.12.1 tell us that, for $\alpha \neq 1$,

$$nE\left(\frac{X_1}{a_n} I\left\{\frac{|X_{n,1}|}{a_n} \leq u\right\}\right) \sim \frac{2-\alpha}{\alpha} \cdot \frac{nL(a_n u)}{a_n (a_n u)^{\alpha-1}}$$

$$= \frac{2-\alpha}{\alpha} \cdot \frac{nU(a_n)}{a_n^2} \cdot \frac{L(a_n u)}{L(a_n)} \cdot u^{1-\alpha} \sim \frac{2-\alpha}{\alpha} u^{1-\alpha} \quad \text{as} \quad n \to \infty,$$

and, similarly, that

$$nE(uI\{X_1 \geq a_n u\} - uI\{X_1 \leq -a_n u\}) = nuP(|X| \geq a_n u)$$

$$\sim u\frac{2-\alpha}{\alpha} \cdot \frac{nL(a_n u)}{(a_n u)^\alpha} \sim \frac{2-\alpha}{\alpha} \cdot u^{1-\alpha} \quad \text{as} \quad n \to \infty,$$

so that

$$nE\, X'_{n,k} \sim \frac{2-\alpha}{\alpha} \cdot u^{1-\alpha} \quad \text{as} \quad n \to \infty.$$

This means that the centerings are constants asymptotically, and can be dismissed; the limit distribution is still a stable one (the convergence to types theorem). When $\alpha > 1$ we know, in addition, from Theorem 3.5 that $E|X| < \infty$, so that we may assume, without restriction, that $E\, X = 0$, or else, center at the mean which amounts to the same. For $\alpha = 1$ the centering causes trouble, mainly because of the integral

$$\int_{|x| \leq a_n u} \frac{L(x)}{x} \, dx.$$

The centering therefore cannot be dispensed with. For example, in the trivial case, when $L(x) \equiv 1$, the integral $\sim \log(a_n u)$.

Except for the exact expression when $\alpha = 1$, the proof of Theorems 3.2 and 3.3, has been completed.    □

*Remark 3.5.* Inspecting (3.15) we find, in view of (3.1), that, asymptotically,

$$\frac{na_n^{2-\alpha}L(a_n)}{a_n^2} \sim 1,$$

so that

$$a_n \sim n^{1/\alpha}\big(L(a_n)\big)^{1/\alpha}.$$

By setting $L^*(n) \sim \big(L(a_n)\big)^{1/\alpha}$, it follows that

$$a_n \sim n^{1/\alpha}L^*(n). \tag{3.16}$$

We also note that $L^* \in \mathcal{SV}$ in view of Lemma A.7.1(e).

For $\alpha = 2$ (3.1) states that $U \in \mathcal{SV}$. One such case is when the variance is finite – $U(\infty) < \infty$ –, in which case $a_n \sim c\sqrt{n}$ as expected (required).

The other case is when the variance is infinite – $U(\infty) = +\infty$. This happens, for example, if $U(x) \sim \log x$ or $U(x) \sim \log \log x$ as $x \to \infty$. In these cases the limit distribution is still normal. However, the normalization will be of the form $a_n \sim \sqrt{n}L^*(n)$, in other words, $\sqrt{n}$ multiplied by some slowly varying function.    □

## 3.3 Two Examples

Let us, as a first illustration, look at the simplest example. Yes, experience from Chapter 6 tells us that we should try the two-sided Pareto distribution.

*Example 3.1.* Let $X, X_1, X_2, \ldots$ be independent random variables with common density

$$f(x) = \begin{cases} \frac{1}{2x^2}, & \text{for} \quad |x| > 1, \\ 0, & \text{otherwise.} \end{cases}$$

The distribution is symmetric, the mean is infinite, and the tails decrease like those of the Cauchy distribution, so we would expect the distribution to belong to the domain of attraction of a symmetric stable distribution with index $\alpha = 1$.

Simple integration shows that

$$P(X > x) = \frac{1}{2x}, \quad P(X < -x) = \frac{1}{2|x|}, \quad P(|X| > x) = \frac{1}{x}, \quad U(x) = x - 1,$$

so that (3.1)–(3.4) are satisfied ($p = 1/2$ and $L(x) = 1$).

The guess that the distribution belongs to the domain of attraction of the standard Cauchy distribution was correct.    □

Next, we consider an interesting boundary case; interesting in the sense that the variance does not exist, but the asymptotic distribution is normal all the same.

*Example 3.2.* Suppose that $X$, $X_1$, $X_2$, ... are independent random variables with common density

$$f(x) = \begin{cases} \frac{1}{|x|^3}, & \text{for} \quad |x| > 1, \\ 0, & \text{otherwise.} \end{cases}$$

The variance is infinite; $\int_1^\infty \frac{x^2}{x^3}\, dx = +\infty$.

The first natural thing to investigate is (3.1):

$$U(x) = \int_{|y| \leq x} y^2 f(y)\, dy = 2 \int_0^x \frac{dy}{y} = 2 \log x,$$

so that $U \in \mathcal{SV}$ as $x \to \infty$, that is, $X$ belongs to the domain of attraction of the normal distribution.

Thus, for a suitable choice of normalizing constants $\{a_n, n \geq 1\}$ (no centering because of symmetry),

$$\frac{S_n}{a_n} \xrightarrow{d} N(0,1) \quad \text{as} \quad n \to \infty.$$

The variance being infinite, (3.16) tells us that

$$a_n \sim \sqrt{n \cdot 2 \log(a_n)}.$$

Since, as a first order approximation, $a_n \approx \sqrt{n}$, we obtain,

$$a_n \sim \sqrt{n \cdot 2 \log \sqrt{n}} = \sqrt{n \log n},$$

suggesting that

$$\frac{S_n}{\sqrt{n \log n}} \xrightarrow{d} N(0,1) \quad \text{as} \quad n \to \infty,$$

which is precisely what was to be shown in Problem 7.8.8.                    □

## 3.4 Two Variations

An interesting feature in Example 3.1 is that the slowly varying function $L \equiv 1$. Distributions for which this happens belong to the domain of *normal* attraction to the normal or more general stable laws. This means that the Pareto distribution in Example 3.1 belongs to the domain of normal attraction of the standard Cauchy distribution. If, as will be seen in the problem section, we modify the density in the example by multiplying with a power of the logarithm, the resulting distribution will belong to the domain of attraction (only).

More formally, a distribution belongs to the *normal* domain of attraction if the normalizing sequence is of the form $cn^{1/\alpha}$, where $c$ is some positive constant. For example, the normal and stable distributions belong to the domain of normal attraction to themselves.

By checking the conditions for attraction one can also show that $X$ belongs to the domain of normal attraction to the normal law, if and only if the variance is finite, and that $X$ belongs to the domain of normal attraction to a stable law with index $\alpha \in (0,2)$ if and only if, for some constant $a > 0$,

$$P(X < -x) \sim c_1 \frac{a^\alpha}{|x|^\alpha}, \quad \text{and} \quad P(X > x) \sim c_2 \frac{a^\alpha}{x^\alpha} \quad \text{as} \quad x \to \infty.$$

Finally, if the condition for belonging to the domain of attraction holds for a *subsequence*, the corresponding distribution is said to belong to the domain of *partial* attraction of the relevant limit distribution. In these cases various kinds of strange behaviors are possible. There exist, for example, distributions that do not belong to their own domain of partial attraction, and there exist distributions that belong to the domain of partial attraction of infinitely many distributions.

Once again, more can be found in [88, 98, 142, 194, 195].

### 3.5 Additional Results

Just as Theorem 7.3.2 is a "random sum central limit theorem", the following one is an Anscombe type random sum *stable* limit theorem.

**Theorem 3.6.** *Let $X_1, X_2, \ldots$ be independent, identically distributed random variables with mean 0, and set $S_n = \sum_{k=1}^n X_k$, $n \geq 1$. Suppose that $\{B_n, n \geq 1\}$ is a sequence of positive normalizing constants, such that*

$$\frac{S_n}{B_n} \xrightarrow{d} Y_\alpha \quad \text{as} \quad n \to \infty,$$

*where $Y_\alpha$ has a stable distribution with index $\alpha \in (1,2]$, and that $\{N(t), t \geq 0\}$ is a family of positive, integer valued random variables, such that, for some $0 < \theta < \infty$,*

$$\frac{N(t)}{t} \xrightarrow{p} \theta \quad \text{as} \quad t \to \infty.$$

*Then*

$$\frac{S_{N(t)}}{B_{N(t)}} \xrightarrow{d} Y_\alpha \quad \text{and} \quad \frac{S_{N(t)}}{B_{[\theta t]}} \xrightarrow{d} Y_\alpha \quad \text{as} \quad t \to \infty.$$

A proof may be modeled after the proof of Theorem 7.3.2, but with some additional care. One first notes that

$$\lim_{\delta \to 0} \lim_{n \to \infty} \frac{B_{[n(1+\delta)]}}{B_n} = \lim_{\delta \to 0} (1 + \delta)^{1/\alpha} = 1,$$

because of the regular variation of the normalizing sequence. This means that Anscombe's condition (7.3.2) is satisfied (cf. [209], p. 77), so that Anscombe's theorem 7.3.1 applies, and the first statement follows.

For the second one we also need the fact that

$$\frac{B_{N(t)}}{B_{[\theta t]}} \xrightarrow{p} 1 \quad \text{as} \quad t \to \infty, \tag{3.17}$$

see [103], Lemma 2.9.a.

**Exercise 3.2.** Complete the details.    □

*Remark 3.6.* In [253] the result is obtained with different methods as a corollary of more general results.    □

As an application of Theorem 3.6 one can prove the following stable analog of Theorem 7.4.1; see [103], Theorem 2.9, or [110], Theorem III.5.3, cf. also [253], Theorem 5.2.

**Theorem 3.7.** *Let* $X_1$, $X_2$, ... *be independent, identically distributed random variables with mean* $\mu > 0$, *and set* $S_n = \sum_{k=1}^{n} X_k$, $n \geq 1$. *Suppose that* $\{B_n,\ n \geq 1\}$ *is a sequence of positive normalizing constants such that*

$$\frac{S_n}{B_n} \xrightarrow{d} Y_\alpha \quad \text{as} \quad n \to \infty,$$

*where* $Y_\alpha$ *has a stable distribution with index* $\alpha \in (1, 2]$. *Furthermore, let* $\{\tau(t),\ t \geq 0\}$ *be the first passage time process, viz.,* $\tau(t) = \min\{n : S_n > t\}$, $t \geq 0$. *Then*

$$\frac{\tau(t) - t/\mu}{\mu^{-1} B_{[t/\mu]}} \xrightarrow{d} -Y_\alpha \quad \text{and} \quad \frac{\tau(t) - t/\mu}{\mu^{-(1+(1/\alpha))} B_{[t]}} \xrightarrow{d} -Y_\alpha \quad \text{as} \quad t \to \infty.$$

The proof follows "the usual pattern", except that taking care of $X_{\tau(t)}$ is significantly harder. In addition to (3.17) we need [103], Lemma 2.9.b:

**Lemma 3.4.** *Under the assumptions of the theorem,*

$$\frac{X_{\tau(t)}}{B_{[t/\mu]}} \xrightarrow{p} 0 \quad \text{as} \quad t \to \infty.$$

**Exercise 3.3.** Prove the lemma and Theorem 3.7.    □

*Remark 3.7.* Note that the limit is $-Y_\alpha$. The reason for this is that general stable distributions are not symmetric (in contrast to the normal case, recall Theorem 7.4.1).    □

Finally, for extensions to convergence rate results and the like we confine ourselves to referring, in chronological order, to [120, 122, 231].

## 4 Infinitely Divisible Distributions

During the process of proving that the stable distributions were the only ones that could appear as limit distributions of normalized partial sums we switched to a triangular array of random variables. Having seen, in Theorem 7.2.4, that the the Lindberg-Lévy-Feller theorem could be extended to arrays, it is not far-fetched to ask for the class of distributions that may appear as possible limit distributions of normalized arrays. Clearly, the stable distributions, in particular the normal ones, must be contained in this, possibly larger, class.

The answer to this question is:

**Theorem 4.1.** *Let $\{(X_{n,j}, 1 \leq j \leq k_n), n \geq 1\}$ be an array of row-wise independent, identically distributed random variables. The class of limit distributions of partial sums $S_n = \sum_{j=1}^{k_n} X_{n,j}$, $n \geq 1$, coincides with the class of infinitely divisible distributions.*

In order to understand the statement, here is a definition of infinite divisibility. Or, the other way around, let the theorem motivate the notion of infinite divisibility.

**Definition 4.1.** *A random variable $X$ has an infinitely divisible distribution iff, for each $n$, there exist independent, identically distributed random variables $X_{n,k}$, $1 \leq k \leq n$, such that*

$$X \stackrel{d}{=} \sum_{k=1}^{n} X_{n,k} \quad \text{for all} \quad n,$$

*or, equivalently, iff*

$$\varphi_X(t) = \left(\varphi_{X_{n,1}}(t)\right)^n \quad \text{for all} \quad n. \qquad \square$$

An inspection of some of the most familiar characteristic functions tells us that, for example, the normal, the Cauchy, the Poisson, the gamma, and the degenerate distributions all are infinitely divisible. The symmetric stable distributions are another easily accessible example. Namely, on the one hand,

$$e^{-|t|^\alpha} = \left(\exp\{-|t/n^{1/\alpha}|^\alpha\}\right)^n,$$

and on the other hand we obtain directly that

$$X_{n,k} = \frac{X_k}{n^{1/\alpha}}, \quad k = 1, 2, \ldots, n,$$

satisfies the definition of infinite divisibility for all $n$. Note that this, in addition, elucidates the switching from $X_k/a_n$ to $X_{n,k}$ in (3.6).

First, some elementary properties.

**Theorem 4.2.** (i) *The characteristic function of an infinitely distribution never vanishes.*
(ii) *If $X$ and $Y$ are independent, infinitely divisible random variables, then, for any $a, b \in \mathbb{R}$, so is $aX + bY$.*
(iii) *Let $0 < p < 1$. If $F_1$ and $F_2$ are infinitely divisible distributions, then $pF_1 + (1-p)F_2$ is* not *necessarily infinitely divisible; convex combinations of infinitely divisible distributions* need not *be infinitely divisible.*
(iv) *If $F_k$, $k \geq 1$, are infinitely divisible, then so is $F = \lim_{k \to \infty} F_k$; the class of infinitely divisible distribution is closed under limits.*

*Proof.* (i): Let $X$ $(F)$ have characteristic function $\varphi$. In order to stay among the real numbers we consider the symmetrized distribution whose characteristic function is $|\varphi|^2$.

We know that $0 \leq |\varphi|^2 \leq 1$, so that

$$|\varphi(t)|^{2/n} \to \varphi^*(t) = \begin{cases} 1, & \text{for} \quad \{t : |\varphi(t)|^2 \neq 0\}, \\ 0, & \text{for} \quad \{t : |\varphi(t)|^2 = 0\}. \end{cases}$$

Considering that $|\varphi(0)|^2 = 1$, and that $|\varphi|^2$ is uniformly continuous, we must have $\varphi^*(t) = 1$ in a neighborhood of the origin, so that, by the continuity theorem for characteristic functions, Theorem 5.9.2, $\varphi^*(t)$ is itself a characteristic function, which, due to continuity, leaves $\varphi^*(t) \equiv 1$, and thus $|\varphi(t)|^2 \neq 0$, and therefore $\varphi(t) \neq 0$ for all $t$, as the only possibility.
(ii): For every $n$ there exist random variables $\{X_{n,k}, 1 \leq k \leq n\}$, and $\{Y_{n,k}, 1 \leq k \leq n\}$, such that

$$X \stackrel{d}{=} \sum_{k=1}^n X_{n,k} \quad \text{and} \quad Y \stackrel{d}{=} \sum_{k=1}^n Y_{n,k},$$

so that

$$aX + bY \stackrel{d}{=} a\sum_{k=1}^n X_{n,k} + b\sum_{k=1}^n Y_{n,k} \stackrel{d}{=} \sum_{k=1}^n (aX_{n,k} + bY_{n,k}),$$

or, alternatively, via characteristic functions,

$$\begin{aligned} \varphi_{aX+bY}(t) &= \varphi_X(at) \cdot \varphi_Y(bt) = \left(\varphi_{X_{n,1}}(at)\right)^n \left(\varphi_{Y_{n,1}}(bt)\right)^n \\ &= \left(\varphi_{X_{n,1}}(at)\varphi_{Y_{n,1}}(bt)\right)^n = \left(\varphi_{aX_{n,1}}(t)\varphi_{bY_{n,1}}(t)\right)^n \\ &= \left(\varphi_{aX_{n,1}+bY_{n,1}}(t)\right)^n. \end{aligned}$$

(iii): Let $X \in \delta(-1)$ and $Y \in \delta(1)$. Then $F = \frac{1}{2}F_X + \frac{1}{2}F_Y$ corresponds to a coin-tossing random variable. Namely,

$$\varphi_F(t) = \frac{1}{2}e^{-it} + \frac{1}{2}e^{it} = \cos t,$$

which, due to (a), shows that $F$ is not infinitely divisible.

(iv): Suppose first that $F_k$, $k \geq 1$, are symmetric, so that the characteristic functions are positive. For each $k$ we then have

$$\varphi_k(t) = \left(\varphi_k^{(n)}(t)\right)^n \quad \text{for all} \quad n.$$

By "the easy half" of the continuity theorem for characteristic functions it follows that

$$\varphi_k(t) \to \varphi(t) \quad \text{as} \quad k \to \infty, \quad \text{for all} \quad t,$$

but also, that *for all* $n$,

$$\varphi_k^{(n)}(t) \to \varphi^{(n)}(t) \quad \text{as} \quad k \to \infty, \quad \text{for all} \quad t.$$

The other half of the same continuity theorem then tells us that $\varphi^{(n)}(t)$ is, indeed, the characteristic function corresponding to some distribution *for every* $n$, which, in turn, shows that

$$\varphi_k(t) = \left(\varphi_k^{(n)}(t)\right)^n \to \left(\varphi^{(n)}(t)\right)^n \quad \text{as} \quad k \to \infty.$$

Joining the displayed formulas, finally shows that

$$\varphi(t) = \left(\varphi^{(n)}(t)\right)^n \quad \text{for all} \quad n,$$

which proves the infinite divisibility of the limit distribution.

For the general case one needs a desymmetrization argument based on the fact that the symmetrized distribution is infinitely divisible (why?), so that, by (a), the characteristic function of the symmetrized distribution never vanishes, so that the logarithm, and therefore $n$th roots are well defined, from which the conclusion follows. We omit the details.

**Exercise 4.1.** Prove that the $U(-1,1)$-distribution is not infinitely divisible. $\quad\square$

The following result characterizes the class of infinitely divisible distributions.

**Theorem 4.3.** *The random variable $X$ has an infinitely divisible distribution iff*

$$\varphi_X(t) = \exp\left\{ i\mu t + \int_{-\infty}^{\infty} \left( e^{itx} - 1 - \frac{itx}{1+x^2} \right) \frac{1+x^2}{x^2} \, dG(x) \right\},$$

*where $\mu \in \mathbb{R}$, and where $G$ is a finite measure. The integrand is defined as $-t^2/2$ for $x = 0$.*

In order to present the basic ideas behind the proof we content ourself with the simpler case of finite variance, for which the characterization is as follows.

**Theorem 4.4.** *Suppose that $X$ is a random variable with finite variance. Then $X$ is infinitely divisible iff*

$$\varphi_X(t) = \exp\left\{ i\mu t + \int_{-\infty}^{\infty} \left( e^{itx} - 1 - itx \right) \frac{1}{x^2} \, dG(x) \right\},$$

*where $\mu \in \mathbb{R}$, and where $G$ is a distribution function up to a multiplicative constant.*

*Remark 4.1.* There are two infinitely divisible distributions for which $G$ is degenerate. If $G$ has point mass $\sigma^2$ at 0 then $X$ is normal with mean $\mu$ and variance $\sigma^2$, and if $G$ has point mass $\lambda$ at 1 and $\mu = 1$, then $X \in \mathrm{Po}(\lambda)$.    □

*Proof.* If $X$ is infinitely divisible, then, for each $n$, there exist random variables $\{X_{n,k}, 1 \leq k \leq n\}$ with characteristic function $\varphi_n$, such that $\varphi_X(t) = (\varphi_n(t))^n$. Moreover, $\varphi_X \neq 0$ so that logarithms are well defined. Thus, by borrowing from the first proof of Theorem 7.2.1 for the second equality, and using the fact that $E X = nE X_{n,1}$ in the last equality, we obtain

$$\log \varphi_X(t) = n \log \varphi_n(t) = n \log \left(1 - (1 - \varphi_n(t))\right)$$

$$= n(\varphi_n(t) - 1) + o(1) = n \int_{-\infty}^{\infty} \left(e^{itx} - 1\right) dF_{X_{n,1}}(x) + o(1)$$

$$= itE X + \int_{-\infty}^{\infty} \left(e^{itx} - 1 - itx\right)\frac{1}{x^2}nx^2\, dF_{X_{n,1}}(x) + o(1)$$

$$= itE X + \int_{-\infty}^{\infty} \left(e^{itx} - 1 - itx\right)\frac{1}{x^2}\, dG_n(x) + o(1) \quad \text{as} \quad n \to \infty,$$

where

$$G_n((c,d]) = nE X_{n,1}^2 I\{c < X_{n,1} \leq d\} \quad \text{for} \quad -\infty < c \leq d < \infty.$$

The measure $G_n$ (which is reminiscent of (3.14)) is a distribution function, except possibly for a scaling factor; viz.,

$$G_n(-\infty) = 0 \quad \text{and} \quad G(+\infty) = nE X_{n,1}^2 = E X^2.$$

By Helly's selection principle, there exists a vaguely convergent subsequence, $\{G_{n_k}, k \geq 1\}$, which, by Theorem 5.8.2, shows that, for any bounded interval, $[A, B]$, where $A, B \in C(G)$,

$$\int_{A}^{B} \left(e^{itx} - 1 - itx\right)\frac{1}{x^2}\, dG_{n_k}(x) \to \int_{A}^{B} \left(e^{itx} - 1 - itx\right)\frac{1}{x^2}\, dG(x) \quad \text{as} \quad k \to \infty.$$

Via estimates of the function $e^{itx}$ from Lemma A.1.2 and the uniform boundedness of the measures $G_{n_k}$ it follows that the integrals $\int_{-\infty}^{A}$ and $\int_{B}^{\infty}$ can be made arbitrarily small for $|A|, B$ sufficiently large, so that, indeed,

$$\log \varphi_X(t) = itE X + \int_{-\infty}^{\infty} \left(e^{itx} - 1 - itx\right)\frac{1}{x^2}\, dG(x).$$

This proves the conclusion, and at the same time identifies $\mu$ as $E X$ (which is why that symbol was chosen).

To prove the converse we recall Remark 4.1 and the Poisson distribution. If, instead $G$ has point mass $\lambda x^2$ at $x$ and $\mu = 0$, then

$$\varphi(t) = \exp\{\lambda(e^{itx} - 1 - itx)\},$$

which is the characteristic function of a linear transformation of a Poisson distribution; if $Y \in \text{Po}(\lambda)$, then $\varphi(t)$ is the characteristic function of $\lambda(Y-x)$.

Exploiting this, let, for $k = 0, \pm 1, \pm 2, \dots, \pm 2^{2n}$, $n \geq 1$, $Y_{n,k}$ be independent, linearly transformed Poisson distributed random variables, corresponding to

$$G_{n,k}\left(\frac{k}{2^n}\right) = G\left(\left(\frac{k}{2^n}, \frac{k+1}{2^n}\right]\right),$$

and set

$$Y_n = \sum_k Y_{n,k},$$

the $G$-measure of which becomes

$$G_n = \sum_k G_{n,k}.$$

Since, as we have just seen, each $Y_{n,k}$ is infinitely divisible, the same can be said for $Y_n$ by Theorem 4.2(ii).

Now, $G_n \xrightarrow{v} G$ as $n \to \infty$, and $G_n(\mathbb{R}) \leq G(\mathbb{R}) < \infty$. By repeating the convergence procedure from the first half of the proof, we conclude that

$$\varphi_{Y_n}(t) \to \exp\left\{i\mu t + \int_{-\infty}^{\infty} \left(e^{itx} - 1 - itx\right)\frac{1}{x^2}\,dG(x)\right\} \quad \text{as} \quad n \to \infty,$$

so that, in this half of the proof, the right-hand side is a characteristic function by the continuity theorem for characteristic functions, and moreover, an infinitely divisible one, thanks to Theorem 4.2(iv).  □

*Remark 4.2.* In words we can say that the class of infinitely divisible distributions with finite variance coincides with limits of compound Poisson distributions.

*Remark 4.3.* A further analysis shows that, in addition to $\mu = EX$, we have $\text{Var}\,X = G(\mathbb{R})$.  □

From Theorem 4.2(iv) we know that limits of infinitely divisible distributions are infinitely divisible. The following theorem provides conditions for convergence.

**Theorem 4.5.** *Let $\{Y_n, n \geq 1\}$ be random variables with finite variances, corresponding to infinitely divisible distributions with characteristic functions*

$$\varphi_n(t) = \exp\left\{i\mu_n t + \int_{-\infty}^{\infty} \left(e^{itx} - 1 - itx\right)\frac{1}{x^2}\,dG_n(x)\right\}.$$

*Then*

$$Y_n \xrightarrow{d} Y \quad and \quad \text{Var}\,Y_n \to \text{Var}\,Y \quad as \quad n \to \infty,$$

*where*

$$\varphi_Y(t) = \exp\left\{i\mu t + \int_{-\infty}^{\infty} \left(e^{itx} - 1 - itx\right)\frac{1}{x^2}\,dG(x)\right\}.$$

*iff*

- $G_n \overset{v}{\to} G$ as $n \to \infty$;
- $G_n(\mathbb{R}) \to G(\mathbb{R})$ as $n \to \infty$;
- $\mu_n \to \mu$ as $n \to \infty$.

*Remark 4.4.* The first two conditions would imply convergence in distribution, had the total mass been equal to 1.                                   □

By further elaborations of the arguments of the proof of the central limit theorem, taking Theorem 4.5 into account one can show that the subclass of infinitely divisible distributions with finite variances coincides with limits of triangular arrays under the assumption of finite variance.

**Theorem 4.6.** *Let* $\{(X_{n,j}, 1 \leq j \leq k_n), n \geq 1\}$ *be an array of row-wise independent random variables with finite variances, set* $S_n = \sum_{j=1}^{k_n} X_{n,j}$ *and* $\mu_n = \sum_{j=1}^{k_n} E X_{n,j}$, *for* $n \geq 1$, *and suppose that*

$$\sup_n s_n^2 = \sup_n \sum_{j=1}^{k_n} \mathrm{Var}\, X_{n,j} < \infty,$$

$$\max_{1 \leq j \leq k_n} \mathrm{Var}\, X_{n,j} \to 0 \quad as \quad n \to \infty.$$

*Then the class of infinitely divisible distributions with characteristic functions*

$$\varphi(t) = \exp\left\{ \mathrm{i}\mu t + \int_{-\infty}^{\infty} \left( \mathrm{e}^{\mathrm{i}tx} - 1 - \mathrm{i}tx \right) \frac{1}{x^2}\, \mathrm{d}G(x) \right\}$$

*coincides with the class of limit distributions of* $S_n$.
  *In addition,* $S_n$ *converges iff (with obvious notation)*

$$G_n \overset{v}{\to} G, \quad G_n(\mathbb{R}) \to G(\mathbb{R}), \quad and \quad \mu_n \to \mu \quad as \quad n \to \infty.$$

**Exercise 4.2.** Note that, and how, the conditions are reminiscent of the Lindeberg conditions. Prove the central limit theorem, Theorem 7.2.1, as a corollary of Theorem 4.6.                                   □

We close this introduction to infinite divisibility by remarking that the proof of Theorem 4.3 follows the same basic strategy as the proof of Theorem 4.4, but that additional care is required at 0 in order to obtain a convergent integral.

One can then show that this, more general, class coincides with the class of infinitely divisible distributions, which extends Theorem 4.6 to Theorem 4.1.

For details and more we refer, e.g., to [19, 21, 48, 71, 88, 97, 98, 142, 177, 194, 195]. A somewhat different source is [25].

# 5 Sums of Dependent Random Variables

The first encounter with dependent random variables that one faces is usually the drawing of balls without replacement from an urn: Suppose that an urn contains balls of two colors, blue and yellow. Drawing balls repeatedly from the urn, and putting $I_k = 1$ if the ball is yellow and 0 if it is blue, produces a sequence of indicator variables. The sum of the first $n$ indicators describes the number of yellow balls obtained after $n$ draws. If we draw with replacement the indicators are independent, if we draw without replacement they are dependent.

There exist many notions of dependence. A fundamental notion is *Markov dependence*. Vaguely speaking, the future then depends on the past only through the present. A random walk is an example. The future depends on where one is *now*, but not on how one got there. Another important dependence concept is *martingale dependence*, which is the topic of the next chapter. There also exist various concepts which are defined via some kind of decay, in the sense that the further two elements are apart in time or index, the weaker is the dependence.

We begin with the simplest such concept, namely *m-dependence*, after which we introduce some of the different concepts of *mixing*.

**Definition 5.1.** *The random variables* $X_1$, $X_2$, ... *are* $m$-dependent *if* $X_i$ *and* $X_j$ *are independent whenever* $|i - j| > m$.    □

*Remark 5.1.* Independence is the same as 0-dependence.[1]    □

*Example 5.1.* If we flip a coin repeatedly and let $A_n$ denote the event that the $n$th toss yields a head, $n \geq 1$, we obtain a sequence of independent events. In Subsection 2.18.2 we discussed this problem in connection with the Borel-Cantelli lemmas. If, however,

$$B_n = \{\text{the } (n-1)\text{th and the } n\text{th toss both yield heads}\}, \quad n \geq 2,$$

then we do not have independent events, and the second Borel-Cantelli lemma is not applicable. However, the events with even indices are independent, and so are those with odd indices. More precisely, $B_i$ and $B_j$ are independent if $|i - j| > 1$. Attaching indicators to the $B$-events, such that $I_n = 1$ whenever $B_n$ occurs and 0 otherwise, we obtain a 1-dependent sequence of random variables.

*Example 5.2. Peak Numbers.* Let $X_1$, $X_2$, ... be independent, $U(0, 1)$-distributed random variables. There is a *peak* at $X_k$ if $X_{k-1}$ and $X_{k+1}$ both are smaller than $X_k$, $k \geq 2$. Let

$$I_k = \begin{cases} 1, & \text{if there is a peak at } X_k, \\ 0, & \text{otherwise.} \end{cases}$$

---

[1] In Swedish this looks prettier: "Oberoende" is the same as "0-beroende".

Then

$$P(I_k = 1) = \frac{1}{3}, \qquad \text{for all} \quad k \geq 2,$$

$$P(I_k = 1, I_{k+1} = 1) = 0, \qquad \text{for all} \quad k \geq 2,$$

$$P(I_k = 1, I_{k+2} = 1) = \frac{2}{15}, \qquad \text{for all} \quad k \geq 2,$$

$$P(I_i = 1, I_j = 1) = \frac{1}{9}, \qquad \text{for} \quad |i - j| > 2, \ i, j \geq 2.$$

The sequence of indicators is 2-dependent. □

**Exercise 5.1.** Go over the example and check the details. □

A common example of $m$-dependent sequences are the so-called $(m+1)$-block factors defined by

$$Y_n = g(X_n, X_{n+1}, \ldots, X_{n+m-1}, X_{n+m}), \quad n \geq 1,$$

where $X_1$, $X_2$, ... are independent random variables, and $g : \mathbb{R}^{m+1} \to \mathbb{R}$. The $B_n$-sequence in the coin tossing example is an example of a 2-block factor.

It was, in fact, long believed that any $m$-dependent sequence must be an $(m+1)$-block factor. However, that this is not necessarily the case has been shown in [1].

In the $m$-dependent case the dependence stops abruptly. A natural generalization would be to allow the dependence to drop gradually. This introduces the concept of *mixing*. There are various variations with different names. Here are some of them.

Let $\mathcal{H}$ and $\mathcal{G}$ be sub-$\sigma$-algebras of $\mathcal{F}$. Following are some *measures of dependence* :

$$\alpha(\mathcal{H}, \mathcal{G}) = \sup_{F \in \mathcal{H}, G \in \mathcal{G}} |P(F \cap G) - P(F)P(G)|,$$

$$\phi(\mathcal{H}, \mathcal{G}) = \sup_{F \in \mathcal{H}, G \in \mathcal{G}} |P(G \mid F) - P(G)|, \qquad \text{for} \quad P(F) > 0,$$

$$\psi(\mathcal{H}, \mathcal{G}) = \sup_{F \in \mathcal{H}, G \in \mathcal{G}} \frac{|P(F \cap G) - P(F)P(G)|}{P(F)P(G)},$$

$$\rho(\mathcal{H}, \mathcal{G}) = \sup_{X \in L^2(\mathcal{H}), Y \in L^2(\mathcal{G})} |\rho_{X,Y}|.$$

The following relations and facts can be shown to hold:

$$\alpha(\mathcal{H}, \mathcal{G}) \leq \frac{1}{4}, \quad \psi(\mathcal{H}, \mathcal{G}) \leq 1, \quad \rho(\mathcal{H}, \mathcal{G}) \leq 1,$$

$$4\alpha(\mathcal{H}, \mathcal{G}) \leq 2\phi(\mathcal{H}, \mathcal{G}) \leq \psi(\mathcal{H}, \mathcal{G}),$$

$$4\alpha(\mathcal{H}, \mathcal{G}) \leq \rho(\mathcal{H}, \mathcal{G}) \leq \psi(\mathcal{H}, \mathcal{G}),$$

$$\rho(\mathcal{H}, \mathcal{G}) \leq 2\sqrt{\phi(\mathcal{H}, \mathcal{G})}.$$

The next step is to extend these notions to random variables. Toward that end, let $X_1$, $X_2$, ... be random variables, and let

$$\mathcal{F}_i^j = \sigma\{X_k,\, i \leq k \leq j\}.$$

By interpreting the index as discrete time, $\mathcal{F}_i^j$ thus contains the information from $i$ o'clock to $j$ o'clock. Corresponding to the above measures of dependence we define the following *mixing coefficients*:

$$\alpha(n) = \sup_{k \in \mathbb{Z}} \alpha(\mathcal{F}_1^k, \mathcal{F}_{k+n}^\infty),$$

$$\phi(n) = \sup_{k \in \mathbb{Z}} \phi(\mathcal{F}_1^k, \mathcal{F}_{k+n}^\infty),$$

$$\psi(n) = \sup_{k \in \mathbb{Z}} \psi(\mathcal{F}_1^k, \mathcal{F}_{k+n}^\infty),$$

$$\rho(n) = \sup_{k \in \mathbb{Z}} \rho(\mathcal{F}_1^k, \mathcal{F}_{k+n}^\infty).$$

These quantities measure the dependence of those portions of the sequence $\{X_k,\, k \geq 1\}$ that are located $n$ "time units" apart. From the above inequalities we can see that some measures of dependence are stronger than others. For independent sequences all of them equal 0 (of course), and *if* the mixing coefficients converge to 0 as $n \to \infty$ we may interpret this as asymptotic independence, in the sense that two events belonging to $\mathcal{F}_1^k$ and $\mathcal{F}_{k+n}^\infty$, respectively, are asymptotically independent as $n \to \infty$.

We call a sequence of random variables $\alpha$-mixing or *strong mixing* if $\alpha(n) \to 0$ as $n \to \infty$, and $\phi$-*mixing*, $\psi$-*mixing*, $\rho$-*mixing*, if $\phi(n) \to 0$, $\psi(n) \to 0$, $\rho(n) \to 0$, respectively, as $n \to \infty$.

An immediate consequence of the inequalities above is that $\psi$-mixing implies $\phi$-mixing, which in turn implies $\alpha$-mixing. This means that if we would like to prove, say, a central limit theorem for $\alpha$-mixing sequences, then it is enough to consider the $\phi$-mixing case if that happens to be more tractable. Conversely, if one wishes to prove some fact in the $\phi$-mixing case and does not have tools enough at ones disposal for $\phi$-mixing, one might try a stronger concept instead (to obtain, at least, a partial result).

A famous problem in this context that has triggered a considerable amount of attention is the *Ibragimov conjecture* [141], which states that a strictly stationary, centered, $\phi$-mixing sequence $X_1$, $X_2$, ..., such that $E\,X_1^2 < \infty$ and $\mathrm{Var}\left(\sum_{k=1}^n X_k\right) \to \infty$ as $n \to \infty$ satisfies the central limit theorem.

A common method of proof is the "big block–small block" technique, which means that the partial sums are split into alternating "smaller" and "bigger" blocks, with the effect that the small blocks can be neglected asymptotically, and the big blocks are asymptotically independent.

Some references are [16, 27, 28, 29, 30, 131, 142, 144, 190, 191, 192, 198]. In particular, Bradley's recent monographs [28, 29, 30] provide an excellent source for the more recent literature and the current state of the art. A further reference for general dependence structures is [128]; results on, e.g., the law of the iterated logarithm may be found in [239]; and [168] is devoted to extremes. Two references on renewal theory in dependent cases are [15, 148].

There also exist analogs of Donsker's theorem, Theorem 7.7.13, in this setting. The first more extensive contribution to this area of research is [199]. A functional version of the Ibragimov conjecture was suggested by Iosifescu [143]. Two further references are [189, 247].

In addition to the above mixing concepts there has been an increased interest in what is called associated sequences, negatively associated sequences, interlaced mixing sequences, and so on, but we stop here, and leave the interested reader to research this further.

# 6 Convergence of Extremes

Paralleling the theory of stable distributions and domains of attraction connected with sums, there exists a theory of *max-stable* distributions and domains of attractions for extremes. Relevant literature for this topic are [19, 21, 71, 91, 100, 123, 168, 207, 208]. Some of the sources also treat dependent cases.

## 6.1 Max-stable and Extremal Distributions

Max-stable distributions are the analog of stable distributions. Extremal distributions correspond to the limit distributions of normalized maxima.

**Definition 6.1.** (i) *A non-degenerate distribution $F$ is* extremal *if it can appear as a limit distribution of standardized partial maxima of independent, identically distributed random variables, that is, if there exist $\{a_n > 0,\ n \geq 1\}$, $\{b_n \in \mathbb{R},\ n \geq 1\}$, and a distribution function $G$, such that, for all $x \in C(F)$,*

$$\left(G(a_n x + b_n)\right)^n \to F(x) \quad as \quad n \to \infty.$$

(ii) *A non-degenerate distribution $F$ is* max-stable *iff there exist $\{a_n > 0,\ n \geq 1\}$ and $\{b_n \in \mathbb{R},\ n \geq 1\}$, such that*

$$\left(F(a_n x + b_n)\right)^n = F(x) \quad for\ all \quad n.$$

(iii) *In terms of random variables: Let $X$, $X_1$, $X_2$, ... be independent, identically distributed random variables with distribution function $G$, and set*

$$Y_n = \max_{1 \leq k \leq n} X_k, \quad n \geq 1.$$

*Then $X$ is* extremal *iff there exist $\{a_n > 0,\ n \geq 1\}$ and $\{b_n \in \mathbb{R},\ n \geq 1\}$, such that*

$$\frac{Y_n - b_n}{a_n} \overset{d}{\to} Y, \quad where \quad F_Y = F,$$

*and $X$ is* max-stable *iff iff there exist $\{a_n > 0,\ n \geq 1\}$ and $\{b_n \in \mathbb{R},\ n \geq 1\}$, such that*

$$Y_n \overset{d}{=} a_n X + b_n \quad for\ all \quad n. \qquad \square$$

In the light of the previous sections of this chapter it is natural to ask for classes of extremal distributions. For sums the conclusion was that only the stable distributions can appear as limit distributions. In this case the "obvious" result should be, and is, the following:

**Theorem 6.1.** *The classes of extremal distributions and max-stable distributions coincide.*

*Proof.* That every max-stable distribution is extremal follows from the definition. The converse follows from the convergence to types theorem. Namely, if $G$ is a limit distribution, then

$$\frac{Y_n - b_n}{a_n} \to G \quad \text{as} \quad n \to \infty,$$

and, by considering blocks,

$$\tilde{Y}_j = \max_{(j-1)n+1 \leq i \leq jn} X_i \quad (\overset{d}{=} Y_n), \quad 1 \leq j \leq k,$$

noticing that

$$Y_{nk} = \max_{1 \leq j \leq k} \tilde{Y}_j,$$

we also have

$$\frac{Y_{nk} - b_n}{a_n} \to G^k \quad \text{as} \quad n \to \infty, \quad \text{for all} \quad k.$$

Theorem 2.1 therefore tells us that there exist constants $A_k > 0$ and $B_k$, such that

$$\max_{1 \leq j \leq k} Z_j \overset{d}{=} A_k Z_1 + B_k,$$

where $\{Z_j\}$ are independent identically distributed random variables with distribution function $G$. This shows that $G$ is max-stable.    □

The next problem is to describe these distributions, preferably explicitly. The following theorem, due to Gnedenko [95], provides the solution; see also [89].

**Theorem 6.2.** *There exist three classes or types of extremal distributions:*

$$\text{Fréchet:} \quad \Phi_\alpha(x) = \begin{cases} 0, & \text{for } x < 0, \\ \exp\{-x^{-\alpha}\}, & \text{for } x \geq 0, \end{cases} \quad \alpha > 0;$$

$$\text{Weibull:} \quad \Psi_\alpha(x) = \begin{cases} \exp\{-(-x)^\alpha\}, & \text{for } x < 0, \\ 1, & \text{for } x \geq 0, \end{cases} \quad \alpha > 0;$$

$$\text{Gumbel:} \quad \Lambda(x) = \exp\{-e^{-x}\}, \quad \text{for } x \in \mathbb{R}.$$

*Remark 6.1.* Note that the Weibull distribution is concentrated on the negative half-axis, whereas "the usual" Weibull distribution, whose distribution function equals $F(x) = 1 - \exp\{-e^{-x^{\alpha}}\}$, $x > 0$, is concentrated on the positive half-axis; for more on this, cf. [71], p. 123.                    □

*Proof.* The instrumental part of the proof is, again, the convergence to types theorem, Theorem 2.1.

Suppose that $G$ is extremal. Considering the usual blocks we have

$$\big(F(a_n x + b_n)\big)^n \to G(x) \quad \text{as} \quad n \to \infty,$$

$$\big(F(a_{nk} x + b_{nk})\big)^{nk} \to G(x) \quad \text{as} \quad n \to \infty, \quad \text{for all} \quad k,$$

for some distribution function $F$. At the same time we also know that

$$\big(F(a_n x + b_n)\big)^{nk} \to \big(G(x)\big)^k \quad \text{as} \quad n \to \infty, \quad \text{for all} \quad k,$$

so that, by the convergence to types theorem, there exist $c_k$ and $d_k$, such that

$$(G(x))^k = G(c_k x + d_k). \tag{6.1}$$

Considering blocks once again it follows that

$$\begin{aligned}(G(x))^{nm} &= \big(G(c_m x + d_m)\big)^n = G\big(c_n(c_m x + d_m) + d_n\big) \\ &= G\big(c_n c_m x + (c_n d_m + d_n)\big),\end{aligned}$$

Combining the two (with $k = nm$) it follows (in the language of distribution functions) from Step (e) in the proof of Theorem 2.1 that

$$c_{nm} = c_n c_m \quad \text{and} \quad d_{nm} = c_n d_m + d_n, \quad n, m \in \mathbb{N}, \tag{6.2}$$

where $c_1 = 1$ and $d_1 = 0$. Changing variables in (6.1), and raising the equation to the power $m$, yields

$$\begin{aligned}(G(x))^{m/n} &= \Big(G\Big(\frac{x - d_n}{c_n}\Big)\Big)^m = G\Big(c_m \frac{x - d_n}{c_n} + d_m\Big) \\ &= G\Big(\frac{c_m}{c_n} x - \frac{c_m d_n}{c_n} + d_m\Big). \end{aligned} \tag{6.3}$$

We now distinguish between the cases $c_k = 1$ for all $k$ and its complement, the latter of which is reduced to the two cases "$d_k = 0$ for all $k$, $G(0) = 0$", and "$d_k = 0$ for all $k$, $G(0) = 1$", respectively.

## The Case $c_k = 1$ for All $k$

In this case (6.3) reduces to

$$(G(x))^r = G(x + \gamma_r), \tag{6.4}$$

with $\gamma_r = d_m - d_n$, for $r = m/n \in \mathbb{Q}$.

Since $0 < G(x) < 1$ for some $x$, it follows that $\gamma_r$ is strictly decreasing in $r$ (since $G^r(x)$ is decreasing in $r$), and, moreover, that, for $r = m/n$ and $s = j/k$,

$$\gamma_{rs} = d_{mj} - d_{nk} = d_m + d_j - d_n - d_k = \gamma_r + \gamma_s. \tag{6.5}$$

Define

$$\gamma(u) = \inf_{0 < r \in \mathbb{Q} < u} \gamma_r \quad \text{for} \quad u \in \mathbb{R}.$$

The function so defined is decreasing in $u$ (for any $u_1 < u_2$ we select rationals $r_1, r_2, r_3$, such that $r_1 < u_1 < r_2 < u_2 < r_3$). Moreover, (6.4) and (6.5) carry over to continuous versions:

$$(G(x))^u = G(x + \gamma(u)) \quad \text{for all} \quad x \in \mathbb{R}, \ u > 0,$$
$$\gamma(uv) = \gamma(u) + \gamma(v).$$

Applying Lemma A.8.2 to the latter yields

$$\gamma(u) = -c \log u,$$

where $c$ must be positive, since the logarithm is increasing, and $\gamma$ is decreasing. This means that

$$G(x) = \big(G(x - c \log u)\big)^{1/u},$$

which, upon setting $u = \exp\{x/c\}$, so that $x - c \log u = 0$, shows that

$$G(x) = \big(G(0)\big)^{\exp\{-x/c\}} = \exp\{\log G(0) e^{-x/c}\}, \quad x \in \mathbb{R}.$$

This proves that the case $c_k = 1$ for all $k$ corresponds to the Gumbel *type* distribution – it is the Gumbel distribution, except for scaling.

## The Case $d_k = 0$ for All $k$ and $G(0) = 0$

This time (6.3) reduces to

$$(G(x))^r = G(\beta_r x), \tag{6.6}$$

where, for any $r = m/n \in \mathbb{Q}$, we define $\beta_r = c_m/c_n$, which is strictly decreasing in $r$, since the distribution only lives on the positive half-axis.

The analog of (6.5) is

$$\beta_{rs} = \frac{c_m c_j}{c_n c_k} = \frac{c_m}{c_n} \cdot \frac{c_j}{c_k} = \beta_r \beta_s,$$

so that, by arguing as above, we find that

$$\beta(u) = \inf_{0 < r \in \mathbb{Q} < u} \beta_r \quad \text{for} \quad u \in \mathbb{R},$$

satisfies the same multiplication rule. Lemma A.8.2 therefore tells us that $\beta(u) = u^{-c}$, where $c > 0$, since $\beta$ is decreasing. Thus

$$G(x) = \left(G(xu^{-c})\right)^{1/u},$$

which, putting $u = x^{1/c}$, so that $xu^{-c} = 1$, yields

$$G(x) = \begin{cases} 0, & \text{for } x < 0, \\ (G(1))^{x^{-1/c}} = \exp\{-x^{-1/c}(\log 1/G(1))\}, & \text{for } x \geq 0, \end{cases}$$

which is a Fréchet type distribution.

## The Case $d_k = 0$ for All $k$ and $G(0) = 1$

This case is completely symmetric: The function $\beta$ will be increasing since the distribution lives on the negative half-axis, which leads to $\beta(u) = (-u)^c$, where $c > 0$, and, finally to

$$G(x) = \begin{cases} (G(-1))^{(-x)^{1/c}} = \exp\{-(-x)^{-1/c}(\log 1/G(-1))\}, & \text{for } x < 0, \\ 1, & \text{for } x \geq 0, \end{cases}$$

which is a Weibull-type distribution. We leave it to the reader to check the details.

## The Case $c_k \neq 1$ for Some $k$

It remains to check how the case "$c_k \neq 1$ for some $k$" is reduced to the two previous ones.

By assumption there exists $k_0 > 1$, such that $c_{k_0} \neq 1$. In view of (6.1) this means that

$$\left(G\left(\frac{d_{k_0}}{1 - c_{k_0}}\right)\right)^{k_0} = G\left(c_{k_0}\frac{d_{k_0}}{1 - c_{k_0}} + d_{k_0}\right) = G\left(\frac{d_{k_0}}{1 - c_{k_0}}\right).$$

We have thus found a point $x_0$, such that $(G(x_0))^{k_0} = G(x_0)$, which forces $G(x_0)$ to be 0 or 1.

This rules out the Gumbel case, since the distribution has its support on the whole real axis.

Suppose first that $G(x_0) = 0$, and let $x^* = \sup\{x : G(x) = 0\}$ be the left end-point of the support of $G$. By translating the distribution if necessary, we may w.l.o.g. assume that $x^* = 0$. The only consequence is that the constants $d_k$ might change.

Now, suppose that there exists some (possibly changed) $d_k > 0$. Then, by letting $x$ be small and negative, we can make $c_k x + d_k > 0$, which contradicts (6.6), since the left-hand side will be equal to 0 and the right-hand side will be positive. If $d_k < 0$ we let, instead, $x$ be small and positive to achieve the same contradiction, except that the roles between the left-hand side and the right-hand side have switched. This forces $d_k$ to be 0 for *every* $k$, which reduces this case to the two previous ones.

The proof of the theorem is complete.                                        □

## 6.2 Domains of Attraction

In analogy with domains of attraction for sums, a distribution $F$ belongs to the domain of attraction of the extremal distribution $G - F \in \mathcal{D}(G)$ – if there exist normalizations $a_n > 0$ and $b_n \in \mathbb{R}$, $n \geq 1$, such that

$$F^n(a_n x + b_n) \to G(x) \quad \text{as} \quad n \to \infty.$$

In terms of random variables: Let $X_1$, $X_2$, ... be independent, identically distributed random variables, and set $Y_n = \max_{1 \leq k \leq n} X_k$, $n \geq 1$. Then $X \in \mathcal{D}(G)$ if there exist normalizations $a_n > 0$ and $b_n \in \mathbb{R}$, $n \geq 1$, such that

$$\frac{Y_n - b_n}{a_n} \xrightarrow{d} G \quad \text{as} \quad n \to \infty.$$

Here is now a theorem that provides necessary and sufficient criteria on the distribution to belong to the domain of attraction to one of the three extremal ones. The result is not exhaustive, there exist other (equivalent) criteria. A good source is [21], Section 8.13.

**Theorem 6.3.** *Let $F$ be a distribution function.*

(a) $F \in \mathcal{D}(\Phi_\alpha)$ *iff* $1 - F \in \mathcal{RV}(-\alpha)$, *in which case one may choose the normalizations* $a_n = \inf\{x : 1 - F(x) \leq 1/n\}$ *and* $b_n = 0$;
(b) $F \in \mathcal{D}(\Psi_\alpha)$ *iff* $x_\infty = \sup\{x : F(x) < 1\} < \infty$, *and* $1 - F^* \in \mathcal{RV}(-\alpha)$, *where* $F^*(x) = F(x_\infty - 1/x)$, *in which case one may choose the normalizations* $a_n = \sup\{x : 1 - F(x_\infty - x) \leq 1/n\}$ *and* $b_n = x_\infty$;
(c) $F \in \mathcal{D}(\Lambda)$ *iff either*

$$\lim_{t \to \infty} \frac{U(tx) - U(t)}{U(ty) - U(t)} = \frac{\log x}{\log y} \quad \text{for all} \quad x, y > 0, \ y \neq 1,$$

*where, $U$ is the inverse of $\frac{1}{1-F}$, or*

$$\frac{V(y + \log x) - V(\log x)}{y} \in \mathcal{SV} \quad \text{as} \quad x \to \infty, \quad \text{for all} \quad y > 0,$$

*where $V$ is the inverse of* $-\log(1 - F)$.

The proofs of the sufficiencies are managable. Let us sketch the first one:

$$\log \big(F(a_n x)\big)^n = n\log(1 - (1 - F(a_n x))) \sim -n(1 - F(a_n x))$$

$$= -\frac{n}{1 - F(a_n)} \cdot \frac{1 - F(a_n x)}{1 - F(a_n)} \to -1 \cdot x^{-\alpha} \quad as \quad n \to \infty.$$

Here $\sim$ stems from Taylor expansion. The limits are consequences of the definition of $a_n$ and the regular variation of the tail $1 - F$.

**Exercise 6.1.** Classify the exponential distribution, the uniform distribution, and the Pareto distribution. □

The expressions in the Gumbel case look more complicated than the others. For the absolutely continuous case there is a swifter condition due to von Mises [185], which we quote without proof.

**Theorem 6.4.** *Suppose that $x_\infty = \sup\{x : F(x) < 1\} = \infty$, and let $f$ be the density of $F$. If*

$$\frac{\mathrm{d}}{\mathrm{d}x}\Big(\frac{1 - F(x)}{f(x)}\Big) \to 0 \quad as \quad x \to \infty,$$

*then $F \in \mathcal{D}(\Lambda)$.*

**Exercise 6.2.** Check that the exponential and normal distributions belong to the domain of attraction of the Gumbel distribution. □

We close by remarking that there also exist local limit theorems, results on large deviations and convergence rates; we refer to [21] and further references give there.

## 6.3 Record Values

The starting point for records (Subsection 2.17.2) was a sequence $X_1, X_2, \ldots$ of independent, identically distributed, continuous random variables. In this subsection we focus on the *record values*, $\{X_{L(n)}, n \geq 1\}$, which have been found to behave like a compressed sequence of partial maxima. This, in turn, makes it reasonable to suggest that there exist three possible limit distributions for $X_{L(n)}$ as $n \to \infty$, namely, some modification(s) (perhaps) of the three extremal distributions, the modification, if necessary, being a consequence of the compression. The following theorem, due to Resnick [206], confirms this.

**Theorem 6.5.** *Suppose that $F$ is absolutely continuous. The possible limit distributions for record values*

$$\Phi(-\log(-\log G(x))),$$

*where $G$ is an extremal distribution and $\Phi$ the distribution function of the standard normal distribution. More precisely, the three classes or types of limit distributions are*

$$\Phi_\alpha^{(R)}(x) = \begin{cases} 0, & for \quad x < 0, \\ \Phi(\alpha \log x), & for \quad x \geq 0, \end{cases} \quad \alpha > 0;$$

$$\Psi_\alpha^{(R)}(x) = \begin{cases} \Phi(-\alpha \log(-x)), & for \quad x < 0, \\ 1, & for \quad x \geq 0 \end{cases} \quad \alpha > 0;$$

$$\Lambda^{(R)}(x) = \Phi(x), \quad for \quad x \in \mathbb{R}.$$

*Remark 6.2.* Note the unfortunate collision between the standard notations $\Phi$ and $\Phi$. □

We only indicate the proof here and refer to the original work or [21, 207] for the full story.

The key ideas are that the order between observations are preserved under non-decreasing transformations; that the exponential distribution is easy to handle; and that, if we associate the random variable $X$ with the distribution function $F$, then $-\log(1 - F(X)) \in \mathrm{Exp}(1)$.

Thus, let us first consider the standard exponential distribution. An elementary argument shows that the increments $X_{L(n)} - X_{L(n-1)}$, $n \geq 2$, are independent, standard exponential random variables, and, hence, that $X_{L(n)} \in \Gamma(n, 1)$ for all $n$. That this is the case can, for example, be seen via the lack of memory property of the exponential distribution and the Poisson process. The central limit theorem therefore tells us that

$$\frac{X_{L(n)} - n}{\sqrt{n}} \xrightarrow{d} N(0, 1) \quad as \quad n \to \infty,$$

in this special case.

The connection to the exponential distribution implies that, in the general case,

$$\frac{-\log(1 - F(X_{L(n)})) - n}{\sqrt{n}} \xrightarrow{d} N(0, 1) \quad as \quad n \to \infty.$$

This fact must be combined with our aim, namely to find normalizing sequences $\{a_n > 0, n \geq 1\}$, and $\{b_n \in \mathbb{R}, n \geq 1\}$, such that

$$\frac{X_{L(n)} - b_n}{a_n} \xrightarrow{d} G \quad as \quad n \to \infty,$$

for some distribution $G$, which is the same as finding $G$, such that

$$P\left(\frac{-\log(1 - F(X_{L(n)})) - n}{\sqrt{n}} \leq \frac{-\log(1 - F(a_n x + b_n)) - n}{\sqrt{n}}\right) \xrightarrow{d} G,$$

as $n \to \infty$. For this to be possible it is necessary that there exists some function $g(x)$, such that

$$\frac{-\log(1 - F(a_n x + b_n)) - n}{\sqrt{n}} \to g(x) \quad as \quad n \to \infty,$$

in which case the limit distribution becomes $G(\cdot) = \Phi(g(\cdot))$.

The analysis of this problem leads to the desired conclusion.

# 7 The Stein-Chen Method

The Stein method is a method due to Stein [237], to prove normal convergence. The Stein-Chen method is an adaptation due to Chen [39], to prove Poisson approximation, primarily of indicator functions. The mode of convergence is variational convergence; recall Definition 5.6.1 and Theorem 5.6.4 (we are dealing with integer valued random variables here). For a full treatment of the method with many applications we refer to the monograph [9].

In a first course one learns that a binomial distribution with a "small" $p$ can be well approximated by a Poisson distribution. However, this requires that the success probabilities in each experiment are the same, and that successive trials are independent.

Suppose now that the probabilities vary; let $X_1, X_2, \ldots, X_n$ be independent random variables, such that $P(X_k = 1) = 1 - P(X_k = 0) = p_k$, $k \geq 1$. Is it then possible to approximate the sum, $S_n$ by $T_n \in \text{Po}(\sum_{k=1}^{n} p_k)$? In particular, what is the variational distance between $S_n$ and $T_n$ as a function of $n$?

Before proceeding we recall the definition of *variational distance* from Definition 5.6.1: If $X$ and $Y$ are random variables, then

$$d(X, Y) = \sup_{A \in \mathcal{R}} |P(X \in A) - P(Y \in A)|.$$

Let $Y_k \in \text{Po}(p_k)$, $k \geq 1$, be independent random variables. The variational distance between $S_n$ and $T_n$ can be conveniently estimated via the following result.

**Lemma 7.1.** (i) *Let $X$ and $Y$ be random variables. Then*

$$d(X, Y) \leq \sum_{k=1}^{n} P(X \neq Y).$$

(ii) *Let $X_1, X_2, \ldots$ and $Y_1, Y_2, \ldots, Y_n$ be sequences of random variables, and set $S_n = \sum_{k=1}^{n} X_k$, and $T_n = \sum_{k=1}^{n} Y_k$, $n \geq 1$. Then*

$$d(S_n, T_n) \leq \sum_{k=1}^{n} P(X_k \neq Y_n).$$

*Proof.* (i): By subtracting and adding $P(\{X \in A\} \cap \{Y \in A\})$ we find that

$$P(X \in A) - P(Y \in A) = P(\{X \in A\} \cap \{Y \in A^c\}) - P(\{Y \in A\} \cap \{X \in A^c\}),$$

which shows that

$$|P(X \in A) - P(Y \in A)| \leq P(X \neq Y) \quad \text{for all} \quad A \subset \mathbb{R}.$$

(ii): The first part, the relation

$$\{S_n \neq T_n\} \subset \bigcup_{k=1}^{n} \{X_n \neq Y_n\},$$

and sub-additivity, together yield

$$d(S_n, T_n) \leq P(S_n \neq T_n) \leq \sum_{k=1}^{n} P(X_k \neq Y_k). \qquad \square$$

A consequence of the lemma is that the problem reduces to estimating the right-most probabilities. Assuming, to begin with, that $X_k$ and $Y_k$ are independent for all $k$ yields

$$\begin{aligned}
P(X_k \neq Y_k) &= P(X_k = 0)P(Y_k \geq 1) + P(X_k = 1)P(Y_k \neq 1) \\
&= (1 - p_k)(1 - e^{-p_k}) + p_k(1 - p_k e^{-p_k}) \\
&\leq p_k(2 - 2p_k + p_k^2) \leq p_k,
\end{aligned}$$

where the first inequality follows from the elementary inequality $e^{-x} \geq 1 - x$. This proves that

$$d(S_n, T_n) \leq \sum_{k=1}^{n} p_k.$$

However, the important observation is that *nothing is assumed about (in)dependence between $X_k$ and $Y_k$* in the formulation of the original problem. In fact, it even turns out that a suitable dependence assumption yields a better rate as compared to independence. With a different terminology: it is advantageous to *couple* the two sequences in some efficient manner.

Let us see how this can be done in two different ways. In [137] the authors assume that

$$\begin{aligned}
P(X_k = Y_k = 1) &= p_k e^{-p_k}, \\
P(X_k = 1, Y_k = 0) &= p_k(1 - e^{-p_k}), \\
P(X_k = Y_k = 0) &= e^{-p_k} - p_k(1 - e^{-p_k}), \\
P(X_k = 0, Y_k = j) &= e^{-p_k} \frac{p_k^j}{j!}, \quad j = 2, 3, \ldots,
\end{aligned}$$

to obtain

$$d(S_n, T_n) \leq 2 \sum_{k=1}^{n} p_k^2.$$

To be precise, this works for $p_k \leq 0.8$, which is enough, since the upper bound $2p_k^2 > 1$ for $p_k > 0.8$.

Serfling [218] improves this bound to

$$d(S_n, T_n) \leq \sum_{k=1}^{n} p_k^2. \qquad (7.1)$$

To achieve this he introduces a sequence $Z_1, Z_2, \ldots, Z_n$ via

$$P(Z_k = 0) = (1 - p_k)e^{p_k}, \quad P(Z_k = 1) = 1 - (1 - p_k)e^{p_k},$$

and then $X_1, X_2, \ldots, X_n$ as

$$X_k = I\{Y_k \geq 1\} + I\{Y_k \geq 0\}I\{Z_k \geq 1\}, \quad k = 1, 2, \ldots, n,$$

where $Y_1, Y_2, \ldots$ are as before. Moreover, it is assumed that $Y_k$ and $Z_k$ are independent for all $k$. Then

$$
\begin{aligned}
P(X_k = 1) &= P(Y_k \geq 1) + P(Y_k = 0)P(Z_k = 1) \\
&= 1 - e^{-p_k} + e^{-p_k}\left(1 - (1 - p_k)e^{p_k}\right) = p_k, \\
P(X_k = 0) &= 1 - p_k,
\end{aligned}
$$

so that $X_k$ is a Bernoulli random variable as required. However, in this case,

$$
\begin{aligned}
P(X_k \neq Y_k) &= P(X_k = 0, Y_k \geq 1) + P(X_k = 1, Y_k \neq 1) \\
&= P(Y_k \geq 2) + P(X_k = 1, Y_k = 0) \\
&= P(Y_k \geq 2) + P(Y_k = 0)P(Z_k = 1) \\
&= 1 - e^{-p_k} - p_k e^{-p_k} + e^{-p_k}\left(1 - (1 - p_k)e^{p_k}\right) \\
&= p_k(1 - e^{-p_k}) \leq p_k^2,
\end{aligned}
$$

which establishes (7.1).

If, in particular, all success probabilities are equal (to $p$), the bound turns into $np^2 = \lambda p$, where $\lambda = np$ is the parameter of the Poisson distribution. In the general case,

$$\sum_{k=1}^{n} p_k^2 \leq \max_{1 \leq k \leq n} p_k \sum_{k=1}^{n} p_k = \lambda \max_{1 \leq k \leq n} p_k.$$

An interpretation of this is that if we consider a sequence of experiments $\{(X_{n,k}, 1 \leq k \leq n), n \geq 1\}$, such that $P(X_{n,k} = 1) = p_{n,k}$, and such that $\sum_{k=1}^{n} p_{n,k} \to \lambda$ as $n \to \infty$, then $S_n = \sum_{k=1}^{n} X_{n,k} \overset{d}{\to} \text{Po}(\lambda)$ as $n \to \infty$ under the additional assumption that $\max_{1 \leq k \leq n} p_{n,k} \to 0$ as $n \to \infty$.

Barbour and Hall [8] use the Stein-Chen method to improve the upper bound (7.1) further as follows.

**Theorem 7.1.** *Suppose that $X_1, X_2, \ldots$ are independent random variables, such that $X_k \in \text{Be}(p_k)$, $k \geq 1$, and set $S_n = \sum_{k=1}^{n} X_k$, $n \geq 1$. Further, let $T_n \in \text{Po}(\lambda_n)$, where $\lambda_n = \sum_{k=1}^{n} p_k$, $n \geq 1$. Then*

$$d(S_n, T_n) \leq \frac{1 - e^{-\lambda_n}}{\lambda_n} \sum_{k=1}^{n} p_k^2 \leq \frac{1}{\lambda_n} \sum_{k=1}^{n} p_k^2. \tag{7.2}$$

Strictly speaking, they prove the first inequality. The second one is trivial, and coincides with Chen's result [39] (without his factor 5). Barbour and Hall also provide a lower bound of the same order of magnitude; cf. also [9], Theorem 2.M, and Corollary 3.D.1.

*Remark 7.1.* In the i.i.d. case the bound (7.2) becomes

$$\frac{1 - e^{-np}}{np} np^2 = \frac{1 - e^{-\lambda}}{\lambda} \lambda p = p(1 - e^{-\lambda}) \le p.$$

This is an improvement over the bound $\lambda p$, but still not very interesting. *The interesting part is unequal success probabilities.*                                                                                                  □

Before we hint on the proof, let us verify the Poisson approximation for $\mu(n)$, the counting process for records:

$$d(\mu(n), \mathrm{Po}(m_n)) \le \frac{\pi^2}{6 \log n},$$

where $m_n = E\,\mu(n) = \sum_{k=1}^n 1/k$, mentioned in connection with Theorem 7.7.5.

Since $p_k = 1/k$, and $\lambda_n = \sum_{k=1}^n 1/k$ as $n \to \infty$, we obtain, utilizing the extreme members in (7.2) and Lemma A.3.1(iii), that

$$d(\mu(n), \mathrm{Po}(m_n)) \le \frac{1}{\sum_{k=1}^n \frac{1}{k}} \sum_{k=1}^n \frac{1}{k^2} \le \frac{1}{\log n} \sum_{k=1}^\infty \frac{1}{k^2} = \frac{\pi^2}{6 \log n},$$

as claimed.

In the discussion so far we have assumed that the indicators are independent. However, the essential feature of the Stein-Chen method is that the method works in many situations where dependence is involved. Once again, we refer to [9] for theory and many applications.

The starting point for the method is the Poisson analog to Stein's equation for normal approximations, that is, the equation

$$E\big(\lambda g(Z + 1) - Zg(Z))\big) = 0,$$

which is satisfied for $Z \in \mathrm{Po}(\lambda)$ and bounded functions $g : \mathbb{Z}^+ \to \mathbb{R}$, where we may set $g(0) = 0$ w.l.o.g. Namely, if $Z \in \mathrm{Po}(\lambda)$, the left-hand side equals

$$e^{-\lambda}\lambda g(1) + e^{-\lambda} \sum_{n=1}^\infty \left( g(n+1)\frac{\lambda^{n+1}}{n!} - g(n)\frac{\lambda^n}{(n-1)!} \right),$$

which equals 0 due to telescoping.

It is, in fact, also possible to prove a converse, namely that any function $h : \mathbb{Z}^+ \to \mathbb{R}$ for which $E\,h(Z) = 0$ must be of the form

$$h(k) = \lambda g(k + 1) - kg(k),$$

for $g$ as above. In other words, this characterizes the Poisson distribution.

Let $X_1, X_2, \ldots, X_n$ be indicators as before, but without any assumption about independence, set $S_n = \sum_{k=1}^n X_k$, and let $T_n \in \mathrm{Po}(\lambda_n)$, where $\lambda_n = \sum_{k=1}^n p_k$.

The central idea is to use the above characterization of the Poisson distribution. Toward this end, let $A \subset \mathbb{R}$, and construct, by recursion, a function $g_{\lambda_n, A} : \mathbb{Z}^+ \to \mathbb{R}$, such that

$$\lambda g_{\lambda_n, A}(k+1) - k g_{\lambda_n, A}(k) = I\{k \in A\} - P(T_n \in A), \quad k \geq 0. \quad (7.3)$$

Inserting $S_n$ and taking expectations yields

$$E\big(\lambda g_{\lambda_n, A}(S_n) - S_n g_{\lambda_n, A}(S_n)\big) = P(S_n \in A) - P(T_n \in A). \quad (7.4)$$

Next, set

$$S_n^{(j)} = \sum_{\substack{k=1 \\ k \neq j}}^n X_k.$$

*In order to give a flavor of the method* we assume, in the following, that the indicators are, indeed, independent.

Conditioning on $X_j$ we obtain, due to independence,

$$E\, S_n g_{\lambda_n, A}(S_n) = \sum_{j=1}^n E\, X_j g_{\lambda_n, A}(S_n^{(j)} + X_j)$$

$$= \sum_{k=1}^n p_j 1 \cdot E\, g_{\lambda_n, A}(S_n^{(j)} + 1) + (1 - p_j) 0 \cdot E\, g_{\lambda_n, A}(S_n^{(j)})$$

$$= \sum_{k=1}^n p_j E\, g_{\lambda_n, A}(S_n^{(j)} + 1),$$

which implies that

$$\left| E\big(\lambda g_{\lambda_n, A}(S_n + 1) - S_n g_{\lambda_n, A}(S_n)\big) \right|$$

$$= \left| \sum_{j=1}^n p_j E\{ g_{\lambda_n, A}(S_n + 1) - g_{\lambda_n, A}(S_n^{(j)} + 1) \} \right|$$

$$\leq \sum_{j=1}^n p_j E \left| g_{\lambda_n, A}(S_n^{(j)} + X_j + 1) - g_{\lambda_n, A}(S_n^{(j)} + 1) \right|$$

$$\leq \sum_{j=1}^n p_j^2 \sup_{1 \leq j \leq n} \left| g_{\lambda_n, A}(j+1) - g_{\lambda_n, A}(j+1) \right| + 0\,,$$

since the only contribution comes from the set $\{X_j = 1\}$.

The final step is to find upper bounds for $\sup_j |g_{\lambda_n, A}|$ and, then, for $\sup_j |g_{\lambda_n, A}(j) - g_{\lambda_n, A}(j)|$. These are, $\min\{1, \sqrt{1/\lambda_n}\}$ and $(1 - e^{-\lambda_n})/\lambda_n \leq$

$1/\lambda_n$, respectively; see [9], Lemma 1.1.1. The uniformity of the bound, and a glance at (7.4), finally establishes (7.2).

The main feature of the Stein-Chen method is that one can handle sums of *dependent* random variables. The assumption that the indicators are independent was used here only toward the end, and *only* in order to illustrate the procedure. The crucial problem is the treatment of the quantity

$$E\, S_n g_{\lambda_n,A}(S_n) = \sum_{k=1}^{n} p_j E\big(g_{\lambda_n,A}(S_n) \mid X_j = 1\big),$$

which, in the general case, introduces some additional term(s) to take care of; once again, we refer to [9] for details.

The final point is how to apply the result. So far we have provided an estimate of the variational distance between the sum of indicators and *one* suitable(?) Poisson distribution. One question remains: Which Poisson distribution is the best one? That is, how shall one construct the approximating Poisson variables?

In the first example we found that independence between $X_1, X_2, \ldots, X_n$ and $Y_1, Y_2, \ldots, Y_n$ provided a poorer estimate than if $X_k$ and $Y_k$ were dependent. Even more so, the two different couplings involving dependence produced different upper bounds for the variational distance. More generally, the problem on how to connect the two sequences introduces *the art of coupling*, which means how to introduce an efficient dependence structure between the two sequences $X_1, X_2, \ldots, X_n$ and $Y_1, Y_2, \ldots, Y_n$ in order to minimize the variational distance. In general this may be tricky and require some ingenuity; see [9]. However, sometimes it suffices to know that a coupling *exists*.

In order to illustrate the wide applicability of the approximation method, we close by mentioning applications to random permutations, random graphs, occupancy problems, extremes, and many more that have been dealt with in [9].

# 8 Problems

1. Prove that if $X$ is strictly stable with index $\alpha \in (0,2)$, and $Y$ is nonnegative and stable with index $\beta \in (0,1)$, then $XY^{1/\alpha}$ is stable with index $\alpha\beta$.

2. Suppose that $X_1, X_2, \ldots$ are independent, strictly stable random variables with index $\alpha \in (0,2)$, and set $S_n = \sum_{k=1}^{n} X_k$, $n \geq 1$. Prove that

$$\frac{\log |S_n|}{\log n} \xrightarrow{p} \frac{1}{\alpha}.$$

♣ In other words, if $\alpha$ is unknown, the ratio $\log n / \log |S_n|$ provides an estimate of the index.

3. Let $X$ be a stable random variables with index $\alpha \in (0, 2)$, suppose that $Y$ is a coin-tossing random variable $(P(Y = 1) = P(Y = -1) = 1/2)$, which is independent of $X$. Show that $XY$ is strictly stable.

4. Suppose that $X_1$, $X_2$, $\ldots$ are independent random variables with common density

$$f(x) = \begin{cases} \frac{c}{|x|^{\alpha+1}(\log |x|)^\gamma}, & \text{for } |x| > \mathrm{e}, \\ 0, & \text{otherwise}, \end{cases}$$

where $\alpha > 0$, $\gamma \in \mathbb{R}$, and where $c$ is a normalizing constant (without interest). Find the (possible) limit distributions for the various values of $\alpha$ and $\gamma$.

5. Let $X_1$, $X_2$, $\ldots$ be independent, symmetric, stable random variables with index $\alpha \in (0, 2]$, and let $a_k \in \mathbb{R}$, $k \geq 1$.
   (a) Prove that $\sum_{k=1}^n a_k X_k$ converges in distribution iff $\sum_{n=1}^\infty |c_n|^\alpha < \infty$.
   (b) Prove that, in fact,

$$\sum_{n=1}^\infty a_n X_n < \infty \text{ a.s.} \quad \Longleftrightarrow \quad \sum_{n=1}^\infty |c_n|^\alpha < \infty.$$

   ♣ Note the special cases $\alpha = 1$, the Cauchy distribution, and $\alpha = 2$ that we have encountered earlier in Chapters 5 and 6.

6. Let $X$, $X_1$, $X_2$, $\ldots$ be independent, identically distributed random variables with partial sums $S_n$, $n \geq 1$. Suppose that $X$ belongs to the domain of attraction of $Y$, that is, suppose that there exist $\{a_n > 0, n \geq 1\}$ and $\{b_n \in \mathbb{R}\}$, such that

$$\frac{S_n - b_n}{a_n} \xrightarrow{d} Y \quad \text{as} \quad n \to \infty.$$

Prove that

$$a_n \to \infty \quad \text{and that} \quad \frac{a_{n+1}}{a_n} \to 1 \quad \text{as} \quad n \to \infty.$$

   ♠ Suppose first that $X$ is symmetric, and check $a_{2n}/a_n$.

7. (a) Let $X$ be standard exponential, which means that the characteristic function equals $\varphi_X(t) = \frac{1}{1-\mathrm{i}t}$. Show that $X$ is infinitely divisible by convincing yourself that

$$\varphi_X(t) = \exp\left\{ \int_0^\infty \left(\mathrm{e}^{\mathrm{i}tx} - 1\right) \frac{\mathrm{e}^{-x}}{x}\, \mathrm{d}x \right\}.$$

   (b) Find the representation for $X \in \mathrm{Exp}(\lambda)$, $\lambda > 0$.
   (c) Check that, if $X \in \Gamma(p, 1)$, then

$$\varphi_X(t) = \exp\left\{ \int_0^\infty \left(\mathrm{e}^{\mathrm{i}tx} - 1\right) x^{p-1} \mathrm{e}^{-x}\, \mathrm{d}x \right\}.$$

8. Show that the geometric distribution is infinitely divisible by exhibiting the canonical representation of the characteristic function.

9. Prove that if $X$ is infinitely divisible with characteristic function

$$\varphi(t) = \exp\left\{ \int_{-\infty}^{\infty} \left( e^{itx} - 1 - \frac{itx}{1+x^2} \right) \frac{1+x^2}{x^2} \, dG(x) \right\},$$

then

$$X \geq 0 \quad \Longleftrightarrow \quad G(0) = 0.$$

10. Let $X_1, X_2, \ldots$ be independent, identically distributed random variables, which are also independent of $N \in \mathrm{Po}(\lambda)$. Show that $X_1 + X_2 + \cdots + X_N$ is infinitely divisible.

11. Show that the partial maxima of independent, identically distributed extremal random variables, properly rescaled, are extremal. More precisely, let $X, X_1, X_2, \ldots$ be independent, identically distributed random variables, and set

$$Y_n = \max\{X_1, X_2, \ldots, X_n\}, \quad n \geq 1.$$

Show that,
(a) if $X$ has a Fréchet distribution, then

$$\frac{Y_n}{n^{1/\alpha}} \overset{d}{=} X;$$

(b) if $X$ has a Weibull distribution, then

$$n^{1/\alpha} Y_n \overset{d}{=} X;$$

(c) if $X$ has a Gumbel distribution, then

$$Y_n - \log n \overset{d}{=} X.$$

12. Let $X_1, X_2, \ldots$ be $\mathrm{Exp}(\theta)$-distributed random variables, let $N \in \mathrm{Po}(\lambda)$, and suppose that all random variables are independent. Show that

$$Y = \max\{X_1, X_2, \ldots, X_N\} \overset{d}{=} V^+ = \max\{0, V\},$$

where $V$ has a Gumbel type distribution.
   ♠ It may help to remember Problem 4.11.25.

13. Suppose that $X_1, X_2, \ldots$ are independent random variables, such that $X_k \in \mathrm{Be}(p_k)$, $k \geq 1$, and set $S_n = \sum_{k=1}^{n} X_k$, $\mu_n = \sum_{k=1}^{n} p_k$, and $s_n^2 = \sum_{k=1}^{n} p_k(1-p_k)$, $n \geq 1$.
   (a) Use Theorem 7.1 to provide an estimate of the closeness in total variation to a suitable Poisson distribution (which one?).
   (b) Compare with the normal approximation from Problem 7.8.1. When is the Poisson approximation better? When is the normal approximation better?

# 10
## Martingales

Martingales are probably the most ingenious invention and generalization of sums of independent random variables with mean 0. They play an important role in probability theory and in statistics. They are also extremely applicable, mathematically tractable, and astonishingly exploitable in purely mathematical contexts. In addition, the theory is extremely elegant and aesthetically appealing.

The term martingale originates in gambling theory.[1] The famous gambling strategy to double one's stake as long as one loses and leave as soon as one wins is called a martingale. Unfortunately though, the gambler will have spent an infinite amount of money on average when he or she, finally, wins. We shall briefly return to this game in Example 3.5.

The first appearance of the term *martingale* in the present context was in 1939 in *Étude Critique de la Notion de Collectif* by Jean Ville; see [249]. The major breakthrough was with the now legendary book *Stochastic Processes* [66] by J. L. Doob, where much of the foundation was coherently proved and described for the first time. Other, more recent books are [186, 252]. A number of textbooks have a chapter devoted to martingales, such as [48, 177, 208].

We open this chapter with a section on conditional expectation, which is an essential concept in the definition of martingales, as well as for most martingale properties. After having provided a number of equivalent definitions of martingales we present a rather extensive selection of examples. This is then followed by the establishment of various properties, such as orthogonality of increments and decomposition theorems.

A major role is played by martingales evaluated at certain random times, called stopping times. After definitions we prove some inequalities and convergence theorems. As for the latter we shall present, not only the traditional proof which is based on so-called upcrossings, but also a different, elegant proof taken from Garsia's monograph [92]. Once a limit has been established it turns

---

[1]The term is also used in non-mathematical contexts, although the origin is somewhat unclear. The traditional example is "a horse's harness".

out that this limit may or may not be a "last" element in the (sub)martingale. This leads to classifications of regular stopping times and regular martingales.

The last part of the chapter is devoted to some applications, such as stopped random walks, after which we close with a section on reversed martingales with an application to the strong law of large numbers and $U$-statistics.

# 1 Conditional Expectation

As mentioned in the introductory text, conditional expectations are cornerstones in martingale theory. Before providing the definition we recall from Definition 2.1.2 that the equivalence class of a random variable $X$ is the collection of random variables that differ from $X$ on a null set.

**Definition 1.1.** *The* conditional expectation $E(X \mid \mathcal{G})$ *of an integrable random variable, $X$, relative to a sub-$\sigma$-algebra $\mathcal{G}$ of $\mathcal{F}$ is any $\mathcal{G}$-measurable, integrable random variable $Z$ in the equivalence class of random variables, such that*

$$\int_\Lambda Z \, dP = \int_\Lambda X \, dP \quad \text{for any} \quad \Lambda \in \mathcal{G}. \qquad \square$$

*Remark 1.1.* Observe that the integrals of $X$ and $Z$ over sets $\Lambda \in \mathcal{G}$ are the same; however, $X \in \mathcal{F}$, whereas $Z \in \mathcal{G}$. $\qquad \square$

The conditional expectation thus satisfies

$$\int_\Lambda E(X \mid \mathcal{G}) \, dP = \int_\Lambda X \, dP \quad \text{for any} \quad \Lambda \in \mathcal{G}, \qquad (1.1)$$

which is called *the defining relation.*

If, in particular, $X$ is the indicator variable of some $\mathcal{F}$-measurable set, $X = I\{A\}$, the defining relation produces

$$\int_\Lambda P(A \mid \mathcal{G}) \, dP = P(A \cap \Lambda) \quad \text{for any} \quad \Lambda \in \mathcal{G}, \qquad (1.2)$$

which means that *the conditional probability $P(A \mid \mathcal{G})$* is any $\mathcal{G}$-measurable random variable in the equivalence class of random variables, satisfying (1.2).

In order to see that conditional expectations exist, we recall that in Section 2.13 we briefly mentioned conditional distributions, and in Proposition 1.4.1, the law of total probability. Combining these we note that if $X$ is a discrete random variable, taking the values $\{x_n, \, n \geq 1\}$, then, for $A \in \mathcal{F}$,

$$P(A) = \sum_{n=1}^{\infty} P(A \mid X = x_n) P(X = x_n).$$

By "randomizing", that is, by replacing $P(X = x_n)$ by the indicator function $I\{X = x_n\}$, we obtain

$$P(A \mid \sigma\{X\}) = P(A \mid X) = \sum_{n=1}^{\infty} P(A \mid X = x_n)I\{X = x_n\},$$

which means that, for a given $\omega \in \Omega$,

$$P(A \mid \sigma\{X\})(\omega) = \begin{cases} P(A \mid X = x_1), & \text{if } X(\omega) = x_1, \\ P(A \mid X = x_2), & \text{if } X(\omega) = x_2, \\ \cdots\cdots & \cdots\cdots\cdots\cdots \\ P(A \mid X = x_n), & \text{if } X(\omega) = x_n, \\ 0, & \text{otherwise.} \end{cases}$$

This coincides with (1.2), since

$$\int_{\Lambda} \left( \sum_{n=1}^{\infty} P(A \mid X = x_n)I\{X = x_n\} \right) dP = \sum_{\{n:x_n \in \Lambda\}} P(A \mid X = x_n)P(X = x_n)$$
$$= P(A \cap \Lambda).$$

More generally, let $\{\Lambda_n, \, n \geq 1\}$ be a partition of $\Omega$, and let $E(X \mid \Lambda_n)$ be the conditional expectation relative to the conditional measure $P(\cdot \mid \Lambda_n)$, so that

$$E(X \mid \Lambda_n) = \int_{\Omega} X(\omega) \, dP(\omega \mid \Lambda_n) = \frac{\int_{\Lambda_n} X \, dP}{P(\Lambda_n)}.$$

Then, for $\Lambda = \sum_{j \in J} \Lambda_j \in \mathcal{G}$, we obtain, noticing the disjointness of the $\Lambda_j$'s,

$$\int_{\Lambda} \left( \sum_{n=1}^{\infty} E(X \mid \Lambda_n)I\{\Lambda_n\} \right) dP = \sum_{j \in J} \sum_{n=1}^{\infty} \int_{\Lambda_j} E(X \mid \Lambda_n)I\{\Lambda_n\}I\{\Lambda_j\} \, dP$$

$$= \sum_{j \in J} \int_{\Lambda_j} E(X \mid \Lambda_j) \, dP = \sum_{j \in J} E(X \mid \Lambda_j)P(\Lambda_j) = \sum_{j \in J} \frac{\int_{\Lambda_j} X \, dP}{P(\Lambda_j)} P(\Lambda_j)$$

$$= \sum_{j \in J} \int_{\Lambda_j} X \, dP = \int_{\cup_{j \in J} \Lambda_j} X \, dP = \int_{\Lambda} X \, dP.$$

This proves that, in this case,

$$E(X \mid \mathcal{G}) = \sum_{n=1}^{\infty} E(X \mid \Lambda_n)I\{\Lambda_n\} \quad \text{a.s.,}$$

or, spelled out, that

$$E(X \mid \mathcal{G}) = \begin{cases} E(X \mid \Lambda_1), & \text{if } \omega \in \Lambda_1, \\ E(X \mid \Lambda_2), & \text{if } \omega \in \Lambda_2, \\ \cdots\cdots & \cdots\cdots\cdots \\ E(X \mid \Lambda_n), & \text{if } \omega \in \Lambda_n, \\ 0, & \text{otherwise.} \end{cases}$$

Another way to express this is that the random variable $E(X \mid \mathcal{G})$ takes on the discrete values $E(X \mid \Lambda_1), E(X \mid \Lambda_2), \ldots, E(X \mid \Lambda_n)$ on the sets $\Lambda_1, \Lambda_2, \ldots, \Lambda_n$, respectively.

Before we turn to existence in the general case, let us prove a.s. uniqueness. Thus, suppose that $Y$ and $Z$ are conditional expectations, satisfying the defining relation. Since both are $\mathcal{G}$-measurable we may choose $\{\omega : Y(\omega) > Z(\omega)\}$ as our $\Lambda \in \mathcal{G}$. The integrals, being equal, yields

$$\int_{\Lambda} (Y - Z)\, \mathrm{d}P = 0,$$

which necessitates $P(\Lambda) = 0$. Similarly for $\{\omega : Y(\omega) < Z(\omega)\}$. This shows that conditional expectations are unique up to null sets (if they exist).

For existence in the general case we need a definition and the Radon-Nikodym theorem as a final preparation.

**Definition 1.2.** *Let $P$ and $Q$ be probability measures. The measure $Q$ is absolutely continuous with respect to $P$ iff*

$$P(A) = 0 \quad \Longrightarrow \quad Q(A) = 0 \quad \text{for all} \quad A \in \mathcal{F}.$$

Notation: $Q \ll P$. □

**Theorem 1.1.** (The Radon-Nikodym theorem)
*Let $(\Omega, \mathcal{F}, P)$ be a probability space, and suppose that $Q$ is a finite measure that is absolutely continuous with respect to $Q \ll P$. Then there exists an $\mathcal{F}$-measurable random variable $X$ with finite mean, such that*

$$Q(A) = \int_A X\, \mathrm{d}P \quad \text{for all} \quad A \in \mathcal{F}.$$

*Moreover, $X$ is $P$-a.s. unique and is written as*

$$X = \frac{\mathrm{d}Q}{\mathrm{d}P}.$$

*It is called the* Radon-Nikodym derivative.

For a proof we refer to the measure theoretic literature. Note, however, that if on the other hand, $P$ and $Q$ are related as in the theorem, then $Q \ll P$. The Radon-Nikodym theorem thus is an "only if" result.

The definition of conditional expectation can now be justified via the following result.

**Theorem 1.2.** *Let $X$ be a random variable with finite expectation, and $\mathcal{G}$ a sub-$\sigma$-algebra of $\mathcal{F}$. Then there exists a unique equivalence class of random variables with finite mean, satisfying the defining relation (1.1).*

*Proof.* Let $P|_{\mathcal{G}}$ be the measure $P$ restricted to $\mathcal{G}$, and set

$$Q(A) = \int_A X \, dP \quad \text{for} \quad A \in \mathcal{G}.$$

From our findings in Section 2.6 we assert that $Q \ll P|_{\mathcal{G}}$ and that the Radon-Nikodym theorem is applicable. The conclusion follows.  □

The equivalence class thus obtained is the Radon-Nikodym derivative, which we henceforth denote $E(X \mid \mathcal{G})$ in the present context.

As a side remark we mention without details that if $X$ is integrable and $\mathcal{G}$ a sub-$\sigma$-algebra of $\mathcal{F}$, then it follows from the defining relation that

$$\int_\Lambda (X - E(X \mid \mathcal{G}))Z \, dP = 0 \quad \text{for all bounded} \quad Z \in \mathcal{G},$$

which, in turn, induces the decomposition

$$X = Y + Z, \tag{1.3}$$

where $Y = E(X \mid \mathcal{G}) \in \mathcal{G}$ and $EYZ = 0$. This means that $Y$ is the "projection" of $X$ onto $\mathcal{G}$, and that $Z$ is the "orthogonal complement".

## 1.1 Properties of Conditional Expectation

We begin by establishing the fact that elementary properties that are known to hold for ordinary (unconditional) expectations remain valid for conditional expectations, except for the fact that they are almost sure properties in the present context. After all, expectations are reals, conditional expectations are functions.

**Proposition 1.1.** *Let $X$ and $Y$ be integrable random variables, $\mathcal{G} \subset \mathcal{F}$, and $a, b, c$ real numbers. Then*

(a)   $E(E(X \mid \mathcal{G})) = E X;$
(b)   $E(aX + bY \mid \mathcal{G}) \stackrel{a.s.}{=} aE(X \mid \mathcal{G}) + bE(Y \mid \mathcal{G});$
(c)   *if* $X \in \mathcal{G}$, *then* $E(X \mid \mathcal{G}) \stackrel{a.s.}{=} X;$
(d)   $E(c \mid \mathcal{G}) \stackrel{a.s.}{=} c;$
(e)   $E(X \mid \{\emptyset, \Omega\}) \stackrel{a.s.}{=} E X;$
(f)   *if* $X \geq 0$ *a.s., then* $E(X \mid \mathcal{G}) \geq 0$ *a.s.;*
(g)   *if* $X \leq Y$ *a.s., then* $E(X \mid \mathcal{G}) \leq E(Y \mid \mathcal{G})$ *a.s.;*
(h)   $|E(X \mid \mathcal{G})| \leq E(|X| \mid \mathcal{G});$
(j)   *if* $X$ *is indpendent of* $\mathcal{G}$, *then* $E(X \mid \mathcal{G}) = E X$ *a.s.*

*Proof.* Every property follows from the defining relation.

The first one follows by setting $\Lambda = \Omega$, and the second one via

$$\int_A E(aX + bY \mid \mathcal{G}) \, dP = \int_A (aX + bY) \, dP = a \int_A X \, dP + b \int_A Y \, dP$$

$$= a \int_A E(X \mid \mathcal{G}) \, dP + b \int_A E(Y \mid \mathcal{G}) \, dP$$

$$= \int_A aE(X \mid \mathcal{G}) + bE(Y \mid \mathcal{G}).$$

Statement (c) follows from the tautology

$$\int_A X \, dP = \int_A X \, dP \quad \text{for any} \quad \Lambda \in \mathcal{G},$$

so that $X$ *being $\mathcal{G}$-measurable!* satisfies the defining relation, and, since any constant is $\mathcal{G}$-measurable, (d) is immediate from (c).

As for (e),

$$\int_A X \, dP = \begin{cases} 0, & \text{for} \quad \Lambda = \emptyset, \\ E\,X, & \text{for} \quad \Lambda = \Omega, \end{cases}$$

that is,

$$\int_A X \, dP = \int_A E\,X \, dP, \quad \text{for all} \quad \Lambda \in \{\emptyset, \Omega\},$$

so that $E\,X$ can be substituted for $E(X \mid \mathcal{G})$.

Next, (f) follows via

$$\int_A E(X \mid \mathcal{G}) \, dP = \int_A X \, dP \geq 0,$$

which, together with (b), applied to the non-negative random variable $Y - X$, yields (g).

Moving down the list, (h) follows via (b)

$$|E(X \mid \mathcal{G})| = |E(X^+ \mid \mathcal{G}) - E(X^- \mid \mathcal{G})|$$
$$\leq E(X^+ \mid \mathcal{G}) + E(X^- \mid \mathcal{G}) = E(|X| \mid \mathcal{G}).$$

To prove (j), we note that $E\,X \in \mathcal{G}$, and that, for any $\Lambda \in \mathcal{G}$,

$$\int_A E\,X \, dP = E\,X P(\Lambda) = E(XI\{\Lambda\}) = \int_\Omega XI\{\Lambda\} \, dP = \int_A X \, dP.$$

The equality between the extreme members tells us that $E\,X$ satisfies the defining relation.  □

*Remark 1.2.* Most of the properties are intuitively "obvious". Consider, for example, (c). If $\mathcal{G}$ is given, then $X$ "is known", so that there is no (additional) randomness, in the sense that $X$ is "constant" on $\mathcal{G}$. And it is well known that the expected value of a constant is the constant itself.  □

Next in line are the conditional counterparts of the monotone convergence theorems, Fatou's lemma and the Lebesgue convergence theorem.

**Proposition 1.2.** *We have*

(a)  *If $X_n \uparrow X$ as $n \to \infty$, then $E(X_n \mid \mathcal{G}) \uparrow E(X \mid \mathcal{G})$ a.s. as $n \to \infty$;*

(b)  *If $X_n \downarrow X$ as $n \to \infty$, then $E(X_n \mid \mathcal{G}) \downarrow E(X \mid \mathcal{G})$ a.s. as $n \to \infty$;*

(c)  *If $\{X_n, n \geq 1\}$ are non-negative and integrable, then*

$$E(\liminf_{n\to\infty} X_n \mid \mathcal{G}) \leq \liminf_{n\to\infty} E(X_n \mid \mathcal{G}) \quad a.s.;$$

(d)  *If $X_n \leq Z \in L^1$ for all $n$, then*

$$E(\limsup_{n\to\infty} X_n \mid \mathcal{G}) \geq \limsup_{n\to\infty} E(X_n \mid \mathcal{G}) \quad a.s.;$$

(e)  *If $|X_n| \leq Y \in L^1$ and $X_n \overset{a.s.}{\to} X$ as $n \to \infty$, then*

$$E(X_n \mid \mathcal{G}) \overset{a.s.}{\to} E(X \mid \mathcal{G}) \quad as \quad n \to \infty.$$

*Proof.* We know from Proposition 1.1(g) that the sequence $E(X_n \mid \mathcal{G})$ is monotone increasing, so the limit $Z = \lim_{n\to\infty} E(X_n \mid \mathcal{G})$ exists. Thus, by the defining relation and monotone convergence (twice),

$$\int_\Lambda Z \, dP = \int_\Lambda \lim_{n\to\infty} E(X_n \mid \mathcal{G}) \, dP = \lim_{n\to\infty} \int_\Lambda E(X_n \mid \mathcal{G}) \, dP$$
$$= \lim_{n\to\infty} \int_\Lambda X_n \, dP = \int_\Lambda \lim_{n\to\infty} X_n \, dP = \int_\Lambda X \, dP.$$

This proves (a), from which (b) follows by changing signs.

To prove (c) we set $Z_n = \inf_{k\geq n} X_n \leq X_n$, and note that $Z_n \overset{a.s.}{\to} \liminf_{n\to\infty} X_n$ monotonically. Thus, by Proposition 1.1(g) and (a), we have

$$E(X_n \mid \mathcal{G}) \geq E(Z_n \mid \mathcal{G}) \uparrow E(\liminf_{n\to\infty} X_n) \mid \mathcal{G},$$

and (c) follows.

For (d) we observe (as in Chapter 2) that $Z - X_n$ is non-negative and integrable, after which we apply (c) and linearity.

Finally, (e) follows by joining (c) and (d) as in the proof of Theorem 2.5.3. We omit the details. $\qquad \square$

**Proposition 1.3.** *If $X$ and $XY$ are integrable, $Y \in \mathcal{G}$, then*

$$E(XY \mid \mathcal{G}) \overset{a.s.}{=} Y E(X \mid \mathcal{G}).$$

*Proof.* First suppose that $X$ and $Y$ are non-negative. For $Y = I\{A\}$, where $A$ is $\mathcal{G}$-measurable, $\Lambda \cap A \in \mathcal{G}$, so that, by the defining relation,

$$\int_\Lambda YE(X\mid\mathcal{G})\,\mathrm{d}P = \int_{\Lambda\cap A} E(X\mid\mathcal{G})\,\mathrm{d}P = \int_{\Lambda\cap A} X\,\mathrm{d}P = \int_\Lambda XY\,\mathrm{d}P,$$

which proves the desired relation for indicators, and hence for simple random variables. Next, if $\{Y_n,\ n \geq 1\}$ are simple random variables, such that $Y_n \nearrow Y$ almost surely as $n \to \infty$, it follows that $Y_n X \nearrow YX$ and $Y_n E(X\mid\mathcal{G}) \nearrow YE(X\mid\mathcal{G})$ almost surely as $n \to \infty$, from which the conclusion follows by monotone convergence. The general case follows by the decomposition $X = X^+ - X^-$ and $Y = Y^+ - Y^-$. □

## 1.2 Smoothing

Many martingale properties and results are proved via iterated, or successive, conditioning. In order to verify such facts we need the following *smoothing lemma*, which turns out to be handy in many situations.

**Lemma 1.1.** *Suppose that $\mathcal{F}_1 \subset \mathcal{F}_2$. Then*

$$E(E(X\mid\mathcal{F}_2)\mid\mathcal{F}_1) = E(X\mid\mathcal{F}_1) = E(E(X\mid\mathcal{F}_1)\mid\mathcal{F}_2)\quad a.s.$$

*Proof.* Since $E(X\mid\mathcal{F}_1) \in \mathcal{F}_2$ the second equality is immediate from Proposition 1.1(c), and it remains to prove the first one. Pick $\Lambda \in \mathcal{F}_1$ and observe that, automatically, $\Lambda \in \mathcal{F}_2$. Applying the defining relation three times, we obtain

$$\int_\Lambda E(E(X\mid\mathcal{F}_2)\mid\mathcal{F}_1)\,\mathrm{d}P = \int_\Lambda E(X\mid\mathcal{F}_2)\,\mathrm{d}P = \int_\Lambda X\,\mathrm{d}P$$
$$= \int_\Lambda E(X\mid\mathcal{F}_1)\,\mathrm{d}P.\qquad \square$$

A proper exploitation of the smoothing lemma and Proposition 1.3 yields the following results. The proofs being similar, we confine ourselves to proving the first one.

**Theorem 1.3.** *Suppose that $Y$ is a random variable with finite variance and that $\mathcal{G}$ is a sub-$\sigma$-algebra of $\mathcal{F}$. Then*

$$E\big(Y - E(Y\mid\mathcal{G})\big)^2 = EY^2 - E\big(E(Y\mid\mathcal{G})\big)^2.$$

*Proof.* By smoothing and Proposition 1.3,

$$E\big(YE(Y\mid\mathcal{G})\big) = E\{E\big(E(YE(Y\mid\mathcal{G})\mid\mathcal{G}\big)\}$$
$$= E\{E(Y\mid\mathcal{G})E(Y\mid\mathcal{G})\} = E\big(E(Y\mid\mathcal{G})^2\big),$$

so that,

$$E\big(Y - E(Y\mid\mathcal{G})\big)^2 = EY^2 + E\big(E(Y\mid\mathcal{G})^2\big) - 2E\big(YE(Y\mid\mathcal{G})\big)$$
$$= EY^2 - E\big(E(Y\mid\mathcal{G})\big)^2.\qquad \square$$

By defining *conditional variance* as the the conditional expectation

$$\mathrm{Var}\,(X \mid \mathcal{G}) = E\big((X - E(X \mid \mathcal{G}))^2 \mid \mathcal{G}\big),$$

another application of the smoothing lemma produces the following:

**Theorem 1.4.** *Let $X$ and $Y$ be random variables with finite variance, let $g$ be a real valued function, such that $E(g(X))^2 < \infty$, and let $\mathcal{G}$ be a sub-$\sigma$-algebra of $\mathcal{F}$. Then*

$$E\big(Y - g(X)\big)^2 = E\,\mathrm{Var}\,(Y \mid \mathcal{G}) + E\big(E(Y \mid \mathcal{G}) - g(X)\big)^2 \geq E\,\mathrm{Var}\,(Y \mid \mathcal{G}),$$

*where equality is obtained for $g(X) = E(Y \mid \mathcal{G})$.*

**Exercise 1.1.** Prove the theorem.                                            □

*Remark 1.3.* Theorem 1.4 has a Pythagorean touch. This is not accidental, in fact, the second moment on the left (the hypotenuse) has been decomposed into the second moment of $Y$ minus its projection, and "the rest" (the two other sides). The inequality in the theorem corresponds to the fact that the shortest distance from a point to a plane is the normal, which in the present context is obtained when $g$ is chosen to be the projection.                □

An important statistical application of the last two results is provided in the following subsection.

## 1.3 The Rao-Blackwell Theorem

Suppose that we are given a family of probability measures $\{P_\theta, \theta \in \Theta\}$, where $\theta$ is some parameter and $\Theta$ is the parameter set. Two examples are the family of exponential distributions, or the family of normal distributions. In the latter case it may be so that the mean is given and the parameter set is the set of possible variances or vice versa, or, possibly, that $\Theta = \{(\mu, \sigma), \mu \in \mathbb{R}, \sigma > 0\}$. If one would like to estimate the unknown parameter $\theta$ it is of obvious interest to do so under minimal "risk". Another feature is *unbiasedness*, which means that the expected value of the estimator $\hat{\theta}$ equals the true value of the parameter; $E\hat{\theta} = \theta$.

A special class of statistics are the *sufficient statistics*, the feature of which is that the conditional distribution of any random variable given that statistic is independent of the actual value of the unknown parameter.

Suppose that $T$ is a sufficient statistic and that $Y$ is an arbitrary unbiased estimator. Then we know that the conditional distribution $(Y \mid T = t)$, more generally, $(Y \mid \sigma\{T\})$, is the same irrespective of the actual true value of the unknown parameter $\theta$.

Theorem 1.3 can now be used to show that the conditional expectation of an unbiased estimator $Y$, given a sufficient statistic, $T$, has smaller variance than $Y$ itself, and is also unbiased.

**Theorem 1.5.** (The Rao-Blackwell theorem)
*Suppose that $T$ is a sufficient statistic for $\theta \in \Theta$, and let $Y$ be an unbiased estimator of $\theta$, such that*

$$E_\theta(Y - \theta)^2 < \infty \quad \text{for all} \quad \theta \in \Theta.$$

*Then*

$$E_\theta(Y - \theta)^2 = E\big(Y - E(Y \mid T)\big)^2 + E_\theta\big(E(Y \mid T) - \theta\big)^2$$

$$\geq \begin{cases} E\big(Y - E(Y \mid T)\big)^2, \\ E_\theta\big(E(Y \mid T) - \theta\big)^2, \end{cases}$$

*so that,*

- *the minimum square loss $E_\theta(Y - \theta)^2$ equals $E\big(Y - E(Y \mid T)\big)^2$;*
- *the expected square loss of $E(Y \mid T)$ is always smaller than that of $Y$;*
- *if $Y$ is unbiased, then so is $E(Y \mid T)$.*

*Remark 1.4.* $E_\theta$ denotes expectation under the hypothesis that the true value of the parameter is $\theta$. The absence of $\theta$ in the expression $E\big(Y - E(Y \mid T)\big)^2$ is a consequence of the sufficiency of the estimator $T$. □

*Proof.* The equality is taken from Theorem 1.3. The first conclusion is a consequence of the first inequality, the second conclusion is a consequence of the second inequality, and the third conclusion follows from the Proposition 1.1(a);

$$E\big(E(Y \mid T)\big) = EY = \theta. \qquad \qquad \square$$

## 1.4 Conditional Moment Inequalities

In order to compare moments or sums of martingales we need conditional versions of several standard inequalities from Chapter 3.

**Theorem 1.6.** (Conditional inequalities) *Let $X$ and $Y$ be random variables, and suppose that $\mathcal{G}$ is a sub-$\sigma$-algebra of $\mathcal{F}$. The following conditional moment inequalities hold almost surely (provided the corresponding moments exist):*

- *Conditional $c_r$*

$$E(|X + Y|^r \mid \mathcal{G}) \leq c_r\big(E(|X|^r \mid \mathcal{G}) + E(|Y|^r \mid \mathcal{G})\big),$$

  *where $c_r = 1$ when $r \leq 1$ and $c_r = 2^{r-1}$ when $r \geq 1$.*
- *Conditional Hölder*

$$|E(XY \mid \mathcal{G})| \leq E(|XY| \mid \mathcal{G}) \leq \|(X \mid \mathcal{G})\|_p \cdot \|(Y \mid \mathcal{G})\|_q.$$

- *Conditional Lyapounov*

$$\|(X \mid \mathcal{G})\|_r \leq \|(X \mid \mathcal{G})\|_p \quad for \;\; 0 < r \leq p.$$

- *Conditional Minkowski*

$$\|((X + Y) \mid \mathcal{G})\|_p \leq \|(X \mid \mathcal{G})\|_p + \|(Y \mid \mathcal{G})\|_p.$$

- *Conditional Jensen*

$$g\big(E(X \mid \mathcal{G})\big) \leq E\big(g(X) \mid \mathcal{G}\big).$$

*Proof.* The proofs are the same as the unconditional ones; they all depart from inequalities for real numbers.    □

**Exercise 1.2.** Please, check the details.    □

Finally we are ready to turn our attention to the theory of martingales.

# 2 Martingale Definitions

The point of departure is the traditional probability space, $(\Omega, \mathcal{F}, P)$ with, additionally, a sequence $\{\mathcal{F}_n, n \geq 0\}$ of increasing sub-$\sigma$-algebras of $\mathcal{F}$ – a *filtration*, – which means that

$$\mathcal{F}_0 \subset \mathcal{F}_1 \subset \cdots \subset \mathcal{F}_n \subset \mathcal{F}_{n+1} \subset \cdots \subset \mathcal{F}.$$

If $n$ is interpreted as (discrete) time, then $\mathcal{F}_n$ contains the information up to time $n$.

We also introduce $\mathcal{F}_\infty = \sigma\{\cup_n \mathcal{F}_n\}$. Recall that the union of a sequence of $\sigma$-algebras is not necessarily a $\sigma$-algebra.

**Definition 2.1.** *A sequence* $\{X_n, n \geq 0\}$ *of random variables is* $\{\mathcal{F}_n\}$-adapted *if* $X_n$ *is* $\mathcal{F}_n$-*measurable for all* $n$. *If* $\mathcal{F}_n = \sigma\{X_0, X_1, X_2, \ldots, X_n\}$ *we call the sequence* adapted, *and we call* $\{\mathcal{F}_n, n \geq 0\}$ *the sequence of* natural $\sigma$-*algebras, or the* natural filtration.

**Definition 2.2.** *A sequence* $\{X_n, n \geq 0\}$ *of random variables is* $\{\mathcal{F}_n\}$-predictable *if* $X_n \in \mathcal{F}_{n-1}$ *for all* $n$. *If* $\mathcal{F}_n = \sigma\{X_0, X_1, X_2, \ldots, X_n\}$ *we call the sequence* predictable.

**Definition 2.3.** *A sequence* $A_n, n \geq 0\}$ *is called an* increasing process *if* $A_0 = 0$, $A_n \nearrow$, *and* $\{A_n\}$ *is predictable (with respect to* $\{\mathcal{F}_n, n \geq 0\})$.    □

Here is now the definition we have been waiting for.

**Definition 2.4.** *An integrable* $\{\mathcal{F}_n\}$*-adapted sequence* $\{X_n\}$ *is called a* martingale *if*

$$E(X_{n+1} \mid \mathcal{F}_n) = X_n \qquad a.s. \quad \text{for all} \quad n \geq 0.$$

*It is called a* submartingale *if*

$$E(X_{n+1} \mid \mathcal{F}_n) \geq X_n \qquad a.s. \quad \text{for all} \quad n \geq 0,$$

*and a* supermartingale *if*

$$E(X_{n+1} \mid \mathcal{F}_n) \leq X_n \qquad a.s. \quad \text{for all} \quad n \geq 0.$$

*We call it an* $L^p$*-martingale (submartingale, supermartingale) if, in addition,* $E|X_n|^p < \infty$ *for all* $n$. *We call it* $L^p$*-bounded if, moreover,* $\sup_n E|X_n|^p < \infty$.

**Definition 2.5.** *A non-negative supermartingale* $\{(X_n, \mathcal{F}_n), n \geq 0\}$, *such that* $E X_n \to 0$ *as* $n \to \infty$ *is called a* potential. □

*Remark 2.1.* The equality and inequalities in the definition are *almost sure* relations. In order to make the material easier to read we shall refrain from repeating that all the time. *We must remember throughout that a statement such as* $X = E(Y \mid \mathcal{G})$ *in reality means* $X \overset{a.s.}{=} E(Y \mid \mathcal{G})$.

Moreover, a statement such as $\{X_n, n \geq 0\}$ is a martingale "does not make sense" without a filtration having been specified. We shall, at times, be somewhat careless on that point. Frequently it is fairly obvious that the natural filtration is the intended one. □

Precisely as in the context of sums of independent random variables it turns out that it is sometimes more convenient to talk about increments.

**Definition 2.6.** *An integrable,* $\{\mathcal{F}_n\}$*-adapted sequence* $\{U_n\}$ *is called a* martingale difference sequence *if*

$$E(U_{n+1} \mid \mathcal{F}_n) = 0 \qquad \text{for all} \quad n \geq 0.$$

*It is called a* submartingale difference sequence *if*

$$E(U_{n+1} \mid \mathcal{F}_n) \geq 0 \qquad \text{for all} \quad n \geq 0,$$

*and a* supermartingale difference sequence *if*

$$E(U_{n+1} \mid \mathcal{F}_n) \leq 0 \qquad \text{for all} \quad n \geq 0.$$ □

*Remark 2.2.* It follows from the definition that a martingale is both a submartingale and a supermartingale, and that switching signs turns a submartingale into a supermartingale (and vice versa).

*Remark 2.3.* Martingales, submartingales and supermartingales are related to harmonic-, subharmonic- and superharmonic functions, which are central objects in the theory of harmonic analysis. Many results in the present context have counterparts in harmonic analysis and there is an important interplay between the two areas, for example, in terms of proof techniques.     □

In order to provide interesting examples we need to establish some properties. However, to have one example in mind already, the sequence of partial sums of independent random variables with mean 0 constitutes a martingale.

## 2.1 The Defining Relation

It follows from the definition of conditional expectations that an equivalent definition of a martingale is that

$$\int_{\Lambda} X_{n+1}\, dP = \int_{\Lambda} X_n\, dP \quad \text{for all } \Lambda \in \mathcal{F}_n, \quad n \geq 0,$$

or that

$$\int_{\Lambda} X_n\, dP = \int_{\Lambda} X_m\, dP \quad \text{for all } \Lambda \in \mathcal{F}_m, \quad 0 \leq m \leq n.$$

Analogously $\{X_n,\, n \geq 0\}$ is a submartingale if

$$\int_{\Lambda} X_{n+1}\, dP \geq \int_{\Lambda} X_n\, dP \quad \text{for all } \Lambda \in \mathcal{F}_n, \quad n \geq 0,$$

or, equivalently,

$$\int_{\Lambda} X_n\, dP \geq \int_{\Lambda} X_m\, dP \quad \text{for all } \Lambda \in \mathcal{F}_m, \quad 0 \leq m \leq n,$$

and a supermartingale if

$$\int_{\Lambda} X_{n+1}\, dP \leq \int_{\Lambda} X_n\, dP \quad \text{for all } \Lambda \in \mathcal{F}_n, \quad n \geq 0,$$

or, equivalently,

$$\int_{\Lambda} X_n\, dP \leq \int_{\Lambda} X_m\, dP \quad \text{for all } \Lambda \in \mathcal{F}_m, \quad 0 \leq m \leq n.$$

Integrability and proper adaptivity must, of course, also be assumed.

## 2.2 Two Equivalent Definitions

We can now prove the statements made prior to the smoothing lemma, thereby obtaining equivalent definitions of martingales, sub- and supermartingales.

**Theorem 2.1.** (i) *An integrable $\{\mathcal{F}_n\}$-adapted sequence $\{X_n\}$ is a martingale iff*

$$E(X_n \mid \mathcal{F}_m) = X_m \quad \text{for all} \quad 0 \leq m \leq n.$$

*It is a submartingale iff*

$$E(X_n \mid \mathcal{F}_m) \geq X_m \quad \text{for all} \quad 0 \leq m \leq n,$$

*and a supermartingale iff*

$$E(X_n \mid \mathcal{F}_m) \leq X_m \quad \text{for all} \quad 0 \leq m \leq n.$$

(ii) *A martingale has constant expectation. A submartingale has non-decreasing expectations. A supermartingale has non-increasing expectations.*

*Proof.* The sufficiency in (i) is, of course, trivial; put $n = m + 1$. To prove the necessity we apply Lemma 1.1.

For martingales,

$$E(X_n \mid \mathcal{F}_m) = E(E(X_n \mid \mathcal{F}_{n-1}) \mid \mathcal{F}_m) = E(X_{n-1} \mid \mathcal{F}_m) = \cdots$$
$$= E(X_{m+1} \mid \mathcal{F}_m) = X_m.$$

For submartingales,

$$E(X_n \mid \mathcal{F}_m) = E(E(X_n \mid \mathcal{F}_{n-1}) \mid \mathcal{F}_m) \geq E(X_{n-1} \mid \mathcal{F}_m) \geq \cdots$$
$$\geq E(X_{m+1} \mid \mathcal{F}_m) \geq X_m,$$

and for supermartingales the conclusion follows by a sign change; recall Remark 2.2.

To prove (ii) we apply part (ii) of the smoothing lemma, according to which

$$E X_m = E(E(X_n \mid \mathcal{F}_m)) = E X_n,$$

for martingales,

$$E X_m \leq E(E(X_n \mid \mathcal{F}_m)) = E X_n,$$

for submartingales, and

$$E X_m \geq E(E(X_n \mid \mathcal{F}_m)) = E X_n,$$

for supermartingales.                                                          □

If we interpret a martingale as a game, part (ii) states that, on average, nothing happens, and part (i) states that the expected state of the game given the past history equals the present state. A poetic way to state this is that in a sense life itself is a martingale. This was formulated by the Nobel Prize winner of 2002, Imre Kertész, in his book *Ad kudarc* [155] with the Swedish title *Fiasko*[2] where he writes as follows about the future:

---

[2]Fiasko is (of course) Swedish for fiasco. The Hungarian title is more toward "failure" without the disastrous connotation of fiasco.

[...] det är inte framtiden som väntar mig, bara nästa ögonblick, framtiden existerar ju inte, den är inget annat än ett ständigt fortgående, ett
tillstädesvarande nu. [...] *Prognosen för min framtid – det är kvalitén
på mitt nu.*[3]

The following, equivalent, definition is expressed in terms of (sub-, super-)
martingale difference sequences.

**Theorem 2.2.** *Let* $\{U_n\}$ *be* $\{\mathcal{F}_n\}$*-adapted and integrable, and set* $X_n = \sum_{k=0}^n U_k$, $n \geq 0$.
(i)  $\{(X_n, \mathcal{F}_n),\ n \geq 0\}$ *is a martingale iff* $\{(U_n, \mathcal{F}_n),\ n \geq 0\}$ *is a martingale
difference sequence, a submartingale iff* $\{(U_n, \mathcal{F}_n),\ n \geq 0\}$ *is a submartingale difference sequence, and a supermartingale iff* $\{(U_n, \mathcal{F}_n),\ n \geq 0\}$ *is a
supermartingale difference sequence.*
(ii)  *A martingale difference sequence has constant expectation* 0*; a submartingale difference sequence has non-negative expectations; a supermartingale difference sequence has non-positive expectations.*

**Exercise 2.1.** Prove the theorem. □

# 3 Examples

We begin by verifying the claim about sums of independent random variables
with mean 0.

*Example 3.1.* Suppose that $Y_1, Y_2, \ldots$ are independent random variables with
mean 0, and set $X_n = \sum_{k=1}^n Y_k$, $n \geq 0$ (with $Y_0 = X_0 = 0$). Moreover,
let $\mathcal{F}_n = \sigma\{Y_0, Y_1, Y_2, \ldots, Y_n\} = \sigma\{X_0, X_1, X_2, \ldots, X_n\}$, $n \geq 0$. Then
$\{(X_n, \mathcal{F}_n),\ n \geq 0\}$ is a martingale (and $\{(Y_n, \mathcal{F}_n),\ n \geq 0\}$ is a martingale
difference sequence). Namely,

$$E(X_{n+1} \mid \mathcal{F}_n) = E(X_n + Y_{n+1} \mid \mathcal{F}_n) = X_n + E(Y_{n+1} \mid \mathcal{F}_n) = X_n + 0 = X_n.$$

*Example 3.2.* If, in particular, $Y_1, Y_2, \ldots$, in addition, are identically distributed, the conclusion in Example 3.1 can be rephrased as *a centered random
walk is a martingale.* □

*Example 3.3.* It is tempting to guess that, for example, the square of an
$L^2$-martingale is a martingale. This is, however, not the case (it is, as we
shall soon find out, a submartingale), but by proper *compensation* one can
exhibit a martingale.

---

[3][...] it is not the future that is expecting me, just the next moment, the future
does not exist, it is nothing but a perpetual ongoing, a present now. [...] *The pre-
diction for my future – is the quality of my present.* Translation (from Swedish) and
italicization, by the author of this book.

As a preliminary example, let $Y_1$, $Y_2$, ... be independent random variables, such that $E Y_k = \mu_k$ and $\operatorname{Var} Y_k = \sigma_k^2$, and set $s_n^2 = \sum_{k=1}^{n} \sigma_k^2$, $n \geq 1$. Once again, $\{\mathcal{F}_n, n \geq 1\}$ are the natural $\sigma$-algebras. Finally, set

$$X_n = \left( \sum_{k=1}^{n} (Y_k - \mu_k) \right)^2 - s_n^2, \quad n \geq 1.$$

Then $\{(X_n, \mathcal{F}_n), n \geq 1\}$ is a martingale.

To see this, we first note that it is no restriction to assume that all means are 0 (otherwise, subtract them from the original summands and rename them). By exploiting the rules for conditional expectations,

$$E(X_{n+1} \mid \mathcal{F}_n) = E\left( \left( \sum_{k=1}^{n} Y_k + Y_{n+1} \right)^2 - s_{n+1}^2 \mid \mathcal{F}_n \right)$$

$$= E\left( \left( \sum_{k=1}^{n} Y_k \right)^2 \mid \mathcal{F}_n \right) + E(Y_{n+1}^2 \mid \mathcal{F}_n) + 2E\left( \left( \sum_{k=1}^{n} Y_k \right) Y_{n+1} \mid \mathcal{F}_n \right) - s_{n+1}^2$$

$$= \left( \sum_{k=1}^{n} Y_k \right)^2 + \sigma_{n+1}^2 + 2\left( \sum_{k=1}^{n} Y_k \right) E(Y_{n+1} \mid \mathcal{F}_n) - s_n^2 - \sigma_{n+1}^2$$

$$= X_n + 2\left( \sum_{k=1}^{n} Y_k \right) \cdot 0 = X_n.$$

If, in particular, $Y_1$, $Y_2$, ... are identically distributed and the mean is 0, then $\{X_n = (\sum_{k=1}^{n} Y_k)^2 - n\sigma_1^2, n \geq 1\}$, is a martingale.

*Example 3.4.* Suppose that $Y_1$, $Y_2$, ... are independent random variables with mean 1, set $X_n = \prod_{k=1}^{n} Y_k$, $n \geq 1$, (with $Y_0 = X_0 = 1$), and let $\{\mathcal{F}_n, n \geq 0\}$ be the natural $\sigma$-algebras. Then $\{(X_n, \mathcal{F}_n), n \geq 0\}$ is a martingale, because in this case

$$E(X_{n+1} \mid \mathcal{F}_n) = E(X_n \cdot Y_{n+1} \mid \mathcal{F}_n) = X_n \cdot E(Y_{n+1} \mid \mathcal{F}_n) = X_n \cdot 1 = X_n.$$

*Example 3.5. Double or nothing.* Set $X_0 = 1$ and, for $n \geq 1$, recursively,

$$X_{n+1} = \begin{cases} 2X_n, & \text{with probability } \frac{1}{2}, \\ 0, & \text{with probability } \frac{1}{2}, \end{cases}$$

or, equivalently,

$$P(X_n = 2^n) = \frac{1}{2^n}, \qquad P(X_n = 0) = 1 - \frac{1}{2^n}.$$

Since

$$X_n = \prod_{k=1}^{n} Y_k,$$

where $Y_1$, $Y_2$, ... are independent, identically distributed random variables which equal 2 or 0, both with probability $1/2$, $X_n$ equals a product of independent, identically distributed random variables with mean 1, so that $\{X_n, n \geq 0\}$ is a martingale.

*Example 3.6.* Suppose that $Y_1$, $Y_2$, ... are independent, identically distributed random variables with a finite moment generating function $\psi$, and set $S_n = \sum_{k=1}^{n} Y_k$, $n \geq 1$. Then

$$X_n = \frac{e^{tS_n}}{(\psi(t))^n} = \prod_{k=1}^{n} \frac{e^{tY_k}}{\psi(t)}, \quad n \geq 1,$$

is a martingale, frequently called *the exponential martingale*, (for $t$ inside the range of convergence of the moment generating function).

This follows from the Example 3.4, since $X_n$ is a product of $n$ independent factors with mean 1.

*Example 3.7.* If $Y_1$, $Y_2$, ... are independent random variables with common density $f$, the sequence of likelihood ratios (Subsection 2.16.4) equals

$$L_n = \prod_{k=1}^{n} \frac{f(Y_k; \theta_1)}{f(Y_k; \theta_0)}, \quad n \geq 0,$$

where $\theta_0$ and $\theta_1$ are the values of some parameter under the null- and alternative hypotheses, respectively, constitutes a martingale of the product type under the null hypothesis;

$$E\left(\frac{f(Y_k; \theta_1)}{f(Y_k; \theta_0)}\right) = \int_{-\infty}^{\infty} \frac{f(y; \theta_1)}{f(y; \theta_0)} f(y; \theta_0)\, dy = \int_{-\infty}^{\infty} f(y; \theta_1)\, dy = 1.$$

*Example 3.8.* One of the applications in Subsection 2.15.1 was the *Galton-Watson process*. Starting with one founding member, $X(0) = 1$, we found that, if

$$X(n) = \# \text{ individuals in generation } n, \quad n \geq 1,$$

and $\{Y_k, k \geq 1\}$ and $Y$ are generic random variables denoting children, then

$$X(2) = Y_1 + \cdots + Y_{X(1)},$$

and, recursively,

$$X(n+1) = Y_1 + \cdots + Y_{X(n)}.$$

Suppose now that the mean number of off-springs equals $m < \infty$, and set $\mathcal{F}_n = \sigma\{X(k), 0 \leq k \leq n\}$. It follows from the reproduction rules that the branching process is Markovian, so that

$$E\big(X(n+1) \mid \mathcal{F}_n\big) = E\big(X(n+1) \mid X(n)\big) = X(n) \cdot m,$$

which implies that, setting $X_n = \frac{X(n)}{m^n}$, $n \geq 1$, we obtain,

$$E(X_{n+1} \mid \mathcal{F}_n) = m^{-(n+1)} E\big(X(n+1) \mid X(n)\big) = m^{-(n+1)} X(n) \cdot m = X_n.$$

This shows that $\{(X_n, \mathcal{F}_n),\, n \geq 0\}$ is a martingale.

*Example 3.9.* A special kind of martingales is constructed as conditional expectations of integrable random variables. Namely, let $Z$ have finite expectation, let $\{\mathcal{F}_n,\, n \geq 0\}$ be a filtration, and set

$$X_n = E(Z \mid \mathcal{F}_n), \quad n \geq 0.$$

Then $X_n \in \mathcal{F}_n$ for all $n$,

$$E|X_n| = E|E(Z \mid \mathcal{F}_n)| \leq E\big(E(|Z| \mid \mathcal{F}_n)\big) = E|Z| < \infty, \tag{3.1}$$

in view of Proposition 1.1(h) and the smoothing lemma, and

$$E(X_{n+1} \mid \mathcal{F}_n) = E\big(E(Z \mid \mathcal{F}_{n+1}) \mid \mathcal{F}_n\big) = E(Z \mid \mathcal{F}_n) = X_n,$$

via another application of the smoothing lemma, which establishes that $\{(X_n, \mathcal{F}_n),\, n \geq 0\}$ is a martingale.

We shall later find that this class of martingales has additional pleasant features.

*Example 3.10.* Any integrable, adapted sequence can be adjusted to become a martingale. To see this, let $\{Y_n,\, n \geq 0\}$ be $\{\mathcal{F}_n\}$-adapted, set $X_0 = Y_0$ and

$$X_n = \sum_{k=1}^{n} \big(Y_k - E(Y_k \mid \mathcal{F}_{k-1})\big), \quad n \geq 1.$$

By smoothing, and the fact that $Y_k \in \mathcal{F}_n$ for $1 \leq k \leq n$,

$$E(X_{n+1} \mid \mathcal{F}_n) = E\Big(\sum_{k=1}^{n+1} \big(Y_k - E(Y_k \mid \mathcal{F}_{k-1})\big) \mid \mathcal{F}_n\Big)$$

$$= \sum_{k=1}^{n} E\big(Y_k - E(Y_k \mid \mathcal{F}_{k-1}) \mid \mathcal{F}_n\big) + E\big((Y_{n+1} - E(Y_{n+1} \mid \mathcal{F}_n)) \mid \mathcal{F}_n\big)$$

$$= \sum_{k=1}^{n} \big(Y_k - E(Y_k \mid \mathcal{F}_{k-1})\big) + E(Y_{n+1} \mid \mathcal{F}_n) - E(Y_{n+1} \mid \mathcal{F}_n)$$

$$= X_n + 0 = X_n,$$

that is, $\{(X_n, \mathcal{F}_n),\, n \geq 0\}$ is a martingale.

As a corollary we find that the partial sums of any adapted, integrable sequence can be decomposed into a martingale + the sum of the conditional expectations:

$$\sum_{k=1}^{n} Y_k = X_n + \sum_{k=1}^{n} E(Y_k \mid \mathcal{F}_{k-1}), \quad n \geq 1.$$

*Example 3.11.* If $\{Y_n, n \geq 0\}$, in addition, are independent with $EY_k = \mu_k$, then conditional expectations reduce to ordinary ones and the decomposition reduces to

$$\sum_{k=1}^{n} Y_k = \sum_{k=1}^{n} (Y_k - \mu_k) + \sum_{k=1}^{n} \mu_k, \quad n \geq 1.$$

*Example 3.12.* Let $\{(Y_n, \mathcal{F}_n), n \geq 0\}$ be a martingale with martingale difference sequence $\{U_n, n \geq 0\}$, suppose that $\{v_k, k \geq 1\}$ is a *predictable* sequence, set $X_0 = 0$, and

$$X_n = \sum_{k=1}^{n} U_k v_k, \quad n \geq 1.$$

Such a sequence $\{X_n, n \geq 0\}$ is called a *martingale transform*, and is in itself a martingale. Namely, recalling Proposition 1.3, we obtain

$$E(X_{n+1} \mid \mathcal{F}_n) = \sum_{k=1}^{n} E(U_k v_k \mid \mathcal{F}_n) + E(U_{n+1} v_{n+1} \mid \mathcal{F}_n)$$
$$= X_n + v_{n+1} E(U_{n+1} \mid \mathcal{F}_n) = X_n + v_{n+1} \cdot 0 = X_n.$$

Typical examples of predictable sequences appear in gambling or finance contexts where they might constitute strategies for future action. The strategy is then based on the current state of affairs. If, for example, $k - 1$ rounds have just been completed, then the strategy for the $k$th round is $v_k \in \mathcal{F}_{k-1}$, the money invested in that round is $U_k \in \mathcal{F}_k$.

Another situation is when $v_k = 1$ as long as some special event has not yet happened and 0 thereafter, that is, the game goes on until that special event occurs. In this case we are faced with a *stopped martingale*, a topic we shall return to in Section 10.8.

*Example 3.13.* Suppose that we are given the setup of Examples 3.1 or 3.2. If $EY_n > 0$ for all $n$, then $\{(X_n, \mathcal{F}_n), n \geq 0\}$ is a submartingale, and if $EY_n < 0$ for all $n$ we have a supermartingale.

To see this we note that $\{X_n - EX_n, n \geq 0\}$ is a martingale, so that, by adding and subtracting the sum of the expectations,

$$E(X_{n+1} \mid \mathcal{F}_n) = E\big((X_n - EX_n + (Y_{n+1} - EY_{n+1}) \mid \mathcal{F}_n\big) + EX_n + EY_{n+1}$$
$$= X_n - EX_n + 0 + EX_n + EY_{n+1} = X_n + EY_{n+1}$$
$$\begin{cases} \geq X_n, & \text{when } EY_n > 0, \\ \leq X_n, & \text{when } EY_n < 0. \end{cases}$$

In particular, in the random walk case, with $EY_1 = \mu$, we have

$$E(X_{n+1} \mid \mathcal{F}_n) = X_n + \mu,$$

and, of course, the analogous conclusion.                                  □

Something between examples and properties is what happens if one adds or subtracts martingales or submartingales, takes the largest of two, and so on.

**Proposition 3.1.** *Let* $a, b \in \mathbb{R}$, *and suppose that* $\{(X_n^{(1)}, \mathcal{F}_n), n \geq 0\}$ *and* $\{(X_n^{(2)}, \mathcal{F}_n), n \geq 0\}$ *are martingales. Then*

- $\{(aX_n^{(1)} + bX_n^{(2)}, \mathcal{F}_n), n \geq 0\}$ *is a martingale;*
- $\{(\max\{X_n^{(1)}, X_n^{(2)}\}, \mathcal{F}_n), n \geq 0\}$ *is a submartingale;*
- $\{(\min\{X_n^{(1)}, X_n^{(2)}\}, \mathcal{F}_n), n \geq 0\}$ *is a supermartingale.*

*Proof.* The first statement is a simple consequence of the linearity of conditional expectation; Proposition 1.1(b):

$$E(aX_{n+1}^{(1)} + bX_{n+1}^{(2)} \mid \mathcal{F}_n) = aE(X_{n+1}^{(1)} \mid \mathcal{F}_n) + bE(X_{n+1}^{(2)} \mid \mathcal{F}_n) = aX_n^{(1)} + bX_n^{(2)}.$$

Next, since $\max\{X_n^{(1)}, X_n^{(2)}\} \geq X_n^{(1)}$ and $\max\{X_n^{(1)}, X_n^{(2)}\} \geq X_n^{(2)}$, it follows that

$$\begin{aligned} E(\max\{X_{n+1}^{(1)}, X_{n+1}^{(2)}\} \mid \mathcal{F}_n) &\geq \max\{E(X_{n+1}^{(1)} \mid \mathcal{F}_n), E(X_{n+1}^{(2)} \mid \mathcal{F}_n)\} \\ &= \max\{X_n^{(1)}, X_n^{(2)}\}, \end{aligned}$$

which proves the second assertion.

The third statement follows similarly with the inequality sign reversed, since that minimum is smaller than each of the individual ones.    □

The analog for submartingales is a bit more delicate, since changing the sign changes the submartingale into a supermartingale, and also because the inequality involved in max and min is not allowed to conflict with the inequality in the definition of submartingales and supermartingales. We leave it to the reader to prove the following results (which, alternatively, might have been called exercises).

**Proposition 3.2.** *Let* $a, b > 0$, *and suppose that* $\{(X_n^{(1)}, \mathcal{F}_n), n \geq 0\}$ *and* $\{(X_n^{(2)}, \mathcal{F}_n), n \geq 0\}$ *are submartingales. Then* $\{(aX_n^{(1)} + bX_n^{(2)}, \mathcal{F}_n), n \geq 0\}$ *and* $\{(\max\{X_n^{(1)}, X_n^{(2)}\}, \mathcal{F}_n), n \geq 0\}$ *are submartingales.*

**Proposition 3.3.** *Let* $a, b > 0$, *and suppose that* $\{(X_n^{(1)}, \mathcal{F}_n), n \geq 0\}$ *and* $\{(X_n^{(2)}, \mathcal{F}_n), n \geq 0\}$ *are supermartingales. Then* $\{(aX_n^{(1)} + bX_n^{(2)}, \mathcal{F}_n), n \geq 0\}$ *and* $\{(\min\{X_n^{(1)}, X_n^{(2)}\}, \mathcal{F}_n), n \geq 0\}$ *are supermartingales.*

**Proposition 3.4.** *Let* $a, b \in \mathbb{R}$, *and suppose that* $\{(X_n^{(1)}, \mathcal{F}_n), n \geq 0\}$ *is a martingale, and that* $\{(X_n^{(2)}, \mathcal{F}_n), n \geq 0\}$ *is a submartingale. Then* $\{(aX_n^{(1)} + bX_n^{(2)}, \mathcal{F}_n), n \geq 0\}$ *is a submartingale for* $b > 0$, *and a supermartingale for* $b < 0$.

We have just seen that changing the sign transforms a submartingale into a supermartingale. Are there other connections of this kind?

The following result, which is a consequence of the conditional Jensen inequality, tells us (a little more than) the fact that a convex function of a martingale is a submartingale.

**Theorem 3.1.** *If* $\{(X_n, \mathcal{F}_n),\ n \geq 0\}$ *is*

(a) *a martingale and* $g$ *a convex function, or*
(b) *a submartingale and* $g$ *a non-decreasing convex function,*

*and, moreover,* $E|g(X_n)| < \infty$ *for all* $n$, *then*

$$\{(g(X_n), \mathcal{F}_n),\ n \geq 0\} \quad \text{is a submartingale.}$$

*Proof.* Suppose that $\{(X_n, \mathcal{F}_n),\ n \geq 0\}$ is a martingale. Then, by convexity,

$$E(g(X_{n+1}) \mid \mathcal{F}_n) \geq g(E(X_{n+1} \mid \mathcal{F}_n)) = g(X_n),$$

which proves the conclusion in that case.

For submartingales, the first inequality remains unchanged, but since $E(X_{n+1} \mid \mathcal{F}_n) \geq X_n$, the final equality becomes a $\geq$-inequality only if $g$ is non-decreasing. $\qquad \square$

Typical martingale examples are the functions $|x|^p$, for $p \geq 1$, $x^+$, $x^-$ and $|x|^p(\log^+ |x|)^r$ for $p, r \geq 1$. Typical submartingale examples are the functions $x^+$ and $(x^+)^p$, $p > 1$. Note that $|x|$ and $x^-$ do not work for general submartingales.

Because of their special importance we collect some such results as a separate theorem (in spite of the fact that it is a corollary).

**Theorem 3.2.** (a) *If* $\{(X_n, \mathcal{F}_n),\ n \geq 0\}$ *is a martingale, then* $\{(X_n^+, \mathcal{F}_n),\ n \geq 0\}$, $\{(X_n^-, \mathcal{F}_n),\ n \geq 0\}$, *and* $\{(|X_n|, \mathcal{F}_n),\ n \geq 0\}$ *are submartingales.*
(b) *If* $\{(X_n, \mathcal{F}_n),\ n \geq 0\}$ *is a martingale, and* $E|X_n|^p < \infty$ *for all* $n$ *and some* $p > 1$, *then* $\{(|X_n|^p, \mathcal{F}_n),\ n \geq 0\}$ *is a submartingale.*
(c) *If* $\{(X_n, \mathcal{F}_n),\ n \geq 0\}$ *is a submartingale, then so is* $\{(X_n^+, \mathcal{F}_n),\ n \geq 0\}$.
(d) *If* $\{(X_n, \mathcal{F}_n),\ n \geq 0\}$ *is a non-negative submartingale, and* $E|X_n|^p < \infty$ *for all* $n$ *and some* $p \geq 1$, *then* $\{(|X_n|^p, \mathcal{F}_n),\ n \geq 0\}$ *is a submartingale.*

*Remark 3.1.* It is not far-fetched to ask for converses. For example, is it true that every non-negative submartingale can be represented as the absolute value of a martingale? It was shown in [94] that the answer is positive. It is also mentioned there that it is *not* true that every submartingale can be represented as a convex function of a martingale. $\qquad \square$

# 4 Orthogonality

Martingales are (i.a.) generalizations of random walks, which have *independent* increments. Martingales do not, but their increments, the martingale differences, are *orthogonal* (provided second moments exist). This also implies a kind of Pythagorean relation for second moments.

**Lemma 4.1.** *Let* $\{(X_n, \mathcal{F}_n), n \geq 0\}$ *be an* $L^2$-*martingale with martingale difference sequence* $\{U_n\}$.
(a) *Then*

$$E U_n U_m = \begin{cases} E U_m^2, & \text{for } n = m, \\ 0, & \text{otherwise.} \end{cases}$$

(b) *For* $m < n$,

$$E U_n X_m = E\big(U_n E(X_n \mid \mathcal{F}_m)\big) = 0,$$
$$E X_n X_m = E\big(X_m E(X_n \mid \mathcal{F}_m)\big) = E X_m^2,$$
$$E(X_n - X_m)^2 = E X_n^2 - E X_m^2,$$
$$E\left(\sum_{k=m+1}^{n} U_k\right)^2 = \sum_{k=m+1}^{n} E U_k^2.$$

(c) *If* $\{(X_n, \mathcal{F}_n), n \geq 0\}$ *is an* $L^2$-*submartingale (supermartingale), then the same hold true with* $=$ *replaced by* $\geq$ $(\leq)$.

*Proof.* This is, once again, an exercise in smoothing, keeping Proposition 1.3 in mind.
(a): There is nothing to prove for the case $n = m$. If $m < n$,

$$E U_n U_m = E(E(U_n U_m \mid \mathcal{F}_m)) = E\big(U_m E(U_n \mid \mathcal{F}_m)\big) = E(U_m \cdot 0) = 0.$$

(b): Similarly,

$$E U_n X_m = E\big(E(U_n X_m \mid \mathcal{F}_m)\big) = E\big(X_m E(U_n \mid \mathcal{F}_m)\big) = E(X_m \cdot 0) = 0,$$
$$E X_n X_m = E\big(E(X_n X_m \mid \mathcal{F}_m)\big) = E\big(X_m E(X_n \mid \mathcal{F}_m)\big) = E X_m^2,$$

which also establishes the next relation, since

$$E(X_n - X_m)^2 = E X_n^2 - 2E X_n X_m + E X_m^2.$$

The last one is just a restatement of the third one, or, alternatively, a consequence of (i).
(c): For submartingales and supermartingales the martingale-equality is replaced by the corresponding inequality at the appropriate place.    □

*Remark 4.1.* In the martingale case, a rewriting of the third relation as

$$E X_n^2 = E X_m^2 + E(X_n - X_m)^2$$

shows that martingales have orthogonal increments, and, in addition, that $E X_n^2 \geq E X_m^2$. Note that the inequaity also is a consequence of the fact that $\{X_n^2, n \geq 1\}$ is a submartingale (Theorem 3.2).

*Remark 4.2.* In particular, if $\{U_n\}$ are independent with finite variances we rediscover the well-known fact that the variance of a sum equals the sum of the variances of the summands.    □

# 5 Decompositions

The analog of centering random walks (whose increments have positive mean) is conditional centering of submartingales.

**Theorem 5.1.** (The Doob decomposition)
*Any submartingale, $\{(X_n, \mathcal{F}_n), n \geq 0\}$, can be uniquely decomposed into the sum of a martingale, $\{(M_n, \mathcal{F}_n), n \geq 0\}$, and an increasing process, $\{(A_n, \mathcal{F}_n), n \geq 0\}$:*

$$X_n = M_n + A_n, \quad n \geq 0.$$

*Proof.* Recalling Example 3.10 we know that any adapted sequence can be adjusted to become a martingale. Using that recipe we can write

$$X_n = M_n + A_n \quad \text{for} \quad n \geq 0,$$

where $M_0 = X_0$, so that $A_0 = X_0 - M_0 = 0$, and

$$M_n = \sum_{k=1}^{n} \left( X_k - E(X_k \mid \mathcal{F}_{k-1}) \right) \quad \text{and} \quad A_n = X_n - M_n, \quad n \geq 1,$$

where, thus $\{(M_n, \mathcal{F}_n), n \geq 0\}$ is a martingale. The next step therefore is to prove that $\{(A_n, \mathcal{F}_n), n \geq 0\}$ is an increasing process.

We already know that $A_0 = 0$. Secondly, $A_n$ is predictable, since

$$A_n = \sum_{k=1}^{n} E(X_k \mid \mathcal{F}_{k-1}) - \sum_{k=1}^{n-1} X_k \in \mathcal{F}_{n-1}.$$

Finally,

$$
\begin{aligned}
A_{n+1} - A_n &= X_{n+1} - M_{n+1} - (X_n - M_n) = (X_{n+1} - X_n) - (M_{n+1} - M_n) \\
&= X_{n+1} - X_n - (X_{n+1} - E(X_{n+1} \mid \mathcal{F}_n)) \\
&= E(X_{n+1} \mid \mathcal{F}_n) - X_n \geq 0,
\end{aligned}
$$

by the submartingale property.

This establishes *existence* of a decomposition, and it remains to prove uniqueness.

Thus, suppose that $X_n = M_n' + A_n'$ is another decomposition. Then, by predictablity,

$$
\begin{aligned}
A_{n+1}' - A_n' &= E(A_{n+1}' - A_n' \mid \mathcal{F}_n) = E(\{(X_{n+1} - X_n) - (M_{n+1}' - M_n')\} \mid \mathcal{F}_n) \\
&= E(X_{n+1} \mid \mathcal{F}_n) - X_n - (M_n' - M_n') = E(X_{n+1} \mid \mathcal{F}_n) - X_n \\
&= A_{n+1} - A_n,
\end{aligned}
$$

which, together with the fact that $A_0 = A_0' = 0$, proves uniqueness of the increasing process, and therefore, since

$$M_n = X_n - A_n = X_n - A_n' = M_n',$$

also of the martingale. □

In developing theories or proving theorems it is, as we have seen a number of times, often convenient to proceed via positive "objects" and then to decompose, $x = x^+ - x^-$, in order to treat the general case. For proving martingale convergence the next decomposition does that for us.

**Theorem 5.2.** (The Krickeberg decomposition)
(i)    For any martingale, $\{(X_n, \mathcal{F}_n), n \geq 0\}$, such that

$$\sup_n E(X_n^+) < \infty,$$

there exist non-negative martingales $\{(M_n^{(i)}, \mathcal{F}_n), n \geq 0\}$, $i = 1, 2$, such that

$$X_n = M_n^{(1)} - M_n^{(2)}.$$

(ii)    For any submartingale $\{(X_n, \mathcal{F}_n), n \geq 0\}$, such that

$$\sup_n E(X_n^+) < \infty,$$

there exist non-negative martingales $\{(M_n^{(i)}, \mathcal{F}_n), n \geq 0\}$, $i = 1, 2$, and an increasing process $\{(A_n, \mathcal{F}_n), n \geq 0\}$, such that

$$X_n = M_n^{(1)} - M_n^{(2)} + A_n.$$

*Proof.* We begin by observing that (ii) follows from (i) with the aid of the Doob decomposition, so we only have to prove the statement for martingales.

The immediate idea is (of course?) the decomposition $X_n = X_n^+ - X_n^-$. However, in view of Theorem 3.1, this decomposes the martingale into the difference of non-negative *sub*martingales.

If the martingale is as given in Example 3.9, that is, of the form $X_n = E(Z \mid \mathcal{F}_n)$ for some integrable random variable $Z$, then

$$X_n = E(Z^+ \mid \mathcal{F}_n) - E(Z^- \mid \mathcal{F}_n), \quad n \geq 0$$

does the job.

In the general case one makes a similar attack, but the details are a bit more sophisticated. Namely, set,

$$Y_{m,n} = E(X_m^+ \mid \mathcal{F}_n) \quad \text{for} \quad m \geq n,$$

and note that $\{(Y_{m,n}, \mathcal{F}_n), 0 \leq n \leq m\}$ is a submartingale for every fixed $m$. The idea is to let $m \to \infty$, thereby obtaining one of the non-negative martingales, (with $+$ replaced by $-$ for the other one).

Since, by Theorem 3.1, $\{(X_n^+, \mathcal{F}_n), n \geq 0\}$ is a submartingale, it follows that $E(X_{m+1}^+ \mid \mathcal{F}_m) \geq X_m^+$, so that, by smoothing,

$$Y_{m+1,n} = E\big(E(X_{m+1}^+ \mid \mathcal{F}_m) \mid \mathcal{F}_n\big) \geq E(X_m^+ \mid \mathcal{F}_n) = Y_{m,n}.$$

Consequently, $Y_{m,n} \uparrow M_n^{(1)}$, say, almost surely as $m \to \infty$ for fixed $n$, where $M_n^{(1)}$ is non-negative and $\mathcal{F}_n$-measurable, since $m \geq n$ is arbitrary..

Moreover, by monotonicity and the $L^1$-boundedness of $X_n^+$,

$$
\begin{aligned}
E\, M_n^{(1)} &= E \lim_{m\to\infty} Y_{m,n} = \lim_{m\to\infty} E(E(X_m^+ \mid \mathcal{F}_n)) \\
&= \lim_{m\to\infty} E\, X_m^+ \leq \sup_m E(X_m^+) < \infty,
\end{aligned}
$$

that is, $M_n^{(1)}$ is integrable. Finally, to verify the martingale property, we exploit monotonicity and smoothing once more:

$$
\begin{aligned}
E(M_{n+1}^{(1)} \mid \mathcal{F}_n) &= E(\lim_{m\to\infty} Y_{m,n+1} \mid \mathcal{F}_n) = \lim_{m\to\infty} E\big(E(X_m^+ \mid \mathcal{F}_{n+1}) \mid \mathcal{F}_n\big) \\
&= \lim_{m\to\infty} E(X_m^+ \mid \mathcal{F}_n) = M_n^{(1)}. \qquad\qquad \square
\end{aligned}
$$

*Remark 5.1.* The Krickeberg decomposition is not unique. This follows from the fact, to be proven later, that $X_n \overset{a.s.}{\to} X_\infty$ as $n \to \infty$, and that therefore

$$
X_n = \big(M_n^{(1)} + E(|X_\infty| \mid \mathcal{F}_n)\big) - \big(M_n^{(2)} + E(|X_\infty| \mid \mathcal{F}_n)\big)
$$

is another decomposition of the desired kind, since the sum of two non-negative martingales is, again a non-negative martingale. This complication will, however, never be a problem in our discussion. $\qquad\qquad \square$

To prove the next decomposition we need the convergence theorem. Since the proof to a large extent is based on the Doob decomposition, and the fact that changing sign in a supermartingale transforms it into a submartingale, we postpone the proof to Problem 17.15. As for the definition of a potential, recall Definition 2.5.

**Theorem 5.3.** (The Riesz decomposition)
*Any supermartingale, $\{(X_n, \mathcal{F}_n),\, n \geq 0\}$, such that*

$$
\inf_n E(X_n) > -\infty,
$$

*can be uniquely decomposed into a martingale, $\{(M_n, \mathcal{F}_n),\, n \geq 0\}$, and a potential, $\{(Z_n, \mathcal{F}_n),\, n \geq 0\}$;*

$$
X_n = M_n + Z_n.
$$

# 6 Stopping Times

Several inequalities in Chapter 3 concerning maximal partial sums were proved by splitting events, such as the maximum being larger than some number, into disjoint "slices" that kept track of when the first passage across that number occurred. At the end of Section 3.7 we found that life would be made easier by introducing a *random index* that kept track of exactly when this first passage occurred. Such indices are examples of the more general concept of *stopping times*, which turn out to have several attractive properties in the martingale context. For example, many martingales can be suitably "stopped" in such a way that the martingale property remains.

Formally we allow stopping times to assume the value $+\infty$ with positive probability, which, vaguely speaking, corresponds to the situation that the kind of stopping alluded to above never occurs.

Set $\mathbb{N} = \{1, 2, \ldots\}$ and $\bar{\mathbb{N}} = \{1, 2, \ldots, \infty\}$.

**Definition 6.1.** *A positive, integer valued, possibly infinite, random variable $\tau$ is called a* stopping time *(with respect to $\{\mathcal{F}_n, n \geq 1\}$) if*

$$\{\tau = n\} \in \mathcal{F}_n \quad \text{for all} \quad n \in \mathbb{N}.$$   □

**Lemma 6.1.** *A positive integer valued, possibly infinite, random variable $\tau$ is a stopping time iff one of the following holds:*

$$\{\tau \leq n\} \in \mathcal{F}_n \quad \text{for all} \quad n \in \mathbb{N},$$
$$\{\tau > n\} \in \mathcal{F}_n \quad \text{for all} \quad n \in \mathbb{N}.$$

*Proof.* Immediate, from the relations

$$\{\tau \leq n\} = \bigcup_{k=1}^{n} \{\tau = k\},$$
$$\{\tau = n\} = \{\tau \leq n\} \setminus \{\tau \leq n-1\},$$
$$\{\tau > n\} = (\{\tau \leq n\})^c.$$   □

The important feature is that stopping times are measurable with respect to "what has happened so far", and, hence, do not depend on the future.

Typical stoppings time (as hinted at in the introductory paragraph) are *first entrance times*, such as the first time a random walk reaches a certain level, the first time a simple, symmetric random walk returns to 0, and so on. Such questions can be answered by looking at what has happened until "now".

Typical random indices that are *not* stopping times are *last exit times*, for example, the last time a simple, symmetric random walk returns to 0. Such questions cannot be answered without knowledge about the future.

The intimate relation between stopping times and time suggests the introduction of a $\sigma$-algebra that contains all information prior to $\tau$, just as $\mathcal{F}_n$ contains all information prior to $n$.

**Definition 6.2.** *The* pre-$\tau$-$\sigma$-algebra $\mathcal{F}_\tau$ *of a stopping time $\tau$ is defined as*

$$\mathcal{F}_\tau = \{A \in \mathcal{F}_\infty : A \cap \{\tau = n\} \in \mathcal{F}_n \quad \text{for all} \quad n\}.$$

Given its name, one would expect that the pre-$\tau$-$\sigma$-algebra is, indeed, a $\sigma$-algebra.

**Proposition 6.1.** *$\mathcal{F}_\tau$ is a $\sigma$-algebra.*

*Proof.* We have to show that $\Omega$, complements, and countable unions of sets in $\mathcal{F}_\tau$ belong to $\mathcal{F}_\tau$.

Clearly $\Omega \in \mathcal{F}_\infty$ and $\Omega \cap \{\tau = n\} = \{\tau = n\} \in \mathcal{F}_n$ for all $n$.

Secondly, suppose that $A \in \mathcal{F}_\tau$. Then, for all $n$,

$$A^c \cap \{\tau = n\} = \{\Omega \setminus A\} \cap \{\tau = n\} = \{\Omega \cap \{\tau = n\}\} \setminus \{A \cap \{\tau = n\}\}$$
$$= \Omega \setminus \{A \cap \{\tau = n\}\} \in \mathcal{F}_n.$$

Finally, if $\{A_k, \ k \geq 1\}$ all belong to $\mathcal{F}_\tau$, then, for all $n$,

$$\left\{\bigcup_k A_k\right\} \cap \{\tau = n\} = \bigcup_k \{A_k \cap \{\tau = n\}\} \in \mathcal{F}_n. \qquad \square$$

**Exercise 6.1.** In the literature one can also find the following definition of the pre-$\tau$-$\sigma$-algebra, namely

$$\mathcal{F}_\tau = \left\{A \in \mathcal{F}_\infty : A = \bigcup_n \{A_n \cap \{\tau \leq n\}\}\right\} \quad \text{where} \quad A_n \in \mathcal{F}_n \quad \text{for all} \quad n.$$

Prove that this definition is equivalent Definition 6.2. $\qquad \square$

Next, some propositions with general facts about stopping times.

**Proposition 6.2.** *Following are some basic facts:*

(a)  *Every positive integer is a stopping time.*
(b)  *If $\tau \equiv k$, then $\mathcal{F}_\tau = \mathcal{F}_k$.*
(c)  $\mathcal{F}_\tau \subset \mathcal{F}_\infty$.
(d)  $\tau \in \mathcal{F}_\tau$.
(e)  $\tau \in \mathcal{F}_\infty$.
(f)  $\{\tau = +\infty\} \in \mathcal{F}_\infty$.

*Proof.* Suppose that $\tau \equiv k$. Then

$$\{\tau = n\} = \begin{cases} \Omega, & \text{when} \quad n = k, \\ \emptyset, & \text{otherwise.} \end{cases}$$

This proves the (a), since $\Omega$ and $\emptyset$ both belong to $\mathcal{F}_n$ for all $n$. As for (b), let $A \in \mathcal{F}_\infty$. Then

$$A \cap \{\tau = n\} = \begin{cases} A, & \text{when} \quad n = k, \\ \emptyset, & \text{otherwise,} \end{cases}$$

and the conclusion follows.

Since $\mathcal{F}_\tau$ is defined as sets in $\mathcal{F}_\infty$ with an additional property it is obvious that $\mathcal{F}_\tau \subset \mathcal{F}_\infty$, which verifies claim (c), and since, for all $m \in \mathbb{N}$,

$$\{\tau = m\} \cap \{\tau = n\} = \begin{cases} \{\tau = n\}, & \text{when } m = n, \\ \emptyset, & \text{otherwise}, \end{cases} \in \mathcal{F}_n \text{ for all } n \in \mathbb{N},$$

it follows that $\{\tau = m\} \in \mathcal{F}_\tau$ for all $m$, which means that $\tau \in \mathcal{F}_\tau$, so that (d) holds true.

Statement (e) follows by joining (c) and (d). Finally, (f) follows from the observation that

$$\{\tau = +\infty\} = \left( \bigcup_n \{\tau = n\} \right)^c \in \mathcal{F}_\infty. \qquad \square$$

The next proposition concerns relations between stopping times.

**Proposition 6.3.** *Suppose that $\tau_1$ and $\tau_2$ are stopping times. Then*

(a)   $\tau_1 + \tau_2$, $\min\{\tau_1, \tau_2\}$, *and* $\max\{\tau_1, \tau_2\}$ *are stopping times;*
(b)   $\tau_M = \min\{\tau, M\}$ *is a bounded stopping time for any $M \geq 1$;*
(c)   *what about $\tau_1 - \tau_2$?*
(d)   *if $\{\tau_k, k \geq 1\}$ are stopping times, then so are $\sum_k \tau_k$, $\min_k\{\tau_k\}$, and $\max_k\{\tau_k\}$;*
(e)   *if $\tau_n$ are stopping times and $\tau_n \downarrow \tau$, then $\tau$ is a stopping time;*
(f)   *if $\tau_n$ are stopping times and $\tau_n \uparrow \tau$, then $\tau$ is a stopping time;*
(g)   *if $A \in \mathcal{F}_{\tau_1}$, then $A \cap \{\tau_1 \leq \tau_2\} \in \mathcal{F}_{\tau_2}$;*
(h)   *if $\tau_1 \leq \tau_2$, then $\mathcal{F}_{\tau_1} \subset \mathcal{F}_{\tau_2}$.*

*Proof.* For all $n$,

$$\{\tau_1 + \tau_2 = n\} = \bigcup_{k=0}^{n} \left\{ \{\tau_1 = k\} \cap \{\tau_2 = n - k\} \right\} \in \mathcal{F}_n,$$

$$\{\min\{\tau_1, \tau_2\} > n\} = \{\tau_1 > n\} \cap \{\tau_2 > n\} \in \mathcal{F}_n,$$

$$\{\max\{\tau_1, \tau_2\} \leq n\} = \{\tau_1 \leq n\} \cap \{\tau_2 \leq n\} \in \mathcal{F}_n,$$

which proves (a) – recall Lemma 6.1. And (b), since $M$ is a stopping time and, obviously, $\tau_M \leq M$. However, the difference between two stopping times may be negative, so the answer to that one is negative. Note also that if the difference equals $n$ then one them must be *larger* than $n$ which violates the measurability condition. (d) follows by induction from (a), and (e) and (f) by monotonicity; $\{\tau = n\} = \cap_k\{\tau_k = n\} \in \mathcal{F}_n$ and $\{\tau = n\} = \cup_k\{\tau_k = n\} \in \mathcal{F}_n$, respectively.

In order to prove (g), we first note that $A \in \mathcal{F}_\infty$. Now, for all $n$,

$$A \cap \{\tau_1 \leq \tau_2\} \cap \{\tau_2 = n\} = \{A \cap \{\tau_1 \leq \tau_2\} \cap \{\tau_1 \leq n\} \cap \{\tau_2 = n\}\}$$

$$\bigcup \{A \cap \{\tau_1 \leq \tau_2\} \cap \{\tau_1 > n\} \cap \{\tau_2 = n\}\}$$

$$= \{A \cap \{\tau_1 \leq n\}\} \cap \{\tau_2 = n\} \in \mathcal{F}_n. \qquad (6.1)$$

This proves that $A \cap \{\tau_1 \leq \tau_2\} \in \mathcal{F}_{\tau_2}$, which establishes (g).

As for (h), the left-most member in (6.1) equals $A \cap \{\tau_2 = n\}$ by assumption, so that, in this case, $A \in \mathcal{F}_{\tau_2}$. This proves that every $A \in \mathcal{F}_\infty$ that belongs to $\mathcal{F}_{\tau_1}$ also belongs to $\mathcal{F}_{\tau_2}$, and, hence, that $\mathcal{F}_{\tau_1} \subset \mathcal{F}_{\tau_2}$.     □

*Remark 6.1.* If $A \in \mathcal{F}_{\tau_2}$, then an analogous argument shows that

$$A \cap \{\tau_1 = n\} = \left(A \cap \{\tau_2 \geq n\}\right) \cap \{\tau_1 = n\},$$

from which it, similarly, would follow that $\mathcal{F}_{\tau_2} \subset \mathcal{F}_{\tau_1}$, and, by extension, that all $\sigma$-algebras would be equal. The reason the argument breaks down is that $A \cap \{\tau_2 \geq n\}$ in general does *not* belong to $\mathcal{F}_{\tau_2}$.     □

Here is another rather natural hypothesis: Martingales have constant expectation, but what about $EX_\tau$ for stopping times $\tau$? And, is it true that $E(X_{\tau_2} \mid \mathcal{F}_{\tau_1}) = X_{\tau_1}$ if $\tau_1 \leq \tau_2$ are stopping times? Our next task is to provide answers to these questions.

# 7 Doob's Optional Sampling Theorem

One question thus concerns whether or not a stopped martingale (submartingale) remains a martingale (submartingale).

The answer is not always positive.

*Example 7.1.* Let $Y, Y_1, Y_2, \ldots$ be independent coin-tossing random variables, that is $P(Y = 1) = P(Y = -1) = 1/2$, set $X_n = \sum_{k=1}^n Y_k$, $n \geq 1$, and let

$$\tau = \min\{n : X_n = 1\}.$$

Since $X_1, X_2, \ldots$ constitutes a centered random walk, we know that it is a martingale (with respect to the natural filtration); in particular, $EX_n = 0$ for all $n$. However, since $X_\tau = 1$ almost surely it follows, in particular, that $EX_\tau = 1 \neq 0$, so that $X_\tau$ cannot be a member of the martingale; $\{X_1, X_\tau\}$ is *not* a martingale. The problem is that $\tau$ is "too large"; in fact, as we shall see later, $E\tau = +\infty$.     □

However, the answer is always positive for a restricted class of martingales, namely those from Example 3.9, and bounded stopping times.

**Theorem 7.1.** (Doob's optional sampling theorem)
*Let $Z \in L^1$, suppose that $\{(X_n, \mathcal{F}_n), n \geq 0\}$ is a martingale of the form $X_n = E(Z \mid \mathcal{F}_n)$, $n \geq 0$, and that $\tau$ is a stopping time. Then $\{(X_\tau, \mathcal{F}_\tau), (Z, \mathcal{F}_\infty)\}$ is a martingale, in particular,*

$$EX_\tau = EZ.$$

*Proof.* That $\{X_n\}$ is indeed a martingale has already been established in Example 3.9. It therefore remains to show that the pair $X_\tau, Z$ satisfies the defining relation

$$\int_A X_\tau \, dP = \int_A Z \, dP. \tag{7.1}$$

To prove this we apply Proposition 1.3 and smoothing: Let $\Lambda \in \mathcal{F}_\tau$. Then, since $I\{\Lambda \cap \{\tau = n\}\} \in \mathcal{F}_n$ for all $n$, we obtain

$$
\begin{aligned}
E\,X_\tau I\{\Lambda\} &= \sum_n E\big(X_\tau I\{\Lambda \cap \{\tau = n\}\}\big) = \sum_n E\big(X_n I\{\Lambda \cap \{\tau = n\}\}\big) \\
&= \sum_n E\big(E(Z \mid \mathcal{F}_n) I\{\Lambda \cap \{\tau = n\}\}\big) \\
&= \sum_n E\big(E(ZI\{\Lambda \cap \{\tau = n\}\} \mid \mathcal{F}_n)\big) \\
&= \sum_n E\big(ZI\{\Lambda \cap \{\tau = n\}\}\big) = E\,ZI\{\Lambda\}.
\end{aligned}
$$

The conclusion now follows upon observing that the equality between the extreme members, in fact, is a rewriting of the defining relation (7.1).    □

**Corollary 7.1.** *Suppose that $\{(X_n, \mathcal{F}_n),\ n \geq 0\}$ is a martingale, and that $\tau$ is a* bounded *stopping time; $P(\tau \leq M) = 1$. Then $\{X_\tau, X_M\}$ is a martingale, and*

$$
E\,X_\tau = E\,X_M.
$$

*Proof.* The conclusion follows from Theorem 7.1 with $Z = X_M$, since

$$
X_n = E(X_M \mid \mathcal{F}_n) \quad \text{for} \quad 0 \leq n \leq M,
$$

and $\tau$ is bounded by $M$.    □

A particularly useful application is the following one.

**Corollary 7.2.** *Suppose that $\{X_n,\ n \geq 0\}$ is a martingale, and that $\tau$ is a stopping time. Then $\{X_{\tau \wedge n}, X_n\}$ is a martingale, and*

$$
E\,X_{\tau \wedge n} = E\,X_n.
$$

*Proof.* Immediate, since $\tau \wedge n$ is bounded by $n$.    □

**Exercise 7.1.** Review Example 7.1; compare $E\,X_\tau$ and $E\,X_{\tau \wedge n}$.    □

An early fact was that martingales have constant expectation. In Corollary 7.1 we have even seen that martingales evaluated at *bounded* stopping times have constant expectation. It turns out that this, in fact, characterizes martingales.

**Theorem 7.2.** *Suppose that $X_1,\ X_2,\ \ldots$ is an $\{\mathcal{F}_n\}$-adapted sequence. Then $\{(X_n, \mathcal{F}_n),\ n \geq 0\}$ is a martingale if and only if*

$$
E\,X_\tau = constant \quad for\ all\ bounded\ stopping\ times \quad \tau. \tag{7.2}
$$

*Proof.* The necessity has just been demonstrated. In order to prove the sufficiency, suppose that (7.2) holds, let $A$ be $\mathcal{F}_m$-measurable, and set

$$\tau(\omega) = \begin{cases} n & \text{if } \omega \in A, \\ m & \text{if } \omega \in A^c, \end{cases} \quad \text{for } 0 \le m < n.$$

Then $E X_\tau = E X_m$, which means that

$$\int_A X_n \, dP + \int_{A^c} X_m \, dP = \int_A X_m \, dP + \int_{A^c} X_m \, dP,$$

which simplifies into

$$\int_A X_n \, dP = \int_A X_m \, dP. \qquad \square$$

Theorem 7.1 can be extended, in a straightforward manner, to cover a
*sequence* of non-decreasing stopping times as follows.

**Theorem 7.3.** *Let $Z \in L^1$, and suppose that $\{(X_n, \mathcal{F}_n), n \ge 0\}$ is a martingale of the form $X_n = E(Z \mid \mathcal{F}_n)$, $n \ge 0$. If $\tau_1 \le \tau_2 \le \ldots \le \tau_k$ are stopping
times, then $\{X_0, X_{\tau_1}, X_{\tau_2}, \ldots, X_{\tau_k}, Z\}$ is a martingale, and*

$$E X_0 = E X_{\tau_1} = E X_{\tau_2} = \cdots = E X_{\tau_k} = E Z.$$

**Exercise 7.2.** Prove Theorem 7.3. $\qquad \square$

An inspection of the proof of Theorem 7.1 (and the corollaries) shows that we
can make the following statements for submartingales.

**Theorem 7.4.** *Suppose that $\{(X_n, \mathcal{F}_n), n \ge 0\}$ is a submartingale, and that
$\tau$ is a bounded stopping time; $P(\tau \le M) = 1$. Then $\{X_\tau, X_M\}$ is a submartingale, and*

$$E X_\tau \le E X_M.$$

*In particular, if $\tau$ is a stopping time, then $\{X_{\tau \wedge n}, X_n\}$ is a submartingale,
and*

$$E X_{\tau \wedge n} \le X_n.$$

*Proof.* The conclusion follows by replacing the third equality in the proof
of the optional sampling theorem with an inequality ($\le$), and the fact that
submartingales have non-decreasing expectations. $\qquad \square$

**Exercise 7.3.** Check the details. $\qquad \square$

# 8 Joining and Stopping Martingales

In this section we shall show how one may join (super)martingales at stopping
times in such a way that the new sequence remains a (super)martingale, and
that a stopped martingale preserves the martingale property in a sense made
precise below.

Our first result, however, is more suited for supermartingales.

**Theorem 8.1.** (i) *Suppose that* $\{(X_n^{(i)}, \mathcal{F}_n), n \geq 0\}$, $i = 1, 2$, *are supermartingales, and that* $\tau$ *is a stopping time such that*

$$X_\tau^{(1)}(\omega) \geq X_\tau^{(2)}(\omega) \quad \text{on the set} \quad \{\tau < \infty\}. \tag{8.1}$$

*Set*

$$X_n(\omega) = \begin{cases} X_n^{(1)}(\omega) & \text{for } n < \tau(\omega), \\ X_n^{(2)}(\omega) & \text{for } n \geq \tau(\omega). \end{cases}$$

*Then* $\{(X_n, \mathcal{F}_n), n \geq 0\}$ *is a supermartingale.*

(ii) *If* $\{(X_n^{(i)}, \mathcal{F}_n), n \geq 0\}$, $i = 1, 2$, *are martingales and equality holds in (8.1), then* $\{(X_n, \mathcal{F}_n), n \geq 0\}$ *is a martingale, and if* $\{(X_n^{(i)}, \mathcal{F}_n), n \geq 0\}$, $i = 1, 2$, *are submartingales and the inequality is reversed in (8.1), then* $\{(X_n, \mathcal{F}_n), n \geq 0\}$ *is a submartingale.*

(iii) *If* $\{(X_n^{(i)}, \mathcal{F}_n), n \geq 0\}$, $i = 1, 2$, *are martingales, and (8.1) holds, then* $\{(X_n, \mathcal{F}_n), n \geq 0\}$ *is a supermartingale.*

*Proof.* (i): By construction,

$$X_n = X_n^{(1)} I\{n < \tau\} + X_n^{(2)} I\{n \geq \tau\},$$

which, first of all, shows that both terms and, hence, $X_n$ is $\mathcal{F}_n$-measurable.

For the transition from $n$ to $n+1$ we single out the event $\{\tau = n+1\}$ and apply (8.1) to obtain

$$X_{n+1}^{(1)} I\{n < \tau\} + X_{n+1}^{(2)} I\{n \geq \tau\}$$
$$= X_{n+1}^{(1)} I\{n + 1 < \tau\} + X_{n+1}^{(1)} I\{n + 1 = \tau\} + X_{n+1}^{(2)} I\{n \geq \tau\}$$
$$\geq X_{n+1}^{(1)} I\{n + 1 < \tau\} + X_{n+1}^{(2)} I\{n + 1 = \tau\} + X_{n+1}^{(2)} I\{n \geq \tau\}$$
$$= X_{n+1}^{(1)} I\{n + 1 < \tau\} + X_{n+1}^{(2)} I\{n + 1 \geq \tau\} = X_{n+1}.$$

Proposition 1.3 and the supermartingale property, $X_n^{(i)} \geq E(X_{n+1}^{(i)} \mid \mathcal{F}_n)$, $i = 1, 2$, now imply that

$$X_n = X_n^{(1)} I\{n < \tau\} + X_n^{(2)} I\{n \geq \tau\}$$
$$\geq I\{n < \tau\} E(X_{n+1}^{(1)} \mid \mathcal{F}_n) + I\{n \geq \tau\} E(X_{n+1}^{(2)} \mid \mathcal{F}_n)$$
$$= E(X_{n+1}^{(1)} I\{n < \tau\} \mid \mathcal{F}_n) + E(X_{n+1}^{(2)} I\{n \geq \tau\} \mid \mathcal{F}_n)$$
$$= E(X_{n+1}^{(1)} I\{n < \tau\} + X_{n+1}^{(2)} I\{n \geq \tau\} \mid \mathcal{F}_n)$$
$$\geq E(X_{n+1} \mid \mathcal{F}_n),$$

which establishes the supermartingale property.

Conclusion (ii) follows similarly by replacing certain inequalities by equalities or by reversed inequalities; recall that switching sign in a supermartingale transforms it into a submartingale, and (iii) is immediate from (i).    □

**Exercise 8.1.** Check the modifications needed for the proofs of (ii) and (iii).    □

A somewhat similar approach leads to the following result.

**Theorem 8.2.** *If* $\{(X_n, \mathcal{F}_n), n \geq 0\}$ *is a (sub)martingale and* $\tau$ *a stopping time, then* $\{(X_{\tau \wedge n}, \mathcal{F}_n), n \geq 0\}$ *is a (sub)martingale.*

*Proof.* Once again we begin by asserting measurability;

$$X_{\tau \wedge n} = X_\tau I\{\tau < n\} + X_n I\{\tau \geq n\} = \sum_{k=1}^{n-1} X_k I\{\tau = k\} + X_n I\{\tau \geq n\} \in \mathcal{F}_n,$$

since each term belongs to $\mathcal{F}_n$.

Now, in the submartingale case,

$$E(X_{\tau \wedge (n+1)} \mid \mathcal{F}_n) = \sum_{k=1}^{n} E(X_k I\{\tau = k\} \mid \mathcal{F}_n) + E(X_{n+1} I\{\tau \geq n+1\} \mid \mathcal{F}_n)$$

$$= \sum_{k=1}^{n} X_k I\{\tau = k\} + I\{\tau > n\} E(X_{n+1} \mid \mathcal{F}_n)$$

$$\geq \sum_{k=1}^{n} X_k I\{\tau = k\} + I\{\tau > n\} X_n$$

$$= \sum_{k=1}^{n-1} X_k I\{\tau = k\} + X_n I\{\tau = n\} + X_n I\{\tau > n\}$$

$$= \sum_{k=1}^{n-1} X_k I\{\tau = k\} + X_n I\{\tau \geq n\} = X_{\tau \wedge n}.$$

As before there is equality instead of inequality in the martingale case.    □

We call the martingales (submartingales, supermartingales) of the kind that have been constructed in Theorems 8.1 and 8.2 *stopped martingales (submartingales, supermartingales)*. The pattern examples, which were hinted at in the introductory lines, are first exit times, first entrance times, first hitting times.

As an illustration, let $\{(X_n, \mathcal{F}_n), n \geq 0\}$ be a supermartingale, and define the stopping time

$$\tau = \min\{n : X_n > t\}, \quad t > 0.$$

In order to apply Theorem 8.1 we let $\{X_n^{(1)}\}$ be the original supermartingale, and $X_n^{(2)} = t$ for all $n$. The new supermartingale is obtained by letting it be equal to the original one before passage has occurred and equal the second one after passage, more precisely,

$$X_0, X_1, X_2, \ldots, t, t, t, t, \ldots.$$

Then (8.1) is satisfied, since $X_\tau^{(1)} \geq t = X_\tau^{(2)}$, so, by Theorem 8.1, the new sequence is also a supermartingale.

The sequence $\{X_{\tau \wedge n}, n \geq 0\}$ in Theorem 8.2 is

$$X_0, X_1, X_2, \ldots, X_\tau, X_\tau, X_\tau, X_\tau, \ldots,$$

whose elements are the original ones as long as $\tau \leq n$, after which they are equal to $X_\tau$.

In both cases the original stretch varies from $\omega$ to $\omega$. The joined martingale and $\{X_{\tau \wedge n}, n \geq 0\}$ coincide up to (but not including) the hitting time, but they take different paths thereafter. Namely, the joined martingale stays at the *constant* level $t$ after passage, whereas the other one stays at the *random* level $X_\tau$ where stopping occurred, a level, which thus varies from realization to realization.

Typical stopped martingales, submartingales and supermartingales are *stopped random walks*. Suppose that $Y_1, Y_2, \ldots$ are independent, identically distributed random variables, and set $X_n = \sum_{k=1}^n Y_k$ as in Examples 3.2 or 3.13. Then

$$\tau = \min\{n : X_n \in A\}, \quad \text{for some} \quad A \subset \mathbb{R},$$

is standard example, typically when $A$ is some interval, such as $(a, \infty)$ for some $a > 0$ or, as in sequential analysis, $(b, a)$, where $b < 0 < a$. In the former case the corresponding supermartingale is a random walk with negative drift (recall Example 3.13), which implies that the probability of ever hitting a positive level is less than one; $P(\tau < \infty) < 1$. This shows that the phrase "on the set $\{\tau < \infty\}$" is important in Theorem 8.1. Since martingales are also supermartingales, random walks with mean 0 also fit the theorem. In this case, $P(\tau < \infty) = 1$. If, instead, we would consider submartingales, then $P(\tau < \infty) = 1$, but then the second submartingale, which is identical to $a$ would be smaller than the first one and thus violate the reversed inequality (8.1).

In the special case of a simple symmetric random walk, that is when the steps are equal to $+1$ and $-1$ with probability $1/2$ each, the random walk is a martingale (recall Example 3.2), with the special property that if the level $a$ is an integer, then $X_\tau = a$ exactly. Thus, if we let $X_n$ be the random walk until the hitting of the integer level $a$ and equal to $a$ ever after, then $X_\tau^{(2)} = a = X_\tau^{(1)}$, so that Theorem 8.1 tells us that $\{(X_n, \mathcal{F}_n), n \geq 0\}$ is a martingale, the realizations of which are

$$X_0, X_1, X_2, \ldots, a, a, a, a, a, \ldots.$$

The realizations of the martingale $\{(X_{\tau \wedge n}, \mathcal{F}_n), n \geq 0\}$ in Theorem 8.2 are the same in this case.

At first sight it might be confusing that this is a martingale, since $EX_0 = 0$ and $E\,a = a$, which would violate the martingale property. The reason for this being in order is that the indices of the $a$'s are random. For any $X_k$ with $k$ fixed we have $E\,X_k = 0$. This is no contradiction, since any such $X_k$ assumes its regular value along the random walk and equals $a$ once the level has been hit, which, in turn, if at all, occurs after a random number of steps.

In this rather simple case, the number of $X$'s before the $a$ run begins is negative binomial – how many times does one have to flip a symmetric coin in order to obtain $a$ more heads than tails? In particular, this shows that $P(\tau < \infty) = 1$.

We shall return to the theory of stopped random walks in Section 10.14.

In the last two sections we have discussed stopped martingales, departing from the question of whether a stopped (sub)martingale remains a (sub)martingale or not, and Doob's optional sampling theorem. There is more to be said about this, and we shall briefly return to the matter in Section 10.15 with some facts about *regular* martingales and *regular* stopping times.

# 9 Martingale Inequalities

Our first result is one that appears under different names in the literature; it seems that the first inequality in the chain of inequalities below is due to Doob and the fact that the left-most one is smaller than the right-most one is due to Kolmogorov. As we shall see after the theorem, in the special case when the martingale is a sum of independent centered random variables, the inequality between the extreme members reduces to the Kolmogorov inequality for sums of independent random variables, Theorem 3.1.6.

**Theorem 9.1.** (The Kolmogorov-Doob inequality)
*Let $\lambda > 0$.*
(i) *Suppose that $\{(X_n, \mathcal{F}_n), n \geq 0\}$ is a submartingale. Then*

$$\lambda P(\max_{0 \leq k \leq n} X_k > \lambda) \leq \int_{\{\max_{0 \leq k \leq n} X_k > \lambda\}} X_n \, dP \leq E\,X_n^+ \leq E|X_n|.$$

(ii) *Suppose that $\{(X_n, \mathcal{F}_n), n \geq 0\}$ is a martingale. Then*

$$\lambda P(\max_{0 \leq k \leq n} |X_k| > \lambda) \leq \int_{\{\max_{0 \leq k \leq n} |X_k| > \lambda\}} |X_n| \, dP \leq E|X_n|.$$

*Proof.* Set

$$\tau = \min\{k : X_k > \lambda\} \quad \text{and} \quad \Lambda = \{\max_{0 \leq k \leq n} X_k > \lambda\} = \{X_{\tau \wedge n} > \lambda\}.$$

Since, by Theorem 7.4, the (sub)martingale property is preserved for the pair $\{X_{\tau \wedge n}, X_n\}$ and, moreover, $\Lambda \in \mathcal{F}_{\tau \wedge n}$, we obtain

$$\lambda P(\Lambda) \le \int_\Lambda X_{\tau \wedge n} \, dP \le \int_\Lambda X_n \, dP \le E X_n^+ \le E|X_n|.$$

This proves (i), from which (ii) follows, since $\{|X_n|, \mathcal{F}_n\}$ is a submartingale (Theorem 3.2).  □

*Example 9.1.* As a particular example, suppose that $Y_1, Y_2, \ldots, Y_n$ are independent random variables with mean 0 and finite variances, and set $X_k = \sum_{j=1}^k Y_j$, $1 \le k \le n$. Then $\{X_k, 1 \le k \le n\}$ is a martingale and, thus, by Theorem 3.2, $\{X_k^2, 1 \le k \le n\}$ a submartingale. An application of Theorem 9.1 therefore yields

$$\lambda P(\max_{0 \le k \le n} X_k^2 > \lambda) \le E(X_n^2),$$

which is the same as

$$P(\max_{0 \le k \le n} |X_k| > \lambda) \le \frac{E(X_n^2)}{\lambda^2},$$

which coincides with the Kolmogorov inequality, Theorem 3.1.6. We also notice that the crucial step there was to prove that $ES_k^2 \le ES_n^2$ for $k \le n$, something that is immediate here because of Theorem 2.1(ii).  □

*Example 9.2.* Let, once again, $Y_1, Y_2, \ldots, Y_n$ be independent random variables with $EY_j = 0$ and $\operatorname{Var} Y_j = \sigma_j^2$. Set, for $k = 1, 2, \ldots, n$, $X_k = \sum_{j=1}^k Y_j$, $s_k^2 = \sum_{j=1}^k \sigma_j^2$ and $Z_k = X_k^2 - s_k^2$. We then know from Example 3.3 that $\{Z_k, 1 \le k \le n\}$ is a martingale.

Suppose, in addition, that there exists $A > 0$, such that $|Y_k| \le A$ for all $k$, define

$$\tau = \min\{k : |X_k| \ge \lambda\},$$

and set $M = \{\max_{1 \le k \le n} |X_k| \ge \lambda\}$. Then, by the optional sampling theorem, Theorem 7.1, $E Z_{\tau \wedge n} = 0$, that is,

$$E X_{\tau \wedge n}^2 = E s_{\tau \wedge n}^2.$$

Moreover, since $s_n^2 \le s_{\tau \wedge n}^2$ on $M^c$ and

$$|X_{\tau \wedge n}| = |X_{\tau \wedge n-1} + Y_{\tau \wedge n}| \le \lambda + A,$$

it follows that

$$s_n^2 P(M^c) \le E s_{\tau \wedge n}^2 = E X_{\tau \wedge n}^2 \le (\lambda + A)^2,$$

which, rewritten, tells us that

$$P(\max_{1 \le k \le n} |X_k| < \lambda) \le \frac{(\lambda + A)^2}{s_n^2}.$$

We have thus obtained a martingale proof of "the other" Kolmogorov inequality, Theorem 3.1.7.  □

**Theorem 9.2.** *If $\{(X_n, \mathcal{F}_n), n \geq 0\}$ is a submartingale, then for $\lambda > 0$,*

$$\lambda P(\min_{0 \leq k \leq n} X_k < -\lambda) \leq E\,X_n - E\,X_0 - \int_{\{\min_{0 \leq k \leq n} X_k < -\lambda\}} X_n \, dP$$

$$\leq E\,X_n - E\,X_0 \leq E\,X_n^+ - E\,X_0.$$

*Proof.* Set

$$\tau = \min\{k : X_k < -\lambda\} \quad \text{and} \quad \Lambda = \{\min_{0 \leq k \leq n} X_k < -\lambda\} = \{X_{\tau \wedge n} < -\lambda\}.$$

Then, via Theorem 7.4 (cf. the proof of Theorem 9.1),

$$E\,X_0 \leq E\,X_{\tau \wedge n} = \int_\Lambda X_{\tau \wedge n} \, dP + \int_{\Lambda^c} X_{\tau \wedge n} \, dP \leq -\lambda P(\Lambda) + \int_{\Lambda^c} X_{\tau \wedge n} \, dP,$$

so that

$$\lambda P(\Lambda) \leq \int_{\Lambda^c} X_n \, dP - E\,X_0 \leq E\,X_n - \int_\Lambda X_{\tau \wedge n} \, dP - E\,X_0,$$

and the conclusion follows. □

Just as the Kolmogorov inequality for sums of independent random variables has a (sub)martingale version, so has the Hájek-Rényi inequality, Theorem 3.1.8; see [43]. The proof below, however, was given by Professor Esseen, on October 1, 1969, during a series of lectures on martingale theory.

**Theorem 9.3.** *Let $\{c_k, 0 \leq k \leq n\}$ be positive, non-increasing real numbers, and suppose that $\{(X_n, \mathcal{F}_n), n \geq 0\}$ is a submartingale. Then, for $\lambda > 0$,*

$$\lambda P(\max_{0 \leq k \leq n} c_k X_k > \lambda) \leq c_0 E\,X_0^+ + \sum_{k=1}^n c_k (E\,X_k^+ - E\,X_{k-1}^+)$$

$$-c_n \int_{\{\max_{0 \leq k \leq n} c_k X_k \leq \lambda\}} X_n^+ \, dP$$

$$\leq c_0 E\,X_0^+ + \sum_{k=1}^n c_k (E\,X_k^+ - E\,X_{k-1}^+) = \sum_{k=0}^{n-1} (c_k - c_{k+1}) E\,X_k^+ + c_n E\,X_n^+.$$

*Proof.* Since $\{\max_{0 \leq k \leq n} c_k X_k \geq \lambda\} = \{\max_{0 \leq k \leq n} c_k X_k^+ \geq \lambda\}$, and since $\{(X_n^+, \mathcal{F}_n), n \geq 0\}$ is also a submartingale, it suffices to prove the theorem for the latter.

Set $Y_0 = c_0 X_0^+$, and, for $1 \leq k \leq n$,

$$Y_k = c_0 X_0^+ + \sum_{j=1}^k c_j (X_j^+ - X_{j-1}^+) = \sum_{j=0}^{k-1} (c_j - c_{j+1}) X_j^+ + c_k X_k^+.$$

Then, $Y_{k+1} = Y_k + c_{k+1}(X_{k+1}^+ - X_k^+)$, $k \geq 0$, so that

$$E(Y_{k+1} \mid \mathcal{F}_k) = Y_k + c_{k+1}E((X_{k+1}^+ - X_k^+) \mid \mathcal{F}_k) \geq Y_k,$$

that is, $\{(Y_n, \mathcal{F}_n), n \geq 0\}$ is a non-negative submartingale, so that, since $Y_k \geq c_k X_k^+$ for all $k$, an application of Theorem 9.1 yields

$$\lambda P(\max_{0 \leq k \leq n} c_k X_k^+ > \lambda) \leq \lambda P(\max_{0 \leq k \leq n} Y_k > \lambda) \leq \int_{\{\max_{0 \leq k \leq n} Y_k > \lambda\}} Y_n \, dP$$

$$\leq \int_{\{\max_{0 \leq k \leq n} c_k X_k^+ k > \lambda\}} \left(\sum_{j=0}^{n-1}(c_j - c_{j-1})X_j^+ + c_n X_n^+\right) dP$$

$$\leq \sum_{j=0}^{n-1} E(c_j - c_{j-1})X_j^+ + c_n E X_n^+ - \int_{\{\max_{0 \leq k \leq n} c_k X_k^+ k \leq \lambda\}} X_n^+ \, dP$$

$$\leq \sum_{j=0}^{n-1} E(c_j - c_{j-1})X_j^+ + c_n E X_n^+. \qquad \Box$$

*Example 9.3.* Let $Y_1, Y_2, \ldots, Y_n$ be independent, identically distributed random variables with mean 0, finite variances, and partial sums $X_1, X_2, \ldots, X_n$. By noticing as before that $\{X_k^2, 1 \leq k \leq n\}$ is a submartingale, an application of Chow's inequality provides another proof of the original Hájek-Rényi inequality, Theorem 3.1.8. $\qquad \Box$

**Exercise 9.1.** Please check this claim. $\qquad \Box$

Next, a famous maximal inequality due to Doob. Whereas the inequalities so far have provided estimates on tail probabilities the Doob inequality relates moments of maxima to moments of the last element in the sequence. We begin, however, with an auxiliary result.

**Lemma 9.1.** *Let $X$ and $Y$ be non-negative random variables. If*

$$P(Y > y) \leq \frac{1}{y} \int_{\{Y \geq y\}} X \, dP,$$

*then*

$$E Y^p \leq \begin{cases} (\frac{p}{p-1})^p E X^p, & when \quad p > 1, \\ \frac{e}{e-1} + \frac{e}{e-1} E X \log^+ X, & when \quad p = 1. \end{cases}$$

*Proof.* We first consider the case $p > 1$. As always, $q$ is defined via the relation $\frac{1}{p} + \frac{1}{q} = 1$. We have

$$E Y^p = p \int_0^\infty y^{p-1} P(Y > y) \, dy \leq p \int_0^\infty y^{p-2} \left(\int_{\{Y \geq y\}} X \, dP\right) dy$$

$$= p \int_\Omega X \left(\int_0^Y y^{p-2} \, dy\right) dP = \frac{p}{p-1} E(XY^{p-1})$$

$$\leq \frac{p}{p-1} \|X\|_p \|Y^{p-1}\|_q = \frac{p}{p-1} \|X\|_p \|Y\|_p^{p-1},$$

where the inequality is a consequence of Hölder's inequality, Theorem 3.2.4.

If $\|Y\|_p = 0$ there is of course nothing to prove. If not, division of the extreme members by $\|Y\|_p^{p-1}$ completes the proof when $p > 1$.

For the case $p = 1$,

$$EY = \int_0^\infty P(Y > y)\,dy \le 1 + \int_1^\infty P(Y > y)\,dy$$

$$\le 1 + \int_1^\infty \left(\frac{1}{y}\int_{\{Y \ge y\}} X\,dP\right)dy = 1 + \int_\Omega X\left(\int_1^Y \left(\frac{1}{y}\,dy\right)\right)dP$$

$$= 1 + E(X\log^+ Y).$$

To finish off we need the following "elementary" inequality

$$a\log^+ b \le a\log^+ a + b/e, \quad a,b > 0, \tag{9.1}$$

the proof of which we postpone for a few lines.

Given this inequality we obtain

$$EY \le 1 + E(X\log^+ X + Y/e) = 1 + EX\log^+ X + EY/e,$$

from which the conclusion follows after some rearranging.

It thus remains to prove the inequality (9.1).

If $0 < a, b \le 1$, the left-hand side equals 0 and there is nothing to prove, and if $a > b$ the inequality is trivial. Thus, suppose that $1 < a < b$. Then

$$a\log^+ b = a\log b \le a\log a + a\log(b/a) = a\log^+ a + b \cdot \frac{\log^+(b/a)}{b/a},$$

so that the proof of (9.1) reduces to showing that

$$\frac{\log c}{c} \le \frac{1}{e}, \quad \text{for} \quad c > 1,$$

which is an easy task.    □

Here is now the promised moment inequality.

**Theorem 9.4.** (Doob's maximal inequality)

(i) *If* $\{(X_n, \mathcal{F}_n), n \ge 0\}$ *is a non-negative submartingale, then*

$$E\left(\max_{0 \le k \le n} X_k\right)^p \le \begin{cases} (\frac{p}{p-1})^p E X_n^p, & \text{when} \quad p > 1, \\ \frac{e}{e-1} + \frac{e}{e-1}E X_n\log^+ X_n, & \text{when} \quad p = 1. \end{cases}$$

(ii) *If* $\{(X_n, \mathcal{F}_n), n \ge 0\}$ *is a martingale, then*

$$E\left(\max_{0 \le k \le n} |X_k|\right)^p \le \begin{cases} (\frac{p}{p-1})^p E|X_n|^p, & \text{when} \quad p > 1, \\ \frac{e}{e-1} + \frac{e}{e-1}E|X_n|\log^+ |X_n|, & \text{when} \quad p = 1. \end{cases}$$

*Proof.* With $Y \longleftrightarrow \max_{0 \le k \le n} X_k$ and $X \longleftrightarrow X_n$ (i) follows immediately from the Kolmogorov inequality and Lemma 9.1, after which (ii) follows from (i) via Theorem 3.1.    □

One might ask whether the additional logarithm when $p = 1$ is really necessary or just a consequence of the method of proof, because of the appearance of the harmonic series. However, it is shown in [101] that a converse inequality holds for a large class of martingales. So, it is necessary. We also refer to [32] and to Subsection 10.16.1 below, where a converse is shown in a law of large numbers context.

## Burkholder Inequalities

As we have noted, martingales are generalizations of sums of independent random variables with mean 0. Increments are not independent, but orthogonal. Several results for sums of independent random variables carry over, as we have seen, to martingales, although sometimes with minor modifications. The Marcinkiewicz-Zygmund inequalities, Theorem 3.8.1 is one such result. It is interesting to note that Marcinkiewicz and Zygmund proved their result for the case $p > 1$ in 1937, and for $p = 1$ in 1938 [180, 182]. For martingales, Burkholder, in his path-breaking 1966 paper [33], proved the result for $p > 1$; Theorem 9 there. For $p = 1$ the result is not generally true. However, it is true if one replaces the martingale by the maximal function [59]. Since one has to limit oneself in order for a book not to be too heavy, we content ourselves by stating the results. Let us, however, mention that Burkholder's result can be proved with methods close to those for sums of independent random variables.

Before stating our results we introduce a piece of notation.

To a martingale $\{(X_n, \mathcal{F}_n), n \geq 0\}$ with martingale difference sequence $\{Y_k, k \geq 0\}$, we associate, the *maximal function* $X_n^* = \max_{0 \leq k \leq n} |X_k|$, the *square function* $S_n(X) = \left(\sum_{k=0}^n Y_k^2\right)^{1/2}$, $n \geq 0$, and the *conditional square function* $s_n(X) = \left(\sum_{k=1}^n E(Y_k^2 \mid \mathcal{F}_{k-1})\right)^{1/2}$, $n \geq 1$.

**Theorem 9.5.** (Burkholder [33], Theorem 9)
(i) *Let $p > 1$. There exist constants $A_p$ and $B_p$, depending only on $p$, such that*
$$A_p \|S_n(X)\|_p \leq \|X_n\|_p \leq B_p \|S_n(X)\|_p.$$

(ii) *Let $p \geq 1$. There exist constants $A_p$ and $B_p$, depending only on $p$, such that*
$$A_p \|S_n(X)\|_p \leq \|X_n^*\|_p \leq B_p \|S_n(X)\|_p.$$

*Remark 9.1.* If the increments are independent with mean 0 we notice immediately that (i) reduces to the Marcinkiewicz-Zygmund inequalities.     □

**Theorem 9.6.** (Burkholder [34], Theorem 21.1)
*Suppose that $\{(X_n, \mathcal{F}_n), n \geq 0\}$ is a martingale with martingale difference sequence $\{Y_k, k \geq 0\}$, let $\Phi$ be a non-negative, convex function, satisfying the growth condition $\Phi(2x) \leq c\Phi(x)$, $x > 0$, for some $c > 0$. Then there exists a constant $C$, depending only on $c$, such that*

$$E\Phi(X_n^*) \le C\big(E\,\Phi(s_n(X)) + E\max_{0 \le k \le n}\Phi(Y_k)\big)$$
$$\le 2C\max\{E\,\Phi(s_n(X)), E\max_{0 \le k \le n}\Phi(Y_k)\}.$$

Particularly useful is the following corollary.

**Corollary 9.1.** *Let $p \ge 2$, and suppose that $\{(X_n, \mathcal{F}_n), n \ge 0\}$ is a martingale, with martingale difference sequence $\{Y_k, k \ge 0\}$. There exists a constant $D_p$, depending only on $p$, such that*

$$\|X_n\|_p \le D_p\big(\|s_n(X)\|_p + \|\max_{1 \le k \le n}|Y_k|\|_p\big)$$

$$\le D_p\big(\|s_n(X)\|_p + \sum_{k=1}^n \|Y_k\|_p\big)$$

$$\le 2D_p \max\big\{\|s_n(X)\|_p, \sum_{k=1}^n \|Y_k\|_p\big\}.$$

If, in particular, the increments are independent with mean 0 and finite variances $\sigma_k^2$, the conditional square function turns into $s_n^2(X) = \sum_{k=1}^n \sigma_k^2$, and we obtain

**Corollary 9.2.** *Let $p \ge 2$, suppose that $Y_1, Y_2, \ldots, Y_n$ are independent with mean 0, finite variances $\sigma_k^2$, that $E|Y_k|^p < \infty$, and set $X_n = \sum_{k=1}^n Y_k$, $n \ge 1$. Then there exists a constant $D_p$, depending only on $p$, such that*

$$E|X_n|^p \le D_p\left(\Big(\sum_{k=1}^n \sigma_k^2\Big)^{p/2} + E\max_{1 \le k \le n}|Y_k|^p\right)$$

$$\le D_p\left(\Big(\sum_{k=1}^n \sigma_k^2\Big)^{p/2} + \sum_{k=1}^n E|Y_k|^p\right)$$

$$\le 2D_p \max\left\{\Big(\sum_{k=1}^n \sigma_k^2\Big)^{p/2}, \sum_{k=1}^n E|Y_k|^p\right\}.$$

*In particular, if $Y_1, Y_2, \ldots, Y_n$ are identically distributed with mean 0 and common variance $\sigma^2$, then*

$$E|X_n|^p \le D_p\big(n^{p/2}\sigma^p + E\max_{1 \le k \le n}|Y_k|^p\big)$$

$$\le D_p\big(n^{p/2}\sigma^p + nE|Y_1|^p\big) \le 2D_p\max\{n^{p/2}\sigma^p, nE|Y_1|^p\}.$$

Observe the similarities to Rosenthal's inequality, Theorem 3.9.1, in the final estimate.

The statements above all contain phrases of the kind "there exist constants $A_p$ and $B_p$ depending only on $p$, such that ...". It should be mentioned that a fair amount of work has been put into finding best constants, sharp constants, and so on. We invite the (interested) reader to search in the literature and on the web. Admittedly, such results may be of interest if one aims at "optimal" results, but if they are used in order to determine finiteness of moments, or just existence of some kind, then the exact numerical value is (usually) of minor importance.

Much more can be said concerning this topic, but let us stop here by mentioning the early very important papers [34, 35, 36], Garsia's monograph [93], and [102], in particular the first half.

# 10 Convergence

One of the most important theorems is the convergence theorem. There exist a few different proofs of this result. The original proof of the convergence theorem is due to Doob [64]. The "standard" one, which is also due to Doob, is based on, what is called, upcrossings (and downcrossings), and will be presented in Subsection 10.10.2. Prior to that we present our favorite proof, which is due to Garsia [92]. The beauty comes from its gradual, stepwise, elegant, approaching of the final target. We close with some hints and references to other approaches.

**Theorem 10.1.** *Suppose that* $\{X_n, n \geq 0\}$ *is a submartingale such that*

$$\sup_n EX_n^+ < \infty.$$

*Then* $X_n$ *converges almost. surely.*

*Remark 10.1.* Since a martingale is also a submartingale it follows in particular that martingales satisfying the required boundedness condition are a.s. convergent.                                                                    □

## 10.1 Garsia's Proof

We follow, essentially, [92].

### $\{X_n\}$ Is an $L^2$-bounded Martingale

Since $\{X_n^2\}$ is an $L^1$-bounded submartingale we know that $E X_n^2 \nearrow$ monotonically and has a limit. From Lemma 4.1 it therefore follows that, for $m < n$,

$$E(X_n - X_m)^2 = E X_n^2 - E X_m^2 \to 0 \quad \text{as} \quad m, n \to \infty,$$

which shows that the sequence $\{X_n\}$ is Cauchy-convergent in $L^2$ and, hence, $L^2$-convergent to some random variable $X$, in particular, convergent in probability.

Since a sequence that converges in probability always contains a subsequence that converges almost surely we know, in particular, that

$$X_{n_k} \overset{a.s.}{\to} X \qquad \text{as} \quad k \to \infty, \tag{10.1}$$

for some subsequence $\{n_k\}$. Moreover, since, once again by Lemma 4.1,

$$E X_{n_k}^2 = E X_{n_0}^2 + \sum_{j=1}^{k} E(X_{n_j} - X_{n_{j-1}})^2,$$

the $L^2$-boundedness implies that also

$$\sum_{k} E(X_{n_k} - X_{n_{k-1}})^2 < \infty. \tag{10.2}$$

Now, since $X_i - X_{n_k} \in \mathcal{F}_i$ for all $i \geq n_k$, and

$$E(X_j - X_{n_k} \mid \mathcal{F}_i) = E(X_j \mid \mathcal{F}_i) - X_{n_k} = X_i - X_{n_k}$$

for all $j \geq i \geq n_k$ it follows that

$$\{X_i - X_{n_k}, \, i \geq n_k\} \quad \text{is a martingale,}$$

so that, by the Kolmogorov inequality and (10.2),

$$\sum_{k} P(\max_{n_k < i \leq n_{k+1}} |X_i - X_{n_k}| > \varepsilon) \leq \frac{1}{\varepsilon^2} \sum_{k} E(X_{n_{k+1}} - X_{n_k})^2 < \infty,$$

for any $\varepsilon > 0$. An application of the first Borel-Cantelli lemma therefore tells us that

$$P(\max_{n_k < i \leq n_{k+1}} |X_i - X_{n_k}| > \varepsilon \text{ i.o.}) = 0 \quad \text{for any } \varepsilon > 0.$$

Combining this with (10.1) finishes the proof.

## $\{X_n\}$ Is a Non-negative $L^2$-bounded Submartingale

Let $X_n = M_n + A_n$, $n \geq 0$, be the Doob decomposition. Then

$$E A_n = E X_n - E M_n = E X_n - E M_0 = E X_n - E X_0$$
$$\leq E X_n \leq \sqrt{E X_n^2} \leq C < \infty,$$

so that $\{A_n\}$ is increasing, $L^1$-bounded, and therefore a.s. convergent.

In order to complete the proof we show that $\{M_n\}$ is $L^2$-bounded, from which the a.s. convergence follows from the previous step.

From the proof of the Doob decomposition we know that the increments of the martingale are

$$M_{n+1} - M_n = X_{n+1} - E(X_{n+1} \mid \mathcal{F}_n),$$

so that, via Lemma 4.1, Theorem 1.3, and the submartingale property,

$$
\begin{aligned}
E\,M_{n+1}^2 - E\,M_n^2 &= E(M_{n+1} - M_n)^2 = E\big(X_{n+1} - E(X_{n+1} \mid \mathcal{F}_n)\big)^2 \\
&= E\,X_{n+1}^2 - E\big(E(X_{n+1} \mid \mathcal{F}_n)^2\big) \le E\,X_{n+1}^2 - E\,X_n^2.
\end{aligned}
$$

Since $X_0 = M_0$ we therefore conclude, via telescoping, that

$$E\,M_n^2 \le E\,X_n^2 < C < \infty,$$

which proves the desired $L^2$-boundedness.

## $\{X_n\}$ Is a Non-negative $L^1$-bounded Martingale

By Theorem 3.1, $\{e^{-X_n}, n \ge 0\}$ is a submartingale, which, moreover, is uniformly bounded. By the previous step this submartingale is a.s. convergent. However, it may converge to 0, in which case $X_n \overset{a.s.}{\to} \infty$ as $n \to \infty$. To see that this is not the case we apply the Kolmogorov inequality to obtain

$$P(\max_{0 \le k \le n} X_k > \lambda) \le \frac{E\,X_n}{\lambda} \le \frac{C}{\lambda} < \infty.$$

Since the upper bound does not depend on $n$ it follows that

$$P(\sup_n X_n > \lambda) \le \frac{C}{\lambda} < \infty,$$

that is, $\sup_n X_n$ is a.s. finite.

## $\{X_n\}$ Is an $L^1$-bounded Martingale

The conclusion is immediate via the Krickeberg decomposition.

## $\{X_n\}$ Is a Martingale, $\sup_n E\,X_n^+ < \infty$

Since

$$E|X_n| = E\,X_n^+ + E\,X_n^- = 2E\,X_n^+ - E\,X_n = 2E\,X_n^+ - E\,X_0,$$

it follows that $\sup_n E\,X_n^+ < \infty$ implies $L^1$-boundedness and the previous step applies.

## $\{X_n\}$ Is an $L^1$-bounded Submartingale

The Doob decomposition yields

$$E\,A_n = E\,X_n - E\,M_n = E\,X_n - E\,M_0 = E\,X_n - E\,X_0,$$

which shows that $\{A_n\}$ is $L^1$-bounded and, as before, a.s. convergent. It follows that $\{M_n\}$ is $L^1$-bounded and, hence, a.s. convergent. The same therefore also holds for $\{X_n\}$.

## $\{X_n\}$ Is a Submartingale, $\sup_n E\,X_n^+ < \infty$

By modifying the computation from the martingale analog, we obtain

$$E|X_n| = E\,X_n^+ + E\,X_n^- = 2E\,X_n^+ - E\,X_n = 2E\,X_n^+ - E\,M_n - E\,A_n,$$
$$= 2E\,X_n^+ - E\,M_0 - E\,A_n \leq 2E\,X_n^+ - E\,M_0,$$

and we conclude, once again, that $\{X_n\}$ is $L^1$-bounded, so that the previous step applies.

The theorem is completely proved. $\qquad\qquad\qquad\qquad\qquad\qquad\qquad\square$

### 10.2 The Upcrossings Proof

Let $\{x_n, n \geq 0\}$ be a sequence of reals, and set, for arbitrary reals $a < b$,

$$\tau_1 = \min\{k \geq 0 : x_k \leq a\},$$
$$\tau_2 = \min\{k \geq \tau_1 : x_k \geq b\},$$
$$\tau_3 = \min\{k \geq \tau_2 : x_k \leq a\},$$
$$\tau_4 = \min\{k \geq \tau_3 : x_k \geq b\},$$
$$\cdots \qquad \cdots\cdots\cdots\cdots\cdots\cdots$$
$$\tau_{2k-1} = \min\{k \geq \tau_{2k-2} : x_k \leq a\},$$
$$\tau_{2k} = \min\{k \geq \tau_{2k-1} : x_k \geq b\},$$

with the convention that the minimum or infimum of the empty set is infinite.

The following, so-called *upcrossings lemma* states that a convergent sequence of real numbers can only oscillate a finite number of times between any two rationals.

**Lemma 10.1.** *The sequence* $\{x_n, n \geq 0\}$ *converges iff*

$$\nu(a,b) = \max\{k : \tau_{2k} < \infty\} < \infty \qquad \text{for all} \quad a, b \in \mathbb{Q}, \ a < b.$$

*Proof.* If $x_n$ does not converge there exist reals $a < b$, such that

$$\liminf_{n\to\infty} x_n < a < b < \limsup_{n\to\infty} x_n,$$

which implies that $\nu(a,b) = \infty$. If, on the other hand, $\nu(a,b) = \infty$ for some $a < b \in \mathbb{Q}$, then we must have infinitely many $x$'s below $a$ and infinitely many $x$'s above $b$, so that

$$\liminf_{n\to\infty} x_n \leq a < b \leq \limsup_{n\to\infty} x_n,$$

in which case $x_n$ cannot converge.

The proof is finished by observing that the rationals are dense in $\mathbb{R}$. $\quad\square$

Upcrossings can equally well be defined for random sequences, and the up-crossings lemma can be extended to such sequences as follows.

**Lemma 10.2.** *The sequence $\{X_n, n \geq 0\}$ converges almost surely iff*

$$\nu(a,b) = \max\{k : \tau_{2k} < \infty\} < \infty \quad a.s. \quad for\ all \quad a, b \in \mathbb{Q},\ a < b.$$

*Proof.* We first note that $\nu(a, b)$ is a random variable, since $\{\nu(a,b) \geq n\} = \{\tau_{2n} < \infty\}$, which is a measurable set.

Next, given $\omega \in \Omega$, it follows from the Lemma 10.1 that $X_n(\omega)$ converges if and only if $\nu(a, b, \omega) < \infty$ for all $a < b \in \mathbb{Q}$, so that

$$\{X_n \text{ converges}\} = \bigcap_{a < b,\ a,b \in \mathbb{Q}} \{\nu(a, b) < \infty\},$$

which is a countable intersection of sets. The conclusion follows. $\qquad\square$

In order to prove the convergence theorem we thus have to prove that $\nu(a, b)$ is finite almost surely for all $a < b \in \mathbb{Q}$. This is achieved by estimating the *expected* number of upcrossings.

**Lemma 10.3.** *Suppose that $\{(X_n, \mathcal{F}_n), n \geq 0\}$ is a submartingale, and let, for $-\infty < a < b < \infty$, $\nu_n(a, b)$ be the number of upcrossings of the sequence $X_1, X_2, \ldots, X_n$.*
(i) *Then*

$$E\,\nu_n(a, b) \leq \frac{E((X_n - a)^+) - E((X_1 - a)^+)}{b - a} \leq \frac{E(X_n^+) + |a|}{b - a}.$$

(ii) *If, in addition,*

$$\sup_n E(X_n^+) = C < \infty,$$

*then*

$$E\,\nu(a, b) \leq \frac{C + a}{b - a}.$$

*Proof.* It suffices to consider the number of crossings of the levels 0 and $b - a$ of the translated sequence $\{(X_n - a)^+, n \geq 0\}$, that is, we may w.l.o.g. assume that $X_n \geq 0$, that $a = 0$ and, hence, that $X_{\tau_j} = 0$ whenever $j$ is odd.

We extend the crossing times $\{\tau_k\}$ with an additional $\tau_0 = 0$. Moreover, since we stop at $\tau_{[\nu_n(a,b)/2]}$ we set $\tau_k = n$ thereafter. Notice that $\tau_n = n$ always.

By telescoping,

$$X_n - X_0 = X_{\tau_n} - X_0 = \sum_{j=1}^{n} (X_{\tau_j} - X_{\tau_{j-1}})$$

$$= \sum_{j \text{ odd}} (X_{\tau_j} - X_{\tau_{j-1}}) + \sum_{j \text{ even}} (X_{\tau_j} - X_{\tau_{j-1}}).$$

In order to take care of the odd terms we observe that the crossing times are a sequence of non-decreasing, bounded (by $n$) stopping times, so that $\{(X_{\tau_k}, \mathcal{F}_{\tau_k}), k \geq 0\}$ is a submartingale by Doob's optional sampling theorem, Theorem 7.1, in particular,

$$E(X_{\tau_{2j+1}} - X_{\tau_{2j}}) \geq 0.$$

As for the even terms,

$$X_{\tau_{2j}} - X_{\tau_{2j-1}} \begin{cases} \geq b - 0 = b, & \text{when} \quad 2j \leq \nu_n(a,b), \\ = X_n - X_{\tau_{2j-1}} \geq 0, & \text{when} \quad 2j = \nu_n(a,b) + 1, \\ = X_n - X_n = 0, & \text{when} \quad 2j > \nu_n(a,b). \end{cases}$$

Putting the above facts together shows that

$$E(X_n - X_0) \geq \sum_{2j \leq \nu_n(a,b)} E(X_{\tau_{2j}} - X_{\tau_{2j-1}}) \geq bE\,\nu_n(a,b),$$

which proves the first inequality in the translated setting. Translating back, the numerator of the upper bound turns into

$$E\big((X_n - a)^+\big) - E\big((X_1 - a)^+\big) \leq E\big((X_n - a)^+\big) \leq E(X_n^+) + |a|.$$

The denominator becomes $b - a$.

This proves (i), from which (ii) is immediate.     □

To prove submartingale convergence it only remains to observe that Lemma 10.3(ii) is applicable, so that, for any rationals $a < b$,

$$E\,\nu_n(a,b) \leq \frac{C + a}{b - a},$$

independently of $n$, so that, by letting $n \to \infty$, it follows that

$$E\,\nu(a,b) < \infty \quad \Longrightarrow \quad P(\nu(a,b) < \infty) = 1,$$

from which almost sure convergence follows from Lemma 10.2.     □

## 10.3 Some Remarks on Additional Proofs

Here we mention some other methods of proof. Only statements are provided. For details we refer to the cited sources.
• A very short, functional analytic proof is given in [166].
• For the following kind of Fatou lemma, which states that if $\{X_n, n \geq 1\}$ is an adapted sequence and $\mathcal{T}_F$ the collection of *finite* stopping times, then

$$E \limsup_{n \to \infty} X_n \leq \limsup_{\tau \in \mathcal{T}_F} E\,X_\tau,$$
$$E \liminf_{n \to \infty} X_n \geq \liminf_{\tau \in \mathcal{T}_F} E\,X_\tau,$$

we refer to [240]. For $L^1$-bounded adapted sequences this has been extended in [38] to

$$E(\limsup_{n\to\infty} X_n - \liminf_{n\to\infty} X_n) \le \limsup_{\tau,\sigma\in\mathcal{T}_B} E(X_\sigma - X_\tau),$$

where now $\mathcal{T}_B$ is the family of *bounded* stopping times.

Since the right-hand side equals 0 for martingales by Theorem 7.2, it follows immediately that $L^1$-bounded martingales are almost surely convergent.
- Another approach is via the following approximation lemma; [6].

**Lemma 10.4.** *Let $Y$ be an $\mathcal{F}_\infty$-measurable random variable, such that $Y(\omega)$ is a cluster point of the sequence $\{X_n(\omega),\ n \ge 1\}$ for every $\omega \in \Omega$. Then there exists a sequence of bounded, non-decreasing stopping times $\tau_n \ge n$, such that*

$$X_{\tau_n} \overset{a.s.}{\to} Y \quad as \quad n \to \infty.$$

If $\{(X_n, \mathcal{F}_n),\ n \ge 0\}$ is a martingale we can use this fact to exhibit two sequences of bounded stopping times, $\{\sigma_k,\ k \ge 1\}$ and $\{\tau_k,\ k \ge 1\}$, with conditions as given in the lemma, such that

$$X_{\sigma_k} \overset{a.s.}{\to} \liminf_{n\to\infty} X_n \quad \text{and} \quad X_{\tau_k} \overset{a.s.}{\to} \limsup_{n\to\infty} X_n \quad \text{as} \quad k \to \infty,$$

which, together with Fatou's lemma and Theorem 7.2, shows that

$$E(\limsup_{n\to\infty} X_n - \liminf_{n\to\infty} X_n) \le \liminf_{k\to\infty} E(X_{\sigma_k} - X_{\tau_k}) = 0,$$

so that $X_n$ converges almost surely as $n \to \infty$.
- *Asymptotic martingales*, abbreviated *amarts*, are defined in terms of *net convergence*: A sequence $\{(X_n, \mathcal{F}_n),\ n \ge 0\}$ is an *amart* iff

$$E X_{\tau_n} \quad \text{converges as} \quad n \to \infty,$$

for every sequence $\{\tau_n,\ n \ge 1\}$ of bounded, non-decreasing stopping times.

The amart convergence theorem can be proved as above for bounded amarts with dominated convergence instead of Fatou's lemma and truncation; truncated amarts remain amarts.

Since martingales have constant expectation for bounded stopping times it follows immediately that martingales are amarts. This also pertains to submartingales.

For more on amarts, see [70] and [106] and references given there.

## 10.4 Some Questions

The theorem suggests (at least) four natural questions:

- Give an example of a non-convergent martingale.

- If $X_n \xrightarrow{a.s.} X_\infty$ (say) as $n \to \infty$, is it true that $X_\infty$ can be interpreted as some kind of last element of the martingale, that is, can we conclude that $\{X_1, X_2, \ldots, X_n, \ldots, X_\infty\}$ is a martingale, in other words, is it true that $X_n = E(X_\infty \mid \mathcal{F}_n)$ for all $n$?
- When, if at all, is there also $L^1$-convergence? $L^p$-convergence?
- What about submartingales?

We shall provide answers to these questions in the following two subsections.

### 10.5 A Non-convergent Martingale

The "immediate" martingale is a sum of independent zero mean random random variables. Let's check if it is convergent or not.

Thus, suppose that $X_n = \sum_{k=1}^{n} Y_k$, where $Y, Y_1, Y_2, \ldots, Y_n$ are i.i.d. coin-tossing random variables, that is, $P(Y = 1) = P(Y = -1) = 1/2$. It is well known (or follows from the law of the iterated logarithm, Theorem 8.1.2) that the random walk $\{X_n, n \geq 1\}$ oscillates and, hence cannot converge. We also note that $E|X_n| \sim \sqrt{\frac{2n}{\pi}}$ as $n \to \infty$, since $\frac{X_n}{\sqrt{n}} \to N(0, 1)$ in (distribution and in) $L^1$ as $n \to \infty$. In particular $\{X_n, n \geq 1\}$ is *not* $L^1$-bounded.

### 10.6 A Central Limit Theorem?

We have extensively dwelt on almost sure convergence of martingales. One might ask whether it is possible that martingales suitably normalized have some asymptotic normal(?) limit. As ever so often, Paul Lévy in the 1930s was the first to prove a result of that kind. A more recent reference is [18]. For further contributions, see [210, 211, 212], the monographs by Hall and Heyde [128], and [239].

Let us only mention that one of the problems is how to tackle conditions on variances. Variances? Conditional variances? Boundedness? Constancy? And so on. In contrast to the simplest case of independent, identically distributed summands where all variances (as well as conditional variances(!)) are the same.

## 11 The Martingale $\{E(Z \mid \mathcal{F}_n)\}$

In this section we shall get acquainted with the subclass of *uniformly integrable* martingales, which have certain additional properties, such as being $L^1$-convergent as one might guess from Chapter 5.

**Theorem 11.1.** *Let $Z \in L^1$, and set*

$$X_n = E(Z \mid \mathcal{F}_n).$$

*Then $\{(X_n, \mathcal{F}_n), n \geq 0\}$ is a uniformly integrable martingale.*

*Proof.* The martingale property of $\{X_n,\ n \geq 0\}\}$ was established in Example 3.9. Checking uniform integrability: Theorem 9.1 and (3.1) together tell us that

$$P(\max_{0 \leq k \leq n} |X_k| > a) \leq \frac{1}{a}E|X_n| \leq \frac{1}{a}E|Z|,$$

so that, by letting $n \to \infty$,

$$P(\sup_n |X_k| > a) \leq \frac{1}{a}E|Z| \to 0 \quad \text{as} \quad a \to \infty,$$

*independently of* $n$, which implies that.

$$E|X_n|I\{|X_n| > a\} = E|E(Z \mid \mathcal{F}_n)|I\{|X_n| > a\}$$
$$\leq E(E(|Z| \mid \mathcal{F}_n)|I\{|X_n| > a\}) = E(E(|Z|I\{|X_n| > a\} \mid \mathcal{F}_n))$$
$$= E(|Z|I\{|X_n| > a\}) \leq E(|Z|I\{\sup_n |X_n| > a\}) \to 0 \quad \text{as} \quad a \to \infty,$$

independently of (and therefore uniformly in) $n$, the convergence to 0 being a consequence of Theorem 2.6.3(ii).

Uniform integrability may, alternatively, be checked directly (that is, without exploiting the Kolmogorov inequality) as follows:

$$E|X_n|I\{|X_n| > a\} \leq E|Z|I\{\sup_n |X_n| > a\}$$
$$= E|Z|I\{\sup_n |X_n| > a, |Z| > \sqrt{a}\} + E|Z|I\{\sup_n |X_n| > a, |Z| \leq \sqrt{a}\}$$
$$\leq E|Z|I\{|Z| > \sqrt{a}\} + \sqrt{a}P(\sup_n |X_n| > a)$$
$$\leq E|Z|I\{|Z| > \sqrt{a}\} + \frac{1}{\sqrt{a}}E|Z| \to 0 \quad \text{as} \quad a \to \infty.$$

Once again, no $n$ is involved at the end.    □

**Corollary 11.1.** *Let* $p \geq 1$. *If* $E|Z|^p < \infty$, *then* $\{(|X_n|^p, \mathcal{F}_n), n \geq 0\}$ *is a uniformly integrable submartingale.*

*Proof.* By Proposition 1.1(h),

$$|X_n|^p = |E(Z \mid \mathcal{F}_n)|^p \leq E(|Z|^p \mid \mathcal{F}_n),$$

so that uniform integrability follows by replacing $Z$ with $|Z|^p$ in the proof of Theorem 11.1, after which an application of Theorem 3.2 establishes the submartingale property.    □

## 12 Regular Martingales and Submartingales

In the first two subsections we provide answers to the last three questions in Subsection 10.10.4, after which we show, by examples, that there exist martingales that are $L^1$-bounded but not uniformly integrable, and therefore a.s. convergent but *not* $L^1$-convergent. A final revisit to the uniformly integrable martingales from Section 10.11 concludes the section.

## 12.1 A Main Martingale Theorem

**Theorem 12.1.** *Suppose that* $\{(X_n, \mathcal{F}_n), n \geq 0\}$ *is a martingale. The following are equivalent:*

(a)  $\{X_n, n \geq 0\}$ *is uniformly integrable.*

(b)  $X_n$ *converges in* $L^1$.

(c)  $X_n \overset{a.s.}{\to} X_\infty$ *as* $n \to \infty$, $X_\infty \in L^1$, *and* $\{(X_n, \mathcal{F}_n), n = 0, 1, 2, \ldots, \infty\}$ *is a martingale* – $X_\infty$ *closes the sequence.*

(d)  *There exists* $Y \in L^1$, *such that* $X_n = E(Y \mid \mathcal{F}_n)$ *for all* $n \geq 0$.

*Remark 12.1.* A martingale satisfying one of these properties is called complete, closable, or *regular* (depending on the literature).    □

*Proof.* We show that (a) $\Rightarrow$ (b) $\Rightarrow$ (c) $\Rightarrow$ (d) $\Rightarrow$ (a).

(a) $\implies$ (b): Uniform integrability implies $L^1$-boundedness and, hence, a.s. convergence and thus also convergence in $L^1$. Note that the $L^1$-convergence is the consequence of a general result concerning a.s. convergence and moment convergence – Theorem 5.5.2.

(b) $\implies$ (c): Convergence in $L^1$ implies $L^1$-boundedness, and, hence, almost sure convergence, and thus also uniform integrability – Theorem 5.5.2. again. We may therefore pass to the limit in the defining relation: For any $\Lambda \in \mathcal{F}_m$,

$$\int_\Lambda X_m \, dP = \int_\Lambda X_n \, dP \to \int_\Lambda X_\infty \, dP \quad \text{as} \quad n \to \infty, \tag{12.1}$$

which proves that $X_\infty$ closes the sequence.

Strictly speaking we have exploited the fact that for every fixed $m$, the sequence $\{X_n I\{\Lambda\}, n \geq m\}$ is a uniformly integrable martingale and, hence, converges almost surely and in $L^1$. The equality in (12.1) is the martingale property, and the arrow denotes convergence of the expectations.

(c) $\implies$ (d): Choose $Y = X_\infty$.

(d) $\implies$ (a): Apply Theorem 11.1.    □

If higher-order moments exist, then additional convergence properties are automatic.

**Theorem 12.2.** *Let* $p > 1$. *Suppose that* $\{(X_n, \mathcal{F}_n), n \geq 0\}$ *is an* $L^p$-*bounded martingale, that is,*

$$\sup_n E|X_n|^p < \infty.$$

*Then,*

(a)  $\{|X_n|^p\}$ *is uniformly integrable;*

(b)  $X_n \to X_\infty$ *a.s. and in* $L^p$;

(c)  $E|X_n|^p \to E|X_\infty|^p$ *as* $n \to \infty$;

(d)  $\{(X_n, \mathcal{F}_n), n = 0, 1, 2, \ldots, \infty\}$ *is a martingale.*

*Proof.* Set $Y_n = \sup_{0 \le k \le n} |X_k|$, and $Y = \sup_n |X_n|$. The $L^1$-boundedness, Fatou's lemma, and the Doob maximal inequality, Theorem 9.4, together yield

$$E\,Y^p \le \liminf_{n \to \infty} E(Y_n)^p \le \liminf_{n \to \infty} \left(\frac{p}{p-1}\right)^p E|X_n|^p$$

$$\le \left(\frac{p}{p-1}\right)^p \sup_n E|X_n|^p < \infty.$$

We have thus exhibited a dominating random variable $Y \in L^p$, which, by Theorem 5.4.4 proves (a).

Since, the martingale, in particular, is $L^1$-bounded it converges almost surely, which together with (a) and Theorem 5.5.4 proves (b), and, together with (a), also (c). Finally, (d) is immediate from Theorem 12.1.   $\square$

*Remark 12.2.* The "problem" for $p = 1$ is that there are $L^1$-bounded martingales that are not uniformly integrable, whereas powers are automatically so when the martingale is $L^p$-bounded. The "reason" for this discrepancy is the additional logarithm in the Doob maximal inequality when $p = 1$.   $\square$

Recalling Corollary 11.1 allows us to formulate the following result.

**Corollary 12.1.** *Suppose that $\{X_n,\, n \ge 0\}$ is a martingale of the form $X_n = E(Z \mid \mathcal{F}_n)$, $n \ge 0$. Then $X_n \to X_\infty$ almost surely and in $L^1$ as $n \to \infty$ and $X_\infty$ closes the martingale. If, in addition, $Z \in L^p$ for some $p > 1$, then $X_n \to X_\infty$ almost surely and in $L^p$ as $n \to \infty$.*

## 12.2 A Main Submartingale Theorem

The version for submartingales differs slightly from the martingale analog, because Theorem 12.1(d) is no longer applicable.

**Theorem 12.3.** *Suppose that $\{(X_n, \mathcal{F}_n),\, n \ge 0\}$ is a submartingale. The following are equivalent:*

(a)   $\{X_n,\, n \ge 0\}$ *is uniformly integrable;*

(b)   $X_n$ *converges in $L^1$;*

(c)   $X_n \xrightarrow{a.s.} X_\infty$ *as $n \to \infty$, $X_\infty \in L^1$, and $\{(X_n, \mathcal{F}_n),\, n = 0, 1, 2, \ldots, \infty\}$ is a submartingale – $X_\infty$ closes the sequence.*

*Proof.* The proofs that (a) $\Rightarrow$ (b) $\Rightarrow$ (c) are the same as for martingales, except that the equality in (12.1) is replaced by an inequality. Moreover, convergence of the expectations follows from almost sure convergence and uniform integrability via Theorem 5.5.4

(c) $\Longrightarrow$ (a): Since, by convexity, $\{(X_n^+, \mathcal{F}_n),\, 0 \le n \le \infty\}$ is a submartingale, it follows from the defining relation that

$$\int_{\{X_n^+ > \lambda\}} X_n^+ \le \int_{\{X_n^+ > \lambda\}} X_\infty^+,$$

and, since, by the Kolmogorov-Doob inequality,

$$P(X_n^+ > \lambda) \leq \lambda^{-1} E X_n^+ \leq \lambda^{-1} E X_\infty^+ \to 0 \quad \text{as} \quad \lambda \to \infty,$$

independently of (and hence uniformly in) $n$, it follows, from Theorem 5.4.1, that $\{X_n^+\}$ is uniformly integrable. This, the fact that $X_n^+ \overset{a.s.}{\to} X_\infty$ as $n \to \infty$, together with an application of Theorem 5.5.2, proves that $E X_n^+ \to E X_\infty^+$ as $n \to \infty$.

Now, since the expectations of $X_n$ and $X_n^+$ converge as $n \to \infty$ the same holds true for $X_n^-$, which, together with almost sure convergence and another application of Theorem 5.5.2, shows that $\{X_n^-\}$ is uniformly integrable.

Finally, since the sum of two uniformly integrable sequences is also uniformly integrable (Theorem 5.4.6), it follows that $\{X_n\}$ is uniformly integrable, which proves (a). □

## 12.3 Two Non-regular Martingales

In this subsection we exhibit two martingales that are $L^1$-bounded but not uniformly integrable, and, hence, not regular, not complete, not closable, depending on the preferred language. This means that they converge almost surely, but *not* in $L^1$, and, hence, do not "fit" Theorem 12.1(d).

*Example 12.1.* Recall the game "double or nothing" from Example 3.5, in which $X_0 = 0$, and, recursively,

$$X_{n+1} = \begin{cases} 2X_n, & \text{with probability } \frac{1}{2}, \\ 0, & \text{with probability } \frac{1}{2}. \end{cases}$$

We proved there that $\{(X_n, \mathcal{F}_n), n \geq 0\}$ was a martingale with mean 1.

Since $P(X_n = 2^n) = 1 - P(X_n = 0) = 2^{-n}$, an application of the first Borel-Cantelli lemma, Theorem 2.18.1, shows, what is intuitively clear, namely, that $X_n \overset{a.s.}{\to} 0$ as $n \to \infty$. However, since $E X_n = 1$ for all $n$ the mean does not converge to 0, so that the martingale is not uniformly integrable.

*Example 12.2.* A *critical Galton-Watson process* is a branching process in which the mean offspring equals 1, which, i.a., implies that the expected number of individuals in every generation equals 1; recall Example 2.15.1. Such processes die out almost surely. Thus, if $X_n = X(n)$ denotes the number of individuals in the $n$th generation, then $\{X_n, n \geq 0\}$ is a martingale with mean 1 (recall Example 3.8). Since $X_n \overset{a.s.}{\to} 0$ and $E X_n = 1$, the martingale is not uniformly integrable. □

## 12.4 Regular Martingales Revisited

Once again, let $X_n = E(Z \mid \mathcal{F}_n)$, $n \geq 0$. The sequence $\{X_n, n \geq 0\}$ is a uniformly integrable martingale, and we know from (the proof of) the main theorem – recall (12.1) that for $\Lambda \in \mathcal{F}_m$,

$$\int_\Lambda Z \, \mathrm{d}P = \int_\Lambda X_n \, \mathrm{d}P \to \int_\Lambda X_\infty \, \mathrm{d}P \quad \text{as} \quad n \to \infty, \quad \text{for all} \quad m.$$

By standard approximation arguments such as "it suffices to check rectangles" one concludes that

$$\int_\Lambda Z \, \mathrm{d}P = \int_\Lambda X_\infty \, \mathrm{d}P \quad \text{for all } \Lambda \in \mathcal{F}_\infty,$$

which, via the defining relation shows that $X_\infty = E(Z \mid \mathcal{F}_\infty)$.

Collecting these facts (and a little more) we obtain the following result.

**Theorem 12.4.** *Suppose that $Z \in L^1$, and set $X_n = E(Z \mid \mathcal{F}_n)$, $n \geq 0$. Then*

$$X_n \to X_\infty = E(Z \mid \mathcal{F}_\infty) \quad \textit{a.s. and in } L^1 \qquad \textit{as} \quad n \to \infty,$$

*in other words $\{X_0, X_1, X_2, \ldots, X_n, X_\infty, Z\}$ is a martingale.*

*In particular, if $Z \in \mathcal{F}_\infty$, then $X_\infty = Z$ a.s.*

*Remark 12.3.* From Theorem 12.1 we know that $Z$ closes the martingale. But so does $X_\infty$. We call $X_\infty$ the *nearest closing* random variable.    □

## 13 The Kolmogorov Zero-one Law

Here is a beautiful application of Theorem 12.4.

Let $Y_1, Y_2, \ldots$ be independent random variables, let $\mathcal{F}_n$ be the natural $\sigma$-algebras, $\mathcal{F}_n = \sigma\{Y_1, Y_2, \ldots, Y_n\}$, and $\mathcal{T}$ the tail-$\sigma$-field.

Let $Z \in \mathcal{T}$ be a random variable, and suppose, to begin with, that $Z \in L^1$. Consider the sequence $\{E(Z \mid \mathcal{F}_n), n \geq 1\}$. Since $\mathcal{T} \subset \mathcal{F}_\infty$ an application of Theorem 12.4 yields

$$E(Z \mid \mathcal{F}_n) \to E(Z \mid \mathcal{F}_\infty) = Z \quad \text{a.s.}$$

The fact that $Z$ is independent of $\mathcal{F}_n$ tells us that $E(Z \mid \mathcal{F}_n) = E Z$, which proves that

$$Z = E Z \quad \text{a.s.}$$

We thus conclude that an integrable random variable that is measurable with respect to the tail-$\sigma$-field must be a.s. constant.

If $Z$ is not integrable we perform the same analysis on, say, $Z/(1 + |Z|)$ with the same conclusion, and since this new random variable is a.s. constant, so is $Z$. Thus:

**Theorem 13.1.** *A random variable that is measurable with respect to the tail-$\sigma$-field of a sequence of independent random variables is a.s. constant.*

Next we apply this result to the special case when $Z$ is the indicator of a set in the tail-$\sigma$-field, thereby rediscovering Theorem 2.10.6.

**Corollary 13.1.** (The Kolmogorov zero-one law)
*Let $Y_1, Y_2, \ldots$ be a sequence of independent random variables, set $\mathcal{F}_n = \sigma\{Y_1, Y_2, \ldots, Y_n\}$, and let $A \in \mathcal{T}$, the tail-$\sigma$-field. Then*

$$P(A) = 0 \quad or \quad 1.$$

# 14 Stopped Random Walks

Random walks were introduced in Section 2.16. As we have already emphasized a number of times, it is frequently more realistic in applications to observe a random walk at *fixed time points*, which means after a *random number of steps*, rather than after a *fixed number of steps*, that is, at a *random time point*. This naturally suggests a theory of *stopped random walks*.

In this section we shall focus on results for stopped random walks pertaining to martingale theory. Throughout our $\sigma$-algebras are assumed to be the natural ones. For more on this topic we refer to [110].

## 14.1 Finiteness of Moments

One application of the Burkholder inequalities is obtained by joining them with Theorem 8.2 in order to prove results about existence of moments of stopped random walks. The conclusion, [110], Theorem I.5.1, runs as follows:

**Theorem 14.1.** *Let $Y, Y_1, Y_2, \ldots$ be independent, identically distributed distributed random variables such that $E|Y|^p < \infty$ for some $p > 0$, with $EY = 0$, when $p \geq 1$, set $X_n = \sum_{k=1}^n Y_k$, and suppose that $\tau$ is a stopping time. Then*

$$E|X_\tau|^p \leq \begin{cases} E\tau \cdot E|Y|^p, & for \;\; 0 < p \leq 1, \\ B_p \cdot E\tau \cdot E|Y|^p, & for \;\; 1 \leq p \leq 2, \\ B_p\big(E\tau^{p/2} \cdot (EY^2)^{p/2} + E\tau \cdot E|Y|^p\big) \\ \quad \leq \begin{cases} 2B_p \max\{E\tau^{p/2} \cdot (EY^2)^{p/2}, E\tau \cdot E|Y|^p\}, \\ 2B_p \cdot E\tau^{p/2} \cdot E|Y|^p, & for \;\; p \geq 2, \end{cases} \end{cases}$$

*where $B_p$ is a numerical constant depending only on $p$.*

*Proof.* If $\tau$ is not sufficiently integrable, there is nothing to prove, so, suppose that the relevant moments of $\tau$ exist whenever they appear.

Let, as usual, $\tau_n = \tau \wedge n$, and observe that it suffices to prove the theorem for $\tau \wedge n$, since $E(\tau \wedge n)^p \leq E(\tau)^p$ by monotonicity, and $E|X_\tau|^p \leq \liminf_{n \to \infty} E|X_{\tau \wedge n}|^p$ by Fatou's lemma.

Secondly, since, for any $p > 0$, $\{|Y_k|^p - E|Y_k|^p, k \geq 1\}$ are independent, identically distributed random variables with mean 0, it follows from Theorem 8.2 that

$$E \sum_{k=1}^{\tau \wedge n} |Y_k|^p = E(\tau \wedge n) \cdot E|Y|^p. \tag{14.1}$$

First, let $0 < p < 1$. Then, by Theorem 3.2.2, the $c_r$-inequalities, and (14.1),

$$E|X_{\tau \wedge n}|^p \leq E \sum_{k=1}^{\tau \wedge n} |Y_k|^p = E(\tau \wedge n) \cdot E|Y|^p.$$

Next, let $1 < p \leq 2$. Then, by Theorem 9.5, the $c_r$-inequalities (note that $p/2 \leq 1$), and (14.1),

$$E|X_{\tau \wedge n}|^p \leq B_p^p E \left( \sum_{k=1}^{\tau \wedge n} Y_k^2 \right)^{p/2} \leq B_p^p E \sum_{k=1}^{\tau \wedge n} |Y_k|^p = B_p^p E(\tau \wedge n) \cdot E|Y|^p.$$

Finally, for $p \geq 2$ we use Theorem 9.1 and independence to obtain

$$E|X_{\tau \wedge n}|^p \leq D_p^p E \left( \sum_{k=1}^{\tau \wedge n} E(Y_k^2 \mid \mathcal{F}_{k-1}) \right)^{p/2} + D_p^p E \sum_{k=1}^{\tau \wedge n} |Y_k|^p$$

$$= D_p^p E \left( (\tau \wedge n) \cdot E Y^2 \right)^{p/2} + D_p^p E(\tau \wedge n) E|Y|^p$$

$$= D_p^p \left( E(\tau \wedge n)^{p/2} (E Y^2)^{p/2} + E(\tau \wedge n) E|Y|^p \right),$$

and what remains of substance follows from Lyapounov's inequality, Theorem 3.2.5.    □

If $EY = \mu \neq 0$ we may apply the Minkowski inequality, Theorem 3.2.6, and Theorem 14.1 to the inequality

$$|X_\tau| \leq |X_\tau - \mu\tau| + |\mu|\tau$$

to obtain the following result.

**Theorem 14.2.** *Let $p \geq 1$, and suppose that $Y, Y_1, Y_2, \ldots$ are independent, identically distributed distributed random variables with $E|Y|^p < \infty$. Set $X_n = \sum_{k=1}^n Y_k$, and suppose that $\tau$ is a stopping time. Then*

$$E|X_\tau|^p \leq B_p' \cdot E\,\tau^p \cdot E|Y|^p,$$

*where $B_p'$ is a numerical constant depending only on $p$.*

## 14.2 The Wald Equations

We have just provided inequalities for moments of stopped sums. But for mean and variance one can frequently obtain *equalities*. Do such equalities hold for the mean and variance of stopped random walks?

From Section 10.3 and the optional sampling theorem we know that

$$E X_{\tau \wedge n} = 0 \quad \text{and that} \quad E\left(X_{\tau \wedge n}^2 - (\tau \wedge n)E Y^2\right) = 0,$$

(under appropriate moment assumptions). One might therefore hope that it is possible to let $n \to \infty$ in order to obtain equalities in these cases. We shall see that this is, indeed, the case in Section 10.15, where we shall also provide more general results concerning when it is possible to let $n \to \infty$ and retain a specific property, such as, for example, allowing the transition $E X_{\tau \wedge n} = 0 \cdots \longrightarrow E X_\tau = 0$.

Let $Y, Y_1, Y_2, \ldots$ be independent, identically distributed random variables with finite mean $\mu$, and let $S_n$, $n \geq 1$, denote the partial sums. From Theorem 2.15.1 we know that $E S_\tau = \mu \cdot E \tau$ if $\tau$ is a (non-negative integer valued) random variable with finite mean, which is *independent* of $Y_1, Y_2, \ldots$, and that an analogous relation holds for $\operatorname{Var} S_\tau$ provided enough integrability is at hand. There are several results of this kind that carry over to stopping times. The following result, the *Wald equations*, is one of them.

**Theorem 14.3.** *Let $Y, Y_1, Y_2, \ldots$ be independent, identically distributed random variables with finite mean, $\mu$, and let $S_n$, $n \geq 1$, denote the partial sums. Suppose that $\tau$ is a stopping time.*

(i) *If $E\tau < \infty$, then*
$$E S_\tau = \mu \cdot E\tau.$$

(ii) *If, moreover, $\operatorname{Var} Y = \sigma^2 < \infty$, then*
$$E(S_\tau - \mu\tau)^2 = \sigma^2 \cdot E\tau.$$

*Proof.* (i): Following the usual pattern, we first have
$$E X_{\tau \wedge n} = E(S_{\tau \wedge n} - \mu(\tau \wedge n)) = 0,$$
so that
$$E S_{\tau \wedge n} = \mu \cdot E(\tau \wedge n) \nearrow \mu \cdot E\tau < \infty \quad \text{as} \quad n \to \infty,$$
by point-wise and monotone convergence.

If the summands are non-negative, then we also have $E S_{\tau \wedge n} \nearrow E S_\tau$ as $n \to \infty$ by point-wise and monotone convergence. In the general case the same follows by point-wise convergence and dominated convergence, since
$$|S_{\tau \wedge n}| \leq \sum_{k=1}^{\tau \wedge n} |Y_k| \overset{a.s.}{\to} \sum_{k=1}^{\tau \wedge n} |Y_k|$$
monotonically. We have thus shown that
$$E S_\tau \leftarrow E S_{\tau \wedge n} = \mu \cdot E(\tau \wedge n) \to \mu E\tau \quad \text{as} \quad n \to \infty.$$

The conclusion follows.

(ii): We exploit the second random walk martingale, Example 3.3. An application of the optional sampling theorem yields
$$E(S_{\tau \wedge n} - \mu(\tau \wedge n))^2 = \sigma^2 \cdot E(\tau \wedge n) \nearrow \sigma^2 \cdot E\tau < \infty \quad \text{as} \quad n \to \infty, \quad (14.2)$$
where convergence follows as in the proof of (i).

In order to take care of the left-hand side we first note that

$$S_{\tau \wedge n} - \mu \cdot (\tau \wedge n) \overset{a.s.}{\to} S_\tau - \mu \cdot \tau \quad \text{as} \quad n \to \infty. \tag{14.3}$$

Secondly, by orthogonality (Section 10.4),

$$
\begin{aligned}
E\big(\{S_{\tau \wedge m} - \mu \cdot (\tau \wedge m)\} &- \{S_{\tau \wedge n} - \mu \cdot (\tau \wedge n)\}\big)^2 \\
&= E\big(S_{\tau \wedge m} - \mu \cdot (\tau \wedge m)\big)^2 - E\big(S_{\tau \wedge n} - \mu \cdot (\tau \wedge n)\big)^2 \\
&= \sigma^2 \big(E\tau \wedge m - (\tau \wedge n)\big) \to 0 \quad \text{as} \quad n, m \to \infty,
\end{aligned}
$$

so that, by (14.3) (and Theorem 5.5.2),

$$E(S_{\tau \wedge n} - \mu \cdot (\tau \wedge n))^2 \to E(S_\tau - \mu \cdot \tau)^2 \quad \text{as} \quad n \to \infty.$$

Joining this with (14.2) finishes the proof. □

*Remark 14.1.* An important pitfall is to falsely conclude that $E(S_\tau - \mu\tau)^2$ is the same as $\operatorname{Var} S_\tau$ (if $\mu \neq 0$); after all, $E(S_n - n\mu)^2 = \operatorname{Var} S_n \dots$. The truth is, by (i), that $\operatorname{Var} S_\tau = E(S_\tau - \mu E \tau)^2$. Moreover, by adding and subtracting $\mu E \tau$, and invoking (ii), we find that

$$
\begin{aligned}
\sigma^2 E \tau &= E(S_\tau - \mu\tau)^2 \\
&= \operatorname{Var}(S_\tau) + E(\mu E \tau - \mu\tau)^2 + 2E(S_\tau - \mu E \tau)(\mu E \tau - \mu\tau) \\
&= \operatorname{Var}(S_\tau) + \mu^2 \operatorname{Var}\tau - 2\mu \operatorname{Cov}(S_\tau, \tau),
\end{aligned}
$$

which can be rewritten as

$$\operatorname{Var}(S_\tau) = \sigma^2 E \tau - \mu^2 \operatorname{Var}\tau + 2\mu \operatorname{Cov}(S_\tau, \tau). \tag{14.4}$$

If, in particular, $\tau$ is *independent* of $Y_1, Y_2, \ldots$, then, by conditioning, $\operatorname{Cov}(S_\tau, \tau) = \mu \operatorname{Var}\tau$, and (14.4) reduces to the well-known

$$\operatorname{Var}(S_\tau) = \sigma^2 E \tau + \mu^2 \operatorname{Var}\tau. \qquad \square$$

## Returning to an Old Promise

We are now in the position to complete the promise given in Example 7.1, namely, to prove that the expected time it takes a symmetric simple random walk to reach the level $+1$ is infinite.

*Example 14.1.* Thus, let $Y, Y_1, Y_2, \ldots$ be independent random variables, such that $P(Y = 1) = P(Y = -1) = 1/2$, set $X_n = \sum_{k=1}^n Y_k$, $n \geq 1$, and let

$$\tau = \min\{n : X_n = 1\}.$$

Assume the contrary, namely that $E\tau < \infty$. It then follows from the first Wald equation, and the fact that $X_\tau = 1$ almost surely, that

$$1 = E\, X_\tau = EY \cdot E\,\tau = 0 \cdot E\,\tau = 0,$$

which is a contradiction. Hence $E\,\tau = +\infty$.

Note, however, that $E\, X_{\tau \wedge n} = 0$ for every fixed $n$ by the optional sampling theorem.

Moreover, the expected time to reach $-1$ is also infinite. And, yet, the minimum of these, that is the time it takes to reach $+1$ or $-1$, equals 1 always; $\min\{n : |X_n| = 1\} = 1$. □

## 14.3 Tossing a Coin Until Success

We begin by showing that the expected number of independent repetitions of zeroes and ones until the first 1 is obtained has mean $1/P(1)$, after which we derive the geometric distribution of this random variable. This is certainly no news, but the procedure illustrates very nicely the martingale and stopping time technique.

Let $Y, Y_1, Y_2, \ldots$ be independent $\mathrm{Be}(p)$-distributed random variables, that is, $P(Y = 1) = 1 - P(Y = 0) = p$, and let

$$\tau = \min\{n : Y_n = 1\}.$$

The claim is that $E\,\tau = 1/p$.

In order to see this we wish to find a suitable martingale to which we can apply $\tau \wedge n$ and then let $n \to \infty$. The natural martingale (with mean 0) is (of course?) $\{(X_n, \mathcal{F}_n),\, n \geq 0\}$, where $X_n = \sum_{k=1}^{n}(Y_k - p)$, $n \geq 1$ (recall Example 3.2). Upon observing that, in fact, $\tau = \min\{n : \sum_{k=1}^{n} Y_k = 1\}$, we can apply the optional sampling theorem, Theorem 7.1, to conclude that $E\, X_{\tau \wedge n} = 0$, that is, that

$$E \sum_{k=1}^{\tau \wedge n} Y_k = pE(\tau \wedge n).$$

Now, without appeal to any general result, we observe that

$$0 \leq \sum_{k=1}^{\tau \wedge n} Y_k \leq \sum_{k=1}^{\tau} Y_k = 1,$$

so that, since $\sum_{k=1}^{\tau \wedge n} Y_k \to \sum_{k=1}^{\tau} Y_k = 1$ as $n \to \infty$, the left-hand side converges to $E \sum_{k=1}^{\tau} Y_k = 1$ by bounded convergence. And, since $\tau \wedge n \nearrow \tau$ as $n \to \infty$, the right-hand side converges to $E\,\tau$ by monotone convergence. Putting the convergences together we find that $E\,\tau = 1/p$ as claimed.

We can, however, do more. Namely, set

$$X_n = \frac{\exp\{s \sum_{k=1}^{n} Y_k\}}{(1 - p + pe^s)^n}, \quad n \geq 1.$$

This defines another martingale (recall Example 3.6). By the optional sampling theorem we therefore know that $E\,X_{\tau \wedge n} = 1$, which, rewritten, tells us that, for $s$ in a neighborhood of 0,

$$E\left(\frac{\exp\{s\sum_{k=1}^{\tau \wedge n} Y_k\}}{(1-p+pe^s)^{\tau \wedge n}}\right) = 1.$$

Since $0 \leq \sum_{k=1}^{\tau \wedge n} Y_k \leq \sum_{k=1}^{\tau} Y_k = 1$, it follows that

$$0 \leq X_{\tau \wedge n} \leq \frac{\exp\{|s|\}}{1-p+pe^{-|s|}},$$

so that the bounded convergence theorem allows us to conclude that

$$E\left(\frac{\exp\{s\sum_{k=1}^{\tau} Y_k\}}{(1-p+pe^s)^{\tau}}\right) = 1,$$

that is, that

$$E\left(\frac{\exp\{s\}}{(1-p+pe^s)^{\tau}}\right) = 1.$$

The change of variable $u = (1-p+pe^s)^{-1}$ then shows that

$$E\,u^{\tau} = \frac{pu}{1-p+pu},$$

which is the probability generating function of a geometric random variable with mean $1/p$. The uniqueness theorem for generating functions, Theorem 4.7.1, therefore allows us to conclude that $\tau \in \text{Ge}(p)$.

A famous, more sophisticated, example is

## 14.4 The Gambler's Ruin Problem

Two persons play a coin tossing game. Player $A$ starts with $a$ Euros, player $B$ with $b$ Euros, both amounts being integers. If the coin turns heads, $A$ wins one Euro from $B$ and if it turns tails $B$ wins one Euro from $A$. The game ends when one player goes broke.

This can be modeled by a simple random walk starting at 0, ending at $a > 0$ or $-b < 0$, whichever is reached first. The traditional, classical solution is to set up a system of difference equations.

The more beautiful way is to use martingales.

Let $p = P(\text{Player } A \text{ wins one round})$, and set $q = 1 - p$. If $Y_k$ is the outcome of round number $k$ with $Y$ as a generic random variable, then

$$P(Y = 1) = 1 - P(Y = -1) = p.$$

The game ends after $\tau$ rounds, where

$$\tau = \min\left\{n : \sum_{k=1}^{n} Y_k = a \text{ or } -b\right\}.$$

Suppose first that $p \neq q$. In this case we can argue with the aid of the moment generating function as in the previous example to obtain

$$E\left(\frac{\exp\{s\sum_{k=1}^{\tau \wedge n} Y_k\}}{(qe^{-s} + pe^s)^{\tau \wedge n}}\right) = 1.$$

By point-wise convergence, and the fact that

$$0 \leq \frac{\exp\{s\sum_{k=1}^{\tau \wedge n} Y_k\}}{(qe^{-s} + pe^s)^{\tau \wedge n}} \leq \frac{\exp\{|s|(a+b)\}}{qe^{-|s|} + pe^{-|s|}} = \exp\{|s|(a+b+1)\},$$

the bounded convergence theorem yields

$$E\left(\frac{\exp\{s\sum_{k=1}^{\tau} Y_k\}}{(qe^{-s} + pe^s)^{\tau}}\right) = 1. \tag{14.5}$$

Let $\alpha = P(\sum_{k=1}^{\tau} Y_k = a) = 1 - P(\sum_{k=1}^{\tau} Y_k = -b)$. Since (14.5) holds for all $s$, we continue our discussion for the particular value $s = \log(q/p)$, which has the special feature that the denominator equals 1 *and therefore vanishes.* Equation (14.5) therefore conveniently reduces to

$$\alpha\left(\frac{q}{p}\right)^a + (1 - \alpha)\left(\frac{q}{p}\right)^{-b} = 1,$$

the solution of which is

$$\alpha = P(\text{Player } A \text{ is the winner of the game}) = \frac{1 - (p/q)^b}{(q/p)^a - (p/q)^b},$$

with $1 - \alpha$ being the probability that player $B$ is the winner.

Before taking care of the symmetric case we present another martingale that does the job. Namely, set

$$X_n = \left(\frac{q}{p}\right)^{T_n}, \quad \text{where} \quad T_n = \sum_{k=1}^{n} Y_k, \quad n \geq 1.$$

Then $\{(X_n, \mathcal{F}_n), n \geq 0\}$ is a martingale of the product type with mean 1. The usual arguments thus lead to $E\,X_{\tau \wedge n} = 1$, and, since

$$0 \leq X_{\tau \wedge n} \leq \left(\frac{q}{p}\right)^{a+b} + \left(\frac{p}{q}\right)^{a+b},$$

via point-wise convergence and bounded convergence, to $E\,X_{\tau} = 1$, which, rewritten, tells us that

$$E\left(\frac{q}{p}\right)^{T_\tau} = 1.$$

With $\alpha = P(\text{Player } A \text{ is the winner of the game})$ as before we obtain

$$\alpha\left(\frac{q}{p}\right)^a + (1 - \alpha)\left(\frac{q}{p}\right)^{-b} = 1,$$

which is the same equation as before, but obtained in a much swifter way. We might say that the second choice of martingale was a better one (but both are interesting as illustrations of the method).

It remains to solve the case $p = q = 1/2$, and for this we consider the martingale from Example 3.2 defined by $X_n = \sum_{k=1}^{n} Y_k$, $n \geq 1$. The usual procedure produces

$$E\, X_{\tau \wedge n} = 0,$$

which, together with the bound

$$|X_{\tau \wedge n}| \leq a + b,$$

and bounded convergence, shows that

$$E\, X_\tau = 0.$$

In this case $\alpha$ is obtained as the solution of the equation

$$\alpha \cdot a + (1 - \alpha) \cdot (-b) = 0,$$

which is

$$\alpha = \frac{b}{a + b}.$$

In particular, if $a = b$, that is, both players start off with the same amount of money, they win with probability $1/2$ each, which is what intuition would suggest.

It is also possible to compute the expected duration of the game.

In the asymmetric case we apply the martingale from Example 3.2, the centered random walk, to the optional sampling theorem, to obtain $E\, X_{\tau \wedge n} = 0$, that is,

$$E \sum_{k=1}^{\tau \wedge n} Y_k = (p - q)E(\tau \wedge n).$$

Since $|\sum_{k=1}^{\tau \wedge n} Y_k| \leq a + |b|$, the corresponding bounded and monotone convergences, respectively, show that

$$E \sum_{k=1}^{\tau} Y_k = (p - q)E\tau.$$

Since the left-hand side is available from the first part of the discussion we are now in the position to find $E\tau$. In fact,

$$E\tau = \frac{E \sum_{k=1}^{\tau} Y_k}{p - q} = \frac{\alpha \cdot a + (1 - \alpha) \cdot (-b)}{p - q}$$

$$= \frac{1}{p - q}\left(a - (a + b)\frac{1 - (p/q)^b}{(q/p)^a - (p/q)^b}\right).$$

In order to find the expected duration of the game in the symmetric case we use the martingale from Example 3.3. Thus, set $X_n = \left(\sum_{k=1}^{n} Y_k\right)^2 - n$, for $n \geq 1$, $(\operatorname{Var} Y = 1)$, so that $\{(X_n, \mathcal{F}_n), n \geq 0\}$ is a martingale with mean 0. Proceeding as before we obtain $E\, X_{\tau \wedge n} = 0$, that is

$$E\left(\sum_{k=1}^{\tau \wedge n} Y_k\right)^2 = E(\tau \wedge n). \tag{14.6}$$

In this case,

$$0 \leq \left(\sum_{k=1}^{\tau \wedge n} Y_k\right)^2 \leq (a+b)^2, \tag{14.7}$$

and the sum converges point-wise to $\sum_{k=1}^{\tau} Y_k$, so that, by bounded convergence, the left-hand side of (14.6) converges to $E\left(\sum_{k=1}^{\tau} Y_k\right)^2$.

The right-hand side of (14.6) converges to $E\,\tau$ by monotone convergence. Moreover, by Fatou's lemma, and (14.6), and (14.7),

$$E\,\tau \leq \liminf_{n \to \infty} E(\tau \wedge n) \leq (a+b)^2 < \infty.$$

Putting our findings together proves that

$$E\left(\sum_{k=1}^{\tau} Y_k\right)^2 = E\,\tau.$$

Inserting the absorption probabilities $\alpha$ and $1 - \alpha$ at the respective ends, we finally obtain

$$E\,\tau = \alpha a^2 + (1-\alpha)b^2 = \frac{b}{a+b} \cdot a^2 + \frac{a}{a+b} \cdot b^2 = ab.$$

## 14.5 A Converse

Theorem 14.1 states, in particular, that if moments of certain orders of the summands and the stopping time exist, then so does the stopped sum. For example, if, for $p \geq 2$, $E|Y|^p < \infty$ and $E\,\tau^{p/2} < \infty$, then $E|X_\tau|^p < \infty$. Theorem 14.3(i) provides a convenient means to compute the expected value of a stopped random walk. However, before applying the theorem we must know that the stopping time has finite mean. On the other hand, sometimes one may have a good grasp on $E\,S_\tau$, and not on $E\,\tau$, and one would like to use the Wald equation in order to find $E\,\tau$. But, once again, we must first know that $E\,\tau$ is finite, which means that we are caught in a catch-22 situation.

A number of "converse results" can be found in [119]. One example, which is to be exploited later in this section, is Theorem 2.2 there (see also [110], Theorem I.5.5).

**Theorem 14.4.** *Let* $Y_1, Y_2, \ldots$ *be independent, identically distributed random variables with partial sums* $S_n$, $n \geq 1$, *and suppose that* $\mu = EY \neq 0$. *Further, let* $\tau$ *be a stopping time.*
(i) *Then*

$$E|S_\tau| < \infty \quad \Longrightarrow \quad E\tau < \infty.$$

(ii) *If, in addition,* $E|Y|^p < \infty$, *then*

$$E|S_\tau|^p < \infty \quad \Longrightarrow \quad E\tau^p < \infty.$$

*Proof.* (i): The proof is based on a beautiful device due to Blackwell [22, 23], and the converse(!) of the strong law of large numbers.

Let $\{\tau_n, n \geq 1\}$ be independent copies of $\tau$, constructed as follows: Let $\tau_1 = \tau$. Restart the random walk at "time" $\tau_1$, and let $\tau_2$ be the corresponding stopping time for that sequence. Technically this means that we consider the sequence $Y_{\tau_1+1}, Y_{\tau_1+2} \ldots$, and let $\tau_2$ play the same role as $\tau_1$ did for the original sequence. Then restart at $\tau_1 + \tau_2$, and so on.

This makes $\{\tau_n, n \geq 1\}$ into a sequence of independent, identically distributed random variables, distributed as $\tau$. Moreover, $\{S_{\tau_1+\tau_2+\cdots+\tau_k}, k \geq 1\}$ is a sequence of partial sums of independent, identically distributed random variables distributed as $S_\tau$.

The strong law of large numbers applied to these sums therefore tells us that

$$\frac{S_{\tau_1+\tau_2+\cdots+\tau_k}}{k} \overset{a.s.}{\to} E S_\tau \quad \text{as} \quad k \to \infty.$$

On the other hand,

$$\frac{S_{\tau_1+\tau_2+\cdots+\tau_k}}{\tau_1 + \tau_2 + \cdots + \tau_k} \overset{a.s.}{\to} EY \quad \text{as} \quad k \to \infty,$$

by the "random index strong law of large numbers", Theorem 6.8.2(iii).

Combining the two yields

$$\frac{\tau_1 + \tau_2 + \cdots + \tau_k}{k} = \frac{\tau_1 + \tau_2 + \cdots + \tau_k}{S_{\tau_1+\tau_2+\cdots+\tau_k}} \cdot \frac{S_{\tau_1+\tau_2+\cdots+\tau_k}}{k}$$

$$\overset{a.s.}{\to} \frac{1}{E S_\tau} \cdot EY \quad \text{as} \quad k \to \infty.$$

By *the converse* of the Kolmogorov strong law of large numbers, Theorem 6.6.1(b), we therefore conclude that $E\tau < \infty$.
(ii): We combine the inequality

$$|\mu\tau| \leq |S_\tau - \mu\tau| + |S_\tau|, \tag{14.8}$$

Theorem 14.1, and the $c_r$-inequality (Theorem 3.2.2), and work ourselves through the powers of 2.

If $1 < p \leq 2$, (14.8), Theorem 14.1, the $c_r$-inequality, and part (i) of the theorem tells us that

$$|\mu|^p E \, \tau^p \le 2^{p-1} B_p E \tau E |Y|^p + 2^{p-1} E |S_\tau|^p < \infty.$$

If $2 < p \le 2^2$ the same procedure yields

$$|\mu|^p E \, \tau^p \le 2^{p-1} 2 B_p E \tau^{p/2} E |Y|^p + 2^{p-1} E |S_\tau|^p < \infty.$$

And so on. $\qquad\qquad\qquad\qquad\qquad\qquad\qquad\qquad\qquad\qquad\qquad\quad$ □

*Remark 14.2.* The proof shows i.a. that the converse of the strong law not only tells us that finite mean is necessary for the strong law to hold, it may also be used to *conclude* that the mean of the summands at hand must be finite.

*Remark 14.3.* The assumption that the mean is not zero cannot be dispensed with; the symmetric coin is a simple counter-example. Remember that we had $S_\tau = 1$, $EY = 0$, and $E\tau = \infty$ in that example.

*Remark 14.4.* A fact that we passed by in the proof was a detailed verification of the fact that $\tau_1, \tau_2, \ldots$ are, indeed, independent, identically distributed random variables. The reason for this being true is that the pre-$\sigma$-algebra is independent of what comes after a stopping time, so that $\mathcal{F}_{\tau_k}$ and $\{Y_n, \, n > \tau_k\}$ are independent for all $k$. In another language this is a consequence of the strong Markov property. $\qquad\qquad\qquad\qquad\qquad\qquad\qquad\qquad$ □

**Exercise 14.1.** Put the details of the previous remark on paper. $\qquad\qquad$ □

*Example 14.2.* If we reconsider tossing a coin until heads appears, Subsection 10.14.3, once more, we know that $EY = p$ and that $S_\tau \, (= X_\tau) = 1$ a.s. Via Theorem 14.4 we obtain $E\tau < \infty$, after which Theorem 14.3 – without resorting to $\tau \wedge n$ and letting $n \to \infty$ – tells us that

$$E\tau = \frac{ES_\tau}{EY} = \frac{1}{p}.$$

Once again, this is really hard artillery for a triviallity, nevertheless, it illustrates a thought. $\qquad\qquad\qquad\qquad\qquad\qquad\qquad\qquad\qquad\qquad\qquad\qquad$ □

# 15 Regularity

Much of the work involved in limit theorems for martingales concerns the asymptotics of moments and conditional moments. We have also seen, in Theorem 12.1, that the closable, or regular martingales have a more pleasant (regular) behavior. We have also seen in Theorem 7.1 that the martingale property is preserved for regular martingales evaluated at stopping times. Finally, in the examples above, and in the work on stopped random walks we had to check expected values or moments of $\tau \wedge n$ as $n \to \infty$.

With this in mind, let us introduce the concept of *regular stopping time*, and collect some general facts for regular martingales and regular stopping times.

**Definition 15.1.** *A stopping time $\tau$ is a* regular stopping time *for the martingale $\{(X_n, \mathcal{F}_n), n \geq 0\}$ if*

$$\{(X_{\tau \wedge n}, \mathcal{F}_n), n \geq 0\} \quad \text{is a regular martingale.} \qquad \square$$

Regular martingales are almost surely convergent. Combining this with the uniform integrability and Theorem 5.5.2 yields the first part of the following result (which we, for simplicity, state under the additional assumption that the stopping time has a finite mean).

**Theorem 15.1.** *Let $\{(X_n, \mathcal{F}_n), n \geq 0\}$ be a martingale, and suppose that $\tau$ is a stopping time with finite mean. The following are equivalent:*

(i)     $\tau$ *is a regular stopping time;*
(ii)    $X_{\tau \wedge n} \to X_\tau$ *almost surely and in $L^1$ as $n \to \infty$.*

*If $\tau$ is regular, then $E\,X_{\tau \wedge n} \to E\,X_\tau$ as $n \to \infty$.*

*Proof.* The equivalence between (i) and (ii) was discussed prior to the statement of the proposition. The convergence of the expected values follows from the fact that

$$|E\,X_{\tau \wedge n} - E\,X_\tau| \leq E|X_{\tau \wedge n} - X_\tau| \to 0 \quad \text{as} \quad n \to \infty. \qquad \square$$

Reviewing Example 14.1 we find that the stopping time defined there is *not* regular. We also note that the main effort in the coin-tossing problem, the Gambler's ruin problem, and the proof of the Wald equations, Theorem 14.3, was to prove regularity. Actually, not quite, it was to verify the last statement in Theorem 15.1, which is somewhat less than proving uniform integrability. However, with some additional arguments we obtain regularity as follows.

**Theorem 15.2.** *Let $Y, Y_1, Y_2, \ldots$ be independent, identically distributed random variables, and let $S_n, n \geq 1$, denote the partial sums. Suppose that $\tau$ is a stopping time relative to a filtration, $\{\mathcal{F}_n, n \geq 1\}$, and that $E\tau < \infty$.*

(i) *If $\mu = E\,Y < \infty$, then $\tau$ is regular for $\{(S_n - n\mu, \mathcal{F}_n), n \geq 0\}$.*
(ii) *If $\sigma^2 = \operatorname{Var} Y < \infty$, then $\tau$ is regular for $\{((S_n - n\mu)^2 - n\sigma^2, \mathcal{F}_n), n \geq 0\}$.*

*Proof.* From the proof of Theorem 14.3 we extract that

$$|X_{\tau \wedge n}| = |S_{\tau \wedge n} - \mu(\tau \wedge n)| \leq \sum_{k=1}^{\tau \wedge n} |Y_k| + \mu(\tau \wedge n) \leq \sum_{k=1}^{\tau} |Y_k| + \mu\tau \in L^1,$$

by Theorems 14.1 or 14.3, that is, we have found an integrable random variable that dominates the sequence $\{X_{\tau \wedge n}, n \geq 1\}$, which therefore, by Theorem 5.4.4, is uniformly integrable. This proves (i).

To prove (ii) we note that in the proof of the second Wald equation we showed that $S_{\tau \wedge n} - \mu(\tau \wedge n) \to S_\tau - \mu(\tau) = X_\tau$ almost surely and in $L^2$, as

$n \to \infty$, which due to Theorem 5.5.2 proves that $\{(S_{\tau \wedge n} - \mu(\tau \wedge n))^2, \, n \geq 1\}$ is uniformly integrable.

Moreover, since $\tau \wedge n \leq \tau$ which is integrable, another application of Theorem 5.5.2 asserts that $\{\tau \wedge n, \, n \geq 1\}$ is uniformly integrable.

And, since the sum (as well as the difference) of two uniformly integrable sequences is uniformly integrable (Theorem 5.4.6), we conclude that

$$\{(S_{\tau \wedge n} - \mu(\tau \wedge n))^2 - \sigma^2(\tau \wedge n), \, n \geq 1\} \quad \text{is uniformly integrable.} \qquad \square$$

By employing the new terminology we may rephrase the results in Sections 10.7 and 10.8 in the following language.

**Theorem 15.3.** *Let $\{(X_n, \mathcal{F}_n), \, n \geq 0\}$ be a martingale.*

- *If $\{(X_n, \mathcal{F}_n), \, n \geq 0\}$ is regular, then any stopping time is regular.*
- *If $\tau$ is regular and $\tau_1 \leq \tau$ is another stopping time, then $\tau_1$ is regular too,*

$$E(X_\tau \mid \mathcal{F}_{\tau_1}) = X_{\tau_1}, \quad \text{and} \quad E\,X_0 = E\,X_{\tau_1} = E\,X_\tau.$$

- *If $\tau$ is regular and $\tau_1 \leq \tau_1 \leq \cdots \leq \tau_k \leq \tau$ are stopping times, then $\{\tau_i, \, 1 \leq i \leq k\}$ are regular,*

$$E(X_{\tau_j} \mid \mathcal{F}_{\tau_i}) = X_{\tau_i} \quad \text{for} \quad 1 \leq i < j \leq k,$$

*and $E\,X_0 = E\,X_{\tau_j} = E\,X_\tau$ for $1 \leq j \leq k$.*

**Exercise 15.1.** Please, verify the details of these statements. $\qquad \square$

The obvious question at this point is whether there are some convenient necessary and sufficient criteria for a stopping time to be regular.

Here is one such result.

**Theorem 15.4.** *Suppose that $\{(X_n, \mathcal{F}_n), \, n \geq 0\}$ is a martingale. The stopping time $\tau$ is regular if and only if*

(i) $\quad E|X_\tau|I\{\tau < \infty\} < \infty$, *and*

(ii) $\quad \{X_n I\{\tau > n\}, \, n \geq 0\}$ *is uniformly integrable.*

*Proof.* Suppose first that $\tau$ is regular. By Fatou's lemma,

$$E|X_\tau|I\{\tau < \infty\} \leq \liminf_{n \to \infty} E|X_{\tau \wedge n}|I\{\tau \wedge n < \infty\}$$
$$= \liminf_{n \to \infty} E|X_{\tau \wedge n}|I\{\tau \leq n\} \leq \sup_n E|X_{\tau \wedge n}| < \infty,$$

since $\{X_{\tau \wedge n}\}$ is uniformly integrable. This verifies (i).

To verify (ii) it suffices to observe that the sequence $\{|X_n I\{\tau > n\}|, \, n \geq 0\}$ is dominated by the uniformly integrable sequence $\{|X_{\tau \wedge n}|, \, n \geq 0\}$, and, therefore, in view of Theorem 5.4.4, is uniformly integrable too.

Conversely, assume that (i) and (ii) are satisfied. Then,

$$
\begin{aligned}
E|X_{\tau\wedge n}|I\{|X_{\tau\wedge n}| > a\} &= E|X_{\tau\wedge n}|I\{\{|X_{\tau\wedge n}| > a\}\cap\{\tau\le n\}\} \\
&\quad + E|X_{\tau\wedge n}|I\{\{|X_{\tau\wedge n}| > a\}\cap\{\tau > n\}\} \\
&= E|X_\tau|I\{\{|X_\tau| > a\}\cap\{\tau\le n\}\} \\
&\quad + E|X_n|I\{\{|X_n| > a\}\cap\{\tau > n\}\} \\
&\le E|X_\tau|I\{\tau < \infty\}I\{|X_\tau|I\{\tau<\infty\} > a\} \\
&\quad + E|X_n|I\{\tau > n\}I\{|X_n|I\{\tau > n\} > a\} \\
&\to 0 \quad\text{as}\quad a\to\infty,
\end{aligned}
$$

by (i) (the tail of a convergent integral) and (ii), respectively.

We have thus shown that $\{X_{\tau\wedge n}, n \ge 0\}$ is uniformly integrable, and, hence, that $\tau$ is regular.   □

*Remark 15.1.* For the Gambler's ruin problem we have shown above that $E\tau < \infty$, so that $X_\tau I\{\tau < \infty\} = X_\tau$, which equals $a$ or $b$ depending on the barrier that is hit. Secondly, $|X_n|I\{\tau > n\} \le \max\{a, |b|\}$ which is a finite constant. An application of Theorem 15.4 shows that $\tau$ is regular for the martingales considered there.   □

Sometimes a sufficient condition is sufficient for ones needs. Here is one that ensures that Theorem 15.4(i) is satisfied.

**Theorem 15.5.** *If* $\{(X_n, \mathcal{F}_n), n \ge 0\}$ *is an* $L^1$*-bounded martingale, then* $E|X_\tau| < \infty$, *and, all the more,* $E|X_\tau|I\{\tau < \infty\} < \infty$.

*The conclusion holds, in particular, for all non-negative martingales.*

*Proof.* The $L^1$-boundedness implies that $X_n \overset{a.s.}{\to} X_\infty$ as $n\to\infty$, so that $X_\tau$ is always well defined. Moreover, since $\{(|X_n|, \mathcal{F}_n), n \ge 0\}$ is a submartingale we can apply the second part of Theorem 7.4 to conclude that $\{X_0, X_{\tau\wedge n}, X_n\}$ is a submartingale. An application of Fatou's lemma therefore yields

$$
E|X_\tau| \le \liminf_{n\to\infty} E|X_{\tau\wedge n}| \le \liminf_{n\to\infty} E|X_n| \le \sup_n E|X_n| < \infty.
$$

The final statement is a consequence of the fact that all non-negative martingales are $L^1$-bounded automatically.   □

Important stopping times are, as has been mentioned before, "the first time something special happens". With the aid of the regularity criteria we can easily establish the following result for first entrance times, or escape times, depending on the angle from which one is observing the process.

**Theorem 15.6.** *Suppose that* $\{(X_n, \mathcal{F}_n), n \ge 0\}$ *is an* $L^1$*-bounded martingale. Then, for* $-\infty < b < 0 < a < \infty$,

$$
\begin{aligned}
\tau_a &= \min\{n : |X_n| > a\}, \quad\text{and} \\
\tau_{b,a} &= \min\{n : X_n < b \text{ or } X_n > a\}
\end{aligned}
$$

*are regular stopping times.*

*Proof.* Because of the $L^1$-boundedness, Theorem 15.5 shows that condition (i) in Theorem 15.4 is satisfied. The fact that $|X_n|I\{\tau > n\} < a$ in the first case, and that $|X_n|I\{\tau > n\} < \max\{a, |b|\}$ in the second case, verifies condition Theorem 15.4(ii).    □

In the following subsection we shall encounter a model where the associated martingale is not $L^1$-bounded, but the stopping time of interest, a one-sided escape time, is regular all the same.

## 15.1 First Passage Times for Random Walks

One application of stopped random walks is the theory of first passage times for random walks. The background was presented in Section 2.16. We have later met the topic again in connection with the three classical limit theorems. Here we shall address questions such as: When is the stopping time finite? What are the appropriate conditions for finiteness of moments? When is (some power of) the stopped sum integrable?

Formally, let $Y, Y_1, Y_2, \ldots$ be independent, random variables with mean $\mu > 0$ and partial sums $S_n = \sum_{k=1}^{n} Y_k$, $n \geq 1$, and first passage times

$$\tau = \tau(t) = \min\{n : S_n > t\}, \quad \text{for some} \quad t > 0.$$

The stopped sum is $S_\tau = S_{\tau(t)}$.

We also recall the important "sandwich inequality" (6.9.3):

$$t < S_\tau = S_{\tau-1} + Y_\tau \leq t + Y_\tau. \tag{15.1}$$

Now, putting $X_n = S_n - n\mu$ for $n \geq 1$ we know that $\{(X_n, \mathcal{F}_n), n \geq 0\}$ is a martingale, so that by Theorem 7.1, $E X_{\tau \wedge n} = 0$, that is,

$$E S_{\tau \wedge n} = \mu \cdot E(\tau \wedge n). \tag{15.2}$$

Suppose, to begin with, that the summands are bounded above; $P(Y \leq M) = 1$ for some $M > 0$. Since $S_{\tau \wedge n} \leq S_\tau$, (15.1) tells us that

$$S_{\tau \wedge n} \leq t + M,$$

which, together with (15.2), and Fatou's lemma shows that

$$E\tau \leq \liminf_{n \to \infty} E(\tau \wedge n) \leq \frac{t + M}{\mu},$$

which is some finite constant.

To remove the restriction we introduce the truncated random walk, whose summands are $Y_k I\{Y_k \leq M\}$, $k \geq 1$. The crucial observation now is that the summands, and therefore the partial sums, of the truncated random walk are *smaller* than those of the original random walk, which means that the level $t$

is reached *more rapidly* for the original random walk than for the truncated one. In other words, the first passage time for the original random walk is *smaller* than that of the truncated random walk, which, in turn had a finite mean.

A reference to Theorem 15.2 shows that $\tau$, in addition, is regular for the martingale $\{(S_n - n\mu, \mathcal{F}_n), n \geq 1\}$.

We have thus established the following result, which because of its importance is given as a theorem.

**Theorem 15.7.** *Let $Y_1$, $Y_2$, ... be independent, random variables with mean $\mu > 0$ and partial sums $S_n$, $n \geq 1$. Let*

$$\tau = \tau(t) = \min\{n : S_n > t\}, \quad for\ some\quad t > 0.$$

*Then*

(i)     $E\tau < \infty;$

(ii)    $\tau$ *is regular for the martingale* $\{(S_n - n\mu, \mathcal{F}_n), n \geq 1\}.$

*Remark 15.2.* The martingale defined in the proof is *not* $L^1$-bounded and that $\tau$ is a *one-sided* escape time. We thus had two reasons for not being able to apply Theorem 15.6. However, $\tau$ turned out to be regular all the same.     □

Returning to (15.1) we note that the "last" summand, $Y_\tau$, plays an important role in this context. The following corollary to Theorem 15.7 provides information about the integrability of $Y_\tau$.

**Theorem 15.8.** *Let $p \geq 1$, and let $Y$, $Y_1$, $Y_2$, ... and $\tau$ be given as before. If $E|Y|^p < \infty$, then*

$$E|Y_\tau|^p \leq E\tau E|Y|^p < \infty.$$

*Proof.* Since

$$|Y_\tau|^p \leq \sum_{k=1}^{\tau} |Y_k|^p,$$

the conclusion follows via the first Wald equation, Theorem 14.3(i).     □

*Remark 15.3.* In this case the last summand must be positive. The theorem therefore remains true with $|Y|$ replaced by $Y^+$.

*Remark 15.4.* As a by-product we point out loudly that $Y_\tau$ does *not* behave like a standard summand; for example, it is always positive. On the one hand this is trivial, on the other it is sometimes forgotten (or overlooked) in the literature.     □

By combining (15.1) and Theorem 15.8 we are in the position to prove

**Theorem 15.9.** *Let $p \geq 1$. In the above setup,*

$$E(S_\tau)^p < \infty \quad \Longleftrightarrow \quad E(Y^+)^p < \infty.$$

*Proof.* "If" is immediate from (15.1) and Theorem 15.8, and "only if" follows from the fact that

$$S_\tau \geq Y_1^+,$$

because if the first summand does not reach the required level we must wait for a larger sum.    □

Whereas the positive tail of the summands is crucial for the integrability of the stopped sum, the negative tail is crucial for the integrability of $\tau$.

**Theorem 15.10.** *Let $p \geq 1$. Under the present setup,*

$$E\,\tau^p < \infty \quad \Longleftrightarrow \quad E(Y^-)^p < \infty.$$

*Proof.* For the sufficiency we exploit Theorem 14.4.

If $E|Y|^p < \infty$ we know that from Theorem 15.9 that $E(S_\tau)^p < \infty$. This, together with Theorem 14.4 proves that $E\,\tau^p < \infty$. The conclusion then follows via the truncation procedure in the proof of Theorem 15.7, because we create smaller summands which results in a larger stopping time which, in turn, is known to have a finite moment of the requested order.

For the necessity we refer to [110], p. 80.    □

**Exercise 15.2.** Check the details.

**Exercise 15.3.** The last two proofs show that integrability of $Y^-$ is irrelevant for the integrability of the stopped sum, and that integrability of $Y^+$ is irrelevant for the integrability of the stopping time. Explain in words why this is "obvious".    □

## 15.2 Complements

The results here concern random walks drifting to $+\infty$. It is known from classical random walk theory that there are three kinds of random walks; those drifting to $+\infty$, those drifting to $-\infty$, and the oscillating ones. In particular, random walks whose increments have positive mean, negative mean, and zero mean, belong to the classes in that order (this can, for example, be seen with the aid of the law of large numbers and the law of the iterated logarithm). Moreover, it can be shown that the hitting times we have discussed in this section are infinite with positive probability when the increments have a negative expected value, and almost surely finite, but with infinite expectation, when the mean is zero. We refer to the classic by Spitzer [234] for more on this. An overview and additional references can be found in [110].

Let us also comment somewhat on, what seems to be almost the same, namely $\nu = \nu(t) = \max\{n : S_n \leq t\}$. If the increments are positive, i.e., for renewal processes, we have

$$\nu + 1 = \tau, \tag{15.3}$$

that is, they are essentially the same. This means that many renewal theoretic results can be obtained with the aid of stopped random walks. As for sequential analysis we recall that the procedure was to continue sampling until $L_n$, the likelihood ratio, escapes from a strip, so that the random sample size becomes

$$\tau_{b,a} = \min\{n : L_n \notin (b, a)\}, \quad \text{where} \quad 0 < b < a < \infty.$$

In order to see that the procedure eventually stops, we recognize that the problem fits into the above framework. In fact, Theorem 15.6 tells us that $\tau_{b,a}$ is regular.

Since, as we have noticed before, a general random walk may well drop below the level $t$ a number of times before finally taking off to $+\infty$, (15.3) does not hold in this case. The mathematical difference, which is an essential one, is that $\nu$ is *not* a stopping time. A distinction between the first passage time and this last exit time can be read off in the following analog of Theorem 15.10. For a nice proof we refer to [147], Theorem 1.

**Theorem 15.11.** *Let $p > 0$. Then*

$$E\,\nu^p < \infty \quad \Longleftrightarrow \quad E(Y^-)^{p+1} < \infty.$$

The "price" for not being a stopping time, we might say, is that an additional moment of $Y^-$ is required in order to obtain the same level if integrability for the last exit time.

## 15.3 The Wald Fundamental Identity

Given a random variable $Y$, let $g$ be the logarithm of the moment generating function:

$$g(u) = \log \psi(u) = \log E \exp\{uY\}, \quad u \in \mathbb{R},$$

and suppose that $g$ is finite in some interval containing 0. (Sometimes in the literature $g$ is called the cumulant generating function. We have, however, kept the more common terminology, namely, that the cumulant generating function is the logarithm of the characteristic function (recall Chapter 4).)

Excluding the trivial case when $g$ is constant, an application of the Hölder inequality shows that $g$ is strictly convex, where it exists. Moreover, $g(0) = 0$, $g'(u) = \psi'(u)/\psi(u)$, so that $E Y = g'(0)$. Checking the second derivative one finds that $\text{Var}\, Y = g''(0)$.

Now, given $Y_1, Y_2, \ldots$, independent, identically distributed random variables with partial sums $S_n$, $n \geq 1$, we can rewrite the exponential Wald martingale, Example 3.6, as

$$X_n = \frac{e^{uS_n}}{(\psi(u))^n} = \exp\{uS_n - ng(u)\}. \tag{15.4}$$

Our next project is to prove that the first passage time across a positive level, $\tau$, is regular (for $\{(X_n, \mathcal{F}_n),\ n \geq 0\}$, where $\{\mathcal{F}_n\}$ are the natural $\sigma$-algebras), and that $E X_\tau = 1$, under the assumption of a positive mean. The proof amounts "of course" to applying Theorem 7.1 and some regularity criterion from Section 10.15.

**Theorem 15.12.** *Let $Y, Y_1, Y_2, \ldots$ be independent, identically distributed random variables, and let $\{(X_n, \mathcal{F}_n),\ n \geq 0\}$ be the Wald martingale as defined in (15.4). Suppose that $E Y \geq 0$, and that $g = \log \psi$ is finite in a neighborhood of 0. Finally, let*

$$\tau = \min\{n : S_n > t\}, \quad t \geq 0.$$

*Then*

(a)   $X_n \overset{a.s.}{\to} 0$ *as* $n \to \infty$;

(b)   $\tau$ *is regular for* $\{(X_n, \mathcal{F}_n),\ n \geq 0\}$;

(c)   $E X_\tau = E \exp\{u S_\tau - \tau g(u)\} = 1$;

(d)   *The same holds for any stopping time* $\tau_* \leq \tau$.

*Proof.* We begin by proving a.s. convergence. Let $G = \{u : g(u) < \infty\}$, and suppose that $2u \in G$. By convexity,

$$g(u) = g\left(\frac{1}{2} \cdot 0 + \frac{1}{2} \cdot 2u\right) < \frac{1}{2}g(0) + \frac{1}{2}g(2u) = \frac{1}{2}g(2u),$$

so that $u \in G$. Therefore, since, by the martingale convergence theorem, the Wald martingale is a.s. convergent, we have

$$X_n \overset{a.s.}{\to} X_\infty \quad \text{as} \quad n \to \infty,$$

which, rewritten and squared, yields

$$\exp\{u S_n - n g(u)\} \overset{a.s.}{\to} X_\infty \quad \text{and} \quad \exp\{2(u S_n - n g(u))\} \overset{a.s.}{\to} (X_\infty)^2$$

as $n \to \infty$. However, since

$$\exp\{2u S_n - n g(u)\} > \exp\{2(u S_n - g(u))\} = \left(\exp\{u S_n - g(u)\}\right)^2$$
$$\overset{a.s.}{\to} (X_\infty)^2 \quad \text{as} \quad n \to \infty,$$

it follows that $(X_\infty)^2 > (X_\infty)^2$, which necessitates that $X_\infty = 0$ a.s.

To prove regularity we wish to apply Theorem 15.4.

From Corollary 7.2 we know that

$$E X_{\tau \wedge n} = E \exp\{u S_{\tau \wedge n} - (\tau \wedge n)g(u)\} = 1.$$

As for $E \tau$, we have already proved finiteness for the case $E Y > 0$. That this also holds when the mean equals 0 follows from general random walk theory (because of the oscillating character of the random walk referred to earlier).

Since the variance is finite this is, in fact, also a consequence of the law of the iterated logarithm, Theorem 8.1.2.

The integrability of $\tau$ is enough for Theorem 15.4(i) in view of Theorem 15.5, since our martingale is non-negative. It thus remains to verify condition Theorem 15.4(ii), that is, that

$$\{X_n I\{\tau > n\}\} = \{\exp\{uS_n - ng(u)\}I\{\tau > n\}\} \quad \text{is uniformly integrable.}$$

However, since $0 \leq X_n I\{\tau > n\} \leq X_n \overset{a.s.}{\to} 0$ as $n \to \infty$, showing uniform integrability is equivalent to showing that

$$E X_n I\{\tau > n\} \to 0 \quad \text{as} \quad n \to \infty, \tag{15.5}$$

(for example by Theorem 5.5.2).

We use what is called *exponential tilting*. This amounts to a change of measure, the tilting meaning that the mean is changed, that "the general direction" (think law of large numbers) of the random walk is changed. Our new measure (on a new probability space) is defined as

$$dP^*(x) = \exp\{ux - g(u)\} \, dP(x),$$

where $P$ is the original measure. The new measure is non-negative. That it is a *probability* measure is seen by integration:

$$\int_{\mathbb{R}} dP^*(x) = \int_{\mathbb{R}} \exp\{ux - g(u)\} \, dP(x) = \exp\{-g(u)\} \, \psi(u) = 1.$$

Moreover, the new mean is non-negative, since, letting a superscript star denote a tilted random variable,

$$EY^* = \int_{\mathbb{R}} x \, dP^*(x) = \int_{\mathbb{R}} x \exp\{ux - g(u)\} \, dP(x)$$

$$= \exp\{-g(u)\} \, \psi'(u) = \frac{EY}{\psi(u)} \geq 0.$$

Moreover, due to independence, the probability measure of $\sum_{k=1}^{n} Y_k^*$ becomes

$$dP_n^*(z_1, z_2, \ldots, z_n) = \exp\left\{u \sum_{k=1}^{n} z_k - ng(u)\right\} \prod_{k=1}^{n} dP_k(z_k).$$

Finally, set $\mathbf{y}_t = \{(y_1, y_2, \ldots, y_n) : \sum_{k=1}^{n} y_k \leq t\}$. Then

$$E X_n I\{\tau > n\} = E \exp\{uS_n - ng(u)\}I\{\tau > n\} = \int_{\mathbf{y}_t} dP_n^*(z_1, z_2, \ldots, z_n)$$

$$= P^*\left(\sum_{k=1}^{n} Y_k^* \leq t\right) = P^*(\tau^* > n) \to 0 \quad \text{as} \quad n \to \infty,$$

since the tilted mean is non-negative, so that, as was pointed out earlier in the proof, $\tau^* < \infty$, $P^*$-a.s.

This verifies condition (ii) in Theorem 15.4, which proves (b), after which (c) and (d) follow from Theorem 15.3.                                                      □

*Remark 15.5.* Note that the first derivation in the Gambler's ruin problem was based on the exponential Wald martingale, and that (14.5) derived there, in fact, is precisely the Wald identity (for that particular stopping time).

*Remark 15.6.* We have i.a. shown that the martingale is *not* regular, but the first passage times are. In other words, we have a stopping time that is regular for a non-regular martingale.

*Remark 15.7.* The tilted measure is also called the *Esscher transform*.      □

The tilting of measures is a common and useful technique. In insurance risk problems where one has a negative drift and wishes to study passages to a positive level one tilts the measure so that it gets a positive drift, employs known results for that case, after which one "factors out" the tilting factor.

Another area is *importance sampling* where one tilts the measure in such a way that sampling (mainly) occurs where most of the original probability measure is concentrated. This technique is also important in *stochastic integration*. For more on this, see [235], in particular, Chapter 13.

# 16 Reversed Martingales and Submartingales

If we interpret $n$ as time, then reversing means reversing time. Traditionally one considers a sequence of *decreasing* $\sigma$-algebras $\{\mathcal{F}_n, \, n \geq 0\}$ and defines the reversed martingale property as

$$X_n = E(X_m \mid \mathcal{F}_n) \quad \text{for all} \quad m < n.$$

Note that the conditioning is on "the future". The more modern way is to let the reversed martingales be defined as ordinary, forward, martingales indexed by the negative integers.

Thus, let

$$\cdots \subset \mathcal{F}_{-(n+1)} \subset \mathcal{F}_{-n} \subset \cdots \subset \mathcal{F}_{-2} \subset \mathcal{F}_{-1} \subset \mathcal{F}_0,$$

and set $\mathcal{F}_{-\infty} = \cap_n \mathcal{F}_n$.

*Remark 16.1.* Note that the intersection of a sequence of $\sigma$-algebras is indeed a $\sigma$-algebra, and remember that the infinite union in the forward case was not necessarily so.                                                                   □

**Definition 16.1.** *An integrable $\{\mathcal{F}_n\}$-adapted sequence $\{X_n \, n \leq 0\}$ is a reversed martingale if*

$$E(X_n \mid \mathcal{F}_m) = X_m \quad \text{for all} \quad m \leq n \leq 0,$$

*or, equivalently,*

$$E(X_n \mid \mathcal{F}_{n-1}) = X_{n-1} \quad \textit{for all} \quad n \leq 0.$$

*It is called a* reversed submartingale *if*

$$E(X_n \mid \mathcal{F}_m) \geq X_m \quad \textit{for all} \quad m \leq n \leq 0,$$

*or, equivalently,*

$$E(X_n \mid \mathcal{F}_{n-1}) \geq X_{n-1} \quad \textit{for all} \quad n \leq 0,$$

*and a* reversed supermartingale *if*

$$E(X_n \mid \mathcal{F}_m) \leq X_m \quad \textit{for all} \quad m \leq n \leq 0,$$

*or, equivalently,*

$$E(X_n \mid \mathcal{F}_{n-1}) \leq X_{n-1} \quad \textit{for all} \quad n \leq 0 \qquad \qquad \square$$

*Remark 16.2.* The important difference between forward martingales and reversed ones is that, whereas the former have a first element but not necessarily a last element, the opposite is true for reversed martingales.          $\square$

Note that if $\{(X_n, \mathcal{F}_n),\ n \geq 0\}$ is a reversed martingale, it follows from the definition that

$$E(X_0 \mid \mathcal{F}_n) = X_n \quad \textit{for all} \quad n \leq 0,$$

which suggests that

**Theorem 16.1.** *Every reversed martingale is uniformly integrable.*

*Proof.* Since $\{(X_k, \mathcal{F}_k)\}$ with $k$ running from $-n$ to $0$, behaves, in reality, as a forward martingale, the proof of Theorem 11.1 carries over directly.          $\square$

**Exercise 16.1.** Write down the details.          $\square$

By the same argument we may consider the Kolmogorov-Doob inequality, Theorem 9.1, and the Doob maximal inequalities, Theorem 9.4. For example, for non-negative reversed submartingales,

$$\lambda P(\max_{-n \leq k \leq 0} X_k > \lambda) \leq E X_0,$$

so that, since the right-hand side does not depend on $n$, we obtain

$$\lambda P(\sup_{n \leq 0} X_n > \lambda) \leq E X_0.$$

For convenience we collect the results in a theorem and urge the reader to check the details.

**Theorem 16.2.** (The Kolmogorov-Doob inequality)
Let $\lambda > 0$.
(i) *Suppose that* $\{(X_n, \mathcal{F}_n), \, n \leq 0\}$ *is a non-negative, reversed submartingale.*
*Then*

$$\lambda P(\sup_{k \leq 0} X_k > \lambda) \leq E\, X_n^+ \leq E\, X_0.$$

(ii) *Suppose that* $\{(X_n, \mathcal{F}_n), \, n \leq 0\}$ *is a reversed martingale. Then*

$$\lambda P(\sup_{k \leq 0} |X_k| > \lambda) \leq E|X_0|.$$

**Theorem 16.3.** (The Doob maximal inequalities)
Let $p \geq 1$.
(i) *If* $\{(X_n, \mathcal{F}_n), \, n \leq 0\}$ *is a non-negative, reversed submartingale, then*

$$E(\sup_{k \leq 0} X_k)^p \leq \begin{cases} (\frac{p}{p-1})^p E\, X_0^p, & when \quad p > 1, \\ \frac{e}{e-1} + \frac{e}{e-1} E\, X_0 \log^+ X_0, & when \quad p = 1. \end{cases}$$

(ii) *If* $\{(X_n, \mathcal{F}_n), \, n \leq 0\}$ *is a reversed martingale, then*

$$E(\sup_{k \leq 0} |X_k|)^p \leq \begin{cases} (\frac{p}{p-1})^p E|X_0|^p, & when \quad p > 1, \\ \frac{e}{e-1} + \frac{e}{e-1} E|X_0| \log^+ |X_0|, & when \quad p = 1. \end{cases}$$

The main theorem for martingales extends to reversed ones, in fact, to *all* reversed martingales, since all of them are uniformly integrable, and, hence, regular.

**Theorem 16.4.** *Suppose that* $\{(X_n, \mathcal{F}_n), \, n \leq 0\}$ *is a reversed martingale. Then*

(a)   $\{X_n, \, n \geq 1\}$ *is uniformly integrable;*
(b)   $X_n \to X_{-\infty}$ *a.s. and in* $L^1$ *as* $n \to -\infty$;
(c)   $\{(X_n, \mathcal{F}_n), \, -\infty \leq n \leq 0\}$ *is a martingale.*

*Proof.* Uniform integrability has already been established in Theorem 16.1.
    The proof of the a.s. convergence – Garsia's proof – is essentially the same as in the forward case (Subsection 10.10.1), except that the first step is easier. Namely, if $\{X_n\}$ is $L^2$-bounded, we first conclude that $X_n$ converges in $L^2$ and in probability as in the forward case. However, since the limit $X_{-\infty} \in \mathcal{F}_{-\infty} \subset \mathcal{F}_n$ for all $n \leq 0$ it follows, for $m \leq n \leq 0$, that

$$E(X_n - X_{-\infty} \mid \mathcal{F}_m) = E(X_n \mid \mathcal{F}_m) - X_{-\infty} = X_m - X_{-\infty},$$

so that $\{X_n - X_{-\infty}, \, n \leq 0\}$ is a martingale, which, by Fatou's lemma is $L^2$-bounded. Thus $\{(X_n - X_{-\infty})^2, \, n \leq 0\}$ is a submartingale, so that an application of the Kolmogorov-Doob inequality, Theorem 16.2, yields

$$P(\sup_{n\leq 0} |X_n - X_{-\infty}| > \varepsilon) \leq \frac{E(X_0 - X)^2}{\varepsilon^2},$$

which shows that $X_n \overset{a.s.}{\to} X_{-\infty}$ as $n \to -\infty$.

The remaining assertions then follow as in the forward case.    □

The following two results correspond to Theorems 12.2 and 12.4, respectively.

**Theorem 16.5.** *Let $p > 1$. Suppose that $\{(X_n, \mathcal{F}_n), n \leq 0\}$ is an $L^p$-bounded reversed martingale. Then*

(a)    $\{|X_n|^p, n \geq 1\}$ *is uniformly integrable;*

(b)    $X_n \to X_{-\infty}$ *a.s. and in $L^p$ as $n \to -\infty$.*

*Proof.* An application of Theorem 16.3 shows that $Y = \sup_n |X_n|$ dominates $X_1, X_2, \ldots$ and has a moment of order $p$. This proves (a). Almost sure convergence follows from the general convergence theorem for reversed martingales, which, together with Theorem 5.5.2 finishes (b).    □

**Theorem 16.6.** *Suppose that $Z \in L^1$, and set $X_n = E(Z \mid \mathcal{F}_n)$, $n \leq 0$. Then*

$$X_n \to X_\infty = E(Z \mid \mathcal{F}_{-\infty}) \qquad a.s. \text{ and in } L^1 \qquad as \quad n \to -\infty,$$

*in other words $\{Z, X_{-\infty}, \ldots, X_n, \ldots, X_{-2}, X_{-1}, X_0\}$ is a martingale.*

*In particular, if $Z \in \mathcal{F}_{-\infty}$, then $X_{-\infty} = Z$ a.s.*

There exist, of course, analogs to the convergence theorem and to Theorem 16.4 for reversed submartingales. We confine ourselves to stating the results (parts (iia) $\Rightarrow$ (iib) $\Rightarrow$ (iic) $\Rightarrow$ (iid) are very much the same as before).

**Theorem 16.7.** *Suppose that $\{(X_n, \mathcal{F}_n), n \leq 0\}$ is a reversed submartingale. Then*

(i) $X_n \overset{a.s.}{\to} X_{-\infty}$ *as $n \to -\infty$, where $-\infty \leq X_{-\infty} < \infty$.*

(ii) *Moreover, the following are equivalent:*

(a)    $\{X_n, n \geq 1\}$ *is uniformly integrable;*

(b)    $X_n \to X_{-\infty}$ *in $L^1$ as $n \to -\infty$;*

(c)    $\{(X_n, \mathcal{F}_n), -\infty \leq n \leq 0\}$ *is a submartingale;*

(d)    $\lim_{-n\to\infty} E X_n > -\infty$.

## 16.1 The Law of Large Numbers

A nice application of reversed martingales is provided by the law of large numbers. Let $Y_1, Y_2, \ldots$ be i.i.d. random variables with finite mean $\mu$, and set $S_n = \sum_{k=1}^n Y_k$, $n \geq 1$. For $n \leq -1$ we define

$$X_{-n} = \frac{S_n}{n}, \quad \text{and} \quad \mathcal{F}_{-n} = \sigma\left\{\frac{S_k}{k}, \ k \geq n\right\}.$$

Since knowing the arithmetic means when $k \geq n$ is the same as knowing $S_n$ and $Y_k$, $k > n$, we have (cf. Problems 1.6.4 and 17.1),

$$\mathcal{F}_{-n} = \sigma\{S_n, Y_{n+1}, Y_{n+2}, Y_{n+3}, \ldots\}.$$

This, together with the fact that the summands are i.i.d. yields

$$\begin{aligned}
E(X_{-n} \mid \mathcal{F}_{-n-1}) &= E\left(\frac{S_n}{n} \mid \sigma\{S_{n+1}, Y_{n+2}, Y_{n+3}, \ldots\}\right) \\
&= E\left(\frac{S_n}{n} \mid \sigma\{S_{n+1}\}\right) = E\left(\frac{S_n}{n} \mid S_{n+1}\right) \\
&= \frac{1}{n}\sum_{k=1}^{n} E(Y_k \mid S_{n+1}) = \frac{1}{n}\sum_{k=1}^{n} \frac{S_{n+1}}{n+1} = \frac{S_{n+1}}{n+1},
\end{aligned}$$

which shows that the sequence of arithmetic means is a reversed martingale $(\{(X_n, \mathcal{F}_n), \ n \leq -1\}$ is a martingale), and, hence, is uniformly integrable. The convergence theorem for reversed martingales therefore tells us that $X_{-n}$ converges almost surely and in $L^1$ as $n \to \infty$, which, rewritten, is the same as

$$\frac{S_n}{n} \to \qquad \text{a.s. and in} \quad L^1 \quad \text{as} \quad n \to \infty.$$

It remains to establish the limit. However, the limit is an element of the tail-$\sigma$-field, and therefore constant. Since the expected value equals $\mu$, the limiting constant must, indeed, be equal to $\mu$.

This proves the strong law of large numbers, including mean convergence.

*Remark 16.3.* Note also the following alternative proof of the martingale property:

$$\frac{S_n}{n} = E\left(\frac{S_n}{n} \mid \mathcal{F}_{-n}\right) = \frac{1}{n}\sum_{k=1}^{n} E(Y_k \mid S_n) = \frac{1}{n} \cdot nE(Y_1 \mid S_n) = E(Y_1 \mid S_n),$$

which exhibits the reversed martingale in the form

$$X_{-n} = \frac{S_n}{n} = E(Y_1 \mid S_n), \quad n \geq 1. \qquad \Box$$

The following result shows that the Doob maximal inequality is sharp for $p = 1$, in the sense that one cannot dispense with the additional logarithmic factor in general. The original proof of the implication (a) $\Longrightarrow$ (b), due to Marcinkiewicz and Zygmund [180], does not use martingale theory. The other implications are due to Burkholder [32].

**Theorem 16.8.** *Let $p \geq 1$. Suppose that $Y, Y_1, Y_2, \ldots$ are independent, identically distributed random variables. The following are equivalent:*

(a)  $E|Y|\log^+ |Y| < \infty$ when $p = 1$ and $E|Y|^p < \infty$, when $p > 1$;

(b)  $E \sup_n \left|\frac{S_n}{n}\right|^p < \infty$;

(c)  $E \sup_n \left|\frac{Y_n}{n}\right|^p < \infty$.

*Proof.* (a) $\Longrightarrow$ (b): Immediate from Theorem 16.3, since the sequence of arithmetic means constitutes a reversed martingale.

(b) $\Longrightarrow$ (c): This implication rests on the fact that

$$\frac{|Y_n|}{n} \le \frac{|S_n|}{n} + \frac{|S_{n-1}|}{n-1} \cdot \frac{n-1}{n},$$

followed by an application of $c_r$-inequalities, Theorem 3.2.2.

(c) $\Longrightarrow$ (a): To begin with, $E|Y|^p < \infty$ (choose $n = 1$). For $p > 1$ there is nothing more to prove, so suppose for the remainder of the proof that $p = 1$.

If the variables are bounded there is nothing more to prove, so suppose they are unbounded. Moreover, via scaling, it is no restriction to assume that $P(|Y| < 1) > 0$.

By Theorem 2.12.1(i), $E|Y| < \infty$ implies that $\sum_{n=1}^{\infty} P(|Y| > n) < \infty$, and hence, by Lemma A.4.1 that

$$A = \prod_{j=1}^{\infty} P(|Y| \le j) < \infty.$$

Now, since

$$P\left(\sup_n \left|\frac{Y_n}{n}\right| > m\right) = \sum_{n=1}^{\infty} P(\{|Y_j| \le jm, 1 \le j \le n-1\} \cap \{|Y_n| > nm\})$$

$$= \sum_{n=1}^{\infty} \prod_{j=1}^{n-1} P(|Y_j| \le jm) P(|Y_n| > nm)$$

$$\ge \sum_{n=1}^{\infty} \prod_{j=1}^{\infty} P(|Y| \le j) P(|Y| > nm)$$

$$= A \sum_{n=1}^{\infty} P(|Y| > nm),$$

another application of Theorem 2.12.1(i) shows that

$$\infty > \sum_{m=1}^{\infty} \sum_{n=1}^{\infty} P\left(\sup_n \left|\frac{Y_n}{n}\right| > m\right) \ge A \sum_{m=1}^{\infty} \sum_{n=1}^{\infty} P(|Y| > nm).$$

An appeal to Theorem 2.12.7 concludes the proof of (a).    □

*Remark 16.4.* In other words, if $E|Y| < \infty$ only, then

$$\sup_n E\left|\frac{S_n}{n}\right| = E|Y_1| < \infty, \quad \text{but} \quad E \sup_n \left|\frac{S_n}{n}\right| = \infty.$$    □

For more on this, see [60, 183], where, in addition, equivalences to moments of randomly indexed arithmetic means are added.

## 16.2 $U$-statistics

Let $X_1$, $X_2$, $\ldots$, $X_n$ be independent, identically distributed random variables. Suppose further that $h : \mathbb{R}^m \to \mathbb{R}$ is a function that is symmetric in its arguments, that is, $h(\sigma(\mathbf{x})) = h(\mathbf{x})$, for any permutation $\sigma$ of the coordinates of $\mathbf{x}$. The $U$-statistic is defined as

$$U_n(\mathbf{X}) = U_n(X_1, X_2, \ldots, X_n) = \frac{1}{\binom{n}{m}} \sum{}^* h(X_{i_1}, X_{i_2}, \ldots, X_{i_m}), \quad n \geq m,$$

where $\sum^*$ denotes that summation extends over all $\binom{n}{m}$ combinations of distinct indices $1 \leq i_1 < i_2 < \cdots < i_m \leq n$. The application one should have in mind is that the *kernel* $h$ is an unbiased estimator of some unknown parameter $\theta$, that is, that

$$E_\theta h(X_1, X_2, \ldots, X_m) = \theta.$$

A preliminary observation is that if $h$ is unbiased, then so is $U_n$.

The connection to the present context is the following: Set, for $n \leq -m$,

$$X_{-n} = U_n, \quad \text{and} \quad \mathcal{F}_{-n} = \sigma\{U_k, k \geq n\}.$$

Then

$$\{(X_n, \mathcal{F}_n), n \leq -m\} \quad \text{is a martingale,}$$

that is $\{(U_n, \mathcal{F}_n), n \geq m\}$ is a reversed martingale.

**Exercise 16.2.** Check this.                                               $\square$

The simplest possible kernel would be $h(x) = x$, so that

$$U_n = \frac{1}{n} \sum_{k=1}^n X_k = \bar{X}_n,$$

in which case the $U$-statistic coincides with the sample mean. The following choice with $m = 2$ relates $U$-statistics to the sample variance.

*Example 16.1.* Let $X_1$, $X_2$, $\ldots$ be independent, identically distributed random variables with mean $\mu$, and finite variance $\sigma^2$, and, let as usual, $\bar{X}_n$ and $s_n^2$ denote sample mean and sample variance, respectively:

$$\bar{X}_n = \frac{1}{n} \sum_{k=1}^n X_k \quad \text{and} \quad s_n^2 = \frac{1}{n-1} \sum_{k=1}^n (X_k - \bar{X}_n)^2.$$

Consider the kernel $h(x_1, x_2) = \frac{1}{2}(x_1 - x_2)^2$.

Set $Y_k = X_k - \bar{X}_n$, and note that $\sum_{k=1}^n Y_k = 0$. Then

$$\sum_{1 \leq i < j \leq n} (X_i - X_j)^2 = \sum_{1 \leq i < j \leq n} (Y_i - Y_j)^2 = \frac{1}{2} \sum_{i,j=1}^n (Y_i - Y_j)^2$$

$$= n \sum_{k=1}^n Y_k^2 = n(n-1)s_n^2,$$

so that

$$U_n = \frac{1}{\binom{n}{2}} \frac{1}{2} n(n-1) s_n^2 = s_n^2.$$

From Chapter 6 we know that $s_n^2 \overset{a.s.}{\to} \sigma^2$ as $n \to \infty$, so that a proper examination of the $U$-statistic should produce the same conclusion ...

From the convergence theorem for reversed martingales we know that $U_n$ converges almost surely and in $L^1$, and from the Kolmogorov zero-one law we know that the limit is almost surely constant. This summarizes into

$$U_n \to E\,U_2 \quad \text{a.s. and in } L^1 \qquad \text{as} \quad n \to \infty.$$

It remains to compute the limit (which ought to be $E\,s_n^2 = \sigma^2$):

$$E\,U_2 = E\left( \sum_{1 \leq i < j = 2} \frac{1}{2}(X_i - X_j)^2 \right) = \frac{1}{2} E(X_1 - X_2)^2 = \sigma^2.$$

This finally tells us that

$$U_n \to \sigma^2 \quad \text{a.s. and in } L^1 \qquad \text{as} \quad n \to \infty,$$

that is, that

$$s_n^2 \to \sigma^2 \quad \text{a.s. and in } L^1 \qquad \text{as} \quad n \to \infty,$$

thereby re-proving almost sure convergence of the sample variance, however, adding $L^1$-convergence, i.e., $E|s_n^2 - \sigma^2| \to 0$ as $n \to \infty$, which is more than the trivial fact that $E\,s_n^2 = \sigma^2$.

Via Theorem 16.5 we may further conclude that if $E(U_2)^r < \infty$ for some $r > 1$, then $U_n \overset{r}{\to} \sigma^2$ as $n \to \infty$, which, restated, tells us that if $E|X|^{2r} < \infty$, then

$$E|s_n^2 - \sigma^2|^r \to 0 \qquad \text{as} \quad n \to \infty,$$

in particular,                                                                                               □

$$E\,s_n^{2r} \to \sigma^{2r} \qquad \text{as} \quad n \to \infty.$$

We close by observing that it follows from the Rao-Blackwell theorem, Theorem 1.5, that if $U$ is a $U$-statistic and $V$ is a sufficient statistic for the unknown parameter $\theta$, then $E(U \mid V)$ is an unbiased estimator for $\theta$, and

$$\text{Var}\,\big(E(U \mid V)\big) \leq \text{Var}\,U,$$

provided the variances are finite, that is, $E(U \mid V)$ is an unbiased estimator with an expected square loss which is smaller than that of $U$.

For an introduction to the theory of $U$-statistics and further examples, we refer to the monograph by Serfling [219].

## 17 Problems

1. Let $Y_1, Y_2, \ldots$ be random variables, and set $X_n = \sum_{k=1}^{n} Y_k$, $n \geq 1$. Show that

$$\sigma\{Y_1, Y_2, \ldots, Y_n\} = \sigma\{X_1, X_2, \ldots, X_n\}.$$

♣ Remember Problem 1.6.4.

2. Suppose that $Y$ is a random variable with finite variance and that $\mathcal{G}$ is a sub-$\sigma$-algebra of $\mathcal{F}$. Prove that

$$\operatorname{Var} Y = E \operatorname{Var}(Y \mid \mathcal{G}) + \operatorname{Var}(E(Y \mid \mathcal{G})).$$

3. Suppose that $\{(X_n, \mathcal{F}_n), n \geq 0\}$ is a martingale. Set $Y = X_{n+1}$ and $\mathcal{G} = \mathcal{F}_n$. Insert this into, that is, verify, the relations in Problem 17.2.

4. Let $\{X(n), n \geq 0\}$ denote the number of individuals in generation $n$ in a Galton-Watson process $(X(0) = 1)$, and suppose that the mean off-spring is finite $= m$.
   (a) Show that $E\,X(n) = m^n$.
      Suppose, in addition, that the off-spring variance is finite $= \sigma^2$.
   (b) Show that
$$\operatorname{Var} X(n) = m^{n-1}\sigma^2 + m^2 \operatorname{Var} X(n-1).$$

   (c) Show that
$$\operatorname{Var} X(n) = \sigma^2 \left(m^{n-1} + m^n + \cdots + m^{2(n-1)}\right).$$

5. Suppose that $\mathcal{G}$ is a sub-$\sigma$-algebra of $\mathcal{F}$, and that $X$ and $Y$ are random variables with the property that

$$E(X \mid \mathcal{G}) = Y \quad \text{and} \quad E(X^2 \mid \mathcal{G}) = Y^2.$$

Show that $X = Y$ a.s.

6. Let $Y$ be an integrable random variable. Prove that

$$\left\{ E(Y \mid \mathcal{G}) : \mathcal{G} \text{ is a sub-}\sigma\text{-algebra of } \mathcal{F} \right\}$$

is uniformly integrable.

7. Let $X$ and $Y$ be independent random variables with mean 0.
   (a) Show that
$$E|X| \leq E|X + Y|.$$

   (b) Suppose, in addition, that $E|X|^r < \infty$ and that $E|X|^r < \infty$ for some $r > 1$. Show that
$$E|X|^r \leq E|X + Y|^r.$$

   ♣ This is Proposition 3.6.5, but in a different context.

8. Let $X$ and $Y$ be independent, identically distributed random variables with finite mean. Show that

$$E(X \mid X + Y) = \frac{X + Y}{2}.$$

9. Suppose that $V \in \operatorname{Exp}(1)$. Find

$$E(V \mid \min\{V, t\}) \quad \text{and} \quad E(V \mid \max\{V, t\}), \quad t > 0.$$

10. Consider Example 3.5 – Double or nothing. Prove that
   • the expected duration of the game is 2;
   • the expected amount of money spent at the time of the first success is infinite;
   • the total gain when the game is over is $+1$.

11. Let $Y_1, Y_2, \ldots$ be an adapted sequence, and let $c_n \in \mathbb{R}$, $n \geq 1$.
   (a) Suppose that $E(Y_{n+1} \mid \mathcal{F}_n) = Y_n + c_n$. Compensate suitably to exhibit a martingale.
   (b) Suppose that $E(Y_{n+1} \mid \mathcal{F}_n) = Y_n \cdot c_n$. Compensate suitably to exhibit a martingale.

12. Toss a symmetric coin repeatedly and set

$$Y_n = \begin{cases} 1, & \text{if the } n\text{th toss is a head,} \\ 0, & \text{otherwise,} \end{cases}$$

so that $S_n = \sum_{k=1}^n Y_k$ equals the number of heads in $n$ tosses. Set

$$X_n = 2S_n - n, \quad n \geq 1.$$

Since $n - S_n$ equals the number of tails in $n$ tosses, it follows that $X_n =$ the number of heads minus the number of tails, that is the excess of heads over tails in $n$ tosses. Show that $\{X_n, n \geq 1\}$ (together with the sequence of natural $\sigma$-algebras) is a martingale.

13. Let $Y, Y_1, Y_2, \ldots$ be independent, identically distributed random variables.
   (a) Suppose that $E|Y|^3 < \infty$. How can $(\sum_{k=1}^n Y_k)^3$ be made into a martingale?
   (b) Suppose that $EY^4 < \infty$. How can $(\sum_{k=1}^n Y_k)^4$ be made into a martingale?

14. Generalize Theorem 9.1 and prove that, for any sub- or supermartingale, $\{(X_n, \mathcal{F}_n), n \geq 0\}$,

$$P(\max_{0 \leq k \leq n} |X_k| > \lambda) \leq \frac{3 \max_{0 \leq k \leq n} E|X_k|}{\lambda}.$$

♣  For martingales and non-negative submartingales the maximal expectation equals the last one, and 3 is superfluous.

15. *The Riesz decomposition*, Theorem 5.3, stated that any supermartingale, $\{(X_n, \mathcal{F}_n), n \geq 0\}$, such that $\inf_n E(X_n) > -\infty$, can be uniquely decomposed into a martingale, $\{(M_n, \mathcal{F}_n), n \geq 0\}$, and a potential, $\{(Z_n, \mathcal{F}_n), n \geq 0\}$;

$$X_n = M_n + Z_n.$$

(a) Prove that there exists a martingale, $\{(V_n, \mathcal{F}_n), n \geq 1\}$, such that

$$X_n = V_n - A_n, \quad n \geq 1.$$

(b) Check that $E A_n \leq -\inf_n E X_n + E V_0$.
(c) Prove/check that $A_n \overset{a.s.}{\to} A_\infty$, say, as $n \to \infty$.
(d) Set

$$M_n = V_n - E(A_\infty \mid \mathcal{F}_n) \quad \text{and} \quad Z_n = E(A_\infty \mid \mathcal{F}_n) - A_n,$$

and prove that this is a decomposition of the desired kind.
(e) Prove that the decomposition is unique.

16. Prove that, if $X_n = M_n - A_n$, $n \geq 1$, is the Doob decomposition of a potential, then

$$X_n = E(A_\infty - A_n), \quad n \geq 1.$$

17. This is an addendum to Problem 6.13.18, where a stick of length 1 was repeatedly randomly broken in the sense that the remaining piece each time was $U(0, Y)$-distributed, where $Y$ was the (random) previous length. Let $Y_n$ denote the remaining piece after the stick has been broken $n$ times, and set $\mathcal{F}_n = \sigma\{Y_k, k \leq n\}$.
(a) Compute $E(Y_n \mid \mathcal{F}_{n-1})$.
(b) Adjust $Y_n$ in order for (a suitable) $\{(X_n, \mathcal{F}_n), n \geq 0\}$ to become a martingale.
(c) Does $X_n$ converge almost surely? In $L^1$?

18. *Another non-regular martingale.* Let $Y, Y_1, Y_2, \ldots$ be independent random variables with common distribution given by

$$P\left(Y = \frac{1}{2}\right) = P\left(Y = \frac{3}{2}\right) = \frac{1}{2},$$

and set $X_n = Y_1 \cdot Y_2 \cdots Y_n$, $n \geq 1$. Show that this produces a martingale of product type with mean one that converges almost surely to 0.

19. In Example 3.7 we found that the likelihood ratios, $\{L_n, n \geq 1\}$, constitute a martingale (of the product type) with mean 1 (if $H_0$ is true).
(a) Consider the log-likelihood ratios $\{\log L_n, n \geq 1\}$. Prove that

$$\log L_n \overset{a.s.}{\to} -\infty \quad \text{as} \quad n \to \infty.$$

♠ Remember Problem 3.10.8.
(b) Prove that $L_n \overset{a.s.}{\to} 0$ as $n \to \infty$.
(c) Is $\{L_n, n \geq 1\}$ closable? Uniformly integrable?

20. Show that a predictable martingale is a.s. constant: Formally, show that if $\{(X_n, \mathcal{F}_n), n \geq 0\}$ is a martingale, such that $X_n \in \mathcal{F}_{n-1}$ for all $n$, then $X_n = X_0$ a.s. for all $n$.

21. (a) Construct a martingale that converges to $-\infty$ a.s. as $n \to \infty$.
♠ Sums of independent random variables with mean 0 are usually a good try.
(b) Construct a non-negative submartingale that converges to 0 a.s. as $n \to \infty$.

22. Prove that a non-negative, uniformly integrable submartingale that converges to 0 a.s. as $n \to \infty$ is identically 0.

23. Let $Y_1, Y_2, \ldots$ be random variables, such that

$$\sup_n |Y_n| \leq Z \in L^1,$$

and suppose that $Y_n \overset{a.s.}{\to} Y_\infty$ as $n \to \infty$. Show that

$$E(Y_n \mid \mathcal{F}_n) \overset{a.s.}{\to} E(Y_\infty \mid \mathcal{F}_\infty) \quad \text{as} \quad n \to \infty,$$

where $\mathcal{F}_n$, $n \geq 1$, and $\mathcal{F}_\infty$ is the usual setup of $\sigma$-algebras.
♣ This generalizes Theorem 12.4.

24. Prove that stopping times with finite mean are regular for martingales with uniformly bounded increments.

25. Let $Y_1, Y_2, \ldots$ be independent standard normal random variables, and set $S_n = \sum_{k=1}^{n} Y_k$, $n \geq 1$. Prove that

   (a) $\{e^{S_n - n/2}, n \geq 1\}$ is a martingale.

   (b) $X_n \overset{a.s.}{\to} 0$ as $n \to \infty$.

   (c) $E(X_n)^r \to 0$ as $n \to \infty$ $\iff$ $r < 1$.

26. *Replacement based on age.* This model was described in Subsection 2.16.6: The lifetimes of some component in a machine are supposed to be independent, identically distributed random variables, $Y_1, Y_2, \ldots$. Replacement based on age means that components are replaced at failure or at some given age, $a$, say, whichever comes first. The inter-replacement times are $W_n = \min\{X_n, a\}$, $n \geq 1$.

   By introducing $Z_n = I\{Y_n \leq a\}$, $n \geq 1$, the quantity $V(t) = \sum_{k=1}^{\tau(t)} Z_k$, where $\tau(t) = \min\{n : \sum_{k=1}^{n} W_k > t\}$ describes the number of components that have been used "at $t$ o'clock". Set $\mu_w = E\,W_1$ and $\mu_z = E\,Z_1$.

   (a) Prove that

   $$E\,\tau(t) < \infty \quad \text{and that} \quad E\,V(t) = E\,\tau(t) \cdot \mu_z < \infty.$$

   (b) Prove that, as $t \to \infty$,

   $$\frac{\tau(t)}{t} \overset{a.s.}{\to} \frac{1}{\mu_w} \quad \text{and that} \quad \frac{V(t)}{t} \overset{a.s.}{\to} \frac{\mu_z}{\mu_w};$$

   $$\frac{E\,\tau(t)}{t} \to \frac{1}{\mu_w} \quad \text{and that} \quad \frac{E\,V(t)}{t} \to \frac{\mu_z}{\mu_w}.$$

   (c) Find explicit expressions for $\mu_w$ and $\mu_z$ and restate the claims.

27. *Continuation.* In practice it might be of interest to determine the value of $a$ in the previous problem in order to minimize some quantity of interest, such as cost. Let us, as an example, consider exponentially distributed lifetimes; $Y_k \in \text{Exp}(\lambda)$, $k \geq 1$.

   (a) Compute $\mu_w$ and $\mu_z$ for that case.

   (b) Restate the conclusions.

   (c) The limit results for $V(t)$ then become

   $$\frac{V(t)}{t} \overset{a.s.}{\to} \frac{1}{\lambda} \quad \text{and} \quad \frac{E\,V(t)}{t} \to \frac{1}{\lambda} \quad \text{as} \quad t \to \infty.$$

   (d) The limits turn out to be *independent of $a$*. Interpret this fact.

28. Let $\{(X_n, \mathcal{F}_n), n \geq 0\}$ be an $L^2$-martingale, and let $S_n(X)$ and $s_n(X)$ be the square function and the conditional square function, respectively, as introduced in connection with Theorem 9.5. Show that

   $$\{X_n^2 - (S_n(X))^2, n \geq 0\}, \quad \{X_n^2 - (s_n(X))^2, n \geq 1\}, \quad \text{and}$$
   $$\{(S_n(X))^2 - (s_n(X))^2, n \geq 1\}$$

   are martingales.

29. Let $Y_1, Y_2, \ldots$ be independent random variables with partial sums $X_n$, $n \geq 1$.

(a) Translate the conclusions of the previous problem into this setting.

(b) The same if, in addition, $Y_1, Y_2, \ldots$ are equidistributed.

30. Let $\{Y_n, n \leq -1\}$ be random variables, such that

$$\sup_n |Y_n| \leq Z \in L^1,$$

and suppose that $Y_n \overset{a.s.}{\to} Y_{-\infty}$ as $n \to -\infty$. Show that

$$E(Y_n \mid \mathcal{F}_n) \overset{a.s.}{\to} E(Y_{-\infty} \mid \mathcal{F}_{-\infty}) \quad \text{as} \quad n \to -\infty,$$

where $\mathcal{F}_n$, $n \leq -1$, and $\mathcal{F}_{-\infty}$ are the usual $\sigma$-algebras.

♣ This generalizes Theorem 16.6.

# A

## Some Useful Mathematics

In this appendix we collect a number of mathematical facts which may or may not be (or have been) familiar to the reader.

### 1 Taylor Expansion

A common method to estimate functions is to use Taylor expansion. Here are a few such estimates.

**Lemma 1.1.** *We have,*

$$e^x \leq 1 + x + x^2 \quad \textit{for } |x| \leq 1, \tag{A.1}$$

$$-\frac{1}{1-\delta}x < \log(1-x) < -x \quad \textit{for } 0 < x < \delta < 1, \tag{A.2}$$

$$|e^z - 1| \leq |z|e^{|z|} \quad \textit{for } z \in \mathbb{C}, \tag{A.3}$$

$$|e^z - 1 - z| \leq |z|^2 \quad \textit{for } z \in \mathbb{C}, \quad |z| \leq 1/2, \tag{A.4}$$

$$|\log(1-z) + z| \leq |z|^2 \quad \textit{for } z \in \mathbb{C}, \quad |z| \leq 1/2. \tag{A.5}$$

*Proof.* Let $0 \leq x \leq 1$. By Taylor expansion, noticing that we have an alternating series that converges,

$$e^{-x} \leq 1 - x + \frac{x^2}{2}.$$

For the other half of (A.1),

$$e^x = 1 + x + x^2 \sum_{k=2}^{\infty} \frac{x^{k-2}}{k!} \leq 1 + x + x^2 \sum_{k=2}^{\infty} \frac{1}{k!}$$

$$= 1 + x + x^2(e - 2) \leq 1 + x + x^2.$$

To prove (A.2), we use Taylor expansion to find that, for $0 < x < \delta < 1$,

$$\log(1-x) = -\sum_{k=1}^{\infty} \frac{x^k}{k} \ge -x - x\sum_{k=2}^{\infty} \frac{\delta^{k-1}}{k}$$

$$\ge -x - x\sum_{k=1}^{\infty} \delta^k = -x - x\frac{\delta}{1-\delta} = -x\frac{1}{1-\delta},$$

which proves the lower inequality. The upper one is trivial.

Next, let $z \in \mathbb{C}$. Then

$$|e^z - 1| \le \sum_{k=1}^{\infty} \frac{|z|^k}{k!} = |z|\sum_{k=0}^{\infty} \frac{|z|^k}{(k+1)!} \le |z|\sum_{k=0}^{\infty} \frac{|z|^k}{k!} = |z|e^{|z|}.$$

If, in addition, $|z| \le 1/2$, then

$$|e^z - 1 - z| \le \sum_{k=2}^{\infty} \frac{|z|^k}{k!} \le \frac{|z|^2}{2}\sum_{k=2}^{\infty} |z|^{k-2} = \frac{|z|^2}{2} \cdot \frac{1}{1-|z|} \le |z|^2,$$

$$|\log(1-z) + z| \le \sum_{k=2}^{\infty} \frac{|z|^k}{k} \le \frac{|z|^2}{2}\sum_{k=2}^{\infty} |z|^{k-2} \le |z|^2. \qquad \square$$

We also need estimates for the tail of the Taylor expansion of the exponential function for imaginary arguments.

**Lemma 1.2.** *For any $n \ge 0$,*

$$\left|e^{iy} - \sum_{k=0}^{n} \frac{(iy)^k}{k!}\right| \le \min\left\{2\frac{|y|^n}{n!}, \frac{|y|^{n+1}}{(n+1)!}\right\}.$$

*Proof.* Let $y > 0$. By partial integration,

$$\int_0^y e^{ix}(y-x)^k \, dx = \frac{y^{k+1}}{k+1} + \frac{i}{k+1}\int_0^y e^{ix}(y-x)^{k+1} \, dx, \quad k \ge 0. \quad (A.6)$$

For $k = 0$ the first formula and direct integration, respectively, yield

$$\int_0^y e^{ix} \, dx = \begin{cases} y + i\int_0^y e^{ix}(y-x) \, dx, \\ \dfrac{e^{iy} - 1}{i}, \end{cases}$$

so that, by equating these expressions, we obtain

$$e^{iy} = 1 + iy + i^2\int_0^y e^{ix}(y-x) \, dx. \quad (A.7)$$

Inserting (A.6) into (A.7) iteratively for $k = 2, 3, \ldots, n-1$ (more formally, by induction), yields

$$e^{iy} = \sum_{k=0}^{n} \frac{(iy)^k}{k!} + \frac{i^{n+1}}{n!} \int_0^y e^{ix}(y-x)^n \, dx, \qquad (A.8)$$

and, hence,

$$\left| e^{iy} - \sum_{k=0}^{n} \frac{(iy)^k}{k!} \right| \le \frac{1}{n!} \int_0^y (y-x)^n \, dx = \frac{y^{n+1}}{(n+1)!}.$$

Replacing $n$ by $n-1$ in (A.8), and then adding and subtracting $\frac{(iy)^n}{n!}$, yields

$$e^{iy} = \sum_{k=0}^{n-1} \frac{(iy)^k}{k!} + \frac{i^n}{(n-1)!} \int_0^y e^{ix}(y-x)^{n-1} \, dx$$

$$= \sum_{k=0}^{n} \frac{(iy)^k}{k!} + \frac{i^n}{(n-1)!} \int_0^y \left( e^{ix} - 1 \right)(y-x)^{n-1} \, dx,$$

so that, in this case, noticing that $|e^{ix} - 1| \le 2$,

$$\left| e^{iy} - \sum_{k=0}^{n} \frac{(iy)^k}{k!} \right| \le \frac{2y^n}{n!}.$$

The proof is finished via the analogous estimates for $y < 0$.           □

Another estimate concerns the integral $\int_0^t \frac{\sin x}{x} \, dx$ as $t \to \infty$. A slight delicacy is that the integral is not absolutely convergent. However, the successive slices $\int_{(n-1)\pi}^{n\pi} \frac{\sin x}{x} \, dx$ are alternating in sign and decreasing in absolute value to 0 as $n \to \infty$, which proves that the limit as $t \to \infty$ exists.

**Lemma 1.3.** *Let $\alpha > 0$. Then*

$$\int_0^t \frac{\sin \alpha x}{x} \, dx \begin{cases} \le \int_0^\pi \frac{\sin x}{x} \, dx \le \pi & \text{for all } t > 0, \\ \to \frac{\pi}{2} & \text{as } t \to \infty. \end{cases}$$

*Proof.* The change of variables $y = \alpha x$ shows that it suffices to check the case $\alpha = 1$.

The first inequality is a consequence of the behavior of the slices mentioned prior to the statement of the lemma, and the fact that $\sin x \le x$.

Since $\frac{1}{x} = \int_0^\infty e^{-yx} \, dy$, and since, for all $t$,

$$\int_0^t \int_0^\infty | \sin x e^{-yx} | dy \, dx \le \int_0^t \frac{|\sin x|}{x} \, dy \le \int_0^t dy = t,$$

we may apply Fubini's theorem to obtain

$$\int_0^t \frac{\sin x}{x} \, dx = \int_0^t \sin x \left( \int_0^\infty e^{-yx} \, dy \right) dx = \int_0^\infty \left( \int_0^t \sin x e^{-yx} \, dx \right) dy.$$

In order to evaluate the inner integral we use partial integration twice:

$$I_t(y) = \int_0^t \sin x\, e^{-yx}\, dx = \left[-\cos x e^{-yx}\right]_0^t - \int_0^t \cos x \cdot y e^{-yx}\, dx$$

$$= 1 - \cos t e^{-yt} - \left[\sin x \cdot y e^{-yx}\right]_0^t - \int_0^t \sin x \cdot y^2 e^{-yx}\, dx$$

$$= 1 - \cos t e^{-yt} - \sin t \cdot y e^{-yt} - y^2 I_t(y),$$

so that

$$I_t(y) = \frac{1}{1+y^2}\left(1 - e^{-yt}(\cos t + y \sin t)\right).$$

Inserting this into the double integral yields

$$\int_0^t \frac{\sin x}{x}\, dx = \int_0^\infty I_t(y)\, dy = \frac{\pi}{2} - \int_0^\infty \frac{1}{1+y^2} e^{-yt}(\cos t + y \sin t)\, dy,$$

so that, finally,

$$\int_0^\infty \frac{\sin x}{x}\, dx = \lim_{t \to \infty} \int_0^t \frac{\sin x}{x}\, dx = \frac{\pi}{2},$$

(since $\int_0^\infty \frac{1}{1+y^2} e^{-yt}(\cos t + y \sin t)\, dy \to 0$ as $t \to \infty$).    □

## 2 Mill's Ratio

The function $e^{-x^2/2}$, which is intimately related to the normal distribution, has no primitive function (expressable in terms of elementary functions), so that integrals must be computed numerically. In many situations, however, estimates or approximations are enough. Mill's ratio is one such result.

**Lemma 2.1.** *Let $\phi(x)$ be the standard normal density, and $\Phi(x)$ the corresponding distribution function. Then*

$$\left(1 - \frac{1}{x^2}\right)\frac{\phi(x)}{x} < 1 - \Phi(x) < \frac{\phi(x)}{x}, \quad x > 0.$$

*In particular,*

$$\lim_{x \to \infty} \frac{x(1 - \Phi(x))}{\phi(x)} = 1.$$

*Proof.* Since $(\phi(x))' = -x\phi(x)$, partial integration yields

$$0 < \int_x^\infty \frac{1}{y^2}\phi(y)\, dy = \frac{\phi(x)}{x} - (1 - \Phi(x)).$$

Rearranging this proves the right-most inequality. Similarly,

$$0 < \int_x^\infty \frac{3}{y^4}\phi(y)\, dy = \frac{\phi(x)}{x^3} - \int_x^\infty \frac{1}{y^2}\phi(y)\, dy,$$

which, together with the previous estimate, proves the left-hand inequality. The limit result follows immediately.    □

*Remark 2.1.* If only at the upper estimate is of interest one can argue as follows:

$$1 - \Phi(x) = \int_x^\infty \frac{y}{y}\phi(y)\,dy < \frac{1}{x}\int_x^\infty y\phi(y)\,dy = \frac{\phi(x)}{x}. \qquad \square$$

## 3 Sums and Integrals

In general it is easier to integrate than to compute sums. One therefore often tries to switch from sums to integrals. Usually this is done by writing $\sum \sim \int$ or $\sum \leq C\int$, where $C$ is some (uninteresting) constant. Following are some more precise estimates of this kind.

**Lemma 3.1.** (i) *For $\alpha > 0$, $n \geq 2$,*

$$\frac{1}{\alpha n^\alpha} \leq \sum_{k=n}^\infty \frac{1}{k^{\alpha+1}} \leq \frac{1}{\alpha(n-1)^\alpha} \leq \frac{2^\alpha}{\alpha n^\alpha}.$$

*Moreover,*

$$\lim_{n\to\infty} n^\alpha \sum_{k=n}^\infty \frac{1}{k^{\alpha+1}} = \frac{1}{\alpha} \quad as \quad n \to \infty.$$

(ii) *For $\beta > 0$,*

$$\frac{n^\beta}{\beta} \leq \sum_{k=1}^n k^{\beta-1} \leq \frac{n^\beta}{\beta} + n^{\beta-1} \leq \left(\frac{1}{\beta}+1\right)n^\beta,$$

*and*

$$\lim_{n\to\infty} n^{-\beta} \sum_{k=1}^n k^{\beta-1} = \frac{1}{\beta}.$$

(iii)

$$\log n + \frac{1}{n} \leq \sum_{k=1}^n \frac{1}{k} \leq \log n + 1,$$

*and*

$$\lim_{n\to\infty} \frac{1}{\log n} \sum_{k=1}^n \frac{1}{k} = 1.$$

*Proof.* (i): We have

$$\frac{1}{\alpha n^\alpha} = \int_n^\infty \frac{dx}{x^{\alpha+1}} = \sum_{k=n}^\infty \int_k^{k+1} \frac{dx}{x^{\alpha+1}} \quad \begin{cases} \leq \sum_{k=n}^\infty \frac{1}{k^{\alpha+1}}, \\[2mm] \geq \sum_{k=n}^\infty \frac{1}{(k+1)^{\alpha+1}}. \end{cases}$$

The proof of (ii) follows the same pattern by departing from

$$\frac{n^\beta}{\beta} = \int_0^n x^{\beta-1}\,dx,$$

however, the arguments for $\beta > 1$ and $0 < \beta < 1$ have to be worked out separately.

The point of departure for (iii) is

$$\log n = \int_1^n \frac{dx}{x}. \qquad \Box$$

**Exercise 3.1.** Finish the proof of the lemma. $\qquad \Box$

*Remark 3.1.* An estimate which is sharper than (iii) is

$$\sum_{k=1}^n \frac{1}{k} = \log n + \gamma + o(1) \quad \text{as} \quad n \to \infty,$$

where $\gamma = 0.5772\ldots$ is *Euler's constant*. However, the corresponding limit coincides with that of the lemma. $\qquad \Box$

## 4 Sums and Products

There is a strong connection between the convergence of sums and that of products, for example through the formula

$$\prod(\cdots) = \exp\Big\{\sum \log(\cdots)\Big\}.$$

One can transform criteria for convergence of sums into criteria for convergence of products, and vice versa, essentially via this connection. For example, if a sum converges, then the tails are small. For a product this means that the tails are close to 1. Here are some useful connections.

**Lemma 4.1.** *For $n \geq 1$, let $0 \leq a_n < 1$. Then*

$$\sum_{n=1}^\infty a_n \text{ converges} \quad \Longleftrightarrow \quad \prod_{n=1}^\infty (1-a_n) \text{ converges.}$$

*Convergence thus holds iff*

$$\sum_{k=m}^n a_k \to 0 \quad \Longleftrightarrow \quad \prod_{k=m}^n (1-a_n) \to 1 \quad \text{as} \quad m,n \to \infty.$$

*Proof.* Taking logarithms, shows that the product converges iff

$$\sum_{n=1}^\infty \log(1-a_n) < \infty.$$

No matter which of the sums we assume convergent, we must have $a_n \to 0$ as $n \to \infty$, so we may assume, without restriction, that $a_n < 1/3$ for all $n$. Formula (A.2) with $\delta = 1/3$ then tells us that

$$-\tfrac{3}{2}a_n \le \log(1 - a_n) \le -a_n. \qquad \square$$

**Lemma 4.2.** *For $n \ge 1$, let $0 \le a_n < \delta < 1$. Then*

$$(1 - a_n)^n \to 1 \quad as \quad n \to \infty \quad \Longleftrightarrow \quad na_n \to 0 \quad as \quad n \to \infty.$$

*Moreover, in either case, given $\delta \in (0,1)$, we have $na_n < \delta(1 - \delta) < 1$ for $n$ large enough, and*

$$(1 - \delta)na_n \le 1 - (1 - a_n)^n \le na_n/(1 - \delta).$$

*Proof.* The sufficiency is well known. Therefore, suppose that $(1 - a_n)^n \to 1$ as $n \to \infty$. Recalling (A.2),

$$1 \leftarrow (1-a_n)^n = \exp\{n\log(1-a_n)\} \begin{cases} \le \exp\{-na_n\}, \\ \ge \exp\{-na_n/(1 - \delta)\}, \end{cases} \quad as \quad n \to \infty,$$

which establishes the preliminary fact that

$$(1 - a_n)^n \to 1 \quad as \quad n \to \infty \quad \Longleftrightarrow \quad na_n \to 0 \quad as \quad n \to \infty.$$

With this in mind we return to the upper bound and apply (A.1). Choose $n$ so large that $na_n < \delta(1 - \delta) < 1$. Then,

$$(1 - a_n)^n \le \exp\{-na_n\} \le 1 - na_n + (na_n)^2 \le 1 - na_n(1 - \delta).$$

For the lower bound there is a simpler way out;

$$(1 - a_n)^n \ge \exp\{-na_n/(1 - \delta)\} \ge 1 - na_n/(1 - \delta).$$

The double inequality follows by joining the upper and lower bounds. $\qquad \square$

## 5 Convexity; Clarkson's Inequality

Convexity plays an important role in many branches of mathematics. Our concern here is some inequalities, such as generalizations of the triangle inequality.

**Definition 5.1.** *A real valued function $g$ is* convex *iff, for every $x, y \in \mathbb{R}$, and $\alpha \in [0,1]$,*

$$g(\alpha x + (1 - \alpha)y) \le \alpha g(x) + (1 - \alpha)g(y).$$

*The function is* concave *if the inequality is reversed.* $\qquad \square$

In words, $g$ is convex if a chord joining two points lies on, or above, the function between those points.

Convex functions always possess derivatives from the left and from the right. The derivatives agree on almost all points. The typical example is $|x|$, which is convex but does not possess a derivative at 0, only left- and right-hand ones.

A twice differentiable function is convex if and only if the second derivative is non-negative (and concave if and only if it is non-positive).

For $x, y \in \mathbb{R}$ the standard triangular inequality states that $|x+y| \leq |x|+|y|$. Following are some analogs for powers.

**Lemma 5.1.** *Let $r > 0$, and suppose that $x, y > 0$. Then*

$$(x+y)^r \leq \begin{cases} 2^r(x^r + y^r), & \text{for } r > 0, \\ x^r + y^r, & \text{for } 0 < r \leq 1, \\ 2^{r-1}(x^r + y^r), & \text{for } r \geq 1. \end{cases}$$

*Proof.* For $r > 0$,

$$(x+y)^r \leq (2\max\{x,y\})^r = 2^r(\max\{x,y\})^r \leq 2^r(x^r + y^r).$$

Next, suppose that $0 < r \leq 1$. Then, since $x^{1/r} \leq x$ for any $0 < x < 1$, it follows that

$$\left(\frac{x^r}{x^r + y^r}\right)^{1/r} + \left(\frac{y^r}{x^r + y^r}\right)^{1/r} \leq \frac{x^r}{x^r + y^r} + \frac{y^r}{x^r + y^r} = 1,$$

and, hence, that

$$x + y \leq (x^r + y^r)^{1/r},$$

which is the same as the second inequality.

For $r \geq 1$ we exploit the fact that the function $|x|^r$ is convex, so that, in particular,

$$\left(\frac{x+y}{2}\right)^r \leq \frac{1}{2}x^r + \frac{1}{2}y^r,$$

which is easily reshuffled into the third inequality.    □

**Lemma 5.2.** *Let $p^{-1} + q^{-1} = 1$. Then*

$$xy \leq \frac{x^p}{p} + \frac{y^q}{q} \quad \text{for } x, y > 0.$$

*Proof.* The concavity of the logarithm, and the fact that $e^x$ is increasing, yield

$$xy = \exp\{\log xy\} = \exp\left\{\frac{1}{p}\log x^p + \frac{1}{q}\log x^q\right\}$$

$$\leq \exp\left\{\log\left(\frac{x^p}{p} + \frac{y^q}{q}\right)\right\} = \frac{x^p}{p} + \frac{y^q}{q}.$$    □

*Remark 5.1.* The numbers $p$ and $q$ are called *conjugate exponents*.

*Remark 5.2.* The case $p = q = 2$ is special, in the sense that the number 2 is the same as its conjugate. (This has a number of special consequences within the theory of functional analysis.) In this case the inequality above becomes

$$xy \le \frac{1}{2}x^2 + \frac{1}{2}y^2,$$

which, on the other hand, is equivalent to the inequality $(x - y)^2 \ge 0$.    □

Clarkson's inequality [50] generalizes the well-known parallelogram identity, which states that

$$|x + y|^2 + |x - y|^2 = 2(|x|^2 + |y|^2) \quad \text{for} \quad x, y \in \mathbb{R},$$

in the same vein as the Lemma 5.1 is a generalization of the triangle inequality. We shall need the following part of [50], Theorem 2.

**Lemma 5.3.** (Clarkson's inequality) *Let $x, y \in \mathbb{R}$. Then*

$$|x + y|^r + |x - y|^r \begin{cases} \le 2(|x|^r + |y|^r), & \text{for } 1 \le r \le 2, \\ \ge 2(|x|^r + |y|^r), & \text{for } r \ge 2. \end{cases}$$

*Proof.* For $r = 1$ this is just a consequence of the triangular inequality. For $r = 2$ it is the parallelogram identity. We therefore assume that $r \ne 1, 2$ in the following.

First, let $1 < r < 2$. If $x = y$, or if one of $x$ and $y$ equals 0, the result is trivial. Moreover, if the inequality is true for $x$ and $y$, then it is also true for $\pm x$ and $\pm y$. We therefore suppose, without restriction, that $0 < y < x$. Putting $a = y/x$ reduces our task to verifying that

$$(1 + a)^r + (1 - a)^r \le 2(1 + a^r) \quad \text{for} \quad 0 < a < 1.$$

Toward that end, set $g(a) = 2(1 + a^r) - (1 + a)^r - (1 - a)^r$. We wish to prove that $g(a) \ge 0$ for $0 < a < 1$. Now,

$$g'(a) = 2ra^{r-1} - r(1 + a)^{r-1} + r(1 - a)^{r-1}$$
$$= r(2 - 2^{r-1})a^{r-1} + r\big((2a)^{r-1} + (1 - a)^{r-1} - (1 + a)^{r-1}\big) \ge 0.$$

Here we have used the fact that $0 < r - 1 < 1$ to conclude that the first expression is non-negative, and Lemma 5.1 for the second one.

Next, let $r > 2$. The analogous argument leads to proving that

$$(1 + a)^r + (1 - a)^r \ge 2(1 + a^r) \quad \text{for} \quad 0 < a < 1,$$

so that with $g(a) = (1 + a)^r + (1 - a)^r - 2(1 + a^r)$,

$$g'(a) = r\big((1+a)^{r-1} - (1-a)^{r-1} - 2a^{r-1}\big), \quad \text{and}$$
$$g''(a) = r(r-1)\big((1+a)^{r-2} - (1-a)^{r-2} - 2a^{r-2}\big)$$
$$= r(r-1)\big([(1+a)^{r-2} + (1-a)^{r-2} - 2] + 2[1 - a^{r-2}]\big) \geq 0,$$

beacuse the first expression in brackets is non-negative by convexity;

$$\frac{1}{2}(1+a)^{r-2} + \frac{1}{2}(1-a)^{r-2} \geq \left(\frac{(1+a)+(1-a)}{2}\right)^{r-2} = 1,$$

and because the second expression is trivially non-negative (since $r - 2 > 0$).

Now, $g''(0) = 0$ and $g''(a) \geq 0$ implies that $g'$ is non-decreasing. Since $g'(0) = 0$, it follows that $g'$ is non-negative, so that $g$ is non-decreasing, which, finally, since $g(0) = 0$, establishes the non-negativity of $g$.    □

*Remark 5.3.* A functional analyst would probably say that (ii) trivially follows from (i) by a standard duality argument.    □

# 6 Convergence of (Weighted) Averages

A fact that is frequently used, often without further ado, is that (weighted) averages of convergent sequences converge (too). After all, by abuse of language, this is pretty "obvious". Namely, if the sequence is convergent, then, after a finite number of elements, the following ones are all close to the limit, so that the average is essentially equal to the average of the last group; the first couple of terms do not matter in the long run. But intuition and proof are not the same. However, a special feature here is that it is unusually transparent how the proof is, literally, a translation of intuition and common sense into formulas.

**Lemma 6.1.** *Suppose that $a_n \in \mathbb{R}$, $n \geq 1$. If $a_n \to a$ as $n \to \infty$, then*

$$\frac{1}{n}\sum_{k=1}^{n} a_k \to a \quad as \quad n \to \infty.$$

*If, in addition, $w_k \in \mathbb{R}^+$, $k \geq 1$, and $B_n = \sum_{k=1}^{n} w_k$, $n \geq 1$, then*

$$\frac{1}{B_n}\sum_{k=1}^{n} w_k a_k \to 0 \quad as \quad n \to \infty.$$

*Proof.* It is no restriction to assume that $a = 0$ (otherwise consider the sequence $a_n - a$, $n \geq 1$). Thus, given an arbitrary $\varepsilon > 0$, we know that $|a_n| < \varepsilon$ as soon as $n > n_0 = n_0(\varepsilon)$. It follows that, for $n > n_0$,

$$\left|\frac{1}{n}\sum_{k=1}^{n} a_k\right| \leq \left|\frac{1}{n}\sum_{k=1}^{n_0} a_k\right| + \frac{n-n_0}{n}\left|\frac{1}{n-n_0}\sum_{k=n_0+1}^{n} a_k\right| \leq \frac{1}{n}\sum_{k=1}^{n_0}|a_k| + \varepsilon,$$

so that

$$\limsup_{n\to\infty}\left|\frac{1}{n}\sum_{k=1}^{n_0}a_k\right|\le\varepsilon,$$

which does it, since $\varepsilon$ can be made arbitrarily small.

This proves the first statement. The second one follows similarly.     □

**Exercise 6.1.** Carry out the proof of the second half of the lemma.     □

*Example 6.1.* If $a_n\to a$ as $n\to\infty$, then, for example,

$$\frac{1}{\log n}\sum_{k=1}^{n}\frac{1}{k}a_k\to a,$$

$$\frac{1}{\log\log n}\sum_{k=1}^{n}\frac{1}{k\log k}a_k\to a,$$

$$n^\alpha\sum_{k=1}^{n}\frac{1}{k^{\alpha+1}}a_k\to\frac{a}{\alpha},\quad \alpha>0.$$     □

Next, an important further development in this context.

**Lemma 6.2.** (Kronecker's lemma) *Suppose that $x_n\in\mathbb{R}$, $n\ge 1$, set $a_0=0$, and let $a_n$, $n\ge 1$, be positive numbers increasing to $+\infty$. Then*

$$\sum_{n=1}^{\infty}\frac{x_n}{a_n}\ \text{converges}\quad\Longrightarrow\quad \frac{1}{a_n}\sum_{k=1}^{n}x_k\to 0\quad\text{as}\quad n\to\infty.$$

*Proof.* The essential tools are partial summation and Lemma 6.1.

Set, $b_0=0$, and, for $1\le n\le\infty$,

$$b_n=\sum_{k=1}^{n}\frac{x_k}{a_k}.$$

Since $x_k=a_k(b_k-b_{k-1})$ for all $k$, it follows by partial summation that

$$\frac{1}{a_n}\sum_{k=1}^{n}x_k=b_n-\frac{1}{a_n}\sum_{k=0}^{n-1}(a_{k+1}-a_k)b_k.$$

Now, $b_n\to b_\infty$ as $n\to\infty$ by assumption, and

$$\frac{1}{a_n}\sum_{k=0}^{n-1}(a_{k+1}-a_k)b_k\to b_\infty\quad\text{as}\quad n\to\infty,$$

by the second half of Lemma 6.1, since we are faced with a weighted average of quantities tending to $b_\infty$.     □

*Example 6.2.* Let $x_n \in \mathbb{R}$, $n \geq 1$. Then

$$\sum_{n=1}^{\infty} \frac{x_n}{n} \text{ converges} \implies \bar{x}_n = \frac{1}{n}\sum_{k=1}^{n} x_k \to 0 \quad \text{as} \quad n \to \infty. \qquad \square$$

The following continuous version is proved similarly.

**Lemma 6.3.** *Suppose that $\{g_n, n \geq 1\}$ are real valued continuous functions such that $g_n \to g$ as $n \to \infty$, where $g$ is continuous in a neighborhood of $b \in \mathbb{R}$. Then, for every $\varepsilon > 0$, there exists $h_0 > 0$, such that*

$$\limsup_{n\to\infty} \left| \frac{1}{2h}\int_{|x-b|<h} g_n(x)\,dx - g(b) \right| < \varepsilon \quad \text{for all} \quad h \in (0, h_0).$$

**Exercise 6.2.** Prove the lemma.

**Exercise 6.3.** State and prove version for weighted averages. $\qquad \square$

## 7 Regularly and Slowly Varying Functions

Regularly and slowly varying functions were introduced by Karamata [152]. Since then, the theory has become increasingly important in probability theory. For more on the topic we refer to [21, 88, 123, 217].

**Definition 7.1.** *Let $a > 0$. A positive measurable function $u$ on $[a, \infty)$ varies regularly at infinity with exponent $\rho$, $-\infty < \rho < \infty$, denoted $u \in \mathcal{RV}(\rho)$, iff*

$$\frac{u(tx)}{u(t)} \to x^\rho \quad \text{as} \quad t \to \infty \quad \text{for all} \quad x > 0.$$

*If $\rho = 0$ the function is* slowly varying *at infinity; $u \in \mathcal{SV}$.* $\qquad \square$

Typical examples of regularly varying functions are

$$x^\rho, \quad x^\rho \log^+ x, \quad x^\rho \log^+ \log^+ x, \quad x^\rho \frac{\log^+ x}{\log^+ \log^+ x}, \quad \text{and so on.}$$

Typical slowly varying functions are the above when $\rho = 0$. Moreover, every positive function with a finite limit as $x \to \infty$ is slowly varying. Regularly varying functions with a non-zero exponent are ultimately monotone.

**Exercise 7.1.** Check that the typical functions behave as claimed. $\qquad \square$

The following lemma contains some elementary properties of regularly and slowly varying functions. The first two are a bit harder to verify, so we refer to the literature for them. The three others follow, essentially, from the definition and the previous lemma.

**Lemma 7.1.** *Let $u \in \mathcal{RV}(\rho)$ be positive on the positive half-axis.*

(a) *If $-\infty < \rho < \infty$, then $u(x) = x^\rho \ell(x)$, where $\ell \in SV$.*
*If, in addition, $u$ has a monotone derivative $u'$, then*

$$\frac{xu'(x)}{u(x)} \to \rho \quad as \quad x \to \infty.$$

*If, moreover, $\rho \neq 0$, then $\mathrm{sgn}\,(u) \cdot u' \in RV\,(\rho - 1)$.*

(b) *Let $\rho > 0$, and set $u^{-1}(y) = \inf\{x : u(x) \geq y\}$, $y > 0$. Then $u^{-1} \in RV\,(1/\rho)$.*

(c) $\log u \in SV$.

(d) *Suppose that $u_i \in RV\,(\rho_i)$, $i = 1, 2$. Then $u_1 + u_2 \in RV\,(\max\{\rho_1, \rho_2\})$.*

(e) *Suppose that $u_i \in RV\,(\rho_i)$, $i = 1, 2$, that $u_2(x) \to \infty$ as $x \to \infty$, and set $u(x) = u_1(u_2(x))$. Then $u \in RV\,(\rho_1 \cdot \rho_2)$. In particular, if one of $u_1$ and $u_2$ is slowly varying, then so is the composition.*

*Proof.* As just mentioned, we omit (a) and (b).

(c): The fact that $\frac{u(tx)}{u(t)} \to x^\rho$ as $t \to \infty$ yields

$$\frac{\log u(tx)}{\log u(t)} = \frac{\log \frac{u(tx)}{u(t)}}{\log u(t)} + 1 \to 0 + 1 \quad as \quad t \to \infty.$$

(d): Suppose that $\rho_1 > \rho_2$. Then

$$\frac{u_1(tx) + u_2(tx)}{u_1(t) + u_2(t)} = \frac{u_1(tx)}{u_1(t)} \cdot \frac{u_1(t)}{u_1(t) + u_2(t)} + \frac{u_2(tx)}{u_2(t)} \cdot \frac{u_2(t)}{u_1(t) + u_2(t)}$$
$$\to x^{\rho_1} \cdot 1 + x^{\rho_2} \cdot 0 = x^{\rho_1} \quad as \quad t \to \infty.$$

If $\rho_1 < \rho_2$ the limit equals $x^{\rho_2}$, and if the exponents are equal (to $\rho$) the limit becomes $x^\rho$.

(e): An application of Lemma 7.2 yields

$$\frac{u(tx)}{u(t)} = \frac{u_1(u_2(tx))}{u_1(u_2(t))} = \frac{u_1(\frac{u_2(tx)}{u_2(t)} \cdot u_2(t))}{u_1(u_2(t))} \to \left(x^{\rho_2}\right)^{\rho_1} = x^{\rho_1 \cdot \rho_2} \quad as \quad t \to \infty. \quad \square$$

*Remark 7.1.* Notice that (c) is contained in (e). $\qquad\square$

In the definition of regular and slow variation the ratio between the arguments of the function is constant. However, the limits remain the same if the ratio converges to a constant.

**Lemma 7.2.** *Suppose that $u \in RV\,(\rho)$, $-\infty < \rho < \infty$, and, in addition, that $u$ is (ultimately) monotone if $\rho = 0$. Moreover, let, for $n \geq 1$, $a_n, b_n \in \mathbb{R}^+$ be such that*

$$a_n, b_n \to \infty \quad and \quad \frac{a_n}{b_n} \to c \quad as \quad n \to \infty \qquad (c \in (0, \infty)).$$

*Then*

$$\frac{u(a_n)}{u(b_n)} \to \begin{cases} 1, & for \quad \rho = 0, \\ c^\rho, & for \quad \rho \neq 0, \end{cases} \quad as \quad n \to \infty.$$

*Proof.* Suppose that $\rho > 0$, so that $u$ is ultimately non-decreasing, let $\varepsilon > 0$, and choose $n$ large enough to ensure that

$$b_n(c - \varepsilon) < a_n < b_n(c + \varepsilon).$$

Then

$$\frac{u((c - \varepsilon)b_n)}{u(b_n)} \leq \frac{u(a_n)}{u(b_n)} \leq \frac{u((c + \varepsilon)b_n)}{u(b_n)},$$

from which the conclusion follows from the fact that the extreme members converge to $(c \pm \varepsilon)^\rho$ as $n \to \infty$.

The case $\rho < 0$ is similar; the inequalities are reversed. In the slowly varying case ($\rho = 0$) the extreme limits are equal to $1$ ($= c^\rho$). $\qquad\square$

Suppose that $u \in RV(\rho)$, where $\rho > -1$. Then, since the slowly varying component is "negligible" with respect to $x^\rho$, it is reasonable to believe that the integral of $u$ is regularly varying with exponent $\rho + 1$. The truth of this fact, which is supported by Lemma 7.1(a) in conjunction with Lemma 3.1, is the first half of the next result.

**Lemma 7.3.** *Let $\rho > -1$.*

(i) *If $u \in RV(\rho)$, then*    $U(x) = \int_a^x u(y)\,dy \in RV(\rho + 1)$.

(ii) *If $\ell \in SV$, then*   $\sum_{j \leq n} j^\rho \ell(j) \sim \frac{1}{\rho+1} n^{\rho+1} \ell(n)$ *as $n \to \infty$.*

There also exists something called *rapid variation*, corresponding to $\rho = +\infty$. A function $u$ is *rapidly varying at infinity* iff

$$\frac{u(tx)}{u(t)} \to \begin{cases} 0, & \text{for } 0 < x < 1, \\ \infty, & \text{for } x > 1, \end{cases} \qquad \text{as}\quad t \to \infty.$$

This means that $u$ increases faster than any power at infinity. The exponential function $e^x$ is one example.

# 8 Cauchy's Functional Equation

This is a well known equation that enters various proofs. If $g$ is a real valued *additive* function, that is,

$$g(x + y) = g(x) + g(y),$$

then it is immediate that $g(x) = cx$ is a solution for any $c \in \mathbb{R}$. The problem is: Are there any other solutions? Yes, there exist pathological ones if nothing more is assumed. However, under certain regularity conditions this is the only solution.

**Lemma 8.1.** *Suppose that $g$ is real valued and additive on an arbitrary interval $I \subset \mathbb{R}$, and satisfies one of the following conditions:*

- $g$ is continuous;
- $g$ is monotone;
- $g$ is bounded.

Then $g(x) = cx$ for some $c \in \mathbb{R}$.

*Proof.* For $x = y$ we find that $g(2x) = 2g(x)$, and, by induction, that

$$g(n) = ng(1) \quad \text{and} \quad g(1) = ng(1/n).$$

Combining these facts for $r = m/n \in \mathbb{Q}$ tells us that

$$g(r) = g(m/n) = mg(1/n) = m\big(g(1)/n\big) = rg(1),$$

and that

$$g(rx) = rg(x) \quad \text{for any} \quad x.$$

The remaining problem is to glue all $x$-values together.

Set $c = g(1)$. If $g$ is continuous, the conclusion follows from the definition of continuity; for any $x \in \mathbb{R}$ there exists, for any given $\delta > 0$, $r \in \mathbb{Q}$, such that $|r - x| < \delta$, which implies that $|g(x) - g(r)| < \varepsilon$, so that

$$|g(x) - cx| \leq |g(x) - g(r)| + c|r - x| \leq \varepsilon + c\delta.$$

The arbitrariness of $\varepsilon$ and $\delta$ completes the proof.

If $g$ is monotone, say non-decreasing, then, for $r_1 < x < r_2$, where $r_2 - r_1 < \delta$,

$$cr_1 = g(r_1) \leq g(x) \leq g(r_2) = cr_2,$$

so that

$$|g(x) - cx| \leq c(r_2 - x) + c(x - r_1) = c(r_2 - r_1) < c\delta.$$

Finally, if $g$ is bounded, it follows, in particular, that, for any given $\delta > 0$, there exists $A$, such that,

$$|g(x)| \leq A \quad \text{for} \quad |x| < \delta.$$

For $|x| < \delta/n$, this implies that

$$|g(x)| = |g(nx)/n| \leq \frac{A}{n}.$$

Next, let $x \in I$ be given, and choose $r \in \mathbb{Q}$, such that $|r - x| < \delta/n$. Then

$$|g(x) - cx| = |g(x - r) + g(r) - cr - c(x - r)| = |g(x - r) - c(x - r)|$$
$$\leq |g(x - r) - c(x - r)| \leq |g(x - r)| + c|x - r|$$
$$\leq \frac{A}{n} + c\frac{\delta}{n} = \frac{C}{n},$$

which can be made arbitrarily small by choosing $n$ sufficiently large. $\qquad\square$

The following lemma contains variations of the previous one. For example, what happens if $g$ is multiplicative?

**Lemma 8.2.** *Let $g$ be a real valued function defined on some interval $I \subset \mathbb{R}^+$, and suppose that $g$ is continuous, monotone or bounded.*

(a) *If $g(xy) = g(x) + g(y)$, then $g(x) = c \log x$ for some $c \in \mathbb{R}$.*

(b) *If $g(xy) = g(x)g(y)$, then $g(x) = x^c$ for some $c \in \mathbb{R}$.*

(c) *If $g(x + y) = g(x)g(y)$, then $g(x) = e^{cx}$ for some $c \in \mathbb{R}$.*

*Remark 8.1.* The relation in (b) is called the *Hamel* equation.    □

*Proof.* (a): A change of variable yields

$$g(e^{x+y}) = g(e^x e^y) = g(e^x) + g(e^y),$$

so that, by Lemma 8.1, $g(e^x) = cx$, which is the same as $g(x) = c \log x$.
(b): In this case a change of variables yields

$$\log g(e^{x+y}) = \log g(e^x e^y) = \log \left(g(e^x) \cdot g(e^y)\right) = \log g(e^x) + \log g(e^y),$$

so that $\log g(e^x) = cx$, and, hence, $g(x) = e^{c \log x} = x^c$.
(c): We reduce to (b) via

$$g(\log xy) = g(\log x + \log y) = g(\log x)g(\log y),$$

so that $g(\log x) = x^c$, and, hence, $g(x) = e^{cx}$.    □

# 9 Functions and Dense Sets

Many proofs are based on the fact that it suffices to prove the desired result on a dense set. Others exploit the fact that the functions under consideration can be arbitrarily well approximated by other functions that are easier to handle; we mentioned this device in Chapter 1 in connection with Theorem 1.2.4 and the magic that "it suffices to check rectangles". In this section we collect some results which rectify some such arguments.

**Definition 9.1.** *Let $A$ and $B$ be sets. The set $A$ is* dense *in $B$ if the closure of $A$ equals $B$; if $\bar{A} = B$.*    □

The typical example one should have in mind is when $B = [0, 1]$ and $A = \mathbb{Q} \cap [0, 1]$:

$$\overline{\mathbb{Q} \cap [0, 1]} = [0, 1].$$

**Definition 9.2.** *Consider the following classes of real valued functions:*

- $C$ = *the continuous functions;*
- $C_0$ = *the functions in $C$ tending to $0$ at $\pm\infty$;*

- $C[a,b]$ = *the functions in $C$ with support on the interval $[a,b]$;*
- $D$ = *the right-continuous, functions with left-hand limits;*
- $D^+$ = *the non-decreasing functions in $D$;*
- $\mathbb{J}_G$ = *the discontinuities of $G \in D$.*

□

**Lemma 9.1.** (i) *If $G \in D^+$, then $\mathbb{J}_G$ is at most countable.*
(ii) *Suppose that $G_i \in D^+$ $i = 1,2$, and that $G_1 = G_2$ on a dense subset of the reals. Then $G_1 = G_2$ for all reals.*

*Proof.* (i): Suppose, w.l.o.g. that $0 \le G \le 1$. Let, for $n \ge 1$,

$$\mathbb{J}_G^{(n)} = \left\{ x : G \text{ has a jump at } x \text{ of magnitude } \in \left( \frac{1}{n+1}, \frac{1}{n} \right] \right\}.$$

The total number of points in $\mathbb{J}_G^{(n)}$ is at most equal to $n+1$, since $G$ is non-decreasing and has total mass 1. The conclusion then follows from the fact that

$$\mathbb{J}_G = \bigcup_{n=1}^{\infty} \mathbb{J}_G^{(n)}.$$

(ii): We first show that a function in $D^+$ is determined by its values on a dense set. Thus, let $D$ be dense in $\mathbb{R}$ (let $D = \mathbb{Q}$, for example), let $G_D \in D^+$ be defined on $D$, and set

$$G(x) = \inf_{\substack{y>x \\ y \in D}} G_D(y). \tag{A.9}$$

To prove that $G \in D^+$ we observe that the limits of $G_D$ and $G$ as $x \to \pm\infty$ coincide and that $G$ is non-decreasing, so that the only problem is to prove right-continuity.

Let $x \in \mathbb{R}$ and $\varepsilon > 0$ be given, and pick $y \in D$ such that

$$G_D(y) \le G(x) + \varepsilon.$$

Moreover, by definition, $G(y) \le G_D(u)$ for all $u \ge y$, so that, in particular,

$$G(y) \le G_D(u) \quad \text{for any} \quad u \in (x,y).$$

Combining this with the previous inequality proves that

$$G(u) \le G(x) + \varepsilon \quad \text{for all} \quad u \in (x,y).$$

The monotonicity of $G$, and the fact that $u$ may be chosen arbitrarily close to $x$, now together imply that

$$G(x) \le G(x+) \le G(x) + \varepsilon,$$

which, due to the arbitrariness of $\varepsilon$, proves that $G(x) = G(x+)$, so that $G \in D^+$.

Finally, if two functions in $D^+$ agree on a dense set, then the extensions to all of $\mathbb{R}$ via (A.9) does the same thing to both functions, so that they agree everywhere. □

**Lemma 9.2.** *Let $G$ and $G_n \in D^+$, $n \geq 1$, and let $J(x) = G(x) - G(x-)$ and $J_n(x) = G_n(x) - G_n(x-)$ denote the jumps of $G$ and $G_n$, respectively, at $x$.*
(i) *Suppose that $G \in D^+ \cap C[a,b]$, where $-\infty < a < b < \infty$. If*

$$G_n(x) \to G(x) \quad as \quad n \to \infty, \quad for\ all \quad x \in D,$$

*then $G_n \to G$ uniformly on $[a,b]$;*

$$\sup_{a \leq x \leq b} |G_n(x) - G(x)| \to 0 \quad as \quad n \to \infty.$$

*Moreover,*

$$\sup_{\substack{a \leq x \leq b \\ x \in \mathbb{J}_G}} |J_n(x)| \to 0 \quad as \quad n \to \infty.$$

(ii) *Suppose that $G \in D^+ \cap C$. If, for some dense subset $D \subset \mathbb{R}$,*

$$G_n(x) \to G(x) \quad as \quad n \to \infty, \quad for\ all \quad x \in D,$$
$$G_n(\pm\infty) \to G(\pm\infty) \quad as \quad n \to \infty,$$

*then*

$$\sup_{x \in \mathbb{R}} |G_n(x) - G(x)| \to 0 \quad as \quad n \to \infty,$$

$$\sup_{\substack{x \in \mathbb{R} \\ x \in \mathbb{J}_G}} |J_n(x)| \to 0 \quad as \quad n \to \infty.$$

(iii) *Suppose that $G \in D^+$. If, for some dense subset $D \subset \mathbb{R}$,*

$$G_n(x) \to G(x) \quad as \quad n \to \infty, \quad for\ all \quad x \in D,$$
$$J_n(x) \to J(x) \quad as \quad n \to \infty, \quad for\ all \quad x \in \mathbb{J}_G,$$
$$G_n(\pm\infty) \to G(\pm\infty) \quad as \quad n \to \infty,$$

*then*

$$\sup_{x \in \mathbb{R}} |G_n(x) - G(x)| \to 0 \quad as \quad n \to \infty,$$

$$\sup_{x \in \mathbb{J}_G} |J_n(x)| \to 0 \quad as \quad n \to \infty.$$

*Proof.* (i): Since $G$ is continuous on a bounded interval it is uniformly continuous. Thus, for any $\varepsilon > 0$, there exists $\delta > 0$, such that

$$\omega_G(\delta) = \sup_{\substack{a \leq x, y \leq b \\ |x-y| < \delta}} |G(x) - G(y)| < \varepsilon.$$

Given the above $\varepsilon$ and the accompanying $\delta$ we let $a = y_0 < y_1 < \cdots < y_m = b$, such that $y_k - y_{k-1} < \delta$ for all $k$. For any $x \in [y_{k-1}y_k]$, $1 \leq k \leq m$, it then follows that

$$G_n(y_{k-1}) - G(x) \le G_n(x) - G(x) \le G_n(y_k) - G(x),$$

so that

$$
\begin{aligned}
|G_n(x) - G(x)| &\le |G_n(y_{k-1}) - G(y_{k-1})| + |G_n(y_k) - G(y_k)| \\
&\quad + |G(y_{k-1}) - G(x)| + |G(y_k) - G(x)| \\
&\le 2 \max_{1 \le k \le m} |G_n(y_k) - G(y_k)| + 2\omega_G(\delta),
\end{aligned}
$$

so that

$$\limsup_{n \to \infty} \sup_{a \le x \le b} |G_n(x) - G(x)| \le 2\omega_G(\delta) \le 2\varepsilon.$$

As for the second statement, noticing that $J(x) = 0$, we obtain

$$
\begin{aligned}
\sup_{x \in J} |G_n(x)| &\le \sup_{x \in J} \left( |G_n(x) - G(x)| + |G(x-) - G_n(x)| \right) \\
&\le 2 \sup_{a \le x \le b} |G_n(x) - G(x)| \to 0 \quad \text{as} \quad n \to \infty.
\end{aligned}
$$

(ii): Since convergence at the infinities is assumed, we have, using (i),

$$
\begin{aligned}
\sup_{x \in \mathbb{R}} |G_n(x) - G(x)| &\le \sup_{|x| > A} |G_n(x) - G(x)| + \sup_{|x| \le A} |G_n(x) - G(x)| \\
&\le \left( G(\infty) - G_n(A) \right) + \left( G_n(-A) - G(-\infty) \right) \\
&\quad + \left( G(\infty) - G(A) \right) + \left( G(-A) - G(-\infty) \right) \\
&\quad + \sup_{|x| \le A} |G_n(x) - G(x)|.
\end{aligned}
$$

Thus, for $\pm A \in C(G)$, we obtain, recalling (i),

$$\limsup_{n \to \infty} \sup_{x \in \mathbb{R}} |G_n(x) - G(x)| \le 2 \left( G(\infty) - G(A) \right) + \left( G(-A) - G(-\infty) \right) + 0,$$

which can be made arbitrarily small by letting $A \to \infty$.

The second statement follows as in (i).

(iii): Assume that the conclusion does not hold, that is, suppose that there exist $\varepsilon > 0$ and a subsequence $\{n_k, k \ge 1\}$, $n_k \nearrow \infty$ as $k \to \infty$, such that

$$|G_{n_k}(x_k) - G(x_k)| > \varepsilon \quad \text{for all} \quad k.$$

The first observation is that we cannot have $x_k \to \pm\infty$, because of the second assumption, which means that $\{x_k, k \ge 1\}$ is bounded, which implies that there exists a convergent subsequence, $x_{k_j} \to x$, say, as $j \to \infty$. By diluting it further, if necessary, we can make it monotone. Since convergence can occur from above and below, and $G_{n_{k_j}}(x_{k_j})$ can be smaller as well as larger than $G(x_{k_j})$ we are faced with four different cases as $j \to \infty$:

- $x_{k_j} \searrow x$,    and    $G(x_{k_j}) - G_{n_{k_j}}(x_{k_j}) > \varepsilon$;

574 A Some Useful Mathematics

- $x_{k_j} \searrow x$, and $G_{n_{k_j}}(x_{k_j}) - G(x_{k_j}) > \varepsilon$;
- $x_{k_j} \nearrow x$, and $G(x_{k_j}) - G_{n_{k_j}}(x_{k_j}) > \varepsilon$;
- $x_{k_j} \nearrow x$, and $G_{n_{k_j}}(x_{k_j}) - G(x_{k_j}) > \varepsilon$.

Choose $r_1, r_2 \in D$, such that $r_1 < x < r_2$. In the first case this leads to

$$\varepsilon < G(x_{k_j}) - G_{n_{k_j}}(x_{k_j}) \leq G(r_2) - G_{n_{k_j}}(x)$$
$$\leq G(r_2) - G(r_1) + G(r_1) - G_{n_{k_j}}(r_1) + J_{n_{k_j}}(x)$$
$$\to G(r_2) - G(r_1) + 0 - J(x) \quad \text{as} \quad j \to \infty.$$

Since $r_1, r_2$ may be chosen arbitrarily close to $x$ from below and above, respectively, the right-hand side can be made arbitrarily close to $0$ if $x \in C(G)$, and arbitrarily close to $J(x) - J(x) = 0$ if $x \in \mathbb{J}_G$, which produces the desired contradiction.

The three other cases are treated similarly:

In the second case,

$$\varepsilon < G_{n_{k_j}}(x_{k_j}) - G(x_{k_j}) \leq G_{n_{k_j}}(r_2) - G(x) \to G(r_2) - G(x) \quad \text{as} \quad j \to \infty,$$

and the contradiction follows from the right-continuity of $G$ by choosing $r_2$ close to $x$.

In the third case,

$$\varepsilon < G(x_{k_j}) - G_{n_{k_j}}(x_{k_j}) \leq G(x-) - G_{n_{k_j}}(r_1) \to G(x-) - G(r_1) \quad \text{as} \quad j \to \infty,$$

after which we let $r_1$ approach $x-$.

Finally,

$$\varepsilon < G_{n_{k_j}}(x_{k_j}) - G(x_{k_j}) \leq G_{n_{k_j}}(x-) - G(r_1)$$
$$\leq -J_{n_{k_j}}(x) + G_{n_{k_j}}(r_2) - G(r_2) + G(r_2) - G(r_1)$$
$$\to -J(x) + G(r_2) - G(r_1) \quad \text{as} \quad j \to \infty,$$

from which the contradiction follows as in the first variant.   $\square$

We close with two approximation lemmas.

**Lemma 9.3.** (Approximation lemma) *Let $f$ be a real valued function such that either*

- $f \in C[a, b]$,   *or*
- $f \in C_0$.

*Then, for every $\varepsilon > 0$, there exists a simple function $g$, such that*

$$\sup_{x \in \mathbb{R}} |f(x) - g(x)| < \varepsilon.$$

*Proof.* Suppose that $f \in C[a, b]$, and set

$$g_1(x) = \begin{cases} \frac{k-1}{2^n}, & \text{for} \quad \frac{k-1}{2^n} \le f(x) < \frac{k}{2^n}, \quad a < x < b, \\ 0, & \text{otherwise,} \end{cases}$$

and

$$g_2(x) = \begin{cases} \frac{k}{2^n}, & \text{for} \quad \frac{k-1}{2^n} \le f(x) < \frac{k}{2^n}, \quad a < x < b, \\ 0, & \text{otherwise.} \end{cases}$$

Then, for $i = 1, 2$,

$$|f(x) - g_i(x)| \le g_2(x) - g_1(x) = \frac{1}{2^n} < \varepsilon,$$

as soon as $n$ is large enough. In addition, $f$ is sandwiched between the $g$-functions; $g_1(x) \le f(x) \le g_2(x)$ for all $x$.

If $f \in C_0$, then $|f(x)| < \varepsilon$ for $|x| > b$, so that $g_1$ and $g_2$ may be defined as above for $|x| \le b$ (that is, $a = -b$) and equal to 0 otherwise. By a slight modification the sandwiching effect can also be retained.     $\square$

**Lemma 9.4.** *Let* $-\infty < a \le b < \infty$. *Any indicator function* $I_{(a,b]}(x)$ *can be arbitrarily well approximated by a bounded, continuous function; there exists* $f_n$, $n \ge 1$, $0 \le f_n \le 1$, *such that*

$$f_n(x) \to I_{(a,b]}(x) \quad \text{for all} \quad x \in \mathbb{R}.$$

*Proof.* Set, for $n \ge 1$,

$$f_n(x) = \begin{cases} 0, & \text{for} \quad x \le a, \\ n(x - a), & \text{for} \quad a < x \le a + \frac{1}{n}, \\ 1, & \text{for} \quad a + \frac{1}{n} < x \le b, \\ 1 - n(x - b), & \text{for} \quad b < x \le b + \frac{1}{n}, \\ 0, & \text{for} \quad x > b. \end{cases}$$

One readily checks that $\{f_n, n \ge 1\}$ does the job.     $\square$

**Exercise 9.1.** Please pursue the checking.     $\square$

# References

In addition to the books and papers that have been cited in the text we also refer to a number of related items that have not been explicitly referred to, but which, nevertheless, may be of interest for further studies.

1. AARONSON, J., GILAT, D., KEANE, M., AND DE VALK, V. An algebraic construction of a class of one-dependent processes. *Ann. Probab.* **17**, 128–143 (1989).

2. DE ACOSTA, A. A new proof of the Hartman-Wintner law of the iterated logarithm. *Ann. Probab.* **11**, 270–276 (1983).

3. DE ACOSTA, A., AND KUELBS, J. Some new results on the cluster set $C(\{S_n/a_n\})$ and the LIL. *Ann. Probab.* **11**, 102–122 (1983).

4. ALEŠKEVIČIENE, A. On the local limit theorem for the first passage time across a barrier. *Litovsk. Mat. Sb.* **XV**, 23–66 (1975).

5. ANSCOMBE, F.J. Large sample-theory of sequential estimation. *Proc. Cambridge Philos. Soc.* **48**, 600–607 (1952).

6. AUSTIN, D.G., EDGAR, G.A., AND IONESCU TULCEA, A. Pointwise convergence in terms of expectations. *Z. Wahrsch. verw. Gebiete* **30**, 17–26 (1974).

7. VON BAHR, B., AND ESSEEN, C.-G. Inequalities for the $r$th absolute moment of a sum of random variables, $1 \leq r \leq 2$. *Ann. Math. Statist.* **36**, 299–302 (1965).

8. BARBOUR, A.D., AND HALL, P. On the rate of Poisson convergence. *Math. Proc. Camb. Phil. Soc.* **95**, 473–480 (1984).

9. BARBOUR, A.D., HOLST, L., AND JANSON, S. *Poisson Approximation.* Oxford Science Publications, Clarendon Press, Oxford (1992).

10. BARNDORFF-NIELSEN O. On the rate of growth of the partial maxima of a sequence of independent identically distributed random variables. *Math. Scand.* **9**, 383–394 (1961).

11. BAUER, H. *Wahrscheinlichkeitstheorie und Grundzüge der Maßtheorie.* De Gruyter, Berlin (1968).

12. BAUM, L.E., AND KATZ, M. On the influence of moments on the asymptotic distribution of sums of random variables. *Ann. Math. Statist.* **34**, 1042–1044 (1963).

578    References

13. BAUM, L.E., AND KATZ, M. Convergence rates in the law of large numbers. *Trans. Amer. Math. Soc.* **120**, 108–123 (1965).

14. VAN BEEK, P. An application of Fourier methods to the problem of sharpening of the Berry-Esseen inequality. *Z. Wahrsch. verw. Gebiete* **23**, 183–196 (1972).

15. BERBEE, H.C.P. *Random Walks with Stationary Increments and Renewal Theory.* Mathematical Centre Tracts **112**, Amsterdam (1979).

16. BERKES, I., AND PHILLIP, W. Limit theorems for mixing sequences without rate assumptions. *Ann. Probab.* **26**, 805–831 (1978).

17. BERRY, A.C. The accuracy of the Gaussian approximation to the sum of independent variates. *Trans. Amer. Math. Soc.* **49**, 122–136 (1941).

18. BILLINGSLEY, P. The Lindeberg-Lévy theorem for martingales. *Proc. Amer. Math. Soc.* **12**, 788–792 (1961).

19. BILLINGSLEY, P. *Probability and Measure*, 3rd ed. Wiley, New York (1995).

20. BILLINGSLEY, P. *Convergence of Probability Measures*, 2nd ed. Wiley, New York (1999).

21. BINGHAM, N.H., GOLDIE, C.M., AND TEUGELS, J.L. *Regular Variation.* Cambridge University Press, Cambridge (1987).

22. BLACKWELL, D. On an equation of Wald. *Ann. Math. Statist.* **17**, 84–87 (1946).

23. BLACKWELL, D. Extension of a renewal theorem. *Pacific J. Math.* **3**, 315–320 (1953).

24. BLUM, J.R., HANSON,D.L., AND ROSENBLATT, J.I. On the central limit theorem for the sum of a random number of independent random variables. *Z. Wahrsch. verw. Gebiete* **1**, 389–393 (1963).

25. BONDESSON, L. *Generalized Gamma Convolutions and Related Classes of Distributions and Densities.* Lecture Notes in Statistics **76**. Springer-Verlag (1992).

26. BOREL, E. Sur les probabilités et leurs applications arithmétiques. *Rend. Circ. Mat. Palermo* **26**, 247–271 (1909).

27. BRADLEY, R.C. Basic properties of strong mixing conditions. In: E. Eberlein and M.S. Taqqu, eds., *Dependence in Probability and Statistics*, 165–192. Birkhäuser, Boston, MA (1986).

28. BRADLEY, R.C. *Introduction to Strong Mixing Conditions, Vol. 1.* Technical Report (March 2003 printing), Department of Mathematics, Indiana University, Bloomington. Custom Publishing of I.U., Bloomington, March 2003.

29. BRADLEY, R.C. *Introduction to Strong Mixing Conditions, Vol. 2.* Technical Report, Department of Mathematics, Indiana University, Bloomington. Custom Publishing of I.U., Bloomington, March 2003.

30. BRADLEY, R.C. *Introduction to Strong Mixing Conditions, Vol. 3.* Technical Report, Department of Mathematics, Indiana University, Bloomington. Custom Publishing of I.U., Bloomington (in preparation).

31. BREIMAN, L. *Probability.* Addison-Wesley, Reading, MA (1968).

32. BURKHOLDER, D.L. Successive conditional expectations of an integrable function. *Ann. Math. Statist.* **33**, 887–893 (1962).

33. BURKHOLDER, D.L. Martingale transforms. *Ann. Math. Statist.* **37**, 1494–1504 (1966).

34. BURKHOLDER, D.L. Distribution function inequalities for martingales. *Ann. Probab.* **1**, 19–42 (1973).

35. BURKHOLDER, D.L., DAVIS, B.J., AND GUNDY, R.F. Integral inequalities for convex functions of operators on martingales. In: *Proc. Sixth Berkeley Symp. Math. Statist. Prob.* **2**, 223–240 (1972).

36. BURKHOLDER, D.L., AND GUNDY, R.F. Extrapolation and interpolation of quasi-linear operators on martingales. *Acta Math.* **124**, 249–304 (1970).

37. CARLEMAN, T. Les fonctions quasi-analythiques. *Collection Borel*, Gauthiers-Villars, Paris (1926).

38. CHACON, R.V. A "stopped" proof of convergence. *Adv. in Math.* **14**, 365–368 (1974).

39. CHEN, L.H.Y. Poisson approximation for dependent trials. *Ann. Probab.* **3**, 534–545 (1975)

40. CHEN, P.-N. Asymptotic refinement of the Berry-Esseen constant. Preprint, National Chiao Tung University, Taiwan (2002).

41. CHEN, R. A remark on the tail probability of a distribution. *J. Multivariate Analysis* **8**, 328–333 (1978).

42. CHERN, H.-H., HWANG, H.-K., AND YEH, Y.-N. Distribution of the number of consecutive records. *Proceedings of the Ninth International Conference "Random Structures and Algorithms" (Poznań, 1999). Random Struct. Alg.* **17**, 169–196 (2000).

43. CHOW, Y.S. A martingale inequality and the law of large numbers. *Proc. Amer. Math. Soc.* **11**, 107–111 (1960).

44. CHOW, Y.S. Delayed sums and Borel summability of independent, identically distributed random variables. *Bull. Inst. Math. Acad. Sinica* **1**, 207–220 (1973).

45. CHOW, Y.S., AND HSIUNG, C.A. Limiting behaviour of $\max_{j \leq n} S_j j^{-\alpha}$ and the first passage times in a random walk with positive drift. *Bull. Inst. Math. Acad. Sinica* **4**, 35–44 (1976).

46. CHOW, Y.S., HSIUNG, C.A., AND LAI, T.L. Extended renewal theory and moment convergence in Anscombe's theorem. *Ann. Probab.* **7**, 304–318 (1979).

47. CHUNG, K.L. On the maximum partial sums of sequences of independent random variables. *Trans. Amer. Math. Soc.* **64**, 205–233 (1948).

48. CHUNG, K.L. *A Course in Probability Theory*, 2nd ed. Academic Press, New York (1974).

49. CHUNG, K.L., AND ERDŐS, P. On the application of the Borel-Cantelli lemma. *Trans. Amer. Math. Soc.* **72**, 179–186 (1952).

50. CLARKSON, J.A. Uniformly convex spaces. *Trans. Amer. Math. Soc.* **40**, 396–414 (1936).

51. COHEN, P.J. *Set Theory and the Continuum Hypothesis*. W. A. Benjamin, Inc., New York (1966).

52. COHN, D.L. *Measure Theory*. Birkhäuser, Boston, MA (1997).

53. COX, D. *Renewal Theory*. Methuen, London (1962).

54. CRAMÉR, H. Über eine Eigenschaft der normalen Verteilungsfunktion. *Math. Z.* **41**, 405–414 (1936).

55. CRAMÉR, H. Sur un nouveau théorème-limite de la théorie des probabilités. *Actual. Sci. Indust.* **736**, 5–23 (1938).

56. CRAMÉR, H. *Mathematical Methods of Statistics*. Princeton University Press, Princeton, NJ (1946).

57. CSÖRGŐ, M., AND RYCHLIK, Z. Weak convergence of sequences of random elements with random indices. *Math. Proc. Camb. Phil. Soc.* **88**, 171–174 (1980).

58. CSÖRGŐ, M., AND RYCHLIK, Z. Asymptotic properties of randomly indexed sequences of random variables. *Canad. J. Statist.* **9**, 101–107 (1981).

59. DAVIS, B. On the integrability of the martingale square function. *Israel J. Math.* **8**, 187–190 (1970).

60. DAVIS, B. Stopping rules for $S_n/n$ and the class $L \log L$. *Z. Wahrsch. verw. Gebiete* **17**, 147–150 (1971).

61. DAVIS, J.A. Convergence rates for the law of the iterated logarithm. *Ann. Math. Statist.* **39**, 1479–1485 (1968).

62. DAVIS, J.A. Convergence rates for probabilities of moderate deviations (1968). *Ann. Math. Statist.* **39**, 2016–2028.

63. DONSKER, M. An invariance principle for certain probability limit theorems. *Mem. Amer. Math. Soc.* **6** (1951).

64. DOOB, J.L. Regularity properties of certain families of chance variables. *Trans. Amer. Math. Soc.* **47**, 455–486 (1940).

65. DOOB, J.L. Renewal theory from the point of view of probability. *Trans. Amer. Math. Soc.* **63**, 422–438 (1948).

66. DOOB, J.L. *Stochastic Processes*. Wiley, New York (1953).

67. DROSIN, M. *The Bible Code*. Simon and Schuster, New York (1997).

68. DUDLEY, R. *Real Analysis and Probability*. Wadsworth & Brooks/Cole, Belmont, CA (1989).

69. DURRETT, R. *Probability: Theory and Examples*. Wadsworth & Brooks/Cole, Belmont, CA (1991).

70. EDGAR, G.A., AND SUCHESTON, L. Amarts: A class of asymptotic martingales. *J. Multivariate Anal.* **6**, 193-221 (1976).

71. EMBRECHTS, P., KLÜPPELBERG, C., AND MIKOSCH, T. *Modelling Extremal Events for Insurance and Finance*. Springer-Verlag, Berlin (1997).

72. ENGLUND, G. Remainder term estimates for the asymptotic normality of the number of renewals. *J. Appl. Probab.* **17**, 1108–1113 (1980).

73. ENGLUND, G. On the coupon collector's remainder term. *Z. Wahrsch. verw. Gebiete* **60**, 381–391 (1982).

74. ERDŐS, P. On a theorem of Hsu and Robbins. *Ann. Math. Statist.* **20**, 286–291 (1949).

75. ERDŐS, P. Remark on my paper "On a theorem of Hsu and Robbins". *Ann. Math. Statist.* **21**, 138 (1950).

76. ESSEEN, C.-G. On the Liapounoff limit of error in the theory of probability. *Ark. Mat., Astr. o. Fysik* **28A** 9, 1-19 (1942).

77. ESSEEN, C.-G. Fourier analysis of distribution functions. A mathematical study of the Laplace-Gaussian law. *Acta Math.* **77**, 1–125 (1945).

78. ESSEEN, C.-G. A moment inequality with an application to the central limit theorem. *Skand. Aktuar. Tidskr.* **XXXIX**, 160–170 (1956).

79. ESSEEN, C.-G. On the Kolmogorov-Rogozin inequality for the concentration function. *Z. Wahrsch. verw. Gebiete* **5**, 210–216 (1966).

80. ESSEEN, C.-G. On the concentration function of a sum of independent random variables. *Z. Wahrsch. verw. Gebiete* **9**, 290–308 (1968).

81. ESSEEN, C.-G. On the remainder term in the central limit theorem. *Arkiv Matematik* **8**, 7–15 (1968).

82. ETEMADI, N. An elementary proof of the strong law of large numbers. *Z. Wahrsch. verw. Gebiete* **55**, 119–122 (1981).

83. FELLER, W. Über den zentralen Grenzwertsatz der Wahrscheinlichkeitsrechnung. *Math. Z.* **40**, 521–559 (1935).

84. FELLER, W. Über den zentralen Grenzwertsatz der Wahrscheinlichkeitsrechnung II. *Math. Z.* **42**, 301–312 (1937).

85. FELLER, W. Fluctuation theory of recurrent events. *Trans. Amer. Math. Soc.* **67**, 98–119 (1949).

86. FELLER, W. An extension of the law of the iterated logarithm to variables without variance. *J. Math. Mech.* **18**, 343–355 (1968).
87. FELLER, W. *An Introduction to Probability Theory and Its Applications, Vol 1*, 3rd ed. Wiley, New York (1968).
88. FELLER, W. *An Introduction to Probability Theory and Its Applications, Vol 2*, 2nd ed. Wiley, New York (1971).
89. FISHER, R.A., AND TIPPETT, L.H.C. Limiting forms of the frequency of the largest or smallest member of a sample. *Proc. Camb. Phil. Soc.* **24**, 180–190 (1928).
90. FRIEDMAN, N., KATZ, M., AND KOOPMANS, L.H. Convergence rates for the central limit theorem. *Proc. Nat. Acad. Sci. USA* **56**, 1062–1065 (1966).
91. GALAMBOS, J. *The Asymptotic Theory of Extreme Order Statistics*, 2nd ed. Krieger, Malabar, FL (1987).
92. GARSIA, A.M. *Topics in Almost Everywhere Convergence.* Markham, Chicago (1970).
93. GARSIA, A.M. *Martingale Inequalities. Seminar Notes on Recent Progress.* W.A. Benjamin, Inc., Reading MA (1973).
94. GILAT, D. Every nonnegative submartingale is the absolute value of a martingale. *Ann. Probab.* **5**, 475–481 (1977).
95. GNEDENKO, B.V. Sur la distribution limite du terme maximum d'une série aléatoire. *Ann. Math.* **44**, 423–453 (1943).
96. GNEDENKO, B.V. On the local limit theorem in the theory of probability. *Uspekhi Mat. Nauk.* **3** (1948) (in Russian).
97. GNEDENKO, B.V. *Theory of Probability*, 4th ed. Chelsea, New York (1967).
98. GNEDENKO, B.V., AND KOLMOGOROV, A.N. *Limit Distributions for Sums of Independent Random Variables*, 2nd ed. Addison-Wesley, Cambridge, MA (1968).
99. GRIMMETT, G.R., AND STIRZAKER, D.R. *Probability Theory and Random Processes*, 2nd ed. Oxford University Press, Oxford (1992).
100. GUMBEL, E.J. *Statistics of Extremes.* Columbia University Press, New York (1958).
101. GUNDY, R.F. On the class $L \log L$, martingales, and singular integrals. *Studia Math.* **XXXIII**, 109–118 (1969).
102. GUNDY, R.F. Inégalités pour martingales à un et deux indices: L'éspace $H^p$. In: *École d'Été de Probabilités de Saint Fluor VIII-1978.* Lecture Notes in Mathematics **774**, 251–334. Springer-Verlag, New York (1980).
103. GUT, A. On the moments and limit distributions of some first passage times. *Ann. Probab.* **2**, 277–308 (1974).
104. GUT, A. Marcinkiewicz laws and convergence rates in the law of large numbers for random variables with multidimensional indices. *Ann. Probab.* **6**, 469–482 (1978).
105. GUT, A. Convergence rates for probabilities of moderate deviations for sums of random variables with multidimensional indices. *Ann. Probab.* **8**, 298–313 (1980).
106. GUT, A. An introduction to the theory of asymptotic martingales. In: Gut, A., and Schmidt, K.D. *Amarts and Set Funtion Processes.* Lecture Notes in Mathematics **284**, 4–49. Springer-Verlag (1983).
107. GUT, A. Complete convergence and convergence rates for randomly indexed partial sums with an application to some first passage times. *Acta Math. Acad. Sci. Hungar.* **42**, 225-232 (1983). Correction *ibid.* **45**, 235–236 (1985).

108. GUT, A. On the law of the iterated logarithm for randomly indexed partial sums with two applications. *Studia Sci. Math. Hungar.* **20**, 63–69 (1985).

109. GUT, A. Law of the iterated logarithm for subsequences. *Probab. Math. Statist.* **7**, 27-58 (1986).

110. GUT, A. *Stopped Random Walks.* Springer-Verlag, New York (1988).

111. GUT, A. Limit theorems for record times. In: *Proc. 5th Vilnius Conf. on Prob. Theory and Math. Stat., Vol. 1*, 490-503. Utrecht, The Netherlands (1990).

112. GUT, A. Convergence rates for record times and the associated counting process. *Stochastic process. Appl.* **36**, 135–151 (1990).

113. GUT, A. *An Intermediate Course in Probability.* Springer-Verlag, New York (1995).

114. GUT, A. Convergence rates in the central limit theorem for multidimensionally indexed random variables. *Studia Math. Hung.* **37**, 401–418 (2001).

115. GUT, A Precise asymptotics for record times and the associated counting process. *Stochastic Process. Appl.* **101**, 233–239 (2002).

116. GUT, A. The moment problem for random sums. *J. Appl. Probab.* **40**, 797–802 (2003).

117. GUT, A. An extension of the Kolmogorov-Feller weak law of large numbers with an application to the St Petersburg game. *J. Theoret. Probab.* **17**, 769–779 (2004).

118. GUT, A., AND JANSON, S. The limiting behaviour of certain stopped sums and some applications. *Scand. J. Statist.* **10**, 281–292 (1983).

119. GUT, A., AND JANSON, S. Converse results for existence of moments and uniform integrability for stopped random walks. *Ann. Probab.* **14**, 1296–1317 (1986).

120. GUT, A., AND SPĂTARU, A. Precise asymptotics in the Baum-Katz and Davis law of large numbers. *J. Math. Anal. Appl.* **248**, 233–246 (2000).

121. GUT, A., AND SPĂTARU, A. Precise asymptotics in the law of the iterated logarithm. *Ann. Probab.* **28**, 1870–1883 (2000).

122. GUT, A., AND STEINEBACH, J. Convergence rates and precise asymptotics for renewal counting processes and some first passage times. In: *Proceedings of the International Conference on Asymptotic Methods in Stochastics, May 23–25, 2002, Carleton University Ottawa. Fields Institute Communications* **44**, 205-227 (2004).

123. DE HAAN, L. *On Regular Variation and Its Application to the Weak Convergence of Extremes.* Mathematical Centre Tracts **32**, Amsterdam (1970).

124. DE HAAN, L., AND HORDIJK, A. The rate of growth of sample maxima. *Ann. Math. Statist.* **43**, 1185–1196 (1972).

125. HAHLIN, L.-O. Double records. *Report U.U.D.M.* 1995:12, Uppsala University (1995).

126. HÁJEK, J., AND RÉNYI, A. Generalization of an inequality of Kolmogorov. *Acta Math. Acad. Sci. Hung.* **6**, 281–283 (1955).

127. HALL, P. Characterizing the rate of convergence in the central limit theorem. *Ann. Probab.* **8**, 1037–1048 (1980).

128. HALL, P., AND HEYDE, C.C. *Martingale Limit Theory and Its Applications.* Academic Press, Cambridge, MA (1980).

129. HALMOS, P.R. *Measure Theory.* Van Nostrand, Princeton, NJ (1950).

130. HARTMAN, P., AND WINTNER, A. On the law of the iterated logarithm. *Amer. J. Math.* **63**, 169–176 (1941).

131. HERRNDORFF, N. Stationary strongly mixing sequences not satisfying the central limit theorem. *Ann. Probab.* **11**, 809–813 (1983).
132. HEWITT, E., AND SAVAGE, L.J. Symmetric measures on Cartesian products. *Trans. Amer. Math. Soc.* **80**, 470–501 (1955).
133. HEYDE, C.C. On a property of the lognormal distribution. *J. Roy. Statist. Soc. Ser. B* **29**, 392–393 (1963).
134. HEYDE, C.C. On the influence of moments on the rate of convergence to the normal distribution. *Z. Wahrsch. verw. Gebiete* **8**, 12–18 (1967).
135. HEYDE, C.C. On the converse to the iterated logarithm law. *J. Appl. Probab.* **5**, 210–215 (1968).
136. HEYDE, C.C. A supplement to the strong law of large numbers. *J. Appl. Probab.* **12**, 173–175 (1975).
137. HODGES, J.L., AND LE CAM, L. The Poisson approximation to the Poisson binomial distribution. *Ann. Math. Statist.* **31**, 737–740 (1960).
138. HOEFFDING, W. Probability inequalities for sums of bounded random variables. *J. Amer. Statist. Assoc.* **58**, 13–30 (1963).
139. HOFFMANN-JØRGENSEN, J. Sums of independent Banach space valued random variables. *Studia Math.* **LII**, 159–186 (1974).
140. HSU, P.L., AND ROBBINS, H. Complete convergence and the law of large numbers. *Proc. Nat. Acad. Sci. USA* **33**, 25–31 (1947).
141. IBRAGIMOV, I.A. Some limit theorems for stationary sequences. *Theor. Prob. Appl.* **7**, 349–382 (1962).
142. IBRAGIMOV, I.A., AND LINNIK, YU.V. *Independent and Stationary Sequences of Random Variables.* Wolters-Noordhoff, Groningen (1971).
143. IOSIFESCU, M. Limit theorems for $\phi$-mixing sequences. A survey. In: *Proc. 5th Conf. on Prob. Theory, Braşov, 1974*, 51-57. Editura Acad. R.S.R. Bucureşti, Romania (1977).
144. IOSIFESCU, M., AND THEODORESCU, R. *Random Processes and Learning.* Springer-Verlag, New York (1969).
145. JACOD, J., AND PROTTER, P. *Probability Essentials*, 2nd ed. Springer-Verlag, Berlin (2003).
146. JAIN, N. Tail probabilities for sums of independent Banach space valued random variables. *Z. Wahrsch. verw. Gebiete* **33**, 155–166 (1975).
147. JANSON, S. Moments for first passage and last exit times, the minimum and related quantities for random walks with positive drift. *Adv. in Appl. Probab.* **18**, 865–879 (1986).
148. JANSON, S. Renewal theory for $m$-dependent variables. *Ann. Probab.* **11**, 558–568 (1983).
149. JANSON, S. Some pairwise independent sequences for which the central limit theorem fails. *Stochastics* **23**, 439–448 (1988).
150. KAHANE, J.-P. *Some Random Series of Functions*, 2nd ed. Cambridge University Press, Cambridge (1985).
151. KALLENBERG, O. *Foundations of Modern Probability,* 2nd ed. Springer-Verlag, New York (2001).
152. KARAMATA, J. Sur une mode de croissance régulière des fonctions. *Mathematica* (Cluj) **4**, 38–53 (1930).
153. KATZ, M. The probability in the tail of a distribution. *Ann. Math. Statist.* **34**, 312–318 (1963).
154. KATZ, M. Note on the Berry-Esseen theorem. *Ann. Math. Statist.* **34**, 1107–1108 (1963).

155. KERTÉSZ, I. *A Kudarc*. (1988). Swedish transl: *Fiasko*. Norstedts Förlag, Stockholm (2000).

156. KESTEN, H. A sharper form of the Doeblin-Lévy-Kolmogorov-Rogozin inequality for concentration functions. *Math. Scand.* **25**, 133–144 (1969).

157. KHAN, R.A. A probabilistic proof of Stirling's formula. *Amer. Math. Monthly* **81**, 366–369 (1974).

158. KHINTHINE, A. Über die diadischen Brüche. *Math. Z.* **18**, 109–116 (1923).

159. KHINTHINE, A. Über einen Satz der Wahrscheinlichkeitsrechnung. *Fund. Math.* **6**, 9–20 (1924).

160. KOLMOGOROV, A.N. Über die Summen durch den Zufall bestimmter unabhängiger Größen. *Math. Ann.* **99**, 309–319 (1928).

161. KOLMOGOROV, A.N. Über das Gesetz des iterierten Logarithmus. *Math. Ann.* **101**, 126–135 (1929).

162. KOLMOGOROV, A.N. Bemerkungen zu meiner Arbeit "Über die Summen zufälliger Größen". *Math. Ann.* **102**, 484–488 (1930).

163. KOLMOGOROV, A.N. *Grundbegriffe der Wahrscheinlichkeitsrechnung*. (1933). English transl: *Foundations of the Theory of Probability*. Chelsea, New York (1956).

164. KREIN, M.G. On one extrapolation problem of A.N. Kolmogorov. *Doklady Akad. Nauk SSSR* **46**, 339–342 (1944) (in Russian).

165. LALLEY, S.P. Limit theorems for first-passage times in linear and nonlinear renewal theory. *Adv. in Appl. Probab.* **16**, 766–803 (1984).

166. LAMB, C.W. A short proof of the martingale convergence theorem. *Proc. Amer. Math. Soc.* **38**, 215–217 (1973).

167. LAMPERTI, J. Wiener's test and Markov chains. *J. Math. Anal. Appl.* **6**, 58–66 (1963).

168. LEADBETTER, M.R., LINDGREN, G., AND ROOTZÉN, H. *Extremes and Related Properties of Random Sequences and Processes*. Springer-Verlag, New York (1983).

169. LE CAM, L. An approximation theorem for the Poisson binomial distribution. *Pacific. J. Math.* **10**, 1181–1197 (1960).

170. LÉVY, P. *Calcul des Probabilités*. Gauthier-Villars, Paris (1925).

171. LÉVY, P. Propriétés asymptotiques de sommes de variables aléatoires indépendantes ou enchaînées. *J. Math. Pures Appl.* **14**, 37–402 (1935).

172. LÉVY, P. *Théorie de l'Addition des Variables Aléatoires*. Gauthier-Villars, Paris (1937). 2nd ed. (1954).

173. LI, D., WANG, X., AND RAO, M.B. Some results on convergence rates for probabilities of moderate deviations for sums of random variables. *Internat. J. Math. & Math. Sci.* **15**, 481–498 (1992).

174. LIN, G.D. On the moment problems. *Statist. Probab. Lett.* **35**, 85-90 (1997). Erratum *ibid.* **50**, 205 (2000).

175. LIN G.D., AND STOYANOV, J.M. On the moment determinacy of the distributions of compound geometric sums. *J. Appl. Probab.* **39**, 545–554 (2002).

176. LINDEBERG, J.W. Eine neue Herleitung des Exponentialgezetzes in der Wahrscheinlichkeitsrechnung. *Math. Z.* **15**, 211–225 (1922).

177. LOÈVE, M. *Probability Theory*, 3rd ed. Van Nostrand, Princeton, NJ (1963).

178. LYAPOUNOV, A.M. Sur une proposition de la théorie des probabilités. *Bull. Acad. Sci. St.-Pétersbourg* (5) **13**, 359–386 (1900).

179. LYAPOUNOV, A.M. Nouvelle forme du théorème sur la limite de probabilités. *Mem. Acad. Sci. St.-Pétersbourg* (8) **12** No 5 (1901).

180. MARCINKIEWICZ, J., AND ZYGMUND, A. Sur les fonctions indépendantes. *Fund. Math.* **29**, 60–90 (1937).

181. MARCINKIEWICZ, J., AND ZYGMUND, A. Remarque sur la loi du logarithm itéré. *Fund. Math.* **4**, 82–105 (1937).

182. MARCINKIEWICZ, J., AND ZYGMUND, A. Quelque théorèmes sur les fonctions indépendantes. *Studia Math.* **VII**, 104–120 (1938).

183. MCCABE, B.J., AND SHEPP, L.A. On the supremum of $S_n/n$. *Ann. Math. Statist.* **41**, 2166–2168 (1970).

184. MEYER, P.A. *Martingales and Stochastic Integrals I.* Lecture Notes in Mathematics **284**, Springer-Verlag (1972).

185. VON MISES, R. La distribution de la plus grande de $n$ valeurs. *Rev. Math. Interbalkanique* **1**, 1–20 (1936). *Selected papers* **II**, 271–294. Amer. Math. Soc., Providence, R.I. (1964)

186. NEVEU, J. *Discrete-parameter Martingales.* North-Holland, Amsterdam (1975).

187. NEVZOROV, V.B. *Records: Mathematical Theory.* Translation of Mathematical Monographs. Amer. Math. Soc. **194**, Providence, RI. (2001).

188. PARTHASARATHY, K.R. *Probability Measures on Metric Spaces.* Academic Press, Cambridge, MA (1967).

189. PELIGRAD, M. Invariance principles for mixing sequences of random variables. *Ann. Probab.* **10**, 968–981 (1982).

190. PELIGRAD, M. Recent advances in the central limit theorem and its invariance principle for mixing sequences of random variables (a survey). In: E. Eberlein and M.S. Taqqu, eds., *Dependence in Probability and Statistics*, 193–223. Birkhäuser, Boston, MA (1986).

191. PELIGRAD, M. On Ibragimov-Iosifescu conjecture for $\phi$-mixing sequences. *Stoch. Proc. Appl.* **35**, 293–308 (1990).

192. PELIGRAD, M. On the asymptotic normality of sequences of weak dependent random variables. *J. Theoret. Probab.* **9**, 703–715 (1996).

193. PETROV, V.V. An estimate of the deviation of the distribution of a sum of independent random variables from the normal law. *Dokl. Akad. Nauk* **160**, 1013–1015 (1965). (In Russian). Engl. transl.: *Sov. Math. Dokl.* **6**, 242–244.

194. PETROV, V.V. *Sums of Independent Random Variables.* Springer-Verlag, New York (1975).

195. PETROV, V.V. *Limit Theorems of Probability Theory.* Oxford Science Publications, Clarendon Press, Oxford (1995).

196. PETROV, V.V. A note on the Borel-Cantelli lemma. *Statist. Probab. Lett.* **58**, 283-286 (2002).

197. PETROV, V.V. A generalization of the Borel-Cantelli lemma. *Statist. Probab. Lett.* **67**, 233-239 (2004).

198. PHILLIP, W. Invariance principles for independent and weakly dependent random variables. In: E. Eberlein and M.S. Taqqu, eds., *Dependence in Probability and Statistics*, 225–268. Birkhäuser, Boston, MA (1986).

199. PHILLIP, W., AND STOUT, W.F. Almost sure invariance principles for sums of weakly dependent random variables. *Mem. Amer. Math. Soc.* **161** (1975).

200. POLLARD, D. *Convergence of Stochastic Processes.* Springer-Verlag, New York (1984).

201. PRABHU, N.U. *Stochastic Processes.* Macmillan, New York (1965).

202. PRATT, J.W. On interchanging limits and integrals. *Ann. Math. Statist.* **31**, 74–77 (1960).

203. PYKE, R., AND ROOT D. On convergence in $r$-mean for normalized partial sums. *Ann. Math. Statist.* **39**, 379–381 (1968).

204. RÉNYI, A. On the asymptotic distribution of the sum of a random number of independent random variables. *Acta Math. Acad. Sci. Hungar.* **8**, 193–199 (1957).

205. RÉNYI, A. On the extreme elements of observations. *MTA III, Oszt. Közl.* **12**, 105–121 (1962). *Collected Works* **III**, 50–66. Akadémiai Kiadó, Budapest (1976).

206. RESNICK, S.I. Limit laws for record values. *Stochastic Process. Appl.* **1**, 67–82 (1973).

207. RESNICK, S.I. *Extreme Values, Regular Variation, and Point Processes.* Springer-Verlag (1987).

208. RESNICK, S.I. *A Probability Path.* Birkhäuser, Boston, MA (1999).

209. RICHTER, W. Limit theorems for sequences of random variables with sequences of random indices. *Theor. Prob. Appl.* **X**, 74–84 (1965).

210. ROOTZÉN, H. On the functional central limit theorem for martingales. *Z. Wahrsch. verw. Gebiete* **38**, 199–210 (1977).

211. ROOTZÉN, H. On the functional central limit theorem for martingales, II. *Z. Wahrsch. verw. Gebiete* **51**, 79–93 (1980).

212. ROOTZÉN, H. Central limit theory for martingales via random change of time. In: A. Gut and L. Holst, eds., *Essays in Honour of Carl Gustav Esseen*, 154–190. Uppsala University (1983).

213. ROSÉN, B. On the asymptotic distribution of sums of independent identically distributed random variables. *Arkiv Matematik* **4**, 323–332 (1962).

214. ROSENTHAL, H.P. On the subspaces of $L^p$ ($p > 2$) spanned by sequences of independent random variables. *Israel J. Math.* **8**, 273–303 (1970).

215. RUSSO, R.R. On the last exit time and the number of exits of partial sums over a moving boundary. *Statist. Probab. Lett.* **6**, 371–377 (1988).

216. SAMORODNITSKY, G., AND TAQQU, M.S. *Stable Non-Gaussian Random Processes.* Chapman & Hall, New York (1994).

217. SENETA, E. *Regularly Varying Functions.* Lecture Notes in Mathematics **508**, Springer-Verlag, Berlin (1976).

218. SERFLING, R.J. A general Poisson approximation theorem. *Ann. Probab.* **3**, 726–731 (1975).

219. SERFLING, R.J. *Approximation Theorems of Mathematical Statistics.* Wiley, New York (1980).

220. SHIGANOV, I.S. Refinement of the upper bound of the constant in the central limit theorem. *J. Sov. Math.* **35**, 2545–2550 (1986).

221. SHORACK, G.A., AND WELLNER, J.A. *Empirical Processes with Applications to Statistics.* Wiley, New York (1986).

222. SHOHAT, J.A., AND TAMARKIN, J.D. *The Problem of Moments.* Amer. Math. Soc., Providence, R.I. (1943).

223. SIEGMUND, D. *Sequential Analysis. Tests and Confidence Intervals.* Springer-Verlag, New York (1985).

224. SKOROHOD, A.V. Limit theorems for stochastic processes. *Theor. Prob. Appl.* **I**, 261–290 (1956).

225. SŁABY, M. The law of the iterated logarithm for subsequences and characterizations of the cluster set of $S_{n_k}/(2n_k \log \log n_k)^{1/2}$ in Banach spaces. *J. Theoret. Probab.* **2**, 343–376 (1989).

226. SLIVKA, J. On the law of the iterated logarithm. *Proc. Nat. Acad. Sci. USA* **63**, 289–291 (1969).

227. SLIVKA, J., AND SEVERO, N.C. On the strong law of large numbers. *Proc. Amer. Math. Soc.* **24**, 729–734 (1970).

228. SLUD, E.V. The moment problem for polynomial forms in normal random variables. *Ann. Probab.* **21**, 2200–2214 (1993).

229. SMITH, W.L. Asymptotic renewal theorems. *Proc. Roy. Soc. Edinburgh Sect A.* **64**, 9–48 (1954).

230. SMITH, W.L. Renewal theory and its ramifications. *J. Roy. Stat. Soc. Ser. B* **20**, 243–302 (1958).

231. SPĂTARU, A. Precise asymptotics in Spitzer's law of large numbers. *J. Theoret. Probab.* **12**, 811–819 (1999).

232. SPITZER, F. A combinatorial lemma and its applications to probability theory. *Trans. Amer. Math. Soc.* **82**, 323–339 (1956).

233. SPITZER, F. A Tauberian theorem and its probability interpretation. *Trans. Amer. Math. Soc.* **94**, 150–169 (1960).

234. SPITZER, F. *Principles of Random Walk,* 2nd ed. Springer-Verlag, New York (1976).

235. STEELE, J.M. *Stochastic Calculus and Financial Applications.* Springer-Verlag, New York (2000).

236. STEIGER, W.L., AND ZAREMBA, S.K. The converse of the Hartman-Wintner theorem. *Z. Wahrsch. verw. Gebiete* **22**, 193–194 (1972).

237. STEIN, C. A bound for the error in the normal approximation to the distribution of a sum of dependent random variables. In: *Proc. Sixth Berkeley Symp. Math. Statist. Prob. 1970* **2**, 583–603 (1972).

238. STRASSEN, V. A converse to the law of the iterated logarithm. *Z. Wahrsch. verw. Gebiete* **4**, 265–268 (1966).

239. STOUT, W.F. *Almost Sure Convergence.* Academic Press, Cambridge, MA (1974).

240. SUDDERTH, W.D. A "Fatou equation" for randomly stopped variables. *Ann. Math. Statist.* **42**, 2143–2146 (1971).

241. TAKÁCS, L. On a probability problem arising in the theory of counters. *Proc. Cambridge Philos. Soc.* **52**, 488–498 (1956).

242. THIELE, T.N. *Almindelig Iakttagelselære: Sandsynlighetsregning og Mindste Kvadraters Methode.* C.A. Reitzel, Copenhagen (1889).

243. THIELE, T.N. *Theory of Observations.* Layton, London (1903). Reprinted in *Ann. Math. Statist.* **2**, 165–307 (1931).

244. THOMAS, D. Hidden messages and the Bible code. *Sceptical Inquirer.* November/December 1997.

245. TORRÅNG, I. Law of the iterated logarithm – cluster points of deterministic and random subsequences. *Probab. Math. Statist.* **8**, 133–141 (1987).

246. TROTTER, H.F. An elementary proof of the central limit theorem. *Arch. Math.* **10**, 226–234 (1959).

247. UTEV, S., AND PELIGRAD, M. Maximal inequalities and an invariance principle for a class of weakly dependent random variables. *J. Theoret. Probab.* **16**, 101–115 (2003).

248. VAN DER VAART, A.W., AND WELLNER, J.A. *Weak Convergence and Empirical Processes.* Springer-Verlag, New York (1996).

249. VILLE, J. *Étude Critique de la Notion de Collectif.* Monographie des Probabilités. Gauthier-Villars, Paris (1939).

250. WALD, A. *Sequential Analysis.* Wiley, New York (1947).

251. WILLIAMS, D. On Rényi's "record" problem and Engel's series. *Bull. London Math. Soc.* **5**, 235–237 (1973).

252. WILLIAMS, D. *Probability with Martingales.* Cambridge University Press, Cambridge (1991).

253. WITTENBERG, H. Limiting distributions of random sums of independent random variables. *Z. Wahrsch. verw. Gebiete* **3**, 7–18 (1964).

254. WITZTUM, D., RIPS, E., AND ROSENBERG, Y. Equidistant letter sequences in the book of Genesis. *Statistical Science* **9**, 429–438 (1994).

# Index

602    Index

# springeronline.com

2nd Edition

## Mathematical Statistics
### J. Shao

This graduate textbook covers topics in statistical theory essential for students preparing for work on a Ph.D. degree in statistics. The new edition makes Chapter 1 a self-contained text for probability theory with emphasis in statistics. Added topics include useful moment inequalities, more discussions of moment generating and characteristic functions, conditional independence, Markov chains, martingales, Edgeworth and Cornish-Fisher expansions, and proofs to many key theorems. A new section in Chapter 5 introduces semiparametric models, and a number of new exercises were added to each chapter.

2003. 591 p. (Springer Texts in Statistics) Hardcover ISBN 0-387-95382-5

2nd Edition

## Monte Carlo Statistical Methods
### C.P. Robert and G. Casella

The second edition has been revised towards a coherent and flowing coverage of these simulation techniques. This is a textbook intended for a second year graduate course, but someone who either wants to apply simulation techniques for the resolution of practical problems or wishes to grasp the fundamental principles behind those methods can also use it.

2004. 680 p. (Springer Texts in Statistics) Hardcover ISBN 0-387-21239-6

2nd Edition

## Mathematics of Financial Markets
### R.J. Elliot and P.E. Kopp

This book presents the mathematics that underpins pricing models for derivative securities. The idealized continuous-time models built upon the famous Black-Scholes theory require sophisticated mathematical tools drawn from modern stochastic calculus. However, many of the underlying ideas can be explained more simply within a discrete-time framework. This is developed extensively in this substantially revised second edition to motivate the technically more demanding continuous-time theory, which includes a detailed analysis of the Black-Scholes model and its generalizations, American put options, term structure models and consumption-investment problems. The mathematics of martingales and stochastic calculus is developed where it is needed.

2004. 340 p. (Springer Finance) Hardcover ISBN 0-387-21292-2

**Easy Ways to Order ▶**   Call: Toll-Free 1-800-SPRINGER • Web: springeronline.com • E-mail: orders-ny@springer-sbm.com
Write: Springer, Dept. S8112, PO Box 2485, Secaucus, NJ 07096-2485 • Visit: Your local scientific
bookstore or urge your librarian to order. Mention S8112 when ordering to guarantee listed prices,
valid until 9/30/05.                                                                 11/04 Promotion #S8112